MICRO- and MACROMECHANICAL PROPERTIES of MATERIALS

Advances in Materials Science and Engineering

Series Editor

Sam Zhang

Aerospace Materials Handbook, *edited by Sam Zhang and Dongliang Zhao*

Biological and Biomedical Coatings Handbook: Applications, *edited by Sam Zhang*

Biological and Biomedical Coatings Handbook: Processing and Characterization, *edited by Sam Zhang*

Hydroxyapatite Coatings for Biomedical Applications, *edited by Sam Zhang*

Nanobiomaterials: Development and Applications, *edited by Dong Yi and Georgia C. Papaefthymiou*

Micro- and Macromechanical Properties of Materials, *Yichun Zhou, Li Yang, and Yongli Huang*

MICRO- and MACROMECHANICAL PROPERTIES of MATERIALS

Yichun Zhou
Li Yang
Yongli Huang

CRC Press is an imprint of the
Taylor & Francis Group, an **informa** business

CRC Press
Taylor & Francis Group
6000 Broken Sound Parkway NW, Suite 300
Boca Raton, FL 33487-2742

First issued in paperback 2017

Version Date: 20130822

ISBN 13: 978-1-138-07233-6 (pbk)
ISBN 13: 978-1-4665-9243-8 (hbk)

Library of Congress Cataloging-in-Publication Data

Zhou, Yichun, 1963-
[Cai liao de hong wei guan li xue xing neng. English]
Micro- and macromechanical properties of materials / author, Yi-Chun Zhou.
pages cm
Includes bibliographical references and index.
ISBN 978-1-4665-9243-8 (hardcover : alk. paper) 1. Materials--Mechanical properties. 2. Microstructure. I. Title.

TA405.Z4713 2014
620.1'1292--dc23 2013028510

Visit the Taylor & Francis Web site at
http://www.taylorandfrancis.com

and the CRC Press Web site at
http://www.crcpress.com

Contents

Foreword

Macro- and micromechanical properties of materials is an important area in materials science because mechanical problems related to structural and functional materials during production, processing, and application still occur. Properties, such as efficient, safe, environmental-friendly and low-powered, and the efficient use of materials and their lifetime prediction are all based on the systematic analysis of their mechanical properties. The ability to analyze the macro- and micromechanical properties of materials is one of the basic skills for all materials science undergraduate and postgraduate students. Yichun Zhou and his colleagues at Xiangtan University, the authors of *Micro- and Macromechanical Properties of Materials*, have written a comprehensive textbook for both students and research scientists.

This book systematically covers macro- and micromechanical properties of metallic and nonmetallic structural materials, and various functional materials, as well as the macro- and microfailure mechanisms under various loading conditions. During the preparation of this book, the authors were committed to introduce structure and function, macroscopic and microscopic scales, scientific theories, and engineering applications. Some cutting-edge research has also been integrated into this book, such as strain gradient theory, scale effects, cross-scale numerical simulation, microfracture mechanics, mechanical properties of smart materials, thin films, and coatings.

This book has three significant features. First, it is an integration between scientific methods and theories. The development of modern materials science has changed rapidly, and it is more useful to learn the fundamentals of scientific methods rather than to learn by rote. This book covers the fundamentals of this discipline and the main theories of the subject while also serving as an introduction to research methods. By combining these two individual areas, the book increases the attractiveness of its contents and also helps to stimulate and innovate the research activities of students.

Second, it is a combination of educational reform and scientific research. Over the last decade, valuable experience has been gained on the subject of mechanical properties of materials by this teaching and research group through undergraduate teaching. Some achievements gained during teaching were also implemented through fostering the materials science talent. Moreover, many research findings conducted by this research group on mechanical properties were incorporated into this book. Introducing some of the latest research techniques and achievements on materials and mechanics has also raised the academic quality of this book.

Third, it is a combination of theories and engineering applications. Materials science is a subject that involves engineering practices and practical applications. This book focuses on the links between scientific theories and engineering applications from various perspectives, from the macro- to the microscale, from theoretical calculation to practical implications, and from materials production to property analysis. Integrating fundamental theories with research findings has enriched the content of this book and has also enabled students to develop their creativity and practical ability.

The authors of this book have long devoted themselves to teaching and research in mechanics and materials science. Editorial director Professor Yichun Zhou, who was the winner of the National Outstanding Youth Science Foundation Award and the National Distinguished Teachers Award, currently serves as the director of the Key Laboratory of Low Dimensional Materials & Application Technology of the Ministry of Education at Xiangtan University. The editorial board's deputy director, Professor Xuejun Zheng, was the winner of the National Outstanding Youth Science Foundation Award. The course entitled "Micro- and Macro-Mechanical Properties of Materials" taught by the authors at the University of Xiangtan was given the National Excellent Course Award in 2005, and their teaching materials were designated the Eleventh Five-Year Project national planning materials for general higher education. The authors then became the backbone of their teaching and research group, Materials and Devices, and in 2007 won the title for being the first national-level teaching/research group.

I believe that the publication of this book will encourage cross-links between mechanics and materials science and will also play a role in nurturing new materials scientists. Therefore, I am delighted to have written the foreword and to recommend this book to readers.

Boyun Huang
Academician of the Chinese Academy of Engineering
Vice-Chairman of the China Association for Science and Technology
President of Central South University

Foreword to the English Language Edition

Micro- and Macromechanical Properties of Materials is a unique textbook, providing a complete and rigorous overview in the research areas of mechanics and materials sciences from the viewpoint of length scales. It combines reviews of both theoretical and applied research, as well as reviews of the most recent studies in the field.

Professor Zhou is a prestigious scientist. He has won research awards from the Chinese Academy of Sciences and the Hunan provincial government in China. He was also granted the Functional Materials Scientist Award by the committee of International Conference on Advances in Functional Materials.

Dr. Changxu Shi, Professor
Member of the Chinese Academy of Sciences
Former Vice-President of the Chinese Academy of Engineering

Dr. Daining Fang, Professor
Vice Dean, College of Engineering, Peking University
Head, Applied Mechanics Key Laboratory of the Ministry of Education, China
Vice President, Theoretical and Applied Mechanics Society of China
Vice Chairman, Asia-Pacific Society of Mechanics of Materials
Chairman, Testing-Machine Branch of the Chinese Society of Instruments

Dr. Wei Pan, Professor
Department of Materials Science and Engineering, Tsinghua University
Director, State Key Laboratory of New Ceramics and Fine Processing
Secretary-General, Chinese Advanced Ceramics Society

Dr. Jingfeng Li, Professor
Changjiang Scholar Professor
Vice Department Chair, Department of Materials Science and Engineering, Tsinghua University

Preface

The development of science, technology, and the national economy generates increasing demands for a vast variety of materials with higher quality and higher performance in more severe operating conditions. Materials efficiency and the service lifetime of materials under certain operating conditions are the most important factors in evaluating material performance. Efficiency, safety, low energy consumption, and environmental friendliness are the ultimate pursuits in new materials development and applied research. According to the statistics of some industrialized countries, losses due to materials failures and the consequent structural failures account for 8%–12% of their national GDP. Moreover, casualties caused by accidents linked directly to these failures are immeasurable. At present, the situation in China is far worse than in the developed Western countries. On the other hand, there is great potential for the development of new materials. Due to the complexity of factors affecting materials performance—the inherent properties of materials, their design, processing status, and operating environment and conditions—it is impossible to design the perfect material at once. Therefore, the recognition of these problems had to rely on countless failures and the gradual accumulation of experience.

In our new economical and harmonious society, we must design and create materials that are efficient, safe, have low power consumption, and are environmental friendly. In order to have a relatively clear prediction of their efficiency and service lifetime, we must have a clear understanding of their mechanical behavior and failure mechanisms during production, processing, and service.

To assist undergraduate students of materials science and engineering in their understanding of the fundamental principles of mechanics, and to subsequently apply their knowledge to research and development of high-performance new materials that are harmonious to our society and environment as well as to improve traditional teaching materials, I have been teaching undergraduate courses on the mechanics of materials since 1996. An appropriate Chinese textbook on the topic is essential in order to teach this course. Initially, we had chosen topics that had the largest impact on the subject from domestic references. As our teaching progressed, we found that the majority of the topics concerned structural materials and their properties on a macroscopic scale. This conformed to the historical development of the subject of materials science in our country, which began with metal materials and their heat treatment, welding, molding, and high-molecular-weight materials. But, in fact, the subject of "materials science and engineering" is a twofold subject.

From a scientific perspective, the relationship between materials structure and their properties involves the exploration of the laws of nature, which is fundamental science. From an engineering perspective, materials were developed to fulfill engineering purposes and to serve economic growth, which is applied science. The main purpose of materials research and development is application, and an appropriate production and processing route is essential to yield materials with the desired properties. Therefore, "materials science and engineering" is a subject that consists of studies of the structure of materials, the production process, and material properties and performance, as well as the relationships between them, and it requires combined knowledge from both scientific and engineering backgrounds.

Materials science has three important features. First, it is a multidisciplinary subject. It involves physics, mechanics, chemistry, metallurgy, and computational science, and these subjects are interlinked. Second, it is a discipline closely linked with practical applications. The aim of materials science is to develop new materials that have improved performance and properties, to have better utilization of materials, and to reduce cost and pollution during their production. Third, materials science is a growing discipline. Unlike physics, chemistry, and mechanics, which are very mature systems, materials science enriches and perfects itself through the development of other disciplines.

In recent years, materials science and technology has been the fastest growing discipline, and it has not only created a large number of new high-performance materials and unprecedented processing methods but has also significantly changed the production of traditional materials. Since the 1990s, materials science and technology has undergone a revolutionary transformation. The trend can be summarized as follows: (1) materials science and engineering has rapidly become a unified discipline and places more emphasis on practical applications; (2) research has focused on the microlevel in order to discover new substances and develop new materials; (3) composite materials have become an important direction of development; (4) research methods have become oriented in the direction of theory–computer simulation–experiment; (5) the trend for innovation in modern materials science and technology has become cheaper and faster, providing better results from a technological innovation perspective; and (6) the trend has been to integrate materials research, development, production, and application.

Based on these characteristics and the attribution and developing trends of the materials science and engineering disciplines, we believe that current teaching materials for structural materials and their macroscopic properties is insufficient. Beginning in 1998, we began to experimentally implement the topics of macro- and micromechanical properties into our course; the course subsequently won the National Excellent Courses Award in 2005. A year later, this book was recognized as the national planning textbook for general higher education in the Eleventh Five-Year project. The backbone of this book is a combination of materials science and mechanics, with three distinct integrated features: the combination of science (fundamental principles) and engineering (applications), across the macroscopic and the microscopic levels, and from structural to functional materials. This book is also a combined effort of all the associated teaching and research staff from the Department of Material and Photoelectronic Physics and the Low Dimension Materials and Application Technology Laboratory (the National Key Laboratory of the Ministry of Education) of Xiangtan University.

The book was arranged in 15 parts: Introduction (written by Yichun Zhou and Shiguo Long), Chapters 1 and 2 (Yichun Zhou), Chapter 3 (Yongli Huang), Chapter 4 (Li Yang, Shiguo Long, and Yichun Zhou), Chapter 5 (Yongli Huang), Chapter 6 (Xuejun Zheng), Chapter 7 (Weiguo Mao), Chapter 8 (Li Yang), Chapter 9 (Shiguo Long and Shangda Chen), Chapter 10 (Jiangyu Li and Shuhong Xie), Chapter 11 (Xuejun Zheng and Yichun Zhou), Chapter 12 (Wenbo Luo), Chapter 13 (Weiguo Mao), and Chapter 14 (Shiguo Long). We would like to acknowledge some of our PhD and master's students, especially Lianghong Xiao, Xuehong Yu, Dan Wu, Sha Zhang, Zhaofeng Zhou, Zengsheng Ma, Jun He, Tonggang Ou, and Bo Wu, for their work in preparing teaching materials and drawings, printing texts, searching the literature, guiding experiments, and helping with tutorial sessions.

During the past five years, there have been detailed discussions with the authors who have written individual chapters about possible modification of equations and graphs. The majority of the content in this book came directly from monographs and papers that we have published. Our research activities have been funded by two of the outstanding youth funds of the National Natural Science Foundation of China (NSFC), two key projects from the NSFC, two projects from National 863 Project in the materials field, and 20 projects from the NSFC and youth projects, a major project of the Ministry of Education. This book is a result of ten years of our teaching experience, under the able guidance of many of our senior professors in the field of mechanics and materials science, including Sirs Changxu Shi, Zhemin Zheng, Jiluan Pan, Hengde Li, and Kezhi Huang; Academicians Yilong Bai, Boyun Huang, and Wei Yang; and Professors Tingqing Yang, Zhuping Duan, Tongyi Zhang, Yonggang Huang, Daining Fang, and Qibin Yang. Many of the chapters are from their earlier publications and theses. A special thank you goes to Academician Boyun Huang for writing the foreword to this book. We would also like to acknowledge the authors whose work has been referenced in this book, the relevant departments who supported this work, and the colleagues and students who contributed to the publication of this book.

Due to various limitations, many topics in this book on macro- and micromechanics were insufficiently covered. In order to improve the content, we welcome feedback from the reader.

Yichun Zhou
Faculty of Materials, Optoelectronics and Physics
Key Laboratory of Low Dimensional Materials and Application Technology, Ministry of Education
Xiangtan University, Xiangtan, Hunan, People's Republic of China

Acknowledgments

We would like to thank Dr. Yugo Ashida, president, and Kenneth MacLean, senior English-language editor, of the JOINLU International Business Services (www.joinlu.com). Their successful effort to have this book published in English is sincerely appreciated. We would like to express our gratitude to the technical translators Dr. Jixi Zhang (Chapters 1, 3, 6, 7, and 9), Dr. Tien-Chien Tsao (Introduction and Chapters 10 and 12), Mr. Jimmy Chu (Chapters 2, 8, and 13), Mr. Cheng Xu (Chapters 4 and 5), Dr. Augustine Chen (Chapter 11), Mr. Xueyang Li (Chapter 11), Dr. Kai Sun (Chapter 14), and Dr. Peng Wang (Foreword and Preface).

We would also like to thank Allison Shatkin, senior acquisitions editor at Taylor & Francis Group, for her help and encouragement and the Chinese publishers of the original work, China Higher Education Press, for their help in the publication of this book.

We would like to acknowledge some of our PhD and master's students—Lianghong Xiao, Xuehong Yu, Dan Wu, Sha Zhang, Zhaofeng Zhou, Zengsheng Ma, Jun He, Tonggang Ou, and Bo Wu—for their work in preparing teaching materials and drawings, printing texts, searching the literature, guiding experiments, and helping with tutorial sessions.

Introduction

Materials are those substances used in the manufacture of goods, devices, components, machinery, or other products and are the basis for mankind's survival and development. The introduction of new materials and applications plays an important role in the progress of human civilization.

Materials, information, and energy are the three pillars of modern civilization. Because of the diversity of materials, there are no uniform standards for classification. However, materials can be grouped into two major areas: (1) structural and functional materials and (2) traditional and new materials.

Structural materials are based on their mechanical properties and are used to make bearing carrier. Structural materials have certain unique physical or chemical properties, such as gloss, thermal conductivity, radiation resistance, corrosion resistance, and antioxidation. Functional materials are also classified mainly by their unique physical and chemical properties, or their biological function. In many cases, a material—such as iron, copper, or aluminum—is both structural and functional. Traditional materials such as steel, cement, and plastics are those that have matured and have been used in mass production and in industrial applications. These materials are known as fundamental materials because of their high value and wide range of use as well as their underlying importance for many pillar industries. New materials (advanced materials) are those that are under development and that have the potential for improved performance. There is no clear boundary between new materials and traditional materials. Traditional materials can become new materials through the use of new technology to improve their technical characteristics and enhance their performance, which substantially brings in added value. New materials can also become traditional materials after long-term production and application. Traditional materials are the foundation for the development of new materials and advanced technologies, and new materials often promote further development of conventional materials.

The origin of mechanics is closely related to human activity. As soon as humans became productive, mechanics inevitably emerged. If a stone could not be moved by hand, for example, a stick could be used as a lever to pry it out. The development of mechanics has gone through several stages. Before the seventeenth century, mechanics was at an early stage of development, concerned mainly with the practical aspects of living. In the period between the seventeenth and nineteenth centuries, mechanics in the classical mechanics stage were represented by Newtonian mechanics. Modern mechanics began after the mid-twentieth century and includes solid mechanics, fluid mechanics, and general mechanics.

Solid mechanics is concerned with the law of displacement, movement, stress, strain, and destruction generated by the internal particles of a deformable solid under the action of external factors such as load, temperature, and humidity.

Mechanical properties of materials refer to the mechanical behavior of materials under external loads. The most common mechanical properties are (1) elastic modulus, which describes the relationship between stress and strain; (2) yield strength, which refers to the minimum stress necessary for the plastic deformation of materials; and (3) hardness, which, not surprisingly, describes a material's softness or hardness. Mechanical properties were the earliest material properties that humans learned to use. In the Stone Age, for example, humans learned how to test the strength and hardness of stone.

The systematic study of the mechanical properties of materials began in the mid-1800s, when metalloscope were used to study the microstructure of materials. Mechanical properties are primarily determined by studying the material's composition and the material's microstructure [4]. With the development of device miniaturization and integration, knowledge of the mechanical properties of structural materials has become increasingly important. At the same time, structural materials and functional materials are being combined, which requires us to know their function and structural strength.

The best way to improve the mechanical properties of structural materials and functional materials is to enhance their micromechanical properties. In addition, materials design is very important in the development of new materials, and to do that, we must also understand the microscopic properties of materials. Therefore, the macro- and micromechanical properties of materials are an important meeting point of materials and mechanics.

MATERIALS SCIENCE AND ENGINEERING AND ITS DEVELOPMENTAL HISTORY AND TRENDS

It can be said that human history is the history of materials development. The continuous development and use of new materials are fundamental for modern human civilization.

The use of materials has undergone seven eras in human history (see Table 1) [3].

From the ancient Stone Age to the Iron Age, the use of metals marked the development of social productivity and the beginning of civilized human society. In the eighteenth century, the advent of the Steel Age caused a global industrial revolution, resulting in a number of economically well-developed powers. The Silicon Age began in 1950, creating the era of information technology that has sparked a revolution with profound impact on the modern world.

The era of steel and silicon led to an understanding of the influence of materials science on social development and progress. Today, scientists specializing in materials research, economists, bankers, business leaders, and leading economic policy makers at the national level are paying close attention to the development of materials research in order to grasp opportunities and to make the right judgments and decisions.

The mistakes Great Britain has made in its technology policies is an example that illustrates this issue. At the advent of the Steel Age, Britain's early development of its steel industry contributed to its dominant position in worldwide steel production and also brought tremendous vitality to its economic development. After World War II, Japan's industrial development required the production of its own low-cost and high-quality steel. In 1952, Japan's annual steel production was only 7 million tons, while Britain annually produced 17 million tons of steel. By 1962, only ten years later, Japan's annual steel output jumped to 27.5 million tons, while Britain's steel production was 20.8 million tons. By 1972, Japan was far ahead, producing 96.9 million tons of steel, compared to

TABLE 1
The Seven Eras of Human Materials Usage

Beginning of Era	Era
ca. 100000 BCE	Stone Age
ca. 3000 BCE	Bronze Age
ca. 1000 BCE	Iron Age
ca. 1 CE	Cement Age
ca. 1800 CE	Steel Age
ca. 1950 CE	Silicon Age
ca. 1990 CE	New Materials Age

Source: Briggs, A., *The Science of New Materials*, Blackwell, Oxford, U.K., 1992.

Britain's 25 million tons. Britain's technology policy promoted the growth of Japan's auto industry, as well as the rapid development of its other major industries that used steel products.

After 1970, the Japanese began to recognize the world's entry into the Silicon Age. In addition to maintaining its dominant position in steel production, Japan also targeted silicon materials and developed its semiconductor industries, which resulted in its worldwide leadership role in the production of household appliances. In contrast, Great Britain ignored the arrival of the silicon era. Because of the absence of corresponding technology policies and strategic vision, nearly 2000 British silicon research scientists flowed into the U.S. Silicon Valley. Consequently, in 1988, the United Kingdom's trade deficit with Japan reached £2.2 billion in information technology products alone, which did not even include products like the silicon-controlled autofocus camera. The result has been, as the British have complained, that "Britain has no semiconductor industry, and the United Kingdom has degraded from a first-rate economic power to the second-tier of economically developed countries, while Japan has developed from a second-tier economically developed country to a first-rate economic power. The attitude of the United Kingdom is like a country that still remains in the Stone Age, but with no progress into the Bronze Age" [4].

It is now recognized that materials development has been a milestone in the history of human evolution, and an important pillar of modern civilization. Further development of science and technology leads to enhancement in the quality of life. Moreover, a rapid increase in world population, the accelerated depletion of resources, and the deteriorating environment have demanded higher levels of sophistication in materials science and technology.

The field of materials science and engineering is now entering into an unprecedented period of innovation and development. New materials are the precursor for high-technology development. New materials research and the scale of industrialization have become important indicators for measuring a country's regional economic development, technological advancement, and national defense capabilities.

"Material" is a term that existed long ago, but the emergence of "materials science" only began in the early 1960s. The Cold War was at its height when the USSR successfully launched Sputnik, the world's first satellite, on October 4, 1957. The U.S. government and the general public were shocked by U.S.S.R's achievement and realized that they were lagging behind in their research into advanced materials. Therefore, in order to catch up, the Americans engaged in an in-depth study of materials and established more than ten material research centers in their universities, achieving significant results through the use of advanced scientific theories and experimental methods. Since then, the term "materials science" has become popular [1].

Materials science is often understood as the study of the relationship between the organization, structure, and the nature of materials, and the exploration of their natural laws. In fact, materials science is an applied science, and the purpose of materials research and development is to practically apply its results to daily life and to promote economic development. A material with practical value must be prepared through rational technological processes and must be mass produced in order to be classified as an engineering material. Therefore, soon after the term "materials science" emerged, "materials science and engineering" was proposed. Materials engineering refers to the study of technologies and the technical problems involved in the process of preparing materials for practical use. Therefore, "materials science and engineering" is a discipline that studies the composition and structure, the synthesis and processing, and the properties and performance of materials, as well as the relationships between them. The image of a tetrahedron on the cover of *Acta Materialia*, a well-known material research journal, best describes the four basic elements of materials science and engineering: composition and structure, synthesis and processing, properties, and performance.

The four-element model of materials science and engineering has two main features [1].

First, there is a special relationship between the properties and the performance of materials, in which the performance of materials represents the behavior of its properties under conditions of use. Environmental conditions such as stress state, ambience, media, and temperature have a great impact on the properties of materials. Some materials perform well under normal circumstances, but when they are placed in a corrosive environmental media, their performance declines significantly. Other materials have excellent performance when their surfaces are smooth, but their performance diminishes greatly when their surfaces have gaps. This behavior is especially prominent for certain high-strength materials: whenever there is a scratch, it may cause catastrophic damage. Therefore, the introduction of environmental factors is very important for engineering materials.

Second, the theory and design of materials has its proper position at the center of the tetrahedron. Because each of these four elements (or a combination of several related elements) has its own corresponding theory, a model can be built based on the theory. Through this model, materials design and technological design are used to enhance the properties and performance of materials, thereby conserving resources, reducing pollution, and cutting down costs. This is the ultimate goal of materials science and engineering.

Materials science has three important attributes [1]:

Interdisciplinary science: Materials science is the result of the integration and interaction between physics, mechanics, chemistry, metallurgy, and computer science.

Practical application: The purpose of materials science is to develop new materials, improve their performance and quality, and rationally use materials, while reducing cost and pollution.

Ongoing development: Unlike physics, chemistry, and mechanics, which are mature sciences, materials science can be enriched and improved through the development of relevant disciplines.

Materials science and technology is one of the fastest growing scientific and technological fields in recent years. It has created many new high-performance materials and unprecedented processing methods but has also dramatically changed the production of traditional materials. Since the 1990s, materials science and technology has undergone revolutionary changes. These developmental trends can be summarized as follows [5]:

- The discovery and development of new substances and new materials at an in-depth and microlevel
- The development of composite materials
- The growing importance of thin film and coating science and technology [6]
- The integration of theory, computer simulation, and experimentation in the development of new materials [7]
- The impact of modern technology and innovation leading to cheaper, faster, and better performance of materials
- The integration of research, development, production, and application in the emergence of new materials
- The integration of materials science and engineering into a unified discipline, with a focus on practical application and development

INTERDEVELOPMENT OF MATERIALS SCIENCE AND SOLID MECHANICS

Mechanics is one of the seven fundamental natural sciences and is closely related to the eight major applied sciences. Solid mechanics studies the deformation, flow, and destruction of a solid material and the structure (components) bearing the force. It is an important branch of mechanics [8].

The infancy of solid mechanics can be traced back to 2000 BC when the Chinese and other ancient civilizations began to construct buildings and design simple travel and hunting tools with embedded mechanical principles. The Zhaozhou stone arch bridge, built in China during the mid-Kaihuang (AD 591–605) of the Sui Dynasty, employed the basic ideas of a modern bar, plate, and shell design. The accumulation of practical experience and the achievements of seventeenth-century physics led to the development of the theory of solid mechanics. The eighteenth century was the developmental period of solid mechanics. Social needs—such as the manufacture of big machines and the construction of large bridges and plants—became the driving force for the advancement of solid mechanics.

Modern solid mechanics refers to the period after World War II. During this period, there were two developments: (1) the finite element method and the computer were widely used and (2) fracture mechanics and composite mechanics emerged as new branches of solid mechanics. After Turner proposed the concept of the finite element method in 1956, a large number of applications in solid mechanics were developed that have solved many complicated problems.

Cracks always exist in the structure of a material, which has motivated the exploration of the crack-tip stress-and-strain fields and the pattern of crack propagation. As early as the 1920s, Griffith first proposed an important concept: the actual strength of glass depends on the crack propagation stress. In 1957, G. R. Irwin introduced the stress intensity factor and its critical values to determine crack propagation. As a result, modern fracture mechanics was born and has grown rapidly since the late 1950s.

Material mechanics is the first developed branch of solid mechanics. It studies the mechanical properties, deformation states, and fracture patterns of a material under external forces. It also provides the basis for the selection of materials, and the sizes of components, in the process of engineering design.

Fracture mechanics, also known as fracture theory, is the latest developed branch in solid mechanics. It studies the crack-tip stress-and-strain fields in engineering structures and analyzes the conditions and patterns of crack propagation.

Solids usually contain cracks. Even without macroscopic cracks, microdefects within the solid (such as pores, grain boundaries, dislocations, and inclusions) develop into macroscopic cracks under the effects of loads and corrosive media, and especially under alternating loading conditions. Therefore, it can be said that fracture theory is crack theory. The fracture toughness and crack propagation rate proposed by fracture theory are important indicators used to predict critical cracking size and component life-span estimation, which have been widely used in the engineering structure. In general, the purpose of fracture mechanics includes the study of crack propagation patterns, the establishment of fracture criteria, and the control and prevention of fracture failure.

Composite material mechanics is the study of the mechanical properties of components in modern composite materials—primarily fiber-reinforced composite materials—and their deformation patterns and design criteria under a variety of different support conditions and external forces. Composite material mechanics is concerned with material design, structural design, and design optimization. It is a new branch of solid mechanics developed in the 1950s.

If elastic mechanics were mainly confined to the nineteenth century, then in the twentieth century, solid mechanics has been greatly developed and substantially expanded, resulting in the emergence of many new branches in the field. Both solid mechanics and materials science play an important role in modern industry. With the development of science and technology, the interrelationship between solid mechanics and materials science has become increasingly evident. Over the years, solid mechanics has established a series of important concepts and methods, such as continuous media, stress, strain, bifurcation, fracture toughness, and the finite element method. These brilliant achievements led to the development of modern civil and construction industry, the machinery manufacturing industry, and the aerospace industry, but also provided the theoretical basis—along with the use of partial differential equations—for a wide range of natural sciences such solid-earth geophysics, nonlinear science, materials science and engineering.

Due to the complexity of the objects involved, modern solid mechanics raises a series of challenging issues that are at the forefront of science. The mechanical properties of materials and the study of the fracture of materials have become the current focus of problems in solid mechanics. In May 2005, the 96th "Young Scientist Forum" was organized by the China Association for Science and Technology (CAST) at Xiangtan, Hunan. The theme of this event was "The challenges posed by the rapid development of materials science on solid mechanics." The forum conducted in-depth discussions on the challenges and opportunities faced by the following two major trends: (1) the development of solid mechanics under the conditions of low-dimensional nanomaterials, smart materials, and multifield coupling and (2) the development of the solid mechanics of materials [9]. The Materials Science and Engineering Committee of the U.S. Department of Energy organized a symposium and invited 19 leading scholars from 7 renowned universities and research institutions for discussions on the future direction of the development of material mechanics. Its research report was published in the prestigious journal *Mechanics of Materials* in 2005 [10]. These discussions show that low-dimensional nanomaterials bring new opportunities for development in the research of solid mechanics.

Theoretical Research

Since the early 1980s, modern manufacturing technology has enabled the manufacture of a variety of new low-dimensional materials with optimum performance, such as zero-dimensional quantum dots, atomic cluster, nanopowders, one-dimensional quantum wires, nanowires, nanotubes, nano-superlattices, quantum two-dimensional arrays, thin films, coatings, and one- or two-dimensional quasicrystals. Most of these materials demonstrate complex structural forms and excellent physical properties. The study of the relationships among their growth patterns, physical properties, and structural forms has attracted scholars from many fields in materials, physics, mechanics, mathematics, and biology. Because of the size effect and the properties of low-dimensional materials—which are very different from corresponding bulk materials and substrate materials—the existing theoretical basis and the experimental methods for research into low-dimensional materials may not be appropriate. The development of low-dimensional materials has brought mechanics scholars into a new realm of nontraditional macro- and microresearch. Therefore, searching for the theories and methods appropriate for the design and prediction of the mechanical properties of low-dimensional materials raises a number of current research challenges: What is the scope of research for the mechanics of low-dimensional materials and nanomaterials? How can researchers propose a new theory that is neither traditional macro-Newtonian mechanics nor traditional microscopic quantum mechanics?

Cross-Scale Numerical Simulations

Research and experience have shown that material properties are not statically dependent on the chemical composition of materials but, to a large extent, dependent on the microstructure of materials. Microstructure refers to the collection and spatial distribution of all lattice defects of nonequilibrium thermodynamics. Its space scales can vary from a few tenths of a nanometer to several meters, and its corresponding temporal scales can range from a few picoseconds to several years. A major goal in the research of materials science has been to understand the quantitative relationship between the macroscopic properties of materials and their microstructure. It is well known that numerical simulation is a powerful tool for materials design and the prediction of material properties. However, it is clearly not feasible, using a single micro- or macroapproach, to study problems of microstructure evolution where there are differences in spatial and temporal scales over several orders of magnitude. For these cross-scale problems, researchers encounter serious issues in mathematics and physics.

Experimental Representations

The challenge of experimental mechanics posed by low-dimensional materials—a new object of research—is how to develop new experimental methods and study new testing technologies that can measure micro- and nanostructures and the mechanical properties of materials. Structural research should emphasize the following: (1) loading methods and deformation measurements of micro- and nanostructures, (2) testing methods for the ultra-high-frequency features of micro- and nanostructures, (3) macroloading environmental control technologies on the microevolution of microzones, (4) the experimental methods and simulation techniques of low-dimensional structures, and (5) measuring methods of microloading and microsensing of microspecimens.

Challenges and Opportunities of Smart Materials, Device Mechanics, and Multifield Coupling Mechanics

Smart devices are extensively used in the rapid development of information technologies and integrated microoptoelectromechanical systems (MOEMS). Smart devices raise scientific issues such as multifield coupling and the structural mechanics of subtle information. When stress, strain, or thermal activity intensively interact with electromagnetic behavior, the laws of mechanics for smart materials, their information technology, and their structural design becomes extremely important. Therefore, solid mechanics faces new challenges in the determination of the constitutive relation of anisotropic smart materials, the analysis of the field under multifield coupling, and the determination of basic physical mechanics parameters.

Direction of Material Solid Mechanics Is Gradually Emerging

Materials science is different from physics, chemistry, and mechanics, which have very mature systems. The system of material solid mechanics is emerging. The difference between actual material strength and current theoretical strength is one to two orders of magnitude. This contradiction has resulted in the establishment of important physical and mechanical theories, such as dislocations and cracks. However, this fundamental contradiction still exists. Today, solid mechanics is not limited to computing small strain and stress, but requires the determination of deformation localization, damage, life span, and fracture. A further problem is how to reasonably configure materials with different properties and functions together, forming composite materials. These composites must practically optimize such factors as specific gravity, stiffness, strength, toughness, and functionality, as well as price, and contribute to the science of materials design. A further step is to develop specific manufacturing and processing techniques, such as plastic forming and particle beam processing technologies, which will achieve rational mechanical understanding and optimization control. At that stage, the entire material and manufacturing industries will transform from so-called kitchen chemistry to resource conserving, energy conserving, and rational, optimized industries.

The composition of materials is important for the development of new materials and the transformation of traditional materials, which include composite materials, surface modification of materials, thin films, and coatings. It makes the material interface problem particularly prominent. The combined effect of residual stress and external load will result in the fracture of materials, because the mismatch of thermal parameters and the mechanics of materials on both sides of the interface will cause residual stress. Therefore, interface mechanics will encounter such problems as micro- and macrodeformation, the theoretical analysis of the entire process from damage to fracture, numerical simulation, and experimental measurement. In particular, the following questions must be asked: Is the concept of interface fracture toughness, and interface strength, suitable? Is it possible to find reasonable parameters representing the entire process of the interface from deformation to damage to fracture?

The Mechanics of Biological and Biomimetic Materials Is Gradually Becoming a Direction of the Discipline

Using the knowledge of solid mechanics, we have been gradually learning to design materials and devices based on inspiration from biology and biotechnology. By drawing from knowledge of biological systems, we can design new materials possessing the stress distribution of the best biological systems, their optimal structural and functional properties, and their best evolution patterns. We can also develop methods for designing hierarchical multiscale new materials by studying the microstructure and stress-and-strain fields of biological materials. We may even boldly predict that microelectromechanical systems (MEMS) or nanoelectromechanical systems (NEMS) already exist in the biological body and that highly sensitive components and new high-performance materials with real social and ecological harmony are biological materials. Therefore, continued development in this area will have a significant impact on the engineering, biological, and medical industries, as well as the materials industry and the military.

New Domains, New Directions, and New Interdisciplinary Approaches Have Pioneered a "New Frontier" of Solid Mechanics

With the rapid development of materials science, traditional solid mechanics has been greatly expanded. New domains, new directions, and new interdisciplinary approaches have pioneered a "new frontier" of solid mechanics. Research on low-dimensional materials has demonstrated that solid mechanics is not limited to the traditional thinking of continuous media but has been extended to the microscopic level of materials, pursuing the nature of the relationship between microstructure and macroscopic properties.

After encountering new materials such as carbon nanotubes and thin films, researchers began to develop new experimental methods and to study new testing techniques that could measure the micro- and nanostructure of materials, and their mechanical properties.

The study of microstructure has evolved. Spatial scales now vary widely from a few tenths of nanometers to several orders of magnitude, and temporal scales range from a few picoseconds to several years, changing the relationship between the macroscopic properties of materials and their corresponding microstructure. This cross-scale research has not only extended the research domain of solid mechanics but has also integrated such disciplines as materials science, physics, mechanics, and mathematics.

The "kitchen cooking" methods of material manufacturing have collapsed, and people are beginning to rationally design materials based on aspects of their microstructure and their micro- and macroproperties. Solid mechanics also plays a very important role in today's information technology and in the rapid development of integrated MOEMS.

INCREASINGLY CLOSE RELATIONSHIP BETWEEN MACROSCOPIC PROPERTIES AND MICROSTRUCTURE OF MATERIALS

In the field of materials science and engineering, there is no unified classification for the levels of material structure, even though there are a variety of classification methods. However, it is possible, based on the spatial scale of objects, to divide materials structure into three levels:

Engineering design level: This scale, corresponding to the macroscopic properties of materials, involves research and design into the processing and performance of bulk materials.

Continuum model scale: When the scale is in millimeters or above, the material is regarded as a continuous medium, and the behavior of single atoms or molecules of the material is not considered.

Microdesign level: When the spatial dimension is in micrometers or below, the scale is atomic- or molecular-level design.

According to a combined spatial and temporal scale, the levels of material structure could be divided into four levels:

Macrolevel: This is the main scope of daily human activity, which uses brute strength or machinery and equipment. The spatial scale of this level is roughly from meters to tens of thousands of kilometers, and the temporal scale is roughly from seconds to years. Today's popular ergonomics uses the human scale as its main reference.

Mesolevel: "Meso" refers to the scale between the "macrolevel" and the "microlevel" and is primarily in the millimeter range. In materials science, grains are the typical representation of materials at the mesolevel, where microstructure features are beginning to emerge. These microstructure features include texture, composition segregation, grain boundary effects, porous absorptions, percolation, and catalysis. Today there are many direct and successful applications of research at the mesolevel to the field of materials engineering, including the refractory materials industry and the metallurgical industry among others.

Microlevel: The microlevel scale is mainly in the micrometers. Over the years, research has focused on the behavior and performance of crystalline and amorphous materials, through the use of optical microscopy, electron microscopy, x-ray diffraction analysis, electron probes, and other techniques. In fact, many new methods have now become conventional in materials science.

Nanolevel: The nanolevel scale varies mainly from nanometers to micrometers, which is roughly equivalent to the collective size of a few dozen to several hundred atoms. At the nanolevel, quantum effects begin to emerge. The object of study can no longer be seen as a "continuum" and can no longer be represented simply by the statistical average volume. The effects of microstructure defects and doping are significantly increased. As a result, materials scientists have had to consider many issues that they have not paid attention to in the past. However, the nanolevel scale has also brought many pleasant surprises to materials scientists.

Over the past decade, the interdevelopment of solid mechanics and materials science has deepened the level of research from the macroscopic into the meso- and even the microscopic [11,12]. New microscopy techniques, the development of nanotechnology, and the research and development of nanocomposite materials and new functional materials reveals the following two typical scientific problems [13–15].

Scale problems: How should we conduct the macro- and microscale transition under different scale levels and quantitatively estimate the impact of a material's microstructure on its macroscopic behavior? The development of nanocomposite materials and composite membrane materials revealed these scale problems. However, current micromechanical theory based on the concept of "average" cannot solve such scale problems, and microfracture mechanics has difficulties giving satisfactory results. We see these scale problems in the two-phase interface, in composite membranes, shear bands, nanoparticle reunion, electromagnetic domain, and micro-crack-tip fields, as well as the analysis of highly nonuniform and localized problems with scales similar to or slightly smaller than that of the mesostructure of composite materials.

Group evolution problems: How should we deal with microstructure and defects as they relate to the interaction and evolutionary dynamics reflected in groups? In macromechanical analysis, the problem of scale and group evolution does not exist. In micro- and macroelastic modulus analysis, the averaged integration method can be used to transition from the micro- to the meso- and the macrolevel, so that the microstructure scale and group evolution do not significantly influence the material's stiffness. However, the microstructure scale and group evolution play a vital role in material strength and toughness. Scale effects significantly distinguish the strength and toughness of nanomaterials from conventional materials, which makes the behavior and reliability of MEMS different from conventional electromechanical systems. The effect of group evolution describes the specific extensions, series, convergences, and localized situations and reveals a variety of unique patterns in the fracture process of materials.

In the 1990s, the science of fracture, known by its theory based on the integration of micro- and macroscales, has become the interdisciplinary research frontier in mechanics and materials science. Since the 1980s, the international community has achieved a significant academic breakthrough in micromechanics, interface mechanics, micro- and mesocomputational mechanics, and the fracture mechanism at the atomic level, which has gradually and effectively guided the design of material toughening and the research of new structural materials. However, this emerging discipline is still in the developmental stage. Current theoretical development is largely confined to the single-factor model of material research and quantitative simulations and is still in a transitional phase from meso- to microscales. Actual defects within materials are very complex. Building a quantitative theory to predict and control fracture patterns from the internal structure and defect distribution of materials still requires arduous effort [10]. Since the early 1950s, it has been widely recognized that further development of antifatigue materials and safety design requires an understanding of the basic process of fatigue from the microscopic level, systematically studying the entire fatigue process at all stages.

In the past, the description of the mechanical behavior of materials was based on the framework of macromechanics. Although such indicators of mechanical properties as strength and toughness can be quantified, they cannot express a scientific relationship with the microstructure of materials. The emergence of fracture mechanics resulted from a large number of brittle fracture accidents of warships and aircraft during and after World War II. The birth of solid mesomechanics is due to the development and extensive application of composite materials, a variety of low-dimensional materials, biomaterials, smart materials, high-speed computer technology, and high-precision measuring technology. In recent years, the integration of mechanics and materials science has promoted the study of solid mechanics from the macroscopic scale through the mesoscopic scale, and gradually down to the microscopic scale, linking the meso- and microstructural design of materials from the qualitative to the semiquantitative and quantitative phases. On the other hand, traditional design formulas can no longer meet the requirements demanded by the variety of new materials with complex microstructures and high costs. In order to effectively enhance the strength, toughness, and other indicators of mechanical properties, it is necessary to use a multicomponent design. However, our understanding is currently still in the qualitative phase. We must learn how to design materials; how to conduct a comprehensive analysis of the mechanical model and explore the fracture mechanism; how to carry out local process parameter design; and how to determine the component ratio of microstructure, form distribution, and process control of the required materials.

Over the years, materials and mechanics scientists, chemists, and physicists have explored these issues from different angles and on different levels of micro-, meso-, and macrostructure. The relatively mature method of materials design is limited to the traditional formula-based designs and preliminary designs based on existing experimental patterns and databases. Materials design methods currently being developed include (1) materials design according to the mechanical properties and the quantitative relationship of tissue elements; (2) materials design based on the integration of traditional methods and computational technology, expert systems, and the use of computer simulation; and (3) materials design that calculates the macroscopic properties of materials from the atomic and electronic level. Each of these materials design methods has its advantages and disadvantages, but improving the design of composite materials to a new stage will only occur when these methods are integrated with each other.

IMPORTANT MECHANICAL PROPERTIES OF FUNCTIONAL MATERIALS AND STRUCTURAL MATERIALS

Smart structures and devices are widely used in modern information technology, new materials, and aerospace and increasingly show their great advantages. Functional materials are one of the core areas of new materials research and account for about 85% of global new materials research. In the twenty-first century, new functional materials play an important role in promoting and supporting the areas of information, biology, energy, the environment, space, and other high-tech areas. Smart devices often use piezoelectric/ferroelectric materials, ferromagnetic materials, and shape-memory alloy materials and usually perform under the electrical–mechanical–magnetic–thermal–coupling load environment. The deformation, vibration, and instability of these structural components and structural systems—in external electric/magnetic fields and in large high-temperature gradient environments—must be considered in their design, and the reliability of their performance is of great concern at home and abroad.

The XXI International Congress of Theoretical and Applied Mechanics (ICTAM04) held at the Warsaw University of Technology, Poland, from August 15 to 21, 2004, declared that smart materials and structures are one of the six most important topics in mechanics today [16]. The mechanics of smart materials include the determination of the constitutive relationship of anisotropic smart materials, field analysis of multifield coupling, and the determination of basic mechanical parameters. Here, we use piezoelectric and ferroelectric materials, for example, to illustrate the problems faced by the mechanics of smart materials.

Piezoelectric materials have been widely used as sensors and actuators in intelligent engineering, electronics technology, laser technology, infrared detection technology, ultrasound technology, and other aspects of engineering. The static and dynamic mechanical properties of piezoelectric materials have been extensively studied, and they have shown that the manufacturing and production of piezoelectric materials will produce a variety of defects such as cracks and microholes. These defects have a great impact on the working life of smart structures and components whose core components are piezoelectric materials and may even cause damage to the entire structure. Because of the close relationship between reliability analysis and the design of smart structures and components, and research into the fracture mechanics of piezoelectric media, it is necessary to establish a proper mechanical model to study the fracture behavior of piezoelectric media.

Just like other issues concerning fractured media, there are many aspects to fracture problems within piezoelectric media due to differences in structural composition and exposure to different loads. The most applicable ones are interface fracture and fracture dynamics. In practical use, smart structures or components containing piezoelectric material as the core component are integrated with other materials and do not independently exist in the device. Therefore, the interface (including interface cracking) is an important part of a structure containing piezoelectric materials. The interface is where defects easily happen, resulting in fracture damage [17]. Regardless of the crack's origin, it often encounters the interface during its expansion process. Therefore, the existence of an interface will inevitably affect the overall mechanical behavior of materials in varying degrees. In addition, the fracture dynamics of piezoelectric media have a definite engineering background in the field of ultrasonic testing, microelectronics and communications, nondestructive testing, and weapons guidance. For example, the piezoelectric accelerometer has undergone much relevant research. In these applications, piezoelectric smart structures or components are often used to conduct surface waves or are subjected to dynamic loads. The crack tip at this time will have a high stress concentration and may lead to further crack propagation, causing damage to the overall structure. Consequently, the analysis of piezoelectric media fracture dynamics has become the focus of attention in recent years.

Ferroelectric materials belong to the category of smart materials. Because of their larger electromechanical function conversion rates and better performance controllability, they are widely used in actuators, microsensors, micropositioning devices, and other smart devices. In recent years, along with the major breakthrough in manufacturing technology for ferroelectric thin films, ferroelectric materials have become the new functional materials for information technology and have gained the attention of researchers [18–20]. Statistics from the 1990s indicate that in electronics applications, ferroelectric ceramics accounted for 60% of the entire worldwide market in high-tech ceramics.

With the extensive application of ferroelectric ceramics, reliability has become particularly important. In a weaker electric field, ferroelectric materials can perform a good linear piezoelectricity. However, when the applied electric field near the coercive field is stronger, material nonlinearity will become very obvious. Ferroelectric thin films, made by the polarization flip features of ferroelectric materials, will produce fatigue and lead to failure after several polarization flips. In addition, the ferroelectric material is a very weak ferroelectric ceramic material: its fracture toughness is only about 1 $MPa/\sqrt{m}$. Therefore, in actual use, fracture behavior caused by electric field loads (electrostrictive fracture) often occurs, as well as fatigue crack propagation caused by alternating electric fields (electrostrictive fatigue). These electrostrictive failure problems limit the wider use of ferroelectric ceramics, as well as further enhancement of the performance of microelectronic devices. This has led to an in-depth study of the constitutive relations of ferromagnetic materials, their electrostrictive failure mechanisms, their reliability analysis and prediction, and a corresponding improvement mechanism [21].

CONTENT OVERVIEW

The content of this book is based on the intersecting of materials science and solid mechanics, and consistently reflects three features: (1) the integration of science (basic theory) and engineering (engineering application), (2) the integration of macro- and microscales, and (3) the integration of materials structure and function.

1. The book's first module deals with basic theory, focusing specifically on the fundamentals of elastoplastic mechanics and micro- and macrofracture mechanics. The second module, from Chapters 3 through 6, stresses engineering application and introduces the conventional mechanical properties of materials and their representation methods, such as basic mechanical properties, hardness, fracture toughness, and residual stress.
2. Micro- and macrointegration: This book emphasizes the macroscopic properties of metal and nonmetal materials, such as the basic mechanical properties of metals, hardness, fracture toughness, fatigue, and creep, as well as the viscoelasticity of polymer materials and ceramic materials. It also covers micromechanical properties such as microscopic damage mechanics analysis, strain gradient theory, scale effects, cross-scale numerical simulation, molecular dynamics simulation of crack tip, and thin-film mechanics.
3. Integration of materials structure and function: This book not only emphasizes the analysis of the mechanical properties of structural materials but also describes the analysis of the mechanical properties of functional materials such as smart materials, thin films, and coating materials.

It includes the latest national and international scientific research results such as strain gradient theory, scale effect, cross-scale numerical simulation, analysis of microfracture mechanics, mechanical properties of smart materials, mechanical properties of thin films, and coating materials. It also contains a series of more than 200 reference articles from the 1990s onward.

The book is divided into four modules and consists of a total of 14 chapters. The first module introduces the study of required fundamental theories of micro- and macromechanical properties of materials, including the basic theory of elastoplasticity in Chapter 1 and the fundamentals of micro- and macrofracture mechanics in Chapter 2. The second module, from Chapters 3 through 6, is written primarily from the engineering point of view and specifically introduces the conventional mechanical properties of materials and their representation methods, such as basic mechanical properties, hardness, fracture toughness, and residual stress. The third module is contained in Chapters 7 and 8, which mainly introduces the mechanical properties of metallic materials most widely used in engineering, specifically including metal fatigue, creep, fatigue and creep interaction, and the mechanical properties of metallic materials in environmental media. The fourth module, from Chapters 9 through 14, is the thematic section. It first introduces the application of computers to the analysis of material properties—the micro- and macrocomputational mechanics of materials—and then focuses on the micro- and macromechanical properties of special materials, including smart materials, thin films, polymers, ceramics, and composite materials. Readers may choose chapters according to their own interest, but we recommend reading through the entirety of the first and second modules.

In order to help readers grasp important concepts, research means, and research methods, there are a relatively large number of exercises, problems, and questions at the end of each chapter. Some of these exercises have standard answers according to the current level of technology, but a considerable number of exercises do not. Some exercises concern an important research topic or direction. Some others, especially the fundamental ones in Chapters 1 and 2, require the reader to repeatedly practice the exercise. Readers should also design exercises among themselves and discuss them with each other. Additionally, the book comes with experimental guidance, including must-do experiments and other selected experiments. The experimental instructions are published by Xiangtan University Press.

REFERENCES

1. Feng D., Shi C. X. and Liu Z. G. *Introduction to Materials Science*. Beijing: Chemical Industry Press, 2002.
2. Tan Y. and Li J. F. *New Materials Outline*. Beijing: Metallurgical Industry Press, 2004.
3. Briggs A. *The Science of New Materials*. Oxford: Blackwell, 1992.
4. Shi D. K. *Material Science Foundation*. Beijing: China Machine Press, 1999.
5. Zhou Y. *Mechanics of Solid Materials* (Vol. 1). Beijing: Science Press, 2005.
6. Freund L. B. and Suresh S. *Thin Film Materials, Stress, Defect Formation and Surface Evolution*. Cambridge: Cambridge University Press, 2003.
7. Raabe D. *Computational Materials Science*. Beijing: Chemical Industry Press, 2002.
8. Zheng Z. M. *Collected Works of Zheng Zhemin*. Beijing: Science Press, 2004: 562–569, 758–766, 809–819.
9. Fang D. N., Zhong Z. and Zhou Y. Challenge of solid mechanics in the development of materials science—The introduction of the 96th forum for young scientists in the Chinese Association of Science and Technology. *Adv Mech*, 2005, 35(3): 461–463.
10. Kassner M. E., Nemat-Nasser S., Suo Z. et al. New directions in mechanics, *Mech Mater*, 2005, 37: 231–259.
11. Huang K. Z. and Xiao J. M. *Damage and Fracture Mehanisms of Materials and Macro-micro-mechanics*. Beijing: Tsinghua University Press, 1999.
12. Huang K. Z. and Wang Z. Q. *Macro-micro-mechanics and Strengthening and Toughening Design of Materials*. Beijing: Tsinghua University Press, 2003.
13. Fan J. H. *Multiscale Analysis for Deformation and Failure of Materials*. Beijing: Science Press, 2008.
14. Fan J. H. and Chen H. B. *Advances in Heterogeneous Material Mechanics*. Lancaster, PA: DEStech Publications, Inc, 2008.
15. Uchic M. D., Dimiduk D. M., Florando J. N. et al. Sample dimensions influence strength and crystal plasticity. *Science*, 2004, 305: 986–989.
16. Gutkowski W. and Kowalewski T. *21st International Congress of Theoretical and Applied Mechanics*. ICTAM04, Warsaw, Poland, 2004.
17. Zheng X. J. and Zhou Y. C. Nano-indentation fracture test of $Pb(Zr_{0.52}Ti_{0.48})O_3$ ferroelectric thin films. *Acta Mater*, 2003, 51(14): 3985–3997.
18. Zhong X. L., Wang J. B., Zheng X. J. et al. Structure evolution and ferroelectric and dielectric properties of $Bi_{3.5}Nd_{0.5}Ti_3O_{12}$ thin films under a moderate temperature annealing. *Appl Phys Lett*, 2004, 85(23): 5661–5663.
19. Zhang J., Tang M. H., Tang J. X. et al. Bilayer model of polarization offset of compositionally graded ferroelectric thin films. *Appl Phys Lett*, 2007, 91(16): 162908.
20. Ye Z., Tang M. H., Zhou Y. C. et al. Electrical properties of V-doped $Bi_{3.15}Nd_{0.85}Ti_3O_{12}$ thin films with different contents. *Appl Phys Lett*, 2007, 90(8): 082905.
21. Yang W. *Mechatronic Reliability*. Beijing: Tsinghua University Press, 2001.

Authors

Yichun Zhou is a professor of mechanics. He received his BS from Xiangtan University in 1985, his MS from the National University of Defense Technology in 1988, and his PhD from the Institute of Mechanics, Chinese Academy of Sciences, in 1994. He currently serves as a vice president at Xiangtan University. He also serves as the editor of *Journal of Materials Science and Technology*, *Transactions of Nonferrous Metals Society of China*, *Materials Focus*, and *Energy Focus*.

Dr. Zhou's research focuses on the synergy of functional coating and films and analytical methods for the evaluation of physical and mechanical properties. His research has been funded by the Distinguished Young Scientists Award from the National Natural Science Foundation of China (NSFC) in 2005, the Major Program Award from NSFC in 2006 and 2011, the Pilot Project of the State High-Tech Development Plan (863 Program) in 2003, and the Major Science and Technology Program Award of Hunan Province in 2009. Dr. Zhou has won the first-place award of the Natural Science Prize of Hunan Province in 2011, the second-place award of the Science and Technology Progress Prize of Hunan Province in 2003, and the first-place award of the Science and Technology Progress Prize of the Chinese Academy of Sciences for his excellence in research in 1996. He has also won China's Distinguished Young Scientists Award as well as the National Famous Teacher Award in 2005. He has been granted 20 patents in China and has published more than 150 scientific papers in major international academic journals, such as *Applied Physics Letters* and *Acta Materialia*.

Dr. Zhou's undergraduate and graduate course titled "Micro- and Macro-Mechanical Properties of Materials" won the National Excellence Higher Education Course Award in 2005. This course, which was published as a textbook in Chinese by Higher Education Press, was selected as one of the national level textbooks in the Eleventh Five-Year Plan in 2006. His book *Solid Mechanics of Materials* (Volumes I and II), published by China Science Publishing, won the Excellent Graduate Course Award of Hunan Province in 2003. His team also won the second-place award in the National Teaching Excellence Prize in 2005 and the first-place award in the Teaching Excellence Prize of Hunan Province in 2004.

Li Yang is an associate professor at Xiangtan University. She received her BS in mechanical engineering from Xiangtan University in 2002 and her PhD in materials science and technology from the same university in 2007. She has been on the faculty of Materials, Optoelectronic and Physics of Xiangtan University as a PhD advisor since 2007. Her research focuses on thermal barrier coatings (TBCs), including (1) developing acoustic emission nondestructive testing methods, (2) designing simulation experiments, and (3) studying failure mechanisms of oxidation, erosion, and CMAS corrosion of TBCs. She has published more than 30 peer-reviewed articles in leading materials science journals, including *Acta Materialia* and *Applied Physics Letters*. She is also the owner and co-owner of 12 Chinese patents and the author of two books.

Yongli Huang received her BS in materials physics in 2003, her MS in materials physics and chemistry in 2006, and her PhD in materials science and engineering in 2013, all from Xiangtan University. Her research focuses on low-dimensional materials, with an emphasis on macromechanical properties and microrelaxation dynamics. Her research is currently supported by four grants from the National Natural Science Foundation of China. She has published ten papers in high-impact journals. Since 2006, as a junior faculty at Xiangtan University, Yongli has been involved in teaching "Micro- and Macro-Mechanical Properties of Materials," a course nationally recognized for excellence, as well as "Solid Mechanics in Materials," a Hunan provincial graduate course also noted for its excellence. She received the Excellence in Undergraduate Program Teaching Award in 2012 and is a very popular teacher in the classroom.

Editorial Committee

1 Fundamentals of Elasto-Plastic Mechanics

A material or structural component does not exist in isolation. It is subjected to the action of surrounding objects, known as the load. How is a load represented mathematically? How is the load or force represented inside the body of a material when it acts on the material? What is the effect of this force? The effect mentioned here is the effect inside the body, that is, the deformation of the object, not the movement of the object as a whole. Our concern here is with a deformable object, not a rigid object. This chapter is intended to illustrate, by using tensors as a mathematical tool, the fundamental concepts and physics ideas of stress, displacement, deformation, strain, and the stress–strain constitutive relationship under the assumption of a continuous medium.

1.1 PREREQUISITES

1.1.1 Objectives of Elasto-Plastic Mechanics

Elasto-plastic mechanics is a branch of solid mechanics, which studies the elasto-plastic deformation and stress and strain states of a deformable solid object subjected to an externally applied load. Elastic mechanics discusses the mechanics of ideal elastic solids during elastic deformation, while plastic mechanics discusses the mechanics of solid materials during plastic deformation.

As an independent branch of solid mechanics, elasto-plastic mechanics has a set of traditional theories and approaches, as well as extensive applications, in many fields of engineering technology. Currently, with the development of modern science and technology, elasto-plastic mechanics is confronted with a series of new demands.

The problems studied by elasto-plastic mechanics are generally the same as those of material mechanics and structural mechanics. However, material mechanics and structural mechanics adopt simplified mathematical models, which can be described by primary theories, while elasto-plastic mechanics employs more accurate mathematical models. Some engineering problems—such as the torsion of a cylinder with noncircular transverse section or stress concentration around the edge of a hole—are not solvable using the theories of material mechanics and structural mechanics. However, these problems can be solved by elasto-plastic mechanics. In other words, some problems can be solved by the approaches of material mechanics and structural mechanics, but the underlying primary theories are defective, and the solutions are not accurate. Elasto-plastic mechanics can evaluate the reliability and accuracy of the solutions obtained by these primary theories.

Elasto-plastic mechanics has two tasks: (1) to construct the theories and approaches to problems that are not solvable in material mechanics and structural mechanics and (2) to evaluate the reliability and accuracy of the solutions obtained by the primary theories.

1.1.2 Fundamental Assumptions of Elasto-Plastic Mechanics

Like material mechanics, elasto-plastic mechanics is also a branch of solid mechanics and makes the same fundamental assumptions concerning deformable solids [1–4].

1.1.2.1 Continuity

Continuity assumes that the matter in the solid object completely fills the space it occupies, leaving no gaps. In fact, matter is made of atoms with empty spaces between neighboring atoms and so is not continuous. However, the size of these empty spaces is minutely small in dimension as compared to the component dimensions and can be ignored. Accordingly, a solid is assumed to be continuous in the whole volume it occupies. When all physical quantities are represented as functions of the coordinates of material points in a solid, infinitesimal analysis with respect to the coordinate increment can be carried out.

1.1.2.2 Homogeneity

According to this assumption, all locations in a solid have identical mechanical properties. As for metals, the mechanical properties of individual crystalline grains that a metal is composed of are not completely identical. However, any arbitrary part of it contains numerous randomly oriented grains, and its mechanical properties are taken to be the statistically averaged values of those of a larger number of grains. As a result, the mechanical properties at any location are homogeneous. In this way, any part taken from a solid, regardless of its position and dimensions, possesses identical mechanical properties.

1.1.2.3 Isotropy

According to this assumption, the mechanical properties of a solid are the same in all directions. For a single crystal in a metal, the properties in different directions are not completely identical. Because the metallic components consist of numerous randomly oriented crystalline grains, the properties in all directions become identical. Materials with such an attribute are called isotropic materials.

In addition, a geometric assumption is made: the deformation of the object is small. In other words, the deformation caused by an applied external load is very small compared to the geometric dimensions of the object. Thus, dimension changes due to deformation can be disregarded. In this way, the geometric dimensions after deformation can be taken to be the ones before deformation. Moreover, the second-order infinitesimal quantities in the expressions for deformation and displacement can be ignored, and the geometric deformation can be expressed as linear functions.

1.1.3 Elasticity and Plasticity

Solid materials deform when subjected to external forces. The deformation before failure generally consists of two stages: the elastic deformation stage and the plastic deformation stage. Depending on material characteristics, some materials exhibit an obvious elastic deformation stage, but an unremarkable plastic deformation stage. Like most brittle materials, fracture occurs immediately after the end of elastic deformation. For materials such as concretes, the elastic stage is unremarkable, while plastic deformation occurs at the very beginning of deformation, and elastic deformation and plastic deformation are always coupled. However, most solid materials display observable elastic and plastic deformation stages. This book focuses on solid materials with apparent elastic and plastic deformation. These materials are generally called elasto-plastic materials.

According to material mechanics, elastic deformation is completely recoverable after the load is removed. Plastic deformation is the residual deformation and cannot recover after unloading. Elasticity and plasticity of solid materials can be illustrated using simple tensile testing. Figure 1.1 shows the well-known stress–strain curve of mild steels under simple tension. The stress at point A, σ_A, is called the proportional limit, and segment OA is a straight line. The stress at point B, σ_0, is called the elastic limit, indicating termination of the elastic deformation stage and the start of the plastic deformation stage. The elastic limit is also called the yield limit. After the stress goes beyond σ_A, the stress–strain curve is no longer a straight line, but the deformation is still elastic. If unloading occurs before the stress reaches point B (σ_0), the stress–strain relationship will be

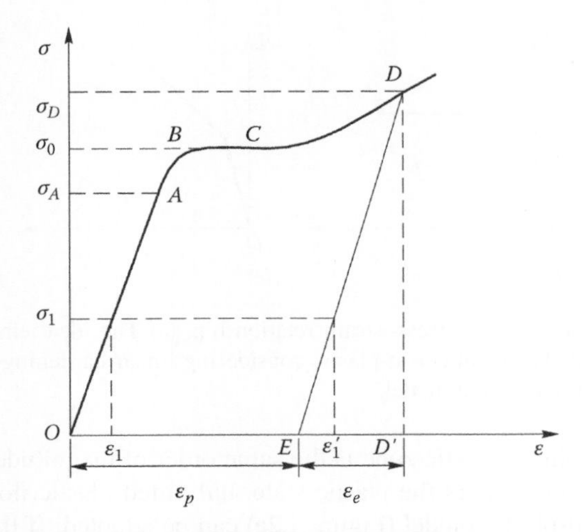

FIGURE 1.1 Stress–strain curve of a mild steel under simple tensile testing.

recovered along the reverse loading path to the original state. σ_0 is called the yield stress. Before the stress reaches yield stress, the material experiences a linear elastic stage (segment *OA*) and a nonlinear elastic stage (segment *AB*). If unloading occurs after the stress goes beyond the yield stress, the stress–strain relationship does not follow the reverse loading path to the original state, and permanent plastic strain remains. Segment *BC* is called the plastic plateau, during which deformation continues at a constant stress and which is generally called plastic flow.

If unloading occurs when the stress reaches σ_D, the stress–strain relationship follows the path *DE* from point *D* to point *E*. Segment *OE* corresponds to plastic deformation, and segment *ED'* is elastic deformation. The total strain ε is divided into two parts: elastic strain ε_e and plastic strain ε_p. Therefore,

$$\varepsilon = \varepsilon_e + \varepsilon_p. \tag{1.1}$$

If the material is unloaded at point *D* and then reloaded, the material exhibits elastic deformation before the stress reaches σ_D and plastic deformation after the stress goes beyond σ_D. This means that the yield stress is enhanced. When stress goes beyond the elastic limit, resistance of the material to deformation is improved. This behavior is called hardening.

In summary, elastic deformation is reversible, the energy stored during deformation can be completely released after unloading, the deformation can be completely resumed to the original state, and one-to-one mapping exists between stress and strain. In the elasto-plastic stage, the strain is not recoverable, and one-to-one mapping between stress and strain does not hold. The strain value depends on the loading history.

Simplification is generally adopted in the theory of plasticity, due to the complexity of elasto-plastic problems. Figure 1.2 illustrates four simplified models. Figure 1.2a represents the ideal elasto-plastic model, Figure 1.2b shows the ideal rigid plastic model, Figure 1.2c is the ideal elasto-plastic model considering linear hardening, and Figure 1.2d shows the ideal rigid plasticity considering linear hardening. These models are represented by the simplified stress–strain curves (Figure 1.1) in terms of the specific features of the problems. For mild steels, hardening does not occur even when the total strain is approximately 10–20 times larger than the elastic strain, so they are generally taken as ideal plastic materials. Although elastic deformation and plastic deformation are at the same order of magnitude, simplified models can still be adopted in the limit equilibrium analysis. For example, when a thick walled cylinder is subjected to an internal pressure, the plastic zone propagates outward from the inner wall, forming an elasto-plastic object with a plastically deformed inner region and an elastically deformed outer region. Due to the constraint of the outer elastic

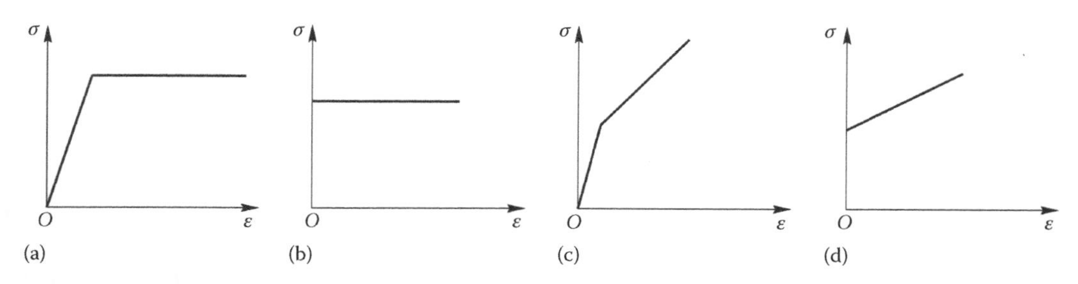

FIGURE 1.2 Simplified models for stress–strain relationship. (a) The ideal elasto-plastic model. (b) The ideal rigid plastic model. (c) The ideal elasto-plastic considering linear hardening model. (d) The ideal rigid plasticity considering linear hardening model.

zone, deformation of the inner plastic zone is the same order of magnitude as the elastic deformation. Once the entire section enters the plastic state, unlimited plastic flow becomes possible. In this case, the ideal elasto-plastic model (Figure 1.2a) can be adopted. If the plastic deformation is not constrained, then elastic deformation can be ignored. In this case, the ideal rigid plastic model (Figure 1.2b) is appropriate. Two simplified models given in Figure 1.2c and d result from the two previous cases, taking linear hardening into consideration and ignoring plastic flow.

1.1.4 Tensor and Summation Conventions [1,2,5]

1.1.4.1 Tensors

Einstein said that any theory that cannot be described by a mathematical language is not yet a science. The state of a nonisolated material must be described by a mathematical language. Therefore, the state of a material is described by a series of physical quantities. Because of the number and variety of these quantities, it is easy to become confused. However, if the physical quantities are expressed by a mathematical language, the problem is very easy. After careful analysis, it is discovered that some physical quantities—such as mass, density, volume, and the kinetic energy of an object—can be completely described by one value. A physical quantity having only magnitude without direction is called a scalar. However, some physical quantities, such as velocity, acceleration, and force, cannot be represented by a single value. They not only have magnitude, but also direction. We have to set up a coordinate system to describe these quantities. For example, as shown in Figure 1.3, a description of the geometric position of point A in a space requires three independent coordinates (x, y, z) in a reference system.

Suppose point A moves to point A' under a force. The displacement has three components u, v, w, respectively, along the x-, y-, and z-directions. Therefore, to represent the displacement of point A requires

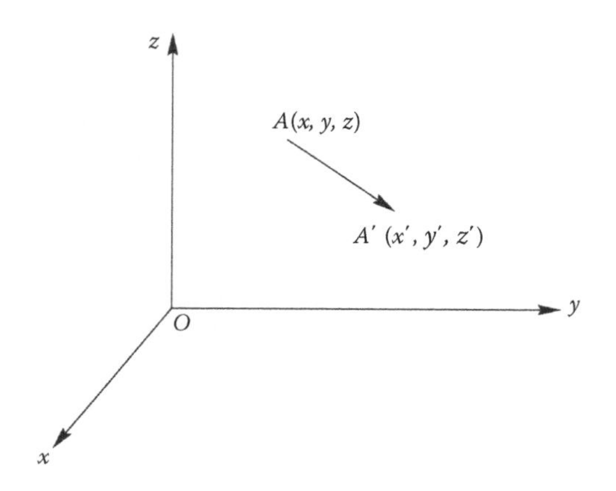

FIGURE 1.3 Vector.

three independent physical quantities (u, v, w) such that $\boldsymbol{u} = \sum_{i=1}^{3} u_i \boldsymbol{e}_i = u\boldsymbol{e}_1 + v\boldsymbol{e}_2 + w\boldsymbol{e}_3$. Such a physical quantity is a set of three independent quantities, called a vector or a first-order tensor. A first-order tensor is a physical quantity with magnitude and direction and is generally represented by a bold font $\boldsymbol{u}$ or by its components (u, v, w).

Some physical quantities cannot be represented by only three components, however. Mathematicians and physicists have found that the number of components is not arbitrarily determined, but follows a rule. By this rule, the number of components is just 3^n. (Ask yourself why this is so. Why is the rule not 4^n or 5^n or the nth power of other values?) In elasto-plastic mechanics, some physical quantities, such as stress (discussed in Section 1.2) and strain (discussed in Section 1.3), are a set of nine independent physical components, represented like this:

$$\begin{bmatrix} \sigma_{11} & \sigma_{12} & \sigma_{13} \\ \sigma_{21} & \sigma_{22} & \sigma_{23} \\ \sigma_{31} & \sigma_{32} & \sigma_{33} \end{bmatrix}.$$

Such physical quantities are called second-order tensors. A second-order tensor maps to a symmetrical 3×3 matrix. Accordingly, the nth-order tensor is a set of 3^n components.

Tensors are generally denoted by indices (or subscripts). The coordinate of point A (x, y, z) can be denoted as x_i ($i = 1, 2, 3$).

A stress tensor

$$\begin{bmatrix} \sigma_{11} & \sigma_{12} & \sigma_{13} \\ \sigma_{21} & \sigma_{22} & \sigma_{23} \\ \sigma_{31} & \sigma_{32} & \sigma_{33} \end{bmatrix}$$

can be denoted as σ_{ij} ($i = 1, 2, 3$; $j = 1, 2, 3$).

And so the number of indices is 1 for a first-order tensor and 2 for a second-order tensor. Accordingly, the number of indices for an nth-order tensor is n. The nth-order tensor can be denoted as $a_{i_1 i_2 \cdots i_n}$ ($i_1 = 1, 2, 3$; $i_2 = 1, 2, 3$; ……; $i_n = 1, 2, 3$).

1.1.4.1.1 Einstein Summation Convention

Only tensors with the same order can carry out addition or subtraction operations. This is similar to addition and subtraction of matrices in linear algebra. Suppose there are two first-order tensors $\boldsymbol{a} = \sum_{i=1}^{3} a_i \boldsymbol{e}_i$ and $\boldsymbol{b} = \sum_{i=1}^{3} b_i \boldsymbol{e}_i$.

Their summation is then

$$\boldsymbol{c} = \sum_{i=1}^{3} c_i \boldsymbol{e}_i = \boldsymbol{a} + \boldsymbol{b}, \quad \text{or} \quad \boldsymbol{c} = \sum_{i=1}^{3} (a_i \boldsymbol{e}_i + b_i \boldsymbol{e}_i) = \sum_{i=1}^{3} (a_i + b_i) \boldsymbol{e}_i.$$

Suppose we have two second-order tensors $\boldsymbol{A} = \sum_{i=1}^{3} \sum_{j=1}^{3} A_{ij} \boldsymbol{e}_i \boldsymbol{e}_j$ and $\boldsymbol{B} = \sum_{i=1}^{3} \sum_{j=1}^{3} B_{ij} \boldsymbol{e}_i \boldsymbol{e}_j$.

Their summation or difference is a new second-order tensor, $\boldsymbol{T} = \sum_{i=1}^{3} \sum_{j=1}^{3} T_{ij} \boldsymbol{e}_i \boldsymbol{e}_j$:

$$\boldsymbol{T} = \boldsymbol{A} \pm \boldsymbol{B} \tag{1.2a}$$

and their components can be expressed as

$$T_{ij} = A_{ij} \pm B_{ij}. \tag{1.2b}$$

The first-order tensor $\boldsymbol{a} = \sum_{i=1}^{3} a_i \boldsymbol{e}_i$ has basis vector $\boldsymbol{e}_i$. Similarly, for the second-order tensor in Equation 1.2a, $\boldsymbol{e}_i\boldsymbol{e}_j$ are called basis tensors or tensor elements. Each basis tensor is a simple concatenation of two basis vectors parallel to coordinate axes, and does not perform any operation, functioning as a unit.

We shall introduce the following Einstein summation convention: when a term of a tensor is expressed by means of indices, the repetition of an index denotes a summation with respect to that index over 1–3. The repeated index (dummy index) is the index for summation, and the summation symbol $\sum$ is omitted. For example, $\boldsymbol{a} = \sum_{i=1}^{3} a_i \boldsymbol{e}_i$ can be simply written as $\boldsymbol{a} = a_i \boldsymbol{e}_i$ and $\boldsymbol{T} = \sum_{i=1}^{3} \sum_{j=1}^{3} T_{ij} \boldsymbol{e}_i \boldsymbol{e}_j$ can be written as $\boldsymbol{T} = T_{ij} \boldsymbol{e}_i \boldsymbol{e}_j$.

And so we can write $\varepsilon_{ii} = \varepsilon_{11} + \varepsilon_{22} + \varepsilon_{33}$, $a_i b_i = a_1 b_1 + a_2 b_2 + a_3 b_3$.

The index that is not repeated in a term is called a free index and can be an arbitrary value from 1 to 3. For example, σ_{ij} and ε_{ij} denote any one of the nine components of stress and strain tensors, respectively.

There are two basic symbols in tensor analysis: the Kronecker symbol δ_{ij} and the permutation symbol (or alternating symbol) e_{rst}. Symbol δ_{ij} is defined as follows

$$\delta_{ij} = \begin{cases} 1, & i = j \\ 0, & i \neq j \end{cases}.$$

δ_{ij} represents nine quantities, but only three of them are not equal to zero. This attribute is very helpful in tensor calculus. For example, $a_i\delta_{ij} = a_1\delta_{1j} + a_2\delta_{2j} + a_3\delta_{3j} = a_j$ (a_1, a_2, or a_3). Symbol e_{rst} is defined as follows

$$e_{rst} = \begin{cases} 1 & \text{when } (r,s,t) = (1,2,3) \text{ or } (2,3,1) \text{ or } (3,1,2) \\ -1 & \text{when } (r,s,t) = (3,2,1) \text{ or } (2,1,3) \text{ or } (1,3,2) \\ 0 & \text{when any two indices are equal} \end{cases}$$

or

$$e_{rst} = \frac{1}{2}(r-s)(s-t)(t-r) \quad (r,s,t = 1,2,3).$$

The definition indicates that e_{rst} contains 27 elements. Three elements with an even permutation of (1, 2, 3) are 1, three elements with an odd permutation are −1, and the remaining elements with repeated indices are 0.

Finally, we discuss the derivative of a tensor. Each component of a tensor is a function of coordinates x_i. The derivative of a tensor is the derivative of each component with respect to the coordinates. In the Cartesian rectangular coordinate system, the derivative of a tensor is still a tensor, but the order is increased by one. For example, for a first-order tensor (vector) V_i, its derivative $\partial V_i/\partial x_j$ is a second-order tensor. The calculation of the derivative of a tensor is the same as for the derivative of a common function.

As a general convention, first-order tensors, second-order tensors, and tensors of order higher than two are represented with a bold font. This book also follows this convention. Although it is very convenient to use tensors to represent the state of a material subjected to external load, it is a little difficult for beginners. In this book, second-order tensors, and tensors of order higher than two, are represented not only by bold font but also by components, so that readers can comparatively study two notations. If readers do not get accustomed to the bold notation at the beginning, they can concentrate on the component notation. Not all first-order tensors (vectors) are written as components, because every undergraduate in science and technology has learned and mastered vectors.

1.2 STRESS

1.2.1 External Forces and Stresses

1.2.1.1 External Forces

There are two external forces acting on a body: body forces and surface forces. Body forces act on the volume elements inside the body and are represented by gravity, magnetic force, and the inertial force of a moving object. The body force is proportional to the mass of the body. However, the force at each point inside a body is usually different. In order to represent the body force at a certain point P, draw a small volume surrounding point P whose volume is ΔV, as shown in Figure 1.4a.

Assume that the body force acting on ΔV is $\Delta \boldsymbol{Q}$. The average body force density is $\Delta \boldsymbol{Q}/\Delta V$. If the surrounding volume shrinks gradually, that is, if ΔV decreases constantly, $\Delta \boldsymbol{Q}$ and $\Delta \boldsymbol{Q}/\Delta V$ (including their directions and magnitudes) change continuously and so does the acting point. Now, let ΔV decrease indefinitely to approach point P, and assume that the body force is continuously distributed. Then $\Delta \boldsymbol{Q}/\Delta V$ tends to a definite limiting value $\boldsymbol{F}$:

$$\lim_{\Delta V \to 0} \frac{\Delta \boldsymbol{Q}}{\Delta V} = \boldsymbol{F}. \tag{1.3}$$

This limiting vector $\boldsymbol{F} = F_i \boldsymbol{e}_i$ is the body force density at point P. Since ΔV is a scalar, the direction of $\boldsymbol{F}$ is the same as the limiting direction of $\Delta \boldsymbol{Q}$. The projections of vector $\boldsymbol{F}$ along coordinate axes x, y, and z are F_1, F_2, and F_3, respectively, are called the body force components of the object at point P. The component is positive if it points to the positive direction of the axis and negative if it points to the negative direction of the axis. Its dimension is [force][length]$^{-3}$.

Surface forces are forces acting on a free body at its bounding surface, such as fluid pressure and contact force. The surface force at each point on the surface of the object is usually different.

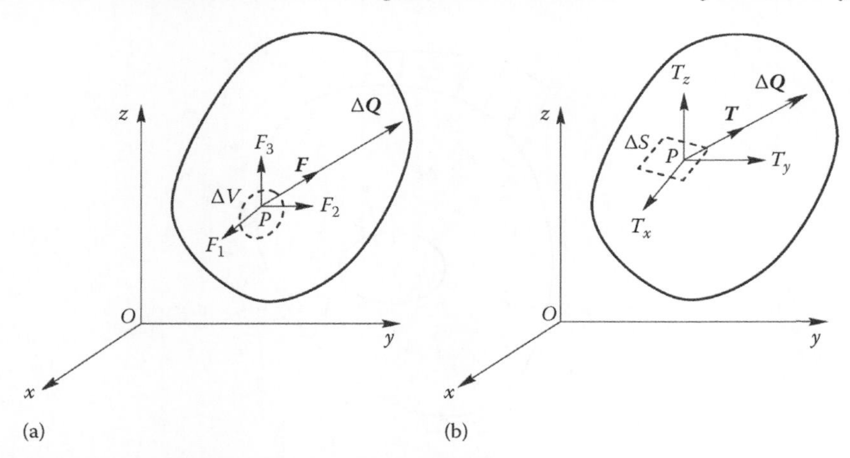

FIGURE 1.4 External force. (a) Body force. (b) Surface force.

In order to represent the surface force at point P on the surface of the object, draw a small area surrounding point P whose area is ΔS, as shown in Figure 1.4b. Suppose the surface force acting on ΔS is $\Delta \boldsymbol{Q}$; then the average density of the surface force is $\Delta \boldsymbol{Q}/\Delta S$. As in the previous case, let ΔS shrink indefinitely to approach point P, and assume that the surface forces are distributed continuously. Then $\Delta \boldsymbol{Q}/\Delta S$ tends to a definite limiting value $\boldsymbol{T}$:

$$\lim_{\Delta S \to 0} \frac{\Delta \boldsymbol{Q}}{\Delta S} = \boldsymbol{T}. \tag{1.4}$$

This limiting vector $\boldsymbol{T} = T_i \boldsymbol{e}_i$ is the surface force density acting on the object at point P. Since ΔS is a scalar, the direction of T is identical to the limiting direction of $\Delta \boldsymbol{Q}$. The projections of vector $\boldsymbol{T}$ along coordinate axes x, y, and z are T_x, T_y, and T_z, respectively, are called the surface force components of the object at point P. The component is positive if it points to the positive direction of the axis and is negative if it points to the negative direction of the axis. Their dimension is [force] [length]$^{-2}$.

1.2.1.2 Stresses

Under an external load, a body deforms and the spacing between molecules changes, resulting in an additional internal force field inside the body. When the internal force field balances the external load, the deformation stops and the body reaches an equilibrium state, as shown in Figure 1.5. Now let us discuss the additional internal force field caused by the external load.

In order to accurately describe the internal force field, Cauchy introduced the concept of stress. Consider a configuration occupied by a body B in an equilibrium state (Figure 1.5). Imagine a closed surface S separating the body into a part exterior to this surface and a part interior to the surface, called the exterior domain and the interior domain, respectively. Point P is an arbitrary point on surface S. Consider a surface element with the area ΔS and with point P as the centroid on the surface S. $\boldsymbol{\nu}$ is the unit vector normal to ΔS with its direction outward from the interior of S, and $-\boldsymbol{\nu}$ is the unit vector normal to ΔS with its direction outward from the exterior of S. $\Delta \boldsymbol{F}$ is the resultant force of the exterior domain acting through the surface element ΔS on the interior domain, and its direction is usually different from the normal vector $\boldsymbol{\nu}$. Assume that as the surface element shrinks down to point P, that is, as $\Delta S \to 0$, the ratio $\Delta \boldsymbol{F}/\Delta S$ tends to a definite limit and that the ratio of the moment of the forces acting on the surface ΔS about any point within the area to ΔS tends to zero. We define

$$\sigma_\nu = \lim_{\Delta S \to 0} \frac{\Delta \boldsymbol{F}}{\Delta S} \tag{1.5}$$

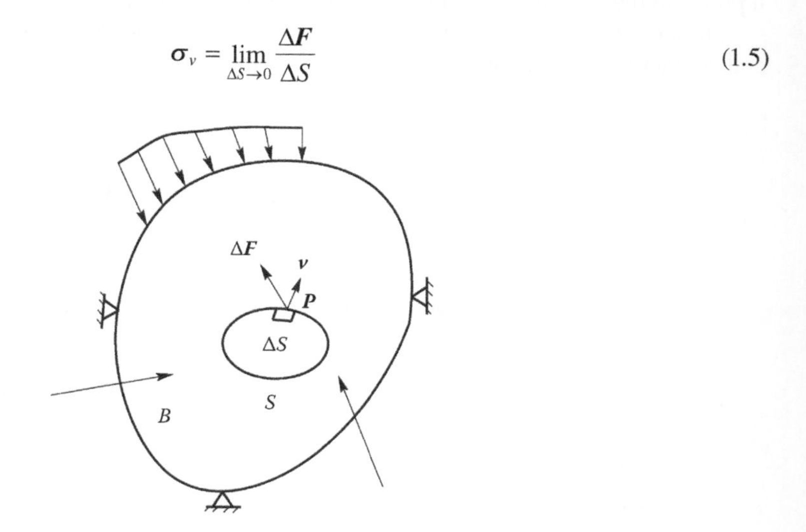

FIGURE 1.5 Stress.

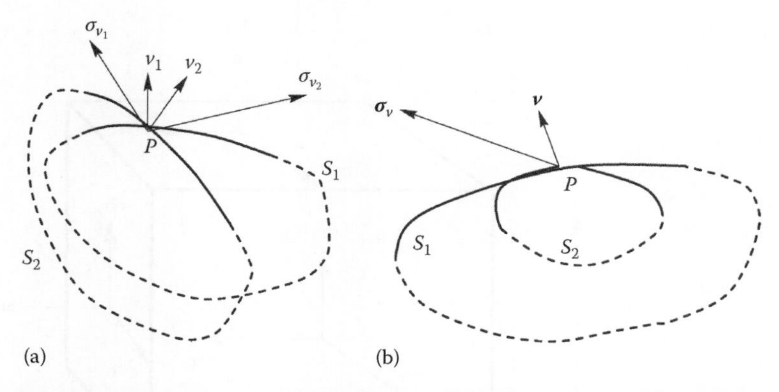

FIGURE 1.6 Dependence of stress on normal vector. (a) Different stress vectors on different surface elements with respective normal vectors at one point. (b) The same stress vector on different surface elements with the same normal vectors at one point.

as the stress vector of the exterior domain acting through the surface element on the interior domain at point P with normal vector $\boldsymbol{\nu}$. It is also the stress vector of the interior domain acting on the surface element at point P, with normal vector $\boldsymbol{\nu}$ exerted by the exterior domain. If ΔS is the original area before deformation, Equation 1.5 gives the engineering stress, or nominal stress, which is generally applied to a small deformation. For a large deformation, ΔS should take the real area of the surface element after deformation, and the corresponding stress is called the true stress. This book only discusses the small deformation, and so the shape change before and after deformation is relatively small. For large deformations, readers can refer to the appropriate references [6].

Compare Equations 1.4 and 1.5. The mathematical definitions and physical dimensions of the stress vector and the surface force vector are identical. The only difference is that the stress is an unknown internal force acting on the interior section of a body, while the surface force is a known external force acting on the exterior surface. When the interior section tends to the exterior surface, the stress also tends to the surface force. The magnitude and direction of vector $\boldsymbol{\sigma}_\nu$ depends not only on the position of point P, but also on the normal vector $\boldsymbol{\nu}$ of the surface element. The stress vectors acting on the surface element at the same point but with different normal vectors are different, as shown in Figure 1.6a. The stress vectors should be the same for the surface elements on different surfaces, as long as they pass through the same point with the same normal direction, as shown in Figure 1.6b. Therefore, the stress vector $\boldsymbol{\sigma}_\nu$ is a function of position $\mathbf{r}$ and the normal vector $\boldsymbol{\nu}$ of a surface passing through point P:

$$\sigma_\nu = \sigma_\nu(\mathbf{r}, \boldsymbol{\nu}). \tag{1.6}$$

Apparently, as long as the stress vectors on the sections passing through point P (with arbitrary directions) are known, the stress state at point P can be determined. However, the sections passing through point P with different orientations are indefinite, and it is impossible to consider them one by one. In this case, how can the stress state at a point be determined?

1.2.1.3 Stress Component

Accordingly, Equation 1.6 is sufficient to describe the stress. However, because the sections passing through point P with different orientations are indefinite, and it is impossible to consider them one by one, Equation 1.6 is obviously not convenient. As we have illustrated in the previous section, we must use a reference coordinate system (such as the Cartesian coordinate system) to discuss the stress state at any point in a body. In the Cartesian coordinate system, a parallelepiped with six planes (or surface elements) parallel to the coordinate planes (called normal cross sections) is taken out in the

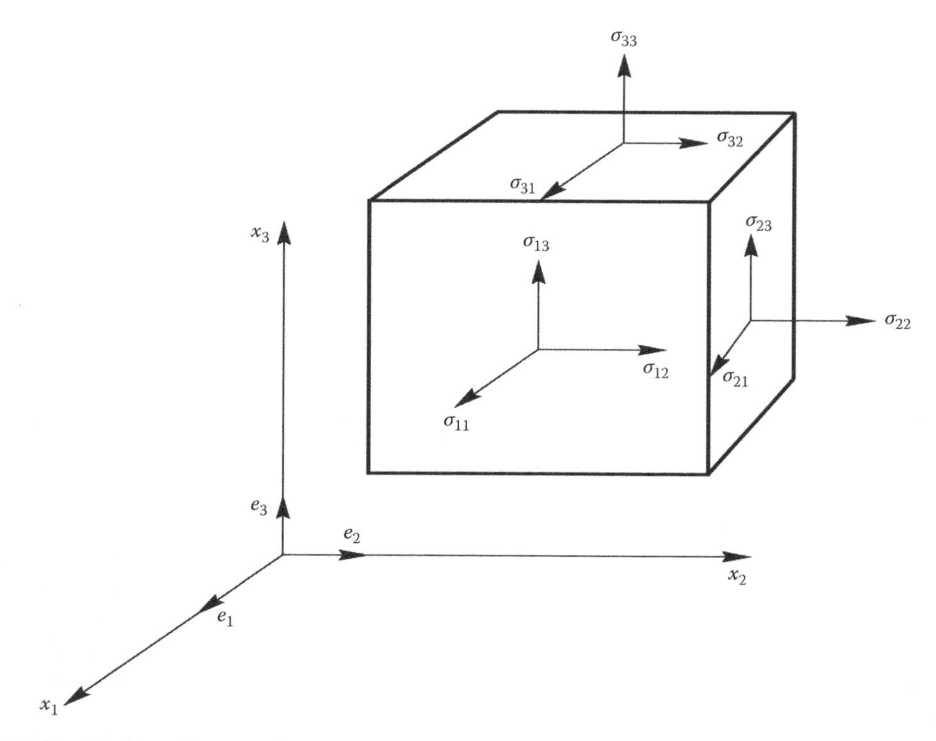

FIGURE 1.7 Stress components in a rectangular coordinate system.

neighboring region of point P, as shown in Figure 1.7. The three surface planes with outward normal directions parallel to the axes x_i (i = 1, 2, 3) are called the positive planes, denoted by dS_i, with the unit normal vectors $\boldsymbol{\nu}_i = \boldsymbol{e}_i$, where $\boldsymbol{e}_i$ are the unit vectors of the coordinate axes. The other three planes with outward normal directions opposite to the coordinate axes are called the negative planes, and the unit normal vectors are denoted by $-\boldsymbol{e}_i$. Decomposing the stress vector $\boldsymbol{\sigma}_{(i)}$ (i = 1, 2, 3) acting on the positive planes dS_i along the positive directions of the coordinate axes, we have

$$\sigma_{(1)} = \sigma_{11}\boldsymbol{e}_1 + \sigma_{12}\boldsymbol{e}_2 + \sigma_{13}\boldsymbol{e}_3 = \sigma_{1j}\boldsymbol{e}_j, \quad \sigma_{(2)} = \sigma_{21}\boldsymbol{e}_1 + \sigma_{22}\boldsymbol{e}_2 + \sigma_{23}\boldsymbol{e}_3 = \sigma_{2j}\boldsymbol{e}_j,$$
$$\sigma_{(3)} = \sigma_{31}\boldsymbol{e}_1 + \sigma_{32}\boldsymbol{e}_2 + \sigma_{33}\boldsymbol{e}_3 = \sigma_{3j}\boldsymbol{e}_j, \tag{1.7a}$$

$$\sigma_{(i)} = \sigma_{ij}\boldsymbol{e}_j. \tag{1.7b}$$

The repeated index j in the previous equations denotes the Einstein summation convention. The nine components in the previous equations can be expressed by a matrix:

$$(\sigma_{ij}) = \begin{pmatrix} \sigma_{11} & \sigma_{12} & \sigma_{13} \\ \sigma_{21} & \sigma_{22} & \sigma_{23} \\ \sigma_{31} & \sigma_{32} & \sigma_{33} \end{pmatrix}. \tag{1.8}$$

The first index i denotes the normal direction of the surface element, called the surface element index. The second index j denotes the direction of stress decomposition, called the direction index. When $i = j$, the stress component is perpendicular to the surface element and is called normal stress.

When $i \neq j$, the stress component is lying on the surface element and is called shear stress. In the Cartesian coordinate system, the nine stress components are denoted by

$$(\sigma_{ij}) = \begin{pmatrix} \sigma_x & \tau_{xy} & \tau_{xz} \\ \tau_{yx} & \sigma_y & \tau_{yz} \\ \tau_{zx} & \tau_{zy} & \sigma_z \end{pmatrix}. \tag{1.9}$$

According to elastic theory, the stress vectors $\boldsymbol{\sigma}_{(-i)}$ (i = 1, 2, 3) acting on the negative surfaces should be decomposed along directions opposite to the coordinate axes. When the element shrinks down toward its centroid to a point, the stresses acting on the positive and negative surfaces have the same magnitude but opposite directions:

$$\sigma_{(-i)} = -\sigma_{(i)} = \sigma_{ij}(-\boldsymbol{e}_j). \tag{1.10}$$

The positive directions of the nine stress components are defined as follows. The stress component acting on the positive surface is positive if it points to the positive direction of the coordinate axis, and the stress component acting on the negative surface is positive if it points to the opposite direction of the coordinate axis. This sign convention agrees with the principle of action and reaction and the traditional concept that "tension is positive and compression is negative" and keeps consistency in mathematical manipulation. However, it should be noted that the positive direction of shear stress defined here is different from that in material mechanics. The stress acting at point P on an arbitrary oblique plane can be represented by σ_{ij}. Therefore, the stress state at a point cannot be described by a scalar or a vector with three components. It must be represented by nine stress components σ_{ij}. Based on the concept of tensors that we discussed in the previous section, the stress state at a point can be described by a second-order tensor $\boldsymbol{\sigma} = \sigma_{ij}\boldsymbol{e}_i\boldsymbol{e}_j$. Readers should not confuse the stress vector (or traction) acting on the oblique plane having normal direction ν in Equation 1.5 with the second-order stress tensor $\boldsymbol{\sigma} = \sigma_{ij}\boldsymbol{e}_i\boldsymbol{e}_j$.

1.2.2 Equations of Equilibrium and Stress Boundary Conditions

1.2.2.1 Equations of Equilibrium

This book is only concerned with the interior effect of a material subjected to an external load or the deformation state. Therefore, the whole body under the external load is defined to be in an equilibrium state. Now we will discuss the static equilibrium state of an infinitesimal volume element. The loads acting on this element consist of the stress components acting on the surfaces, and the body force $\boldsymbol{F}$. The following static equilibrium problems discuss the force equilibrium conditions of the volume element along the three coordinate axes and the moment equilibrium conditions with respect to the three coordinate axes.

Select a Cartesian coordinate system as the frame of reference, and consider an infinitesimal rectangular cube in the neighboring region of point P with edge lengths $\mathrm{d}x_1$, $\mathrm{d}x_2$, $\mathrm{d}x_3$ (see Figure 1.8), which is called an infinitesimal volume element. The body forces F_i(i = 1, 2, 3) act at the centroid C of the infinitesimal volume element. Assume σ_{ij} are the stress components acting at the centroids of the three negative surfaces (with normals opposite to the coordinate axes). The stress components acting at the centroids of the three positive surfaces (with normals parallel to the coordinate axes) have increments with respect to those on the negative surfaces. By expanding Taylor series and ignoring the higher-order infinitesimal quantities, these stress components can be expressed by the stress components on the negative surfaces and their first-order derivatives. For example, the stress component σ_{11} on the negative plane evolves to $\sigma_{11} + (\partial\sigma_{11}/\partial x_1)\,\mathrm{d}x_1 + \cdots$ on the positive plane, which is $\mathrm{d}x_1$ away from the negative plane.

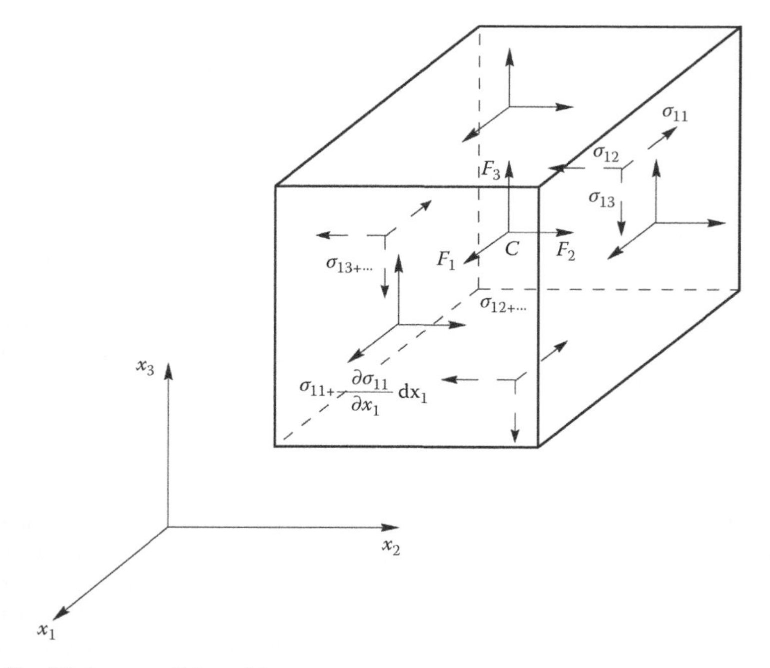

FIGURE 1.8 Equilibrium condition of force.

Therefore, the forces acting on this volume are shown in Figure 1.8. The force equilibrium condition of the volume along direction x_1 can be expressed as follows

$$\left(\sigma_{11}+\frac{\partial\sigma_{11}}{\partial x_1}\mathrm{d}x_1\right)\mathrm{d}x_2\mathrm{d}x_3-\sigma_{11}\mathrm{d}x_2\mathrm{d}x_3+\left(\sigma_{21}+\frac{\partial\sigma_{21}}{\partial x_2}\mathrm{d}x_2\right)\mathrm{d}x_3\mathrm{d}x_1-\sigma_{21}\mathrm{d}x_3\mathrm{d}x_1$$

$$+\left(\sigma_{31}+\frac{\partial\sigma_{31}}{\partial x_3}\mathrm{d}x_3\right)\mathrm{d}x_1\mathrm{d}x_2-\sigma_{31}\mathrm{d}x_1\mathrm{d}x_2+F_1\mathrm{d}x_1\mathrm{d}x_2\mathrm{d}x_3=0.$$

After elimination, and dividing by the volume, the limit value when the volume tends to point x_1, x_2, x_3 is obtained:

$$\frac{\partial\sigma_{11}}{\partial x_1}+\frac{\partial\sigma_{21}}{\partial x_2}+\frac{\partial\sigma_{31}}{\partial x_3}+F_1=0. \tag{1.11a}$$

Similarly, the force equilibrium conditions along x_2 and x_3 can be derived as follows

$$\frac{\partial\sigma_{12}}{\partial x_1}+\frac{\partial\sigma_{22}}{\partial x_2}+\frac{\partial\sigma_{32}}{\partial x_3}+F_2=0, \tag{1.11b}$$

$$\frac{\partial\sigma_{13}}{\partial x_1}+\frac{\partial\sigma_{23}}{\partial x_2}+\frac{\partial\sigma_{33}}{\partial x_3}+F_3=0. \tag{1.11c}$$

The three aforementioned equations can be written concisely in indicial notation:

$$\sigma_{ji,j}+F_i=0. \tag{1.12}$$

The aforementioned equations are called differential equilibrium equations, or simply equilibrium equations, and describe the relationship between the first-order derivatives of stress components and the body force components. The general formulae for the equilibrium equations are as follows

$$\frac{\partial \sigma_x}{\partial x}+\frac{\partial \tau_{yx}}{\partial y}+\frac{\partial \tau_{zx}}{\partial z}+F_x=0,\quad \frac{\partial \tau_{xy}}{\partial x}+\frac{\partial \sigma_y}{\partial y}+\frac{\partial \tau_{zy}}{\partial z}+F_y=0,$$
$$\frac{\partial \tau_{xz}}{\partial x}+\frac{\partial \tau_{yz}}{\partial y}+\frac{\partial \sigma_z}{\partial z}+F_z=0. \tag{1.13}$$

Next, we discuss the moment equilibrium of the infinitesimal volume element. Consider the moment of all forces about the axis passing through the centroid C and parallel to the x_3 axis. The resultant moment is zero for the stress components or body force components with the acting line passing through point C or the acting direction parallel to this axis. Therefore, the equilibrium equation of moment can be written as

$$\left(\sigma_{12}+\frac{\partial \sigma_{12}}{\partial x_1}\mathrm{d}x_1\right)\mathrm{d}x_2\mathrm{d}x_3\cdot\frac{\mathrm{d}x_1}{2}+\sigma_{12}\mathrm{d}x_2\mathrm{d}x_3\cdot\frac{\mathrm{d}x_1}{2}$$
$$-\left(\sigma_{21}+\frac{\partial \sigma_{21}}{\partial x_2}\mathrm{d}x_2\right)\mathrm{d}x_1\mathrm{d}x_3\cdot\frac{\mathrm{d}x_2}{2}-\sigma_{21}\mathrm{d}x_1\mathrm{d}x_3\cdot\frac{\mathrm{d}x_2}{2}=0$$

Ignoring the higher-order terms, we only need to consider two terms:

$$\sigma_{12}\mathrm{d}x_2\mathrm{d}x_3\cdot\mathrm{d}x_1-\sigma_{21}\mathrm{d}x_3\mathrm{d}x_1\cdot\mathrm{d}x_2=0.$$

Thus,

$$\sigma_{12}=\sigma_{21}. \tag{1.14a}$$

Similarly, the resultant moment about the axis passing through the centroid and parallel to the x_1 or x_2 direction is zero, giving

$$\sigma_{23}=\sigma_{32},\quad \sigma_{31}=\sigma_{13}. \tag{1.14b}$$

In a general form, we have

$$\sigma_{ij}=\sigma_{ji}. \tag{1.15}$$

This is the theorem of conjugate shear stresses or the symmetric property of stress tensors.

From Equation 1.8 or 1.9, we conclude that the complete description of the stress state at any point needs nine stress components. From the symmetry in Equation 1.15, only six components are independent. Therefore, six stress components are required to describe the stress state at a point.

1.2.2.2 Stress Boundary Conditions

In Section 1.2.1, we illustrated that the external forces acting on a body consist of body forces and surface forces. The effect of the body force on the equilibrium state of internal forces has been represented by Equation 1.12. How does the surface force influence the internal forces of a body? This section will

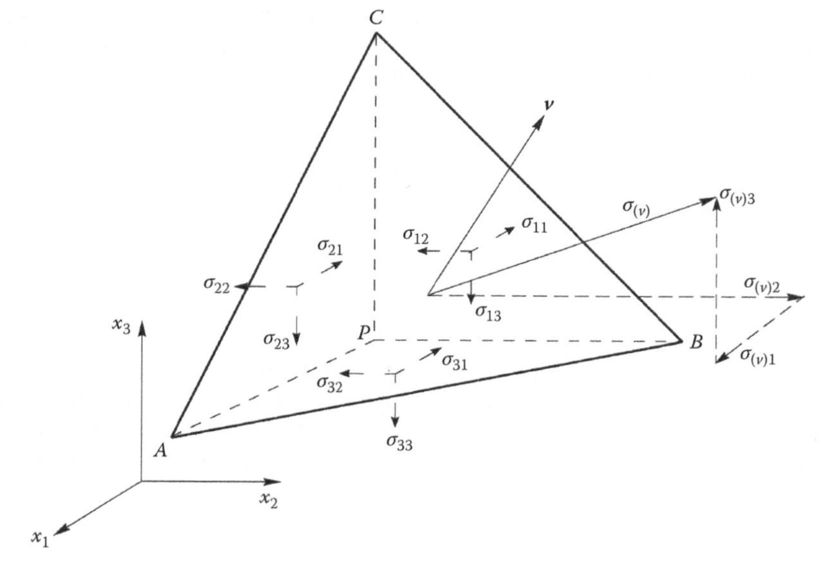

FIGURE 1.9 Stress on tetrahedron.

analyze this type of stress boundary condition. First, we derive the expression for the stress on an arbitrary oblique plane using the principle of equilibrium. Consider the tetrahedron *PABC* in Figure 1.9. It consists of three negative planes and an oblique plane with the outward normal vector

$$\boldsymbol{v} = v_1\boldsymbol{e}_1 + v_2\boldsymbol{e}_2 + v_3\boldsymbol{e}_3 = v_i\boldsymbol{e}_i, \tag{1.16a}$$

where

$$v_i = \cos(\boldsymbol{v}, \boldsymbol{e}_i) \tag{1.16b}$$

are the cosines between the normal direction of the oblique plane, $\boldsymbol{\nu}$, and the unit vectors of the coordinates axes, $\boldsymbol{e}_i$, called direction cosines. Suppose the area of the oblique plane ΔABC is $\mathrm{d}S$. Then the areas of three negative planes are as follows

$$\begin{aligned} \mathrm{d}S_1 &= \text{the area of } \Delta PBC = v_1\mathrm{d}S, \\ \mathrm{d}S_2 &= \text{the area of } \Delta PCA = v_2\mathrm{d}S, \\ \mathrm{d}S_3 &= \text{the area of } \Delta PAB = v_3\mathrm{d}S. \end{aligned} \tag{1.17}$$

The volume of the tetrahedron is

$$V = \frac{1}{3}\mathrm{d}h\mathrm{d}S, \tag{1.18}$$

where $\mathrm{d}h$ is the perpendicular distance from point P to the oblique plane of ΔABC.

According to Equation 1.5, the stress acting on the oblique plane with normal direction ν is a vector $\boldsymbol{\sigma}_{(\nu)} = \sigma_{(\nu)j}\boldsymbol{e}_j$. Now we discuss how to represent $\boldsymbol{\sigma}_{(\nu)} = \sigma_{(\nu)j}\boldsymbol{e}_j$ via the stress vectors of the three negative planes. The equilibrium condition for the forces satisfies

$$-\boldsymbol{\sigma}_{(1)}\mathrm{d}S_1 - \boldsymbol{\sigma}_{(2)}\mathrm{d}S_2 - \boldsymbol{\sigma}_{(3)}\mathrm{d}S_3 + \boldsymbol{\sigma}_{(v)}\mathrm{d}S + \boldsymbol{F}\left(\frac{1}{3}\mathrm{d}h\mathrm{d}S\right) = 0, \tag{1.19}$$

where $\boldsymbol{\sigma}_{(i)}$ is the stress vector on the ith plane (see Equation 1.7b).

The first four terms are the forces on the negative planes and the oblique plane, and the fifth term is the resultant force of the body force. As $dS \to 0$, dh tends to be zero; thus the term of the body force can be ignored when compared to the first four terms.

Substituting Equation 1.17 into Equation 1.19, and dividing by the common factor dS, we obtain

$$\boldsymbol{\sigma}_{(v)} = v_1\boldsymbol{\sigma}_{(1)} + v_2\boldsymbol{\sigma}_{(2)} + v_3\boldsymbol{\sigma}_{(3)} = v_i\boldsymbol{\sigma}_{(i)}.$$

From Equation 1.7a,

$$\boldsymbol{\sigma}_{(v)} = v_i\sigma_{ij}\boldsymbol{e}_j = \sigma_{(v)j}\boldsymbol{e}_j.$$

Therefore, the stress vector on the oblique plane with the normal direction $\boldsymbol{\nu}$ can be expressed as

$$\sigma_{(v)j} = v_i\sigma_{ij}. \tag{1.20}$$

That is,

$$\begin{aligned} \sigma_{(v)1} &= v_1\sigma_{11} + v_2\sigma_{21} + v_3\sigma_{31} \\ \sigma_{(v)2} &= v_1\sigma_{12} + v_2\sigma_{22} + v_3\sigma_{32} \\ \sigma_{(v)3} &= v_1\sigma_{13} + v_2\sigma_{23} + v_3\sigma_{33} \end{aligned} \tag{1.21}$$

in which $\sigma_{(v)1}\sigma_{(v)2}$, and $\sigma_{(v)3}$ are the components of $\boldsymbol{\sigma}_{(v)}$ in the x_1x_2, and x_3 directions, respectively. Generally, they are not the normal stress or shear stress on the oblique plane. This is the famous Cauchy's formula, also called the stress formula for the oblique plane. In fact this is the equilibrium condition for the infinitesimal tetrahedron.

Cauchy's formula has two important applications.

1. Calculating a variety of stresses acting on the oblique plane
   In terms of the direction cosine of the oblique plane ν_i and the stress components σ_{ij} of the normal planes, the three components of the stress vector on the oblique plane along the coordinate axes $\sigma_{(v)j}$ can be calculated using Equation 1.21. The magnitude of the stress vector on the oblique plane (also called the full stress) can be calculated as

$$\sigma_v \equiv \left|\boldsymbol{\sigma}_{(v)}\right| = \sqrt{\sigma^2_{(v)1} + \sigma^2_{(v)2} + \sigma^2_{(v)3}}. \tag{1.22}$$

   The direction of the stress vector is

$$\begin{aligned} \cos(\boldsymbol{\sigma}_{(v)}, x_1) &= \frac{\sigma_{(v)1}}{\sigma_v} \\ \cos(\boldsymbol{\sigma}_{(v)}, x_2) &= \frac{\sigma_{(v)2}}{\sigma_v} \\ \cos(\boldsymbol{\sigma}_{(v)}, x_3) &= \frac{\sigma_{(v)3}}{\sigma_v} \end{aligned} \tag{1.23}$$

The normal stress on the oblique plane, $\boldsymbol{\sigma}_n$, is the component of $\boldsymbol{\sigma}_{(v)}$ along the normal direction of the oblique plane:

$$\boldsymbol{\sigma}_n = \sigma_n \boldsymbol{v}$$

$$\begin{aligned} \sigma_n &\equiv |\boldsymbol{\sigma}_n| = \boldsymbol{\sigma}_{(v)} \cdot \boldsymbol{v} = \sigma_{ij} v_i v_j \\ &= \sigma_x l^2 + \sigma_y m^2 + \sigma_z n^2 + 2\tau_{xy} lm + 2\tau_{yz} mn + 2\tau_{zx} nl \end{aligned} \tag{1.24}$$

where
- $l = v_1$, $m = v_2$, $n = v_3$ are direction cosines
- σ_x, σ_y, σ_z denote the normal stresses on the upright planes perpendicular to the x, y, z axes, respectively

They are the simplified notations of σ_{xx}, σ_{yy}, σ_{zz}, respectively.

The shear stress on the oblique plane, τ, is the component of $\boldsymbol{\sigma}_{(v)}$ in the oblique plane:

$$\begin{aligned} \boldsymbol{\tau} &= \boldsymbol{\sigma}_{(v)} - \boldsymbol{\sigma}_n \\ \tau &\equiv |\boldsymbol{\tau}| = \sqrt{\sigma_v^2 - \sigma_n^2} \end{aligned} \tag{1.25}$$

It should be pointed out that the decomposed component of τ along the coordinate axes is not the shear component in the oblique plane.

2. Defining the force boundary conditions
If the oblique plane is the boundary of a body and the surface force $\boldsymbol{T}$ is given, Cauchy's formula can be adopted as the force boundary conditions of the unknown stress field:

$$\begin{aligned} \boldsymbol{T} &= \boldsymbol{\sigma}_{(v)} = v_i \sigma_{ij} \boldsymbol{e}_j, \quad \text{or,} \\ T_j &= v_i \sigma_{ij} \end{aligned} \tag{1.26}$$

in which T_j are the components of surface traction $\boldsymbol{T}$ along the coordinate axes. The component notations for the force boundary conditions in Equation 1.26 can be expressed as

$$T_x = \sigma_x l + \tau_{yx} m + \tau_{zx} n, \quad T_y = \tau_{xy} l + \sigma_y m + \tau_{zy} n, \quad T_z = \tau_{xz} l + \tau_{yz} m + \sigma_z n. \tag{1.27}$$

1.2.3 Principal Stresses and Principal Directions

In general, normal stress and shear stress are present on the infinitesimal area element of a body subjected to external loads. These stresses change with the normal direction of the area element. When the infinitesimal area element rotates, its normal direction v changes, and the magnitudes and directions of the normal stress σ_v and the shear stress τ_v change correspondingly. During the process of $\boldsymbol{v}$ changing its direction, normal stress and shear stress may sometimes vanish on the infinitesimal area element. In this case, we call the normal direction $\boldsymbol{v}$ of the area element the principal direction, the corresponding stress σ_v is called the principal stress, and the plane of the area element is called the principal plane.

Let us describe this problem with mathematical language and find a normal direction $\boldsymbol{v}$ satisfying

$$\boldsymbol{\sigma}_{(v)} = v_i \sigma_{ij} \boldsymbol{e}_j = \sigma_v \boldsymbol{v} = \sigma_v v_j \boldsymbol{e}_j. \tag{1.28}$$

Equating the corresponding components gives

$$v_i\sigma_{ij} - \sigma_v v_j = 0. \tag{1.29}$$

Replacing indices using δ_{ij} gives

$$v_i(\sigma_{ij} - \sigma_v\delta_{ij}) = 0 \quad (j = 1, 2, 3). \tag{1.30}$$

This is a set of three linear homogeneous algebraic equations of v_i.

The necessary and sufficient condition for the existence of nonzero solutions is that the determinant of the coefficients is equal to zero; that is,

$$\begin{vmatrix} \sigma_{11}-\sigma_v & \sigma_{12} & \sigma_{13} \\ \sigma_{21} & \sigma_{22}-\sigma_v & \sigma_{23} \\ \sigma_{31} & \sigma_{32} & \sigma_{33}-\sigma_v \end{vmatrix} = 0.$$

We obtain a cubic equation in σ_v after expanding, called the characteristic equation:

$$\sigma_v^3 - J_1\sigma_v^2 + J_2\sigma_v - J_3 = 0, \tag{1.31a}$$

in which

$$J_1 = \sigma_{11} + \sigma_{22} + \sigma_{33} = \sigma_{ii} = \sigma_x + \sigma_y + \sigma_z \tag{1.31b}$$

is the sum of the elements on the main diagonal of the stress matrix, called the trace of the stress tensor $\boldsymbol{\sigma}$, denoted as tr$\boldsymbol{\sigma}$, where

$$\begin{aligned} J_2 &= \begin{vmatrix} \sigma_{22} & \sigma_{23} \\ \sigma_{32} & \sigma_{33} \end{vmatrix} + \begin{vmatrix} \sigma_{11} & \sigma_{13} \\ \sigma_{31} & \sigma_{33} \end{vmatrix} + \begin{vmatrix} \sigma_{11} & \sigma_{12} \\ \sigma_{21} & \sigma_{22} \end{vmatrix} \\ &= \frac{1}{2}(\sigma_{ii}\sigma_{jj} - \sigma_{ij}\sigma_{ij}) = \frac{1}{2}\left(J_1^2 - \sigma_{ij}\sigma_{ij}\right) \\ &= \sigma_x\sigma_y + \sigma_y\sigma_z + \sigma_z\sigma_x - \tau_{xy}^2 - \tau_{yz}^2 - \tau_{zx}^2 \end{aligned} \tag{1.31c}$$

is the sum of the second-order cofactors of the stress matrix.

$$\begin{aligned} J_3 &= \begin{vmatrix} \sigma_{11} & \sigma_{12} & \sigma_{13} \\ \sigma_{21} & \sigma_{22} & \sigma_{23} \\ \sigma_{31} & \sigma_{32} & \sigma_{33} \end{vmatrix} = e_{ijk}\sigma_{1i}\sigma_{2j}\sigma_{3k} \\ &= \frac{1}{3}\sigma_{ij}\sigma_{jk}\sigma_{ki} + J_1\left(J_2 - \frac{1}{3}J_1^2\right) \\ &= \sigma_x\sigma_y\sigma_z + 2\tau_{xy}\tau_{yz}\tau_{zx} - \sigma_x\tau_{yz}^2 - \sigma_y\tau_{zx}^2 - \sigma_z\tau_{xy}^2 \end{aligned} \tag{1.31d}$$

is the determinant of the stress matrix, denoted *det*$\boldsymbol{\sigma}$.

The magnitudes of J_1, J_2, and J_3 do not depend on the choice of coordinate system; therefore, they are called the first, second, and third invariants of a stress tensor, respectively. Three Eigen roots of the characteristic equation (1.31) are the principal stresses, ordered by their magnitude, and called the first principal stress σ_1, the second principal stress σ_2, and the third principal stress σ_3.

Substituting the three principal stresses σ_k into Equation 1.30, and $\nu_i\nu_i = 1$ as the conditions of the unit vector ν, we can solve three Eigen directions ν^k which are the principal directions. The three planes with normal directions ν^k are the principal planes. It can be proved that the three principal planes are mutually perpendicular [1,2]. There are only normal stresses and no shearing stresses on the principal planes. The magnitudes and directions of the principal stresses do not depend on the choice of the coordinate system.

1.2.4 Spherical and Deviatoric Stress Tensors

Generally, the stress state at a point can be decomposed into two parts: the first part is an isotropic tensile (or compressive) stress tensor $\sigma_m \boldsymbol{I}$ and the second part is a tensor denoted by $\boldsymbol{S}$:

$$\sigma = \sigma_m \boldsymbol{I} + \boldsymbol{S} \quad \text{or} \quad \sigma_{ij} = \sigma_m \delta_{ij} + S_{ij} \tag{1.32a}$$

where

$$\sigma_m \boldsymbol{I} = \begin{bmatrix} \sigma_m & 0 & 0 \\ 0 & \sigma_m & 0 \\ 0 & 0 & \sigma_m \end{bmatrix}, \tag{1.32b}$$

$$\boldsymbol{S} = \begin{bmatrix} \sigma_x - \sigma_m & \tau_{xy} & \tau_{xz} \\ \tau_{yx} & \sigma_y - \sigma_m & \tau_{yz} \\ \tau_{zx} & \tau_{zy} & \sigma_z - \sigma_m \end{bmatrix}, \tag{1.32c}$$

$$\sigma_m = \frac{1}{3}(\sigma_x + \sigma_y + \sigma_z) = \frac{1}{3}(\sigma_1 + \sigma_2 + \sigma_3). \tag{1.32d}$$

The first term $\sigma_m \boldsymbol{I}$ at the right-hand side of Equation 1.32a is the spherical stress tensor, and the second term $\boldsymbol{S}$ is the deviatorical stress tensor. $\boldsymbol{I}$ is a unit tensor, represented by the components δ_{ij}.

Like the stress tensor, the deviatorical stress tensor also has three invariants, which can be calculated similarly to Equation 1.31b through d. The three invariants of the deviatorical stress tensor are

$$I_1' = S_{kk} = 0, \tag{1.33a}$$

$$I_2' = \frac{1}{2}\left(I_1' - S_{ij}S_{ij}\right) = -\frac{1}{2}S_{ij}S_{ij}, \tag{1.33b}$$

$$I_3' = \frac{1}{3}S_{ij}S_{jk}S_{ki} + I_1'\left(I_2' - \frac{1}{3}I_1'^2\right) = \frac{1}{3}S_{ij}S_{jk}S_{ki}. \tag{1.33c}$$

Since I_2' is always negative, the following quantities are alternatively defined:

$$J_2' = -I_2' = \frac{1}{2} S_{ij} S_{ij}$$

$$J_3' = I_3' = \frac{1}{3} S_{ij} S_{jk} S_{ki} \tag{1.33d}$$

J_2' can also be expressed by the stress components:

$$\begin{aligned} J_2' &= \frac{1}{6}\left[(\sigma_{11}-\sigma_{22})^2+(\sigma_{22}-\sigma_{33})^2+(\sigma_{33}-\sigma_{11})^2\right]+\sigma_{12}^2+\sigma_{23}^2+\sigma_{31}^2 \\ &= \frac{1}{6}\left[(\sigma_1-\sigma_2)^2+(\sigma_2-\sigma_3)^2+(\sigma_3-\sigma_1)^2\right]. \end{aligned} \tag{1.34}$$

Here, J_2' is a parameter related to the fourth strength theory, which has important applications in plastic mechanics. Here, we introduce a definition that has been used many times, $\bar{\sigma} = \sqrt{3J'}$.

1.3 STRAIN

1.3.1 Deformation and Strain

1.3.1.1 Description of Displacement

When a body is subjected to an external load, the location of each material point in the body changes, and displacement occurs. If the relative positions of all points after deformation are identical to the original ones before deformation, the body only generates rigid movement and rotation. This type of displacement is called the rigid displacement. If the original relative positions of points change after displacement, then deformation occurs, including volume change and shape distortion. Strain analysis studies the deformation of a stressed body, called the strain state, but it does not discuss rigid displacement.

The displacement of an arbitrary point in a body can be represented by the displacement components u, v, w along the x-, y-, z-directions, respectively. Therefore, once the displacement of each point in a body is known, the deformation state is determined. Because the displacement of each point in a body is generally different, the displacement components u, v, w should be the functions of the coordinates:

$$u = u(x, y, z), \quad v = v(x, y, z), \quad w = w(x, y, z).$$

1.3.1.2 Description of Strain

In order to describe the deformation of an arbitrary point P in a body, consider three mutually perpendicular infinitesimal line elements PA, PB, and PC with lengths dx, dy, and dz respectively, as shown in Figure 1.10.

After the body deforms under external loading, the lengths of the three line elements change and so do the angles between them. The relative changes in lengths of these line elements are called the normal strains at point P, denoted by ε. The normal strains of line elements PA, PB, PC along the x-, y-, z-directions are represented by ε_x, ε_y, ε_z, respectively. The normal strain is positive if the line element elongates and is negative if the line element shortens. The change in angle between two line elements is called the shear strain at point P, denoted by γ. The angle change between line elements

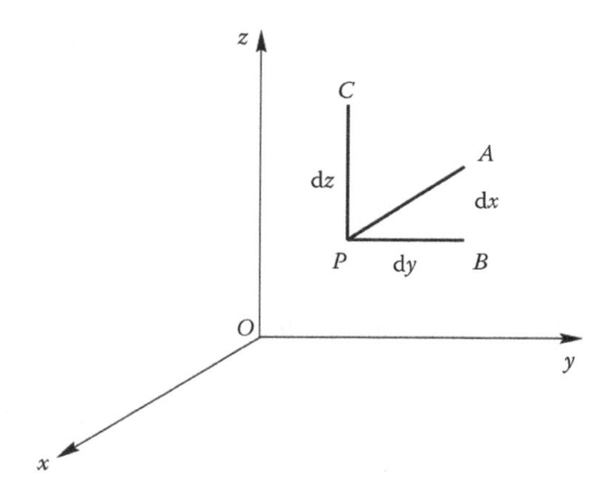

FIGURE 1.10 Description of strain.

PA and PB along the x- and y-axes is represented by γ_{xy}. Similarly, the angle change between line elements PB and PC in the y- and z-directions is denoted by γ_{yz}, and the angle change between line elements PC and PA on the z- and x-axes is denoted by γ_{zx}. The shear strain is positive if the angle between two line elements decreases and negative if the angle increases.

1.3.1.3 Geometric Equations

In order to describe deformation we set up a Cartesian coordinate system, as shown in Figure 1.11. The geometry is represented by B before deformation and by B' after deformation. For convenience, we call configuration B the geometric state before deformation and configuration B' the geometric state after deformation. For an arbitrary line element $\overrightarrow{PQ}$ before deformation, the position vectors of its two ends $P(a_1, a_2, a_3)$ and $Q(a_1 + \mathrm{d}a_1, a_2 + \mathrm{d}a_2, a_3 + \mathrm{d}a_3)$ are

$$\overrightarrow{OP} = \boldsymbol{a} = a_i\boldsymbol{e}_i, \ \overrightarrow{OQ} = \boldsymbol{a} + \mathrm{d}\boldsymbol{a} = (a_i + \mathrm{d}a_i)\boldsymbol{e}_i.$$

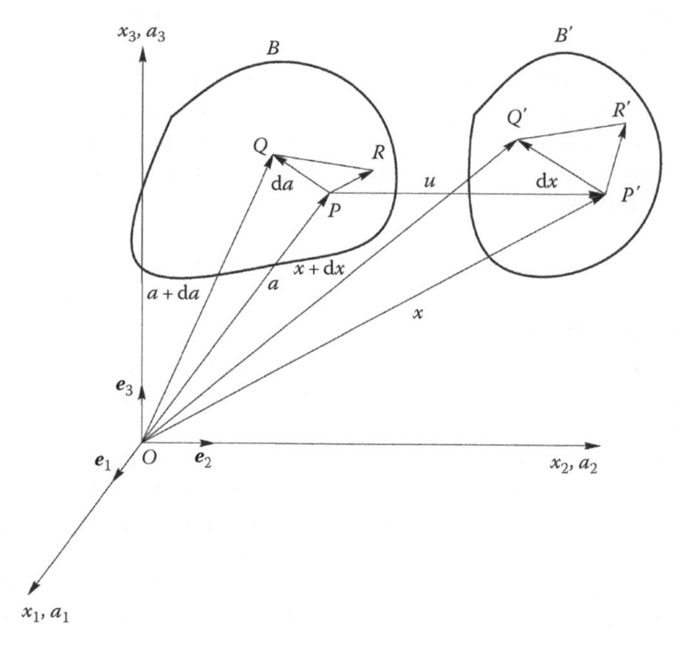

FIGURE 1.11 Configurations B and B'.

Therefore, the line element $\overrightarrow{PQ}$ can be represented by

$$\overrightarrow{PQ} = \overrightarrow{OQ} - \overrightarrow{OP} = \mathrm{d}\boldsymbol{a} = \mathrm{d}a_i\boldsymbol{e}_i = \mathrm{d}a_1\boldsymbol{e}_1 + \mathrm{d}a_2\boldsymbol{e}_2 + \mathrm{d}a_3\boldsymbol{e}_3.$$

After deformation, points P and Q move to P' and Q', and the corresponding position vectors are

$$\overrightarrow{OP'} = \boldsymbol{x} = x_i\boldsymbol{e}_i, \quad \overrightarrow{OQ'} = \boldsymbol{x} + \mathrm{d}\boldsymbol{x} = (x_i + \mathrm{d}x_i)\boldsymbol{e}_i.$$

Here, x are functions of a_i, and $x_m = x_m(a_i)$. Thus, the line element after deformation can be represented by

$$\overrightarrow{P'Q'} = \overrightarrow{OQ'} - \overrightarrow{OP'} = \mathrm{d}\boldsymbol{x} = \mathrm{d}x_i\boldsymbol{e}_i = \mathrm{d}x_1\boldsymbol{e}_1 + \mathrm{d}x_2\boldsymbol{e}_2 + \mathrm{d}x_3\boldsymbol{e}_3.$$

Before and after deformation, the squares of the lengths of line elements $\overrightarrow{PQ}$ and $\overrightarrow{P'Q'}$, respectively, are

$$\mathrm{d}s_0^2 = \mathrm{d}\boldsymbol{a}\cdot\mathrm{d}\boldsymbol{a} = \mathrm{d}a_i\mathrm{d}a_i = \delta_{ij}\mathrm{d}a_i\mathrm{d}a_j = \mathrm{d}a_1^2 + \mathrm{d}a_2^2 + \mathrm{d}a_3^2, \tag{1.35a}$$

$$\mathrm{d}s^2 = \mathrm{d}\boldsymbol{x}\cdot\mathrm{d}\boldsymbol{x} = \mathrm{d}x_m\mathrm{d}x_m = \mathrm{d}x_1^2 + \mathrm{d}x_2^3 + \mathrm{d}x_3^2. \tag{1.35b}$$

Because x are functions of a_i, that is, $x_m = x_m(a_i)$, we have

$$\mathrm{d}x_m = \frac{\partial x_m}{\partial a_i}\mathrm{d}a_i = \frac{\partial x_m}{\partial a_1}\mathrm{d}a_1 + \frac{\partial x_m}{\partial a_2}\mathrm{d}a_2 + \frac{\partial x_m}{\partial a_3}\mathrm{d}a_3. \tag{1.36}$$

Substituting into Equation 1.35b gives

$$\mathrm{d}s^2 = \frac{\partial x_m}{\partial a_i}\frac{\partial x_m}{\partial a_j}\mathrm{d}a_i\mathrm{d}a_j. \tag{1.37}$$

Subtracting Equation 1.35a gives the change in the square of the length of line element after deformation:

$$\mathrm{d}s^2 - \mathrm{d}s_0^2 = 2E_{ij}\mathrm{d}a_i\mathrm{d}a_j, \tag{1.38}$$

where

$$E_{ij} = \frac{1}{2}\left(\frac{\partial x_m}{\partial a_i}\frac{\partial x_m}{\partial a_j} - \delta_{ij}\right). \tag{1.39}$$

From Figure 1.11, $x_m(a_i) = a_m + u_m(a_i)$. After derivation,

$$\frac{\partial x_m}{\partial a_i} = \delta_{mi} + \frac{\partial u_m}{\partial a_i}. \tag{1.40}$$

This book studies the case of small deformations, in which the displacement is much smaller than the minimum dimension of the deformed object. In this case, the first-order derivative of the displacement component is much less than 1:

$$\left|\frac{\partial u_i}{\partial a_j}\right| << 1, \quad \left|\frac{\partial u_j}{\partial x_j}\right| << 1.$$

Ignoring the higher-order infinitesimal quantities gives

$$\frac{\partial u_i}{\partial a_j} = \frac{\partial u_i}{\partial x_k}\frac{\partial x_k}{\partial a_j} = \frac{\partial u_i}{\partial x_k}\left(\delta_{kj} + \frac{\partial u_k}{\partial a_j}\right) \approx \frac{\partial u_i}{\partial x_k}\delta_{kj} = \frac{\partial u_i}{\partial x_j}.$$

We remind readers again that the repeated index in the above equation represents the summation operation, taking Equation 1.40 into consideration. Therefore, when describing the small deformation of a body, the distinction between coordinates a_i and x_i is ignored. For the small deformation case, Equation 1.39 can be simplified as

$$E_{ij} \approx \varepsilon_{ij} = \frac{1}{2}\left(\frac{\partial u_i}{\partial x_j} + \frac{\partial u_j}{\partial x_i}\right). \tag{1.41}$$

Here, ε_{ij} are the components of the Cauchy strain tensor or the small strain tensor. $\boldsymbol{\varepsilon}$ is a second-order symmetric tensor, with six independent components. In Cartesian coordinates, the general formulae are

$$\begin{aligned}
&\varepsilon_{11} = \frac{\partial u_1}{\partial x_1}, \quad \varepsilon_{12} = \varepsilon_{21} = \frac{1}{2}\left(\frac{\partial u_1}{\partial x_2} + \frac{\partial u_2}{\partial x_1}\right), \quad \varepsilon_{22} = \frac{\partial u_2}{\partial x_2}, \\
&\varepsilon_{23} = \varepsilon_{32} = \frac{1}{2}\left(\frac{\partial u_2}{\partial x_3} + \frac{\partial u_3}{\partial x_2}\right), \quad \varepsilon_{33} = \frac{\partial u_3}{\partial x_3}, \quad \varepsilon_{31} = \varepsilon_{13} = \frac{1}{2}\left(\frac{\partial u_3}{\partial x_1} + \frac{\partial u_1}{\partial x_3}\right).
\end{aligned} \tag{1.42}$$

This is a set of linear differential equations, called the strain–displacement formulae or geometric equations. In terms of Equation 1.42, the strain components can either be obtained from the derivatives of the displacement components, or the displacement components can be obtained by integrating the strain components.

The strain tensor ε is used to determine the change in length of the line element and the change of angle between line elements. First, the change in length of the line element is analyzed.

Before deformation, the unit vector for the direction of line element $\overrightarrow{PQ}$ is

$$\boldsymbol{v} = \frac{\mathrm{d}\boldsymbol{a}}{\mathrm{d}s_0} = \frac{\mathrm{d}a_i}{\mathrm{d}s_0}\boldsymbol{e}_i = v_i\boldsymbol{e}_i, \tag{1.43}$$

where $v_i = \mathrm{d}a_i/\mathrm{d}s_0$ are the direction cosines of line element $\overrightarrow{PQ}$.

Define $\lambda_v = \mathrm{d}s/\mathrm{d}s_0$ as the change in length of the line element before and after deformation, called the length ratio.

From Equations 1.38, 1.41, and 1.43, we have

$$\lambda_v = \frac{\mathrm{d}s}{\mathrm{d}s_0} = (1 + 2\varepsilon_{ij}v_iv_j)^{1/2} \approx 1 + \varepsilon_{ij}v_iv_j. \tag{1.44}$$

Here, the condition for the small deformation $2\varepsilon_{ij}\nu_i\nu_j << 1$ is used. Generally, the engineering normal strain ε_ν for the line element in the direction of $\boldsymbol{\nu}$ is defined as the relative change in length before and after deformation:

$$\varepsilon_\nu = \frac{\mathrm{d}s - \mathrm{d}s_0}{\mathrm{d}s_0} = \lambda_\nu - 1$$

Substituting Equation 1.44 gives

$$\varepsilon_\nu = \varepsilon_{ij}\nu_i\nu_j. \tag{1.45a}$$

By expanding,

$$\varepsilon_\nu = \varepsilon_{11}\nu_1\nu_1 + \varepsilon_{22}\nu_2\nu_2 + \varepsilon_{33}\nu_3\nu_3 + 2\varepsilon_{12}\nu_1\nu_2 + 2\varepsilon_{23}\nu_2\nu_3 + 2\varepsilon_{31}\nu_3\nu_1. \tag{1.45b}$$

When the values of $\boldsymbol{\nu}$ are $\boldsymbol{e}_i$ (i = 1, 2, 3), we obtain from Equation 1.45a

$$\varepsilon_x = \varepsilon_{11} \quad \varepsilon_y = \varepsilon_{22} \quad \varepsilon_z = \varepsilon_{33}. \tag{1.45c}$$

Therefore, the three diagonal components of the strain tensor ε_{ij} are the engineering normal strains of the line elements parallel to the coordinate axes. The normal strain is positive if the line element elongates and negative if the line element shortens.

Next, we discuss the direction of a line element. After deformation, the unit vector of the direction of line element $\overrightarrow{P'Q'}$ is

$$\boldsymbol{\nu}' = \frac{\mathrm{d}\mathbf{x}}{\mathrm{d}s} = \frac{\mathrm{d}x_i}{\mathrm{d}s}\boldsymbol{e}_i = \nu_i'\boldsymbol{e}_i. \tag{1.46}$$

The direction cosines can be denoted by

$$\nu_i' = \frac{\mathrm{d}x_i}{\mathrm{d}s} = \frac{\partial x_i}{\partial a_j}\frac{\mathrm{d}a_j}{\mathrm{d}s_0}\frac{\mathrm{d}s_0}{\mathrm{d}s} = \frac{\partial x_i}{\partial a_j}\nu_j\frac{1}{\lambda_\nu}.$$

Using Equation 1.40, the direction cosines of an arbitrary line element after deformation can be represented by the displacement

$$\nu_i' = \left(\delta_{ji} + \frac{\partial u_i}{\partial a_j}\right)\nu_j\frac{1}{\lambda_\nu}. \tag{1.47a}$$

Using Equations 1.44 and 1.45, and ignoring the second-order infinitesimal quantities, we have

$$\frac{1}{\lambda_\nu} = \frac{1}{1+\varepsilon_\nu} \approx 1 - \varepsilon_\nu. \tag{1.47b}$$

Substituting this equation into Equation 1.47a, and ignoring the second-order infinitesimal quantities, we obtain the direction cosines of the line element after deformation:

$$\nu_i' = \nu_i + \frac{\partial u_i}{\partial a_j}\nu_j - \nu_i\varepsilon_\nu. \tag{1.48a}$$

According to Equation 1.48a, the direction cosines of a line element after deformation can be calculated from the displacement gradient components $\partial u_i/\partial a_j$ and the normal strain ε_v of the line element. For example, consider the line element that is parallel to coordinate axis a_1 before deformation. Its unit vector and direction cosines are

$$\boldsymbol{v} = \boldsymbol{e}_1, \quad \text{or} \quad v_1 = 1, \quad v_2 = v_3 = 0.$$

Substituting into Equation 1.45a gives

$$\varepsilon_v = \varepsilon_{ij} v_i v_j = \varepsilon_{11}.$$

From Equation 1.48a, the direction cosines after deformation are

$$v_1' = 1 + \frac{\partial u_1}{\partial a_1} - \varepsilon_{11} \approx 1, \quad v_2' = \frac{\partial u_2}{\partial a_1}, \quad v_3' = \frac{\partial u_3}{\partial a_1}. \tag{1.48b}$$

The sum of the squares of these three components is not strictly equal to 1. However, in the small deformation case, the discrepancy is only a second-order infinitesimal quantity, which is tolerable. Therefore, the unit vector after deformation is

$$\boldsymbol{e}_1' = \boldsymbol{e}_1 + \frac{\partial u_2}{\partial a_1}\boldsymbol{e}_2 + \frac{\partial u_3}{\partial a_1}\boldsymbol{e}_3.$$

Assuming that the angle between $\boldsymbol{e}_1'$ and $\boldsymbol{e}_2$ is $(\pi/2) - \theta_2$,

$$\cos\left(\frac{\pi}{2} - \theta_2\right) = \boldsymbol{e}_1' \cdot \boldsymbol{e}_2 = \frac{\partial u_2}{\partial a_1}.$$

When θ_2 is very small,

$$\cos\left(\frac{\pi}{2} - \theta_2\right) = \sin\theta_2 \approx \theta_2 \approx \frac{\partial u_2}{\partial a_1}. \tag{1.48c}$$

Similarly,

$$\cos\left(\frac{\pi}{2} - \theta_3\right) = \boldsymbol{e}_1' \cdot \boldsymbol{e}_3 = \frac{\partial u_3}{\partial a_1} \approx \theta_3. \tag{1.48d}$$

Equation 1.48c and d indicate that the line elements perpendicular to axes a_2 and a_3 rotate an angle $\partial u_2/\partial a_1$ about a_2 before deformation and an angle $\partial u_3/\partial a_1$ about a_3 after deformation. Similarly, the line elements along the axes a_2 and a_3 also rotate after deformation, and the corresponding rotation angles and their directions are shown in Figure 1.12. This figure intuitively illustrates the rotations of three line elements along the coordinate axes after deformation.

Finally, we discuss the change in angle between two line elements. As shown in Figure 1.11, we consider two arbitrary line elements $\overrightarrow{PQ}$ and $\overrightarrow{PR}$ before deformation, with respective unit vectors $\boldsymbol{\nu}$ and $\boldsymbol{t}$ and direction cosines v_i and t_i. The cosine of the angle between $\overrightarrow{PQ}$ and $\overrightarrow{PR}$ is

$$\cos(\boldsymbol{v}, \boldsymbol{t}) = \boldsymbol{v} \cdot \boldsymbol{t} = v_i t_i.$$

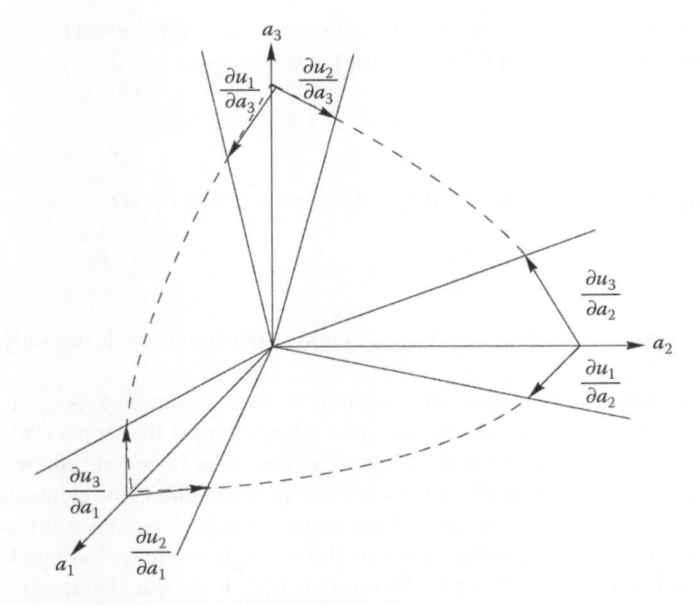

FIGURE 1.12 Rotation of line element after deformation.

After deformation, the two line elements change to $\overrightarrow{P'Q'}$ and $\overrightarrow{P'R'}$, with respective unit vectors $\boldsymbol{\nu}'$ and $\boldsymbol{t}'$ and direction cosines v_i' and t_i'. Using Equation 1.47a, the cosine of the angle between $\overrightarrow{P'Q'}$ and $\overrightarrow{P'R'}$ is

$$\cos(\boldsymbol{\nu}',\boldsymbol{t}') = \boldsymbol{\nu}'\cdot\boldsymbol{t}' = v_i't_i' = \left(\delta_{mn} + \frac{\partial u_n}{\partial a_m} + \frac{\partial u_m}{\partial a_n} + \frac{\partial u_i}{\partial a_m}\frac{\partial u_i}{\partial a_n}\right)v_m t_n \frac{1}{\lambda_\nu}\frac{1}{\lambda_t}.$$

Using Equations 1.39 and 1.41, the previous equation can be written as

$$\cos(\boldsymbol{\nu}',\boldsymbol{t}') = (\nu_m t_m + 2\varepsilon_{mn}\nu_m t_n)\frac{1}{\lambda_\nu\lambda_t}. \tag{1.49}$$

This equation gives the change in angle between line elements after deformation. Substituting Equations 1.47b and 1.41 into Equation 1.49, and ignoring the second-order infinitesimal quantities, Equation 1.49 is simplified as

$$\boldsymbol{\nu}'\cdot\boldsymbol{t}' = \cos(\boldsymbol{\nu}',\boldsymbol{t}') = (1-\varepsilon_\nu-\varepsilon_t)\boldsymbol{\nu}\cdot\boldsymbol{t} + 2\nu_i\varepsilon_{ij}t_j.$$

If line elements $\overrightarrow{PQ}$ and $\overrightarrow{PR}$ before deformation are perpendicular, that is, $\boldsymbol{\nu}\cdot\boldsymbol{t}=0$, and letting θ be the decrease of the right angle between the line elements after deformation, the previous equation gives

$$\theta \approx \cos\left(\frac{\pi}{2}-\theta\right) = \cos(\boldsymbol{\nu}',\boldsymbol{t}') = 2\varepsilon_{ij}\nu_i t_j = 2\varepsilon_{\nu t}. \tag{1.50}$$

Generally, the decrease of the right angle between two perpendicular line elements is defined as the engineering shear strain $\gamma_{\nu t}$:

$$\gamma_{\nu t} = 2\varepsilon_{\nu t} = 2\varepsilon_{ij}\nu_i t_j. \tag{1.51}$$

If $\boldsymbol{\nu}$, $\boldsymbol{t}$ are unit vectors in the directions of the coordinate axes, for example, $\nu_i = 1$, $t_j = 1 (i \neq j)$, and the remaining direction cosines are 0, the previous equation gives

$$\gamma_{ij} = 2\varepsilon_{ij}, \quad i \neq j. \tag{1.52}$$

Because the angle between $\boldsymbol{\nu}$ and $\boldsymbol{t}$ is identical to that between $\boldsymbol{t}$ and $\boldsymbol{\nu}$,

$$\varepsilon_{ij} = \varepsilon_{ji}. \tag{1.53}$$

Like the stress tensor, the strain tensor is also a symmetric tensor. It has only six independent components.

From this discussion, the geometric interpolation of the six components ε_{ij} of a strain tensor is clear. When $i = j$, ε_{ij} represents the normal strain of a line element that is parallel to the coordinate axis i, and it is positive if the line element elongates and negative if the line element shortens. When $i \neq j$, double ε_{ij} represents the shear strain between two perpendicular line elements that are parallel to the coordinate axes i and j. It is positive if the right angle becomes an acute angle (decrease in angle) and negative if the right angle becomes an obtuse angle (increase in angle).

From Equations 1.45a, 1.46, 1.49, and 1.50 through 1.52, it is clear that the strain tensor $\boldsymbol{\varepsilon}$ gives all necessary information about the deformation state of a body.

1.3.2 Principal Strains and Principal Directions

In the discussion of stress state at a point, we have introduced the principal planes, the principal stresses, and the principal directions. Now we discuss the principal strains and the principal directions for the strain state at a point. If the shear strain on a plane is zero, the plane is called a principal plane. The normal direction of the principal plane is called the principal strain direction, and the normal strain on the principal plane is called the principal strain. The line element with unit normal $\boldsymbol{\nu}$ becomes the line element with unit normal $\boldsymbol{\nu}'$ after deformation, while the normal strain in direction $\boldsymbol{\nu}$ is calculated by Equation 1.45a: $\varepsilon_\nu = \varepsilon_{ij}\nu_i\nu_j$. By a careful analysis of Equation 1.48a, we discover that $\boldsymbol{\nu}'$ can be written as $\boldsymbol{\nu}' = \boldsymbol{\nu} + \varepsilon_{ij}\nu_j\boldsymbol{e}_i$. In order to solve the principal strain, we assume that the direction $\boldsymbol{\nu}'$ after deformation of the line element, which has a unit normal $\boldsymbol{\nu}$ before deformation, is identical to the direction $\boldsymbol{\nu}$; that is, only normal strain exists. In this case, $\boldsymbol{\nu}' = \boldsymbol{\nu} + \varepsilon_\nu\boldsymbol{\nu}$.

Comparing $\boldsymbol{\nu}' = \boldsymbol{\nu} + \varepsilon_{ij}\nu_j\boldsymbol{e}_i$ and $\boldsymbol{\nu}' = \boldsymbol{\nu} + \varepsilon_\nu\boldsymbol{\nu}$, we have

$$\varepsilon_{ij}\nu_j\boldsymbol{e}_i = \varepsilon_\nu\nu_i\boldsymbol{e}_i, \quad \text{or} \quad \varepsilon_{ij}\nu_j = \varepsilon_\nu\nu_i. \tag{1.54}$$

That is,

$$(\varepsilon_{ij} - \varepsilon_\nu\delta_{ij})\nu_j = 0. \tag{1.55}$$

Letting the determinant of the coefficients be zero, we obtain the characteristic equation for the principal strains:

$$\varepsilon_\nu^3 - I_1\varepsilon_\nu^2 + I_2\varepsilon_\nu + I_3 = 0, \tag{1.56}$$

where the coefficients are

$$I_1 = \varepsilon_{ii} = \varepsilon_{11} + \varepsilon_{22} + \varepsilon_{33}, \tag{1.57a}$$

$$I_2 = \frac{1}{2}(\varepsilon_{ii}\varepsilon_{jj} - \varepsilon_{ij}\varepsilon_{ij}) = (\varepsilon_{11}\varepsilon_{22} + \varepsilon_{22}\varepsilon_{33} + \varepsilon_{33}\varepsilon_{11}) - \left(\varepsilon_{12}^2 + \varepsilon_{23}^2 + \varepsilon_{31}^2\right), \tag{1.57b}$$

$$I_3 = e_{ijk}\varepsilon_{1i}\varepsilon_{2j}\varepsilon_{3k} = \varepsilon_{11}\varepsilon_{22}\varepsilon_{33} + 2\varepsilon_{12}\varepsilon_{23}\varepsilon_{31} - \left(\varepsilon_{11}\varepsilon_{23}^2 + \varepsilon_{22}\varepsilon_{31}^2 + \varepsilon_{33}\varepsilon_{12}^2\right). \tag{1.57c}$$

These three quantities are independent of the choice of coordinate axes; therefore, they are called the first, second, and third invariants.

Consider a right parallelepiped with edge lengths of dx_1, dx_2, dx_3 in the principal directions. The relative volume change after deformation (ignoring the higher-order infinitesimal quantities) is

$$\begin{aligned} \varepsilon_v &= \frac{dV' - dV}{dV} \\ &= \frac{(1+\varepsilon_{11})dx_1(1+\varepsilon_{22})dx_2(1+\varepsilon_{33})dx_3 - dx_1dx_2dx_3}{dx_1dx_2dx_3} \\ &\approx \varepsilon_{11} + \varepsilon_{22} + \varepsilon_{33} = I_1. \end{aligned} \tag{1.58}$$

Thus, the first invariant represents the volume change of per unit volume due to deformation, also called the volumetric strain.

Like the stress tensor, the strain tensor can be decomposed into a spherical strain tensor and a deviatoric strain tensor:

$$\boldsymbol{\varepsilon} = \frac{1}{3}\varepsilon_{kk}\mathbf{I} + \boldsymbol{\varepsilon}' \quad \text{or} \quad \varepsilon_{ij} = \frac{1}{3}\varepsilon_{kk}\delta_{ij} + \varepsilon'_{ij}, \tag{1.59}$$

where

$$\left(\frac{1}{3}\varepsilon_{kk}\delta_{ij}\right) = \left(\varepsilon_m\delta_{ij}\right) = \begin{pmatrix} \varepsilon_m & 0 & 0 \\ 0 & \varepsilon_m & 0 \\ 0 & 0 & \varepsilon_m \end{pmatrix}$$

is called the spherical strain tensor, representing the volume expansion or contraction. It does not introduce shape distortion. ε_m is the average normal strain.

From Equation 1.59,

$$\left(\varepsilon'_{ij}\right) = \left(\varepsilon_{ij} - \frac{1}{3}\varepsilon_{kk}\delta_{ij}\right) = \begin{pmatrix} \varepsilon_{11} - \varepsilon_m & \varepsilon_{12} & \varepsilon_{13} \\ \varepsilon_{21} & \varepsilon_{22} - \varepsilon_m & \varepsilon_{23} \\ \varepsilon_{31} & \varepsilon_{32} & \varepsilon_{33} - \varepsilon_m \end{pmatrix}$$

is called the deviatoric strain tensor. It is easy to see that $\varepsilon'_{ii} = 0$; that is, the deviatoric strain tensor does not cause volume change, it only causes shape distortion. In addition, the equivalent strain $\bar{\varepsilon} = \sqrt{\frac{2}{3}\varepsilon'_{ij}\varepsilon'_{ij}}$ is introduced here.

1.4 STRESS–STRAIN RELATIONSHIP

In order to describe the relationship between strain components and displacement components, we introduced the concept of the stress state at an arbitrary point in a deformed body, the equilibrium differential equations of stress components, the geometric equations, and the equations of compatibility.

These equations are irrelevant to material properties and are applicable to any continuous medium. We know that stresses inside a body lead to deformation, and the degree of deformation is described by strains. Hence, a certain relationship exists between stresses and strains. We call this relationship a constitutive relationship, which is described by constitutive equations. For a specific material, the constitutive relationship is determined by its material properties. That is, the constitutive relationship of a material is inherent to the material.

1.4.1 Hooke's Law of Isotropic Elastic Material

First, we assume that the material is isotropic; that is, the properties in all directions are identical. Next, we discuss the relationship between stress components and strain components for an isotropic elastic material in the condition of linear elasticity. Consider an infinitesimal right parallelepiped with all six planes parallel to the coordinate planes. If a normal stress σ_{xx} is applied in the direction of the x-axis, then the normal strain in the x-direction is

$$\varepsilon_{xx} = \frac{\sigma_{xx}}{E},$$

where E is the elastic modulus of the material.

During tension, the transverse cross-sectional area of the specimen decreases constantly with elongation in the tensile direction. During compression, the transverse cross-sectional area of the specimen increases constantly with contraction in the compressive direction. The degree of increase or decrease in the transverse cross-sectional area is obviously related to the degree of unidirectional tension or compression. When the deformation is very small (in the linear elastic stage), we assume that this relationship is linear. That is, when the tensile (or compressive) strain in the direction of the x-axis is ε_{xx}, the compressive (or tensile) degree in the y-direction and in the z-direction should be identical. Thus, the lateral strains ε_{yy} and ε_{zz} are

$$\varepsilon_{yy} = \varepsilon_{zz} = -v\frac{\sigma_{xx}}{E},$$

where v denotes the proportionality factor of a linear relationship called the transverse deformation coefficient or the Poisson ratio.

Similarly, if only normal stress σ_{yy} is applied in the y-direction, then

$$\varepsilon_{yy} = \frac{\sigma_{yy}}{E}, \quad \varepsilon_{zz} = \varepsilon_{xx} = -v\frac{\sigma_{yy}}{E}.$$

If only normal stress σ_{zz} is exerted in the z-direction, we have

$$\varepsilon_{zz} = \frac{\sigma_{zz}}{E}, \quad \varepsilon_{xx} = \varepsilon_{yy} = -v\frac{\sigma_{zz}}{E}.$$

Under the condition of linear elasticity, we can apply the superposition principle. If three normal stresses, σ_{xx}, σ_{yy}, σ_{zz} are applied simultaneously along three coordinate axes, then the total strain along each coordinate axis is the linear summation of strains generated by individual stresses in this direction:

$$\begin{aligned}
\varepsilon_{xx} &= \frac{1}{E}\left[\sigma_{xx} - v(\sigma_{yy} + \sigma_{zz})\right] \\
\varepsilon_{yy} &= \frac{1}{E}\left[\sigma_{yy} - v(\sigma_{xx} + \sigma_{zz})\right]. \\
\varepsilon_{zz} &= \frac{1}{E}\left[\sigma_{zz} - v(\sigma_{xx} + \sigma_{yy})\right]
\end{aligned} \tag{1.60a}$$

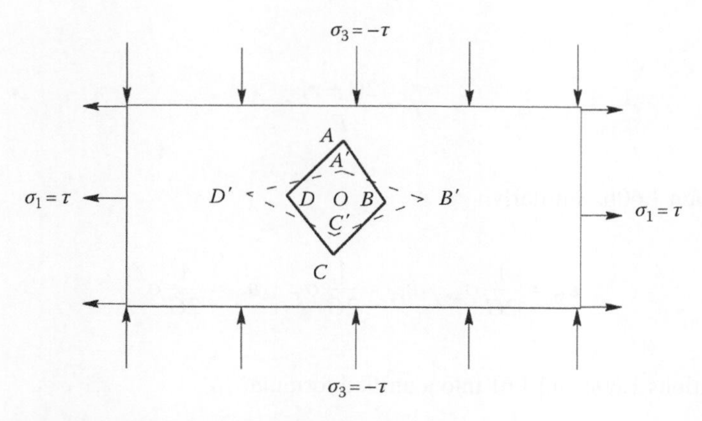

FIGURE 1.13 Deformation caused by the pure shear stress state.

In the stress state of pure shear, a linear relationship exists between the shear stress τ_{xy} and the shear strain γ_{xy}. Deriving from Hooke's law,

$$\tau_{xy} = G\gamma_{xy}, \tag{1.60b}$$

where $G = E/2(1 + \nu)$ is the shear modulus. In fact, for the stress state of pure shear shown in Figure 1.13, the square $ABCD$ with side length $\sqrt{2}a$ transforms to a parallelogram $A'B'C'D'$.

The principal stresses are $\sigma_1 = \tau$, $\sigma_2 = 0$, $\sigma_3 = -\tau$. From Equation 1.60, the relative elongation of the horizontal diagonal in Figure 1.13 is

$$\varepsilon_1 = \frac{1}{E}(\sigma_1 - \nu\sigma_3) = \frac{1+\nu}{E}\tau.$$

The relative contraction of the vertical diagonal is

$$\varepsilon_3 = \frac{1}{E}(\sigma_3 - \nu\sigma_1) = -\frac{1+\nu}{E}\tau.$$

In triangle $A'OB'$, $OB' = a(1 + \varepsilon_1)$, $OA' = a(1 + \varepsilon_3)$.
Therefore,

$$A'B' = \sqrt{a^2(1+\varepsilon_1)^2 + (1+\varepsilon_3)^2 a^2} = a\sqrt{2(1+\varepsilon_1+\varepsilon_2+\varepsilon_3)+\varepsilon_1^2+\varepsilon_2^2}$$

From $\varepsilon_1 + \varepsilon_3 = 0$, we have $A'B' \approx AB = \sqrt{2}a$.
If the change of right angle is γ, then

$$\angle OA'B' = \frac{\pi}{4} + \frac{\gamma}{2}, \quad \tan\left(\frac{\pi}{4} + \frac{\gamma}{2}\right) = \frac{OB'}{OA'} = \frac{a(1+\varepsilon_1)}{a(1+\varepsilon_3)}.$$

Considering that

$$\tan\left(\frac{\pi}{4} + \frac{\gamma}{2}\right) = \frac{1+\dfrac{\gamma}{2}}{1-\dfrac{\gamma}{2}} \approx 1+\gamma, \quad \frac{1+\varepsilon_1}{1+\varepsilon_3} \approx 1+\varepsilon_1-\varepsilon_3,$$

we have

$$\gamma = \frac{2(1+\nu)}{E}\tau. \tag{1.60c}$$

This is Equation 1.60b. Similarly,

$$\varepsilon_{xy} = \frac{1}{2G}\sigma_{xy} \quad \varepsilon_{yz} = \frac{1}{2G}\sigma_{yz} \quad \varepsilon_{zx} = \frac{1}{2G}\sigma_{zx}. \tag{1.61}$$

Writing Equations 1.60a and 1.61 into a unified formula,

$$\varepsilon_{ij} = \frac{1+\nu}{E}\sigma_{ij} - \frac{\nu}{E}\delta_{ij}\Theta, \quad \text{or} \quad \varepsilon = \frac{1+\nu}{E}\sigma - \frac{\nu}{E}\Theta\boldsymbol{I}, \tag{1.62}$$

where $\Theta = \sigma_{xx} + \sigma_{yy} + \sigma_{zz} = \sigma_{kk}$ is the first invariant of the stress tensor. Equation 1.62 is called Hooke's law of isotropic elastic materials. Summation of the three equations in 1.60 gives

$$e = \frac{1-2\nu}{E}\Theta, \tag{1.63a}$$

where $e = \varepsilon_{xx} + \varepsilon_{yy} + \varepsilon_{zz} = \varepsilon_{ii}$ is the volumetric strain. Equation 1.63a can be written as

$$\sigma_m = Ke, \tag{1.63b}$$

where $K = E/3(1 - 2\nu)$ is the bulk modulus, and $\sigma_m = (1/3)\sigma_{kk}$ is the normal mean stress.

Equation 1.63b indicates that the volumetric strain is proportional to the mean normal stress. This is Hooke's law in volume form for isotropic elastic materials. The relationship between the deviatoric stress tensor S and the deviatoric strain tensor $\boldsymbol{\varepsilon}'$ is

$$\boldsymbol{S} = 2G\varepsilon', \quad \text{or} \quad S_{ij} = 2G\varepsilon'_{ij}. \tag{1.63c}$$

Let $\lambda = E\nu/(1 + \nu)(1 - 2\nu)$ and $\mu = E/2(1 + \nu) = G$. Equation 1.62 can be written as

$$\sigma = \lambda e\mathbf{I} + 2\mu\varepsilon, \quad \text{or} \quad \sigma_{ij} = \lambda e\delta_{ij} + 2\mu\varepsilon_{ij}. \tag{1.64}$$

In the aforementioned elastic relationships of isotropic materials, we defined five elastic constants: E, ν, G, λ, K. Only two of them are independent and we generally use E, ν or λ, G or K, G. The conversions between them can be found in Reference [1]. For a given engineering material, E and ν can be measured by a uniaxial tension test, and G can be measured by a torsion test of a thin-walled tube. K can be measured using a hydrostatic pressure test. As an ideal limit case, if $\nu = 1/2$, then the bulk modulus is $K = \infty$. This material is said to be noncompressible, and the associated shear modulus $G = E/3$. Plastic mechanics often uses the noncompressible assumption. The material with a negative Poison ratio has not yet been discovered. The experimentally measured value for ν is in the range of $0 < \nu < 1/2$. For metallic materials, the value of the Poisson ratio has insignificant influence on stress and strain and frequently takes a value of approximately 0.3.

Since the 1990s, a new name has been proposed for metamaterials. Metamaterials are typically man-made microstructural materials, with many properties that are quite different from the conventional materials, such as negative Poisson's ratio. Readers can refer to the appropriate references [7–9].

Many engineering components—such as dams, channels, thick-walled tubes, cylinders, and thin plates subjected to in-plane loading—can be simplified as two-dimensional plane problems: plane-stress problems and plane-strain problems.

In plane-stress problems, we study objects like thin plates, which deform elastically. Assume that a thin plate with thickness h (much less than the dimensions of the other two directions) is subjected to a surface force. The force acts on the side of the plate, is parallel to the plate plane, and does not change along the thickness direction. Now, take the mid-section plane at $z = 0$ as the xoy plane, and take the line perpendicular to the mid-section plane as the z-axis. Since the two surfaces at $z = \pm(h/2)$ are without stress, $\sigma_z = \sigma_{zy} = \sigma_{zx} = 0$. In this way, the stress–strain relationship for the plane-stress state can be expressed by

$$\varepsilon_z = -\frac{V}{E}(\sigma_x + \sigma_y), \quad \varepsilon_{xy} = \frac{1}{2G}\sigma_{xy} = \frac{1+v}{E}\sigma_{xy} \tag{1.65}$$

As a typical plane-strain problem, we consider a long cylinder with an equal transverse section. The axis of the cylinder is parallel to the z-axis, and the dimension in the axis-direction is much larger than the dimensions in the other two directions. Assume that the side of the cylinder is subjected to a surface force, which is perpendicular to the z-axis and does not change in the z-direction. If the cylinder is indefinitely long or rigid constraint is applied at two ends (the length of the cylinder being the limiting distance), then the displacement in the axis direction is constrained, and any transverse section can be regarded as a symmetric plane. As a result, any point inside the cylinder can displace only in the x- and y-directions, but cannot displace in the z-direction. That is, the displacement components u and v are irrelevant to the coordinate z of the point and are functions of x and y only, and the displacement component w is zero. Therefore, the strain components are

$$\varepsilon_x = \frac{\partial u}{\partial x}, \quad \varepsilon_y = \frac{\partial v}{\partial y}, \quad \varepsilon_{xy} = \frac{1}{2}\left(\frac{\partial v}{\partial x} + \frac{\partial u}{\partial y}\right), \quad \varepsilon_z = \varepsilon_{yz} = \varepsilon_{zx} = 0\,. \tag{1.66}$$

From $\varepsilon_z = 0$, we have $\sigma_z = \nu(\sigma_x + \sigma_y)$. From Equation 1.62, we obtain the stress–strain relationship of the plane-strain state:

$$\varepsilon_x = \frac{1}{E}\left[(1-v^2)\sigma_x - v(1+v)\sigma_y\right], \quad \varepsilon_y = \frac{1}{E}\left[(1-v^2)\sigma_y - v(1+v)\sigma_x\right], \quad \varepsilon_{xy} = \frac{1}{2G}\sigma_{xy}. \tag{1.67}$$

Now introduce symbols

$$E' = \frac{E}{1-v^2}, \quad v' = \frac{v}{1-v}.$$

The constitutive equations of the plane-strain state can be written as

$$\varepsilon_x = \frac{1}{E'}(\sigma_x - v'\sigma_y), \quad \varepsilon_y = \frac{1}{E'}(\sigma_y - v'\sigma_x), \quad \varepsilon_{xy} = \frac{1}{2G}\sigma_{xy}. \tag{1.68}$$

In this way, the constitutive relationships for plane problems can be unified as in Equation 1.68, where

$$E' = \begin{cases} E & \text{plane stress} \\ \dfrac{E}{1-\nu^2} & \text{plane strain} \end{cases}$$

$$\nu' = \begin{cases} \nu & \text{plane stress} \\ \dfrac{\nu}{1-\nu} & \text{plane strain} \end{cases}$$

Consequently, the mathematical manipulation of the two types of plane problems is identical.

1.4.2 Elastic Strain Energy Function

When subjected to an external load, an elastic body unavoidably deforms. Simultaneously, the potential energy of the external force changes. When the external load is applied slowly, the kinetic energy of the body can be ignored. If the dissipation of other energies (such as heat energy) can be ignored, the change in potential energy of the external force completely converts to strain energy (a type of potential energy) stored in the deformed body [1,2].

For the infinitesimal element in the body as shown in Figure 1.14, the stress can be viewed as an external force acting on its surface. We calculate the work done by all surface stresses to the infinitesimal element. Generally, the stress–strain relationship is nonlinear, as shown in Figure 1.14a.

The coupled stresses σ_{11} acting on the two sides of the infinitesimal element only do work on the elongation of the infinitesimal element caused by the normal strain ε_{11}. This work is

$$\mathrm{d}A_1 = \int_0^{\varepsilon_{11}} \sigma_{11}\mathrm{d}x_2\mathrm{d}x_3 \cdot \mathrm{d}\varepsilon_{11}\mathrm{d}x_1 = \int_0^{\varepsilon_{11}} \sigma_{11}\mathrm{d}\varepsilon_{11}\mathrm{d}V.$$

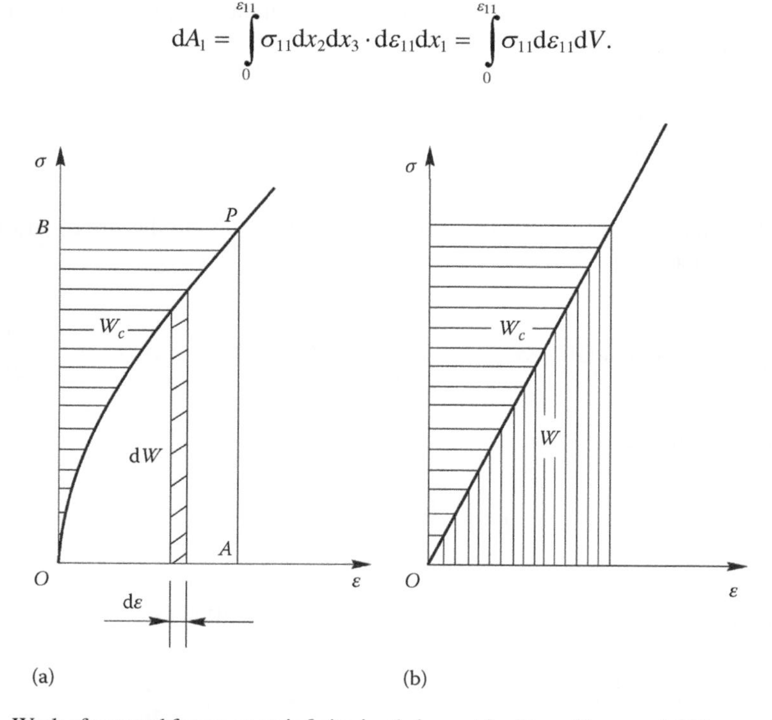

FIGURE 1.14 Work of external forces on an infinitesimal element for (a) nonlinear and (b) linear elastic materials.

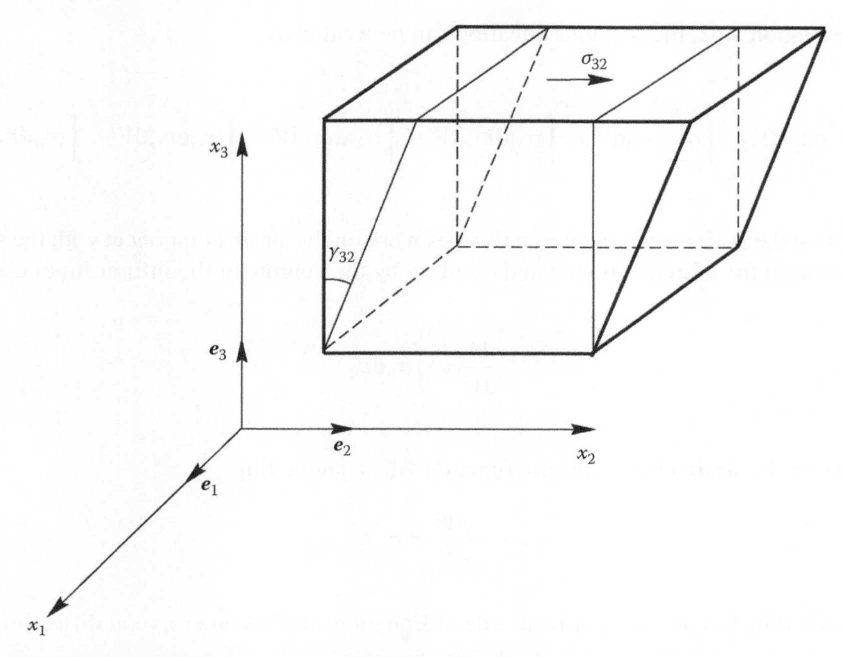

FIGURE 1.15 Work of the shear stress σ_{32} on the distortion of an infinitesimal element.

Similar results can be obtained for other normal stresses. Therefore, the total work done by the normal stresses on the elongation of the infinitesimal element is

$$\int_0^{\varepsilon_{11}} (\sigma_{11} \mathrm{d}\varepsilon_{11}) \mathrm{d}V + \int_0^{\varepsilon_{22}} \sigma_{22} \mathrm{d}\varepsilon_{22} \mathrm{d}V + \int_0^{\varepsilon_{33}} \sigma_{33} \mathrm{d}\varepsilon_{33} \mathrm{d}V.$$

Next, we discuss the work done by shear stresses, which cause distortion of the infinitesimal element. For the right parallelepiped shown in Figure 1.15, we consider the effect of shear stress σ_{32} acting on face 3. An increment of σ_{32} leads to an infinitesimal shear strain $\mathrm{d}\gamma_{32}$, due to the reduction in angle between the y-axis and the z-axis. A simple analysis shows that the infinitesimal displacement due to a reduction in angle between the y-axis and the z-axis, caused by the force, is $\mathrm{d}\gamma_{32} \cdot \mathrm{d}x_3$. Therefore, the work done by the force $\sigma_{32} \mathrm{d}x_1 \mathrm{d}x_2$ is

$$\mathrm{d}A_2 = \int_0^{\gamma_{32}} \sigma_{32} \mathrm{d}x_1 \mathrm{d}x_2 \cdot \mathrm{d}\gamma_{32} \mathrm{d}x_3 = \int_0^{\gamma_{32}} \sigma_{32} \mathrm{d}\gamma_{32} \mathrm{d}V.$$

Similar results can be obtained for other shear stresses. Therefore, the work done by the shear stresses on the distortion of an infinitesimal element is

$$\int_0^{\gamma_{12}} \sigma_{12} \mathrm{d}\gamma_{12} \mathrm{d}V + \int_0^{\gamma_{32}} \sigma_{32} \mathrm{d}\gamma_{32} \mathrm{d}V + \int_0^{\gamma_{13}} \sigma_{13} \mathrm{d}\gamma_{13} \mathrm{d}V.$$

Using Equation 1.52, the previous equation can be written as

$$\int_0^{\varepsilon_{12}}\sigma_{12}\mathrm{d}\varepsilon_{12}\mathrm{d}V+\int_0^{\varepsilon_{21}}\sigma_{21}\mathrm{d}\varepsilon_{21}\mathrm{d}V+\int_0^{\varepsilon_{32}}\sigma_{32}\mathrm{d}\varepsilon_{32}\mathrm{d}V+\int_0^{\varepsilon_{23}}\sigma_{23}\mathrm{d}\varepsilon_{23}\mathrm{d}V+\int_0^{\varepsilon_{13}}\sigma_{13}\mathrm{d}\varepsilon_{13}\mathrm{d}V+\int_0^{\varepsilon_{31}}\sigma_{31}\mathrm{d}\varepsilon_{31}\mathrm{d}V$$

indicating that the stress component σ_{ij} only does work on the strain component with the same indices ε_{ij}. Summing the aforementioned and dividing by the volume of the infinitesimal element dV,

$$\frac{\mathrm{d}A}{\mathrm{d}V}=\int_0^{\varepsilon_{ij}}\sigma_{ij}\mathrm{d}\varepsilon_{ij}. \tag{1.69}$$

Introducing the strain energy density function $W(\varepsilon_{ij})$, and letting

$$\frac{\partial W}{\partial\varepsilon_{ij}}=\sigma_{ij}, \tag{1.70}$$

the integrated function on the right-hand side of Equation 1.69 becomes a total differential:

$$\sigma_{ij}\mathrm{d}\varepsilon_{ij}=\frac{\partial W}{\partial\varepsilon_{ij}}\mathrm{d}\varepsilon_{ij}=\mathrm{d}W. \tag{1.71}$$

Equation 1.69 becomes

$$\frac{\mathrm{d}A}{\mathrm{d}V}=\int_0^{\varepsilon_{ij}}\mathrm{d}W=W(\varepsilon_{ij})-W(0), \tag{1.72}$$

where $W(0)$ and $W(\varepsilon_{ij})$ are the strain energy densities before and after deformation, respectively. In general, the original state before deformation is taken as a reference state; thus, $W(0)=0$. The aforementioned equation indicates that (1) the strain energy density stored during deformation is equal to the work per unit volume done by the external force and (2) the strain energy density after deformation is only relevant to the original state and the final deformation state and is irrelevant to the deformation history. This type of function, which only depends on the state and is irrelevant to the history of the state, is called a state function in thermodynamics.

Equation 1.70 is called Green's formula. It indicates that the strain energy is another expression of the constitutive equation of an elastic material. When the specific function of $W(\varepsilon_{ij})$ is given, the stress–strain relationship can be completely determined by Equation 1.70.

When W is second or higher-order differentiable with respect to ε_{ij}, from

$$\frac{\partial}{\partial\varepsilon_{kl}}\left(\frac{\partial W}{\partial\varepsilon_{ij}}\right)=\frac{\partial}{\partial\varepsilon_{ij}}\left(\frac{\partial W}{\partial\varepsilon_{kl}}\right)$$ and Equation 1.70, we have

$$\frac{\partial\sigma_{ij}}{\partial\varepsilon_{kl}}=\frac{\partial\sigma_{kl}}{\partial\varepsilon_{ij}}. \tag{1.73}$$

This is called general Green's formula.

For the linear elastic materials, $W = 1/2\sigma_{ij}\varepsilon_{ij}$. For the isotropy elastic materials, $W = \left(\frac{1}{2}\lambda + G\right)I_1^2 - 2GI_2$. Here, I_1 and I_2 are respective the first and second invariants of a strain tensor in Equation 1.57. Similar to Equation 1.69, another state function, $W_c = \int_0^{\sigma_{ij}} \varepsilon_{ij} d\sigma_{ij}$, named complementary strain energy or complementary energy for simplify, as shown in Figure 1.14b.

We have mentioned that the deformation of a body can be divided into two parts: volume change and shape change. Correspondingly, strain energy can also be divided into two parts. It is easy to understand that the isotropic mean normal stress (hydrostatic stress) causing the volume change is $\sigma_m = 1/3(\sigma_x + \sigma_y + \sigma_z)$, and the corresponding mean normal strain is $\varepsilon_m = 1/3(\varepsilon_x + \varepsilon_y + \varepsilon_z)$. That is, the following stress state does not introduce shape change of the infinitesimal element

$$\sigma_{ij} = \begin{bmatrix} \sigma_m & 0 & 0 \\ 0 & \sigma_m & 0 \\ 0 & 0 & \sigma_m \end{bmatrix}.$$

Therefore, the strain energy per unit volume due to the volumetric change (volume-related strain energy) is

$$u_V = \frac{3}{2}\sigma_m\varepsilon_m = \frac{\sigma_m^2}{2K} = \frac{1}{18K}(\sigma_x + \sigma_y + \sigma_z)^2. \tag{1.74}$$

The stress state causing shape change is the deviatoric stress tensor S_{ij}:

$$S_{ij} = \begin{bmatrix} \sigma_x - \sigma_m & \tau_{xy} & \tau_{xz} \\ \tau_{yx} & \sigma_y - \sigma_m & \tau_{yz} \\ \tau_{zx} & \tau_{zy} & \sigma_z - \sigma_m \end{bmatrix}.$$

The strain energy per unit volume due to the shape change (distortion energy) is

$$u_d = \frac{1}{2}S_{ij}\varepsilon'_{ij} = \frac{1}{2G}J_2. \tag{1.75}$$

The total strain energy is

$$W = u_V + u_d = \frac{1}{18K}I_1^2 + \frac{1}{2G}J_2. \tag{1.76}$$

1.4.3 Yield Function and Yield Surface

1.4.3.1 Yield Function

As shown on the stress–strain curve of a material under simple tension or compression (Figure 1.1), when the stress goes beyond σ_0, the stress–strain relationship no longer follows Hooke's law. σ_0 is the yield stress under simple tension. In such a simple stress state, the yield stress can be immediately detected from the simple tension (or compression) curve. When the boundary between elastic deformation and plastic deformation is not apparent, σ_0 can be determined by a certain rule, and its value can be used in engineering design. For a complicated stress state, the analysis in Section 1.2.1 indicates that six stress components are needed to completely describe the stress state at a point. In the elastic range, the stress–strain relationship obeys Hooke's law. If the stress at this point is large, Hooke's law no longer holds true. Then we face the following problems: How do we judge whether this point yields? What is the stress–strain relationship after yielding? Does the material yield when one stress component reaches a critical value, when each stress component reaches its critical value, or when a certain combination of six stress components reaches a critical value?

Here are two examples demonstrating the complexity in defining the yield condition.

Example 1.1

Consider a thin-walled tube subjected to an internal pressure p, an axial tensile force q, and a circumferential torque T. The stress state in the tube wall can be regarded as a plane-stress state with sufficient accuracy. When the external loads change, the combination of the internal stresses changes. If only the axial tensile force q and torque T are present, the circumferential stress, axial stress, and shear stress, respectively, are

$$\sigma_{\varphi} = 0, \quad \sigma_z = \frac{q}{2\pi a h}, \quad \tau_{\varphi z} = \frac{T}{2\pi a^2 h}$$

where

a is the average radius

h is the wall thickness

$h \ll a$

Example 1.2

If only the axial tensile force q and the internal pressure p are present, the corresponding stresses are

$$\sigma_{\varphi} = \frac{pa}{h}, \quad \sigma_z = \frac{q}{2\pi a h}, \quad \tau_{\varphi z} = 0.$$

These two examples indicate that under the simultaneous action of circumferential stress, axial stress, and shear stress, the stress state of a certain point in the tube wall can enter the plastic state, and thus we can obtain the yield condition under this stress state. Whether or not the stress state reaches the yield condition needs to be determined by experimentation. In general, the yield condition under a complicated stress state is determined by experimentation. Moreover, if the yield conditions under a variety of combinations of internal stresses are determined only by experimentation, then the number of experiments will be remarkable. At the same time, theoretical analysis requires us to obtain the analytical expression for the yield condition. Consequently, it is necessary to experimentally develop the yield theory [1–3].

The boundary of the original elastic state under a complex stress state is called the yield criterion. In general, it is a function of stress σ_{ij}, strain ε_{ij}, time t, and temperature T. Therefore, the yield criterion can be written as the following function:

$$\Phi(\sigma_{ij}, \varepsilon_{ij}, t, T) = 0. \tag{1.77}$$

For the cases in which the time effect can be ignored and the temperature is fixed, time t and temperature T have no influence on the yield criterion, and the plastic state, after yielding. Thus, the variables of Φ will not include t and T. The material is in an elastic state before yielding, and there exists a one-to-one mapping between stress and strain, so ε_{ij} in Φ can be represented by σ_{ij}. Consequently, the yield criterion is only a function of the stress components. We write this function as

$$F(\sigma_{ij}) = 0. \tag{1.78}$$

Advanced mathematics has established the concept of n-dimensional spaces, and we will illustrate the yield criterion via the language of n-dimensional space. If we take the six stress components σ_{ij} as coordinate axes, we can construct a six-dimensional stress space. In this space, the equation $F(\sigma_{ij}) = 0$ represents a surface surrounding the origin of the coordinates. The origin represents a zero-stress state or the original state before deformation. We call this surface the yield surface. When the point of stress σ_{ij} is inside of this surface, $F(\sigma_{ij}) < 0$, and the material is in an elastic state. When the point of stress is on the surface, $F(\sigma_{ij}) = 0$, and the material begins to yield.

When isotropy is assumed, the yield criterion is irrelevant to the choice of coordinate axes. Although the magnitudes of the six stress components depend on the choice of coordinate system, the three principal stresses of a stress state are independent of the choice of coordinate system. Thus, Equation 1.78 can be written as

$$F(\sigma_1,\sigma_2,\sigma_3) = 0, \tag{1.79}$$

and the yield criterion can be represented by a surface in the principal stress space σ_1, σ_2, σ_3.

Experimentation has discovered that hydrostatic stress has no influence on the yield condition and the plastic state. Thus, yield criterion is only related to the invariants of the deviatoric stress components. Equation 1.79 can also be written as

$$f(S_{ij}) = 0. \tag{1.80}$$

In this way, the yield function is a function of the deviatoric stress components and can be discussed in the principal stress space of σ_1, σ_2, σ_3, which is a Cartesian coordinate system. The principal stress space is a three-dimensional space that gives the geometric shape of yield function. Understanding the intuitive geometric shape is helpful for understanding the yield surface.

1.4.3.2 Π Space

We can represent the stress state of a point by a geometric method. When the directions of the principal stresses are known, we can select the principal stresses σ_1, σ_2, σ_3 as coordinate axes and represent the stress state of a material point as a point in the principal stress space. Because the principal stress space is three dimensional, we can obtain a relatively intuitive geometric image.

In the principal stress space, the stress state of a point can be described by a vector $\overrightarrow{OP}$ (Figure 1.16). Letting $\boldsymbol{i}, \boldsymbol{j}, \boldsymbol{k}$ be the unit vectors of the three coordinate axes in the principal stress space, we have

$$\overrightarrow{OP} = \sigma_1\boldsymbol{i} + \sigma_2\boldsymbol{j} + \sigma_3\boldsymbol{k} = (S_1\boldsymbol{i} + S_2\boldsymbol{j} + S_3\boldsymbol{k}) + (\sigma_m\boldsymbol{i} + \sigma_m\boldsymbol{j} + \sigma_m\boldsymbol{k}) = \overrightarrow{OQ} + \overrightarrow{ON}. \tag{1.81}$$

Here, $\sigma_m = 1/3(\sigma_1 + \sigma_2 + \sigma_3)$ is the mean normal stress. From Figure 1.16, vector $\overrightarrow{OQ}$ is the deviatoric vector of the principal stresses, and the vector $\overrightarrow{ON}$ is equally inclined to the σ_1, σ_2, σ_3 axes, which is normal to the following plane passing through the origin:

$$\sigma_1 + \sigma_2 + \sigma_3 = 0. \tag{1.82}$$

The aforementioned plane, on which the mean normal stress is zero, is called the π plane. The π plane is an isoclinic plane in the principal stress space. Since the three components S_1, S_2, S_3 of $\overrightarrow{OQ}$ satisfy

$$S_1 + S_2 + S_3 = 0, \tag{1.83}$$

the deviatoric stress vector $\overrightarrow{OQ}$ is always on the π plane.

In order to discuss the yield surface in the principal stress space, we introduce the line Λ. Line Λ passes through the origin O and is equally inclined to the three coordinate axes (54°44′). Its direction cosines are $\nu_1 = \nu_2 = \nu_3 = 1/\sqrt{3}$, represented by the vector $\overrightarrow{ON}$ shown in Figure 1.16. The stress state of

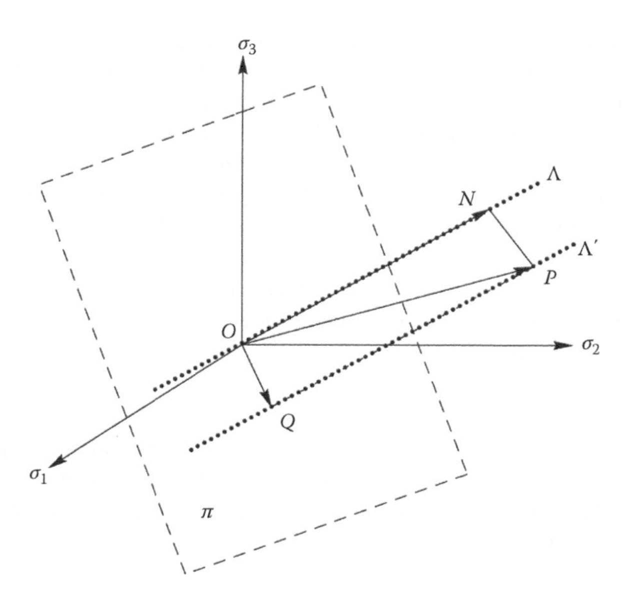

FIGURE 1.16 Principal stress state.

any point on line Λ has only a spherical stress tensor, but not a deviatoric stress tensor. This stress state is called the hydrostatic stress state, and it has no influence on the yielding of a material. The vector $\overrightarrow{OQ}$ represents the deviatoric stress tensor of the stress state at point P and leads to the yielding of material.

1.4.3.3 Yield Surface

Now draw a line Λ' passing though point P and parallel to line Λ. Obviously, the decomposed vector on the plane π of the stress vector of any point on line Λ' is identical to the decomposed vector $\overrightarrow{OQ}$ on the plane π of the stress vector $\overrightarrow{OP}$. That is, they all have deviatoric stress tensors identical to point P. The differences in their stress states are the decomposed vectors on line Λ, indicating that their spherical stress tensors are different. The yielding of a material is dependent only on the deviatoric stress tensor; so if point P yields, then all points on line Λ' yield. Therefore, in the principal stress space, the yield surface is the surface of a right cylinder with line Λ as the axis. The generatrix line is parallel to line Λ and perpendicular to plane π, as shown in Figure 1.17a.

If the stress state of a point is inside the yield surface, the point is in the elastic state. If the stress state of a point is on the yield surface, the point begins to yield.

Examining the yielding from a physics viewpoint, the yield locus on the π plane has the following important characteristics:

1. The yield locus is a closed curve surrounding the origin of the coordinates. At the origin, all stress components are zero, and the material obviously does not yield. Therefore, the yield locus does not pass through the origin, but surrounds the origin. The interior of the yield locus represents the elastic state, while the exterior of the yield locus represents the plastic state. If the locus is not closed, then some stress states with very high stress values exist even though the material does not yield. Apparently, this is impossible.
2. (a) The yield locus must intersect with an arbitrary position vector starting from the origin of the coordinates, and (b) the intersection occurs only once. (a) holds true apparently. As for (b), because the initial yield occurs, it is impossible to yield again at a stress state with higher stress components. That is why the initial yield takes place only once.
3. The yield locus is symmetric with respect to the three coordinate axes and their perpendicular lines.

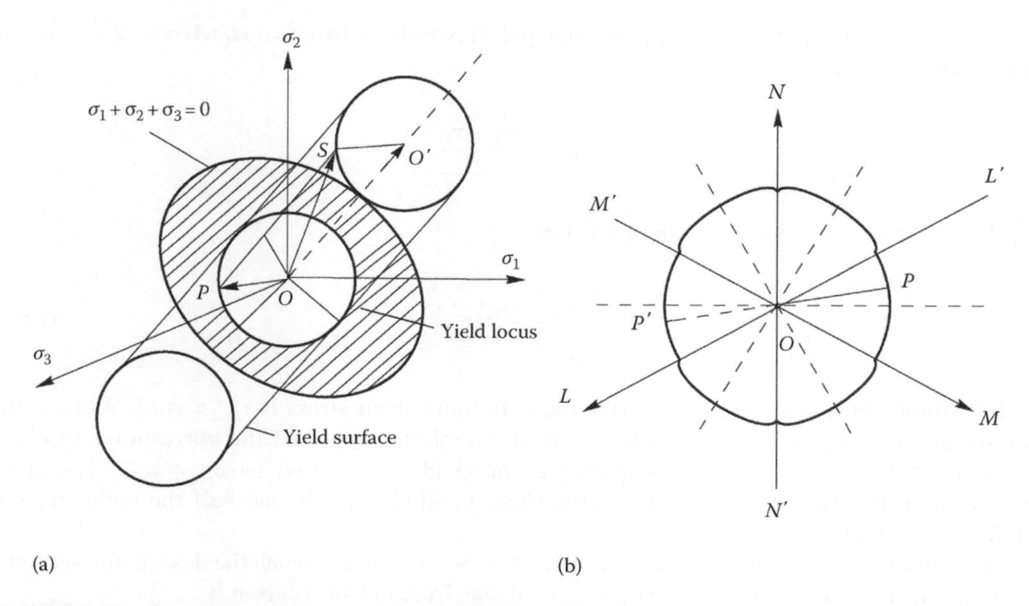

FIGURE 1.17 Symmetry of (a) yield surface and (b) yield locus.

Because the π plane is equally inclined with respect to the three principal stresses, the projections of the three principal stress axes σ_1, σ_2, σ_3 on the π plane oL, oM, and oN are three axes forming mutual 120° angles with each other, as shown in Figure 1.17b. Because the initial yield of a material is isotropic, if σ_1, σ_2, σ_3 are the yield stresses, then σ_1, σ_2, σ_3 are also yield stresses. Thus, the yield locus must be symmetric with respect to LL', the projection of the σ_1 axis on the π plane. Similarly, the yield locus must also be symmetric to MM' and NN'.

Another assumption is made regarding the yield of a material: the magnitudes of the positive yield stress and the negative yield stress are equal. Therefore, the line passing through point o must intersect with the yield locus at points that are equidistant from the origin. Thus, the yield locus is not only symmetric to the three axes noted as LL', MM', NN', but also symmetric to the isogonal lines of the three axes.

In summary, the yield locus has 6 symmetric lines, which divide the yield locus into 12 identical fan-shaped zones with a central angle of 30°. As long as we know any 1 of the 12 zones, we can obtain the entire yield locus based on the yield locus symmetry.

1.4.4 Two General Yield Criteria

In the previous section, we learned that the yield criterion is the surface of a cylinder in the space of principal stresses. In this section, we will discuss two general yield conditions, also called yield criteria. These two criteria—which have been validated by extensive engineering practice—are appropriate for practical metallic materials and are convenient in engineering applications. One is the Tresca yield criterion, and the other is the von Mises yield criterion [1,3].

1.4.4.1 The Tresca Yield Criterion

The Tresca yield criterion is also called the maximum shear stress yield criterion [1,3]. According to this criterion, when the maximum shear stress reaches a critical value

$$\tau_{\max} = \tau_y, \tag{1.84a}$$

the material yields. τ_y is the shear yield stress of the material, which can be determined by experimentation.

When the descending sequence of the principal stresses is known, that is, when $\sigma_1 > \sigma_2 > \sigma_3$, the maximum shear stress is [1]

$$\tau_{\max} = \frac{\sigma_1 - \sigma_3}{2}.$$

Hence, Tresca yield criterion can be written as

$$\frac{\sigma_1 - \sigma_3}{2} = \tau_y. \tag{1.84b}$$

For simple tension $\sigma_1 > 0$, $\sigma_2 = \sigma_3 = 0$, the maximum shear stress is $\tau_{\max} = \sigma_1/2$. When yielding occurs, $\sigma_1 = \sigma_s$. σ_s is the yield stress under simple tension, and the corresponding maximum shear stress is $\tau_{\max} = \sigma_s/2$. Comparing to the yield criterion, we have $\tau_y = \sigma_s/2$. Therefore, according to the Tresca yield criterion, the shear yield stress τ_y is one-half the yield stress σ_s under simple tension.

Substituting $\tau_y = \sigma_s/2$ into $\sigma_1 - \sigma_3/2 = \tau_y$ gives $\sigma_1 - \sigma_3 = \sigma_s$. Hence, when the descending sequence of principal stresses is known (when $\sigma_1 > \sigma_2 > \sigma_3$), the Tresca yield criterion is

$$\frac{\sigma_1 - \sigma_3}{2} = \tau_y = \frac{\sigma_s}{2} \tag{1.84c}$$

or

$$\sigma_1 - \sigma_2 = \sigma_s.$$

When the sequence of principal stresses is unknown, the criterion can be expressed as

$$|\sigma_1 - \sigma_2| = \sigma_s \quad |\sigma_2 - \sigma_3| = \sigma_s \quad |\sigma_3 - \sigma_1| = \sigma_s. \tag{1.85a}$$

As long as any of the equations in (1.85a) holds true, the material starts to yield. The aforementioned conditions can be written in a unified form:

$$\tau_{\max} = \frac{1}{2}\max\{|\sigma_1 - \sigma_2|, \quad |\sigma_2 - \sigma_3|, \quad |\sigma_3 - \sigma_1|\} = \tau_y = \frac{\sigma_s}{2}. \tag{1.85b}$$

As long as any term of $1/2|\sigma_1 - \sigma_2|, 1/2|\sigma_2 - \sigma_3|, 1/2|\sigma_3 - \sigma_1|$ in Equation 1.85b reaches τ_y, the material yields; if all three terms are less than τ_y, the material is in the elastic state. It can be seen from the criterion (1.84c) that the expression only contains the maximum principal stress σ_1 and the minimum principal stress σ_3, but does not contain the middle principal stress σ_2. Obviously, in the Tresca yield criterion, the middle principal stress does not influence the yielding condition and is not considered in the expression.

From the unified form Equation 1.85b of Tresca's criterion, the yield surface of Tresca's criterion is a right hexagonal prism perpendicular to the plane π, as shown in Figure 1.18a. This right hexagonal prism is usually called Tresca's hexagonal prism. The yield locus of the projection of the yield surface on the π plane is a regular hexagon, see Figure 1.18b. In general, this regular hexagon is called Tresca's regular hexagon.

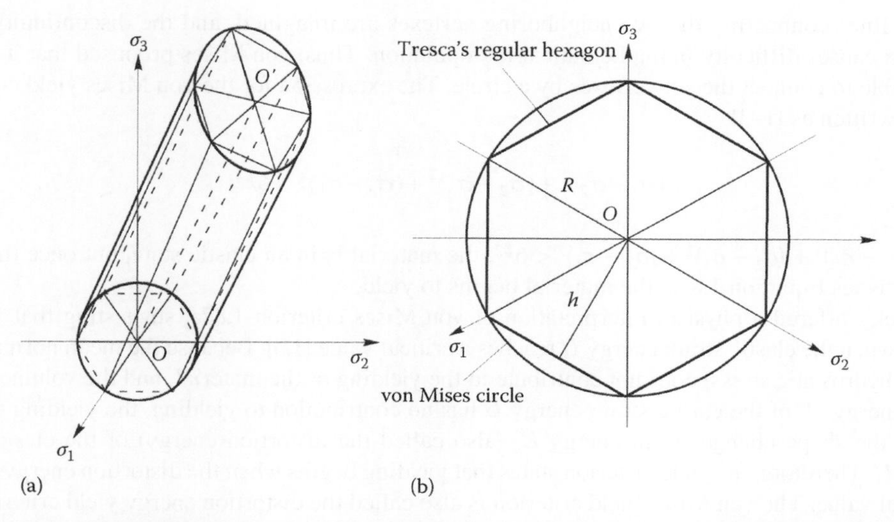

FIGURE 1.18 (a) Right hexagonal prism of the Tresca yield criterion. (b) The regular hexagon on the π plane.

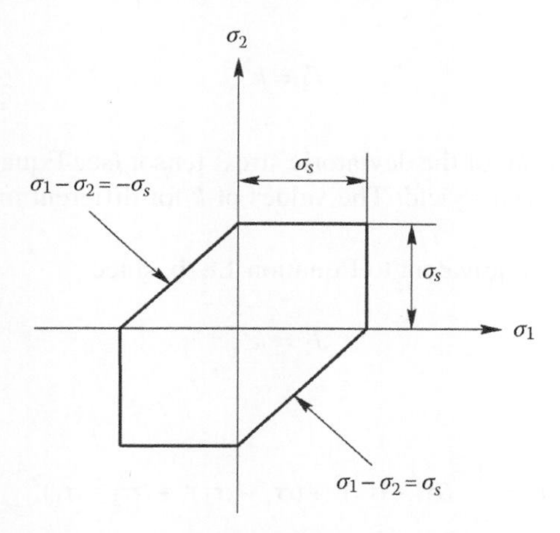

FIGURE 1.19 Irregular hexagon on the $\sigma_1 - \sigma_2$ plane of Tresca yield criterion of the plane-stress state.

For the plane-stress state, $\sigma_3 = 0$, the criterion can be simplified as

$$|\sigma_1 - \sigma_2| = \sigma_s \quad |\sigma_2| = \sigma_s \quad |\sigma_1| = \sigma_s. \tag{1.86}$$

The projection of this equation on the $\sigma_1 - \sigma_2$ plane is an irregular hexagon, as shown in Figure 1.19. This irregular hexagon is usually called Tresca's irregular hexagon, and it is the intersection of Tresca's hexagonal prism with the $\sigma_1 - \sigma_2$ plane.

The mathematical expression of the Tresca yield criterion is very simple and agrees with experimental results relatively well. However, when applying this criterion, the sequence of the principal stresses must be predetermined in order to calculate the maximum shear stress τ_{max}. However, it is relatively difficult to apply this criterion because in general, the sequence of the principal stresses is unknown and varies with the load.

1.4.4.2 The von Mises Yield Criterion

The von Mises yield criterion was initially proposed to simplify the calculation. It should be pointed out that the six vertexes of Tresca's regular hexagon on the π plane are obtained by experiment, but the lines connecting the two neighboring vertexes are imagined, and the discontinuity of the hexagon causes difficulty in mathematical manipulation. Thus, von Mises proposed that it is more reasonable to connect the six vertexes by a circle. The expression for the von Mises yield condition can be written as [1–3]

$$(\sigma_1-\sigma_2)^2+(\sigma_2-\sigma_3)^2+(\sigma_3-\sigma_1)^2=6k^2. \tag{1.87a}$$

If $(\sigma_1-\sigma_2)^2+(\sigma_2-\sigma_3)^2+(\sigma_3-\sigma_1)^2<6k^2$, the material is in an elastic state, but once the stress state satisfies Equation 1.87a, the material begins to yield.

Hencky offered a physical interpretation of von Mises criterion 1.87a, suggesting that yielding begins when the elastic strain energy U reaches a critical value [1,3]. Because the mean normal stress σ_m (the hydrostatic stress) does not contribute to the yielding of the material, and the volume-related strain energy U_v of the elastic strain energy U has no contribution to yielding, the yielding quantity is only the shape change strain energy U_d (also called the distortion energy) of the elastic strain energy U. Therefore, the yield criterion states that yielding begins when the distortion energy reaches a critical value. The von Mises yield criterion is also called the distortion energy yield criterion.

Because the distortion energy $U_d=(1/2G)J_2'$, or $J_2'=2GU_d$, the von Mises yield criterion can be written as

$$J_2'=k^2, \tag{1.87b}$$

where J_2' is the second invariant of the deviatoric stress tensor (see Equation 1.34) and k is a parameter to characterize the material's yield. The values of k for different materials can be determined by experimental testing.

Equation 1.87a is actually equivalent to Equation 1.87b, since

$$J_2'=k^2,$$

while

$$J_2'=\frac{1}{6}\left[(\sigma_1-\sigma_2)^2+(\sigma_2-\sigma_3)^2+(\sigma_3-\sigma_1)^2\right],$$

thus

$$\frac{1}{6}\left[(\sigma_1-\sigma_2)^2+(\sigma_2-\sigma_3)^2+(\sigma_3-\sigma_1)^2\right]=k^2,$$

and

$$(\sigma_1-\sigma_2)^2+(\sigma_2-\sigma_3)^2+(\sigma_3-\sigma_1)^2=6k^2.$$

When yielding occurs under simple tension, $\sigma_1=\sigma_s$, $\sigma_2=\sigma_3=0$. Substituting into the previous equation gives $k=\left(1/\sqrt{3}\right)\sigma_s$.

When yielding occurs under pure shear, $\sigma_1=-\sigma_3=\tau_y$, $\sigma_2=0$, substituting into the previous equation gives $k=\tau_y$.

Comparing the results under simple tension and pure shear, we have

$$\tau_y=\frac{1}{\sqrt{3}}\sigma_s.$$

According to this criterion, the shear yield stress τ_y is $1/\sqrt{3} \approx 0.577$ of the yield stress σ_s under simple tension.

Substituting $k = \left(1/\sqrt{3}\right)\sigma_s$ into the criterion, we have

$$J_2' = \frac{1}{6}\left[(\sigma_1-\sigma_2)^2+(\sigma_2-\sigma_3)^2+(\sigma_3-\sigma_1)^2\right] = k^2 = \frac{1}{3}\sigma_s^2,$$

or

$$(\sigma_1-\sigma_2)^2+(\sigma_2-\sigma_3)^2+(\sigma_3-\sigma_1)^2 = 2\sigma_s^2. \tag{1.87c}$$

Apparently, the yield surface represented by Equation 1.87c is a circular cylinder perpendicular to the π plane, as shown in Figure 1.18a. This circular cylinder is usually called the von Mises circular cylinder. Its yield locus on the π plane is a circle, as shown in Figure 1.18b.

It can be proved that the von Mises circular cylinder is outscribed to Tresca's regular hexagonal prism and that, on the π plane, von Mises circle is the circumcircle of Tresca's regular hexagon, and that the radius of this circle is $r = \sqrt{2/3}\,\sigma_s$.

In the case of plane stress, $\sigma_3 = 0$. From Equation 1.87c, the criterion can be simplified as

$$\sigma_1^2 - \sigma_1 \quad \sigma_2 + \sigma_2^2 = \sigma_s^2, \tag{1.88}$$

or

$$\left(\frac{\sigma_1}{\sigma_s}\right)^2 - \left(\frac{\sigma_1}{\sigma_s}\right)\left(\frac{\sigma_2}{\sigma_s}\right) + \left(\frac{\sigma_2}{\sigma_s}\right)^2 = 1.$$

This equation represents an ellipse in the plane $\sigma_1 - \sigma_2$; see Figure 1.20. This ellipse is usually called von Mises ellipse, and it is the intersection of von Mises circular cylinder with the plane $\sigma_1 - \sigma_2$. Experimental results show that the von Mises yield criterion is closer to the experimental results than the Tresca yield criterion [1,3].

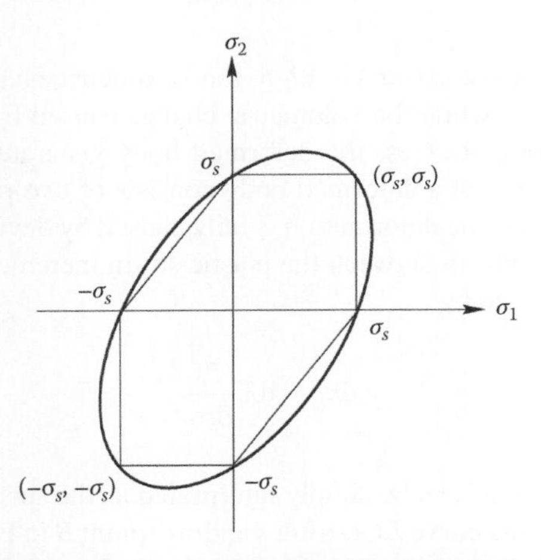

FIGURE 1.20 Irregular hexagon on the plane $\sigma_1 - \sigma_2$ of the Tresca yield criterion and the ellipse on the plane $\sigma_1 - \sigma_2$ of von Mises yield criterion in the case of plane stress.

1.4.5 Incremental Theory of Plasticity

When the stress state of a point in the deformed body meets the yield criterion, the material begins plastic deformation, and the elastic constitutive relationship is no longer applicable to the point. We need to develop the constitutive equations for plastic deformation, which describe the relationship between plastic stress and strain or between the stress increment and the strain increment.

From Figure 1.1 in Section 1.1.3, we see that the important characteristics of the plastic stress–strain relationship are nonlinearity and nonuniqueness. Nonlinearity means that the stress–strain relationship is not linear. Nonuniqueness means that the strain cannot be uniquely determined by stress. As we know, when the load changes, stress also changes, and the point in the stress space, representing the stress state, moves correspondingly. The locus of the movement of the stress point is called the stress path, and this activity is called the stress history. In the plastic deformation stage, the strain state depends not only on the stress state, but also on the entire stress history. Strain is a function of stress and stress history.

From Figure 1.1 and Equation 1.1, the total strain is the sum of elastic strain and plastic strain when the stress state of a point enters the plastic state under uniaxial tension. For a three-dimensional stress state, there is no mutual influence among the stress components. Therefore, the total strain ε_{ij} can be divided into two parts, the elastic strain ε_{ij}^e and the plastic strain ε_{ij}^p:

$$\varepsilon_{ij} = \varepsilon_{ij}^e + \varepsilon_{ij}^p. \tag{1.89}$$

The elastic strain obeys Hooke's law. The plastic strain, which is the permanent residual strain after unloading, is the difference between the total strain and the elastic strain. The plastic strain is unchanged during unloading, and only changes after the load is further increased.

The aforementioned analysis indicates that plastic strain depends on the loading path. Therefore, we have to discuss the characteristics of stress change and strain change. The total strain is calculated as the integral or summation of the infinitesimal strain increment over the whole loading history. That is why plastic theory is characterized by incrementation. In the following, we will discuss the incremental theory (or flow theory) of plasticity.

When the external load has an infinitesimal increment, the strain correspondingly has an infinitesimal increment $d\varepsilon_{ij}$. From Equation 1.89, $d\varepsilon_{ij}$ is the sum of the elastic strain increment $d\varepsilon_{ij}^e$ and the plastic strain increment $d\varepsilon_{ij}^p$:

$$d\varepsilon_{ij}^p = d\varepsilon_{ij} - d\varepsilon_{ij}^e. \tag{1.90}$$

When hydrostatic stress is not extremely high, the volumetric change of the deformed body is caused by elastic deformation, while the volumetric change caused by plastic deformation is zero. Under the action of deviatoric stresses, the deformed body generates distortion, but the volume does not change. The distortion of a deformed body consists of two parts, elastic deformation and plastic deformation. That is, plastic deformation is only caused by deviatoric stresses [1,3]. It can be strictly proved that the relationship between the plastic strain increment $d\varepsilon_{ij}^p$ and the loading function $f = f(\sigma_{ij}, k)$ can be expressed by

$$d\varepsilon_{ij}^p = d\lambda \frac{\partial f}{\partial \sigma_{ij}}. \tag{1.91}$$

(Note: The "loading function" can be simply interpreted as the surface in the three-dimensional stress state corresponding to the curve *BCD* after yielding [point *B* in Figure 1.1]. The detailed interpretation of this concept goes beyond the scope of this book. Readers should refer to Reference [1], where k denotes the extent of hardening.)

In the plastic loading stage, $d\lambda > 0$. In the unloading and elastic stage, $d\lambda = 0$. Equation 1.91 is called the rule of plastic flow. In addition, the material is incompressible in the plastic state, so the volumetric strain of the plastic deformation is zero:

$$d\varepsilon_x^p + d\varepsilon_y^p + d\varepsilon_z^p = 0, \tag{1.92a}$$

or

$$d\varepsilon_{ii}^p = 0, \tag{1.92b}$$

because

$$d\varepsilon_m = \frac{1}{3}(d\varepsilon_x + d\varepsilon_y + d\varepsilon_z) = \frac{1}{3} d\varepsilon_{ii}^e = d\varepsilon_m^e. \tag{1.93}$$

The deviatoric strain increments are

$$d\varepsilon_x' = d\varepsilon_x - d\varepsilon_m \quad d\varepsilon_y' = d\varepsilon_y - d\varepsilon_m \quad d\varepsilon_z' = d\varepsilon_z - d\varepsilon_m,$$

or

$$d\varepsilon_{ij}' = d\varepsilon_{ij} - d\varepsilon_m \delta_{ij}. \tag{1.94}$$

The corresponding deviatoric stress increment can be expressed as

$$dS_{ij} = d\sigma_{ij} - \frac{1}{3} d\sigma_{kk} \delta_{ij}. \tag{1.95}$$

In the elastic stage, according to Hooke's law

$$\begin{aligned} d\varepsilon_x'^e &= \frac{1}{3}\left(2d\varepsilon_x^e - d\varepsilon_y^e - d\varepsilon_z^e\right) = \frac{1+\nu}{3E}\left(2d\sigma_x - d\sigma_y - d\sigma_z\right) \\ &= \frac{1}{3}\cdot\frac{1}{2G}\left(2d\sigma_x - d\sigma_y - d\sigma_z\right) = \frac{1}{2G} dS_x \end{aligned}$$

$$d\varepsilon_y'^e = \frac{1}{3}\cdot\frac{1}{2G}\left(2d\sigma_y - d\sigma_z - d\sigma_x\right) = \frac{1}{2G} dS_y$$

or

$$\left.\begin{aligned} d\varepsilon_x'^e &= \frac{1}{2G} dS_x \\ &\cdots\cdots \\ d\varepsilon_{xy}^e &= \frac{1}{G} d\tau_{xy} \\ &\cdots\cdots \end{aligned}\right\} \tag{1.96a}$$

or

$$\frac{dS_x}{d\varepsilon_x'^e} = \frac{dS_y}{d\varepsilon_y'^e} = \frac{dS_z}{d\varepsilon_z'^e} = \frac{d\tau_{xy}}{d\varepsilon_{xy}^e} = \frac{d\tau_{yz}}{d\varepsilon_{yz}^e} = \frac{d\tau_{zx}}{d\varepsilon_{zx}^e} = 2G. \tag{1.96b}$$

This equation shows that the deviatoric stress increment is proportional to the corresponding deviatoric strain increment and with the proportional constant $2G$ in the elastic stage.

Next we discuss the expression for plastic strain incremental theory. It is based on the assumption that in any infinitesimal time increment during plastic deformation, the plastic strain increment is proportional to the instantaneous deviatoric stress component

$$\frac{d\varepsilon_x^p}{S_x} = \frac{d\varepsilon_y^p}{S_y} = \frac{d\varepsilon_z^p}{S_z} = \frac{d\varepsilon_{xy}^p}{\tau_{xy}} = \frac{d\varepsilon_{yz}^p}{\tau_{yz}} = \frac{d\varepsilon_{zx}^p}{\tau_{zx}} = d\lambda, \tag{1.97a}$$

or

$$d\varepsilon_{ij}^p = d\lambda S_{ij}, \tag{1.97b}$$

where $d\lambda$ is a proportional constant that is a nonnegative scalar and can vary with the loading history. If the loading function is selected as von Mises yield function of the ideal elasto-plastic material, Equation 1.97b can be obtained from Equation 1.91 [1,3].

Merging Equations 1.96a and 1.97b, we have

$$\left.\begin{array}{lll} d\varepsilon_x' = \dfrac{1}{2G}dS_x + d\lambda S_x & d\varepsilon_y' = \dfrac{1}{2G}dS_y + d\lambda S_y & d\varepsilon'_z = \dfrac{1}{2G}dS_z + d\lambda S_z \\ d\gamma_{xy} = \dfrac{1}{G}d\tau_{xy} + d\lambda\tau_{xy} & d\gamma_{yz} = \dfrac{1}{G}d\tau_{yz} + d\lambda\tau_{yz} & d\gamma_{zx} = \dfrac{1}{G}d\tau_{zx} + d\lambda\tau_{zx} \end{array}\right\}, \tag{1.98a}$$

or

$$\begin{aligned} d\varepsilon_{ij}' &= \frac{1}{2G}dS_{ij} + d\lambda S_{ij} \\ d\varepsilon_{KK} &= \frac{1-2v}{E}d\sigma_{KK} \end{aligned} \tag{1.98b}$$

Equation 1.98b is called the Prandtl–Reuss relationship.

It can be seen from Equation 1.98a that even if σ_{ij} and $d\sigma_{ij}$ are given, $d\lambda$ cannot be determined, and so $d\varepsilon_{ij}'$ cannot be determined. On the contrary, if σ_{ij} and $d\varepsilon_{ij}$ are given, $d\sigma_{ij}$ can be determined. In fact,

$$dW = S_{ij}d\varepsilon_{ij}' = S_{ij}\left(\frac{dS_{ij}}{2G} + d\lambda S_{ij}\right)$$

$$= \frac{1}{2G}dJ_2' + d\lambda 2J_2' = 2\tau_s^2 d\lambda.$$

Thus,

$$d\lambda = \frac{dW}{2\tau_s^2}. \tag{1.99}$$

Here, $dJ_2' = 0$ when an ideal plastic material is employed. Therefore, if σ_{ij} and $d\varepsilon_{ij}$ are given, then S_{ij}, $d\varepsilon_{ij}'$, and $d\lambda$ can be determined, and $d\sigma_{ij}$ can be calculated from Equation 1.98a.

1.4.6 Total Strain Theory of Plasticity

In the incremental theory of plasticity, we have obtained the relationship between the increments of plastic strain components and the stress components. In order to obtain the relationship between the total plastic strain component and the stress component, we need to integrate Equation 1.98a over the loading path. Consequently, the relationship between total stress and total strain must depend on the loading path. While the total strain theory (or deformation theory) attempts to formulate the constitutive model of total strain independent of the loading path, we know that such a theory is generally incorrect, because plastic strain is generally dependent on the loading path. However, theoretically, it is possible to integrate along the loading path and in some special cases, we can obtain from the integral—regardless of the strain history—the explicit total-stress total-strain relationship. Next, we will illustrate such a case.

Under proportional loading, where the stress components at a point increase proportionally and are proportional to a common factor, the incremental theory can be transformed into the total strain theory. In practice, if σ_{ij}^0 is a nonzero reference stress state at time t_0, the instantaneous stress state at an arbitrary time t is t_0, where k is a monotonically increased function with time. Then

$$S_{ij} = kS_{ij}^0, \quad \bar{\sigma} = k\bar{\sigma}^0,$$

where $\bar{\sigma}$ and $\bar{\sigma}^0$ are the effective stresses at time t and t_0, respectively. Equation 1.97b becomes

$$d\varepsilon_{ij}^p = \frac{2}{3}\frac{d\bar{\varepsilon}}{\bar{\sigma}^0} S_{ij}^0, \tag{1.100}$$

where $\bar{\varepsilon}$ is the equivalent strain at time t. Integrating both sides of the previous equation gives

$$\varepsilon_{ij}^p = \frac{3}{2}\frac{\bar{\varepsilon}}{\bar{\sigma}} S_{ij}, \quad \bar{\varepsilon} = \bar{\varepsilon}(\bar{\sigma}). \tag{1.101}$$

After expansion,

$$\left.\begin{aligned} &\varepsilon_x^p = \frac{\bar{\varepsilon}}{\bar{\sigma}}\left[\sigma_x - \frac{1}{2}(\sigma_y + \sigma_z)\right] \varepsilon_y^p = \frac{\bar{\varepsilon}}{\bar{\sigma}}\left[\sigma_y - \frac{1}{2}(\sigma_z + \sigma_x)\right] \varepsilon_z^p = \frac{\bar{\varepsilon}}{\bar{\sigma}}\left[\sigma_z - \frac{1}{2}(\sigma_x + \sigma_y)\right] \\ &\gamma_{xy}^p = \frac{3\bar{\varepsilon}}{\bar{\sigma}}\tau_{xy} \quad \gamma_{yz}^p = \frac{3\bar{\varepsilon}}{\bar{\sigma}}\tau_{yz} \quad \gamma_{zx}^p = \frac{3\bar{\varepsilon}}{\bar{\sigma}}\tau_{zx} \end{aligned}\right\}. \tag{1.102}$$

Thus, plastic strains ε_{ij}^p are only functions of the instantaneous stress state. The constitutive Equations 1.102 of total strain theory are called the Henky–Ilyushin equations. It should be noted that Equation 1.101 is the constitutive equation of physically nonlinear theory. When it is applied to the elasto-plastic process, the stresses at each point must be proportionally loaded. As we know, when unloading occurs, the constitutive relationship does not obey Equation 1.101, because the plastic deformation is unchanged. Therefore, during unloading, the change of strain components and the change of stress components follow Hooke's law.

The stresses and strains after unloading can be calculated by letting the change of external loading be the assumed load acting on the body. The stresses and strains are calculated according to elastic theory and are subtracted from the stresses and strains before unloading. This is the rule governing the calculation of stress and strain during unloading. Apparently, after the external load is completely removed, the remaining stress is called residual stress and the remaining strain is called residual strain. In engineering practice, proportional loading is a good approximation. Thus, the total strain theory can solve many practical problems and obtain satisfactory results.

EXERCISES

1.1 What are the objectives and contents of elasto-plastic mechanics? What are the similarities and differences between elasto-plastic mechanics and material mechanics? What are the basic assumptions of elasto-plastic mechanics?

1.2 As shown in Figure 1.21, a triangle section dam with specific gravity γ is subject to the pressure of a liquid with specific gravity γ_1. The solution for the stress components is

$$\left.\begin{aligned} \sigma_x &= ax + by \\ \sigma_y &= cx + dy - \gamma y \\ \sigma_{xy} &= -dx - ay \end{aligned}\right\}.$$

Determine the coefficients a, b, c, and d in terms of the surface conditions on the vertical and oblique edges.

1.3 Derive Equation 1.19 and the Cauchy's formulae using the force equilibrium conditions in the x-, y-, z-directions.

1.4 For the rectangular plate shown in Figure 1.22, side AB is subject to a surface force perpendicular to the boundary, and side CD is a free surface. Suppose the stress components can be expressed by $\left.\begin{aligned} \sigma_x &= qx^2y - (2/3)qy^3 \\ \sigma_y &= (1/3)qy^3 - c_1y + c_2 \\ \tau_{xy} &= -qxy^2 + c_1x \end{aligned}\right\}$, and the volumetric force is zero. Determine the constants c_1 and c_2, and plot the distribution of surface forces on side AB and side BC.

1.5 Prove that (1) the three principal stress directions are mutually perpendicular and (2) the three principal stresses σ_1, σ_2, σ_3, must be real numbers.

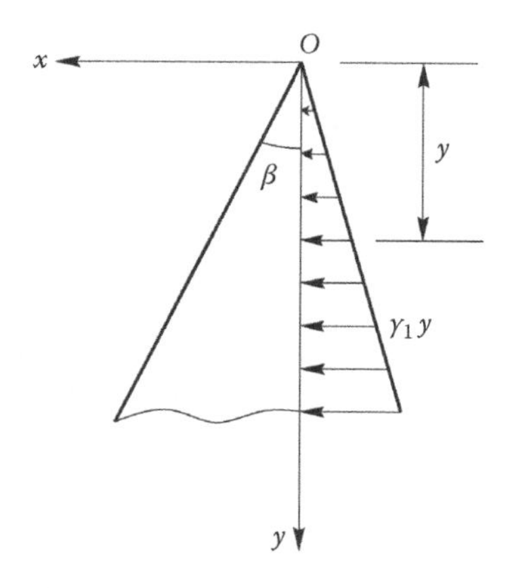

FIGURE 1.21 Figure for Exercise 1.2.

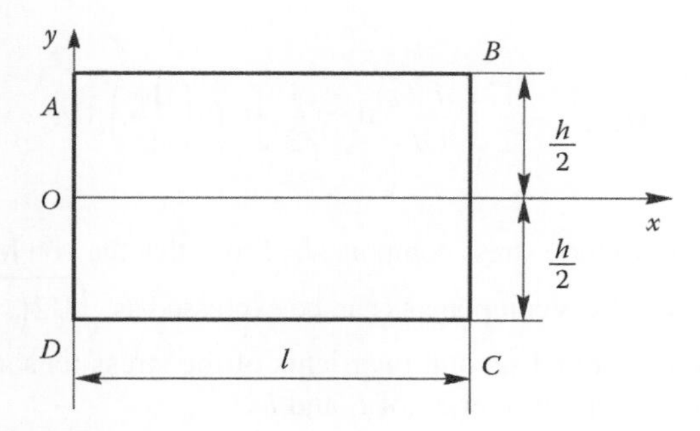

FIGURE 1.22 Figure for Exercise 1.4.

1.6 Judge the truth or falsity of the following statements and briefly explain the following:
1. If the displacements of a point in a body u, v, w are all zero, then the strains at this point must be $\varepsilon_x = \varepsilon_y = \varepsilon_z = 0$.
2. On a line with the x-coordinate being a constant, if $u = 0$, then $\varepsilon_x = 0$ on this line.
3. On a line with the y-coordinate being a constant, if $u = 0$, then $\varepsilon_x = 0$ on this line.
4. The stresses satisfying the equilibrium differential equations and stress boundary conditions must be the correct solution (suppose all boundary conditions are stress boundary conditions).

1.7 What type of surface becomes spherical after homogeneous deformation?

1.8 Place a small object into a high-pressure vessel with hydrostatic pressure of $p = 0.45$ N/mm². The measured volumetric strain is $e = -3.6 \times 10^{-5}$. Assuming that the Poisson ratio is $v = 0.3$, determine the elastic modulus E of the object.

1.9 The normal strains at a point in the directions of 0°, 60°, 90° are $\varepsilon_{0^\circ} = -100 \times 10^{-6}$, $\varepsilon_{60^\circ} = 50 \times 10^{-6}$, and $\varepsilon_{90^\circ} = 150 \times 10^{-6}$, respectively. Calculate the principal strains, the maximum shear strain, and the principal stresses at this point ($E = 2.1 \times 10^5$ N/mm², $v = 0.3$).

1.10 Derive the volume-related deformation strain energy density W_v and the distortion strain energy density W_d. Respectively, they are

$$W_v = \frac{1}{6}\sigma_{ii}\varepsilon_{jj} = \frac{1}{18K}(\sigma_{ii})^2$$

$$W_d = \frac{1}{2}\sigma'_{ij}\varepsilon'_{ij} = \frac{1}{4G}\left(\sigma'_{ij}\sigma'_{ij}\right) = \frac{1}{4G}\left[\sigma_{ij}\sigma_{ij} - \frac{1}{3}(\sigma_{ii})^2\right].$$

1.11 Using the formula of elastic strain energy $W = (1/2)\sigma_{ij}\varepsilon_{ij}$, derive the strain energy formulae in material mechanics for the tension of a rod, the bending of a rod, and the torsion of a cylinder. Respectively, these are

$$U_{tension} = \frac{1}{2}\int_0^l \frac{N^2(x)}{EA}dx = \frac{1}{2}\int_0^l EA\left(\frac{du}{dx}\right)^2 dx,$$

$$U_{bending} = \frac{1}{2}\int_0^l \frac{M^2(x)}{EI}dx = \frac{1}{2}\int_0^l EI\left(\frac{d^2\omega}{dx^2}\right)^2 dx,$$

and

$$U_{torsion} = \frac{1}{2}\int_0^l \frac{M^2(z)}{GI_P}dz = \frac{1}{2}\int_0^l GI_P\left(\frac{d\phi}{dz}\right)^2 dz.$$

1.12 Assume s_1, s_2, s_3 are deviatoric stress components. Prove that the von Mises yield criterion represented by the deviatoric stress components can be expressed as $\sqrt{3/2\left(s_1^2+s_2^2+s_3^2\right)}=\sigma_s$.

1.13 Suppose I_1, I_2 are the first and second invariants of the stress tensor, respectively. Represent the von Mises yield condition in terms of I_1 and I_2.

1.14 A thin-walled tube with a radius of 50 mm and a thickness of 3 mm keeps a stress state of $\tau_{z\theta}/\sigma_z = 1$. The tensile yield limit of the material is 40 kg/mm^2. Calculate the axial load P and torque M_s when yielding occurs in the tube.

1.15 For Exercise 1.14, calculate the ratio of plastic strain increments in the following two conditions:

1. Uniaxial tensile stress state, $\sigma_1 = \sigma_s$.
2. Pure shear stress state, $\tau = \sigma_s/\sqrt{3}$.

1.16 The stress–strain curve of a material is given as $\varepsilon = (\sigma/E) + ((\sigma - \sigma_s)/E_1)$. A thin-walled cylinder made of this material is subjected to tension and torsion. Calculate the axial strain ε_z and the shear strain $\gamma_{\theta z}$ using the incremental theory of plasticity for the following loading paths:

1. Load from the beginning to $\sigma_z = \sigma_s$ along the z-axis, keeping this stress value and increasing the shear stress until $\tau_{\theta z} = \sigma_s/\sqrt{3}$.
2. Load from the beginning to $\tau_{\theta z} = \sigma_s/\sqrt{3}$, keeping the stress $\tau_{\theta z}$, and increasing axial stress σ_z until it reaches σ_s.
3. Axial stress and shear stress increase proportionally with the ratio $\sqrt{3}:1$ until $\sigma_z = \sigma_s$, $\tau_{\theta z} = \sigma_s/\sqrt{3}$.

1.17 Discussion: Why is the number of components of a tensor the nth power of 3 and not the nth power of 4, the nth power of 5, or the nth power of other numbers? What are the answers for two-dimensional plane problems, one-dimensional problems, and zero-dimensional problems? Can this conclusion be simply extended? Are there similar problems?

1.18 We propose the concept of "stress" to describe the internal "force" of a body subjected to a load. Is the stress inside a body measurable? Do we have other, more scientific methods, to describe it?

REFERENCES

1. Zhou Y. *Mechanics of Solid Materials* (Vol. 1, 2). Beijing: Science Press, 2005.
2. Lu M. and Luo X. *Fundamentals of Elasticity Theory* (Vol. 1, 2). Beijing: Qinghua University Press, 2001.
3. Wang R., Xiong Z., and Huang W. *Fundamentals of Mechanics of Plasticity*. Beijing: Science Press, 1998.
4. Xu Z. *Elasticity Theory* (Vol. 1, 2). Beijing: People's Education Press, 1982.
5. Huang K., Xue M., and Lu M. *Tensor Analysis*. Beijing: Qinghua University Press, 1986.
6. Huang K. *Non-Linear Mechanics of Continuum*. Beijing: Science Press, 1989.
7. Shelby R. A., Smith D. R., and Schultz S. Experimental verification of a negative index of refraction, *Science*, 2001, 292: 77–79.
8. Pendry J. B., Schurig D., and Smith D. R. Controlling electromagnetic fields. *Science*, 2006, 312: 1780–1782.
9. Zhou X. M., Hu G. K., and Lu T. J. Elastic wave transparency of a solid sphere coated with metamaterials. *Phys. Rev. B.*, 2008, 77: 024101.

2 Basis of Macro- and Microfracture Mechanics

The "fracture" of materials or components is evident everywhere: in the production process and in life. For example, during the Wenchuan Earthquake in Sichuan, China, on May 12, 2008, uncountable houses collapsed instantaneously, and nearly 100,000 people lost their lives. In daily life, we see bridge collapses, train derailments, and air crashes, as well as minor and serious traffic accidents. There are too many examples to list.

As college students and as researchers, what should we learn from these natural disasters and terrible accidents? Why are there so many destructive incidents? How can we prevent these destructive incidents from occurring? Why do materials or components become fractured? Are there identifiable patterns that characterize these destructive incidents, which seem to have no common features? This chapter will describe, from various destructive incidents, how to extract the fundamental theory, namely, the concept of "crack," upon which the engineering and scientific problems of the fracture are analyzed. This chapter mainly describes the fundamental laws of macro- and microfracture mechanics from three aspects: macrofracture mechanics, microfracture mechanics, and nanofracture mechanics.

2.1 ANALYSIS OF MACROFRACTURE MECHANICS

Because accidents happen we must ask: "why do materials or components become fractured?" From Chapter 1, we learned that materials and components do not exist in isolation but are applied with a certain load, which can result in deformation. When deformation or stress increases to a certain level, fracture occurs. This is the theoretical basis of the *mechanics of materials*. When designing engineering components based on the traditional strength theory in the *mechanics of materials*, the requirements are that

$$\sigma \le [\sigma], \quad [\sigma] = \begin{cases} \dfrac{\sigma_s}{k}, & \text{plastic material} \\[2ex] \dfrac{\sigma_b}{k}, & \text{brittle material} \end{cases} \tag{2.1}$$

that is, the working stress of the component σ must be lower than or equal to the material's allowable stress $[\sigma]$. In Equation 2.1, σ_s is the yield strength of the material, σ_b is the tensile strength of the material, and k is the safety factor. In general, the typical value is $1.3 \sim 2.0$. If the stress to the component caused by the external load σ is accurately calculated, and the stress measured on the selected specimen σ_s (or σ_b) accurately represents the resistance to the fracture of the component inside the material, the k value can be lowered as appropriate.

"Having various engineering components meet the requirements in Equation 2.1" is the method used by the traditional design. However, since the end of World War II, testing and experimentation around the world has shown that components designed according to the traditional strength theory occasionally fracture accidentally at low stress. For example, in the United States in the 1950s, the pressure housing of the solid fuel engine on the Polaris missile was designed and completely accepted using the traditional strength theory, but low-stress brittle fracture occurred unexpectedly during the launch process. In this incident, when the fracture occurred, the stress was much lower than the allowable stress of the material $[\sigma]$.

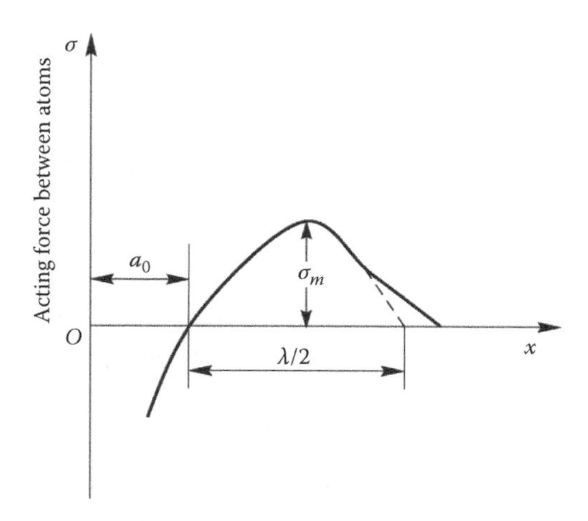

FIGURE 2.1 Acting force between atoms is a function of the spacing between atoms.

Moreover, low-stress brittle fracture often occurs in daily life. Products like glass and ceramics fracture when applied with a small external force. These facts clearly underline the limitations of the traditional strength design theory, and it is clear that the theoretical basis of the *mechanics of materials* is ill-founded from the viewpoint of engineering. What should we do then?

In the following, we describe the limitations of the strength theory in the *mechanics of materials* from the viewpoint of science. From *general physics*, we learn that inter-reaction exists between atoms: when the spacing is small, atoms mutually repel each other; when the spacing becomes large, they mutually attract each other. The reason for material fracture is that the external force must overcome the bonding force between the atoms. As shown in Figure 2.1, when the spacing increases as the stress rises, the stress overcomes the acting force between the atoms and reaches a maximum value σ_m on one point. This maximum value is the theoretical fracture strength σ_t.

Assuming that the variation of the acting force between the atoms, with respect to the atomic spacing x, can be represented approximately by a sine wave with a wavelength of λ. Then,

$$\sigma = \sigma_t \sin\frac{2\pi x}{\lambda}. \tag{2.2}$$

Material fracture is a process in which the atoms perpendicular to the tensile stress are separated due to the application of tensile stress. In this process, in order to make the fracture occur, sufficient energy must be provided to create two new surfaces. When the density of the surface energy is 2γ, which means that the energy used in the formation of the fracture surface through the external work is at least equal to 2γ—that is, the external work is equal to the surface energy of the fracture surface—then

$$\int_0^{\lambda/2} \sigma_t \sin\frac{2\pi x}{\lambda}\,dx = \frac{\lambda\sigma_t}{\pi} = 2\gamma. \tag{2.3}$$

Assuming that the material has small deformation, $\sin x \approx x$ applies, and it follows Hooke's law:

$$\sigma = E\frac{x}{b}, \tag{2.4}$$

where

b is the atomic spacing in the balanced state
E is the elastic modulus

As a result, from Equations 2.2 through 2.4, the theoretical fracture strength of the material (σ_t) may be calculated approximately, using the following equation:

$$\sigma_t = \left(\frac{E\gamma}{b} \right)^{1/2}. \tag{2.5}$$

Substituting the typical values of the material ($E = 10^{10}$ Pa, $\gamma = 10^{-4}$ J/cm^2, and $b = 3 \times 10^{-10}$ cm), the theoretical fracture strength of the material is calculated and we obtain $\sigma_t = 3 \times 10^4$ MPa. At present, the steel with the highest strength has a theoretical fracture strength of 4500 MPa. Many experimental results show that the actual fracture strength of the material is less than the theoretical value by 1–3 orders of magnitude. What is the reason for this?

Whether in engineering or in the sciences, severe challenges appear in attempting to resolve the problem of fracture. For example, when the engineering design is correct, why does the material or component become fractured when the load is within the designed range? Even though the estimate is strictly based on up-to-date atomic theory, the theoretical value seriously deviates from the actual value. The only possible explanation is that "traditional strength theory is built on an incorrect basis." Where is the problem? The problem is that we think too ideally. In 1913, Inglis first pointed out that various defects inevitably exist in materials, such as microcracks, cavities, notches, scratches, and so on [1,2]. There is a local area with high stress (or high strain) near the tip. The stress in this area is several times that of the stress in the area far away from the tip, becoming the "original point" of the fracture. He therefore came up with the concept of "crack." Griffith, a U.K. scientist, is the pioneer of macrofracture mechanics. In 1920 and 1924, he published two papers in which a basic framework of the brittle fracture theory [1,2] was established. The concept of "crack" is one of the greatest contributions to the natural sciences in the twentieth century.

This section describes the macrofracture problem of materials, closely based on the concept of "crack," and outlines the main framework of macrofracture mechanics with regard to linear elastic-plastic fracture mechanics and elastic fracture mechanics.

2.1.1 Cracking Mode and the Elastic Stress Field near the Crack Tip

Although we have learned the concept of "crack," the situations in which fracture occur are diversified, with no common pattern. We cannot manufacture materials that are difficult to fracture, or prevent them from fracturing, until we know the fracture pattern of materials. To know the fracture pattern of materials, we should pay closer attention to the keyword "crack." What common patterns do the cracks have? Can the diversified cracks be classified? Of course! Through persistent effort, scientists have discovered that, in the stress field plane problem, the stress and strain field near the crack can be classified into three basic modes, based on the relationship between the position of the crack and the direction of stress.

1. Opening-mode crack (Mode I)
   The crack plane is perpendicular to the stress σ, and the crack tip opens when applying σ. The direction of crack growth is also perpendicular to σ. This type of crack is called the opening-mode (Mode I) crack, as shown in Figure 2.2a. Figure 2.2b is its left view. The longitudinal crack on the cylindrical vessel as shown in Figure 2.2c and the transverse crack on the plate under tension as shown in Figure 2.2d are Mode I cracks.
2. Sliding-mode crack (Mode II)
   The crack plane is parallel to the shear stress τ, and the crack grows in a sliding manner when applying τ. The direction of crack growth keeps a certain angle with τ. This type of crack is called a sliding mode (Mode II) or an in-plane shear mode crack, as shown in Figure 2.3a. Figure 2.3b is its left view. The bolt shown in Figure 2.3c and the crack at the joint of the plates are Mode II cracks. Figure 2.3d shows the diagrams of applying load.

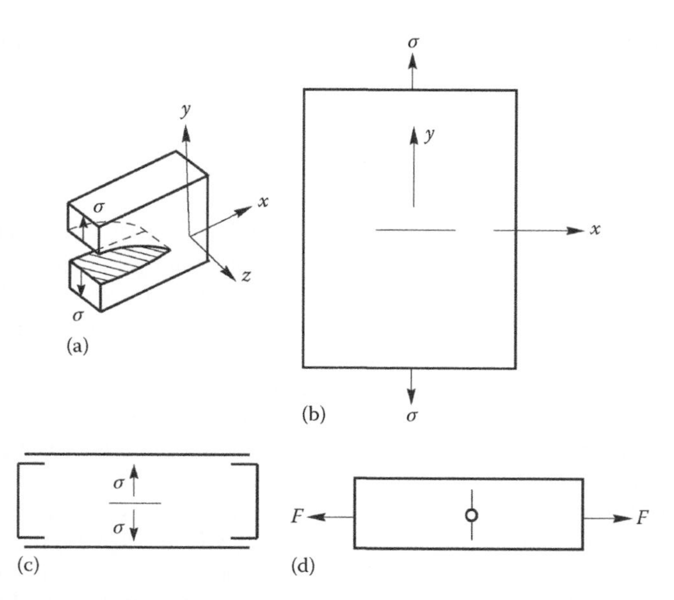

FIGURE 2.2 Mode I crack. (a) Front view. (b) Left view. (c) Longitudinal crack in a cylindrical vessel. (d) Transversal crack in a stretched plate.

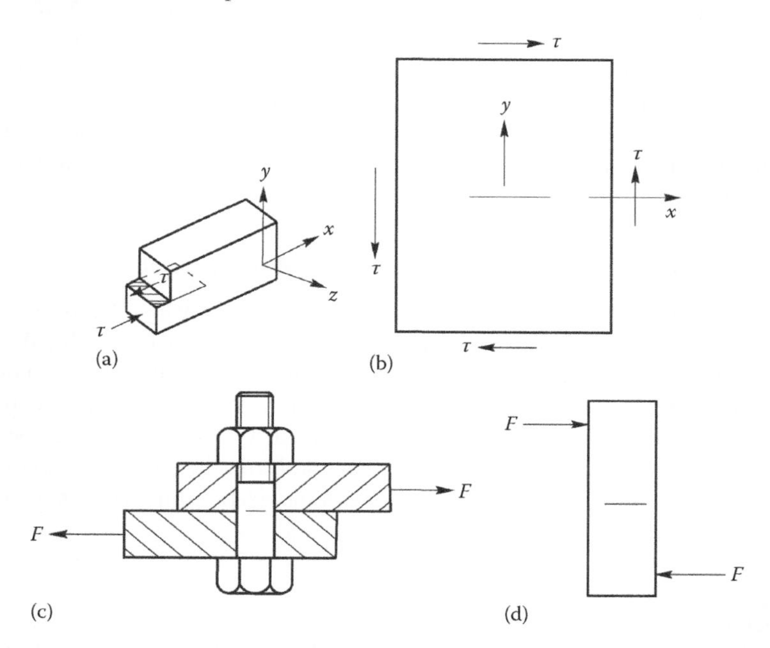

FIGURE 2.3 Mode II crack. (a) Front view. (b) Left view. (c) Crack at the interface of a bolt. (d) Its force distribution.

3. Tearing mode crack (Mode III)
   The crack plane is parallel to the shear stress τ, and the crack displaces (it does not slide) along the original crack plane when applying τ. However, the direction of crack growth is perpendicular to τ. This type of crack is called a tearing mode (Mode III), an antiplane shear mode, or a twist mode crack, as shown in Figure 2.4a. Figure 2.4b is its left view. For example, the cloth is sheared with scissors and then torn. This is Mode III crack growth. In an engineering application, the radial crack of the twisted cylinder is also a Mode III crack. Figure 2.4c and d show the diagrams of applying load.

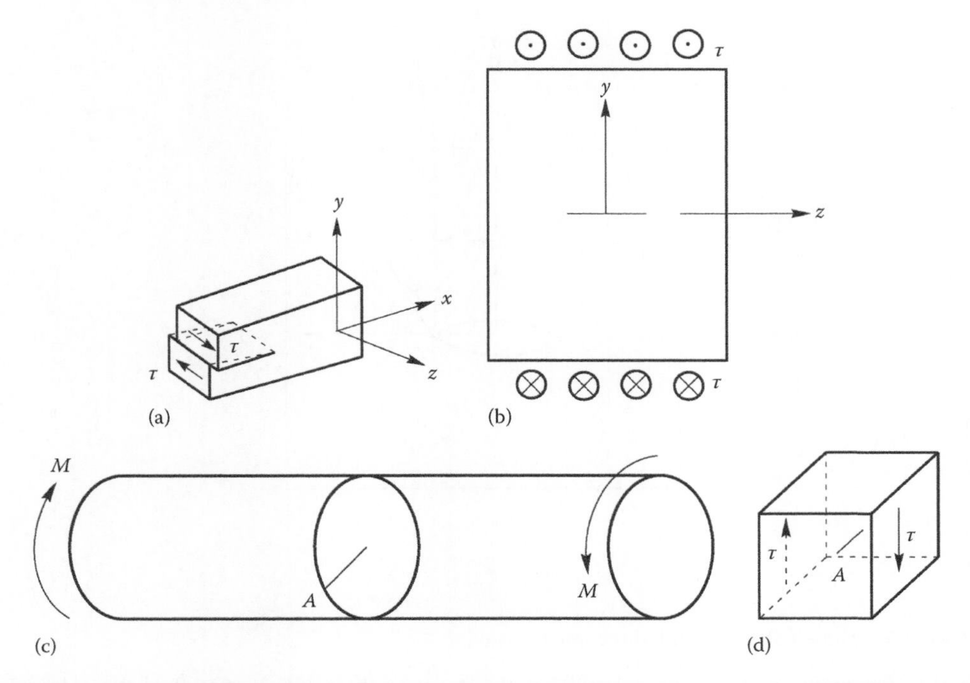

FIGURE 2.4 Mode III crack. (a) Front view. (b) Left view. (c) Radial crack in a torsion cylinder. (d) Its force distribution in the cubic unit.

When a material deforms, displacement occurs on all mass points. Mode I and Mode II cracks do not have deformation in the thickness direction, namely, $u \neq 0$, $v \neq 0$, $w = 0$. However, in a Mode III crack, this is not the case. In the deformation, $u = v = 0$, $w \neq 0$ applies, where u, v, w are the displacements in x, y, z direction, respectively.

If the inside crack is also applied with normal stress and shear stress, or the crack keeps a certain angle with the normal stress (e.g., oblique crack of the thin-walled container), then Mode I and Mode II (or Mode I and Mode III) cracks exist simultaneously. This is called a mixed-mode crack.

The opening-mode crack (Mode I) is the most dangerous, which tends to result in low-stress brittle fracture. Even if the actual crack is a mixed-mode crack, it is often taken to be an opening-mode crack. This is the simple and safe practice. Therefore, when studying fracture mechanics, the focus is on Mode I cracking. In the following, we describe the elastic stress field near the crack tip when a Mode I crack is under bidirectional tension and unidirectional tension.

As shown in Figure 2.5, there is a $2a$ long central crack in the plate without boundary. A bidirectional tensile stress σ is applied at infinity. This is a question regarding the plane. As the crack goes through the plate thickness, each xoy plane has the same stress state; which is to say, the stress and strain do not change with the thickness. If the plate is thin, it is a question regarding the plane stress; if the plate is thick, it is a question regarding the plane strain.

For Mode I cracking under bidirectional tensile stress, its boundary condition is as follows:

1. In a location where $y = 0$ and $-a < x < a$ apply, it is $\sigma_y = 0$. Because the crack has a cavity inside, there is no stress. That is, in the location where $|x| < a$ applies, all stress components are zero.
2. In a location where $y = 0$ and $|x| > a$ apply, it is $\sigma_y > \sigma$. Moreover, the closer x is to a, the higher is σ_y. The crack tip has intensive stress, so the closer it is to the crack tip, the more intensive is the stress.
3. In a location where $x \to \pm\infty$ applies, it is $\sigma_y = \sigma$, $\sigma_x = \sigma$. At a position far away from the crack, as the stress intensive effect generated by the crack disappears, the stress is equal to the external stress.

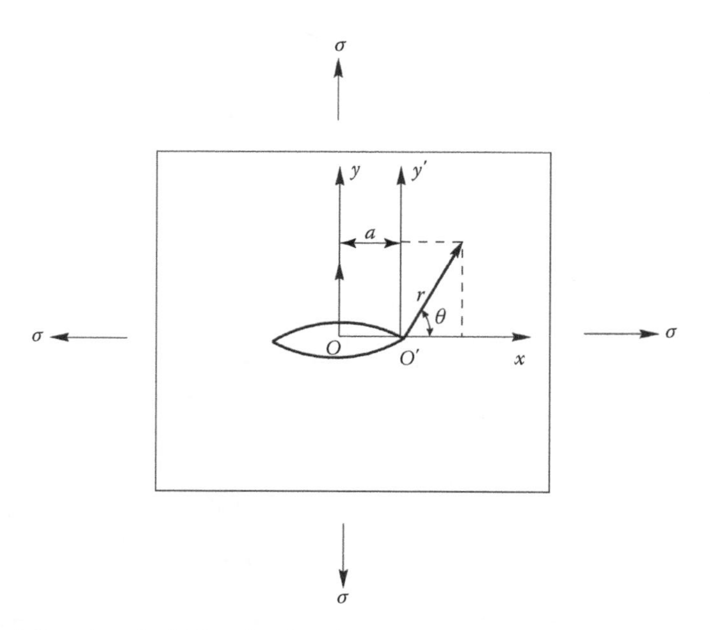

FIGURE 2.5 Mode I crack under bidirectional tension.

In the aforementioned boundary condition, the stress on the position of (r, θ) in the polar coordinate system is as follows (the detailed derivation is beyond the scope of this book) [2]:

$$\sigma_x = \frac{\sigma\sqrt{\pi a}}{\sqrt{2\pi r}}\cos\frac{\theta}{2}\left(1-\sin\frac{\theta}{2}\sin\frac{3}{2}\theta\right)$$

$$\sigma_y = \frac{\sigma\sqrt{\pi a}}{\sqrt{2\pi r}}\cos\frac{\theta}{2}\left(1+\sin\frac{\theta}{2}\sin\frac{3}{2}\theta\right), \tag{2.6}$$

$$\tau_{xy} = \frac{\sigma\sqrt{\pi a}}{\sqrt{2\pi r}}\cos\frac{\theta}{2}\sin\frac{\theta}{2}\cos\frac{3}{2}\theta$$

where the origin of the coordinate system is in the right tip of the crack, r is the distance to the crack tip, and θ is the angle with the axis x on the coordinate point. It should be pointed out that the range of θ here is from 0 to π, and the three stress components, σ_x, σ_y, and τ_{xy}, are all symmetrical with respect to both the x and y axes shown in Figure 2.5, which has not been specified in Reference [2] or other textbooks. Readers are urged to carefully analyze Equation 2.6 and check whether interesting things can be found.

In Equation 2.6, each stress component has the item $\sigma\sqrt{\pi a}$. For easy analysis, if $K_{\mathrm{I}} = \sigma\sqrt{\pi a}$ applies, the previous equation may be changed to

$$\left.\begin{aligned}
\sigma_x &= \frac{K_{\mathrm{I}}}{\sqrt{2\pi r}}\cos\frac{\theta}{2}\left(1-\sin\frac{\theta}{2}\sin\frac{3\theta}{2}\right)\\
\sigma_y &= \frac{K_{\mathrm{I}}}{\sqrt{2\pi r}}\cos\frac{\theta}{2}\left(1+\sin\frac{\theta}{2}\sin\frac{3\theta}{2}\right)\\
\tau_{xy} &= \frac{K_{\mathrm{I}}}{\sqrt{2\pi r}}\cos\frac{\theta}{2}\sin\frac{\theta}{2}\cos\frac{3\theta}{2}
\end{aligned}\right\}. \tag{2.7a}$$

The expression of the corresponding strain and displacement is as follows:

$$\left.\begin{aligned}
\varepsilon_x &= \frac{1}{2G(1+\nu')}\frac{K_{\mathrm{I}}}{\sqrt{2\pi r}}\cos\frac{\theta}{2}\left[(1-\nu')-(1+\nu')\sin\frac{\theta}{2}\sin\frac{3\theta}{2}\right]\\
\varepsilon_y &= \frac{1}{2G(1+\nu')}\frac{K_{\mathrm{I}}}{\sqrt{2\pi r}}\cos\frac{\theta}{2}\left[(1-\nu')+(1+\nu')\sin\frac{\theta}{2}\sin\frac{3\theta}{2}\right]\\
\gamma_{xy} &= \frac{1}{G}\frac{K_{\mathrm{I}}}{\sqrt{2\pi r}}\cos\frac{\theta}{2}\sin\frac{\theta}{2}\cos\frac{3\theta}{2}\\
u &= \frac{K_{\mathrm{I}}}{G(1+\nu')}\cdot\sqrt{\frac{r}{2\pi}}\cos\frac{\theta}{2}\left[(1-\nu')+(1+\nu')\sin^2\frac{\theta}{2}\right]\\
v &= \frac{K_{\mathrm{I}}}{G(1+\nu')}\cdot\sqrt{\frac{r}{2\pi}}\sin\frac{\theta}{2}\left[2-(1+\nu')\cos^2\frac{\theta}{2}\right]
\end{aligned}\right\} \tag{2.7b}$$

where ν' corresponds to the definition of the plane stress and plane strain (see Section 1.4.1).

Now, let us carefully analyze Equation 2.6 or Equation 2.7a. One of the common and important methods used in our analysis is to check what will happen in the extreme case. From Equation 2.6, we know that the stress state varies with (r, θ), and the value inside the brackets is always positive for the stress components σ_x, σ_y. Therefore, the positive/negative sign of σ_x, σ_y depends on cos $\theta/2$. When $0 \leq \theta \leq 2\pi$ applies, cos $\theta/2$ is always positive. In other words, in the application of bidirectional tensile stress, the normal stress—whether in the x-direction or in the y-direction—is tensile stress. Based on Equation 2.1, when the tensile stress reaches a certain value, it is possible that the material has fractured.

Now, let us analyze the variation of σ_x, σ_y with r. When r is large, σ_x, σ_y are small. However, when r becomes small, σ_x, σ_y become large. When $r \to 0$ applies, σ_x, σ_y tend to infinity (i.e., for items containing $1/\sqrt{r}$, when $r \to 0$ applies, this item tends to infinity, and so it is called a singular item). We call this situation the singularity of the crack tip stress field. What does "tending to infinity" mean? It means that as long as $\sigma \neq 0$ applies, the tensile stress on the crack tip will exceed the allowable stress of the material $[\sigma]$, and the material will certainly fracture. In fact, it is impossible that no force is applied on the material or that no crack exists inside the material. As a result, the conclusion is that the true material or component does not exist.

Where is the problem? It seems that we have come to a dead end. Obviously, "the singularity of the crack tip stress field" does not obey the physical law. The appearance of the singularity is reasonable, because we take the true "crack" that exists and that has a certain width, as a zero-width crack in a mathematical conception. Therefore, due to the "singularity" of the crack tip, it is obvious that Equation 2.1 of the traditional strength theory cannot be applied. Therefore, we must come up with a new concept or idea to replace the traditional strength theory. We may reason that there is only a single load factor due to no crack (i.e., geometrical factor) in the traditional strength theory, while the component containing the crack has not only a load factor but also a crack (i.e., geometrical factor). Therefore, can we reason that a "generalized load" exists that contains both load and geometry? From the stress field of the crack tip, we find that $K_{\mathrm{I}} = \sigma\sqrt{\pi a}$ is the item that represents the intensity of the area stress field near the crack tip. $K_{\mathrm{I}} = \sigma\sqrt{\pi a}$ is called the stress intensity factor (SIF), which functions as a "generalized load."

For the unidirectional tension, we can obtain the stress component in a similar manner as follows:

$$\left.\begin{aligned}\sigma_x &= \frac{K_{\mathrm{I}}}{\sqrt{2\pi r}}\cos\frac{\theta}{2}\left(1-\sin\frac{\theta}{2}\sin\frac{3\theta}{2}\right)-\sigma\\ \sigma_y &= \frac{K_{\mathrm{I}}}{\sqrt{2\pi r}}\cos\frac{\theta}{2}\left(1+\sin\frac{\theta}{2}\sin\frac{3\theta}{2}\right)\\ \tau_{xy} &= \frac{K_{\mathrm{I}}}{\sqrt{2\pi r}}\cos\frac{\theta}{2}\sin\frac{\theta}{2}\cos\frac{3\theta}{2}\end{aligned}\right\}. \tag{2.8}$$

Based on a comparison between Equations 2.6 and 2.8, we know that, for the stress field near the crack tip, the only difference between the unidirectional tension and the bidirectional tension is one constant item $-\sigma$ in σ_x. The singular item of the stress field is absolutely the same. When r is small, the singular item is much larger than the additional item ($-\sigma$). Therefore, in general, Equation 2.6 can be used as the stress field calculation equation for unidirectional tension.

For Mode II and Mode III cracks, the stress component also has $1/\sqrt{r}$ singularity on the crack tip.

2.1.2 Stress Intensity Factor

For Mode I, Mode II, and Mode III cracks, the total solution expression of the stress component can be represented as follows [2]:

$$\sigma_{ij} = \frac{K_m}{\sqrt{2\pi}}(r^{-1/2})\sigma_{ij}(\theta)\; O(r^0)+\cdots \tag{2.9}$$

As r is very small in the crack tip area, the first item in the aforementioned equation is much larger than the following items. After all items following r to the zero power are omitted, we get

$$\sigma_{ij} = \frac{K_m}{\sqrt{2\pi}}(r^{-1/2})\sigma_{ij}(\theta). \tag{2.10}$$

This equation represents the stress solution of the area near the crack tip (crack tip solution or asymptotic solution for short), where $\bar{\sigma}_{ij}(\theta)$ is a function of θ and is called the angular distribution function. K_m is the intensity of the stress field of the area near the crack tip; m is I, II, III (i.e., K_{I}, K_{II}, K_{III}), respectively, representing the intensity of the tip stress field of Mode I, Mode II, and Mode III cracks (stress intensity factor or K factor for short).

In linear elastic fracture mechanics, because the intensity of the crack tip stress field is mainly represented by K_m, it can establish the fracture criterion (also called the K criterion) when the generalized load reaches one critical value—namely, $K_{\mathrm{I}} = K_{\mathrm{IC}}$ or $K_m = K_{mC}$ (m = I, II, III)—to resolve the problem of brittle fracture in engineering practice. Therefore, it is common to pay more attention to solving the stress intensity factor K_m.

K_m is an item whose value depends on factors such as the nature of the external load and crack and elastic body geometry. Its general equation is

$$\left.\begin{aligned} K_{\mathrm{I}} &= \alpha\sigma\sqrt{\pi a} \\ K_{\mathrm{II}} &= \beta\tau\sqrt{\pi a} \\ K_{\mathrm{III}} &= \gamma\tau_l\sqrt{\pi a} \end{aligned}\right\}, \tag{2.11}$$

where

α, β, γ are, respectively, called the geometrical factor of Mode I, Mode II, and Mode III cracks
σ is tensile stress
τ and τ_l are, respectively, the in-plane shear stress and the out-plane shear stress

Identifying the stress intensity factor is an important task in linear elastic fracture mechanics. For Mode I, Mode II, and Mode III cracks, the key to identifying the stress intensity factor is to identify the crack geometry factor. Generally, it is rather complex to identify the crack geometry factor.

The methods for identifying the stress intensity factor mainly include the analytical method, the numerical method, and the test method. In the case of simple geometry, the analytical method may be used. At present, the formulas for the stress intensity factor of different components at different loading conditions have been compiled into the handbooks for reference in engineering [3].

To determine whether the material will become fractured, the "generalized load" (the stress intensity factor) must be identified, and the critical value $K_m = K_{mC}$ (m = I, II, III) must also be identified. The fracture criterion and critical value will be discussed in detail in Section 2.1.4.

2.1.3 Plastic Correction under Small-Scale Yielding

Now that we have discussed the elastic stress field near the crack tip, we know that a stress singularity exists on the crack tip. When the singularity is infinitely close to the crack tip ($r \to 0$), the stresses σ_x, σ_y, τ_{xy} tend to infinity. However, we know from Chapter 1 that for a general metal material—even one that has ultrahigh strength—when the stress near the crack tip reaches a certain level, plastic deformation will occur. This means that around the crack tip there is always an area where plastic deformation occurs. If the induration of the material is not taken into consideration, the stress will maintain at a constant level. Within the plastic area of the crack tip, the material no longer obeys the elastic law. Therefore, when studying the stress intensity factor K_{I} of a Mode I crack tip, the theory and method of linear elastic fracture mechanics—which assumes that the material is in the fully linear elastic state—is, in principle, not applicable to the plastic area. However, when the plastic area size is much smaller than the crack size, in the so-called small-scale yielding situation, the wide area around its plastic area is still an elastic area. Therefore, after proper correction, the conclusions of linear elastic fracture mechanics can still be applied and used approximately.

From the *mechanics of materials*, we know that under unidirectional tension, the material generates yield and plastic deformation as long as its stress reaches the yield point σ_s. However, in the complex stress state, two theories are generally adopted to establish the yielding condition for the plastic material: the Tresca's yield criterion and the Mises's yield criterion (see Chapter 1). For the component containing cracks, even if the external load is undergoing unidirectional tension, the area near its crack tip is in the complex stress state.

2.1.3.1 Plastic Area of Crack Tip under Small-Scale Yielding

Here, we use the Mode I crack as an example in discussing the plastic area of the crack tip. We know that for a Mode I crack, the stress component of the area near the crack tip is determined by Equation 2.7a. From Chapter 1, or the *mechanics of materials*, we know that the calculation equation of the main stress is

$$\left.\begin{array}{ll} \left.\begin{array}{l}\sigma_1 \\ \sigma_2\end{array}\right\} = \dfrac{\sigma_x+\sigma_y}{2} \pm \sqrt{\left(\dfrac{\sigma_x-\sigma_y}{2}\right)^2 + \tau_{xy}^2} & \text{plane stress} \\ \sigma_3 = \begin{cases} 0 \\ \nu(\sigma_1+\sigma_2) \end{cases} & \text{plane strain} \end{array}\right\}. \tag{2.12}$$

By substituting Equation 2.7a into Equation 2.12, the main stress of the area near the crack tip is obtained:

$$\left.\begin{array}{ll} \sigma_1 = \dfrac{K_{\mathrm{I}}}{\sqrt{2\pi r}}\cos\dfrac{\theta}{2}\left(1+\sin\dfrac{\theta}{2}\right) & \\ \sigma_2 = \dfrac{K_{\mathrm{I}}}{\sqrt{2\pi r}}\cos\dfrac{\theta}{2}\left(1-\sin\dfrac{\theta}{2}\right) & \\ \sigma_3 = \begin{cases} 0 & \text{plane stress} \\ 2\nu\dfrac{K_{\mathrm{I}}}{\sqrt{2\pi r}}\cos\dfrac{\theta}{2} & \text{plane strain} \end{cases} \end{array}\right\}. \tag{2.13}$$

After the expression for the main stress is obtained, the shape and size of the plastic area on the crack tip can be determined by using the yield criterion.

1. *Shape of plastic area in Tresca's yield criterion*
   For the plane stress, when $\sigma_3 = 0$ applies, then $\sigma_1 = \sigma_s$.

   Applying Equation 2.13, $\dfrac{K_{\mathrm{I}}}{\sqrt{2\pi r}}\cos\dfrac{\theta}{2}\left(1+\sin\dfrac{\theta}{2}\right) = \sigma_s$, and

$$r(\theta) = \frac{1}{2\pi}\left(\frac{K_{\mathrm{I}}}{\sigma_s}\right)^2 \cos^2\frac{\theta}{2}\left(1+\sin\frac{\theta}{2}\right)^2. \tag{2.14}$$

   This is the boundary equation of the plastic area on the tip of the Mode I crack, represented with polar coordinates, in the state of plane stress. In the extension line of the crack, when $\theta = 0$ applies,

$$r_0 = \frac{1}{2\pi}\left(\frac{K_{\mathrm{I}}}{\sigma_s}\right)^2. \tag{2.15}$$

   After Equation 2.14 is divided by r_0 on both sides, a dimensionless equation is obtained:

$$\frac{r(\theta)}{r_0} = \cos^2\frac{\theta}{2}\left(1+\sin\frac{\theta}{2}\right)^2. \tag{2.16}$$

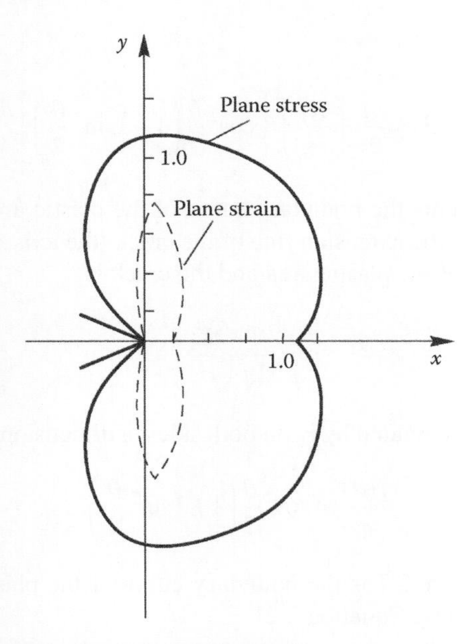

FIGURE 2.6 Plastic boundary curve in Mises' yield criterion.

The solid line in Figure 2.6 is the boundary curve of the plastic area, which is drawn based on the dimensionless Equation 2.16.

For the plane strain, when $\sigma_3 = 2\nu \dfrac{K_{\text{I}}}{\sqrt{2\pi r}} \cos\dfrac{\theta}{2}$ applies,

$$\frac{K_{\text{I}}}{\sqrt{2\pi r}} \cos\frac{\theta}{2}\left(1+\sin\frac{\theta}{2}\right) - 2\nu \frac{K_{\text{I}}}{\sqrt{2\pi r}} \cos\frac{\theta}{2} = \sigma_s,$$

and

$$r(\theta) = \frac{1}{2\pi}\left(\frac{K_{\text{I}}}{\sigma_s}\right)^2 \cos^2\frac{\theta}{2}\left(1-2\nu+\sin\frac{\theta}{2}\right)^2. \tag{2.17}$$

This is the boundary equation of the plastic area on the tip of the Mode I crack, represented with polar coordinates, in the state of plane strain. Similarly, after Equation 2.17 is divided by r_0 on both sides, a dimensionless equation is obtained

$$\frac{r(\theta)}{r_0} = \cos^2\frac{\theta}{2}\left(1-2\nu+\sin\frac{\theta}{2}\right)^2. \tag{2.18}$$

If $\nu = 0.33$ applies, the boundary curve of the dimensionless plastic area represented by the previous equation is the dotted line shown in Figure 2.6.

2. *Shape of plastic area in Mises's yield criterion*

For the plane stress, it is

$$\frac{K_{\text{I}}^2}{2\pi r}\left[\cos^2\frac{\theta}{2}\left(1+3\sin^2\frac{\theta}{2}\right)\right] = \sigma_s^2,$$

or

$$r(\theta) = \frac{1}{2\pi}\left(\frac{K_{\mathrm{I}}}{\sigma_s}\right)^2 \left[\cos^2\frac{\theta}{2}\left(1+3\sin^2\frac{\theta}{2}\right)\right]. \tag{2.19}$$

Equation 2.19 represents the boundary curve of the plastic area on the crack tip in the state of plane stress. On the extension line of the crack (the axis x with $\theta = 0$), the distance between the boundary of the plastic area and the crack is

$$r_0 = \frac{1}{2\pi}\left(\frac{K_{\mathrm{I}}}{\sigma_s}\right)^2. \tag{2.20}$$

After Equation 2.19 is divided by r_0 on both sides, a dimensionless equation is obtained:

$$\frac{r(\theta)}{r_0} = \cos^2\frac{\theta}{2}\left(1+3\sin^2\frac{\theta}{2}\right). \tag{2.21}$$

The solid line in Figure 2.7 is the boundary curve of the plastic area, which is drawn based on the dimensionless Equation 2.21.

For the plane strain, it is

$$\frac{K_{\mathrm{I}}^2}{2\pi r}\left[\frac{3}{4}\sin^2\theta + (1-2v)^2\cos^2\frac{\theta}{2}\right] = \sigma_s^2,$$

or

$$r(\theta) = \frac{1}{2\pi}\left(\frac{K_{\mathrm{I}}}{\sigma_s}\right)^2 \cos^2\frac{\theta}{2}\left[(1-2v)^2 + 3\sin^2\frac{\theta}{2}\right]. \tag{2.22}$$

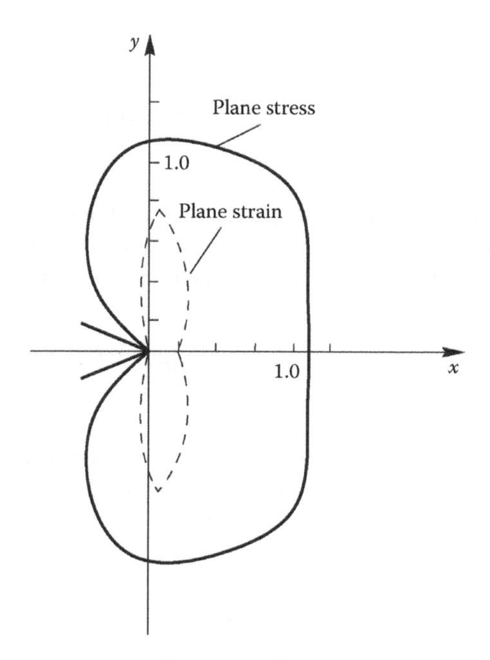

FIGURE 2.7 Plastic boundary curve in Mises' yield criterion.

Equation 2.22 represents the boundary curve of the plastic area of the crack tip in the state of plane strain.

Similarly, after Equation 2.22 is divided by r_0 on both sides, a dimensionless equation is obtained:

$$\frac{r(\theta)}{r_0} = \cos^2\frac{\theta}{2}\left[(1-2\nu)^2 + 3\sin^2\frac{\theta}{2}\right]. \tag{2.23}$$

If $\nu = 0.33$ applies, the boundary curve of the dimensionless plastic area represented by this equation is the dotted line shown in Figure 2.7.

Generally, the size of the plastic area on the crack tip is represented by the plastic area size on the extension line of the crack r_0. r_0 is called the plastic area size. From the previous analysis, we get

$$r_0 = \begin{cases} \dfrac{1}{2\pi}\left(\dfrac{K_\mathrm{I}}{\sigma_s}\right)^2 & \text{plane stress} \\ \dfrac{1}{2\pi}\left(\dfrac{K_\mathrm{I}}{\sigma_s}\right)^2 (1-2\nu)^2 & \text{plane strain} \end{cases} \tag{2.24}$$

Therefore, the plastic area in the state of plane strain is much smaller than that in the state of plane stress. Along axis x ($\theta = 0$), the value $r(\theta)$ in the state of plane strain is much less than that in the state of plane stress. If $\nu = 0.33$ applies, $r(\theta)$ (plane strain) $= 0.12r(\theta)$ (plane stress), because the elastic restraint in the z-direction causes the material on the crack tip to be under tensile stress in three directions, in the state of plane strain. At this time, plastic deformation does not tend to occur.

2.1.3.2 Interaction between Stress State and Plastic Area

Pure plane stress or plane strain only appears in the ideal situation. For the common plate, the crack tip on the middle part of the plate thickness is in the state of plane strain, where the plastic area is small. When it is closer to the plate surface, as the elastic restraint becomes small, σ_3 goes down. It gradually becomes the state of plane stress with increasing plastic area. The variation of the entire plastic area with the plate thickness is shown in Figure 2.8.

The previous paragraphs point out the influence of the stress state to the plastic area size. On the other hand, the plastic area size also influences the stress state. When large displacement occurs on the material of the plastic area, the material in other places must compensate. When the plastic area size is basically equal to the plate thickness (Figure 2.9a), free yielding is possible in the plate thickness direction. If the plastic area size is small (Figure 2.9b), free yielding is not possible in the

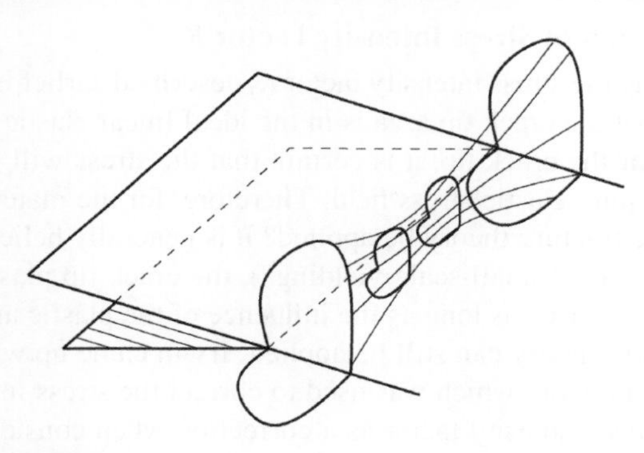

FIGURE 2.8 Shape of plastic area.

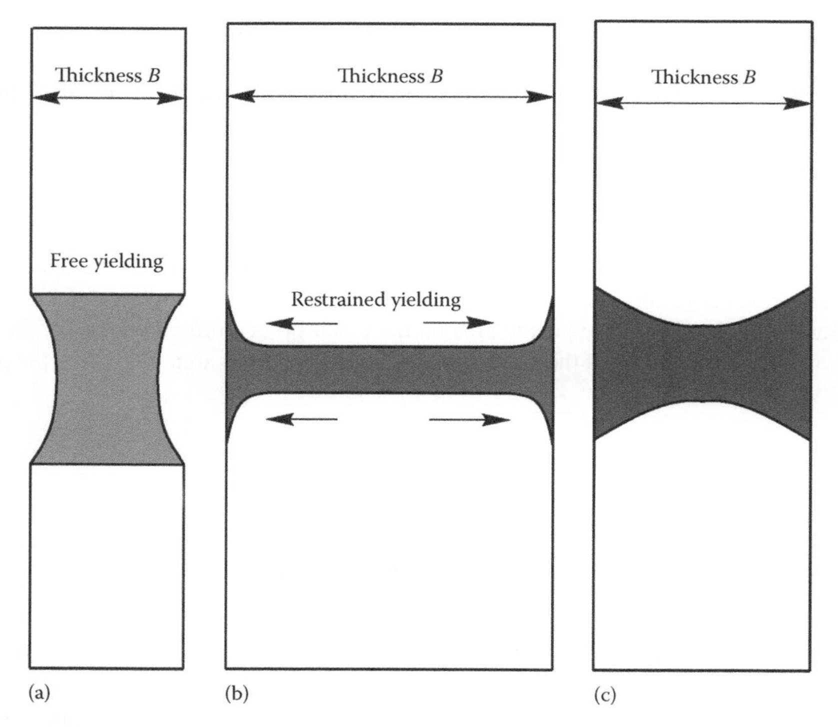

FIGURE 2.9 Influence of plastic zone size to stress state.

thickness direction, because the restraint of the surrounding material keeps ε_z at zero. The result is that the small plastic area is in the state of plane strain, while the large plastic area accelerates the development of the state of plane stress.

For the stress state, the ratio between the plastic area size r_0 and the plate thickness B is an important parameter. If the plastic area size and the plate thickness have the same order of magnitude (r_0/B tends to 1), the plane stress continues to develop. To keep most parts of the thickness in the state of plane strain, the aforementioned ratio must be much less than one (the plane stress area on the surface is only a small part of the thickness). The test proves that if $r_0/B = 0.025$ applies, the crack is a typical plane strain type. Because the plastic area size r_0 is proportional to $(K_I/\sigma_s)^2$, when the fracture toughness test is performed on the material with low-yield stress and high toughness (to ensure that the specimen is in the state of plane strain), the specimen thickness should be large. On the contrary, for the material with high-yield stress and low toughness, the specimen thickness should be small.

2.1.3.3 Plastic Correction of Stress Intensity Factor K_I

The method for calculating the stress intensity factor K_I described earlier is built on the linear elastic theory, which assumes that the crack tip area is in the ideal linear elastic stress field. In fact, when the plastic area exists near the crack tip, it is certain that the stress will become loose; and so the crack stress field is not a pure elastic stress field. Therefore, for the material with plastic deformation, can the linear elastic fracture theory be applied? It is generally believed that, when the plastic area on the crack tip is small ("small-scale yielding"), the crack tip plastic area is surrounded by the wide elastic area. In this case, as long as the influence of the plastic area is taken into consideration, linear elastic fracture theory can still be applied. Irwin came up with one simple and practical "effective crack size" method, which was used to correct the stress intensity factor K_I to obtain the so-called effective stress intensity factor as a correction when considering the influence of the plastic area [4–6].

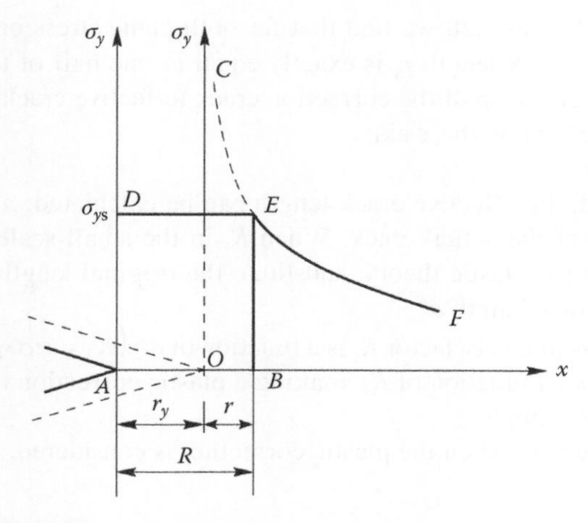

FIGURE 2.10 Plastic correction of crack length.

1. *Irwin's effective crack size*

   Assuming that the material is an ideal elastic-plastic material (for its stress–strain relationship, see Figure 1.2a), after the stress becomes loose, the plastic area near the crack tip has a size of $R = AB$ on the x-axis. The actual stress distribution pattern is shown by the solid line DEF in Figure 2.10. This proves that $R = 2r_0$ applies for the plane stress or for the plane strain. Readers should complete this proof. If it is too hard, refer to Reference [2].

   To make the solution of the linear elastic theory $\sigma_y|_{\theta=0} = K_I/\sqrt{2\pi r}$ still apply, assume that the crack tip is moved rightward, to the point O, to change the actual elastic-plastic stress field into a virtual elastic stress field. Here, the curve of the elastic stress σ_y, substituted with the dotted line, happens to overlap with the elastic part of the curve of the elastic-plastic stress, represented by the solid line at the boundary E of the plastic area. When the point O is the tip of the assumed crack, at the point where $r = R - r_y$ applies, $\sigma_y(r)|_{\theta=0} = \sigma_{ys}$ applies. From Equation 2.7a, we obtain

$$\sigma_y(r)\big|_{\theta=0} = \frac{K_I}{\sqrt{2\pi r}} = \frac{K_I}{\sqrt{2\pi(R-r_y)}} = \sigma_{ys}. \tag{2.25}$$

   Therefore,

$$r_y = R - \frac{1}{2\pi}\left(\frac{K_I}{\sigma_{ys}}\right)^2. \tag{2.26}$$

   For the state of plane stress, $R = (K_I/\sigma_{ys})^{2/\pi}$, $\sigma_{ys} = \sigma_s$, and

$$r_y = \frac{1}{\pi}\left(\frac{K_I}{\sigma_s}\right)^2 - \frac{1}{2\pi}\left(\frac{K_I}{\sigma_s}\right)^2 = \frac{1}{2\pi}\left(\frac{K_I}{\sigma_s}\right)^2. \tag{2.27}$$

   For the state of plane strain, $R = (K_I/\sigma_s)^{2/\pi}$, $\sigma_{ys} = \sqrt{2\sqrt{2}}\sigma_s$, and

$$r_y = \frac{1}{2\sqrt{2}\pi}\left(\frac{K_I}{\sigma_s}\right)^2 - \frac{1}{2\pi}\left(\frac{K_I}{\sqrt{2\sqrt{2}}\sigma_s}\right)^2 = \frac{1}{4\sqrt{2}\pi}\left(\frac{K_I}{\sigma_s}\right)^2. \tag{2.28}$$

From Equations 2.27 and 2.28 we find that for both plane stress or plane strain, the correction value of the crack length r_y is exactly equal to one half of the plastic area size R. In other words, the crack tip of the correction crack (effective crack) is located just on the center of the plastic area on the x-axis.

2. *K factor correction*

After r_y is obtained, the effective crack length can be calculated: $a^* = a + r_y$, where a is the original length of the actual crack. When K_I in the small-scale yielding condition is calculated by using the elastic theory, substitute the original length of the actual crack a with the effective crack length a^*.

Because the stress intensity factor K_I is a function of a^* $\left(K_I = \alpha\sigma\sqrt{\pi a^*}\right)$, and $a^* = a + r_y$ applies, and r_y is also a function of K_I, making a plastic correction to the crack tip's stress intensity factor K_I is complex.

For an ordinary crack, when the plastic correction is considered, the K_I expression may be changed to

$$K_I = \alpha\sigma\sqrt{\pi a^*} = \alpha\sigma\sqrt{\pi(a + r_y)}.$$

After substitution of r_y's Equations 2.27 and 2.28 under the plane stress and plane strain, into the previous equation, and simplifying, we obtain

Plane stress:

$$K_I = \alpha\sigma\sqrt{\pi a}\frac{1}{\sqrt{1 - \frac{\alpha^2}{2}\left(\frac{\sigma}{\sigma_s}\right)^2}}. \tag{2.29}$$

Plane strain:

$$K_I = \alpha\sigma\sqrt{\pi a}\frac{1}{\sqrt{1 - \frac{\alpha^2}{4\sqrt{2}}\left(\frac{\sigma}{\sigma_s}\right)^2}}. \tag{2.30}$$

Therefore, after the influence of the plastic area is considered, K_I increases to a certain extent. This increase is

$$M_P = \frac{1}{\sqrt{1 - \frac{\alpha^2}{2}\left(\frac{\sigma}{\sigma_s}\right)^2}} \quad \text{plane stress.} \tag{2.31}$$

$$M_P = \frac{1}{\sqrt{1 - \frac{\alpha^2}{4\sqrt{2}}\left(\frac{\sigma}{\sigma_s}\right)^2}} \quad \text{plane strain.} \tag{2.32}$$

Generally, M_P is called the plastic correction factor.

It should be noted that the aforementioned analysis is only applicable to "small-scale yielding." When the crack tip plastic size is smaller than the crack length and the component size by one order of magnitude or above, the linear elastic-plastic fracture theory after the plastic correction can be used. For "large-scale yielding" on the crack tip area, or for general yielding, the elastic fracture theory must be used.

2.1.4 Fracture Criterion and Fracture Toughness

2.1.4.1 Stress Intensity Factor Fracture Criterion

The stress intensity factor K is a "generalized load," a parameter that describes the intensity of the stress field near the crack tip. Whether the crack growth becomes unstable depends on the K value. Therefore, the K factor can be used to establish the fracture criterion (also called the K criterion) and $K = K_C$. This means that, for the elastic body containing the crack under an external load, when the K factor of the crack tip reaches the critical value K_C, the crack growth will become unstable, and the cracked body becomes fractured.

For a Mode I crack under plane strain, its crack criterion is

$$K_{\mathrm{I}} = K_{\mathrm{I}C}, \tag{2.33}$$

where K_{I} is the stress intensity factor of the Mode I crack, which is a function of the load that the cracked component bears and the geometry and size of the crack. $K_{\mathrm{I}C}$ is the K_{I} critical value under plane strain. It is a constant of the material, called the material plane strain fracture toughness. It can be determined by experiment (refer to Chapter 5).

In principle, for Mode II, Mode III, and mixed cracks, the corresponding fracture criterion can be established by using Equation 2.33. However, it is difficult to measure $K_{\mathrm{II}C}$ and $K_{\mathrm{III}C}$. At present, the mixed fracture criterion is usually used to establish the relationship between $K_{\mathrm{II}C}$, $K_{\mathrm{III}C}$, and $K_{\mathrm{I}C}$.

After the fracture criterion is established, we can solve the problem of cracked component fracture, which cannot be solved within conventional strength design. However, it must be noted that when the "K criterion" is used for the fracture analysis, (1) the position, shape, and size of the defect is first discovered by using nondestructive testing (NDT) technology—ultrasonic detection, magnetic particle and fluorescent detection, etc.—and then (2) the defect is simplified into a crack model for analysis. In the case of a design component, the possible maximum crack size is estimated as the basis for fracture prevention. On the other hand, the fracture toughness $K_{\mathrm{I}C}$ value of the material is measured accurately.

The "K criterion" can be used to solve the following problems:

1. *Identifying the critical load of the cracked component.* If the geometrical factor, the crack size, and the material's fracture toughness value of the component are known, the "K criterion" can be used to identify the critical load of the cracked component.
2. *Identifying the tolerated size of the crack.* When the load, the material's fracture toughness value, and the cracked body's geometry are given, the "K criterion" can be used to identify the tolerated size of the crack (the crack size when the crack growth becomes unstable).
3. *Identifying the safety level of the cracked component.*
4. *Selecting and assessing the material.* For the traditional design concept, select and assess the material based on the yield limit σ_s or the strength limit σ_b. For alternating stress, it is the endurance limit. However, from the viewpoint of fracture prevention, a material with high $K_{\mathrm{I}C}$ should be selected. In the general case, the higher is σ_s, the lower $K_{\mathrm{I}C}$ becomes. Therefore, when we select and assess the material, we should consider both aspects.

2.1.4.2 Crack Propagation Energy Criterion

1. *Crack propagation resistance R*

   Let us study the energy relationship in the crack propagation process, so as to more clearly reveal the physical meaning of fracture toughness. Obviously, the crack propagation needs energy, and the crack surface area will increase. The crack surface energy intensity is γ, and when two new surfaces are formed during crack propagation, a surface energy of 2γ per unit area is needed. For a metal material, plastic deformation occurs before crack propagation. This process also consumes the energy, called the plastic deformation work. Assume that the plastic deformation work consumed per unit area for the crack propagation is γ_p. For a metal material, γ_p is much larger than γ, such that $\gamma_p = 10^2\gamma - 10^4\gamma$. In general, the energy consumed per unit area for the crack propagation is represented by R:

$$R = 2\gamma + \gamma_p. \tag{2.34}$$

   Obviously, R is the resistance of the crack propagation. As the crack grows, γ is unchanged (it is the unit area energy); however, it is possible for γ_p to go up. This may be related to the size of the plastic area on the crack tip and the deformation. Therefore, as the crack grows, R also goes up continuously, or it soon stabilizes. The resistance curve (R–Δa curve) is shown in Figure 2.11. It is related not only to K_{IC}/σ_s and the material itself, but also to the specimen size. Typically, as the crack grows under plane stress, R goes up sharply, as shown by the curve *ABCD* in Figure 2.11, under the conditions that the specimen thickness B is much less than $(K_{IC}/\sigma_s)^2$ and under the plane strain, where $B \geq 2.5(K_{IC}/\sigma_s)^2$. (Why is the latter requirement imposed? You can make the analysis yourself. If you feel that it is too difficult, please refer to Reference [2].) After the crack grows a little, R tends to saturation, as shown by the curve *AEF*. This is also the resistance curve of most brittle materials. (From the resistance curve of the plane stress and plane strain, which requirements are imposed on the specimen when the fracture toughness is measured?)

   However, for an intermetallic compound like TiAl and Ti3Al+Nb, its plane strain curve is shown by *ABCD*, similar to the tough material's resistance curve [7].

2. *Crack propagation power G_I*

   The crack propagation needs power. Assume that the power provided by the system for the crack propagation is G_I per unit area. Then, in the crack propagation process, $G_I \geq R$ applies. Assume that that the energy (potential energy) of the entire system—the specimen

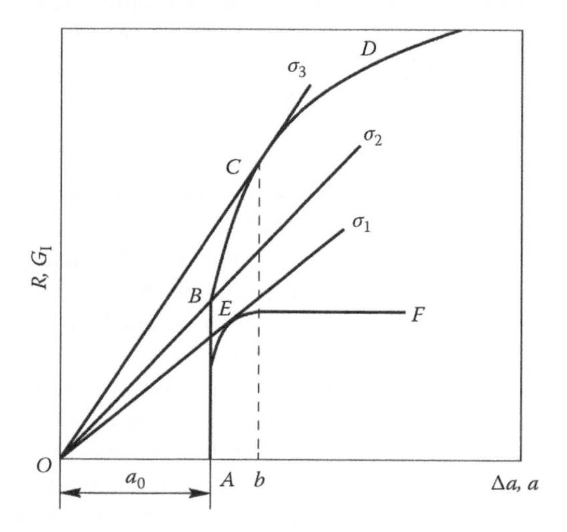

FIGURE 2.11 Resistance curve and power curve.

and the tester constitute one system—is represented by U. The energy consumed for the crack propagation ΔA area is just equal to the power provided by the system; that is, $R\Delta A = G_I\Delta A$. This means that the system's potential energy drops $-\Delta U$. In other words, because the energy required for the crack propagation comes from the system's potential energy, the potential energy drops during the crack propagation, and $G_I\Delta A = -\Delta U$. In the extreme condition, the following equation applies:

$$G_I = -\frac{\partial U}{\partial A}, \tag{2.35}$$

where G_I is the system's energy drop rate per unit area for the crack propagation (called the system's energy release rate). It is the power of the crack propagation. The subscript "I" represents a Mode I crack. For a through crack with a length of a, $\mathrm{d}A = B\mathrm{d}a$, where B is the specimen thickness. For the unit thickness, $B = 1$, and

$$G_I = -\frac{\partial U}{\partial a}. \tag{2.36}$$

G_I is the drop rate of the system's potential energy per unit length of the crack propagation, called the crack propagation power.

For the cracked specimen under an external force F, it is extended by $\mathrm{d}\delta$, and the work of the external force is $\mathrm{d}W = P\mathrm{d}\delta$. From Equation 1.69 in Chapter 1, we know that when the specimen is extended under unidirectional tension, the increment of the elastic strain energy is

$$\mathrm{d}E = \sigma \cdot \varepsilon \cdot \frac{V}{2} = \frac{F}{A} \cdot \frac{\mathrm{d}\delta}{L} \cdot \frac{V}{2} = F\frac{\mathrm{d}\delta}{2}, \tag{2.37}$$

where

L is the length of the specimen
A is the sectional area of the specimen
V is the volume of the specimen

The energy consumed in the crack propagation process is the energy that the system is to provide or $G_I\mathrm{d}A$ (the energy required for unit area of the crack propagation is G_I). Obviously, in the crack propagation process, the work increment of the external force $\mathrm{d}W$ causes the internal strain energy to increase by $\mathrm{d}\Omega$, but it also realizes the crack propagation $\mathrm{d}W = \mathrm{d}\Omega + G_I\mathrm{d}A$. Therefore,

$$G_I = -\frac{\partial(\Omega - W)}{\partial A}. \tag{2.38}$$

After comparing this with Equation 2.35, we know that

$$U = \Omega - W. \tag{2.39}$$

For a specimen with constant displacement, δ = constant, $\mathrm{d}\delta = 0$, and $\mathrm{d}W = 0$ apply, so that

$$G_I = -\frac{\partial \Omega}{\partial A} = -\frac{\partial \Omega}{\partial a} \ (B = 1). \tag{2.40}$$

This equation shows that, as the crack grows, the originally stored elastic strain energy is released. When the released elastic strain energy $-d\Omega$ is equal to or larger than the energy consumed for the crack propagation Rda, the crack can grow. Under the condition of constant displacement, G_I is called the strain energy release rate. However, as the crack grows under the condition of constant load or tension, the stored elastic strain energy increases, instead of being released. The work increment of the external force dW, after the strain energy increment dE is deducted, is used for the crack propagation. In this case, G_I is not called the strain energy release rate. When $dW - dE \geq G_I da$, the crack can grow.

3. *Relationship between G_I and K_I*

G_I is a power for the crack propagation, while K_I is a generalized load. They both represent the "load" contributing to the crack propagation, and therefore, some relationship should exist between them. Below we discuss the relationship between G_I and K_I.

Consider the crack model [8] shown in Figure 2.12. Assume that the board is fixed at both ends. Consequently, when the crack grows, the board's strain energy goes down. Obviously, the strain energy released during the crack propagation should be equal to the work required to force the grown crack to return to its original state with respect to the value. As a result, we can change the question, "how do we calculate the decrement of the strain energy?" to the question, "how do we calculate the work?"

From the analysis in Figure 2.12c, we know that the work ΔW needed to force the crack to close is

$$\Delta\bar{W} = 4B\int_0^{\Delta a}\frac{1}{2}\sigma_y(r,0)v(r,\pi)\mathrm{d}x, \tag{2.41}$$

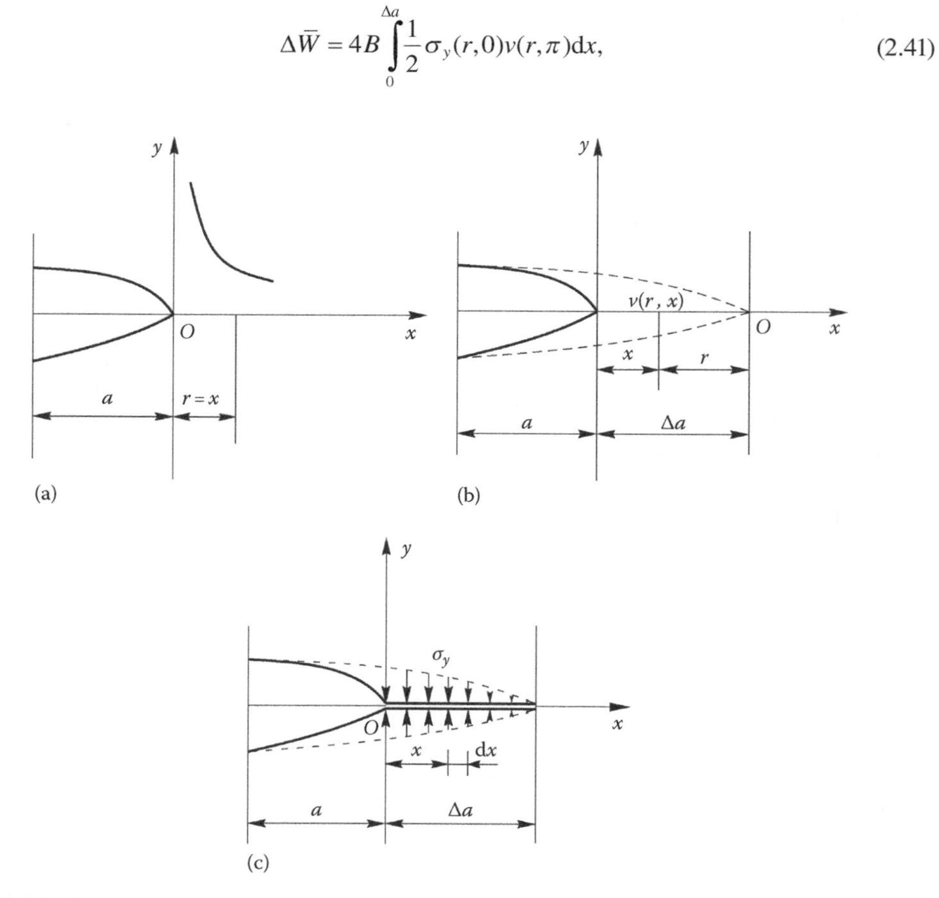

FIGURE 2.12 Diagram of force causing the crack to close.

where $\sigma_y(r,0) = K_I/\sqrt{2\pi x}$, and $v(r, \pi)$ is the displacement of all points on the crack surface while the crack is closing. Its value can be represented by substituting $\theta = \pi$, $r = \Delta a - x$ into Equation 2.7b:

$$v(r,\pi) = \frac{(\kappa+1)K_I}{2G}\sqrt{\frac{\Delta a - x}{2\pi}}, \tag{2.42}$$

where, under plane strain, $\kappa = 3 - 4\nu$, and under plane stress, $\kappa = (3 - \nu)/(4 + \nu)$. After substituting $\sigma_y(r, 0)$ and $v(r, \pi)$ into Equation 2.41, we obtain

$$\Delta\bar{W} = \frac{(\kappa+1)(1+\nu)}{4E}\Delta A K_I^2, \tag{2.43}$$

Where $\Delta A = 2B\Delta a$.

The strain energy released by the system when the crack grows should be equal to the work required to force the crack to return to its original state with respect to the value, so

$$-\Delta\Omega = \Delta\bar{W} = \frac{(\kappa+1)(1+\nu)}{4E}\Delta A K_I^2. \tag{2.44}$$

After substituting Equation 2.44 into Equation 2.40, we obtain the relationship between the crack propagation power G_I and the stress intensity factor K_I:

$$G_I = \frac{(\kappa+1)(1+\nu)}{4E}K_I^2 \quad B = 1. \tag{2.45}$$

This equation can be changed to

$$G_I = \frac{K_I^2}{H}, \tag{2.46}$$

where

$$H = \begin{cases} E & \text{Plane stress} \\ \dfrac{E}{1-\nu^2} & \text{Plane strain} \end{cases}$$

Hence, it is proved that, for Mode II and Mode III cracks, the similar relationship is

$$G_{II} = \frac{K_{II}^2}{H} \quad G_{III} = \frac{(1+\nu)K_{III}^2}{E}. \tag{2.47}$$

In Equation 2.7a, for the through crack in the center, $K_I^2 = \sigma^2\pi a$; that is, $G_I = \sigma^2\pi a/H$. The power curves (G_I–a curve) under different external stresses σ are straight lines passing through the origin, as shown by the straight lines *OE*, *OB*, and *OC* in Figure 2.11.

2.1.4.3 Fracture Toughness and Critical Fracture Stress

Obviously, the crack grows only when $G_I \geq R$. Figure 2.11 shows that as the crack grows, both R and G_I go up. However, if dR/da is larger than dG_I/da, $G_I < R$ applies after a certain distance is reached. At this time, it will stop growing, and the component will not become fractured. If constant stress σ_2 is applied, the power curve is OB, intersecting with the resistance curve $ABCD$ of the toughness material (or plane stress) at point B. Below point B, $G_I > R$, and the crack grows. However, above point B, $G_I < R$, and the crack stops growing. If constant stress σ_3 is applied, the power curve is OC and is tangent with the resistance curve. As the crack grows, G_I is always much larger than (or equal to) R, and the crack can keep growing until the specimen becomes fractured. The point of tangency C of the power curve and the resistance curve corresponds to the critical state of the crack's unstable propagation. When the gradient of the power curve dG_I/da is equal to the gradient of the resistance curve dR/da, we can obtain the coordinate of the critical point C (the point of tangency). By making $dG_I/da = dR/da$, we can obtain the crack length a_c (Ob) and the external stress σ_3 to which the critical point C corresponds. After substituting them in Equation 2.46, we can obtain the critical power that results in unstable crack propagation $G_{IC} = \sigma_3^2 a_c/H$. It is equal to the critical resistance of the unstable crack propagation $R_C = 2\gamma + \gamma_{PC}$.

When the specimen does not meet the plane strain condition, its resistance curve shape is related to the specimen thickness B, so that the critical resistance $R_c = G_{IC}$ is also related to the thickness. Once the specimen meets the plane strain condition—for example, when $B > 2.5(K_{IC}/\sigma_s)^2$—the resistance curve no longer changes with the specimen thickness. Its shape is shown by the curve AEF in Figure 2.11. Many experiments show that, under plane strain, the critical crack length to which the critical point C corresponds (the point of tangency of the resistance curve and the power curve) is $a_c = 1.02a_0$, where a_0 is the original crack length. On the critical point, the crack's relative growth $\Delta a/a_0$ is 2%. That is, under plane strain, the crack will propagate unstably after growing by 2% relatively, leading to final fracture. In this case, the critical crack propagation resistance $R_C = G_{IC}$ is a minimum stable value. It is the material constant, also called the material's fracture toughness, because it is the metric for evaluating the capability of the material to resist unstable crack propagation:

$$G_{IC} = R_C = 2\gamma + \gamma_{PC}. \tag{2.48}$$

Under plane strain, both G_{IC} and K_{IC} are the metrics for evaluating the capability of the material to resist unstable crack propagation and are both called the fracture toughness. They are correlated by Equation 2.46, under the plane strain:

$$G_{IC} = (1-\nu^2)K_{IC}^2/[E = 2\gamma + \gamma_P], \tag{2.49}$$

where γ_P is γ_{PC} in Equation 2.48. The $G_{IC} = R_C$ measure, under plane stress, is not the material constant (it is related to the specimen thickness). Therefore, only $G_{IC} = R_C$ and K_{IC} measured under plane strain is the material constant, which is unrelated to the specimen thickness, and is called the material's fracture toughness.

From Figure 2.11, we know that as the crack grows, the power is much larger than the resistance, once the crack propagation power $G_I \geq R_C = G_{IC}$ (resistance at the critical point) applies. Even when the external stress does not go up, the crack grows, until the specimen or component becomes fractured. When $G_I = (1-\nu^2)K_I^2/E$ applies, $G_I \geq G_{IC}$ is equivalent to $K_I \geq K_{IC}$. That is, the crack propagates unstably, and the mechanical criterion of the specimen fracture is

$$G_I \geq R_{IC} = R_C \quad K_I \geq K_{IC}. \tag{2.50}$$

2.1.5 Elastic-Plastic Fracture Mechanics

Linear elastic-plastic fracture mechanics is subject to the small-scale yielding condition. It is actually impossible to meet the small-scale yielding condition for low-strength materials that have high toughness and withstand large plastic deformation and whose crack tip is passivated before becoming cracked. In this case, the size of the plastic area on the crack tip is also closed and even exceeds the crack size. Such a fracture is the large-scale yielding fracture. The other problem is that, due to the presence of high local stress, the material in this area is in the general yielding state. In the plastic area with such high strain, even a short crack can keep growing, which will finally lead to fracture. Such a fracture is called the general yielding fracture. Both the large-scale yielding fracture and the general yielding fracture are within the scope of elastic-plastic fracture mechanics.

2.1.5.1 Crack Opening Displacement

In 1961, Wells introduced the COD theory [9]. His experiment and analysis showed that, after a load is applied on the cracked body, the plastic area near the crack tip will cause the surface of the crack tip to open. This extent of this opening is called the COD, generally represented by δ. Wells thought that when the COD δ reached the material's critical value δ_c, that the crack would begin to grow unstably. This is the COD criterion of elastic-plastic fracture mechanics: $\delta = \delta_c$.

Dugdale [10] and Barenblatt [11] came up with an assumption that the plastic area on the crack tip showed an independent wedge strip feature. This model is called the D–B strip plastic area model and is applicable to both small-scale yielding and large-scale yielding. It can be used to analyze the elastic-plastic fracture problem of an infinite board with a through crack in the center, under even tension stress. The D–B model assumes that under even tensile stress σ from a far field, the crack tip area with a crack length of $2a$ extends in a wedge strip form along both edges of the crack line. The strip length is $2R$, as shown in Figure 2.13a. The material of the plastic area is an ideal plastic material. The entire crack and plastic area are still surrounded by a wide elastic area. The joint surface of the plastic area and the elastic area is applied with an even bonding force σ_s, which Brenblatt called the cohesive force [11]. The relationship between the cohesive force and the opening displacement is shown in Figure 2.13b. When $\delta = \delta_c$ applies, the bonding becomes fractured and the cohesive force goes to zero.

Assuming that the strip plastic area is cut, the crack length is $2L = 2(a + R)$. As the media other than the strip yield area shows linear elasticity, the superposition principle can be applied. Hence, the crack tip's stress intensity factor K_I consists of two parts: one is the solution of the even tensile

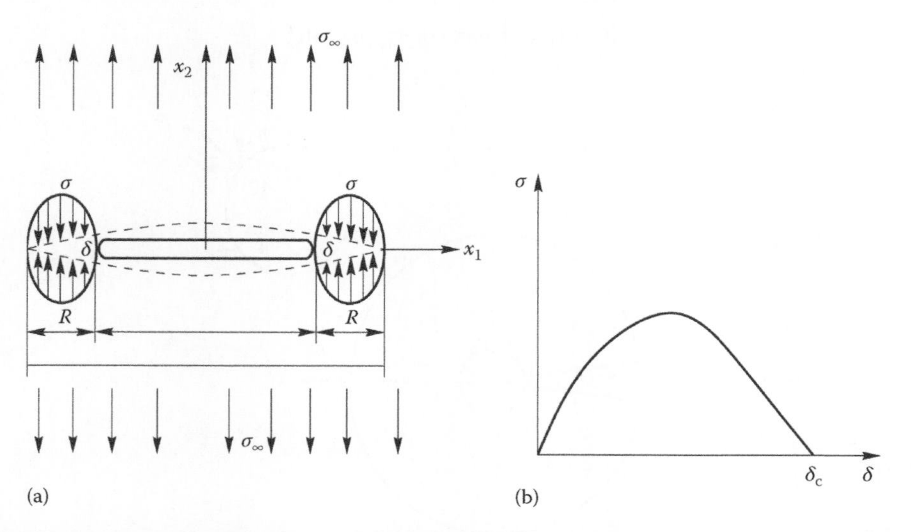

FIGURE 2.13 Cohesive force model. (a) D–B strip plastic area. (b) The cohesive force-opening displacement curve.

stress σ from a far field and the other one is the solution of the cohesive force. Through the derivation, we obtain the crack opening displacement [2]:

$$\delta(a) = \frac{8\sigma_s a}{\pi E} \ln\left[\sec\left(\frac{\pi}{2}\frac{\sigma}{\sigma_s}\right)\right]$$

$$R = a\left[\sec\left(\frac{\pi}{2}\frac{\sigma}{\sigma_s}\right) - 1\right] \tag{2.51}$$

Under small-scale yielding (SSY), take σ/σ_s as the small quantity. Expand the previous two equations with $\left(\frac{\pi}{2}\frac{\sigma}{\sigma_s}\right)$ as the small quantity to obtain the SSY solution:

$$R_{SSY} = \frac{\pi^2}{8}\left(\frac{\sigma}{\sigma_s}\right)^2 \quad a = \frac{\pi^2}{8}\left(\frac{K_I}{\sigma_s}\right)^2, \tag{2.52}$$

$$\delta_{SSY} = \frac{\pi a}{E\sigma_s} \quad \sigma^2 = \frac{G}{\sigma_s}, \tag{2.53}$$

where $K_I = \sigma\sqrt{\pi a}$ is the stress intensity factor (SIF) when the strip plastic area is not considered, and G is the energy release rate when the strip plastic area is not considered. After comparing Equation 2.51 with Equation 2.53, we find

$$\frac{\delta}{\delta_{SSY}} = \frac{8}{\pi^2}\left(\frac{\sigma_s}{\sigma}\right)^2 \ln\left[\sec\left(\frac{\pi}{2}\frac{\sigma}{\sigma_s}\right)\right]. \tag{2.54}$$

This equation can be used to identify the precision of the small-scale yielding solution.

2.1.5.2 *J* Integral Theory

Rice came up with *J* integral theory [12,13] in 1968, which brought vitality to the study of elastic-plastic fracture mechanics. Consider the two-dimensional problem shown in Figure 2.14 and define the *J* integral as

$$J = J_1 = \int_\Gamma (\omega n_1 - n_\alpha \sigma_{\alpha\beta} u_{\beta,1}) \mathrm{d}\Gamma. \tag{2.55}$$

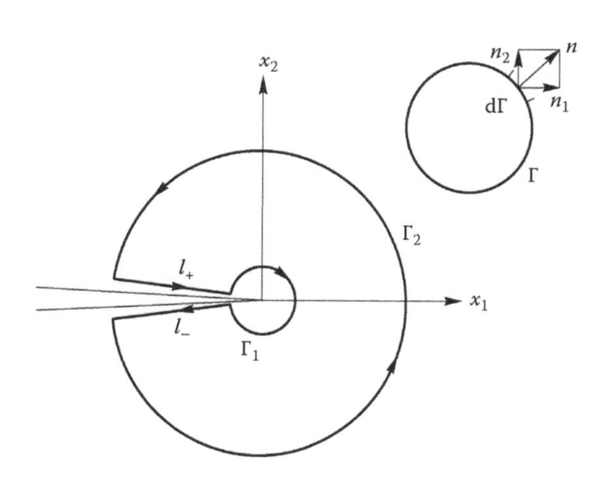

FIGURE 2.14 *J* integral.

where Γ is a contour on the $x_1 - x_2$ plane, Γ_1 and Γ_2 take the anticlockwise rotation as the positive direction, and l_- and l_+ take the direction from left to right as the positive direction. $\omega = \int_0^{\varepsilon_{ij}} \sigma_{ij} \mathrm{d}\varepsilon_{ij}$ is the strain energy intensity of any point on the loop Γ. It is proved that, [1] under the following assumptions, we can obtain Equation 2.56: (1) The hyper elasticity constitutive relationship $\sigma_{ij} = \partial\omega/\partial\varepsilon_{ij}$ applies and the material is even along x_i, (2) there is no physical effect, (3) the material is quasistatic, (4) there is no singular point in the Γ area, and (5) there is small deformation.

$$J_i = \int_\Gamma (\omega n_i - n_j \sigma_{jk} u_{k,i}) \mathrm{d}\Gamma = 0, \quad (i, j = 1, 2, 3). \tag{2.56}$$

Hence the special cases of taking J as J_i include

$$\int_{-\Gamma_1 - l_- + \Gamma_2 + l_+} (\omega n_1 - n_\alpha \sigma_{\alpha\beta} u_{\beta,1}) \mathrm{d}\Gamma = 0, \quad (\alpha, \beta = 1, 2). \tag{2.57}$$

For the contour section l_- and l_+ along the upper and lower crack shores, $n_1 = 0$ applies. If we further assume that the crack surface is free ($\sigma_{\alpha\beta}\, n_\alpha = 0$ applies on l_+ and l_-), their sections do not contribute to the J integral. Therefore, we obtain

$$J = \int_{\Gamma_1} (\omega n_1 - n_\alpha \sigma_{\alpha\beta} u_{\beta,1}) \mathrm{d}\Gamma = \int_{\Gamma_2} (\omega n_1 - n_\alpha \sigma_{\alpha\beta} u_{\beta,1}) \mathrm{d}\Gamma, \tag{2.58}$$

and the J integral value is unrelated to the integral path.

Now, we analyze the J integral of the linear elastic body. For the linear elastic body, the several prerequisite conditions under which J integral conservation are established exist naturally. Therefore, J integral theory can also be used to analyze the problem of the linear elastic plane crack. After substituting the stress and strain field of a Mode I crack tip area under plane strain into the strain energy intensity, we obtain

$$\omega = \frac{1}{2}\sigma_{ij}\varepsilon_{ij} = \frac{K_\mathrm{I}^2(1+\nu)}{2\pi r E}\left[\cos^2\frac{\theta}{2}\left(1 - 2\nu + \sin^2\frac{\theta}{2}\right)\right]. \tag{2.59}$$

If we take the circle with the crack tip as the center and r as the radius of the integral loop Γ, we obtain

$$\int_\Gamma \omega \mathrm{d}y = \int_{-\pi}^{\pi} \omega r \cos\theta \mathrm{d}\theta = \frac{K_\mathrm{I}^2(1+\nu)(1-2\nu)}{4E}. \tag{2.60}$$

After substituting the stress field and displacement field of the crack tip into the other part of the J integral, we obtain

$$\int_\Gamma T_i \frac{\partial u_i}{\partial x_1} \mathrm{d}s = \int_{-\pi}^{\pi} \left(T_1 \frac{\partial u_1}{\partial x_1} + T_2 \frac{\partial u_2}{\partial x_1}\right) r \mathrm{d}\theta = -\frac{K_\mathrm{I}^2(1+\nu)(3-2\nu)}{4E}. \tag{2.61}$$

Finally, we obtain

$$J = \frac{1-\nu^2}{E} \quad K_\mathrm{I}^2 = \frac{K_\mathrm{I}^2}{H} = G_\mathrm{I}. \tag{2.62}$$

This equation shows, in the linear elastic state, the relationship between the J integral and the stress intensity factor K_I and the crack growth energy release rate G_I. In addition, through the analysis, we can also obtain the relationship between the J integral and the COD. Therefore, J is a widely applicable parameter in fracture mechanics. In the linear elastic state, $J = J_{IC}$ still applies. What is more, it is fully equivalent to the stress intensity factor criterion and the energy criterion. To call J_{IC} the fracture toughness, experimentation is required.

2.2 ANALYSIS OF MICROFRACTURE MECHANICS

Microfracture mechanics includes meso- and nanofracture mechanics. Generally, micromechanics is applied at the micron level. In this case, the English word "micron" is well correlated with the micromechanics. Its theoretical framework is also called damage mechanics. Nanomechanics goes further to the nanoscopic level. Its object of study may be nanocrystals or nanomaterials, but in most cases, study is made on the mechanical behavior of common solid materials at the nanolevel [1].

Damage mechanics involves the process from the presence of microdefects on the raw material or component to its growth into a macrocrack. Fracture mechanics involves the process from the macrocrack to the final fracture. Damage mechanics has gradually evolved. In 1958, Kachanov first introduced a theory that described the continuous performance variation in the material damage process using continuous variables [14]. Rabotnov, his student, later spread this theory and laid a firm foundation for damage mechanics [15]. In 1977, Janson and Hult came up with a new term "damage mechanics" [16]. At present, damage mechanics has become the frontier research area of solid mechanics, materials science, and condensed matter physics [17–22].

This section describes the main framework of microfracture mechanics from the concept and classification of damage, isotropic damage, and anisotropic damage. Although nanofracture mechanics attracts the attention of many researchers and many achievements have been made in this area, this book does not offer a systematic description due to the area's immaturity. If you are interested in this subject, please see References [1,23–27].

2.2.1 Basic Concept and Classification of Damage [18]

Damage mechanics is the study of the evolutionary pattern of the microdefect or damage field, and its influence on a material's mechanical performance, when the deformable solids containing various continuously distributed microdefects inside the material are applied with external factors such as a load. The quantitative level of damage mechanics concerns microdefects like microcracks and microholes to the material's volume element fracture (macrocracking). As various materials have greatly different component and minimum grain sizes, their microsize and volume element levels also differ somewhat.

We explain the concept of damage from the description of the one-dimensional damage state. Consider a straight rod pulled evenly, as shown in Figure 2.15. We reason that the main mechanism of material damage is the reduction of the effective area due to the microdefect. Assuming that the sectional area in the undamaged state is A and that the effective loaded area is reduced to $\tilde{A}$ after damage, the physical meaning of the continuity ψ is the ratio between the effective loaded area and the sectional area in the undamaged state:

$$\psi = \frac{\tilde{A}}{A}. \tag{2.63}$$

Obviously, the continuity ψ is a dimensionless scalar field variable. $\psi = 1$ corresponds to an ideal state of the material without any defects, while $\psi = 0$ corresponds to the state of a fully fractured material without any bearing capacity.

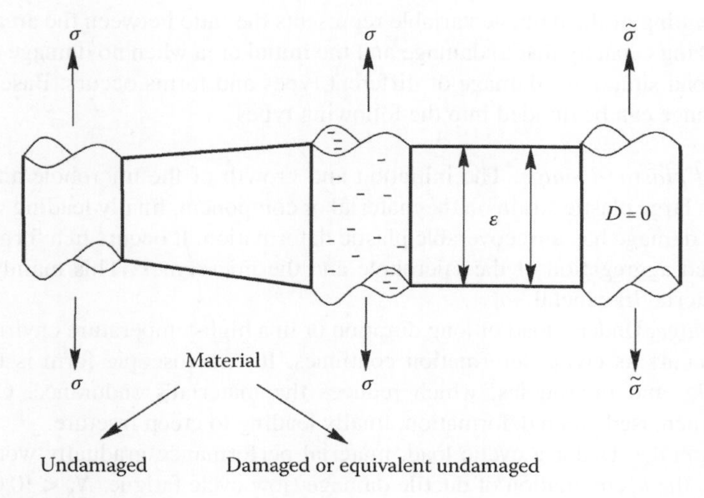

FIGURE 2.15 Strain equivalence.

The ratio between the external load F and the effective loaded area $\tilde{A}$ is defined as the effective stress $\tilde{\sigma}$:

$$\tilde{\sigma} = \frac{F}{\tilde{A}} = \frac{\sigma}{\psi}, \tag{2.64}$$

where $\sigma = F/A$ is the Cauchy's stress. The continuity goes down monotonically. Assuming that the material becomes fractured when ψ reaches one critical value ψ_c, the material's fracture condition is represented as

$$\psi = \psi_c. \tag{2.65}$$

The experiment shows that, for most metal materials, $0.2 \leq \psi_c \leq 0.8$. Even for the same material, ψ_c is not a constant. It is also related to the load [28,29].

In 1963, Rabotnov, a famous dynamicist, suggested the use of a damage variable D to describe the damage [15], when he studied the creep constitutive equation of a metal. D is defined as

$$D = 1 - \psi = 1 - \frac{\tilde{A}}{A}. \tag{2.66}$$

When the material is in a fully undamaged state, D in = 0 applies. For the material without any bearing capacity, $D = 1$ applies. From Equations 2.63 and 2.66, we obtain

$$D = \frac{A - \tilde{A}}{A}. \tag{2.67}$$

Therefore, the relationship between the effective stress $\tilde{\sigma}$ and the damage variable D is

$$\tilde{\sigma} = \frac{\sigma}{\psi} = \frac{\sigma}{1 - D}. \tag{2.68}$$

The physical meaning of the damage variable represents the ratio between the area when the material loses its bearing capacity due to damage and the initial area when no damage occurs.

In different load situations, damage of different types and forms occurs. Based on the loading process, the damage can be divided into the following types:

1. *Tough and plastic damage.* The initiation and growth of the microhole and microcrack results in a large plastic strain on the material or component, finally leading to plastic fracture. Such damage has unrecoverable plastic deformation. It occurs in a form of initiation, growth, and aggregation of the microhole and the microcrack. This mainly occurs on a plastic material like metal.
2. *Creep damage.* Under a load of long duration or in a high-temperature environment, creep damage occurs as creep deformation continues. Its macroscopic form is the growth of microcracks and microholes, which reduces the material's endurance. Creep damage results in increased creep deformation, finally leading to creep fracture.
3. *Fatigue damage.* Under a cyclic load, material performance gradually worsens. In each load cycle, the accumulation of ductile damage (low cycle fatigue: $N_R < 10{,}000$) or brittle damage (high cycle fatigue: $N_R > 10{,}000$) shortens the material's life, leading to fatigue fracture.
4. *Dynamic damage.* Under a dynamic load such as an impact load, many microcracks initiate and grow inside the material. These microcracks have large quantities, but generally do not grow fully, because the load time is very short (typically several microseconds). However, when one section is full of microcracks, fracture occurs.

2.2.2 Example: One-Dimensional Creep Damage

Let us use an example to better our understanding of the damage concept: an analysis of Kachanov's one-dimensional creep damage model [14,18]. The fracture behavior of a metal is different at high temperatures and under high and low loads. When the load is high, the specimen becomes longer and the sectional area becomes smaller, leading to a monotonic increase in stress, until ductile fracture occurs on the material. The corresponding meso mechanism is that the microhole growth in the metal crystalline grain results in transgranular fracture. When the load is low, the specimen becomes slightly longer, and the sectional area basically remains unchanged. However, microcracks and microholes still appear on the grain boundary inside the material. They become larger as time goes by, finally becoming macrocracks, which leads to intergranular brittle fracture.

Assume that the specimen's initial sectional area before loading is A_0, the sectional area after the loading is A, its effective loaded area is $\tilde{A} = A(1 - D)$, the nominal stress is σ_0, and the Cauchy's stress is σ. The effective stresses $\tilde{\sigma}$ are respectively defined as

$$\sigma_0 = \frac{P}{A_0}, \quad \sigma = \frac{P}{A} \quad \tilde{\sigma} = \frac{P}{\tilde{A}} = \frac{P}{A(1-D)} = \frac{\sigma}{1-D}. \tag{2.69}$$

Ignoring the elastic deformation and considering the damage situation, the creep law (for the creep concept, refer to Chapter 7) is assumed to be

$$\frac{\mathrm{d}\varepsilon}{\mathrm{d}t} = B\tilde{\sigma}^n, \tag{2.70}$$

where

ε is the total strain

B and n are the material constants

In the undamaged situation, $\tilde{\sigma} = \sigma$, and Equation 2.70 is often called Norton's law. When studying creep damage, we must establish the damage evolution process. That is, we must establish the relationship between the damage evolution rate $\mathrm{d}D/\mathrm{d}t$ and the correlated dynamical variables. For some simple cases, we can assume that the evolution rate equation also has the form of an exponential function:

$$\frac{\mathrm{d}D}{\mathrm{d}t} = C\tilde{\sigma}^{\nu} = C\left(\frac{\sigma}{1-D}\right)^{\nu}, \tag{2.71}$$

where C and ν are the material constants.

Assuming that the nominal stress σ_0 remains unchanged and based on the material's incompressible volume condition $AL = A_0L_0$, we obtain the effective stress

$$\tilde{\sigma} = \frac{\sigma}{1-D} = \frac{\sigma_0 A_0}{A(1-D)} = \frac{\sigma_0 L}{L_0(1-D)} = \frac{\sigma_0}{1-D}\exp\varepsilon. \tag{2.72}$$

where $\varepsilon = \ln L/L_0$ is used, and L_0 and L are, respectively, the length of the specimen before and after the damage. We discuss the creep fracture of the metal material in three situations as follows.

2.2.2.1 Undamaged Ductile Fracture

When damage is not considered (i.e., $D \equiv 0$), Equation 2.72 is simplified into

$$\tilde{\sigma} = \sigma_0 \exp\varepsilon. \tag{2.73}$$

After substituting it into Equation 2.70, we obtain

$$\frac{\mathrm{d}\varepsilon}{\mathrm{d}t} = B\sigma_0^n \exp(n\varepsilon). \tag{2.74}$$

After integrating this equation, by using the initial condition $\varepsilon(0) = 0$, we obtain

$$\varepsilon(t) = -\frac{1}{n}\ln\left(1 - nB\sigma_0^n t\right). \tag{2.75}$$

The condition of ductile creep fracture is $\varepsilon \to \infty$, so we obtain the time for the ductile creep fracture:

$$t_{RH} = \frac{1}{nB\sigma_0^n}. \tag{2.76}$$

2.2.2.2 Damaged Brittle Fracture without Deformation

When deformation is not considered (i.e., $\varepsilon \equiv 0$), $A = A_0$ applies and the effective stress in Equation 2.72 is simplified to

$$\tilde{\sigma} = \frac{\sigma_0}{1-D}. \tag{2.77}$$

After substituting it into Equation 2.71, we obtain

$$\frac{\mathrm{d}D}{\mathrm{d}t} = C\sigma_0^{\nu}(1-D)^{-\nu}. \tag{2.78}$$

After integrating this equation, by using the initial condition $D(0) = 0$, we obtain

$$D = 1 - \left[1-(1+\nu)C\sigma_0^{\nu}t\right]^{1/(\nu+1)}. \tag{2.79}$$

The condition of damage brittle fracture is $D = D_c = 1$, so we obtain the time for the brittle fracture:

$$t_{RK} = \frac{1}{(1+\nu)C\sigma_0^{\nu}}. \tag{2.80}$$

This expression was derived by Kachanov in 1958 [14].

2.2.2.3 Damage and Deformation Considered Simultaneously

Similar to the definition of logarithmic strain

$$\mathrm{d}\varepsilon = \frac{\mathrm{d}L}{L} = -\frac{\mathrm{d}A}{A}, \tag{2.81}$$

damage is defined in the following form [30]:

$$\mathrm{d}D = -\frac{\mathrm{d}A_n}{A_n}, \tag{2.82}$$

where A_n is the assumed effective loaded area, which is defined as

$$\tilde{\sigma} = \frac{F}{A_n}. \tag{2.83}$$

Hence, the effective stress in Equation 2.72 is changed to

$$\tilde{\sigma} = \sigma_0 \exp(\varepsilon + D). \tag{2.84}$$

From Equations 2.70, 2.71, and 2.84, we obtain the following control equation of the effective stress $\tilde{\sigma}$:

$$\frac{\mathrm{d}\tilde{\sigma}}{\tilde{\sigma}\mathrm{d}t} - B\tilde{\sigma}^{n} - C\tilde{\sigma}^{\nu} = \frac{\mathrm{d}\sigma_0}{\sigma_0\mathrm{d}t}. \tag{2.85}$$

By giving the loading history $\sigma_0(t)$ at random, we can obtain the variation process $\tilde{\sigma}(t)$ of the effective stress from the previous equation. For example, for the Heaviside loading history shown in Figure 2.16, on section 0–1, it is

$$\frac{1}{\tilde{\sigma}}\mathrm{d}\tilde{\sigma} = \frac{1}{\sigma_0}\mathrm{d}\sigma_0, \tag{2.86}$$

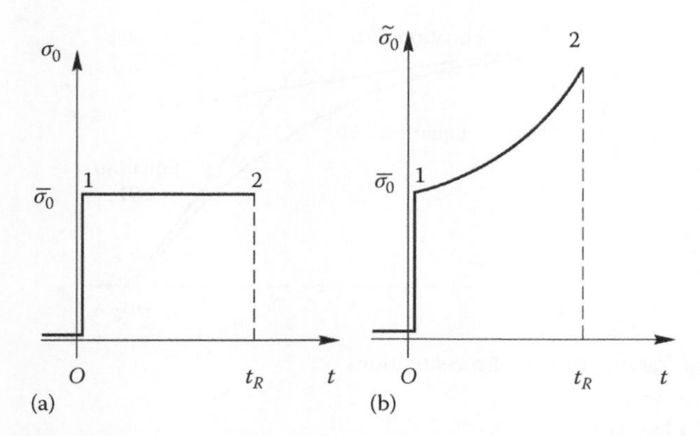

FIGURE 2.16 (a) Heaviside loading history $\sigma_0(t)$. (b) Effective stress $\tilde{\sigma}$ (t).

and we obtain

$$\tilde{\sigma}_0 = \bar{\sigma}_0. \tag{2.87}$$

This equation shows that, in the transient loading process, there is no creep deformation or damage development. From point 1 to point 2, Equation 2.85 is simplified to

$$\frac{\mathrm{d}\tilde{\sigma}}{\tilde{\sigma}\mathrm{d}t} - B\tilde{\sigma}^n - C\tilde{\sigma}^\nu = 0. \tag{2.88}$$

After integrating this equation, by using the initial condition in Equation 2.87, we obtain

$$t = \int_{\bar{\sigma}_0}^{\tilde{\sigma}} (B\tilde{\sigma}^{n+1} + C\tilde{\sigma}^{\nu+1})^{-1}\mathrm{d}\tilde{\sigma}. \tag{2.89}$$

From the previous equation and the condition that $\tilde{\sigma} \to \infty$, we obtain the fracture time when both damage evolution and creep deformation are considered:

$$t_R = \int_{\bar{\sigma}_0}^{\infty} (B\tilde{\sigma}^{n+1} + C\tilde{\sigma}^{\nu+1})^{-1}\mathrm{d}\tilde{\sigma}. \tag{2.90}$$

If $C = 0$ applies, it is the fracture time without considering the damage, which is same as t_{RH} in Equation 2.76. If $B = 0$ applies, it is the fracture time without considering the creep deformation:

$$t_R = \frac{1}{\nu C\bar{\sigma}_0^\nu}. \tag{2.91}$$

Because different damage definitions are used, t_{RK} in Equations 2.80 and 2.91 are a little different. When $B > 0$ and $C > 0$, we can obtain the numeric integral result of the fracture time, as shown in Figure 2.17. From this figure, we know that when the stress is high, we can use Equation 2.76, and when the stress is low, we can use Equation 2.91. When the stress is moderate, we should consider both damage and creep deformation.

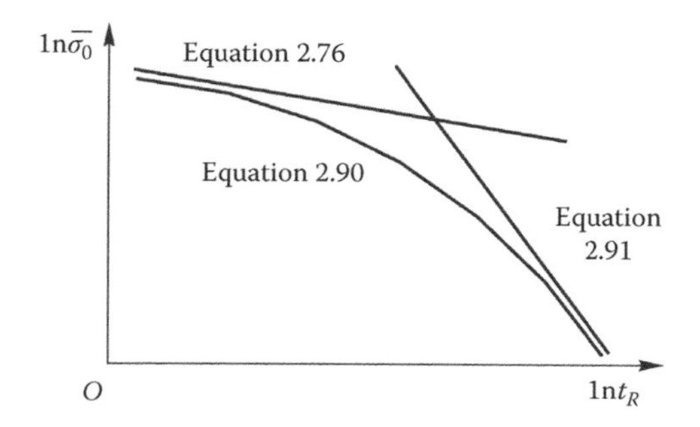

FIGURE 2.17 Creep fracture time in three situations.

2.2.3 Isotropic Damage

2.2.3.1 Definition of Isotropic Damage

Based on the aforementioned description of the one-dimensional damage state, we will now analyze the problem of the damage field. First, we analyze the isotropic damage. In many problems, the damage distribution and its influence on material performance does not differ too much in all directions. For such problems, we can assume that the influence of the damage in all directions is the same. Such problems are thus concerned with isotropic damage, in which the damage variable can be described by one scalar, generally represented by the variable D [17]:

$$D = \frac{\delta S_D}{\delta S}, \tag{2.92}$$

where

δS is one section area in the micelle

δS_D is the damaged area (defect) on the section in question, as shown in Figure 2.18

From Equation 2.92, we find that the variation range of the damage variable D is $0 \le D \le 1$. When $D = 0$ applies, $\delta S_D = 0$ is established, which means that the section is not damaged. When $D = 1$ applies, $\delta S_D = \delta S$ is established, which means that the section is full of damage (defects) and the material becomes fully fractured. In fact, when $D < 1$ applies, fracture or damage has occurred.

For the complex stress state, when damage is represented by one scalar, the effective stress $\tilde{\sigma}_{ij}$ can be defined using Equation 2.68:

$$\tilde{\sigma}_{ij} = \frac{\sigma_{ij}}{1-D} = \sigma_{ij}\frac{\delta S}{\delta S - \delta S_D}, \tag{2.93}$$

where σ_{ij} is a component of the Cauchy's stress tensor.

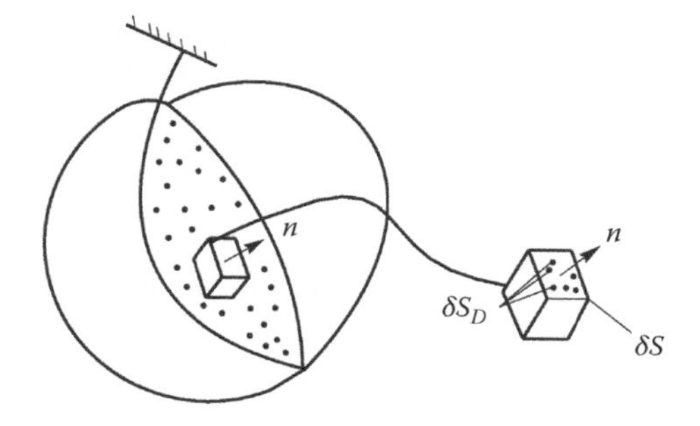

FIGURE 2.18 Definition of damage variable.

2.2.3.2 Strain Equivalence Principle

In the medium containing damage, it is difficult to analyze each defect form and damage mechanism at the mesoscale in order to identify the effective loaded area. To indirectly measure the damage, Lemaitre [31], in 1971, introduced an important strain equivalence hypothesis. The strain equivalence principle can be described by saying that the deformation behavior of the damaged material ($D \neq 0$) can only be expressed by the effective stress. In other words, the constitutive relationship of the damaged material ($D \neq 0$) can use that of the undamaged material ($D \neq 0$). We only need to substitute the nominal stress in the undamaged material constitutive relationship—the so-called Cauchy's stress σ_{ij}—with the effective stress after damage $\tilde{\sigma}_{ij}$, as shown in Figure 2.15. For example, the one-dimensional elastic constitutive equation can be represented as $\varepsilon_e = \sigma/E$ when it is undamaged. After substituting the nominal stress with the effective stress, we obtain the one-dimensional elastic constitutive equation after damage:

$$\varepsilon_e = \frac{\tilde{\sigma}}{E} = \frac{\sigma}{E(1-D)}. \tag{2.94}$$

2.2.3.3 Promotion of Effective Stress Concept

All of the aforementioned effective stress concepts are established on the basis of isotropic damage. However, many experiments show that damage under tension and compression often differs to a large extent. Under a cyclic load, the material usually shows different tensile and compressive elastic modulus. These phenomena are related to the closure effect of the crack. When the stress perpendicular to the crack is compressive stress, the crack surface still has a certain bearing capacity. Considering these issues, we should make some corrections to the effective stress, so that it shows different performance under tension and compression.

In the one-dimensional problem, the effective stress can be corrected as

$$\begin{aligned} \tilde{\sigma} &= \frac{\sigma}{1-D} \qquad \sigma \geq 0 \\ \tilde{\sigma} &= \frac{\sigma}{1-hD} \qquad \sigma < 0 \end{aligned}, \tag{2.95}$$

where h is the crack closure factor, typically $0 \leq h < 1$.

For the three-dimensional problem, how can we judge whether the stress is tensile stress or compressive stress? Reviewing the description regarding the principal stress in Chapter 1, we use it to solve this problem. In a principal stress space, we have

$$\sigma_{ij} = \begin{bmatrix} \sigma_1 & 0 & 0 \\ 0 & \sigma_2 & 0 \\ 0 & 0 & \sigma_3 \end{bmatrix}. \tag{2.96}$$

Hence, we can determine whether the principal stresses $\sigma_1, \sigma_2, \sigma_3$ are tensile stresses or compressive stresses. If we define

$$\begin{aligned} <\sigma_{ij}> &= \begin{bmatrix} <\sigma_1> & 0 & 0 \\ 0 & <\sigma_2> & 0 \\ 0 & 0 & <\sigma_3> \end{bmatrix} \\ <-\sigma_{ij}> &= \begin{bmatrix} <-\sigma_1> & 0 & 0 \\ 0 & <-\sigma_2> & 0 \\ 0 & 0 & <-\sigma_3> \end{bmatrix}, \end{aligned} \tag{2.97}$$

we obtain

$$\sigma_{ij} = <\sigma_{ij}> - <-\sigma_{ij}>,$$

where the symbol $<>$ represents $<x> = x$ $(x \geq 0)$; $<x> = 0$ $(x < 0)$. In this case, based on Equation 2.95, we can obtain the effective stress

$$\tilde{\sigma}_{ij} = \frac{<\sigma_{ij}>}{1-D} - \frac{<-\sigma_{ij}>}{1-hD}, \tag{2.98}$$

where the factor h represents the closure effect of the microcrack and microhole, which depends on the shape and intensity of the microdefect. We conclude that it is a material constant.

In the one-dimensional problem, when the material obeys the elastic constitutive relationship, we obtain

$$\begin{aligned} \sigma &= (1-D)E\varepsilon_e \quad \sigma \geq 0 \\ \sigma &= (1-hD)E\varepsilon_e \quad \sigma < 0. \end{aligned} \tag{2.99}$$

Assuming that Young's modulus E is known, we want to measure the elastic modulus after tension damage $\tilde{E}_t$ and the elastic modulus after compression damage $\tilde{E}_c$. Based on $\tilde{E}_t = E(1-D)$ and $\tilde{E}_c = E(1-hD)$, we obtain

$$\frac{\tilde{E}_c}{\tilde{E}_t} = \frac{1-hD}{1-D}. \tag{2.100}$$

Assuming that the damage is determined by the instant tension modulus $D = 1 - (\tilde{E}_t/E)$, the h value can be obtained using the following equation:

$$h = \frac{E - \tilde{E}_c}{E - \tilde{E}_t}. \tag{2.101}$$

2.2.3.4 Measurement of Toughness Damage

The toughness damage is also called the plastic damage. In metals, it generally refers to the damage caused in a large deformation process. In the metal molding process, we must consider this damage. After the molding, the damage will continue to influence the material's performance.

Toughness damage shows the physical phenomena of initiation, growth, and aggregation of the hole. The method of measuring the plastic damage is indirect measurement of the elastic modulus after damage. As the elastic modulus of the damaged material goes down, measure the unloaded modulus using the continuous unloading method, then calculate the damage variable, as shown in Figure 2.19.

Applying the strain equivalence principle, we obtain

$$\tilde{\sigma} = \frac{\sigma}{1-D} = E\varepsilon_e, \tag{2.102}$$

$$\sigma = E(1-D)\varepsilon_e = \tilde{E}\varepsilon_e. \tag{2.103}$$

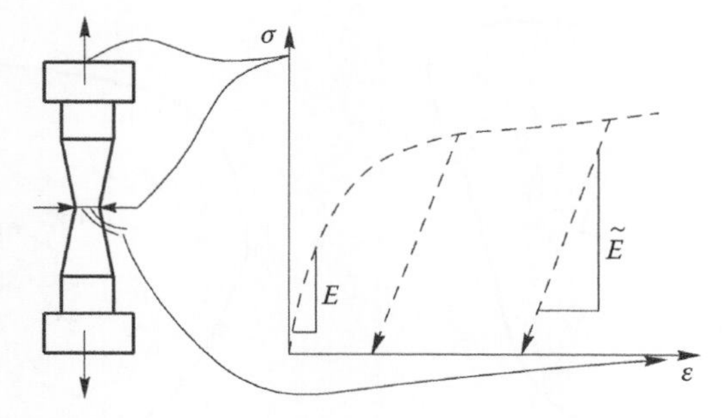

FIGURE 2.19 Measurement of plastic damage.

$\tilde{E}$ is the elastic modulus after damage:

$$\tilde{E} = \frac{\sigma}{\varepsilon_e}. \tag{2.104}$$

$$D = 1 - \frac{\tilde{E}}{E}. \tag{2.105}$$

However, it is usually difficult to measure $\tilde{E}$, because (1) the damage is local, so it is necessary to use a small strain gauge for the measurement and (2) a small nonlinearity exists in the elastic area, even when unloaded.

2.2.4 Anisotropic Damage

The isotropic damage theory is established on the basis of a homogenous and isotropic material and isotropic damage. For some engineering materials, if analysis is performed based on this theory, large differences from the actual condition may exist. These materials are often initially anisotropic, heterogeneous, and inhomogenous, or the damage evolution shows obvious anisotropic features. For such materials, we must further develop the corresponding anisotropic damage theory.

The physical mechanism of damage is mainly microholes and microcracks. They have directionality, which means that they do not exhibit the same performance in all directions. Therefore, it is not enough to use one scalar to accurately describe these anisotropic damages; rather, it is necessary to introduce a vector or a tensor to act as the damage variable. The material's anisotropic damage characteristics are the growth of the microholes distributed inside the material, and the rising deterioration of the mechanical performance of the material in all directions, to different extents. When using mechanical variables to define material damage, we will encounter the following two problems: (1) Definition of the damage variable. These damage variables should have definite mathematical and physical meanings in order to reflect the material's damage characteristics. (2) Quantization of the damage variable. We will discuss how to define the damage variable based on the concept that hole initiation results in a reduced net-loaded area of the material in the following.

Murakami defined the damage variable as a second-order tensor [32]. To define the damage state, Murakami introduced the concept of "virtual (reference) undamaged configuration," as shown in Figure 2.20. Consider any surface element *PQR* in a "current (real) damage configuration," where the line element *PQ*, *PR*, and the surface element *PQR* are represented by the vectors $\mathrm{d}\boldsymbol{x}$, $\mathrm{d}\boldsymbol{y}$, and $\mathrm{d}\boldsymbol{A}$, respectively. Due to damage influence, the load applied on the surface element $\mathrm{d}\boldsymbol{A} = \boldsymbol{\nu}\mathrm{d}A$ goes down. Assuming that this is equivalent to the undamaged (virtual) situation, the surface element $\mathrm{d}\boldsymbol{A}$ goes down, becoming $\mathrm{d}\boldsymbol{A}^* = \boldsymbol{\nu}^*\mathrm{d}A^*$ in the virtual undamaged configuration. Of course, the line elements $\mathrm{d}\boldsymbol{x}$ and $\mathrm{d}\boldsymbol{y}$ are also changed to $\mathrm{d}\boldsymbol{x}^*$ and $\mathrm{d}\boldsymbol{y}^*$, as shown in Figure 2.20b. We define the tensor $(\delta_{ij} - D_{ij})$ as the reduction of the surface element $\mathrm{d}\boldsymbol{A}^*$ due to the damage:

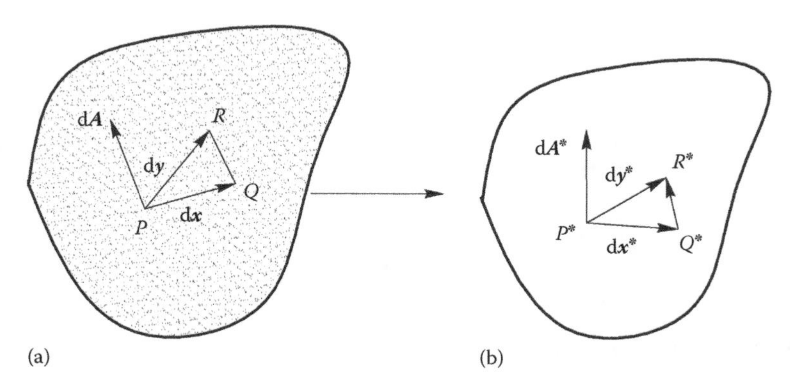

FIGURE 2.20 Damaged state in virtual (reference) undamaged configuration. (a) Current (real) damaged configuration. (b) Virtual (reference) undamaged configuration.

$$dA_i^* = (\delta_{ij} - D_{ij})dA_j. \tag{2.106}$$

The second-order tensor D_{ij} represents the internal variable of the material's anisotropic damage state, called the damage tensor.

To further study the nature of the tensor $(\delta_{ij} - D_{ij})$ [18] in Equation 2.106, we can assume that the tensor D_{ij} is symmetrical ($D_{ij} = D_{ji}$) and that there are always three orthogonal principal directions n_i (i = 1, 2, 3), where the principal value D_i is the real variable. Therefore, the damage tensor can be represented by (D_1, D_2, D_3). In the current (real) damage configuration B_t, and the virtual (reference) undamaged configuration B_f, take a group of principal coordinate systems $ox_1x_2x_3$ and $o^*x_1x_2x_3$ of the tensor D_{ij}, whose coordinate axes, respectively, pass the point P, Q, R and P^*, Q^*, R^*, as shown in Figure 2.21. Hence, we obtain two tetrahedrons $oPQR$ and $o^*P^*Q^*R^*$, respectively, composed of the surface elements PQR, $P^*Q^*R^*$, and the side surfaces perpendicular to the axes x_1, x_2, x_3. It is easy to obtain [17,18]

$$\mathbf{dA}^* = \boldsymbol{n}_1 dA_1^* + \boldsymbol{n}_2 dA_2^* + \boldsymbol{n}_3 dA_3^*, \tag{2.107}$$

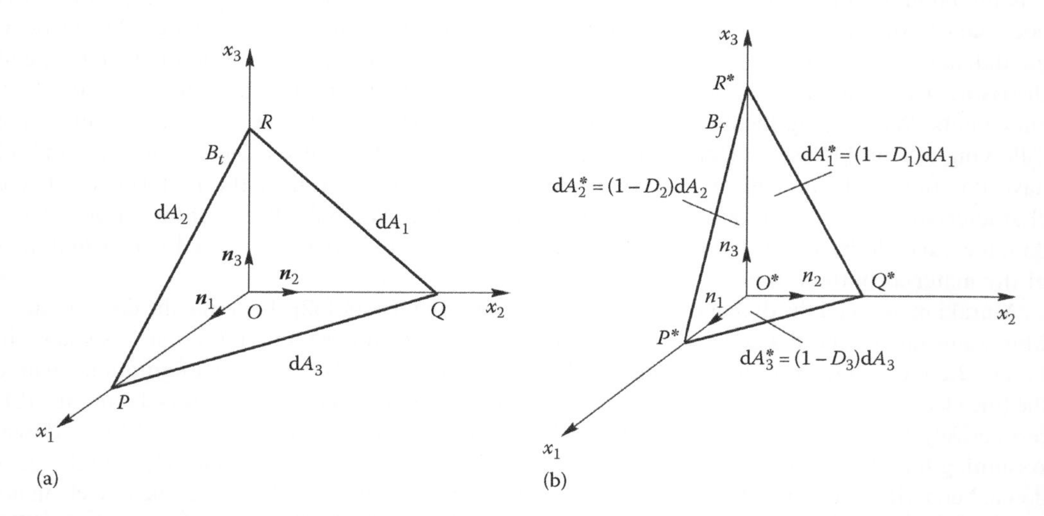

FIGURE 2.21 Geometrical explanation of damage tensor D_{ij}. (a) Instant damage configuration. (b) Virtual undamaged configuration.

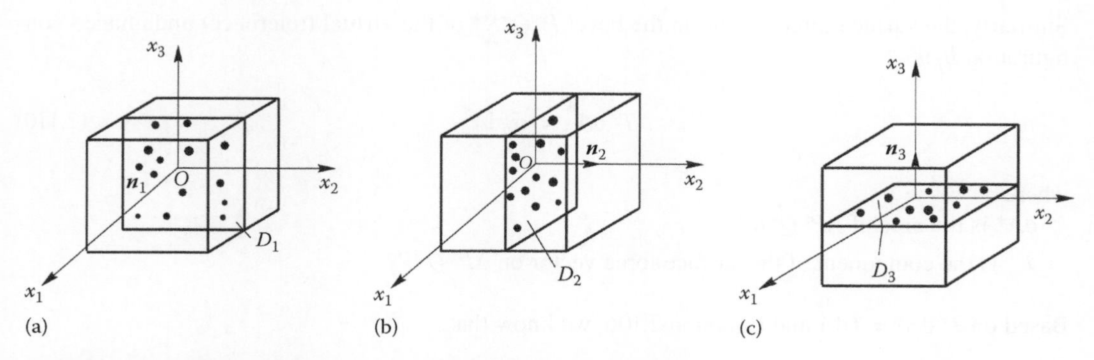

FIGURE 2.22 Reduction of area on the principal plane of damage tensor.

where

$$dA_1^* = (1-D_1)dA_1 \quad dA_2^* = (1-D_2)dA_2 \quad dA_3^* = (1-D_3)d\,A_3. \tag{2.108}$$

$dA_i (i = 1, 2, 3)$ and $dA_i^* \ (i = 1, 2, 3)$, respectively, represent the area of the three side surfaces of the tetrahedron in B_t and B_f, as shown in Figure 2.21. The damage variable (D_1, D_2, D_3) represents the reduction of the effective loaded area on the three principal planes of D_{ij} in B_t and B_f, as shown in Figure 2.22.

In conclusion, the material damage caused by the microcracks and microholes can be described by the reduction of the net-loaded area. No matter how the microdefects are distributed, the damage state can be represented by a second-order symmetrical tensor D_{ij}. The principal value of the damage tensor D_i represents the area density of the gap or crack in the damaged principal plane.

To establish the relationship between the net stress tensor (also called the effective stress tensor) $\tilde{\sigma}_{ij}$ and the Cauchy's stress tensor σ_{ij}, we analyze the tetrahedrons $oPQR$ and $o^*P^*Q^*R^*$, as shown in Figure 2.23. In the current (real) damage configuration B_t, the surface force vector on the bevel PQR is obtained based on Equations 1.20 and 1.26:

$$T_i dA = \sigma_{ij} dA_j, \tag{2.109}$$

where

dA is the area of ΔPQR

T_i is the component of the surface force vector on ΔPQR

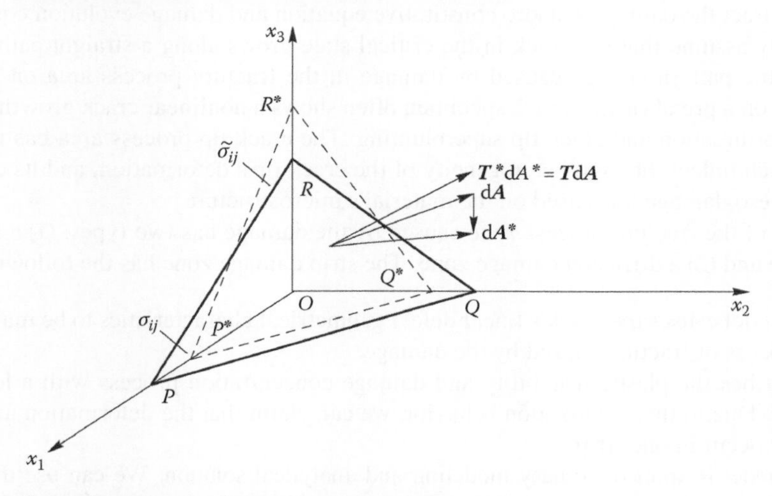

FIGURE 2.23 Reduction of net loaded area on any section.

Similarly, the surface force vector on the bevel $P^*Q^*R^*$ of the virtual (reference) undamaged configuration B_f is

$$T_i^* \mathrm{d}A^* = \tilde{\sigma}_{ij} \mathrm{d}A_j^*, \tag{2.110}$$

where

$\mathrm{d}A^*$ is the area of $\Delta P^*Q^*R^*$

T_i^* is the component of the surface force vector on $\Delta P^*Q^*R^*$

Based on $\boldsymbol{T}^*\mathrm{d}A^* = \boldsymbol{T}\mathrm{d}A$ and Equation 2.106, we know that

$$\sigma_{ij}\nu_j = \tilde{\sigma}_{ij}(\nu_j - D_{jk}\nu_k). \tag{2.111}$$

From Equation 2.111, we can obtain the component of the effective stress tensor. However, the effective stress tensor defined above is an asymmetrical tensor. Because it is a little difficult to use the asymmetrical tensor initiation damage evolution equation and the damage constitutive equation, the symmetrization of these effective stresses is necessary. One method is to take the symmetrical part of $\sigma_{ij}\left(\sigma_{ij}^*\right)$ to substitute σ_{ij} [17,18]. As a result, when the stress tensor and the principal damage tensor overlap in the principal direction, it is easy to obtain the component in its principal coordinate system:

$$\tilde{\sigma}_1^* = \frac{\sigma_1}{1-D_1}, \quad \tilde{\sigma}_2^* = \frac{\sigma_2}{1-D_2}, \quad \tilde{\sigma}_3^* = \frac{\sigma_3}{1-D_3}. \tag{2.112}$$

In fact, the aforementioned equations are the descriptions of three-dimensional effective stress in the traditional Kachanov–Rabotnov theory [17,18].

2.2.5 Interaction of Damage and Fracture [1]

The fracture process area of the macrocrack contains one meso-damage zone, where the damage grows and the substances separate and are respectively controlled by the damage evolution equation and the critical damage condition. Meso-damage mechanics applies continuous medium mechanics to study the solid material containing the meso-damage structure and uses the homogenization method to extract the damage's macro constitutive equation and damage evolution equation.

We usually assume that the crack in the critical state grows along a straight path and pay less attention to the pattern change caused by damage in the fracture process area on the crack tip. Experiments on a prefabricated crack specimen often shows a nonlinear crack growth pattern, such as crack tip bifurcation and crack tip superblunting. The crack tip process area has many damage patterns, which reflect the high heterogeneity of the crack tip's deformation, and its close relationship to the meso damage law based on the material's microstructure.

Modeling of the fracture process zone caused by the damage has two types: (1) a linear or strip damage zone and (2) a diffusion damage zone. The strip damage zone has the following features:

1. This model causes the crack's linear defect geometrical characteristics to be maintained in the process of fracture caused by the damage.
2. It describes the plastic instability and damage concentration process with a localization feature. Due to this localization behavior, we can claim that the deformation and damage mainly occur in one strip.
3. This model is applied for easy modeling and analytical solution. We can use the bridging mechanics model to study the crack growth problem and estimate the strength and toughness.

We may introduce a damage field variable D in the fracture process area to describe the loss of the stress bearing capacity of the material unit. D varies from 0 (no damage) to 1 (fully damaged). For microhole damage, D is the hole volume percentage. In many actual situations, material damage instability may occur when $D = D_c \leq 1$ applies.

As the simplest damage evolution law, the following equation can be used to describe the shear yield damage due to plastic flow:

$$\dot{D} = A\dot{\bar{\varepsilon}}^{p}. \tag{2.113}$$

Here, the damage rate $\dot{D}$ is proportional to the equivalent plastic strain rate $\dot{\bar{\varepsilon}}^{P}$. The amplitude factor A can be roughly represented as $Dc/\bar{\varepsilon}_C^P$, where $\bar{\varepsilon}_C^P$ is the logarithmic plastic strain when the specimen is damaged due to tension.

Equation 2.113 is established only in the extreme condition of isotropic damage and no hydrostatic stress. However, hydrostatic stress is, generally, the main parameter that restrains many damage mechanisms (such as damage nucleation at the two-phase particle and microhole growth). Rice and Tracy [33] gave the quantitative law of hydrostatic stress on the damage evolution influence. They demonstrated that the damage evolution rate should be proportional to $\exp(3\sigma_m/2\bar{\sigma})$, where σ_m is the averaged hydrostatic stress, and $\bar{\sigma}$ is the J_2 flow stress. If the influence of hydrostatic stress on the damage anisotropy is considered at the same time, Equation 2.113 can be generalized as

$$\dot{D} = Af(\theta)\exp\left(\frac{3\sigma_m}{2\bar{\sigma}}\right). \tag{2.114}$$

Besides the Rice–Tracy factor $\exp(3\sigma_m/2\bar{\sigma})$, the aforementioned equation also includes an orientation function $f(\theta)$. θ is the angle between the material's mesostructure orientation—such as a macromolecule material's initial tension direction or a composite material's fiber direction—and the principal direction of the maximum deformation rate. $f(\theta)$ represents the influence of damage anisotropy. Normalize $f(\theta)$ in the special case of damage isotropy to make it the unit value.

We can perform value simulation on the damage process of the crack tip, using the aforementioned damage model; add the correction that accommodates the damage evolution in the calculation and simulate the damage evolution in the unit average display format:

$$\dot{\Omega} = \frac{1}{V^e}\int_{V^e} \dot{D}\mathrm{d}v, \tag{2.115}$$

where

V^e is the unit volume in the reference configuration

$\dot{D}$ is given by Equation 2.114

The integral in Equation 2.115 can be obtained by calculating the weighted mean $\dot{D}$ value of all Gaussian integral points in the unit. The result, simulated by the finite element value, shows the five crack tip damage patterns in Figure 2.24: superblunting, bifurcation, tribifurcation, wedge split, and blunt split.

2.2.6 Nanofracture Mechanics [1]

Understanding the nature of the solid fracture process cannot be realized without combining micromechanics with nanomechanics. The origin and evolution of the four basic elements of the mesofracture process (hole, microcrack, interface failure, and deformation localization strip) cannot be

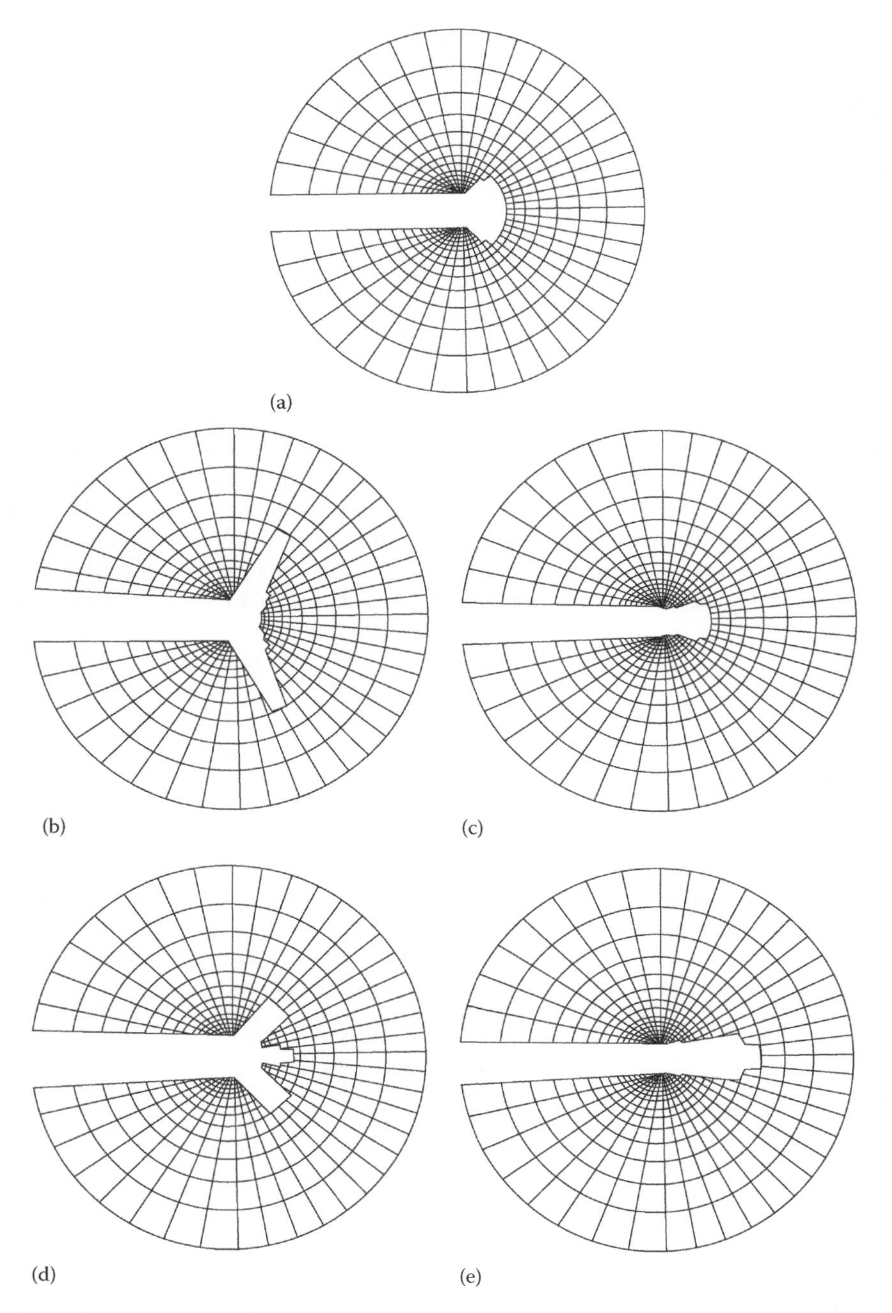

FIGURE 2.24 Numerical simulation of crack tip pattern caused by damage: (a) Superblunting, (b) wide bifurcation, (c) blunt split, (d) bifurcation, and (e) wedge split. (From Yang, W., *Macro and Microscopical Fracture Mechanics*, National Defense Industry Press, Beijing, China, 1995; Taken from Figure 7.3.)

described clearly at a level other than the nanoscale. In this case, its fracture state equation can realize closure based on the physical failure rule (such as the atomic binding force curve). An in-depth study from the macroscale to the mesoscale and then to the nanoscale results in a conceptual breakthrough from phenomenological epistemology to the damage mechanism, and to fracture physics. In recent years, it has been possible to analyze the microscopic nature of material fracture due to a number of factors: the in-depth study of micromechanics, the emergence of the continuous medium/particle nested configuration concept, the development of the large-scale calculation mean, and the breakthrough of high-resolution electron microscopy and single atom detection technologies, such as the scanning tunneling microscope and the atomic force microscope.

Nanofracture mechanics embodies the in-depth transformation from traditional solid mechanics to materials physics and condensed matter physics. It subsumes the contiguous medium hypothesis of macromechanics and goes directly to the atomic level. It discusses fracture behavior of solids at the mesoscale, by studying the motion of particles under the potential function.

Meso-fracture mechanics considers that separation of the lattice close-packed planes results in material cleavage fracture and dislocation movement of the close-packed planes along the lattice direction, resulting in toughness caused by "crack tip passivation." By applying molecular dynamics, the Monte Carlo method, and atomic transposition technology, we can simulate the fine structure at the nanospatial scale and atomic motion at the femtosecond to picosecond timescale, as well as the nanoprocess of further fracture.

For numerical simulation of the crack tip's nanofracture process, refer to Chapter 9.

EXERCISES

2.1 What are the types of cracks? What are their characteristics?

2.2 Describe the difference between toughness fracture and brittle fracture. In the engineering component, why is brittle fracture most dangerous?

2.3 For annealed pure iron, $\gamma = 2$ J/m^2, $E = 2 \times 10^5$ MPa, and $d = 2.5 \times 10^{-8}$ cm apply. Determine its theoretical fracture strength σ_{th}.

2.4 If one sheet has one 3 mm crack inside and $d = 3 \times 10^{-8}$ cm applies, determine the fracture stress of its brittle fracture ($\sigma_{th} = 0.1E = 2 \times 10^4$ MPa).

2.5 For one material, $E = 2 \times 10^{11}$ N/m^2 and $\gamma = 8$ N/m apply. Calculate the minimum crack length of this material under a tensile stress of 7×10^7 N/m^2 ($K_I = \sigma\sqrt{\pi a}$).

2.6 Linear elastic fracture mechanics establishes a new fracture criterion. Please illustrate its application in engineering, as follows:

Example 1: one steel sheet containing a 16 mm long central through crack, under stress of 350 MPa perpendicular to the crack plane.

1. If the material's yield strength is 1400 MPa, determine the plastic area size and the crack tip's effective stress intensity factor value.
2. If the material's yield strength is 385 MPa, determine the plastic area size and the crack tip's effective stress intensity factor value.
3. Please compare and discuss the significance of performing plastic correction to the stress field's intensity factor in the aforementioned two situations.

Example 2: Assume that there are two very wide alloy steel plates with a yield strength of 415 MPa, fracture toughness K_{IC} of 132 MPa$\cdot$m$^{1/2}$, and thickness of 100 and 260 mm, respectively. If both plates are under tensile stress of 300 MPa, and assuming that both plates have a 46 mm long central through crack, does the crack in both plates grow?

2.7 Describe the relationship between the stress state and the plastic area.

2.8 Describe the relationship between fracture toughness and critical fracture stress.

2.9 How did Kachanov define the continuous defect variable in 1958? What difference exists between it and the current damage variable definition?

2.10 Which are the different types of damage? What features do all types of damage mainly have?

2.11 How are proper damage variables selected? What are the main principles when selecting damage variables?

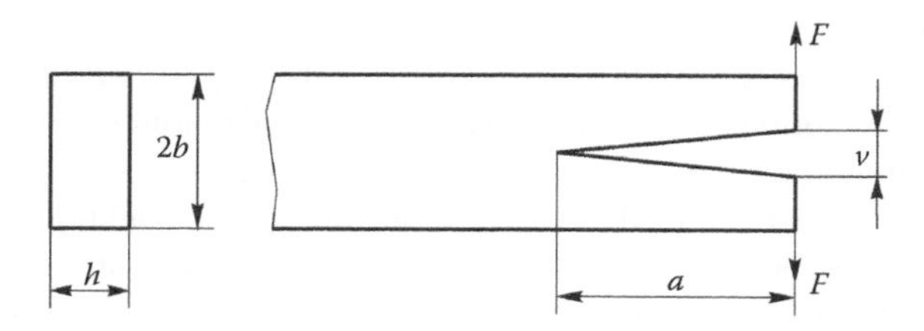

FIGURE 2.25 Figure for Exercise 2.15.

2.12 What is the effective stress when damage exists? What is the strain equivalence principle?

2.13 If the crack closure effect is considered, how should the effective stress be corrected?

2.14 What are the main anisotropic characteristics of damage? In various types of damage, what do the maximum damage surfaces (crack surface, hole surface) have in common, with respect to their orientation?

2.15 On a plane cracked body as shown in Figure 2.25, if b is much less than a and the relative displacement on the end is v, please calculate the strain energy release rate G_I.

2.16 What is the "virtual undamaged configuration"? What function does it have? How did Murakami define the second-order symmetrical damage tensor? What physical meaning does it have?

2.17 In a Mode I plane crack problem as shown in Figure 2.5, assume that in the state of plane strain and that $2a = 2.5$ cm. For $\nu = 0.3$, $\sigma/E = 0.1\%$, and $\sigma/E = 0.5\%$, calculate the opening displacement $2\nu^*$ in the crack center, and explain whether such a large displacement can be easily observed.

2.18 For one pressure container with wall thickness h, radius R, and material yield limit of $\sigma_s = 1000$ N/mm², and fracture toughness $K_{IC} = 1200\,\text{N}\cdot\text{mm}^{-\frac{3}{2}}$, assume that the container has a through crack whose length direction is the container's axis direction, with a length of $2a = 3.8$ mm. Determine the critical pressure P_{IC} when brittle fracture occurs on the container.

2.19 Based on your understanding, what relationship exists between macro- and microfracture mechanics and between micro- and nanofracture mechanics? How can macro-, micro-, and mesofracture mechanics be combined? Please make your analysis both from theory and from experiment.

REFERENCES

1. Yang W. *Macro and Micro Fracture Mechanics*. Beijing: National Defense Industry Press, 1995.
2. Zhou Y. C. *Materials Mechanics of Solids (Vol. 2)*. Beijing: Science Press, 2005.
3. Chinese Aeronautical Establishment. *Stress Intensity Factors Handbook*. Beijing: Science Press, 1981.
4. Ding S. D. and Sun L. M. *Fracture Mechanics*. Beijing: Machine Press, 1997.
5. Irwin G. R. *"Fracture Dynamics"*. *Fracturing of Metals*. Cleveland: ASM, 1948.
6. Hong Q. C. *Elementary Engineering Fracture Mechanics*. Shanghai: Shanghai Jiao Tong University Press, 1987.
7. Wu Y, Qiao L. J., Chen Q. Z. et al. *Fracture and Environmental Fracture*. Beijing: Science Press, 2000.
8. Bortman Y. and Bank-Sills L. An extended weight function method for mixed-mode elastic crack analysis. *J Appl Mech*, 1983, 50: 907–909.
9. Wells A. A. Application of fracture mechanics at and beyond general yielding. *Brit Weld J*, 1963, 10: 563–570.
10. Dugdale D. S. Yielding in steel sheets containing slits. *J Mech Phys Solids*, 1960, 8: 100–108.
11. Barenblatt G. I. The mathematical theory of equilibrium cracks in brittle fracture. *Adv Appl Mech*, 1962, 7: 55–129.
12. Rice J. R. and Bosengren G. F. Plane strain deformation near a crack tip in power-law hardening material. *J Mech Phys Solids*, 1968, 16: 1–12.

13. Rice J. R. A path independent integral and the approximate analysis of strain concentration by notches and cracks. *J Appl Mech*, 1968, 35: 379–386.
14. Kachanov L. M. On the time to failure under creep condition. *Izv Akad Nauk USSR Otd Tekhn Nauk*, 1958, 8: 26–31.
15. Rabotnov Y. N. On the equation of state for creep. In: *Proceedings of the Institution of Mechanical Engineers, Conference Proceedings*, 1963, pp. 307–315.
16. Janson J. and Hult J. Fracture mechanics and damage mechanics, a combined approach. *J de Mech Appl*, 1977, 1(1): 59–64.
17. Li Z. X. *Damage Mechanics and Its Application*. Beijing: Science Press, 2002.
18. Yu S. W. and Feng X. Q. *Damage Mechanics*. Beijing: Tsinghua University Press, 1997.
19. Bai Y. L., Ling Z, Luo L. M. et al. Initial development of microdamage under impact loading. *ASME Trans J Appl Mech*, 1992, 59: 622–627.
20. Bai Y. L., Xia M. F., Ke F. J. et al. Dynamic function of damage and its implications. *Int J Key Eng Mat*, 1998, 145–149: 411–420.
21. Xing X. S. New advances in non-equilibrium statistical physics principle. *Chin Sci Bull*, 2000, 45(12): 1235–1242.
22. Ha K. F. *Fundamentals of Fracture Physics*. Beijing: Science Press, 2000.
23. Yang W., Ma X. L., Wang H. T. et al. Advances in nanomechanics. *Adv Mech*, 2002, 32(2): 161–174.
24. Yang W., Ma X. L., Wang H. T. et al. Advances in nanomechanics (continued). *Adv Mech*, 2003, 33(2): 175–186.
25. Yang W., Tan H. L., and Guo T. F. Evolution of crack tip process areas. *Model Simul Mater Sci Eng*, 1994, 2(3a): 767–782.
26. Guo Y. F., Wang C. Y., and Wang Y. S. The effect of stacking fault or twin formation on bcc-iron crack propagation. *Phil Mag Lett*, 2004, 84(12): 763–770.
27. Guo Y. F., Wang C. Y., and Zhao D. L. Atomistic simulation of crack cleavage and blunting in bcc-Fe. *Mater Sci Eng A*, 2003, 349(1–2): 29–35.
28. Zhou Y. C., Long S. G., and Liu Y. W. Thermal failure mechanism and failure threshold of SiC particle reinforced metal matrix composites induced by laser beam. *Mech Mater*, 2003, 35(10): 1003–1020.
29. Long S. G. and Zhou Y. C. Thermal fatigue of particle reinforced metal-matrix composite induced by laser heating and mechanical load. *Compos Sci Tech*, 2005, 65(9): 1391–1400.
30. Hult J. Introduction and general overview. In: D. Krajcinovic and J. Lmaitre. eds. *Continuum Damage Mechanics: Theory and Application*. Wien-New York: Springer-Verlag, 1989.
31. Lemaitre J. Evaluation of dissipation and damage in metals submitted to dynamic loading. In: *Proceedings of ICM-1*, Kyoto, 1971.
32. Murakami S. Mechanical modeling of material damage. *J Appl Mech*, 1988, 55: 280–286
33. Rice J. R. and Tracey D. M. On the ductile enlargement of voids in triaxial stress fields. *J Mech Phys Solids*, 1969, 17: 201–213.

3 Basic Mechanical Properties of Materials

In the previous two chapters, we established the theoretical fundamentals for the study and analysis of the micro- and macromechanical properties of materials. We have discussed physical quantities such as elastic modulus, elastic limit, and yield strength. These physical quantities are called the basic mechanical properties of a material and include the mechanical properties under tension, compression, torsion, bending, and shearing. These basic physical quantities must be fully understood in the development, production, and application of a material. These properties are primarily measured through the appropriate mechanical test. For example, the tensile test can be used to determine many important mechanical properties such as elasticity, strength, plasticity, strain hardening, and toughness [1,2]. Beginning with the tensile properties of a material, this chapter introduces the basic mechanical properties under tension, compression, torsion, and bending [3–6] at room temperature in the atmosphere and the corresponding experimental tests to measure these mechanical properties [7].

3.1 BASIC MECHANICAL PROPERTIES OF MATERIALS

3.1.1 Mechanical Properties of a Material under Tension

3.1.1.1 Tensile-Test Diagram

Figure 3.1 shows the commonly used standardized tensile specimen [8]. The rod segment between marks m and n is called the gauge section, and its length l is called the gauge length. For a test specimen with a circular section of diameter d (Figure 3.1a), the required gauge length is usually $l = 10d$ or $l = 5d$.

For a test specimen with a rectangular cross section of area A (Figure 3.1b), the required gauge length is usually

$$l = 11.3\sqrt{A} \quad \text{or} \quad l = 5.56\sqrt{A}.$$

The tensile test is usually carried out at a strain rate of $\dot{\varepsilon} \leq 10^{-1}$/s (why should the strain rate be considered?). Because the tensile loading rate is relatively low, it is generally called the static tensile test.

Usually, a tensile testing machine has an automatic recording or drawing device to record or draw a curve of load F versus elongation $\Delta l(\Delta l = l - l_0)$ of the specimen under the load. Here, l is the gauge length of the specimen after loading. Figure 3.2 shows a tensile-test diagram of low-carbon steel.

3.1.1.2 Stress–Strain Curve

From Figure 3.2, dividing the tensile force F by the original cross-sectional area A_0 gives stress $\sigma = F/A_0$, which is also called engineering stress or nominal stress. Dividing Δl by the original gauge length l gives strain $\varepsilon = \Delta l/l$, also known as engineering strain or nominal strain. In this way, we can get the curve to represent the relationship between σ and ε (Figure 3.3), which is called the stress–strain curve or the σ–ε curve.

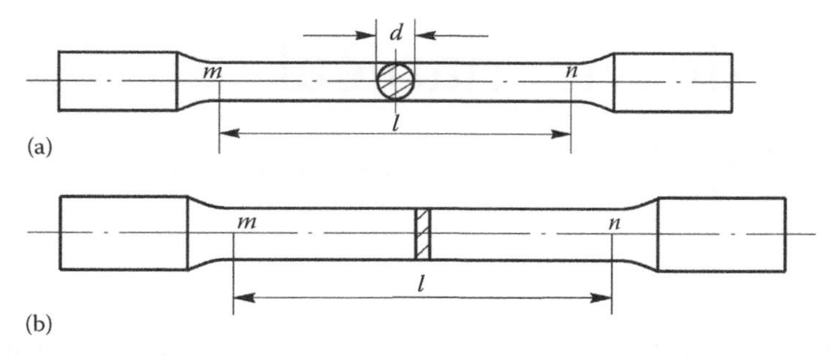

FIGURE 3.1 Generally used standardized tensile specimens. (a) Tensile specimen with a circular cross section. (b) Tensile specimen with a rectangular cross section.

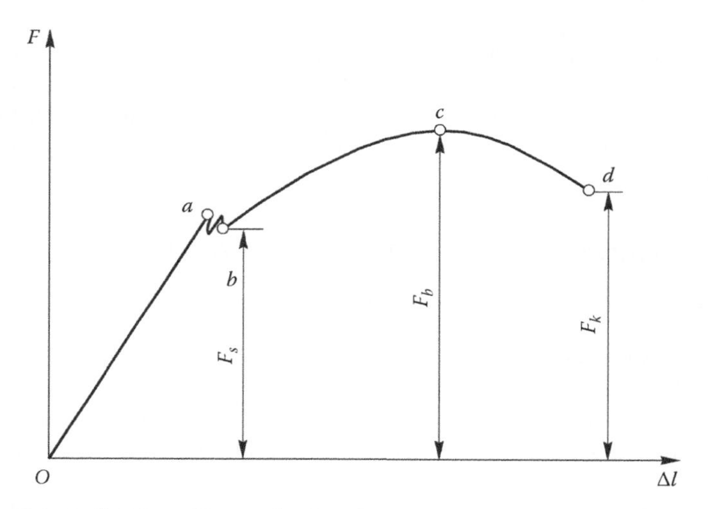

FIGURE 3.2 Tensile-test diagram of low-carbon steel.

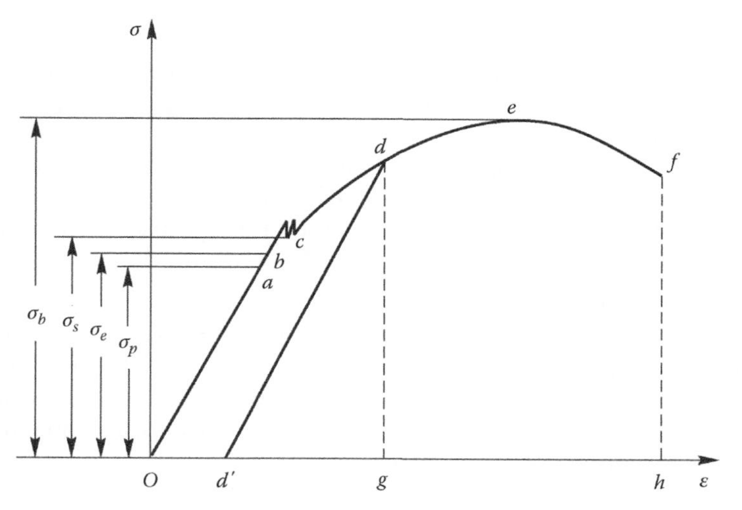

FIGURE 3.3 Tensile stress–strain curve of mild steel.

In Section 1.1.3, we described in detail the characteristics of the stress–strain relationship shown in Figure 3.3. Here, we briefly review these characteristics. The tensile process of the specimen can be divided into four stages (readers are advised to carry out the tensile test to understand the four stages):

1. *Elastic stage*: In the initial stage of tension, the relationship between σ and ε follows the straight line Oa. The stress σ_p, corresponding to the highest point a of the line, is called the proportional limit. After the stress goes beyond the proportional limit, from point a to point b, the relationship between σ and ε no longer follows a straight line, but deformation can be completely recovered upon removal of the load. This type of deformation is called elastic deformation. The stress σ_e at point b is called the elastic limit and corresponds to the maximum stress below which only elastic deformation occurs. Typically, the proportional limit and the elastic limit are not strictly distinguished, but they are different for some materials.
2. *Yield stage*: When the stress exceeds the elastic limit and reaches a certain value, it drops suddenly and remains almost unchanged with only slight fluctuations. However, the strain increases significantly and forms a serrated plateau on the σ–ε curve. The phenomenon in which the stress remains unchanged but the strain increases significantly is called yield or flow. In the yield stage, the lowest stress that is relatively stable among the fluctuating stresses is called the yield point or the yield strength and is denoted by σ_s, as shown in Figure 3.3 midpoint c.
3. *Hardening stage*: After the yield stage, the tensile force must be increased in order to continue to increase the deformation. That is, the material enhances the ability to resist deformation. This stage is known as the strain-hardening stage. The stress σ_b corresponding to the highest point e in the strain-hardening stage is the maximum stress that the material can withstand, known as the ultimate tensile strength.
4. *Necking stage*: After the stress reaches the ultimate tensile strength, the transverse dimensions in a local region decrease suddenly, as shown in Figure 3.4. We call this phenomenon necking. The local deformation in the necking region leads to a rapid increase in elongation. Because the cross-sectional area of the necking region rapidly decreases, the tensile force acting on the specimen significantly reduces, and the specimen finally ruptures at point f.

We consider the situation after unloading. After the specimen is pulled to point d in the hardening stage and then unloaded gradually, the stress–strain curve during unloading changes along a straight line dd' back to d', and dd' is nearly parallel to Oa. This is generally known as the unloading law. After the load is completely removed, $d'g$ in the σ–ε curve represents the disappeared elastic deformation. Od' represents the permanent plastic deformation, which does not disappear after unloading.

If the specimen is loaded again shortly after unloading, the σ–ε curve first follows line $d'd$, which is nearly parallel to Oa, and then follows the path def after reaching the point d. It can be seen that reloading extends the elastic stage and decreases the plastic deformation. This phenomenon shows that at room temperature, preloading the material to the hardening stage, followed by unloading, leads to a higher proportional limit but lower plasticity in subsequent loading. This phenomenon is known as cold working or work hardening. In engineering, cold working is frequently applied to improve the load-bearing capacity of components in the elastic deformation range.

Different materials have different chemical compositions and microstructures. Under the same experimental conditions, they exhibit different stress–strain responses. Figure 3.5 displays several typical stress–strain curves.

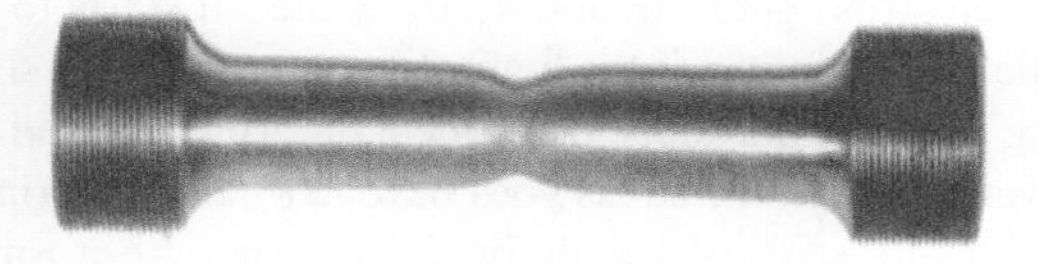

FIGURE 3.4 Necking during uniaxial tension. (Taken from Figure 7.151; http://course.cau-edu.net.cn/course/z0125/ch07/se06/slide/slide01.html)

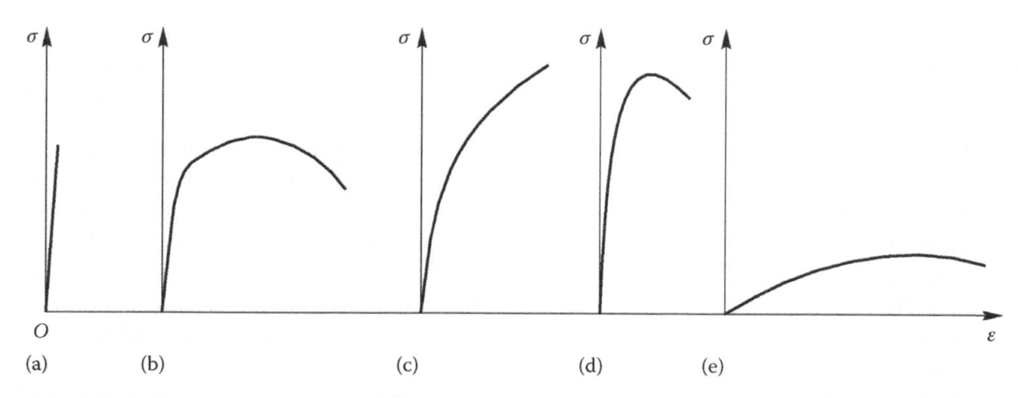

FIGURE 3.5 Typical stress–strain curves of brittle materials (a), ductile materials (b, d), and materials with low ductility (c, e).

In engineering practice—depending upon whether plastic deformation occurs before tensile fracture—materials can be classified into two categories: brittle materials and ductile materials. Brittle materials do not produce significant plastic deformation before tensile fracture. Ductile materials not only produce uniform elongation before tensile fracture but also incur necking with a large amount of plastic deformation (see Figure 3.5b and d). Low-carbon steels show large tensile plastic deformation prior to fracture and are thus materials with high ductility. Materials that undergo uniform tensile elongation before fracture, with a small amount of plastic deformation, are materials with low ductility (see Figure 3.5c and e).

3.1.1.3 Tensile Properties

The tensile properties of a material, also known as their mechanical properties, are represented by the values of critical points in the stress–strain curve, which indicate the change in the nature of the deformation process. The mechanical properties of a material can be classified into two categories: strength properties, reflecting the resistance of the material to plastic deformation, and fracture; plasticity properties, reflecting the material's plastic deformation capability.

3.1.1.3.1 Strength Properties

1. *Yield strength*: In principle, the yield strength of a material should be interpreted as the stress at the initiation of plastic deformation. However, it is difficult to use such a yield criterion for continuous yielding materials. In engineering, the grains of a polycrystalline material have different orientations and they cannot simultaneously begin plastic deformation. When only a few grains experience plastic deformation, it is very difficult to detect plastic deformation in the stress–strain curve. Only when most grains undergo plastic deformation can the effect of macroscopic plastic deformation be observed. Therefore, the stress indicating the beginning of plastic deformation will depend on the sensitivity of the testing equipment. In engineering, the yield strength is determined in terms of a certain amount of residual plastic deformation.

   There are generally three types of yield strength in engineering:

   a. *Proportional limit*: The previously mentioned σ_p is the proportional limit or the maximum stress in the stress–strain curve below which the stress and strain satisfy a linear relationship. When stress goes beyond σ_p, the material starts to yield.
   b. *Elastic limit*: The elastic limit σ_e is the highest stress value in the stress–strain curve below which unloading does not introduce permanent deformation, and the deformation can be elastically recovered. When the stress goes beyond σ_e, the material begins to yield.
   c. *Yield strength*: Yield strength is defined in terms of a small amount of residual deformation. Generally, the yield strength is taken as the stress resulting in 0.2% residual deformation, denoted by $\sigma_{0.2}$.

2. *Ultimate tensile strength*: The limit of a material's load-bearing capacity is denoted by the ultimate tensile strength. In a tensile test, the stress σ_b corresponding to the maximum load F_b is the ultimate tensile strength. (Why do we not take the fracture stress as tensile strength?)

$$\sigma_b = \frac{F_b}{A_0} \tag{3.1}$$

For brittle materials and materials that do not incur necking, the tensile load is the maximum fracture load. Therefore, the ultimate tensile strength represents the fracture strength. For ductile materials that experience necking, the ultimate tensile strength represents the resistance to the largest homogeneous deformation and also represents the limit of a material's load-bearing capacity under static tension.

3. *True fracture strength*: The tensile fracture load F_K divided by the actual fracture cross-sectional area A_K is called the true fracture strength S_K:

$$S_K = \frac{F_K}{A_K}, \tag{3.2}$$

where S_K is the true stress, characterizing the resistance of a material to fracture. Therefore, it is sometimes called the true fracture stress.

3.1.1.3.2 Ductility Properties

A material's capacity for plastic deformation, or its ductility, can be represented by the elongation ratio δ and the reduction of area ψ.

1. Elongation ratio: The total elongation ratio after specimen fracture is called the limit elongation ratio, denoted by δ_K and expressed in percent as

$$\delta_K = \frac{l_K - l_0}{l_0} \times 100\%, \tag{3.3}$$

where l_K is the gauge length after fracture.

2. Reduction of area: The total reduction of the cross-sectional area after specimen fracture is called the limit reduction of area, denoted by ψ_K and expressed in percent as

$$\psi_K = \frac{A_0 - A_K}{A_0} \times 100\%, \tag{3.4}$$

where A_K is the minimum cross-sectional area at the fracture point.

3.1.1.4 True Stress–True Strain Curve

Through extensive experimentation, we know that stress in the tensile stress–strain curve begins to decrease when it reaches a certain point, such as the highest point b in Figure 3.6. We have previously illustrated that necking occurs at this moment with drastic reduction in the cross-sectional area. However, in Figures 3.2 and 3.3, the stress is defined by $\sigma = P/A_0$ (here P is the tensile load and A_0 is the original cross-sectional area of the specimen), in which the change in cross-sectional area is not taken into consideration. In order to truly reflect the characteristics of stress and strain in a material, alternative definitions of stress and strain are given as follows:

$$S = \frac{F}{A} = \frac{F}{A_0} \cdot \frac{A_0}{A} = \frac{\sigma}{1-\psi} = \sigma(1+\varepsilon), \tag{3.5}$$

$$e = \int_{l_0}^{l} \frac{\mathrm{d}l}{l} = \ln\frac{l}{l_0} = \ln(1+\varepsilon) = \ln\frac{1}{1-\psi}, \tag{3.6}$$

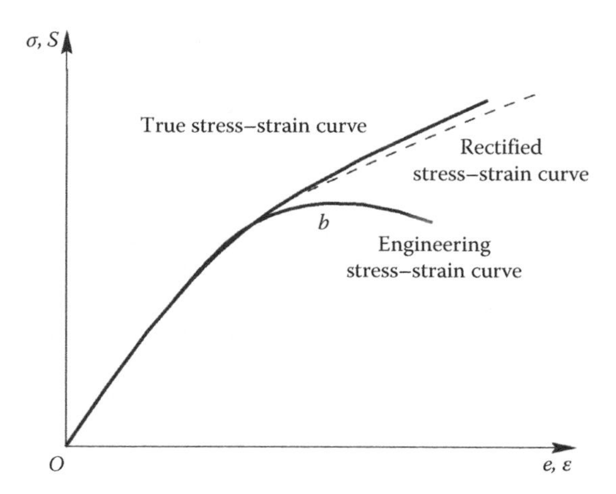

FIGURE 3.6 True stress–true strain curve.

where A and l are the instantaneous cross-sectional area and the gauge length, respectively. The defined stress S and strain e are called true stress and true strain, respectively. When the specimen is subjected to tension, the cross-sectional area reduces, resulting in a larger true stress than the engineering stress $\sigma = P/A_0$, at the point with the minimum cross-sectional area. Additionally, an increase in gauge length leads to a smaller true strain than the engineering strain at the same elongation Δl (see Equation 3.5). Figure 3.6 plots the true stress–strain curve. It can be seen that the true stress–true strain curve continues to rise until the specimen fractures.

It can be seen from Equation 3.4 that the strain in the elastic deformation stage is relatively small, generally less than 1%. Thus, the contraction in the transverse direction is small and the true stress–true strain curve almost coincides with the engineering stress–strain curve (see Figure 3.6). From the initiation of plastic deformation to point b (the homogenous plastic deformation stage), the true stress is higher than the engineering stress. The difference between them increases with the increase in strain. At point b, we have

$$\begin{cases} S_b = \dfrac{\sigma_b}{1-\psi_b} = \sigma_b(1+\varepsilon_b) \\ e_b = \ln(1+\delta_b) = \ln\left(\dfrac{1}{1-\psi_b}\right) \end{cases}. \tag{3.7}$$

After necking, plastic deformation concentrates in the necking zone, and the specimen's cross-sectional area decreases drastically. Although the engineering stress decreases with the increase in strain, true stress increases with the strain. Therefore, the true stress–true strain curve and the engineering stress–strain curve exhibit different trends.

In addition, deformation before necking takes place under unidirectional stress. After necking, the stress state of the necking zone changes from the uniaxial to the triaxial stress state. Correspondingly, the uniaxial stress decreases, as shown by the rectified curve in Figure 3.6.

3.1.1.5 Important Parameters

The elastic behavior of various materials is different, embodied in the discrepancy in elastic constants. In engineering, the general elastic constants include E, v, G, and K. In the next section, we will demonstrate their physical meanings using the generalized Hooke's law (see Chapter 1).

E, the elastic modulus, under uniaxial loading, can be expressed as

$$E = \frac{\sigma_x}{\varepsilon_x}. \tag{3.8}$$

Thus, it reflects the capacity of a material to resist normal strain.

ν, the Poison ratio, under uniaxial loading, can be expressed as

$$\nu = -\frac{\varepsilon_y}{\varepsilon_x}. \tag{3.9}$$

It reflects the ratio of longitudinal normal strain to transversal normal strain.

G, the shear modulus, under pure shear, can be expressed as

$$G = \frac{\tau_{xy}}{\gamma_{xy}}. \tag{3.10}$$

It reflects the capacity of a material to resist shear strain.

K, the bulk modulus, can be expressed as

$$K = \frac{E}{3(1-2\nu)}. \tag{3.11}$$

Notice that only two constants are independent for an isotropic material. Therefore, we need two equations to correlate the aforementioned four constants:

$$E = 2G(1+\nu) \tag{3.12}$$

and

$$E = 3K(1-2\nu). \tag{3.13}$$

3.1.2 Mechanical Properties of Materials under Compression

3.1.2.1 Compression-Test Diagram and Stress–Strain Curve

Usually, the compression specimens are cylinders but can sometimes be cubes or prisms. In order to prevent buckling, a specimen's height to diameter ratio h_0/d_0 should be 1.5–2.0. (Suggestions for readers to think about and derive: What is buckling? How is buckling represented by a mathematical language? Why is there such a requirement?) A specimen's height to diameter ratio h_0/d_0 has a significant influence on testing results. The larger h_0/d_0 is, the lower is the compressive strength. In order to compare the testing results of compressive strength, specimen values of h_0/d_0 must be identical. For specimens with different geometric shapes, the values of $h_0/\sqrt{A_0}$ should be identical.

Like a tensile test, a compressive test is also conducted using the universal material testing system. The compressive test usually measures the relationship between pressure and compressive deformation, namely, the F–Δh curve. Figure 3.7 shows two typical F–Δh curves of mild steels and cast irons. The dashed part of curve b keeps rising, indicating the good plasticity of mild steels. Therefore, one cannot measure the strength parameters of mild steels by compression test.

Referring to the processing of the tensile-test diagram in the last section, we can obtain the stress–strain curve corresponding to the compressive-test diagram, as shown in Figure 3.8. It can be seen that mild steels under compression exhibit elastic limit, proportional limit, and yield limit.

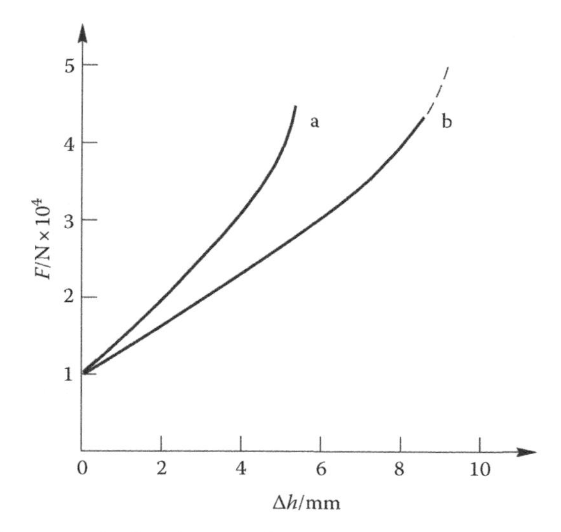

FIGURE 3.7 Compressive curve. (a) Cast iron and (b) mild steel.

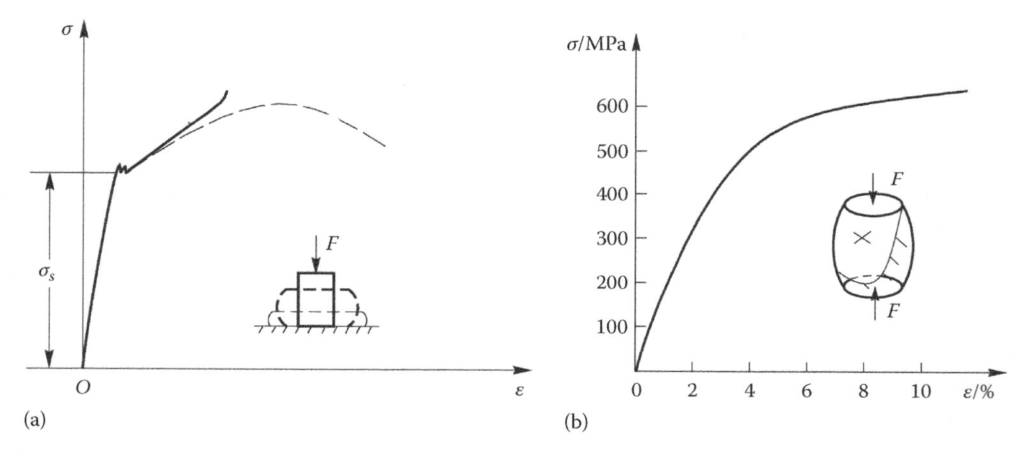

FIGURE 3.8 Curves under compression. (a) Curve of mild steel under compression. (b) Curve of cast iron under compression.

The test shows that the yield limit of mild steel under compression is close to that under tension, except that the yielding phenomenon under compression is not as obvious as that under tension. With an increase in pressure, the specimen changes from a barrel shape to a flattened biscuit and becomes more and more flattened; however, compressive fracture will not occur, so the ultimate compressive strength cannot be obtained. Generally, the yield limit is taken as the characteristic value of the compressive strength of mild steel. It can be seen from the compressive curve of cast iron that the specimen abruptly fractures at a relatively small deformation. The normal of the fracture surface is at an angle of approximately 45°–55° with respect to the specimen axis (readers should pay close attention to this angle), indicating that the fracture is caused by the relative movement of the upper parts and the lower parts along the glide plane. The ultimate compressive strength is 4–5 times that of the ultimate tensile strength.

Materials with good ductility will not fracture when subjected to compressive deformation. Therefore, for these materials, a compression test can only measure elastic modulus, proportional limit, elastic limit, and yield strength but cannot measure the limit of compressive strength. For low-ductility or brittle materials that are prone to fracture under tension, a compressive test can measure the mechanical properties in the ductile state.

3.1.2.2 Important Mechanical Properties

Compression can be viewed as tension in the opposite direction. Therefore, the mechanical properties and formulae defined in tensile testing can also be applied in compressive testing. Actually, the compressive test is different from the tensile test. For example, the specimen in compressive testing undergoes shortening instead of elongation, and the transverse section experiences expansion instead of contraction.

Next, we will introduce some general mechanical properties in compressive testing:

σ_{bc}, or the compressive strength, is expressed as

$$\sigma_{bc} = \frac{F_{bc}}{A_0}. \tag{3.14}$$

ε_{ck}, or the relative compressive ratio, can be calculated as

$$\varepsilon_{ck} = \frac{h_0 - h_k}{h_0} \times 100\%. \tag{3.15}$$

ψ_{ck}, or the relative expansion of area, is expressed as

$$\psi_{ck} = \frac{A_k - A_0}{A_0} \times 100\%, \tag{3.16}$$

where

F_{bc} is the load when the specimen fractures under compression

h_0 and h_k are, respectively, the original height of the specimen and the height at the moment of fracture

A_0 and A_k are, respectively, the original cross-sectional area and the cross-sectional area at the moment of fracture

3.1.3 Mechanical Properties of Materials under Torsion

3.1.3.1 Stress and Strain under Torsion

3.1.3.1.1 Characteristics of Stress

The stress state of a material in a torsion test is pure shear. Shear stresses distribute in two perpendicular sections along the longitudinal and transverse directions. The principal stresses σ_1 and σ_3 are at an angle of approximately 45° with respect to the longitudinal axis and are equal to the shear stress in magnitude. σ_1 is a tensile stress, σ_3 is a compressive stress with the same magnitude, and $\sigma_2 = 0$ (see Figure 3.9). Therefore, if fracture occurs in the transversal section, it is caused by shear. However, if fracture is caused by the maximum normal stress, the fracture surface exhibits a spiral shape, at an angle of 45° with respect to the longitudinal axis.

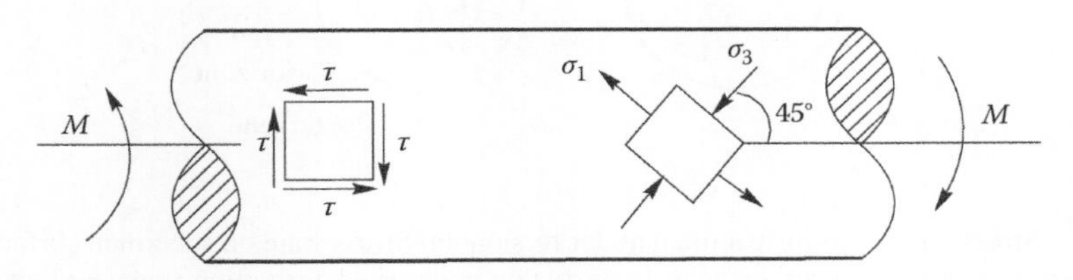

FIGURE 3.9 Stress state under torsion.

In order to illustrate the features of stress and deformation under torsion, we compare the tensile test and the torsion test.

	Tension	Torsion
a.	$\sigma_1 = \sigma_{max}; \sigma_2 = \sigma_3 = 0$	$\sigma_1 = -\sigma_3; \sigma_2 = 0$
b.	$\tau_{max} = \dfrac{\sigma_1 - \sigma_3}{2} = \dfrac{\sigma_1}{2}$	$\tau_{max} = \dfrac{2\sigma_1}{2} = \sigma_1$
c.	$\varepsilon_{max} = \varepsilon_1; \varepsilon_2 = \varepsilon_3 = -v\varepsilon_1$	$\varepsilon_{max} = \varepsilon_1 = -\varepsilon_3; \varepsilon_2 = 0$

According to the law of conservation of volume,

$$\Delta = \frac{\Delta V}{V} = \frac{1-2v}{E}(\sigma_1 + \sigma_2 + \sigma_3)$$

when $\Delta = 0$, and thus $v = 1/2$; consequently, $\varepsilon_2 = \varepsilon_3 = -1/2\varepsilon_1$.

According the generalized Hooke's law,

$$\varepsilon_1 - \varepsilon_3 = \frac{\sigma_1 - \sigma_3}{E}(1+v) \quad \text{and} \quad \tau_{max} = \frac{1}{2}(\sigma_1 - \sigma_3) = G\gamma = \frac{E}{2(1+v)}\gamma.$$

Thus,

$$\text{d.} \quad \gamma_{max} = \varepsilon_1 - \varepsilon_3 = \frac{3}{2}\varepsilon_1 \qquad \gamma_{max} = \varepsilon_1 - \varepsilon_3 = 2\varepsilon_1 .$$

3.1.3.1.2 Analysis of Stress and Strain

Based on the previous analysis, we can obtain the distributions of stress and strain in a round rod with uniform diameter under torsion [9]. On the transverse section, there is only shear stress, but no normal stress (Figure 3.10). In the elastic deformation stage, the shear stress at each point on the

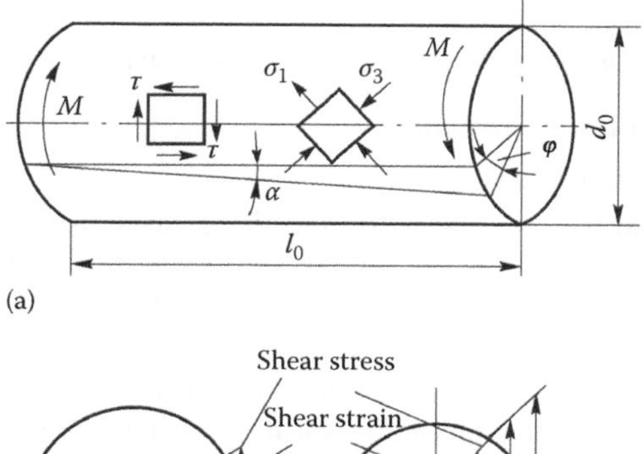

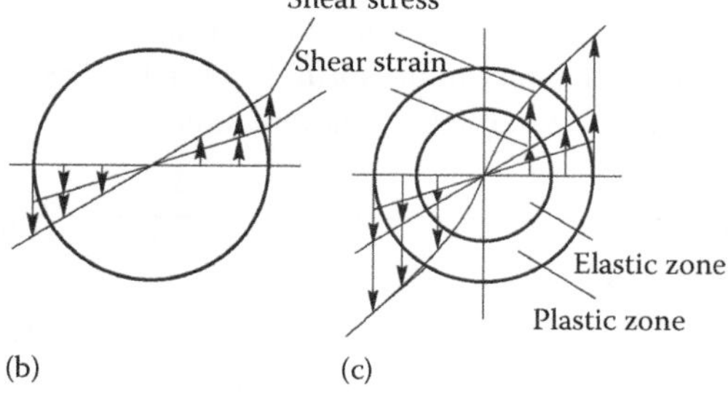

FIGURE 3.10 Stress and strain in specimen under torsion. (a) Stress state on specimen surface, (b) distributions of shear stress and shear strain on transverse section in elastic deformation stage, and (c) shear stress on transverse section in elastoplastic deformation stage.

transverse section is normal to the radial direction, and its magnitude is proportional to the distance from this point to the center. The shear stress is zero at the center and reaches its maximum at the surface (see Figure 3.10b). After plastic deformation occurs in the surface layer, the shear strain at each point is still proportional to the distance from the point to the center, but the shear stress reduces due to plastic deformation, as shown in Figure 3.10c. On the surface of the round rod, maximum shear stress occurs perpendicular to the specimen axis, normal stress reaches maximum at an angle of 45° to the specimen axis, and normal stress is equal to shear stress (see Figure 3.10a).

According to the mechanics of materials, in the elastic deformation range, shear stress on the surface of a round rod can be calculated as

$$\tau = \frac{M}{W}, \tag{3.17}$$

where

M is the torque
W is the torsion coefficient for the section

For a solid round rod, $W = \pi d_0^3/16$; for a hollow round rod, $W = \pi d_0^3\left(1 - d_1^4/d_0^4\right)/16$, where d_0 is the outer diameter and d_1 is the inner diameter.

The shear strain on the surface of a round rod due to the action of shear stress is

$$\gamma = \tan\alpha = \frac{\varphi d_0}{2l_0} \times 100\%, \tag{3.18}$$

where

α is the rotation angle of an arbitrary line on the surface of a rod parallel to the specimen axis due to the application of τ (see Figure 3.10a)
φ is the angle of twist
l_0 is the length of the rod

3.1.3.2 Mechanical Properties in Torsion Testing

A torsion test is carried out by a torsion testing machine, using cylindrical (solid or hollow) specimens. The torsion specimen is shown in Figure 3.11. Sometimes, short specimens with a gauge length of 50 mm are adopted.

During the test, when the torque is increased, the sections at the two ends of the specimens within the gauge length continuously rotate each other, resulting in an increase of the twist angle φ.

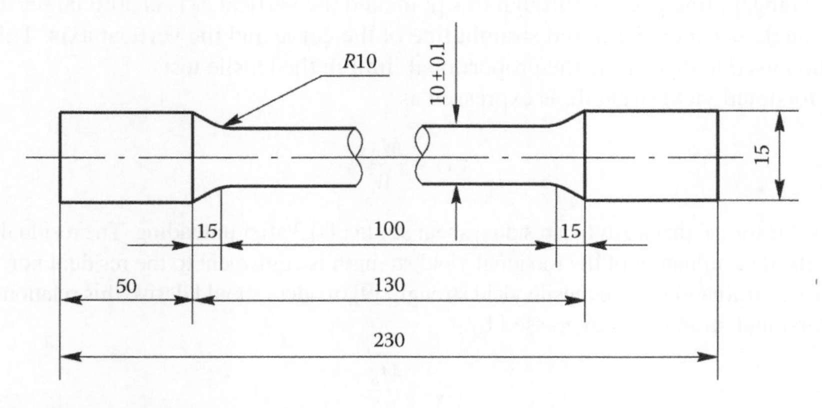

FIGURE 3.11 Torsion specimen.

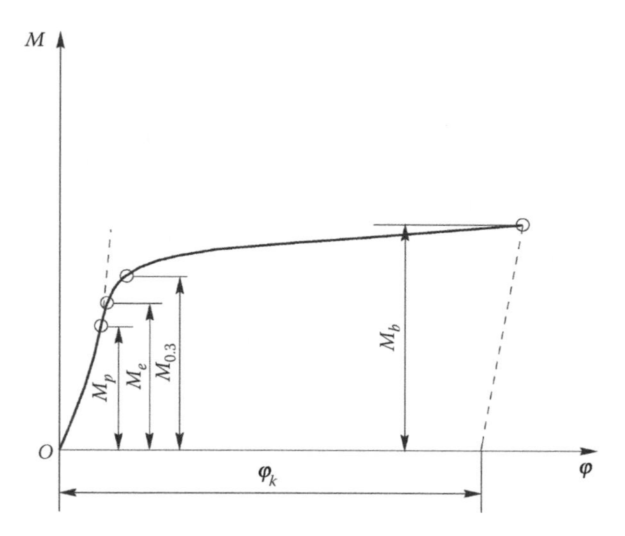

FIGURE 3.12 Torsion-test diagram.

The M–φ curve can be obtained using the drawing apparatus of the torsion testing machine and is called the torsion-test diagram, as shown in Figure 3.12.

This curve is very similar to the true stress–strain curve measured in the tensile test. The reason for this similarity is that the specimen shape does not change, and the deformation is always uniform under torsion. Even in the plastic deformation stage, torque increases with the increase in deformation, until the specimen fractures.

In terms of the torsion diagram (Figure 3.12) and Equations 3.16 and 3.17, we obtain the generally used mechanical properties in a torsion test:

G, the shear modulus, is expressed as

$$G = \frac{\tau}{\gamma} = \frac{32Ml_0}{\pi\varphi d_0^4}. \tag{3.19}$$

τ_p, the torsional proportional limit, is expressed as

$$\tau_p = \frac{M_p}{W}, \tag{3.20}$$

where M_p is the torque when the torque–twist angle curve begins to deviate from the initial straight line. M_p can be determined by the torque at the point in the curve at which the tangent of the angle between the tangent line passing through this point and the vertical axis is 50% larger than the tangent of the angle between the initial straight line of the curve and the vertical axis. This is similar to the method used to determine the proportional limit in the tensile test.

$\tau_{0.3}$, the torsional yield strength, is expressed as

$$\tau_{0.3} = \frac{M_{0.3}}{W}, \tag{3.21}$$

where $M_{0.3}$ is the torque that leads to a residual shear strain of 0.3 after unloading. The residual shear strain of 0.3% for the determination of the torsional yield strength is equivalent to the residual normal strain of 0.2% for the determination of the tensile yield strength [9] (readers should derive this relationship).

τ_b, the torsional strength, is expressed by

$$\tau_b = \frac{M_b}{W}, \tag{3.22}$$

where M_b is the maximum torque before the specimen ruptures. It should be pointed out that τ_b is calculated by the formula in the elastic deformation state. It can be seen from Figure 3.10c that τ_b is larger than the true torsional strength and is thus called the conditional torsional strength. Taking plastic deformation into consideration, the true torsional strength τ_k in the plastic state is calculated as

$$\tau_k = \frac{4}{\pi d_0^3}\left[3M_k + \theta_k\left(\frac{\mathrm{d}M}{\mathrm{d}\theta}\right)_k\right], \tag{3.23}$$

where

M_k is the maximum torque before rupture
θ_k is the relative twist angle of the unit length before rupture, $\theta_k = \mathrm{d}\varphi/\mathrm{d}l$
$(\mathrm{d}M/\mathrm{d}\theta)_k$ is the slope $\tan\alpha$ of the tangent line at $M = M_k$ in the M–θ curve, as shown in Figure 3.13

If the last part of the M–θ curve is approximately parallel to the horizontal axis, then $(\mathrm{d}M/\mathrm{d}\theta)_k = 0$. In this case, Equation 3.22 can be simplified as

$$\tau_k = \frac{12M_k}{\pi d_0^3}. \tag{3.24}$$

The true torsional strength τ_k can also be measured directly using a thin-walled tubular specimen. Since the tubular wall is very thin, it can be observed that shear stress in the cross section is approximately equal. Therefore, when a thin-walled tubular specimen ruptures, the shear stress is the true torsional strength τ_k, which can be calculated by

$$\tau_k = \frac{M_k}{2\pi t r^2}, \tag{3.25}$$

where

M_k is the torque before rupture
r is the average value of the inner and outer radii
t is the wall thickness
$2\pi t r^2$ is the resist torsional coefficient for a section of the thin-walled tubular specimen

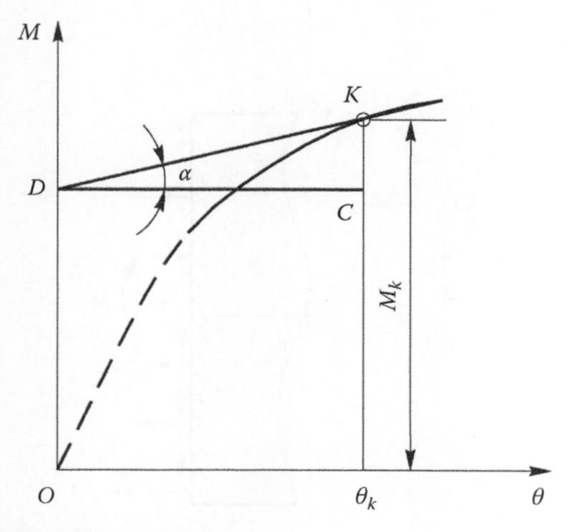

FIGURE 3.13 Schematic illustration of $(\mathrm{d}M/\mathrm{d}\theta)_k$.

The plastic deformation under torsion can be represented by the residual torsional shear strain γ_k, which can be calculated as

$$\gamma_k = \frac{\varphi_k d_0}{2l_0} \times 100\%, \tag{3.26}$$

where φ_k is the relative twist angle in the gauge length l_0 before the specimen ruptures. The total shear strain under torsion is the sum of the plastic shear strain and the elastic shear strain. For materials with high ductility, the elastic shear strain is very small, and therefore the plastic shear strain in Equation 3.25 is approximately equal to the total shear strain.

3.1.4 Mechanical Properties of Material under Bending

3.1.4.1 Stress and Deformation under Bending

A rod with bending as its main deformation is called a beam. In the general case, shearing and bending coexist in a beam. Therefore, on a section of the beam, shear stress and normal stress coexist, as shown in Figure 3.14.

According to the mechanics of materials, in the elastic deformation range the normal stress under bending in the cross section of a beam can be expressed as

$$\sigma = \frac{My}{I_z}, \tag{3.27}$$

where

M is the bending moment

I_z is the cross-sectional area's moment of inertia

y is the distance to the neutral axis (see details in Figure 3.15)

At $y = y_{max}$ (the points furthest away from the neutral axis on the cross section), the normal stress due to bending reaches the maximum value

$$\sigma_{max} = \frac{My_{max}}{I_z} = \frac{M}{I_z/y_{max}}, \tag{3.28}$$

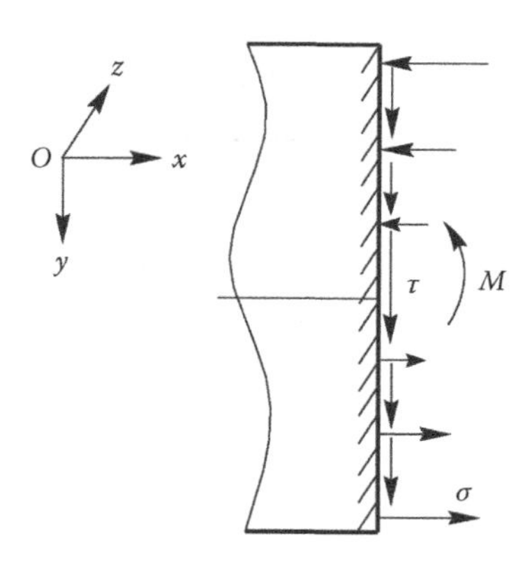

FIGURE 3.14 Stress diagram in beam.

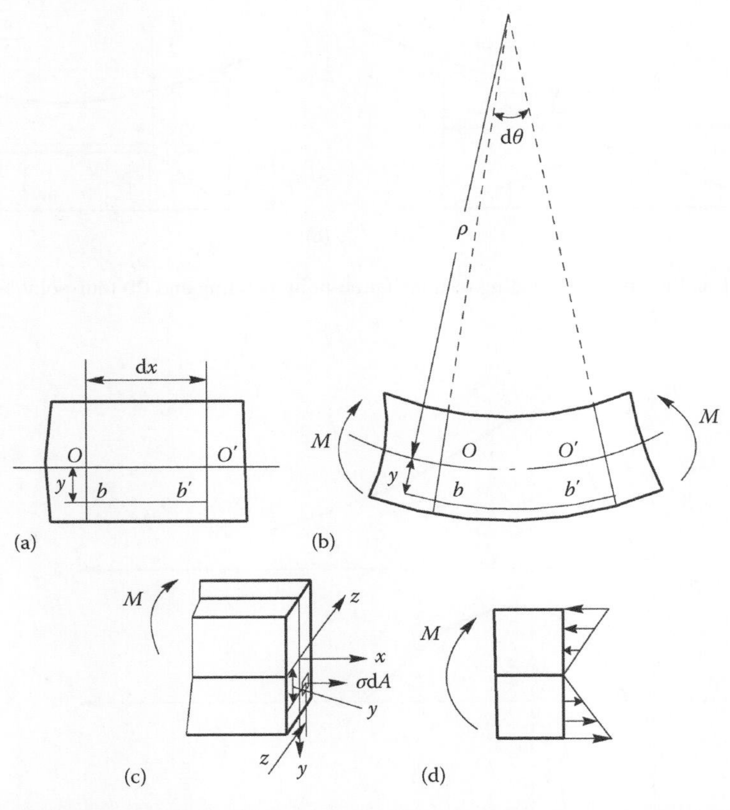

FIGURE 3.15 Deformation and stress distribution in a segment of beam: The segment (a) before and (b) after deformation, (c) the coordinate system, and (d) the normal stress distribution on the section.

where the ratio of I_z/y_{max} is only related to the shape and size of the cross section, called the bending coefficient for the section, and denoted by W_z:

$$W_z = \frac{I_z}{y_{max}}. \tag{3.29}$$

Therefore, the maximum bending normal stress is

$$\sigma_{max} = \frac{M}{W_z}. \tag{3.30}$$

After the beam bends, the displacement of the cross section's centroid in the direction perpendicular to the axis of the beam is called the deflection, denoted by ω. A bending test usually obtains the deflection of the middle point of the gauge length, denoted by $\omega_{L/2}$, where L is the gauge length for the specimen. Then, the related mechanical properties can be determined through the $F-\omega_{L/2}$ curve.

3.1.4.2 Mechanical Properties Measured in a Bending Test

Specimens with a rectangular or circular cross section are adapted in the bending test. The specimen is placed on the supports with a specified span. A concentrated load (three-point bending) or two equivalent loads (four-point bending) are applied, as shown in Figure 3.16. In the test, deflection is measured at the span center of the specimen, and the $F-\omega_{L/2}$ curve, called the bending-test diagram, is drawn. Figure 3.17 shows the bending-test diagrams of three different materials.

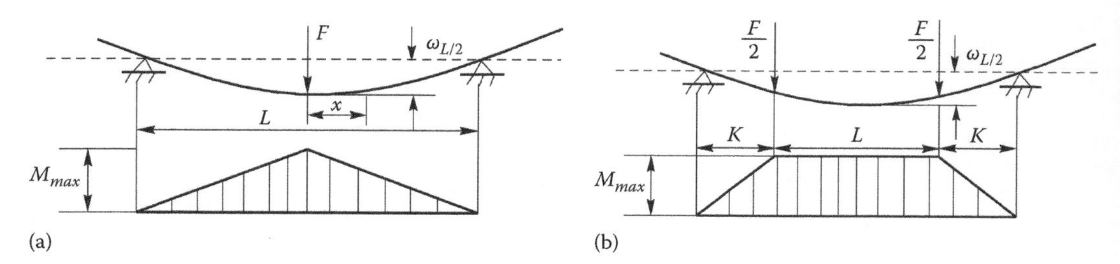

FIGURE 3.16 Loading mode in bending test. (a) Three-point bending and (b) four-point bending.

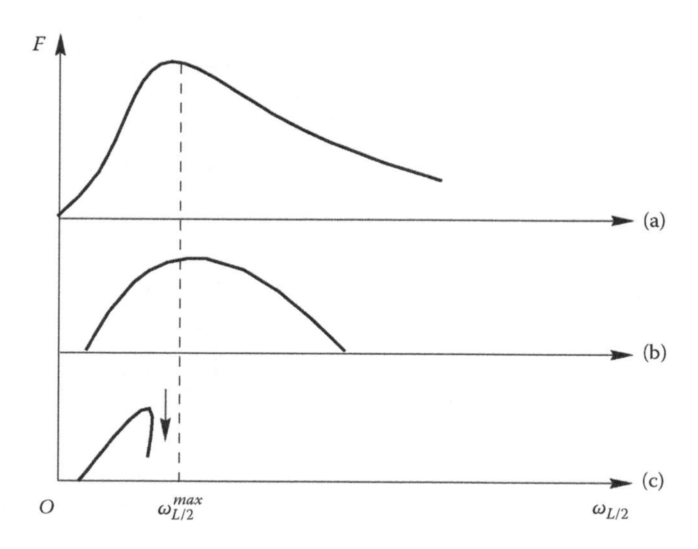

FIGURE 3.17 Typical bending-test diagrams. (a) Ductile material, (b) medium ductile material, and (c) brittle material.

For materials with high ductility, the bending test does not rupture the specimen. The last portion of the F–$\omega_{L/2}$ curve can be very long, as shown in Figure 3.17a. Therefore, a bending test is not suitable for measuring the strength of ductile materials, and analyzing the testing results is very complicated. Instead of using a bending test, the mechanical properties of ductile materials are measured by a tensile test.

Usually, for brittle materials, only the bending strength before fracture σ_{bb} is measured. From Equation 3.29, we have

$$\sigma_{bb} = \frac{M_b}{W}, \tag{3.31}$$

where

M_b is the bending moment before fracture
W is the section's bending coefficient

For specimens with a circular cross section, $W = \pi d^3/32$, d is the diameter; for specimens with a rectangular cross section, $W = bh^2/6$, where b and h are the width and height of a specimen, respectively.

From the initial straight-line portion of the $F–f_{max}$ curve, the flexural modulus of elasticity can be calculated. For the specimen with a rectangular cross section, the flexural modulus of elasticity is

$$E_b = \frac{ml^3}{4bh^3}, \tag{3.32}$$

where

m is the slope of the initial straight-line portion of the $F–\omega_{L/2}$ curve
L is the span

Because the previous section covers the stress and deformation related to shearing, it is not discussed separately here.

3.2 MEASUREMENT OF THE BASIC MECHANICAL PROPERTIES OF MATERIALS

Generally, a material exhibits different mechanical performance in different stress states. The same material can be "soft" in one state—prone to plastic deformation and ductile rupture—but can become "hard" in another stress state, where the material becomes difficult to deform and exhibits brittle fracture. In order to distinguish the softness or hardness of a material in different stress states, and to help the tester select an appropriate testing method to accurately measure the required mechanical properties, we need to understand the concept of "softness (or viscous) coefficient of the stress state." [3]

The maximum normal stress and the maximum shear stress play different roles during the deformation and fracture of a material. In general, maximum shear stress causes plastic deformation, which results in ductile fracture, while maximum normal stress leads to brittle fracture. Therefore, at the macroscopic level, we can use the ratio of the maximum shear stress τ_{max} to the maximum normal stress σ_{max} in different stress states to predict the tendency of deformation and fracture under the applied load. This ratio

$$\alpha = \frac{\tau_{max}}{\sigma_{max}} = \frac{\sigma_1 - \sigma_3}{2\left[\sigma_1 - \nu(\sigma_2 + \sigma_3)\right]}, \tag{3.33}$$

is called the flexibility coefficient of the stress state.

The larger the value of α, the larger the portion of the maximum shear stress, indicating that the stress state is softer and the material is more prone to plastic deformation and ductile fracture. The smaller the value of α, the harder the stress state, and the material is more prone to brittle fracture. Table 3.1 gives the flexibility coefficients of stress state under the commonly used loading methods for materials with $\nu = 0.25$.

TABLE 3.1
Flexibility Coefficients of Stress State under Different Loading Methods

Loading Method	Principal Stress σ_1	σ_2	σ_3	Softness Coefficient α
Nonuniform triaxial tension	σ	$(8/9)\sigma$	$(8/9)\sigma$	0.1
Uniaxial tension	σ	0	0	0.5
Torsion	σ	0	$-\sigma$	0.8
Uniform biaxial compression	0	$-\sigma$	$-\sigma$	1
Uniaxial compression	0	0	$-\sigma$	2
Nonuniform triaxial compression	$-\sigma$	$-(7/3)\sigma$	$-(7/3)\sigma$	4

In Table 3.1, nonuniform triaxial tension and nonuniform triaxial compression represent two extreme conditions. Under uniaxial tension, $\alpha = 0.5$, "softness" and "hardness" are medium, since, it has the most extensive application in the mechanical testing of materials.

3.2.1 Measurement of the Tensile Properties of Materials

3.2.1.1 Tensile Test

If not specified, a tensile test is the method used to measure the mechanical properties of smooth specimens with slow uniaxial loading rates at room temperature in the atmosphere. We have introduced the shapes and dimensions of commonly used tensile specimens, as shown in Figure 3.18. If the smooth cylindrical specimen is adopted and the gauge length l_0 is much larger than the diameter d_0 (in general, $l_0 > 5d_0$), then, according to Saint Venant's principle, the stress distribution in the cross section in the middle part of the specimen is homogeneous, and the uniform axial loading in the middle part of the specimen can be implemented. A cylindrically shaped specimen is conveniently used to measure the radial strain, and the machining is relatively easier. The dimensions of the specimen and the tolerances in machining are regulated by the Chinese national standard GB6397-1976.

When measuring the mechanical properties of sheets and strips, plate-like specimens can be adopted, as shown in Figure 3.18b. However, the gauge length l_0 of the specimen should satisfy the following equation:

$$l_0 = 11.3\sqrt{A_0} \quad \text{or} \quad l_0 = 5.56\sqrt{A_0},$$

where A_0 is the original cross-sectional area of the specimen. This regulation corresponds to $l_0 = 10d_0$ or $l_0 = 5d_0$ for a cylindrical specimen.

Before yielding, the regulated tensile loading rate is $d\sigma/dt = 1 \sim 10$ MPa/s. The preparation of specimens, the requirements of the testing apparatus, operational procedures, and data processing should follow the technical regulations of the national standard. The testing results are valid only when the tensile test strictly follows the national standard, and the mechanical properties measured by different laboratories and staff can be compared.

Usually, the tensile testing machine has automatic recording or drawing equipment that records or draws the curve between the load F acting on the specimen and the elongation Δl (see the tensile-test diagram in Figure 3.2). The tensile curve of annealed mild steel is shown in Figure 3.19. Dividing the load by the original cross-sectional area gives the engineering stress σ, $\sigma = F/A_0$; diving the elongation by the original gauge length gives the engineering strain ε, $\varepsilon = \Delta l/l_0$. Figure 3.20 shows the engineering stress–strain curve, simply called the stress–strain curve or the tensile curve. Comparison of Figures 3.19 and 3.20 indicates that they have the same or similar shape, but different scales for coordinate axes and different physical meanings.

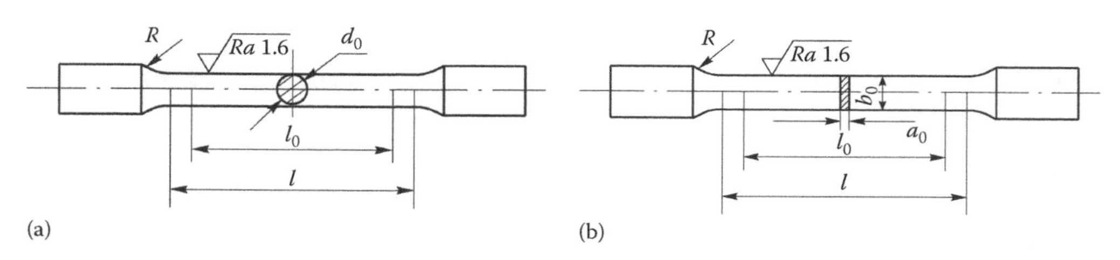

FIGURE 3.18 Commonly used tensile specimens. (a) Standardized cylindrical tensile specimen. (b) Plate-like tensile specimen.

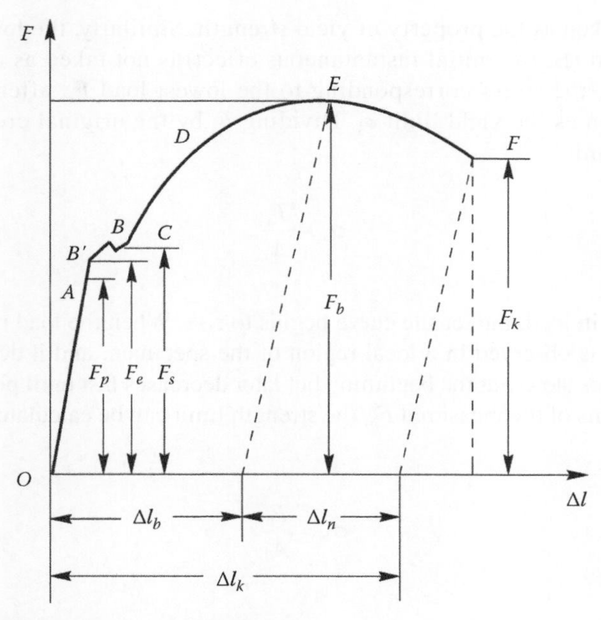

FIGURE 3.19 Tensile diagram of annealed mild steel.

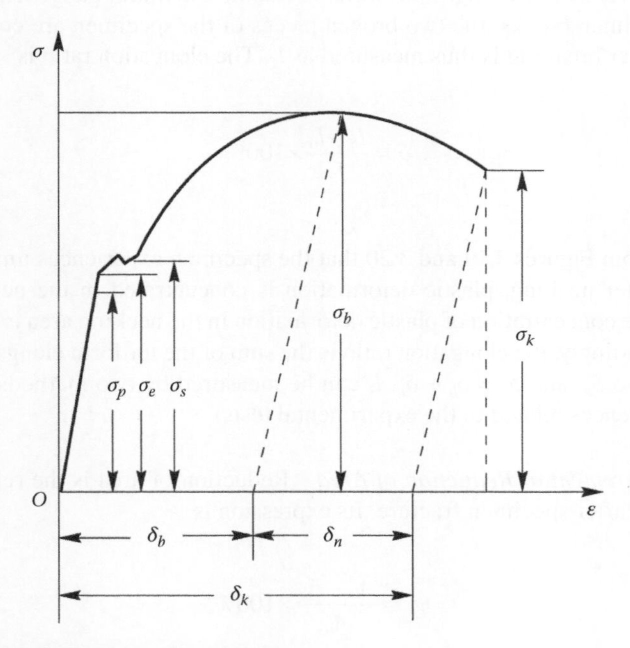

FIGURE 3.20 Engineering stress–strain curve.

3.2.1.2 Measurement of Material Properties

3.2.1.2.1 Measurement of Strength Properties

See Figure 3.19. The yield stage (segment B' to C) of mild steel displays a serrated shape. In this stage, the stress corresponding to the maximum load at point B' is called the upper yield limit. Because it is considerably influenced by the deformation rate and specimen shape, it is generally not taken as the property of yield strength. Similarly, the lowest point due to the first decrease of load (B, the initial instantaneous effect) is not taken as the property of yield strength. In general, the stress corresponding to the lowest load F_s, after the initial instantaneous effect, is taken as the yield limit σ_s. Dividing F_s by the original cross-sectional area A_0 gives the yielding limit

$$\sigma_s = \frac{F_s}{A_0}.$$

With the increase in load, the tensile curve begins to rise. When the load reaches the maximum value of F_b, necking is observed in a local region of the specimen, and it develops quickly. After that, the load decreases slowly at the beginning but later decreases fast until point F, at which specimen fractures. In terms of the measured F_b, the strength limit can be calculated using the following equation:

$$\sigma_b = \frac{F_b}{A_0}.$$

3.2.1.2.2 Measurement of Ductile Properties

3.2.1.2.2.1 Measurement of Elongation Ratio Assume the initial gauge length of the specimen is l_0. After the specimen breaks, the two broken pieces of the specimen are connected tightly, and the gauge length after breaking is thus measured as l_k. The elongation ratio is

$$\delta = \frac{l_k - l_0}{l_0} \times 100\%.$$

It can be seen from Figures 3.19 and 3.20 that the specimen experiences uniform elongation Δl_b before necking. After necking, plastic deformation is concentrated in the necking area, and the elongation due to the concentration of plastic deformation in the necking area is Δl_n. Therefore, $\Delta l_k = \Delta l_b + \Delta l_n$. Correspondingly, the elongation ratio is the sum of the uniform elongation ratio δ_b and the local elongation ratio δ_n, and $\delta_k = \delta_b + \delta_n$. l_k can be measured by two methods (please refer to the accompanying references related to the experimental tests).

3.2.1.2.2.2 Measurement of Reduction of Area Reduction of area is the relative contraction of the fracture surface after specimen fracture. Its expression is

$$\psi_k = \frac{A_0 - A_k}{A_0} \times 100\%,$$

where

A_k is the minimum cross-sectional area of the fracture surface
A_0 is the original cross-sectional area of the specimen

It is convenient to measure ψ_k for the circular section. We only need to measure the minimum diameter of the fracture surface; then we can calculate A_k and ψ_k. ψ_k is not influenced by the specimen's gauge length but is slightly influenced by the specimen's original diameter.

3.2.2 Measurement of the Compressive Properties of Materials

3.2.2.1 Uniaxial Compressive Test

The softness coefficient of the stress state under uniaxial compression is high. Thus, the compressive test is used to measure the mechanical properties of brittle materials such as cast iron, bearing alloy, concrete, brick, stone, and the like. Since the stress state under compression is relatively soft, the mechanical behavior (which cannot be displayed under tension, torsion, and bending) can be obtained under compression. Ductile materials under compression only undergo compressive deformation without rupture, and the compressive curves rise constantly, as in curve 1 shown in Figure 3.21. For this reason, the compressive test is rarely performed for ductile materials. If the compressive test is required for a ductile material, the purpose is only to evaluate its adaptability to the machining process.

Curve 2 in Figure 3.21 is the compressive curve of a brittle material. In terms of the compressive curve, we can obtain the properties of compressive strength and ductility. For materials with low ductility and brittle materials, in general, we only measure the compressive strength σ_{bc}, shortening the ratio ε_{ck} and expansion ratio of the fracture surface ψ_{ck}:

$$\sigma_{bc} = \frac{F_{bc}}{A_0},$$

$$\varepsilon_{ck} = \frac{h_0 - h_k}{h_0} \times 100\%,$$

$$\psi_{ck} = \frac{A_k - A_0}{A_0} \times 100\%,$$

where σ_{bc} is the nominal compressive strength (with the assumption that the cross-sectional area does not change during compression). If the influence of the specimen's cross-sectional area is taken into consideration, we obtain the true compressive strength (F_k/A_k). Since $A_k > A_0$, the true compressive strength is less than or equal to the nominal compressive strength.

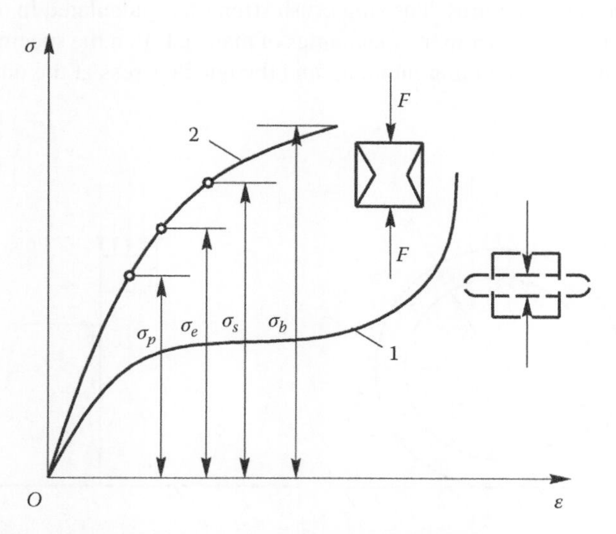

FIGURE 3.21 Compressive load-deformation curve. 1. Ductile material. 2. Brittle material.

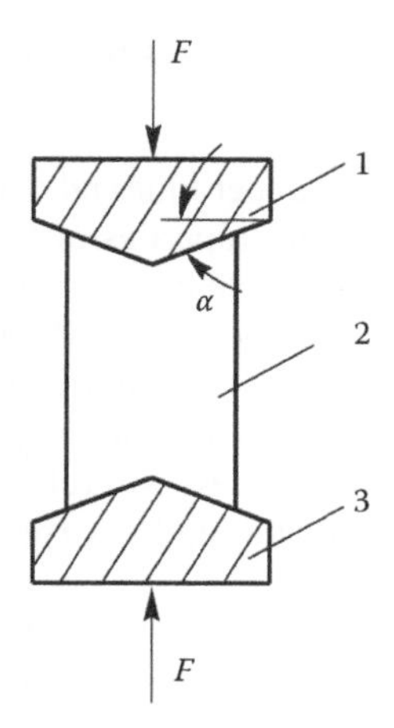

FIGURE 3.22 Reducing the friction at the specimen ends: the shape of rams and specimen. 1: upper ram, 2: specimen, and 3: lower ram.

During a compressive test, a huge friction force exists between the specimen ends and the upper and lower rams, which influences not only the testing results but also the mode of rupture. In order to reduce the influence of friction force, the two ends of the specimen must be smooth, flat, parallel to each other, and lubricated with lubricating oil or graphite powder. The specimen ends can also be machined into a concave and conic surface, and the tilting angle of the conical surface is equal to the friction angle: $\tan\alpha = f$, where f is the friction coefficient. At the same time, the rams should be changed to the corresponding cones, as shown in Figure 3.22.

3.2.2.2 Ring Crush Strength Test

There are a lot of tubular products in the ceramic materials industry. During research, development, and quality examination, a ring crush strength test is frequently adopted. This test is also frequently employed in the quality examination of power metallurgical products. The test adopts circular ring specimens, with the shape and loading mode shown in Figure 3.23.

During the test, the specimen is placed between the upper and lower rams, and pressure is applied until the specimen fractures. The ring crush strength is calculated in terms of the compressive load at rupture. It is known from the mechanics of materials that the specimen's I-I cross section is subjected to the maximum bending moment, and the tensile stress at the outer surface on the I–I

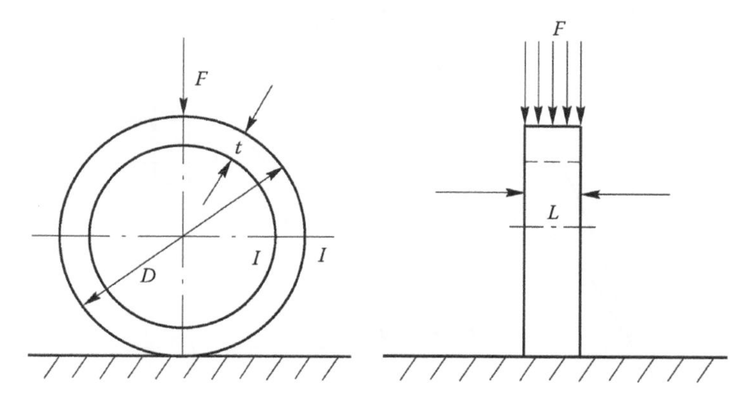

FIGURE 3.23 Illustration of ring crush strength test.

cross section is the maximum. When the specimen fractures, the maximum tensile stress on the I–I section is the ring crush strength, which can be calculated using the following equation:

$$\sigma_r = \frac{1.908P_r(D-t)}{2Lt^2}, \tag{3.34}$$

where

P_r is the load when the specimen crushes
D is the outer diameter of the ring
t is the specimen's wall thickness
L is the specimen's width

Please note that the specimen must keep its circularity (the surface of the cylindrical specimen must be smooth), there must be no scratches on the surface, and the wall thickness should be uniform.

3.2.3 Measurement of the Mechanical Properties of Materials under Torsion

3.2.3.1 Characteristics of the Torsion Test

The torsion test is one of the important tests for mechanical properties. It has the following characteristics:

1. The softness coefficient of the stress state under torsion is relatively high. Thus, the torsion test can be used to measure the properties of plastic deformation and strength of materials that exhibit brittle normal fracture under tensile test (such as tool steels quenched and tempered at low temperature).
2. When a cylindrical specimen is subjected to torsion, plastic deformation within the entire gauge length is always uniform, the cross section and the gauge length are primarily unchanged, and the necking that occurs under static tension does not take place. Therefore, a torsion test can accurately measure the resistance of materials with high ductility to deformation and the capacity for deformation, which are difficult to measure using a uniaxial tensile test or a compressive test.
3. A torsion test can definitely distinguish the fracture mode of a material: normal fracture or shear fracture. For a ductile material, shear fracture caused by shear stress is indicated if the fracture surface is perpendicular to the specimen axis and the fracture surface is flat with the rotating trace of plastic deformation. For a brittle material, the shear fracture surface is at an angle of approximately 45° to the specimen axis and exhibits a helical shape. If the shearing strength in the axial direction is lower than that in the transverse direction—such as wood or alloying plates with serious strip-like segregation—the fracture under torsion can display a laminated fracture surface. Therefore, in terms of the features of fracture surface under torsion, we can judge the causes of fracture and the relative magnitudes of torsional strength and tensile (or compressive) strength. Using this feature, we can analyze some testing results very well, such as the influence of carbon content in low-temperature tempering martensite on the toughness of carbon steels.
4. In a torsion test, the torsional shear stress distributes nonuniformly on the cross section, and it reaches maximum on the surface and reduces toward the center. Therefore, it is very sensitive to surface defects. In engineering, a torsion test is usually used to investigate or examine the surface quality of heat treatment or the effect of a variety of surface hardening processes.
5. In a torsion test, the specimen is subjected to a relatively large shear stress. Therefore, it is widely used to investigate the problems related to the non-simultaneous initiation of plastic deformation, such as elastic lag and dissipation.

In summary, a torsion test can measure all of the mechanical properties of shear deformation and fracture of ductile materials and brittle materials. In addition, it has the non-matching advantages over the other mechanical testing methods. Therefore, torsion testing is widely used in research and product examination. However, the features of torsion testing can become disadvantages in some cases. For example, since the shear stress reaches the maximum on the surface and reduces toward the center of the section, the central part is still in elastic deformation, while plastic deformation occurs in the surface layer. Therefore, it is difficult to accurately measure the moment of the surface layer to initiate plastic deformation. Thus, a torsion test cannot accurately measure the resistance of a material to microplastic deformation.

3.2.3.2 Torsion Test

The torsiometer is mounted on the specimen surface to measure the angle of twist. If the two sections pinned by the torsiometer rotate away from each other, the torsiometer can measure the angle of twist φ, as shown in Figure 3.24.

Within the torsional proportional limit of a material, the equation for the angle of twist is

$$\varphi = \frac{Ml_0}{GI_p}, \tag{3.35}$$

where

M is the torque

I_p is the area inertia moment of the circular cross section

Using the incremental method, the load is increased step by step. If the increment of the angle of twist $\Delta\varphi$ is primarily identical for each equal incremental torque ΔM, this verifies Hooke's law in shear. In terms of the incremental of the angle of twist $\Delta\varphi_i$ at each step, the shear elastic modulus can be calculated using the following expression:

$$G_i = \frac{\Delta M l_0}{\Delta\varphi_i I_p}, \tag{3.36}$$

where the subscript i denotes the loading step ($i = 1, 2, \ldots, n$).

The aforementioned torsiometer can only measure small twist angles. After the material yields, it has to be removed. The automatic drawing apparatus of the testing machine continues to record the M–φ curves, as shown in Figure 3.25.

When the torque reaches a value of M_p, the shear stress at the cross-sectional edge reaches the shear yield limit τ_s. After the torque goes beyond M_p, the distribution of shear stress on the cross section is no

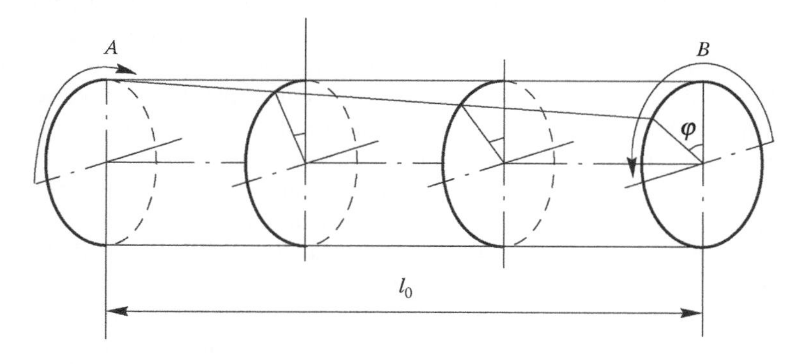

FIGURE 3.24 Illustration of torsion.

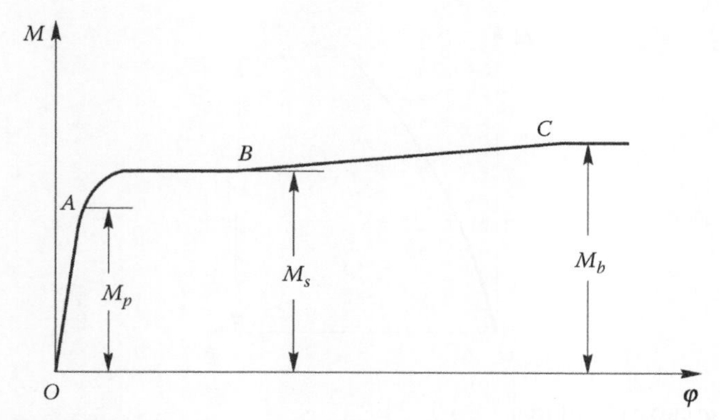

FIGURE 3.25 Curve of mild steel.

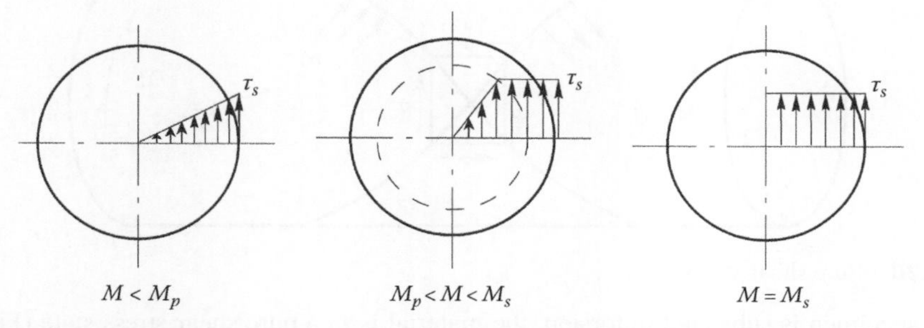

FIGURE 3.26 Distributions of shear stress in a circular shaft of mild steel at different torques.

longer linear (see Figure 3.26). In the cylinder's outer area, the material yields and forms a ring-shaped plastic zone, at the same time that the M–φ diagram changes from a straight line to a curve.

With continuous torsional deformation of the specimen, the plastic zone propagates to the circle center and the M–φ curve rises slightly until point B, after which the curve tends to flatten. The minimum torque M_s indicated by the torqueometer's dial gauge is the torque when the whole specimen yields. At this moment, the plastic zone occupies almost the whole cross section (see Figure 3.26). τ_s is approximately equal to

$$\tau_s = \frac{3}{4} \cdot \frac{M_s}{W_p}, \tag{3.37}$$

where W_p is the torsional modulus of the specimen's cross section.

With the continuous increase of deformation, the material further hardens until point C in the M–φ curve, and the specimen fractures. The maximum torque M_b can be read from the torqueometer's dial gauge. Similar to Equation 3.36, we have

$$\tau_b = \frac{3}{4} \cdot \frac{M_b}{W_p}. \tag{3.38}$$

The M–φ curve for cast iron is far different from that of mild steel, as shown in Figure 3.27. From the beginning of torsion to fracture, it is approximately a straight line. Therefore, the torsional strength can be approximately calculated according to the elastic deformation

$$\tau_b = \frac{M_b}{W_p}. \tag{3.39}$$

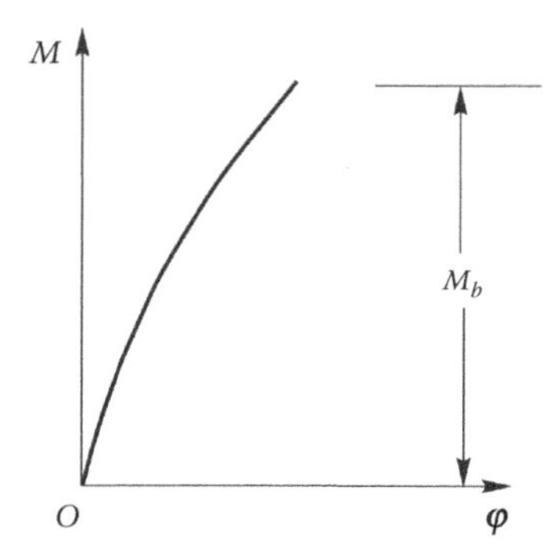

FIGURE 3.27 M–φ curve of cast iron.

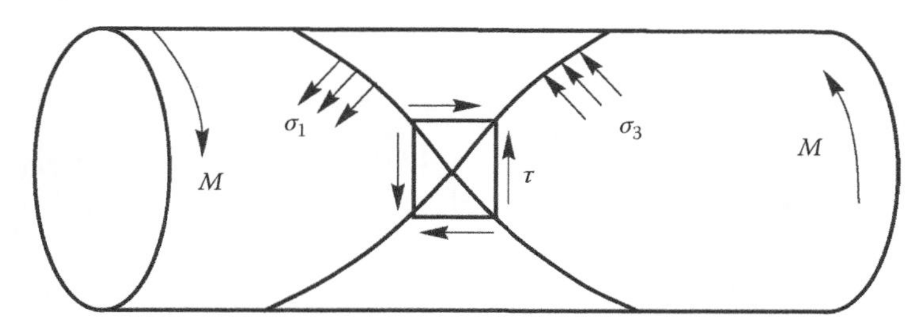

FIGURE 3.28 Pure shear stress.

When the specimen is subjected to torsion, the material is in a pure shear stress state (Figure 3.28). On the helical surface at an angle of ±45° to the axis of the shaft, the principal stresses are $\sigma_1 = \tau$ and $\sigma_3 = -\tau$, respectively. For mild steel, the tensile strength is larger than the torsional strength, and thus shear fracture occurs in the cross section. For cast iron, the tensile strength is lower than the shear strength, thus tensile fracture occurs in the direction perpendicular to σ_1.

3.2.4 Measurement of Material Bending Properties

A static bending test has the following characteristics:

1. From the viewpoint of specimen tension, the stress state in a bending test is similar to that in a tensile test. However, the geometry of the specimen as a whole in a bending test is simpler than that in a tensile test. Thus, it is suitable to measure the fracture strength and ductility of brittle materials such as cast irons, tool steels, hard alloys, and even ceramic materials. For polymer materials, a bending test is also frequently employed to measure the bending strength and modulus, in order to select ingredients and to control the quality of products.
2. For annealing, normalizing, and tempering structural carbon steels or alloy steels, due to their good ductility, they frequently cannot be broken during test and the last part of the F–$\omega_{L/2}$ curve can be very long. For this type of material, a tensile test should be employed to measure the resistance to fracture.
3. Similar to a torsion test, the bending test has a nonuniform stress distribution, and stress reaches the maximum value on the surface. So the bending test can sensitively detect the surface defects of a material to evaluate the product quality after carburization or surface quenching.

As previously mentioned, the bending test has two loading modes: three-point bending and four-point bending. In a four-point bending test, the specimen between the two loading points is subjected to an equal bending moment. Thus, the specimen generally fractures at the site where the microstructural defects are located. It can reflect the material properties well, and the experimental results are relatively accurate. However, we must pay attention to the balance of loading in a four-point bending test. In a three-point bending test, the specimen always fractures at the site with the maximum bending moment. A three-point bending test is relatively simple and is frequently employed.

3.2.5 Measurement of Material Shear Properties

For components subjected to shear loading, the shear test is often performed to simulate practical service conditions and to provide data on the shear strength of materials for design. This is particularly important for components like rivets and pins. The general shear tests include the simple shear test, the double shear test, and the punch shear test.

3.2.5.1 Simple Shear Test

The shear test is used to measure the shear strength of plates and wires. Thus, shear test specimens are cut from sheet metals or wire rods. In the test, the specimen is fixed on the pedestal and pressure is applied to the upper ram until shear breaking of the specimen occurs along the shear plane *m–m*, as shown in Figure 3.29. At this moment, the maximum shear stress on the shear plane is the material's shear strength. In terms of the maximum load F_b and the original cross-sectional area A_0 of the specimen before shear fracture, the material's shear strength can be calculated as

$$\tau_b = \frac{F_b}{A_0}. \tag{3.40}$$

Figure 3.29 illustrates the loading and deformation of a specimen in a simple shear test. The external loads acting on the two lateral faces have the same magnitudes and the opposite directions and are very close to the acting line, causing the two parts of the specimen to move relatively along the shear plane (*m–m*). Therefore, shear stress is generated in the shear plane. Because of the accompanying squeezing and bending when the specimen is subjected to shearing, the distribution of shear stress is complex. However, in a simple shear test, it is usually assumed that shear stress distributes uniformly in the shear plane. The shear test cannot measure the shear proportional limit and the shear yield strength. If these properties are required, the previously introduced torsion test should be employed.

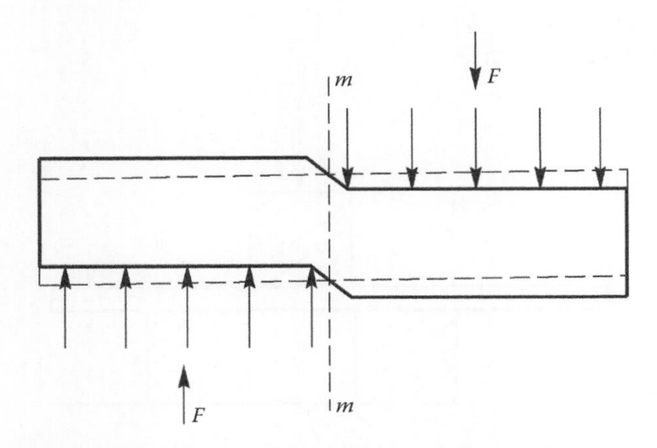

FIGURE 3.29 Illustration of loading and deformation of specimen in simple shear.

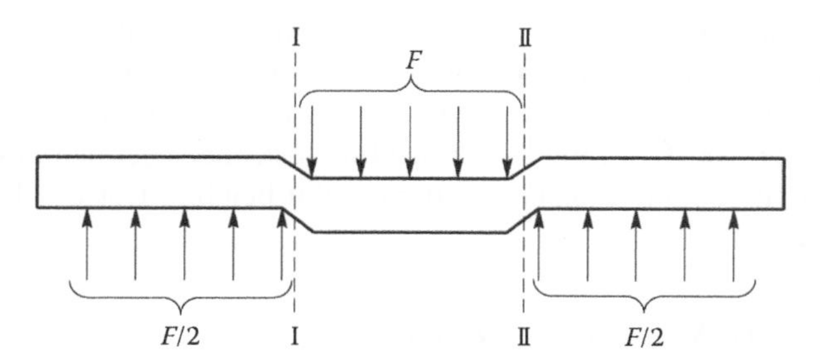

FIGURE 3.30 Illustration of double shear test.

3.2.5.2 Double Shear Test

The double shear test is the most commonly used shear test. In this test, the specimen is mounted inside the compressive or tensile shear cutter and the load is applied. The specimen is subjected to shear on the I-I and II-II sections at the same time (see Figure 3.30). Assuming that the load when the specimen breaks is P_b, the shear strength is

$$\tau_b = \frac{P_b}{2A_0}. \tag{3.41}$$

A cylindrical specimen is used in the double shear test, and the length of the shearing part cannot be too long. The reason is that during the shearing process, two shear planes are subjected to shearing and the specimen is also subjected to bending. In order to reduce the influence of bending, the ratio of the length of the shearing part over the specimen diameter should not exceed 1.5.

The hardness of the lining ring should not be lower than 700HV30 (see Chapter 4). The shear speed in general is 1 mm/min, and the fastest speed should not exceed 10 mm/min. If the specimen undergoes obvious bending deformation after shear fracture, the testing result is invalid.

3.2.5.3 Punch Shear Test

The shear strength of a thin metal sheet is measured using a punch shear test, as schematically illustrated in Figure 3.31. The load at the breaking of specimen is P_b, and the fracture surface is a cylindrical surface. Thus, the shear strength is

$$\tau_b = \frac{P_b}{\pi d_0 t}, \tag{3.42}$$

where

d_0 is the diameter of the punching hole
t is the thickness of the metal sheet

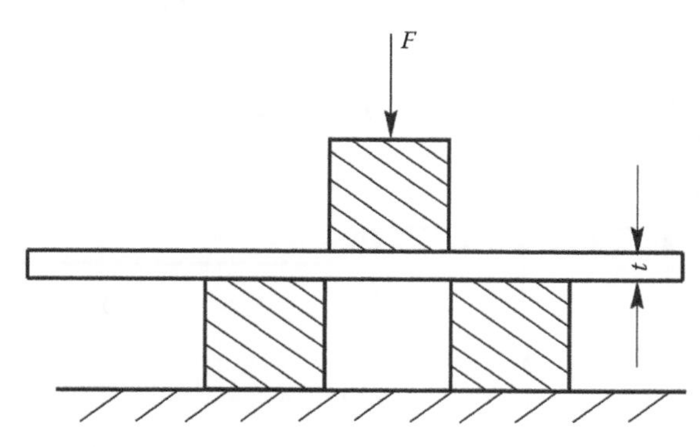

FIGURE 3.31 Schematic illustration of punch shear test.

3.2.6 New Advances in the Application of Testing Methods to the Basic Mechanical Properties of Materials

In an era of rapid development of innovative technology and new materials, testing methods for the basic mechanical properties of materials, including tension, compression, torsion, bending, and shear, are assigned a new mission. We take the conventional tensile test as an example. With the development of new technology, the invention of micro-tensile machines with lower load and higher accuracy provides great help to scientific researchers. It is extensively used to measure the mechanical properties of materials at the microscale. In the work of Choi et al. [10–13], a micro-tensile test was used to measure a variety of mechanical properties of low-dimensional materials such as films, fiber, MEMS, and the like. Many researchers combined a micro-tensile test with other testing methods and obtained very good results. Mili [14] applied acoustic emission monitoring in the micro-tension of glass fiber bundles and characterized the glass fiber strength very well. Qin [15] studied the strain-hardening properties of Cu/TiN polycrystalline thin films by in situ XRD stress analysis, combined with a tensile test. Compression, torsion, and bending tests—the basic mechanical properties testing methods—also find better applications in the new era. Xin et al. [16] investigated the buckling of carbon nanotubes under axial compression by molecular dynamics simulation. Yerranmalli [17] investigated the failure behavior of fiber reinforced polymer composites under combined compression–torsion loading. Using scanning electron microscopy together with the bending testing method, Morrison et al. [18–20] studied the fatigue behavior of Zr-based bulk metallic glass, the damage behavior of ceramic films, and the mechanical properties of TiO_2 nanofibers.

With the development of new equipment, new technology, and new materials, the traditional basic mechanical properties testing methods are no longer restricted to the measurement of simple, basic mechanical properties. Combined with innovative measuring methods, they can also be applied more extensively in other fields.

EXERCISES

3.1 What mechanical properties can be measured by the tensile test? What is the basic requirement for the tensile specimen?

3.2 What is the difference between a tensile-test diagram and the stress–strain curve? What is the difference between the stress–strain curve and the true stress–true strain curve? How are tensile properties determined in terms of the stress–strain curve?

3.3 How is the reduction of area measured?

3.4 How is the yield strength enhanced?

3.5 The following data are measured in the tensile test of normalized 60 Mn steel with an original diameter of 10 mm ($d = 9.9$ mm is the diameter of the specimen at the end of the yield plateau):

$$F/\text{kN}\ 39.5, 43.5, 47.6, 52.9, 55.4, 54.0, 52.4, 48.0, 43.1$$

$$d/\text{mm}\ 9.91, 9.87, 9.81, 9.65, 9.21, 8.61, 8.21, 7.41, 6.78$$

1. Plot the stress–strain curve.
2. Plot the uncorrected and corrected true stress–true strain curves.
3. Solve σ_s, σ_b, ε_b, ψ_b, ψ_k.

3.6 $\tau_{0.2}$ is not adopted as the measure of torsional yield strength, while $\sigma_{0.2}$ is adopted as the measure of tensile yield strength. What is the reason for this?

3.7 Why can the torsion test approximately judge the relative magnitudes of τ_k and S_k of a material? Can the features of stress causing fracture be analyzed in terms of the features of fracture surface in a torsion test?

3.8 A torsion test is performed using a specimen of GCrl5 steel (quenched and tempered at 200°C) with $d_0 = 0.9$ mm, $l_0 = 75$ mm. The experimental data are recorded as follows:

$$M/(\text{N}\cdot\text{m}): 139.7, 186.3, 217.7, 254.0, 269.7, 283.4, 293.0, 305.0$$

$$(\varphi/^\circ): 11.5, 15.7, 26.3, 34.3, 46.0, 59.6, 78.7, 98.1, 117.5$$

where $M_p = 139.7$ N · m, $M_{0.3} = 217.7$ N · m, $M_k = 305.0$ N · m

1. Plot the τ–γ curve.
2. Solve τ_p, $\tau_{0.3}$, τ_k, G.

3.9 Which materials are suitable for a bending test? Which two loading modes are adopted in the bending test? Where is the location of the maximum bending moment? Illustrate with a diagram.

3.10 The hull for a rotating wheel is made of cast iron. The technical requirement of the component is that the bending strength is larger than 400 MPa. A three-point bending test is performed using specimens of φ30 mm × 340 mm. The testing results are given as follows. Is the technical requirement satisfied?
Group 1: $d = 30.2$ mm, $L = 300$ mm, $P_{bb} = 14.2$ kN
Group 2: $d = 32.2$ mm, $L = 300$ mm, $P_{bb} = 18.3$ kN

3.11 Compare the characteristics of the uniaxial tensile test, the torsion test, the bending test, the compressive test, and the shear test. How is the appropriate testing method selected to measure the mechanical properties in terms of the practical application of a material?

3.12 The testing material is cast iron. The specimen diameter is $d = 30$ mm, and the initial gauge length is $h_0 = 45$ mm. In the compressive test, the specimen fractures when the load is 485 kN. The gauge length after testing is $h = 40$ mm. Calculate the compressive strength and the shortening ratio.

3.13 The dimension of a specimen in a double shear test is φ12.3 mm. The specimen fractures when the load reaches 21.45 kN. Calculate the shear strength of the material.

3.14 A hole must be punched with a specified diameter on a metal sheet using a puncher. Which mechanical property should be known in order to determine the shearing load? Which mechanical test should be used to measure it?

3.15 We need to evaluate the mechanical properties of the following materials under static loading. The testing methods include uniaxial tension, uniaxial compression, bending, torsion, and hardness. For the given materials, select one or two of the most appropriate testing methods. Materials: mild steel, cast iron, high-carbon tool steel (quenched and tempered at low temperature), structural ceramic, glass, and thermoplastic material.

REFERENCES

1. Zhang S., Sun D., Fu Y. Q. et al. Toughness measurement of ceramic thin films by two-step uniaxial tensile method. *Thin Solid Films*, 2004, 469–470: 233–238.
2. Li X. P., Kasai T., Nakao S. et al. Measurement for fracture toughness of single crystal silicon film with tensile test. *Sensor Actuat A*, 2005, 119: 229–235.
3. Zheng X. *The Mechanical Properties of Materials,* 2nd edn. Xi'an, China: Northwestern Polytechnical University Press, 2000.
4. Shan H. *Mechanics of Materials (I)*. Beijing: Higher Education Press, 1999.
5. Jiang W., Zhao S., Wang C. et al. *The Mechanical Properties of Engineering Materials*. Beijing: Beijing University of Aeronautics and Astronautics Press, 2000.
6. Liu H. *Introduction to Mechanics of Materials*. Beijing: Higher Education Press, 1997.
7. Pan X. and He Y. *The Principles and Methods of Experiments in Mechanics of Materials*. Harbin, China: Harbin Engineering University Press, 1995.

8. Liang X., Li J., and Zhang Z. *Assembly of National Standards for Experimental Testing Methods of Mechanical Properties and Process Properties of Metallic Materials*. Beijing: Chinese Standard Publishing House, 1996.
9. Wei W. *The Test of Mechanical Properties of Metals*. Beijing: Science Press, 1980.
10. Choi H. W. and Lee K. R. Fracture behavior of diamond-like carbon films on stainless steel under a micro-tensile test condition. *Diam Relat Mater*, 2006, 15(1): 38–43.
11. Hua T. and Xie H. M. A new micro-tensile system for measuring the mechanical properties of low-dimensional materials-fibers and films. *Polym Test*, 2007, 26(4): 513–518.
12. Yang Y. and Yao N. Deformation and fracture in micro-tensile tests of freestanding electrodeposited nickel thin films. *Script Mater*, 2008, 58(12): 1062–1065.
13. Modlinski R., Puers R., and Wolf I. D. AlCuMgMn micro-tensile samples: Mechanical characterization of MEMS materials at micro-scale. *Sensor Actuat A-Phys*, 2008, 143(1): 120–128.
14. Mili M. R., Moevus M., and Godin N. Statistical fracture of E-glass fibres using a bundle tensile test and acoustic emission monitoring. *Compos Sci Technol*, 2008, 68(7–8): 1800–1808.
15. Qin M. and Ji V. Determination of proof stress and strain-hardening exponent for thin film with biaxial residual stresses by in-situ XRD stress analysis combined with tensile test. *Surf Coat Technol*, 2005, 192(2–3): 139–144.
16. Xin H., Han Q., and Yao X. H. Buckling of defective single-walled and double-walled carbon nanotubes under axial compression by molecular dynamics simulation. *Compos Sci Technol*, 2008, 68(7–8): 1809–1814.
17. Yerramalli C. S. and Waas A. M. A failure criterion for fiber reinforced polymer composites under combined compression–torsion loading. *Int J Solids Struct*, 2003, 40(5): 1139–1164.
18. Morrison M. L. and Buchanan R. A. Four-point-bending-fatigue behavior of the Zr-based Vitreloy 105 bulk metallic glass. *Mat Sci Eng A*, 2007, 467(1–2): 190–197.
19. Carneiro J. O., Alpuim J. P., and Teixeira V. Experimental bending tests and numerical approach to determine the fracture mechanical properties of thin ceramic coatings deposited by magnetron sputtering. *Surf Coat Tech*, 2006, 200(8): 2744–2752.
20. Lee S. H., Tekmen C., and Sigmund W. M. Three-point bending of electrospun TiO_2 nanofibers. *Mat Sci Eng A*, 2005, 398(1–2): 77–81.

4 Material Hardness and the Size Effect

Even as children we understand that things such as cake, cotton, and sponge are soft, while stones, metals, and bricks are hard. The most basic method for detecting hardness or softness is by the sense of touch. As materials scientists, however, we need to quantify the amount of hardness in order to have a systematic method of deciding where a hard material can be used, and where a soft one is more applicable. While the sense of touch does enable us to distinguish the soft from the hard, how do we use the mathematical language to distinguish them? This chapter will provide mathematical descriptions of hardness and describe various test equipment developed for measuring hardness. We will first focus on the commonly used macroscopic hardness testing methods, macrostatic indentation, macrodynamic indentation, and macroscratch tests for measuring macroscopic static indentation hardness, dynamic indentation hardness, and macroscopic scratch indentation such as Brinell hardness, Rockwell hardness, and Vickers hardness. At the same time, we will introduce methods for measuring microhardness and nanohardness in order to satisfy our need to know the hardness of microsized and nano-sized materials. Lastly, we will briefly discuss an interesting hardness phenomenon that occurs with nano-sized materials called the "size effect," from which we can gain valuable insight into the fundamental physics behind the macroscopic and microscopic mechanical properties of materials.

4.1 INTRODUCTION TO MATERIAL HARDNESS

4.1.1 Definition of Hardness

How do we describe the feeling of "soft" or "hard"? First, we will name it "hardness" so that we can quantify it. The higher the hardness value is, the harder the material will be, and vice versa. How do we define this "hardness" value? Currently, there is no overall consensus on the definition. From a practical point of view, it can be defined as "a measure of the degree that a material is capable of resisting deformation produced by another material." In deformation mechanics, hardness could be defined as "the ability to resist elastic deformation, plastic deformation, and fracture," or "the ability of a material to resist residual deformation and fracture." However hardness is defined, it is always measured by applying a certain amount of pressure on the surface of the solid sample being tested by using a harder object of definite shape and size, that is, an indenter. The measured hardness value not only depends on the elastic modulus, yield strength, tensile strength, and other mechanical properties but also is heavily influenced by the test instrument and the test conditions. Therefore, the hardness is a measure of bulk characteristics from the combined effects of local mechanical properties under a specific set of conditions. For example, if we indent and apply the same force on the surfaces of copper and iron with indenters of the same material properties and then measure the depth of penetration from each surface, we will find that the indentation depth taken from copper is a lot deeper than that of iron. This indicates that copper's deformation resisting strength is greater than that of iron, which means that copper is softer than iron and has a lower hardness value. In reality, this comparison also reflects the differences in elastic and plastic anti-resistance capabilities between the two metals.

4.1.2 Material Hardness Testing

The hardness of a material is determined by hardness testing. In order to compare the hardness between a wild variety of materials, various testing methods and hardness standards were developed. These include static indentation hardness, dynamic or rebound hardness, and scratch hardness. Static indentation hardness is determined by applying pressure on the sample surface using a ball—or diamond-shaped indenter. For a given applied load, a hardness value can be obtained from the area and depth of the indentation. Dynamic or rebound hardness uses a standardized object of a given weight and shape and drops it upon the sample surface from a given height. The height of the rebound is then measured and a hardness value is determined from that measurement. Scratch hardness testing applies a normal load on a tip with a radius of curvature. The tip is then dragged across the surface resulting in a scratch that is used to determine hardness. Depending on the applied load of the hardness tests, hardness values are separated into different categories: macrohardness (over 10 N in Japan, the United States, Russia, and over 2 N in the EU), microhardness (below 10–2 N but above 10 mN), and nanohardness (lower than 700 mN). During nanohardness tests, materials are observed to have increasing hardness with a decrease in size, showing signs of "size effect" [1–3] (see Section 4.9). These tests spurred new research into the accurate determination of hardness values, materials properties, and the relationship between the size and hardness of nanomaterials.

Different hardness measurements are not directly comparable due to different testing methods, testing conditions, and material responses to each test. Nevertheless, hardness values are widely used as a standard for quality assurance of mechanical performance in machine manufacturing, aerospace, and other modern industries. A number of factors make this possible: the simplicity of the testing method, the minimum requirement in sample geometry, and the relatively small damage area, as well as the relationship [4–6] between hardness and elastic modulus, shear modulus, tensile strength, and other materials' properties. Hardness measurements are the most common testing methods in materials testing and play a very important role in materials science research.

4.2 BRINELL HARDNESS

4.2.1 Brinell Hardness Measurement Method and Principles

The Brinell hardness test uses a hardened steel or carbide ball with a diameter D as an indenter tip, pressing into the sample surface with a load P. After a set time, the load is removed and an indentation is left on the surface of the sample (Figure 4.1). From the indentation diameter d, or indentation depth h, the area of the indentation A can be calculated. The average stress of the sample surface during the indentation P/A is defined as the Brinell hardness with units in kgf/mm^2. When the load and tip diameter is kept constant, a bigger indentation area will result in a lower Brinell hardness value, because that suggests the material has a lower resistance to deformation. Conversely, a higher Brinell hardness value implies a higher resistance to deformation by the material. Brinell hardness is commonly denoted as HB without units. For example, HB = 45 means the hardness is 45 kgf/mm^2. As seen in Figure 4.1, the indentation surface takes the shape of a half hemisphere, and the indentation area is $A = \pi Dh$. The Brinell hardness equation defines HB mathematically in the following formula:

$$\mathrm{HB} = \frac{F}{A} = \frac{F}{\pi Dh} = \frac{2P}{\pi D[D - \sqrt{D^2 - d^2}\,]}. \tag{4.1}$$

The Brinell hardness value could be represented in different ways, depending on the indenter material. When the indenter is made of hardened steel, the Brinell hardness is denoted by HBS. When the indenter is made of tungsten carbide, the hardness value is denoted by HBW. The hardness value is represented by a number preceding the indenter type. The numbers after the indenter type represent the testing conditions: the indenter diameter, load, and hold time. For example,

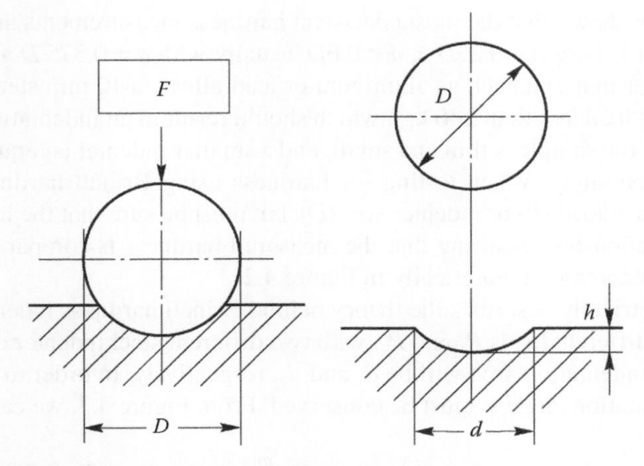

FIGURE 4.1 Diagram of the Brinell hardness measurement technique.

150 HBS10/3000/30 represents a material with 150 Brinell hardness tested using a 10 mm diameter steel tip with 3000 kgf (29.4 kN) over 30 s. As another example, 500 HBW5/750 means the material has an hardness value of 500 tested by a 5 mm tungsten carbide tip with 750 kgf (7.355 kN) over 10 ~ 15 s. A typical indentation time of 10 ~ 15 s is assumed if not explicitly denoted.

4.2.2 Other Measurement Considerations Related to Brinell Hardness

Under ideal conditions, repeated indentation on the same surface with the same indenter with increasing load would yield an increase in the indentation area, thus resulting in a constant Brinell hardness measurement. The difference in the effect of changing the load and the indenter tip should also be constant, in order to ensure adequate comparison between the hardness of different materials. In practice, the measured Brinell hardness from different applied loads is often different when testing on the same material. Figure 4.2 shows the results of the hardness value obtained from the same annealed steel using different applied loads. At small loads, the hardness value (HB) increases as the load (*P*) increases. At a certain critical loading range, the hardness hits a maximum and becomes independent of the load applied. If the load is further increased past this critical range, the hardness value will begin to decrease as the load increases. Therefore, for consistent hardness measurement, one must make the measurement using a load within the region shown in Figure 4.2: a region that is flat and independent of the applied load.

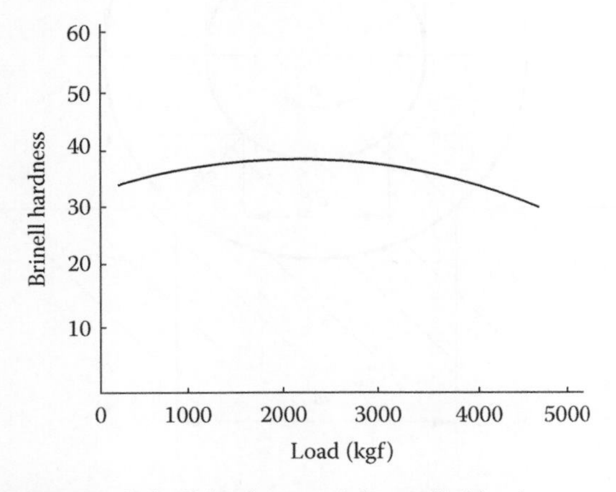

FIGURE 4.2 Relationship between Brinell hardness and the applied load.

Experiments have shown that the most consistent hardness measurements are obtained when the indentation diameter is between $0.25D < d < 0.6D$, usually with $d = 0.375\ D$ and penetration angle at $\Phi = 44°$. For softer material such as aluminum or lead alloys, a 10 mm steel indenter should be used with an applied load less than 240 kgf, which should result in an indentation diameter between $0.25D < d < 0.6D$. If the sample is thin and small, and a smaller indenter is required, the load should be reduced correspondingly. When testing for hardness using Brinell hardness indentation, one cannot use an arbitrary load (P) or indenter size (D), but must be sure that the load and indenter size have the correct relationship, ensuring that the measured hardness is comparable with other tests. This relationship is expressed theoretically in Figure 4.2.

Figure 4.3 geometrically describes the theory behind Brinell hardness indentation. For the same material under two different loads P_1 and P_2, with two different steel indenter diameters D_1 and D_2, the diameter of the indentation area will be d_1 and d_2, respectively. In order to have the same hardness value, the indentation angle φ must be conserved. From Figure 4.3, we can see that

$$d_1 = D_1 \sin\frac{\varphi}{2}. \tag{4.2}$$

Substituting Equation 4.2 into the Brinell hardness formula, we obtain

$$\mathrm{HB}_1 = \frac{F_1}{D_1^2}\frac{2}{\pi\left(1-\sqrt{1-\sin^2\frac{\varphi}{2}}\right)}. \tag{4.3}$$

Similarly, we may do the same for d_2:

$$\mathrm{HB}_2 = \frac{F_2}{D_2^2}\frac{2}{\pi\left(1-\sqrt{1-\sin^2\frac{\varphi}{2}}\right)}. \tag{4.4}$$

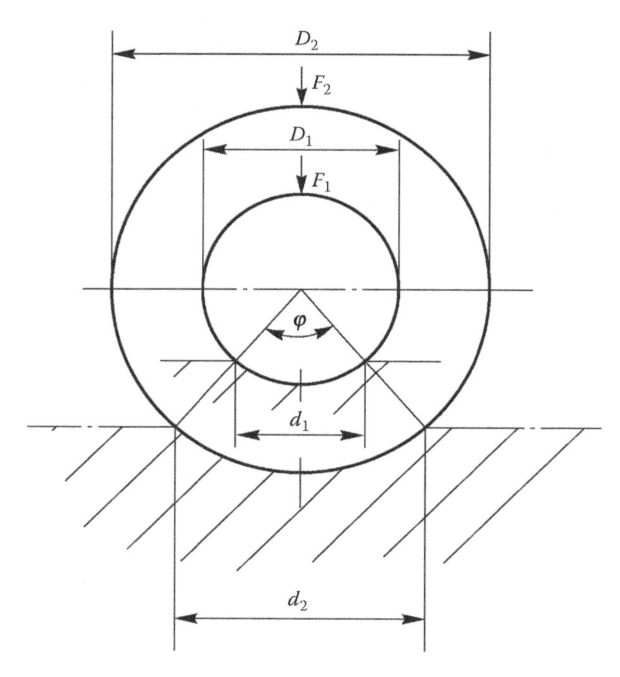

FIGURE 4.3 Geometric representation of indentation.

TABLE 4.1
Brinell Hardness (HB) and Relevant *K* Value

Material	HB	*K*
Cast steels	<140	10
	≥140	30
Copper alloys	<35	5
	35 ~ 130	10
	>130	30
Light metals and alloys	<35	1.25(2.5)
	35 ~ 80	5(10)(15)
	>80	10(15)
Lead, tin		1(1.25)

Therefore, for the same material with the same hardness, the following relation is obtained:

$$\frac{P_1}{D_1^2} = \frac{P_2}{D_2^2} = \frac{F}{D^2} = K. \tag{4.5}$$

As long as the ratio of the load to the square of the indentation diameter is constant, the Brinell hardness value is constant for the same material, making comparison between different materials possible. Different values of *K* are often used to ensure that the indentation angle is consistent for different materials and that the indenter size stays within $0.25D < d < 0.6D$ of the indentation diameter. National Standard GB231–84 describes seven different *K* values for Brinell hardness measurements of different materials, as shown in Table 4.1.

During Brinell hardness testing, the sample thickness should be at least 10 times larger than the indentation depth. If the sample thickness allows, a 10 mm indenter is preferred. Then, depending on the material being tested, consult Table 4.1 to choose the correct *K* value and choose the best load *P* for the test. After indentation, one must make sure that the indenter size and the indentation diameter fall within the $0.25D < d < 0.6D$ range; otherwise, another *K* should be chosen and the test redone.

During load application, the sample and the indenter should be perpendicular to ensure an even application of the load. Avoid any shock or vibration during the indentation. Loading time is also a function of the material being tested. The softer the material is, the more plastic deformation it will receive, and thus a longer loading time is required. For harder metals and alloys, the loading time is typically 10–15 s. Softer metals usually need 30 s of loading time. For soft materials with HB <35, a loading time of 60 s is often necessary.

4.2.3 Characteristics of Brinell Hardness Testing and Its Applications

Due to the large indenter size, load, and indentation area, the Brinell hardness value reflects the average bulk material characteristic and is not affected by local heterogeneity of the material. Therefore, Brinell hardness testing has good reproducibility with a small standard deviation of error, making it especially useful for grey cast irons and bearing alloys with large grain sizes and phases. Brinell hardness is most accurate for soft alloys such as pure aluminum, lead, and tin.

Because the indentation area is so large, the Brinell hardness test is not suitable for testing surface hardness or thin samples. If repeated testing is required during mass production, the limitation in minimum load time during hardness testing could increase the labor time and cost. For materials with very high hardness values, the accuracy of testing could be compromised due to deformation of the indenter tip. Therefore, steel indenter tips should only be used for materials with hardness value HB < 450; tungsten alloy tips could test materials with hardness as high as 650 HB.

4.3 ROCKWELL HARDNESS

4.3.1 Rockwell Hardness Testing Method and Principles

The Rockwell hardness test uses an indenter with a certain load to cause an indentation on the surface and directly measures the indentation depth as a representation of the material's hardness. The deeper the indentation is, the lower the hardness value becomes. To guarantee good contact between the indenter and the sample surface, a Rockwell hardness test preloads the indenter with 10 kgf (98.07 N), before applying the actual load. How much of the actual load is applied depends on the material hardness. The Rockwell hardness formula is defined as follows:

$$\mathrm{HR} = K - \frac{h_1 - h_2}{S}. \tag{4.6}$$

HR is the unit for Rockwell hardness and K is a constant describing the indenter tip. A diamond cone-shaped indenter tip has a K value of 100; a steel tip has a K value of 130; and h_2 is the indentation depth (mm) resulting from the preload of 10 kgf; h_1 is the indentation depth after the main load has been removed; and S is another constant with the value of 0.002 mm, which defines a unit depth for indentation.

Rockwell hardness test typically uses two different types of indenter: a diamond cone indenter with 120° angle and a steel indenter with a diameter of $\varphi = 1.588$ mm. By using different indenters with different loads, different Rockwell test standards are formed. Most widely used are the A, B, and C standards as shown in Table 4.2, with C being the most common. The hardness values of each standard are denoted by HRA, HRB, and HRC. For example, a Rockwell hardness of 60 made with the A standard would have units of 60 HRA.

Figure 4.4 describes the testing procedure using the Rockwell hardness C standard. As shown by Figure 4.4a, a 10 kgf preload was applied with an indentation depth of h_2. At this point, the dial is shown to be zeroed. A main load of 140 kgf is then applied, causing an indentation depth of h_3, and the dial is turned counterclockwise to the relevant hardness value (see Figure 4.4b). Under the main load, the surface of the metal both plastically and elastically deforms. After the load is removed,

TABLE 4.2
Common Rockwell Hardness Testing Standards and Applications

Standard	Hardness Unit	Indenter Type	Preload F_0(kgf)	Main Load F(kgf)	Hardness Range	Applications
A	HRA	120° angle diamond cone	10	50	20 ~ 88	Carbide, hardened steel sheet, thin layer of hardened steel surface
B	HRB	ϕ1.588 mm steel ball		90	25 ~ 100	Low-carbon steel, copper alloys, ferrite, soft cast iron
C	HRC	120° diamond cone		140	20 ~ 70	Hardened steel, hard cast iron, pearlite

Source: Han, D.W., *The Hardness of Metals and Its Testing Method*, Hunan Science & Technology Press, Changsha, China, 1983; Taken from Tables 3.1 through 3.3.

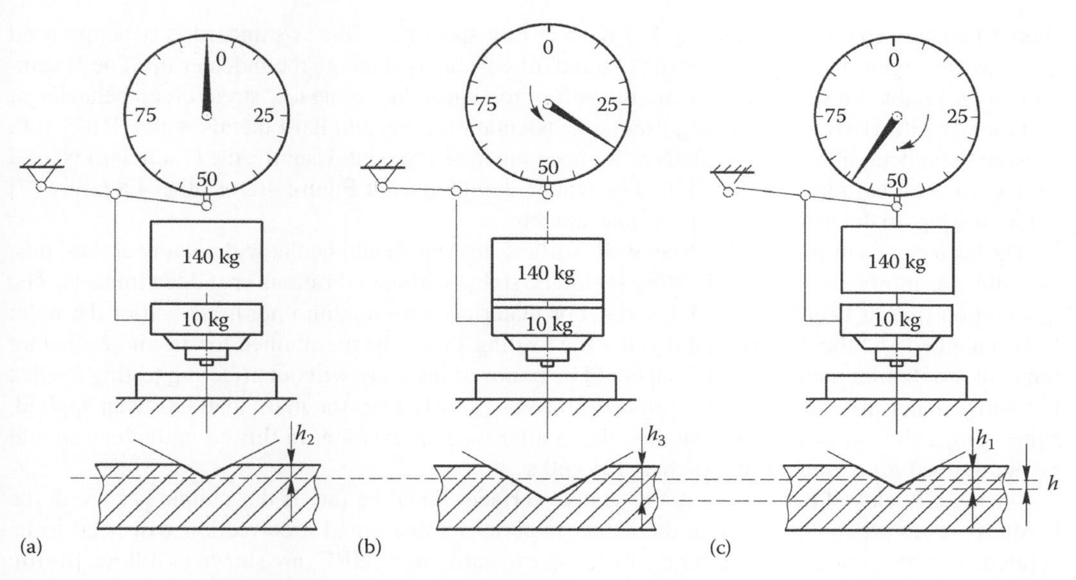

FIGURE 4.4 Rockwell hardness testing procedure. (a) A 10 kgf preload was applied with an indentation depth of h_2. (b) A main load of 140 kgf is then applied, causing an indentation depth of h_3. (c) Uninstall the main load, pointer to a reversal, ending with a depth of h_1.

the elastic deformation is recovered and the dial shows a corresponding decrease (see Figure 4.4c). The indentation depth at this point is defined as h_1. The final indentation depth after the test is $h = h_1 - h_2$. The constant K is defined as the highest hardness value for the Rockwell hardness test. As the indentation depth is increased by 0.002 mm, the hardness value is reduced by one unit. The amount of hardness unit that remains after the depth subtraction from K is the Rockwell hardness of the material. Therefore, the smaller the indentation depth, the less the value is subtracted from K, and the higher is the hardness value. For example, HRC and HRA both have K values of 100. When $h = h_1 - h_2 = 0.08$ mm, that is, 40 times 0.002 mm, the hardness value is calculated as

$$\mathrm{HRC(A)} = 100 - \frac{0.08\,\mathrm{mm}}{0.002\,\mathrm{mm}} = 60. \tag{4.7}$$

Rockwell hardness is defined as a unitless quantity and has a linear relationship with h. This property enables Rockwell hardness charts to be made and installed directly onto the testing machine, so that after the load has been removed, the hardness values can be read directly from the machine. The K value for diamond cone indenters and steel ball indenters are 100 and 130, respectively. The diamond cone indenter is usually used on hard materials with the HRC(A) standard, and it will therefore rarely reach the 0.2 mm depth and give a hardness of zero. The steel ball indenter is often used to test soft metals, where the indentation depth could exceed 0.2 mm, causing negative hardness values when K is defined as 100. To avoid that situation, the K for steel ball indenters is defined as 130, changing the Rockwell hardness formula to the following:

$$\mathrm{HRB} = 130 - \frac{h}{0.002\,\mathrm{mm}}. \tag{4.8}$$

Note the linear relationship between hardness and h is still maintained, and so the hardness value can still be read directly from the testing machine.

For Rockwell hardness testing, the sample thickness should be eight times the indentation depth. The sample's back surface should not have visible scratches or deformations. The correct standard should be determined before testing. For hard or thin specimens, the A standard is typically used because the larger load required for the C standard will likely damage the indenter tip. The B standard is unsuitable for materials that are too soft or too hard, due to the low stress creep behavior of soft metals with HRB < 25, causing excessive indentation time, and hard metals with HRB > 100, causing the indentation to be too shallow for accurate measurement. Usually, the C standard is used for metals with hardness HRB > 100. The reader should consult Figure 4.2 or other literature [7] before testing, to determine the appropriate standard.

The surface of samples used in Rockwell hardness testing should be flat without any cracks, pits, or visible machining marks. The loading should be stable without vibrations or sudden impacts. The indentation time is usually around 4 ~ 6 s. For materials with medium hardness—after the main load is applied and the dial has stabilized—the loading should be maintained for around 2 s before removal. For harder materials, the load could be removed instantly without affecting testing results. For softer materials, the dial will continue to increase slowly after the main load has been applied, suggesting a slow increase in indentation depth after load application. In this case, the load should be maintained for a longer time, such as 10 ~ 60 s.

During a Rockwell hardness test, the sample surface should be flat. If the sample is curved, the hardness result will underestimate the actual material hardness and a correction will need to be applied. The correction formulae for the difference in hardness (ΔHRC) are shown as follows [8–10]:

Cylindrical surface:

$$\frac{\Delta HRC = 0.06(100 - HRC')^2}{D}. \tag{4.9}$$

Spherical surface:

$$\frac{\Delta HRC = 0.012(100 - HRC')^2}{D}. \tag{4.10}$$

HRC is the hardness value tested on the surface of the curved surface, and D is the diameter of the sphere or cylinder.

4.3.2 Advantages and Disadvantages of Rockwell Hardness and Its Application

Rockwell hardness testing is simple and quick to perform, with a hardness value that can be read directly from the testing machine. Therefore, it is efficient and widely used in materials testing during mass production. In addition, the loading conditions for Rockwell hardness are smaller compared to Brinell hardness and are less likely to damage the sample surface, making it usable as a quality control for semi-finished or finished products. Because of preloading, Rockwell hardness can reduce the variation in test results due to uneven surface roughness. Moreover, Rockwell hardness testing uses both diamond and steel indenter tips and allows a wide range of loading to be applied for different samples. Therefore, Rockwell hardness testing can be applied for a very wide range of materials of different hardness.

The disadvantages of Rockwell hardness testing come from its inability to compare material hardness between different standards. Because the indentation area is very small, it is also very sensitive to localized heterogeneity in the material, causing it to have bad repeatability over different areas of the material. Therefore, Rockwell hardness is not suitable for testing inhomogeneous material properties.

4.3.3 Rockwell Surface Hardness

The Rockwell surface hardness test was developed based on the conventional Rockwell hardness testing. The basic principles behind the two tests are nearly identical. However, because

TABLE 4.3
Rockwell Surface Hardness Standard, Test Conditions, and Applications

Standard	Hardness Unit	Indenter Type	Preload F_0/kgf	Main Load F/kgf	Hardness Range	Applications
15N	HR15N	120° diamond	3	12	70 ~ 90	Carburized steel, nitride steel,
30N	HR30N	cone		27	42 ~ 86	thin steel plate, cutting edge,
45N	HR45N			42	20 ~ 77	the edge of some parts, surface coatings
15T	HR15T	ϕ1.588 mm	3	10	67 ~ 93	Low-carbon steel, aluminum,
30T	HR30T	steel ball		27	29 ~ 82	and other copper alloy sheet
45T	HR45T			42	1 ~ 72	

conventional Rockwell hardness testing uses a fairly heavy load, it is not suitable for thin materials or materials whose surfaces have been heat treated or chemically heat treated. Rockwell surface hardness is defined using the same principles and indenters: 120° diamond cones and 1.588 mm diameter steel balls, but lowering the loading conditions for Rockwell hardness. Instead of the conventional 10 kgf preload, surface hardness uses 3 kgf. The main load is also reduced from 60, 100, and 150 kgf down to 15, 30, and 45 kgf. The formula for calculating the hardness is still the same as Equation 4.6, but S is redefined from 0.002 to 0.001 mm, and K is still constant at 100.

Similar to conventional Rockwell hardness, the Rockwell surface hardness value is still defined as HR. However, to differentiate the two tests, additional symbols are used after the HR to denote a surface hardness test. For example, when doing a surface hardness test using diamond tip with 15 kgf main load, the hardness should be represented as HR15N (note: the "N" does not denote Newton, but denotes a diamond indenter tip). If a 30 or 45 kgf force is used, the hardness notations are correspondingly HR30N or HR45N. When using a 1.588 mm steel ball indenter with 15, 30, or 45 kgf, the hardness is denoted as HR15T (note: "T" stands for steel ball tip), HR30T, or HR45T. For example, a hardness of 45HR30N means a hardness of 45 by a Rockwell surface hardness test using a diamond cone indenter with a 30 kgf load. With these notations, Rockwell surface hardness has a total of six different standards as described in Table 4.3 [7].

4.4 VICKERS HARDNESS

4.4.1 Vickers Hardness Principles and Methods

Vickers hardness testing uses a 136° diamond pyramid shaped indenter to apply a static load into the sample surface. After the static load has been applied over a certain amount of time, the load is removed, and the diagonal length of the diamond indent is measured to calculate the indentation area in relation to the load applied, giving the Vickers hardness value in units of kgf/mm^2.

The indentation area is measured by the diagonals of the diamond-shaped indent as shown by Figure 4.5. When the diamond indent is pressed into the surface, the area is defined as

$$S = \frac{d^2}{2\sin 68^\circ}. \tag{4.11}$$

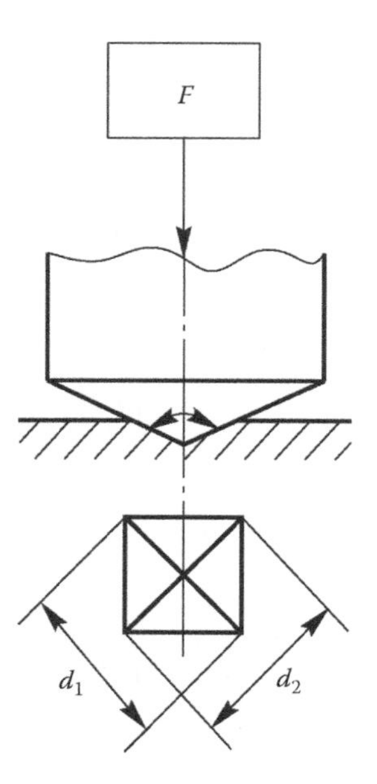

FIGURE 4.5 Vickers hardness testing principle.

By definition, the Vickers hardness value will be defined as

$$\text{HV} = \frac{P}{S} = \frac{2P \sin 68^\circ}{d^2} = 1.854\frac{F}{d^2}. \tag{4.12}$$

where

HV is the symbol for Vickers hardness (the units are kgf/mm^2, but it is usually shown as unitless)
F is the load applied (kgf)
S is the indentation area (mm^2)
d is the average diagonal length of the indent (mm)

During Vickers hardness testing, six different loads are possible: 5, 10, 20, 30, 50, and 100 kgf. By plugging in the known quantities of P and d, HV can be determined. Vickers hardness notation is identical to Brinell hardness notation. For example, 640 HV30/20 denotes a hardness of 640 with 30 kgf for 20 s.

Vickers hardness was developed on the basis of Brinell hardness and Rockwell hardness testing. Vickers hardness improved the indenter material and indenter tip design. By using a pyramid shape with a constant surface angle α, F/d^2 will always remain constant for the same hardness. If the applied load is changed, the shape of the indentation will also stay the same. From Equation 4.11, the indentation area has a linear relationship with d^2. Therefore, Vickers hardness can use any load when testing a homogenous material and still get the same hardness value. The 136° indenter angle is chosen because it is more comparable with Brinell hardness. For Brinell hardness testing, when $d = 0.375D$, the hardness is most unaffected by the loading conditions, and the indentation angle is 44°, making the outer angle of the ball indenter equivalent to 136°. Vickers hardness uses a pyramidal indenter with an angle of 136°, causing each indent to have an indentation angle of 44°C. Both Vickers hardness and Brinell hardness calculate the area of the indent to reflect the hardness value, and if the tests are done at the same indentation angle ($\alpha = 44°$), the resistance to plastic deformation on the surface will be the same for both tests. Because of the similarity in the indentation angle, the results from Vickers hardness and Brinell hardness testing will be very similar for a wide range of hardness values. For example, for hardness values less than 400, HV is roughly the same as HB.

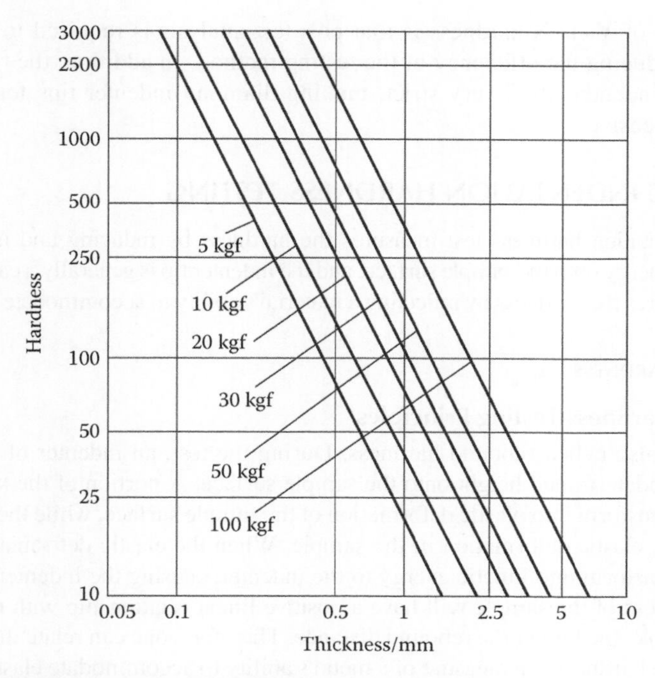

FIGURE 4.6 Relationship between load, hardness, and minimum thickness.

Vickers hardness testing is especially suited for hardened surfaces or thin samples. The hardened surface and sample surface should have a minimum thickness of 1.5 d. If the hardened surface thickness cannot be determined, begin with the minimum load and increase the load slowly. If a dramatic drop in hardness is seen upon a load increase, then a smaller load should be used until two consecutive loads yield the same hardness value. If a prediction of the hardness is available, Figure 4.6 can be consulted to pick the proper load. If a sample thickness is large enough, a higher load should be used to minimize the surface effects and the error in determining the diagonal of the indentation area, thus increasing the accuracy. For materials with HV > 500, the indentation load should not exceed 50 kgf in order to avoid damaging the indenter tip.

Similar to Brinell hardness and Rockwell hardness, loading during the Vickers hardness test should be stable and without impact. The loading time for steels are typically 10 ~ 15 s and 30 ± 2 s.

4.4.2 Vickers Hardness Characteristics and Applications

Vickers hardness testing uses a pyramid-shaped diamond indenter tip and it will yield the same hardness value regardless of the applied load. Therefore, unlike Brinell hardness, Vickers hardness testing is not restricted by the applied load P, nor by the indenter size D. Vickers hardness has a wide range of applications over both soft and hard materials, and it does not share the Rockwell hardness disadvantage of having multiple standards that make comparison of hardness values difficult. Vickers hardness is also very good at determining the hardness of thin samples and sample surfaces, and it is therefore often used to test the hardness of surface layers or instrument components. It is especially suited for testing samples with various surface hardening treatments such as nitriding, carburizing, vanadizing, boriding, and a variety of surface coatings. In addition, because the indentation from Vickers hardness leaves a clear diamond shape, the accuracy of measurement is greatly increased. When material hardness is less than 450 HV, Vickers hardness will share the same hardness value as Brinell hardness.

A disadvantage of Vickers hardness is that post-test analysis is required to obtain a hardness value, therefore reducing the efficiency of the testing method. In addition, the requirement for the Vickers hardness indenter tip is very strict, making diamond indenter tips for Vickers hardness machines very expensive.

4.5 DYNAMIC INDENTATION HARDNESS TESTING

The dynamic indentation hardness test measures the hardness by inducing and measuring a certain amount of kinetic energy onto the sample surface, and the indenter tip is generally a carbide ball or sphere. This method measures the hardness by reflecting a material's ability to accommodate elastic deformation.

4.5.1 Shore Hardness

4.5.1.1 Shore Hardness Testing Principles

Shore hardness is also called rebound hardness. During the test, an indenter of a certain weight is dropped from a predetermined height onto the sample surface. A portion of the kinetic energy from the indenter will transform into plastic deformation of the sample surface, while the rest will be instantaneously stored as elastic deformation in the sample. When the elastic deformation is recovered, it will once again transform into kinetic energy to the indenter, causing the indenter to rebound off the surface. The hardness of the sample will have a positive linear relationship with the rebound height: the harder the sample, the higher the rebound distance. Therefore, one can relate the rebound height to the hardness. Shore hardness is a measure of a metal's ability to accommodate elastic deformation and reflects the ratio between its elastic deformation and plastic deformation at a fixed energy. Shore hardness is defined by the initial height h_0 and the rebound height h, as shown in the following formula:

$$\mathrm{HS} = \frac{Kh}{h_0}. \tag{4.13}$$

where

HS is the shore hardness value and is unitless

K is the shore hardness constant: For C type shore hardness (optical measurement using a ruler scale), $K = 10^4/65$; for D type shore hardness (height measurement using a dial indicator), $K = 140$

No posttest analysis is needed for shore hardness, as the hardness value can be directly read from the ruler or dial.

Shore hardness testing requires the sample surface to have a roughness of better than 8. The sample surface should be completely flat with an even thickness of over 2 mm, which will reduce the chance of the bulk sample plastically deforming under impact. For testing of thin samples, a bulk piece of steel over 400 g can be soldered to the back of the sample to provide the necessary structural support for a shore hardness test. Details concerning the experimental method of shore hardness testing can be found in the literature [11].

4.5.1.2 Shore Hardness Characteristics and Applications

A shore hardness tester is a portable and lightweight machine. Its mobility, simple design, ease of operation, and high testing efficiency makes it especially suited for the on-site testing of large work pieces such as machine crankshafts, guides, or gears. The disadvantage of this method is the lower accuracy in comparison to static hardness testing, with a higher standard deviation between the same tests, as well as the inability to compare hardness values between materials with severely different K values. Moreover, the shore hardness test is unable to ensure a complete vertical drop and adequate surface roughness of big machine parts, increasing the testing error.

4.5.2 Brinell Hammer Test

4.5.2.1 Brinell Hammer Testing Principles

The Brinell hammer test uses a human hammering force to calculate the hardness of a material. To calculate the hardness value, it compares the indentation area of a quenched steel ball bearing and a standardized rod hammering into the sample surface. It uses the symbol HBO to denote hardness. Under the same loading conditions, the indentation area of a steel ball and a rod are S_O and S_B, respectively. From the Brinell hardness principle, the sample hardness is HBO = P/S_O, and the standardized rod hardness is HBB = P/S_B, giving the equation

$$\text{HBO} = \text{HBB} \cdot \frac{S_B}{S_O} = \text{HBB} \cdot \frac{D\left(D - \sqrt{D^2 - d_B^2}\right)}{D\left(D - \sqrt{D^2 - d_o^2}\right)}. \quad (4.14)$$

The equation can be simplified to

$$\text{HBO} = \text{HBB} \cdot \frac{d_B^2}{d_o^2}. \quad (4.15)$$

where

D is defined as the indenter diameter
d_o is the indentation diameter
d_B is the indentation diameter from the standardized rod

During Brinell hammer testing, the hardness of the standardized rod (HBB) is a known quantity. The hardness can therefore be determined by obtaining the ratio of the indentation diameters.

4.5.2.2 Brinell Hammer Hardness Testing Characteristics and Applications

The Brinell hammer test has the advantage of small size, simple design, cheap production, and ease of transport. The disadvantage is that the elasticity of a standardized rod might not be applicable to the elasticity of the sample. In certain cases where the difference in hardness of the material is dramatic, the test will not satisfy the linear *P–D* relationship, increasing the error of the testing method.

The Brinell hammer test excels in testing large machinery, forgings, racks, or small parts already assembled or stocked that are hard to determine with conventional hardness testing methods. For example, materials in steel mills, such as tempered steel parts, cast parts, copper, aluminum, magnesium, lead, and tin alloys, are all tested in stock using the Brinell hammer test. However, because the indenter head is a hardened steel ball, the Brinell hammer test is not suitable for testing materials with high hardness values.

4.6 SCRATCH TESTING FOR MATERIALS' HARDNESS

4.6.1 Testing Principles and Theoretical Formulae

4.6.1.1 Mohs Hardness

Mohs hardness measures hardness by testing a material's resistance to surface scratches. There are 10 different standards for Mohs hardness, categorized into 10 levels from soft to hard (see Table 4.4). If a material's hardness cannot be expressed using the scratch of a material level n but can only be expressed using the scratch of a material level $n + 1$, then the hardness of the material lies between the two standard materials, $((n + 1)/2)$. The application for Mohs hardness has expanded with time, and the numbers of standard levels have also increased. For Mohs hardness results for pure metals, see Table 4.5.

TABLE 4.4
Mohs Hardness

Material	Hardness	Material	Hardness	Material	Hardness
Talc	1	Apatite	5	Topaz	8
Plaster	2	Feldspar	6	Corundum	9
Calcite	3	Quartz	7	Diamond	10
Fluorite	4				

Source: Han, D.W., *The Hardness of Metals and Its Testing Method*, Hunan Science & Technology Press, Changsha, China, 1983; Taken from Table 7.1.

TABLE 4.5
Mohs Hardness for Metals

Metal	Mohs Hardness	Metal	Mohs Hardness	Metal	Mohs Hardness
Cesium	0.2	Cerium	2.5	Palladium	4
Sodium	0.4	Gold	2.5	Platinum	4.3
Potassium	0.5	Zinc	2.5	Nickel	5
Lead	1.5	Magnesium	2.6	Manganese	6
Germanium	1.5	Aluminum	2.7	Molybdenum	6
Tin	1.8	Antimony	2.9	Iridium	5 ~ 6
Bismuth	1.8 ~ 1.9	Copper	3	Tungsten	6.5 ~ 7.5
Cadmium	2	Iron	3	Tantalum	7
Calcium	2.2 ~ 2.5	Silver	4	Chromium	9

Source: Han, D.W., *The Hardness of Metals and Its Testing Method*, Hunan Science & Technology Press, Changsha, China, 1983; Taken from Table 7.2.

Because the Mohs hardness variation at high hardness levels is rather large for the 10 standard hardness system, the standards have now increased to 15 levels [12].

4.6.1.2 Martens Scratch Hardness

When a standardized indenter tip is depressed into the sample surface and then moved, it will leave a scratch of a certain width and depth. Martens scratch hardness testing uses that as a way to reflect the hardness of the sample. The hardness value is represented by the load that is needed to achieve a scratch of a certain width and depth on a sample surface. Using a diamond cone with a 90° angle, the formula for the load necessary to achieve a 10 μm wide scratch is as follows:

$$H_m \approx \frac{F}{b}. \tag{4.16}$$

where

F is defined as the vertical loading (g)
b is the scratch width (mm)
H_m has units of gf/mm

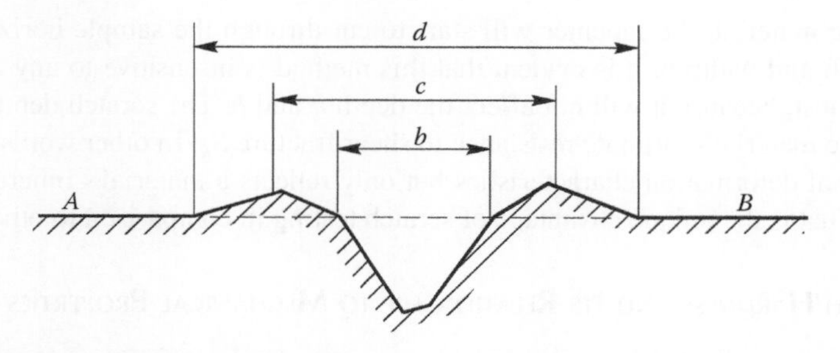

FIGURE 4.7 Cross section of scratched surface.

Research has shown that the Martens scratch hardness test is more accurately represented by the square of the width of the scratch [10], and the equation should be written as follows:

$$H_m \approx \frac{F}{b^2}, \tag{4.17}$$

H_m in this case having units of gf/mm^2.

Because a large amount of plastic deformation is necessary to induce a scratch onto the sample surface, there will be ridges on the sample near the scratch (see Figure 4.7). To measure the scratch width, one should not measure outside of the ridge c, nor from the top of ridge d, but only the width of the scratch b.

4.6.2 Scratch Process and Analysis

When an indenter is depressed into the sample surface, the indentation depth is defined as h_c (see Figure 4.8). When the tip is moved across the surface, a shear stress P_h is applied in addition to the normal stress P. The sample surface is under a total stress of P_d, which is the vector sum of P_h and P. The angle between the normal stress to the indenter cone surface and P_d vector is defined as α. The indenter cone will slip up when P_h is increased so that α becomes greater than the friction angle and overcomes the dynamic friction between the indenter and the sample. After the cone slips up from the indentation, the area of contact between the sample surface is reduced, and P_d is consequently increased enough to cause plastic deformation and strain hardening. When the pressure is increased beyond the yield

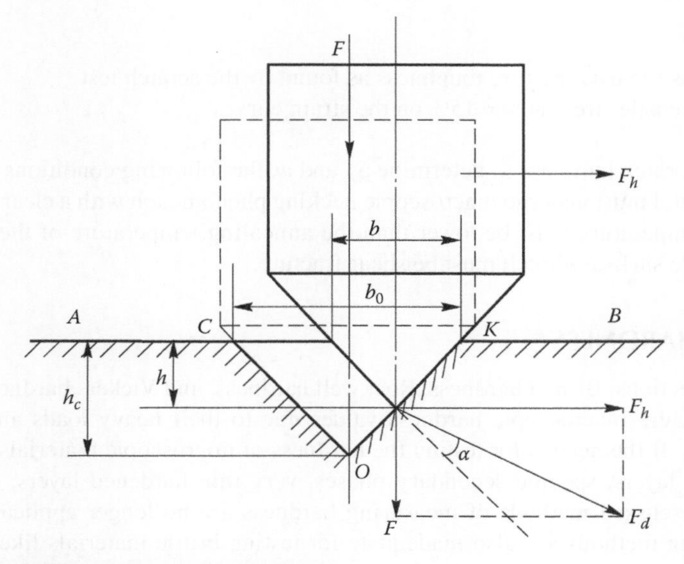

FIGURE 4.8 Scratch hardness testing principles.

strength σ_k of the material, the indenter will start to cut through the sample horizontally, causing a scratch of depth h and width b. It is evident that this method is insensitive to any deformation hardening before the test, because it will not affect the depth *h* and *b*. The scratch depth h and width b is only related to the material's ultimate resistance to shear fracture S_K. In other words, this test does not reflect any external deformation characteristics but only reflects a material's inherent microstructure characteristics. That is the unique advantage of scratch testing in comparison to other hardness tests.

4.6.3 Scratch Hardness and Its Relationship to Mechanical Properties

Scratch hardness can reflect other mechanical properties such as fracture toughness and reduction of area. From the discussion in scratch testing theory, there is a strong relationship between scratch hardness and true fracture toughness S_K. This relationship for lead, copper, and iron can be described by the following formula [10]:

$$S_K = 4.02\left(\frac{1}{b_{50}} - 22.5\right), \tag{4.18}$$

where

b_{50} denotes the scratch width from a 90° angle cone indenter under a 50 g load
S_K is in units of kgf/mm^2

If the Rockwell hardness indenter, a 120° diamond tip, is used under 8 kgf, then the formula becomes

$$S_K = \frac{86}{b} - 136, \tag{4.19}$$

S_K is still in units of kgf/mm^2.

When using the Rockwell hardness diamond indenter for scratch hardness, the reduction in area and the scratch width share the following relationship:

$$\psi = \left(15 + \frac{S_K - S_{II}}{1.30 - 0.004S_K}\right)\%. \tag{4.20}$$

where

S_K is defined as the true fracture toughness as found by the scratch test
S_{II} is the true tensile stress at $\psi = 15\%$ on the strain curve

When using scratch hardness to determine S_K and ψ, the following conditions must be satisfied: (1) the sample metal must undergo macroscopic necking phenomenon with a clear reduction in area, (2) the testing temperature must be lower than the annealing temperature of the sample material, and (3) the sample surface scratch must be shear fracture.

4.7 MICROHARDNESS

In the previous sections, Brinell hardness, Rockwell hardness, and Vickers hardness were discussed as a way to measure macroscopic hardness values due to their heavy loads and relatively large indentation areas. If the need is for testing the hardness of microscopic materials—such as certain grains, diffusion layers, specific secondary phases, very thin hardened layers, or the like—then the previously discussed methods of measuring hardness are no longer applicable. Furthermore, these three testing methods are also inadequate for testing brittle materials like ceramics, due to the ease of fracture under heavy loading. Microhardness testing methods provided the solution for these applications and are now widely used in industry. In the twenty-first century, advancements in microtechnology and nanotechnology are accelerating in the fields of materials science and electric science, which increases the demand for more and more microhardness testing [13]. Microhardness testing is any hardness test method that requires the aid of a microscope [14]. The loads used are often very small (maximum of 10 or 2 N, minimum 10 mN), resulting in very small indents.

Depending on the different indenters used, microhardness testing is usually separated into Vickers microhardness, Knoop microhardness, and Double-cone microhardness. The most common are the Vickers and Knoop microhardness tests, and these are the two that will be focused in this chapter.

4.7.1 Microhardness Testing Principles

During microhardness testing, a pyramid-shaped diamond indenter with a 136° angle or a 172.5° Knoop indenter is depressed into the sample surface on the relevant features, held for a specific time, and, after the load is removed, the indentation cross diameter (d) is measured to obtain the indentation area (the Knoop test measures the projected area). Calculation obtains the relationship between the indentation area to the applied load, and a hardness value is acquired from that relationship using the hardness charts.

When a 136° indenter tip is used, the hardness values are represented as HV, because the governing principle for such a hardness test is exactly the same as the Vickers hardness test with the exception of smaller loading conditions. Therefore, it is necessary for the load to be defined after the HV symbol. For example, 340 HV0.1 denotes a hardness of 340 from a Vickers hardness test under a load of 0.1 kgf. Similarly, 340 HV0.05 denotes a hardness of 340 from a Vickers hardness test performed with a load of 0.05 kgf.

Using an indenter tip with an angle of 172.5° is denoted with HK, and the equation for hardness is

$$\mathrm{HK} = 14.21\frac{F}{l^2}, \tag{4.21}$$

where

HK denotes Knoop hardness (units of kgf/mm^2, but often shown as unitless)
F is the applied load during testing (kgf)
l is the length of the diagonal of the indentation area in mm

4.7.2 Knoop Hardness Characteristics

As seen from Figure 4.9, the Knoop hardness indenter is an elongated pyramid shape with an angle of 172.5°, with the indenter angle at the short diagonal side being 130°. Therefore, the long diagonal of the indent is seven times as long as the short diagonal, making it possible to just measure the longer

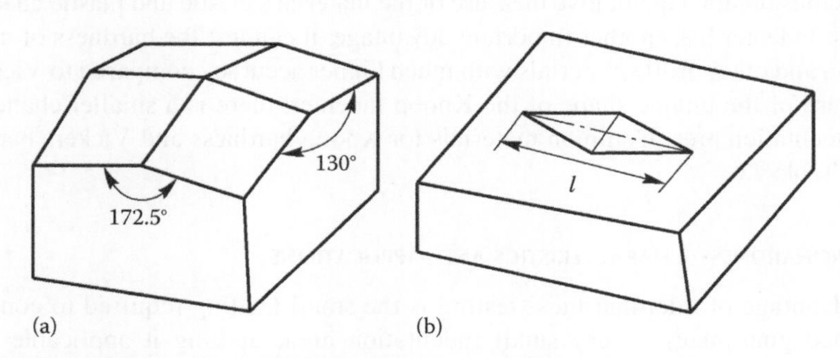

FIGURE 4.9 Knoop indenter geometry. (a) Indenter shape. (b) Indentation shape.

TABLE 4.6
Knoop Hardness and Vickers Hardness Comparison

Material	Heat Treatment	HK	HV
1.1% carbon steel	775°C water quench 150°C tempering	700 ~ 800	810
	775°C water quench 350°C tempering	550 ~ 580	580
	775°C water quench 450°C tempering	460 ~ 500	455
	850°C annealing	280	240
0.1% carbon steel	800°C water quench 150°C tempering	600 ~ 700	690
0.2% carbon steel	Cold work	260	240
QBc2.25	Solution treatment and aging	496	400
	Solution treatment and cold work	175	180
H65	Cold drawn	145	140
	Annealing	80	80
Industrial pure nickel	Annealing	125	125
Hard aluminum	Age hardening	120	125
Pure aluminum	Cold work	35	30
	Annealing	24	19
Pure zinc	As cast	34	28

Source: Han, D.W., *Handbook for Technology of Hardness Determination of Metals*, Central South University Press, Changsha, China, 2003; Taken from Table 7.5.

diagonal length, and increasing the accuracy and precision of the measurement. If the same load is applied for HV and HK, HK will have a shallower indent, making it more applicable for determining the hardness of thin layers. Because Knoop hardness uses a pyramidal indenter to press into the sample surface, it can also test the nonelastic recovery effect on microhardness. When the load is removed from the sample, the indentation will usually be reduced in size and changed in shape, due to the elastic recovery of the material. This elastic recovery is dependent on both the material characteristics and the shape of the indenter. When the load is removed (because of the special design of the Knoop indenter), elastic recovery will affect the short diagonal, but the effect on the long diagonal length is essentially negligible. Since Knoop hardness is calculated independently of elastic recovery, it has a different physical meaning compared to Vickers hardness. In addition, Knoop hardness can measure the ratio of the short diagonal (with elastic recovery) and the long diagonal (without elastic recovery) and thus obtain a qualitative measure of the material's elastic and plastic characteristics.

The Knoop indenter has another important advantage: it can test the hardness of materials such as glass, agate, and other brittle materials with much higher accuracy compared to Vickers hardness testing. Because of the unique shape of the Knoop indenter, there is a smaller chance of fracture around the indentation area. Common materials for Knoop hardness and Vickers hardness testing are shown in Table 4.6.

4.7.3 Microhardness Characteristics and Applications

The major advantage of microhardness testing is the small loading required to conduct the test. The small load guarantees a very small indentation area, making it applicable for hardness measurements in localized areas of interest, without causing excessive damage to the material.

Another advantage of a microhardness test is its high sensitivity. Microhardness testing is already widely used in metallurgy, machine manufacturing, precision instrumentation, and other industrial sectors. It is becoming increasingly popular in materials science research, metallography, metallurgy, and materials physics research as well.

As a method for quality control testing, microhardness is especially suited for determining the hardness of heat-treated surface layers, diffusion layers, thin oxide films, or thin metals such as spring wires, miniature bearings, pivots, foils, and the like. In addition, microhardness testing is also very common in determining the hardness of different phases in the same bulk material, identifying the nature of the different phases (e.g., chromium carbides have dramatically different hardness than the iron chromium matrix in binary alloys). For carbide research, when alloy compositions and production process changed, microhardness testing is used to determine the hardness of titanium, manganese, tungsten, and other carbide formers to analyze their effect in bulk alloy materials. In metallurgy and mechanical behavior research, microhardness is useful for the study of intergranular segregation, aging, diffusion, phase transformations, the chemical composition of inhomogeneities, brain-boundary effects, and impurities affecting lattice distortions on a microscopic level. It is also useful for studying the causes of variation in hardness and composition of solid–solution alloys of different elements. Lastly, the indentations themselves can serve as coordinate markings prior to other materials' analysis such as electron microscopy, probes, and high temperature alloy surface analysis.

4.8 NANOHARDNESS

With modern surface engineering (vapor deposition, sputtering deposition, ion implantation, high-energy surface modification, and thermal spraying) advances in electronics, MEMS, bio-engineering, and medical materials, the thickness of surface layers that can be modified is becoming thinner and thinner. The macrohardness and microhardness techniques described in previous chapters cannot test thicknesses less than 10 μm under the best conditions even with the aid of electron microscopes and so are not adequate for nanotechnology applications. Therefore, to satisfy that need, nanohardness testing methods have been developed. The most common nanohardness testing methods are nanoindentation and nanoscratch hardness.

4.8.1 NANOINDENTATION

Nanoindentation works on the principle of depressing an indenter tip into the material surface with a predetermined pressure and by analyzing the relationship between the indentation depth, the applied load, and the indenter geometry, in order to calculate the indentation area during the loading and unloading phases. This method is also called depth hardness testing.

Elastic–plastic materials are used in this chapter to describe the principles and methods behind nanoindentation. When a material is subjected to applied pressure from the indenter tip, it undergoes both elastic and plastic deformations. The deformation of an elastic–plastic material under loading can be described as a summation of the elastic deformation and the plastic deformation. Therefore, the total deformation depth h can be defined as the sum of the elastic depth h_e and the plastic depth h_p. From nondimensional analysis, it is found that the load is proportional to the square of the indentation depth whether the material is plastic, elastic, or elastic–plastic [15,16]. Therefore, during the loading phase, the relationship between the load P and the square of the indentation depth h follows a linear relationship for elastic–plastic materials. This suggests a parabolic relationship between load and depth during the loading phase. (This is confirmed by nondimensional analysis, but often not observed experimentally. Discussion on why this phenomenon occurs is encouraged.) During the unloading phase, plastic deformation becomes permanent and does not recover, while elastic deformation will recover. Therefore, the relationship between the load and the indentation depth is also parabolic (see Figure 4.10). For this description, h is the final indentation depth, P_{max} is the maximum load, and h_{max} is the maximum indentation depth.

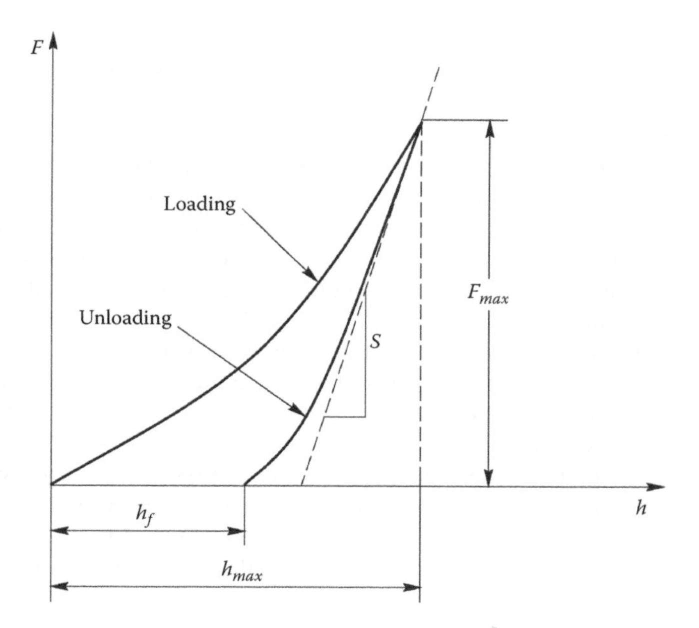

FIGURE 4.10 Relationship between load and indentation depth.

FIGURE 4.11 Depression and elevation during indentation.

In order to determine the hardness values from loading versus displacement curves, the elasticity of the material and indentation area must be clearly defined. Because the material around the indenter also deforms during an indent, it will cause a depressed curvature during loading, and an elevated curvature (due to recovery) during unloading (see Figure 4.11). This depression and elevation cannot easily be taken into account in the indentation area calculation using h_f and must be dealt with empirically. This is a major difference between nanoindentation and static macroindentation methods.

Currently, the most widely used method for determining the indentation area is called the Oliver–Pharr method [17]. This method utilizes the relationship between the load and the displacement at the top of the unloading curve to determine experimental constants relevant to that specific loading test:

$$P = B(h - h_f)^m. \tag{4.22}$$

B and m are the empirical constants from the exponential fit. With these constants known, the elastic modulus can be determined by taking the derivative of Equation 4.21 with respect to depth:

$$S = \left[\frac{dP}{dh}\right]_{h=h_{max}} = Bm(h_{max} - h_f)^{m-1}. \tag{4.23}$$

By extrapolating the unloading curve with a correction factor, the effective indentation depth h_e can be determined:

$$h_e = h - \varepsilon \frac{F}{S}. \tag{4.24}$$

Here, ε is a constant dependent on the indenter geometry. For a ball indenter or a pyramidal Berkovich indenter, $\varepsilon = 0.75$; for a conical indenter, $\varepsilon = 0.72$; and when the indenter tip is flat, $\varepsilon = 1$. After this correction factor, the error in testing can be minimized. When using nanoindentation methods to test for material hardness, the hardness value can be obtained using the following formula:

$$H = \frac{F_{max}}{A}, A = Kh_e^2. \tag{4.25}$$

where

H is the hardness value, with units of GPa
F_{max} is the maximum load with units of μN
A is the projected area of the effective indentation depth with units of μm^2
h_e is the effective indentation depth with units of μm
K is a constant dependent on material and indenter tip

4.8.2 Nanoscratch Hardness

Nanoscratch hardness testing operates on the same principle as conventional scratch testing. It utilizes the normal and tangential load applied by the indenter at a fixed speed to create a scratch on the sample surface and determines the hardness by measuring the final scratch depth and width. The only difference is that nanohardness uses a smaller load, usually between 1 ~ 100 mN.

Nanoscratch hardness can be calculated in three different ways: (1) determining the ratio between the normal force applied by the indenter using the axial projection of the indenter with the sample surface area, (2) determining the normal force applied by the indenter with the total area of contact between the indenter and the sample, and (3) determining the total work applied by the indenter with the total deformation area from the scratch. The most common method is the first one, described by the following formula:

$$\mathrm{HP} = \frac{F(N)}{S_{proj}(\mathrm{mm}^2)}. \tag{4.26}$$

This can be simplified to have the hardness value be a function of the applied pressure and the scratch width:

$$\mathrm{HP} = \frac{FA}{b^2}. \tag{4.27}$$

where

A is defined as a constant related to the indenter geometry
F is the applied load
b is the width of the scratch

Nanoscratch hardness is designed to study materials' resistance to normal pressure between 1 ~ 100 mN. It is usually used in tandem with nanoindentation hardness. The normal force and depth of the scratch test is recorded using high-resolution pressure meters, which also (1) precisely record the movement of the sample stage that guarantees the precision of the scratch made by the indenter and (2) measure the magnitude of the shear stress with high resolution. This makes nanoscratch hardness applicable for research in friction properties, fracture properties, and bond strength between substrates with films and coatings [18,19]. Chapter 11 will further discuss the role of nanoscratch hardness testing in thin film materials. Currently, nanohardness techniques are already widely used for thin film, spray coating, and surface modification research and are becoming increasingly popular in the electronic, MEMS, bio-engineering, and medical instrumentation fields. It is currently the most effective testing method for determining the mechanical properties of materials on the micro- and nanoscale [20,21].

4.9 SIZE EFFECT IN MATERIALS AND HARDNESS

4.9.1 Size Effect

With the advancement of nanotechnology, it is now possible to obtain accurate material characteristics on the submicro- and nanolevel. Experiments have shown repeatedly that hardness testing at very small levels will yield different results than macrohardness testing. This new phenomenon is termed to be the "size effect" and describes the unique physical properties exhibited by materials at the nanoscale. A wealth of experimental data from the past 10 years has shown that when a material is subjected to nonuniform plastic deformation at the nanoscale, it shows a strong size-effect dependence. Two typical experiments illustrate the size-effect phenomenon: (1) when the indentation depth of a metal or ceramic is reduced from 10 μm down to 1 μm, the hardness will be doubled. Figure 4.12 shows the experimental result by Stelmashenko et al. [22] in 1993 for nanoindentation of pure tungsten. (2) Fleck et al. [23] also observed in copper torsion experiments that the dimensionless torsion hardness increased three times for a copper wire diameter of 12 μm in comparison to a diameter of 170 μm (see Figure 4.13). Furthermore, Ma et al. [1] conducted nanoindentation experiments on thin nickel films for various loads and generated loading versus indentation curves as shown in Figure 4.14a, as well as obtaining the relationship between hardness and indentation depth as shown in Figure 4.14b.

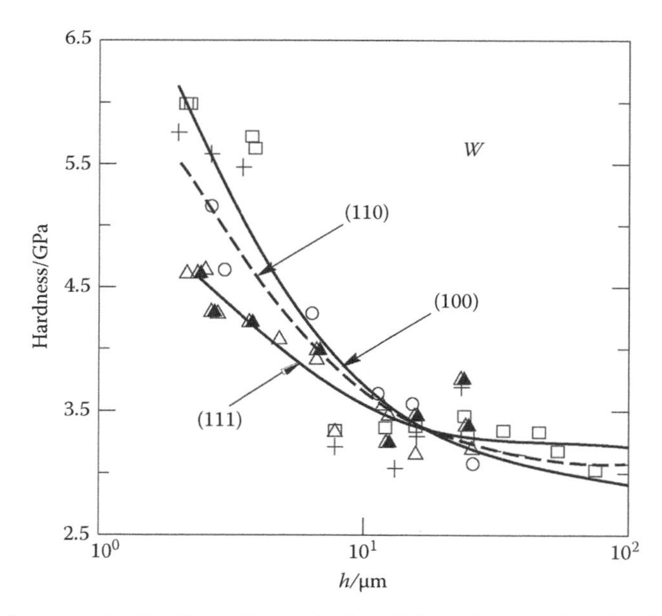

FIGURE 4.12 Hardness vs. depth. (From Stelmashenko, N.A. et al., *Acta Metall. Mater.*, 41, 2855, 1993; Taken from Figure 9.)

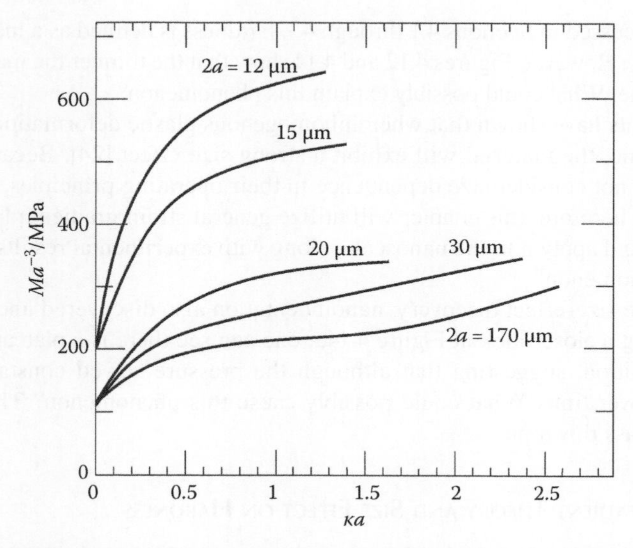

FIGURE 4.13 Thin copper wire torsion vs. diameter. M = torsion; k = unit angle; 2a = wire diameter. (From Fleck, N.A. et al., *Acta. Metall. Mater.*, 42, 475, 1994; Adapted from Figure 7.)

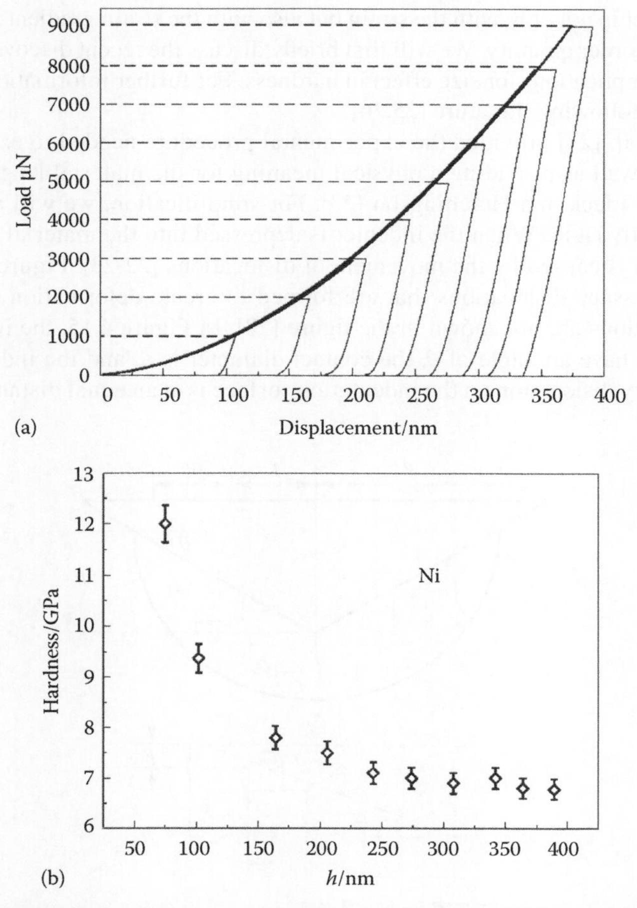

FIGURE 4.14 Nanoindentation result of nickel thin films. (a) Loading vs. depth curve. (b) Hardness vs. depth relationship. (From Ma, Z.S. et al., *J Appl. Phys.*, 103, 043512, 2008; Taken from Figures 3 and 4.)

From the results discussed in Sections 4.1 through 4.7, hardness is defined as a material constant independent of geometry. However, Figures 4.12 and 4.14 show that the thinner the material, the harder the material will become. What could possibly explain this phenomenon?

Many experiments have shown that when inhomogenous plastic deformation is applied to materials at the nanoscale, the material will exhibit a strong size effect [24]. Because traditional macrohardness tests do not consider size dependence in their operating principles, they cannot explain this phenomenon. Therefore, this chapter will utilize general strain gradient plasticity theory at the macroscopic level and apply it to the nanoscale, along with experimental results, in order to analyze the size-effect phenomenon.

In addition to the size-effect discovery, nanoindentation also discovered another intriguing phenomenon. By taking a closer look at Figure 4.14a, one can see there is a plateau during each maximum loading condition, suggesting that although the pressure stayed constant, the deformation is still increasing over time. What could possibly cause this phenomenon? The reader is strongly encouraged to discuss this topic.

4.9.2 Strain Gradient Theory and Size Effect on Hardness

In the description of classical plasticity theory as discussed in Chapter 1, the stress at a point should only be dependent on the strain and strain history; therefore, materials' property is a simple first-order quantity. But recent experiments have shown that, under certain conditions, the stress at a point has a relationship not only with the strain but also with the strain gradient at that point as well, making it a second-order quantity. We will first briefly discuss the recent discoveries in strain gradient theory and its implications for size effect in hardness. For further information on strain gradient theory, consult the following literature [25,26].

Nix and Gao et al. [27] provided the experimental procedure needed to examine the effect of strain gradient, as well as providing a physical meaning for the material length l as described in the experiments of Fleck and Hutchingson [23]. For simplification, we will assume the nanoindenter tip is perfectly rigid. When the indenter is depressed into the material surface, permanent plastic deformation is caused by the movement of dislocations [22,23]. Figure 4.15 illustrates the geometrically necessary dislocations that are formed to create deformation during indentation. The stored dislocations are not shown in the figure [28]. In Figure 4.15, the indenter surface and the sample surface have an angle of θ, the contact diameter is a, and the indentation depth is h. Assuming that every dislocation on the indentation surface is at an equal distance from each other,

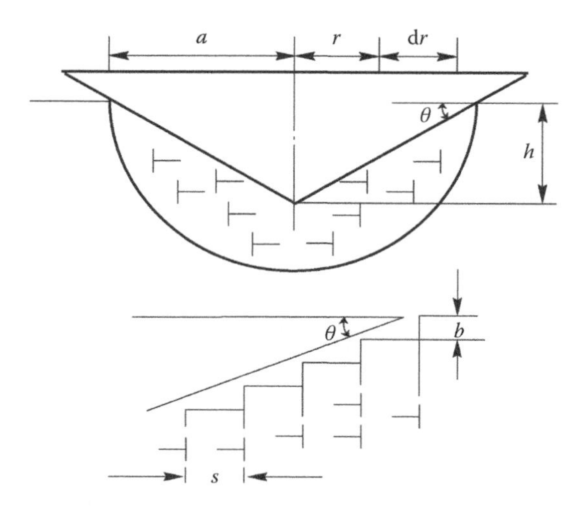

FIGURE 4.15 Geometry of dislocation steps during deformation.

it is easy to calculate the width between the surface deformation steps, as seen in Figure 4.15. If λ is the total length of the dislocation, then the following equation holds:

$$d\lambda = 2\pi r\frac{dr}{s} = 2\pi r\frac{h}{ba}dr. \tag{4.28}$$

Integrating the equation yields the following equation:

$$\lambda = \int_0^a \frac{h}{ba}2\pi r dr = \frac{\pi ha}{b}, \tag{4.29}$$

$$\tan\theta = \frac{h}{a} = \frac{b}{s}, \quad s = \frac{ba}{h}. \tag{4.30}$$

Assuming all the dislocations exist in a half-hemisphere of volume V, then $V = (2/3)\pi a^3$, and a dislocation density of

$$\rho_G = \frac{\lambda}{V} = \frac{3h}{2ba^2} = \frac{3}{2bh}\tan^2\theta. \tag{4.31}$$

Nix and Gao et al. [27] used the relationship between dislocation density and the Taylor factor to describe the material shear resistance:

$$\tau = \alpha\mu b\sqrt{\rho_T} = a\mu b\sqrt{\rho_S + \rho_G}, \tag{4.32}$$

where

ρ_T is the total dislocation density
ρ_s is the statistically stored dislocation density
ρ_G is the geometrically necessary dislocation density
μ is the shear modulus
b is the Burgers vector
α is an experimental constant on the order of 1

The strain gradient is only a function of the geometrically necessary dislocation density and can be defined as

$$\eta = \rho_G b. \tag{4.33}$$

From this formulation, η is explained as the curvature of the bending caused by extra dislocations, analogous to the unit bending angle in a torsion problem.

For the majority of elastic materials, the uniaxial tensile stress–strain relationship can be written in its power form:

$$\sigma_{eq} = \sigma_{ref}\varepsilon_{eq}^N, \tag{4.34}$$

where

N $(0 < N < 1)$ is defined as plastic hardening
σ_{ref} is the reference stress

For polycrystalline materials, the tensile flow stress σ_{eq} is usually $M = 3.06$ times larger than the shear flow stress τ. During uniaxial tension, the strain gradient becomes zero, and Equation 4.33 reflects only the hardening caused by the statistically stored dislocations, and so ρ_s can be determined from the uniaxial stress–strain relationship. Equation 4.31 will give the hardening caused by the strain gradient:

$$\sigma_{eq} = \sigma_{ref}\sqrt{\varepsilon^{2N} + l\eta}, \tag{4.35}$$

$$l = M^2\alpha^2\left(\frac{\mu}{\sigma_{ref}}\right)^2 b, \tag{4.36}$$

where l is the intrinsic material length in strain gradient plasticity theory, for elastic metals, and is on the order of microns. This quantity is estimated by Fleck et al. [25] by thin copper wire torsion tests and Stolken and Evans et al. [29] by thin nickel film bending tests. It should also be mentioned that Nix and Gao et al. [27] used $M = \sqrt{3}$ in Equation 4.31 by utilizing the Von Mises's criteria for various solids.

Tensile flow stress can also be related to hardness through the following equations:

$$\sigma_{eq} = \sqrt{3}\tau, \quad H = 3\sigma_{eq}. \tag{4.37}$$

From the aforementioned relationship, a hardness ratio can be deduced:

$$\frac{H}{H_0} = \sqrt{1 + \frac{h^*}{h}}. \tag{4.38}$$

$H_0 = 3\sqrt{3}\alpha\mu b\sqrt{\rho_s}$ does not take into account the hardness from the strain gradient:

$$h^* = \frac{81}{2}b\alpha^2\tan^2\theta\left(\frac{\mu}{H_0}\right)^2. \tag{4.39}$$

Equation 4.37 agrees well with the microindentation data found by McElhaney et al. [29] on single crystal and cold-worked polycrystalline copper. Figure 4.16 shows the theoretical relationship between indentation depth and hardness for (111) single crystal copper using Equation 4.37 [29]. From Figure 4.16, we can see that the calculated data from Equation 4.37 fits very well with the microindentation results done on (111) single crystal copper by McElhaney et al. On the other hand, classical plasticity theory will generate a horizontal line, suggesting complete independence of hardness and indentation depth as shown in Figure 4.16.

4.9.3 Relationship between Free Surface and Size Effect

In the previous chapter, we used strain gradient theory to predict the size effect of materials on the micron and submicron levels. It used a phenomenological description of the relationship between geometrically necessary dislocation density and the Taylor dislocation model [30] to explain the effect and produced predictions that are consistent with experimental data. However, it has been shown that when the indentation depth becomes very shallow ($h < 100$ nm), nanoindentation data will deviate from predictions made by the strain gradient theory.

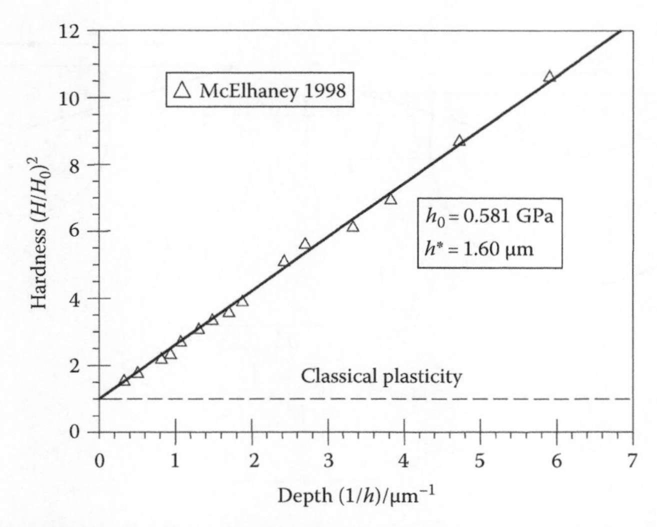

FIGURE 4.16 Relationship between indentation depth and hardness of (111) single crystal copper and Equation 4.35. (From McElhaney, K.W. et al., *J. Mater. Res.*, 13, 1300, 1998; Adapted from Figure 7.)

The most basic building blocks of materials are atoms (including protons and other charged particles), and the characteristics of all known elements are included in the table of elements. Different atoms can create different crystal structures or substructures by utilizing ionic, covalent, or metallic bonding between their charged particles to create a variety of materials such as metals and ceramics [31]. When indentation hardness testing is used, the indenter tip causes deformation on the material surface, causing bonding in the affected area to undergo changes without affecting the atomic structure of the material. This means that the size-effect phenomenon is on a scale greater than the atomic scale.

Although the definition of materials' quantities seem to differ in different physical descriptions, the total mechanical behavior of materials such as hardness, elastic modulus, and yield strength, all depend on the sum of the bond energy per unit volume on a fundamental level. Therefore, any deformation can cause a change in strain energy and coordination number of atoms near defects or surfaces, which in turn may cause a big change in the atomic bonding energy, bond length, and energy density near those defects and interfaces. These changes are highly sensitive to external factors such as pressure and temperature [32].

When an indentation is deep, the surface effect does not influence the measurement, and the hardness measurement is mainly affected by the diffusion of the internal dislocation (see Figure 4.17a). Inversely, when an indentation is shallow, there are very few dislocations, so the effect of different atomic bonding energies on the surface layer plays a much more important role in influencing the hardness deformed area (see Figure 4.17b). Therefore, in explaining the size effect of nano-sized materials, the free energy of the surface must be taken into account.

Graça et al. [33] made modifications to the Nix–Gao model by incorporating the surface free energy effect on the size effect's influence on indentation hardness. Graça et al. [33] assumed that surface free energy is independent of the dislocation densities and arrived at the following relationship:

$$H = H_0 + \Delta H_{dislocations} + \Delta H_{SFE} = H_0\sqrt{1+\frac{h^*}{h}} + \left\{\kappa\frac{E_S}{h}\right\}_{SFE} \tag{4.40}$$

where

$\Delta H_{dislocations}$ is the influence of dislocations as found by Nix–Gao
ΔH_{SFE} is the effect of the surface free energy on the hardness
κ is a constant related to the indenter geometry
E_s is the surface free energy, around 2.1 J/m^2 [34]

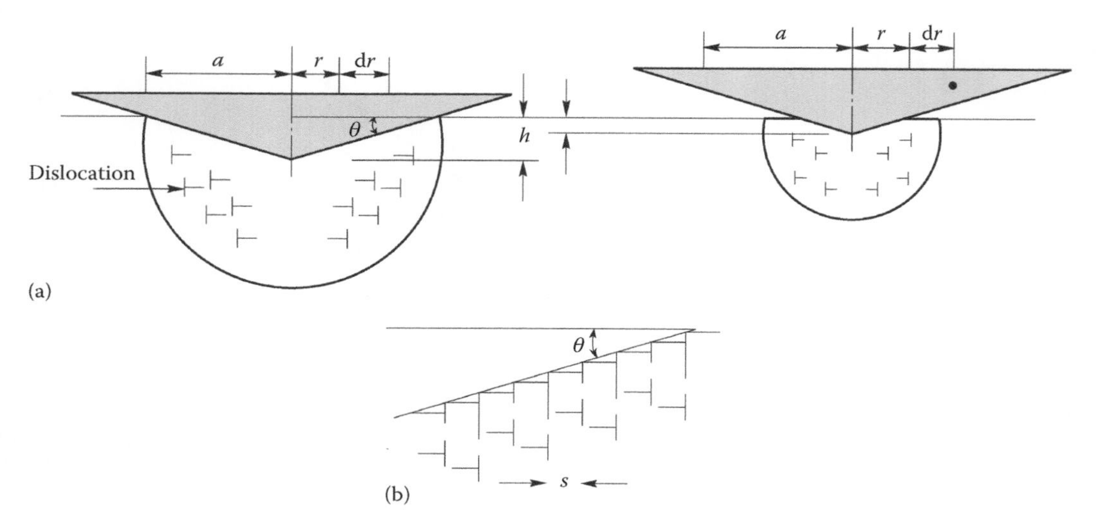

FIGURE 4.17 Dislocations near indentation. (a) Deep indent. (b) Shallow indent. (From Ma, Z. et al., *J. Appl. Phys.*, 103, 043512, 2008; Taken from Figure 5.)

From Equation 4.39, the GCV model includes both the effect of dislocations and the surface free energy.

If the surface free energy is ignored for hardness, the Nix–Gao model (Equation 4.37) will predict indentation hardness as shown by Figure 4.18. From the figure, one can see that the Nix–Gao model agrees very closely to experimental data with $R > 0.98$. Therefore, when the indentation depth is larger than 100 nm, the Nix–Gao model is adequate in predicting the size-effect influence on hardness. However, when the indentation depth is less than 100 nm, the surface effect can no longer be ignored.

As shown in Figure 4.19, both the GVC (4.39) and Nix–Gao (4.37) models have been used to predict experimental data. The effect of κ is only pronounced when the sample is less than 100 nm and when the surface free energy has a big influence on the hardness. For the Berkovich indenter,

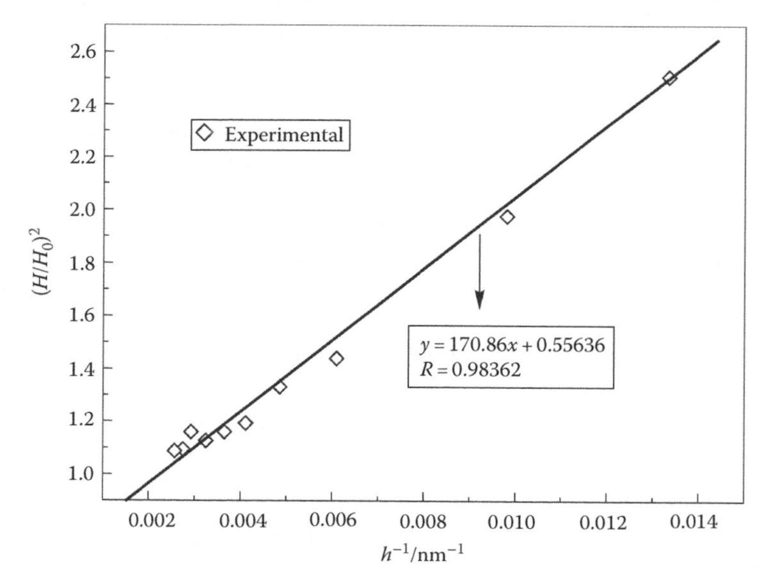

FIGURE 4.18 Nix–Gao model in comparison to experimental data. (From Ma, Z. et al., *J. Appl. Phys.*, 103, 043512, 2008; Adapted from Figure 6.)

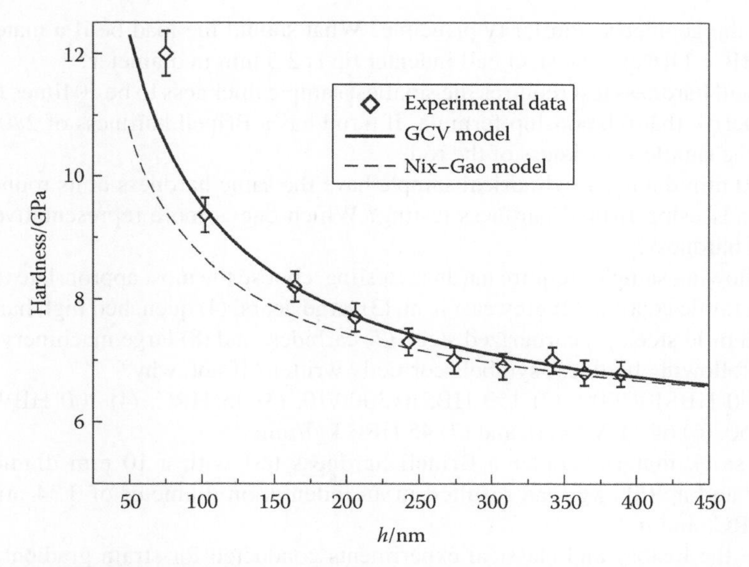

FIGURE 4.19 Experimental data for thin film nickel in comparison to GCV and Nix–Gao model. (From Ma, Z. et al., *J. Appl. Phys.*, 103, 043512, 2008; Taken from Figure 7.)

κ is taken as 0.157. For samples that are greater than 100 nm thick, both the Nix–Gao and GCV models can predict the experimental data within experimental error. However, when the indentation depth is smaller than 100 nm, the Nix–Gao model no longer fits well into the experimental data, but GCV can still predict the hardness fairly well. This means that the surface free energy has a large influence on the size-effect phenomenon and becomes even more pronounced as the indentation depth gets smaller and smaller. Therefore, GCV is the preferred model in predicting the size effect and hardness.

In conclusion, strain gradient plasticity theory can explain the majority of questions regarding the size-effect phenomenon, until the sample thickness gets to the nanoscale. This is mainly due to the imperfections of the testing method and the testing material. Any small difference in surface roughness, defects, localized hardening, oxidation, and other free surface effects could greatly affect the size effect in hardness. These variables are currently very difficult to control and are the main obstacle in obtaining a clear picture of this phenomenon. This is a major area of research and discussion for materials scientists.

EXERCISES

4.1 What is hardness? What are the characteristics of hardness testing? What are the common methods for hardness testing?

4.2 Explain the meaning of the following abbreviations: HB, HRA, 100HV, and HS.

4.3 Compare and contrast Brinell hardness, Vickers hardness, and Rockwell hardness testing principles. Also explain their advantages and disadvantages and their applications.

4.4 The Shore hardness test and the Brinell hammer hardness test are both dynamic hardness testing methods. Are their operating principles the same? Why or why not?

4.5 What is the relationship between hardness values from different hardness tests on the same material? Why is it so? If two materials have a hardness of 200 HBS10/1000 and 45 HRC, respectively, which one is harder?

4.6 What is the usefulness of microhardness testing?

4.7 What is the geometric similarity principle? What should the load be if a material is a black metal (HB < 140) and the steel ball indenter tip is 2.5 mm in diameter?

4.8 The Brinell hardness test requires the smallest sample thickness to be 10 times the indentation depth. Derive that relationship formula. If a rod has a Brinell hardness of 280 HBS10/3000, what is the smallest thickness of the rod?

4.9 Will a 10 mm diameter cylindrical sample have the same hardness at its rounded surface as its flat ends using Brinell hardness testing? Which one is more representative of the actual material hardness?

4.10 If the following samples require hardness testing, choose the most appropriate testing method:
(1) Ceramic coating, (2) grey cast iron, (3) large gears, (4) quenched high hardness part, (5) annealed mild steel, (6) carburized steel, (7) carbides, and (8) large machinery.

4.11 Are the following hardness symbols correctly written? If not, why?
(1) 100 HBS10/3000, (2) 150 HBS10/3000/10, (3) 15 HRC, (4) 100 HBW0.05, (5) 500 HRS5/750, (6) 640 HV30/20, and (7) 45 HRC kgf/mm^2.

4.12 A steel stock material under a Brinell hardness test with a 10 mm diameter steel ball indenter and a 3000 kg load resulted in an indentation diameter of 3.34 mm. What is its HBS, HRC, and σ_b?

4.13 Describe the history and classical experiments conducted for strain gradient plastic theory and size effect.

4.14 Give examples of the size-effect phenomenon and how it is experimentally observed.

4.15 Thin wire torsion tests show a clear size-effect influence. Does the size-effect also influence the results of thin wire uniaxial tensile testing? Why or why not?

4.16 Derive Equations 4.34 and 4.37 from first principles in detail.

4.17 Describe the motivation behind using strain gradient plasticity theory to explain microscopic mechanisms.

4.18 When the indentation depth is very small, both Nix–Gao and GCV cannot satisfactorily explain experimental data. Give specifics on why that is and suggestions for practical improvements of testing methods.

4.19 Discussion: What is the implication of the plateau seen on the loading vs indentation depth curve of Figure 4.14a?

4.20 Discussion: From nondimensional analysis found in Reference [18], the loading curve is proportional to the square of the indentation depth during nanoindentation tests. Affirm this hypothesis by experiment. If this relationship is not seen, what are the reasons why this is not seen?

REFERENCES

1. Ma Z. S., Long S. G., Pan Y. et al. Indentation depth dependence of the mechanical strength of Ni films. *J Appl Phys*, 2008, 103: 043512.
2. Chang S. Y. and Chang T. K. Grain size effect on nanomechanical properties and deformation behavior of copper under nanoindentation test. *J Appl Phys*, 2007, 101: 033507.
3. Durst K., Franke O., Bohner A. et al. Indentation size effect in Ni-Fe solid solutions. *Acta Materialia*, 2007, 55: 6825–6833.
4. Chung H. Y., Weinberger M. B., Yang J. M. et al. Correlation between hardness and elastic moduli of the ultraincompressible transition metal diborides RuB_2, OsB_2, and ReB_2. *Appl Phys Letts*, 2008, 92: 261904.
5. Ramamurty U., Jana S., Kawamura Y., and Chattopadhyay K. Hardness and plastic deformation in a bulk metallic glass. *Acta Materialia*, 2005, 53: 705–717.
6. Yang R., Zhang T. H., Jiang P. et al. Experimental verification and theoretical analysis of the relationships between hardness, elastic modulus, and the work of indentation. *Appl Phys Lett*, 2008, 92: 231906.
7. Li J. L. Introduction of main technical contents in new national standard of metallic rockwell hardness test. *Phys Exam Test*, 2005, 23(1): 37–39.
8. Shu D. L. *Mechanical Properties of Engineering Materials*. Beijing: China Machine Press, 2003.

9. Han D. W. *Handbook for Technology of Hardness Determination of Metals*. Changsha, China: Central South University Press, 2003.
10. Han D. W. *The Hardness of Metals and Its Testing Method*. Changsha, China: Hunan Science & Technology Press, 1983.
11. Liang X. B., Li J. L., and Zhang Z. W. *National Standard Collection. Metal Mechanics and Method for Process Properties Test*. Beijing: China Standard Press, 1996.
12. Tabor D. *The Hardness of Metals*. Oxford: Clarendon Press, 1951.
13. Yu W. L. Study of the latest development and trend of hardness measurement technology. *Phys Test Chem Anal* (Part A: Physical Testing), 2003, 39(8): 401–405
14. Yang D. and Li F. X. *Microhardness Testing*. Beijing: Metrology Press, 1988.
15. Cheng Y. T. and Cheng C. M. Scaling approach to conical indentation in elastic-plastic solids with work hardening. *J Appl Phys*, 1998, 84(3): 1284–1291.
16. Cheng Y. T. and Cheng C. M. Relationship between hardness, elastic modulus, and the work of indentation. *Appl Phys Lett*, 1998, 73(5): 614–616.
17. Oliver W. C. and Pharr G. M. An improved technique for determining hardness and elastic modulus using load and displacement sensing indentation experiments. *J Mater Res*, 1992, 7(6): 1564–1583.
18. Lee K. M, Yeo C. D., and Polycarpou A. A. Mechanical property measurements of thin-film carbon overcoat on recording media towards 1 Tbit/in^2. *J Appl Phys*, 2006, 99: 08 G906.
19. Chang S. Y., Tsai H. C., Chang J. Y. et al. Analyses of interface adhesion between porous SiOCH low-k film and SiCN layers by nanoindentation and nanoscratch tests. *Thin Solid Films*, 2008, 516(16): 5334–5338.
20. Cole D. P., Bruck H. A., and Roytburd A. L. Nanoindentation studies of graded shape memory alloy thin films processed using diffusion modification. *J Appl Phys*, 2008, 103: 064315.
21. Soh M. T., Discher-Cripps A. C., and Savvides N. Nanoindentation of plasma-deposited nitrogen-rich silicon nitride thin films. *J Appl Phys*, 2006, 100: 024310.
22. Stelmashenko N. A., Walls A. G., Brown L. M. et al. Microindentations on W and Mo oriented single crystals; an STM study. *Acta Metal Mater*, 1993, 41: 2855–2865.
23. Fleck N. A. Strain gradient plasticity: Theory and experiment. *Acta Metal Mater*, 1994, 42(2): 475–487.
24. Stolken J. S. and Evans A. G. A microbend test method for measuring the plasticity length scale. *Acta Materialia*, 1998, 46: 5109–5115.
25. Huang K. C. and Huang Y. *The Constitutive Relation of Solid*. Beijing: Tsinghua University Press, 1999.
26. Zhou Y. C. *Solid Mechanics in Materials (Part II)*. Beijing: Science Press, 2005.
27. Nix W. D. and Gao H. Indentation size effects in crystalline materials: A law for strain gradient plasticity. *J Mech Phys Solids*, 1998, 46: 411–425.
28. Ashby M. F. The deformation of plastically non-homogeneous alloys. *Phil Mag*, 1970, 21: 399–424.
29. McElhaney K. W, Vlassak J. J., and Nix W. D. Determination of indenter tip geometry and indentation contact area for depth-sensing indentation experiments. *J Mat Res*, 1998, 13: 1300–1306.
30. Taylor G. I. The mechanism of plastic deformation of crystals. Part I—Theoretical. *Proc. Roy Soc London A*, 1934, 145: 362–38
31. Feng D., Shi C. X., and Liu Z. G. *Introduction to Materials Science*. Beijing: Chemical Industry Press, 2005: 17–46.
32. Sun C. Q. Size dependence of nanostructures: Impact of bond order deficiency. *Prog Solid State Chem*, 2007, 35: 1–159.
33. Graça S., Colaço R., and Vilar R. Indentation size effect in nickel and cobalt laser clad coatings. *Surf Coat Technol*, 2007, 202(3): 538–548.
34. Jäger I. L. Surface free energy-a possible source of error in nanohardness? *Surf Sci*, 2004, 565: 173–179.

5 Testing of Material Fracture Toughness

Parents provide their children with a plastic bowl and not a ceramic bowl at mealtime because the ceramic bowl is fragile. Glass and chinaware break to pieces when they fall to the ground, but things like iron are undamaged even when violently handled. Why? Experientially we understand that some materials are brittle, while others have good toughness. But can we describe "brittleness" or "toughness" in a mathematical language? What methods can we use to measure brittle things, or tough things, in a quantitative manner? In Chapter 2, we established the theoretical basis of material damage, introduced some important concepts such as the stress intensity factor and fracture toughness, and obtained the material damage criterion.

In the material damage criterion, fracture toughness is a performance indicator deemed by fracture mechanics to reflect a material's resistance to unstable crack growth. It is a requisite and important parameter for the design of a material or a component. This chapter introduces the testing technology of material fracture toughness [1–4], including (1) testing of the plane strain's fracture toughness in thick components, (2) testing of the plane stress's fracture toughness in thin components, (3) testing of the surface crack's fracture toughness in cracks that lie on the component's surface, (4) testing of the component's J integral critical value, and (5) the crack-opening displacement critical value that represents its fracture performance.

5.1 TESTING OF PLANE STRAIN FRACTURE TOUGHNESS K_{IC}

5.1.1 Common Fracture Toughness Measurement Method and K_{IC} Representation

As described in Chapter 2, the dominant term of the stress field inside the crack tip of a Mode I crack (opening mode crack) is

$$\sigma_x = \sigma_y = \frac{K_I}{\sqrt{2\pi r}}, \tag{5.1}$$

then,

$$K_I = Y\sigma\sqrt{a}, \tag{5.2}$$

where

r is the distance from the crack tip

Y is an item related to the crack shape, sample type, and loading method

For a crack across the center of an infinite plate, $Y = \sqrt{\pi}$ applies. From Equation 5.1, we know that for each point on the crack tip (the coordinate r is known) the stress field near the crack tip depends solely on the stress intensity factor K_I. When K_I reaches the critical value K_{IC}, unstable crack growth occurs, leading to sample fracture. In this case, K_{IC} is the material's fracture toughness. If the sample is thick enough, it is in the plane strain state. This K_{IC} is called the plane strain's fracture toughness. Based on Equation 5.2, when $\sigma = \sigma_C$, the unstable crack growth is in the critical state

$$K_{IC} = Y\sigma_C\sqrt{a}. \quad (5.3)$$

Because the plane strain's fracture toughness K_{IC} is the material constant, it is unrelated to the loading method and the sample type under certain conditions. Therefore, in principle, K_{IC} measured on different types of samples should be the same as long as the sample meets the plane strain conditions.

Using the linear elasticity mechanics method for specimens that meet the plane strain conditions, one can obtain

$$K_I = \sigma\sqrt{\pi a}\cdot f, \quad (5.4)$$

where f is a function value related to the specimen size, also called the correction factor. When the unstable crack growth begins, the fracture toughness K_{IC} can be calculated based on Equation 5.4 while the input values of σ, a, and f are the corresponding critical ones. Refer to the GB4161-1984 National Standard issued in 1984. The common K_{IC} representations are described next:

1. Three-point bending specimen (S:W = 4:1, as shown in Figure 5.1):

$$K_I = \frac{FS}{BW^{3/2}} f\left(\frac{a}{W}\right), \quad (5.5)$$

$$f\left(\frac{a}{W}\right) = 2.9\left(\frac{a}{W}\right)^{1/2} - 4.6\left(\frac{a}{W}\right)^{3/2} + 21.8\left(\frac{a}{W}\right)^{5/2} - 37.6\left(\frac{a}{W}\right)^{7/2} + 38.7\left(\frac{a}{W}\right)^{9/2} \quad (5.6)$$

$$0.25 \le \frac{a}{W} \le 0.75,$$

where
F is the load applied on the specimen, which becomes F_Q when it is in the critical state (for its determination method, see Section 5.1.3)
S is the span
B is the specimen thickness
W is the specimen height
a is the crack length

Here, the correction factor $f(a/W)$ is a function of a/W.

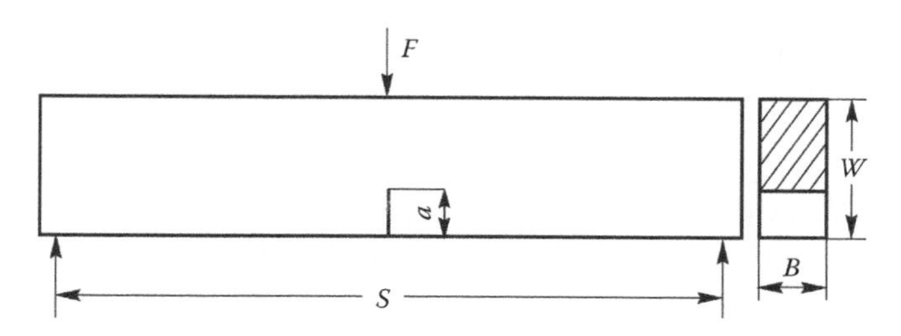

FIGURE 5.1 Three-point bending specimen.

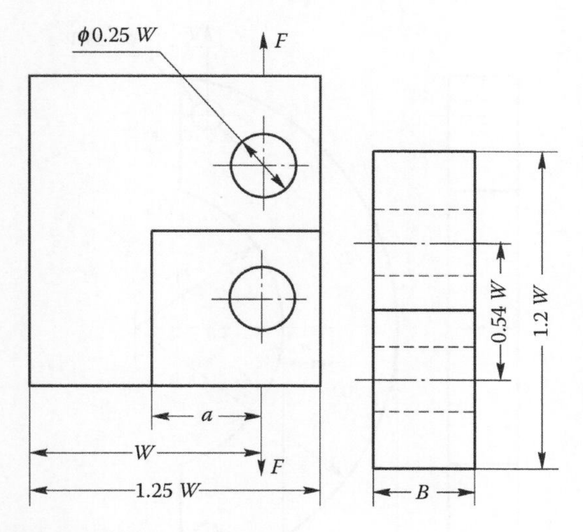

FIGURE 5.2 Compact tension specimen.

2. Compact tension specimen (as shown in Figure 5.2):

$$K_I = \frac{F}{BW^{1/2}} f\left(\frac{a}{W}\right), \tag{5.7}$$

$$f\left(\frac{a}{W}\right) = 29.6\left(\frac{a}{W}\right)^{1/2} - 185.5\left(\frac{a}{W}\right)^{3/2} + 655.7\left(\frac{a}{W}\right)^{5/2} - 1017\left(\frac{a}{W}\right)^{7/2} + 638.9\left(\frac{a}{W}\right)^{9/2},$$

$$0.30 \le \frac{a}{W} \le 0.70 \tag{5.8}$$

where the meanings of F, B, W, and a are shown in Figure 5.2.

3. C-shaped specimen (as shown in Figure 5.3):

$$K_I = \frac{F}{BW^{1/2}} f\left(\frac{a}{W}\right)\left(1 + 1.54\frac{x}{W} + 0.50\frac{a}{W}\right)\left[1 + 0.22\left(1 - \frac{a}{W}\right)^{1/2}\right]\left(1 - \frac{r_1}{r_2}\right) \tag{5.9}$$

$$f\left(\frac{a}{W}\right) = \left(\frac{a}{W}\right)^{1/2} \begin{bmatrix} 18.23 - 106.2\frac{a}{W} + 379.7\left(\frac{a}{W}\right)^2 \\ -582.0\left(\frac{a}{W}\right)^3 + 369.1\left(\frac{a}{W}\right)^4 \end{bmatrix}, \tag{5.10}$$

$$0.45 \le \frac{a}{W} \le 0.55; \quad 0 < \frac{x}{W} \le 0.5; \quad 0 < \frac{r_1}{r_2} \le 1.0$$

where the meanings of P, B, W, a, r, r_2 are shown in Figure 5.3.

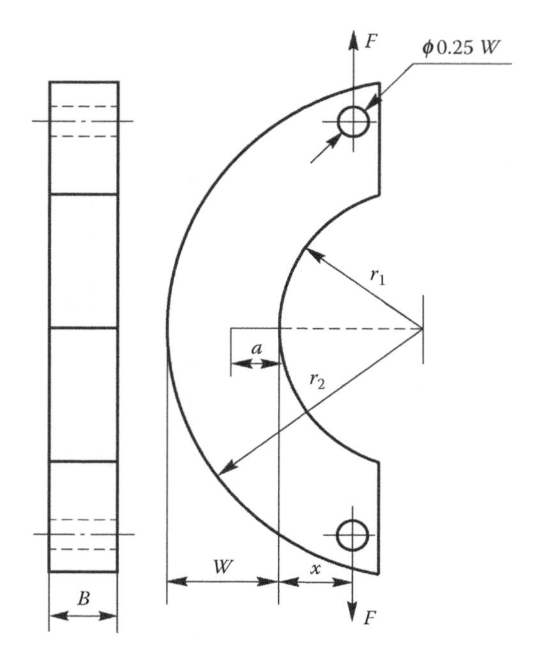

FIGURE 5.3 C-shaped specimen.

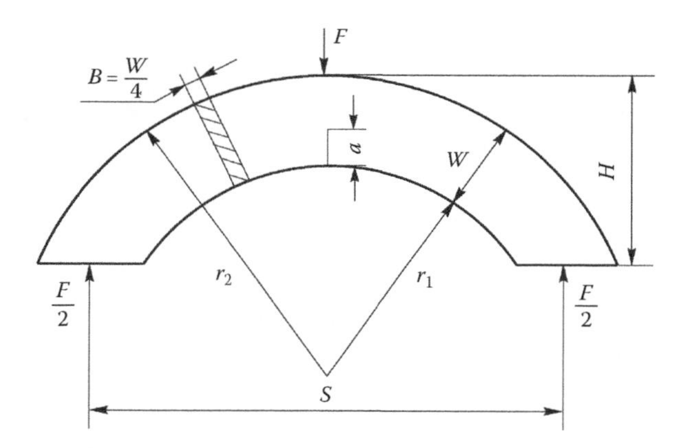

FIGURE 5.4 Arch three-point bending specimen.

4. Arch three-point bending specimen (as shown in Figure 5.4):

$$K_I = \frac{FS}{4BW^{3/2}} f\left(\frac{a}{W}\right), \tag{5.11}$$

where the meanings of P, B, W, S, a are shown in Figure 5.4.

5. Circular compact tension specimen (as shown in Figure 5.5):

$$K_I = \frac{F}{BW^{1/2}} f\left(\frac{a}{W}\right), \tag{5.12}$$

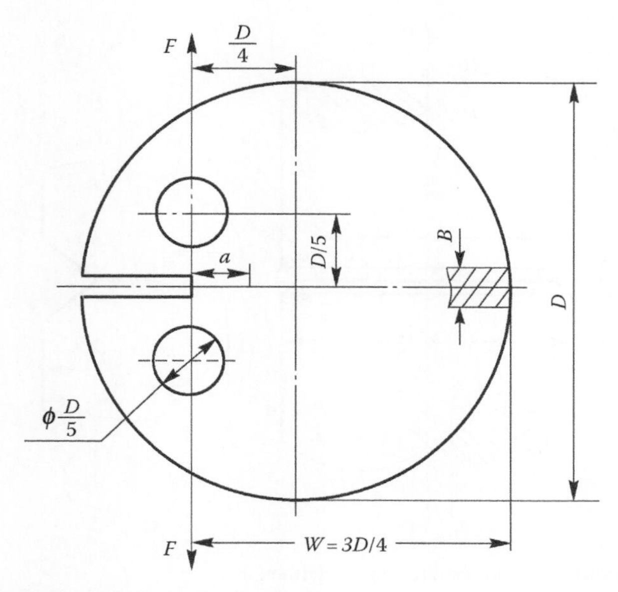

FIGURE 5.5 Circular compact tension specimen.

$$f\left(\frac{a}{W}\right)=29.6\left(\frac{a}{W}\right)^{1/2}-162\left(\frac{a}{W}\right)^{3/2}$$
$$+492.6\left(\frac{a}{W}\right)^{5/2}-663.4\left(\frac{a}{W}\right)^{7/2}+405.6\left(\frac{a}{W}\right)^{9/2}, \tag{5.13}$$

where the meaning of P, B, W, a are shown in Figure 5.5.

6. Circumferentially notched tension specimen (as shown in Figure 5.6):

$$K_I=\frac{F}{D^{3/2}}f\left(\frac{d}{D}\right), \tag{5.14}$$

where the meanings of P, D, d are shown in Figure 5.6. The correction factor $f(d/D)$ is a function of d/D.

In the six specimens, the three-point bending specimen and compact tension specimen are standard specimens for determining the material's K_{IC}. When testing, the three-point bending specimen needs a simple clamp, while the compact tension specimen needs a special clamp and is difficult to process. For specimens with different thicknesses, different clamps are required. However, the compact tension specimen is material-saving. For a large specimen of medium-strength steel, this becomes more obvious. In a pressure container, the most dangerous case is when the crack grows along the thickness (radial) direction under circumferential tensile stress. In this case, the C-shaped specimen and the arch three-point bending specimen are not only easy to process but also fully use the pipe's wall thickness. As a result, it can easily meet the mechanical condition of small-scale yielding, and we can obtain an effective K_{IC} value. For a component such as a tie bar, the preference is to use a circular compact tension specimen and a circumferentially notched tension specimen to determine its K_{IC} value.

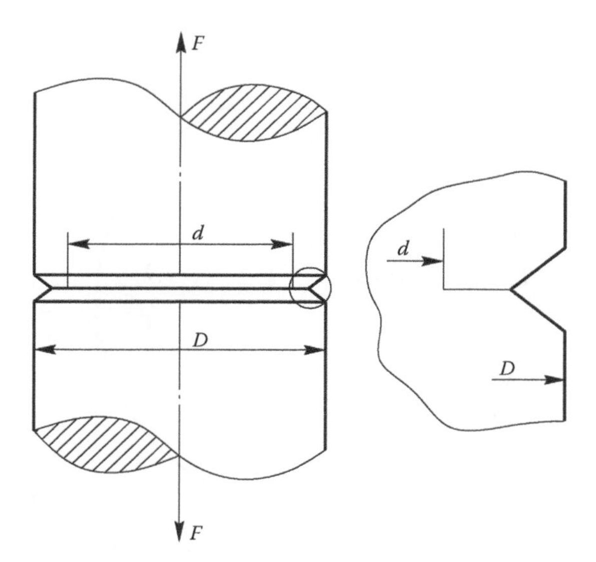

FIGURE 5.6 Circumferentially notched tension specimen.

5.1.2 Requirements on Specimen Size

Generally, the material's critical stress intensity factor K_{IC} is related to the specimen thickness B, crack length a, and ligament width $(W - a)$. We can obtain a stable K_{IC} value only when the specimen size meets the mechanical conditions of plane strain and small-scale yielding. In this case, the fracture toughness K_{IC} represents the material's inherent characteristics, which are unrelated to the specimen size.

1. Requirements of plane strain condition on thickness
Only when the specimen is thick enough can it can generate large enough constraint in the z-direction so that the strain component in the z-direction ε_z is equal to zero, leading to a plane strain state. For a through-crack specimen, the plastic zone on the tip of the crack on the surface layer is always in the plane stress state. Many tests have shown that, for the same material, the thickness of the plane stress layer (the width of the tensioned edge or shear lip) remains basically unchanged. When the specimen thickness goes up, the width of the tensioned edge seldom changes. Therefore, when the specimen is thick enough, the plane stress layer in the thickness direction occupies a small proportion, while the major part of the crack tip is in the plane strain state. In this case, the entire specimen is approximately in the plane strain condition so that we can measure a stable K_{IC} value.
Which thickness ensures that the specimen meets the plane strain condition? At present, we cannot make a precise conclusion in theory. Based on the results of several tests, we can basically measure and obtain a stable K_{IC} value using the "burst" load (to be discussed in Section 5.1.3). Therefore, it is recommended that the specimen thickness is

$$B \geq 2.5\left(\frac{K_{IC}}{\sigma_s}\right)^2. \tag{5.15}$$

2. Requirements of small-scale yielding condition on crack length
It should be noted that even for an ideal linear elastic body (absence of small-scale yielding) when the stress component's approximate representation is used to substitute for the precise solution, error exists.

For the common three-point bending and compact tension specimen, when $r/a = 0.02$ applies, the deviation between the stress field given by the single parameter K_I, and the precise value is approximately 6%–7%. Because a large or a small plastic zone exists on the crack tip, the radius of the plastic zone r_y cannot tend to zero. The r value when K_I is approximately established corresponds to the junction of the crack tip's plastic zone and the wide elastic zone. From Equation 2.28 in Chapter 2, we obtain the radius of the plane strain's plastic zone,

$$r_y = \frac{1}{4\sqrt{2}\pi}\left(\frac{K_{IC}}{\sigma_s}\right)^2. \tag{5.16}$$

To establish K_{IC}'s approximate deviation of ≤10% for a three-point bending and compact tension specimen used in the standard test, $r_y/a \le 0.02$ must apply, and so

$$a \ge 50\ r_y \approx 2.5\left(\frac{K_{IC}}{\sigma_s}\right)^2. \tag{5.17}$$

3. Requirements related to ligament size
   The ligament size, also called the ligament width $(W - a)$, has a great influence on the value of the stress intensity factor K_I. If the ligament width is too small, the back surface will lose the constraint effect on the crack's plastic deformation so that the entire ligament yields in the loading process. The cracked specimen is no longer approximately taken as an elastic body. In this case, linear elastic theory does not apply and a K_I approximate value does not exist. Therefore, the specimen must have a ligament size large enough to meet the conditions of small-scale yielding, ensuring that the back surface has a sufficient constraint effect on the plastic deformation of the crack tip. The requirement for the ligament width is

$$W - a \ge 2.5\left(\frac{K_{IC}}{\sigma_s}\right)^2. \tag{5.18}$$

5.1.3 Determination of Critical Load

After determining the specimen's form, size, and corresponding K_I representation, the key issue during testing is how to determine the critical load, that is, the load under which the crack begins unstable growth. In unstable crack growth, Equation 5.4 becomes

$$K_{IC} = \sigma_C\sqrt{\pi a_C}\cdot f\left(\frac{a_C}{W}\right). \tag{5.19}$$

In theory, under the condition of plane strain, there should not be a process of slow growth from the front edge to unstable growth (also called subcritical growth). In actual testing, a subcritical growth of 2% a (equal to the plastic zone's equivalent growth) is allowed. Therefore, $a_C = a_0$ (initial averaged crack length) is used in the calculation. The only issue is how to determine σ_C, the critical load.

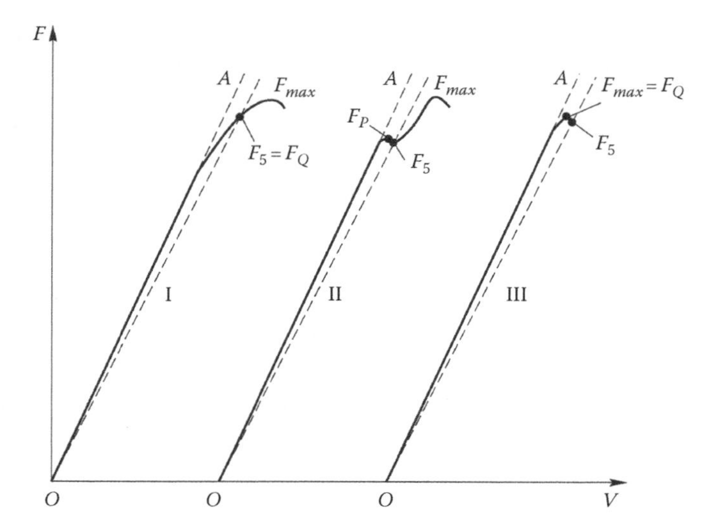

FIGURE 5.7 Three typical F–V curves.

In common K_{IC} testing, the curve of the obtained load F versus the opening displacement V is roughly divided into three types, as shown in Figure 5.7. The critical load is determined according to different types of curves under certain conditions. The value obtained is F_Q in the critical load condition. Next we will discuss them.

1. When testing is performed on a specimen with sufficient thickness, it often happens that a Mode III curve is obtained. In the loading process, the front edge of the crack does not grow. When the load reaches the maximum value, sudden brittle fracture occurs on the specimen. Most fractures are even. In this case, the maximum load can be taken as F_Q.
2. When testing is performed on a specimen with small thickness, a Mode II curve is obtained. Such a curve has one obvious "burst" platform. This is because, in the loading process, the specimen's center layer is in the plane strain state and grows first, while the surface layer is in the plane stress state and does not grow. Therefore, the crack growth of the center layer is drawn by the surface layer in a short period of time. In the testing process, when this specimen reaches the "burst" load, we can often hear a clear "burst" sound. In this case, the "burst" load F_P can be taken as F_Q.
3. When testing is performed on a specimen with minimum limit thickness, it often happens that a Mode I curve is obtained. In this case, the fracture toughness cannot be calculated by using the maximum load, because when the fracture toughness is lower than the maximum load, the specimen's crack has grown gradually. Because the front edge of the crack has a relatively small part in the plane strain state, the crack's initial "burst" growth is small and too imperceptible to be found. For such a specimen, the so-called conditional value can be determined from the F–V curve only by using a certain engineering hypothesis. If the test curve of the three-point bending specimen is a Mode I curve as in Figure 5.7, then, from the coordinate's origin O, make a secant line OF_5 with a gradient 5% smaller than that of the curve's initial tangent line OA. The intersection point of the secant line OF_5, and the curve, corresponds to the load F_5. When $(F_{max}/F_5) < 1.1$ applies, the load F_5 is taken as F_Q. For the theoretical principle, see the relevant references.

If the test curve is a Mode II or a Mode III curve, the load F_5 can be obtained using the same graphing method. However, as you can see from the Mode III curve in Figure 5.7, the peak value of the curve F_{max} appears at the left of F_5 and is larger than F_5. In this case, this load F_{max}

should be taken as F_Q. It should be noted that this method of calculating F_Q using the secant line with a 5% reduced gradient does not apply to all specimens with various forms and different crack lengths.

5.1.4 Testing the Plane Strain Fracture Toughness K_{IC}

A three-point bending specimen is taken as an example in describing the steps taken to test the plane strain's fracture toughness K_{IC}.

5.1.4.1 Preparation of Specimen

1. Sampling direction: Metallic materials, including forged, plate, pipe, or bar, have anisotropy to different extents. This becomes more obvious in the fracture toughness value, and so the fracture toughness is related to the specimen orientation. Generally, the specimen orientation is represented by two letters. The first letter represents the crack surface's normal direction, while the second letter represents the crack's growth direction. Figure 5.8 shows a rolled plate, from which fracture toughness specimens of six different orientations can be cut. In the figure, *L* is the rolling direction (longitudinal), *T* is transverse direction, and *S* is the plate thickness direction. For example, *L–T* represents that the crack that goes through the plate grows along the transverse direction, while the *S–L* specimen represents that the crack that is parallel to the plate grows along the longitudinal direction. Research shows that the K_{IC} value is highest along the *L–S* orientation, while the K_{IC} value is lowest along the *S–L* orientation. When sampling among the actual components, the specimen's crack orientation should be identical to the component's most dangerous crack direction. In a pressure container, the most dangerous case is the internal and external surface cracks that grow along the thickness (radial) direction under circumferential stress. Therefore, a three-point bending, or an arch three-point bending specimen, along the *C–R* orientation (*C* is tangential, *R* is radial) can be taken. Figure 5.9 shows a three-point bending specimen along the *L–S* orientation.

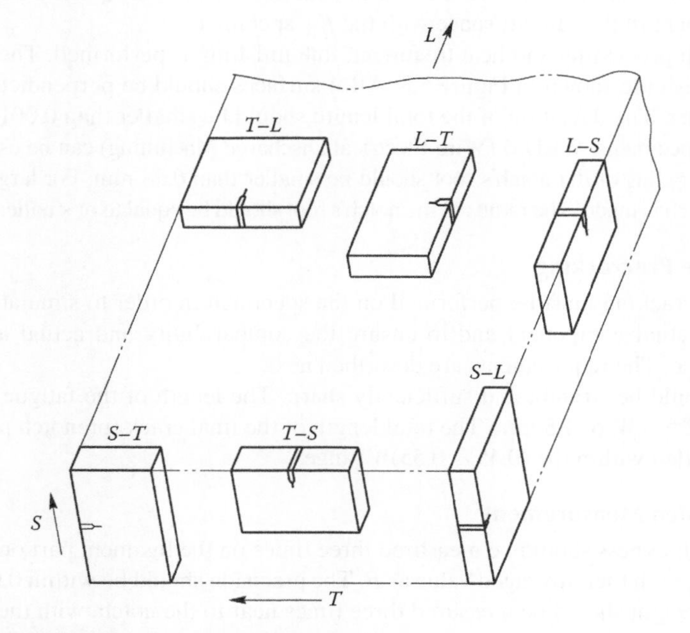

FIGURE 5.8 Plate specimen's cracked surface orientation.

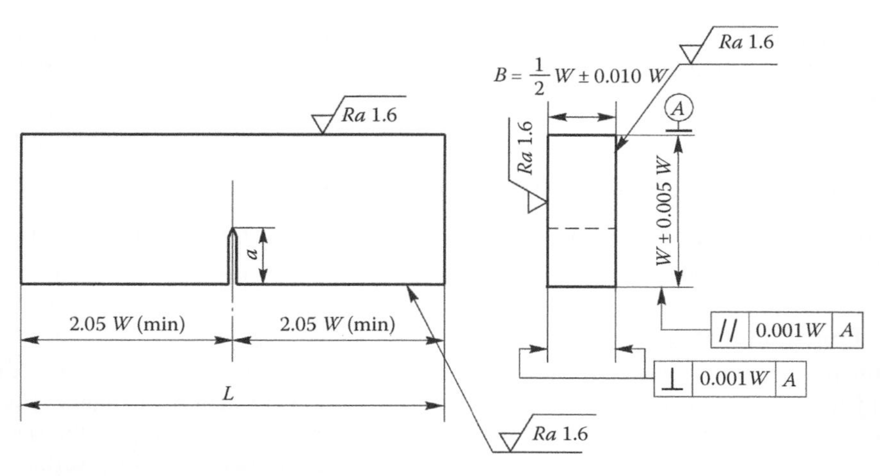

FIGURE 5.9 Three-point bending specimen.

2. The specimen thickness B can be selected on the basis of the ratio between the material's K_{IC} estimated value and σ_s, or the ratio between σ_s and E. Based on the condition of the plane strain and small-scale yielding,

$$B \geq 2.5\left(\frac{K_{IC}}{\sigma_s}\right)^2, \quad a = W - a = B.$$

The ratio among the specimen's thickness B, height W, and span S, is

$$B : W : S = 1 : 2 : 8.$$

3. Two to three routine tension specimens are prepared from the material in the same furnace used for measuring routine mechanical indicators like σ_s. The specimens must undergo heat treatment in the same furnace with the K_{IC} specimen.
4. After rough processing and heat treatment, fine finishing is performed. The final size and surface finish are shown in Figure 5.8. All A surfaces should be perpendicular or parallel to each other. The deviation of the total length should be smaller than 0.001 W.
5. For small specimens, WEDM (Wire Electrical Discharge Machining) can be used to make the notch. The radius of the notch's root should be smaller than 0.08 mm. For large specimens, a chevron notch is made. The radius of the notch's root should be equal to or smaller than 0.25 mm.

5.1.4.2 Fatigue Precracking

Fatigue load precracking must be performed on the specimen in order to simulate the sharp crack existing in the actual component and to ensure the comparability and actual application of the obtained K_{IC} value. The requirements are described next.

The crack should be straight and sufficiently sharp. The length of the fatigue crack should not be smaller than 2.5% W, or 1.5 mm. The total length of the final crack (prenotch plus fatigue crack) should be controlled within the $(0.45 \sim 0.55)W$ range.

5.1.4.3 Specimen Measurement

The specimen's thickness should be measured three times on the ligament part on the front edge of the fatigue crack, with their averaged value as B. The precision should be within 0.02 mm or 0.1% B. The specimen's height should be measured three times near to the notch, with their averaged value as W. The precision should be within 0.02 mm or 0.1% W.

5.1.4.4 Test Procedure

In the bending test, special support rollers must be used. Attention should be paid to the following:

1. The support rollers should roll freely to minimize the error caused by friction between the specimen and the support.
2. The line of the load action should pass the center of the span (center-to-center spacing between two support rollers), with a deviation of less than 1% of the span.
3. The span error should be within 0.5% of the nominal length.
4. The crack ends should be placed on the central line between two support rollers, with a deviation of less than 1% of the span.
5. The specimen should be perpendicular to the support roller's axis, with a deviation of less than 2°.
6. An overall blade or paste blade should be made on the specimen in advance. The blades should be parallel, with spacing symmetrical to the cracked surface. When a clamp-type extensometer is installed, the blade should match the extensometer's groove.
7. The load sensor and clamp-type extensometer are connected to the strain meter to amplify the output signal, then respectively connected to the load and displacement terminals on the X–Y recorder.
8. The X–Y recorder's full range is selected carefully so that the gradient of the initial elastic part is $0.7 \sim 1.5$, and the drawn graph has proper size. After the full range is selected, calibration should be performed on the load sensor and extensometer.
9. The loading rate should be kept even. If the estimated K_{IC} is approximately 2500 N/mm$^{3/2}$, the load continues to rise until fracture occurs within 0.5–2.5 min. In the loading process, the initial load and the fracture load should be recorded on the curve.
10. The test temperature and fracture appearance should be recorded.

5.1.4.5 K_Q Calculation

1. Determine F_Q from the recorded F–V curve.
2. Measure five crack lengths using the reading microscope, a_1, a_2, a_3, a_4, a_5, as shown in Figure 5.10. Take the averaged value of the middle three readings as the effective crack length, with an error of less than 0.5%.
3. Calculate a/W and find $f(a/W)$ based on the measured a and W.
4. Substitute F_Q, B, W, and $f(a/W)$ into Equation 5.5 to calculate K_Q.

If the aforementioned conditions are followed strictly in the test process, K_Q is the plane strain's fracture toughness K_{IC}.

5.2 TESTING OF SURFACE CRACK'S FRACTURE TOUGHNESS K_{IE}

In engineering practice, brittle fracture is caused by growth of surface cracks that do not go through the plate thickness in most cases. For example, the explosion of the Polaris missile's engine housing in the launch test and the subsequent damage that frequently occurred in hydraulic tests performed on the high-strength steel housing in the accident analysis are all caused by low-stress brittle fracture due to the growth of surface cracks. Most parts on the front edge of the surface crack are parallel to the plate width W, as shown in Figure 5.11. Therefore, the elastic constraint in the z-direction is borne by the plate width, so the deepest edge of the surface crack is in the three-direction tension stress state of the most dangerous plane strain. This is the crack growth that needs special attention in one plane strain condition in fracture mechanics. The front edge of the crack is the key zone in the fracture process. This zone is basically in the

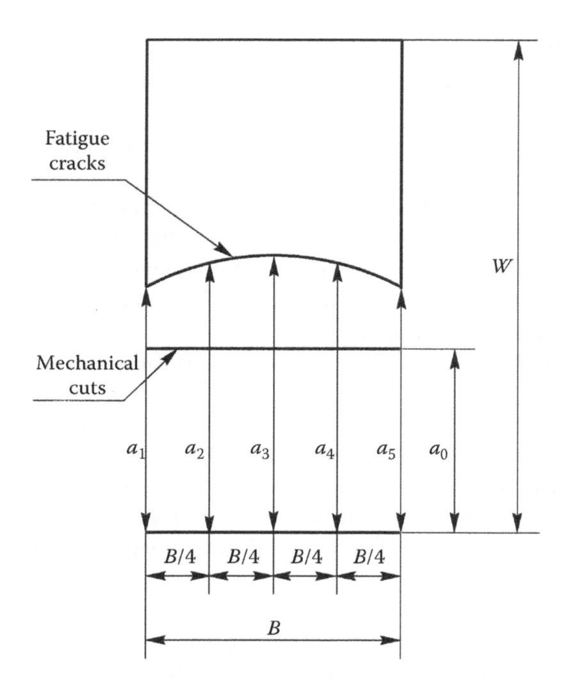

FIGURE 5.10 Crack length measurement.

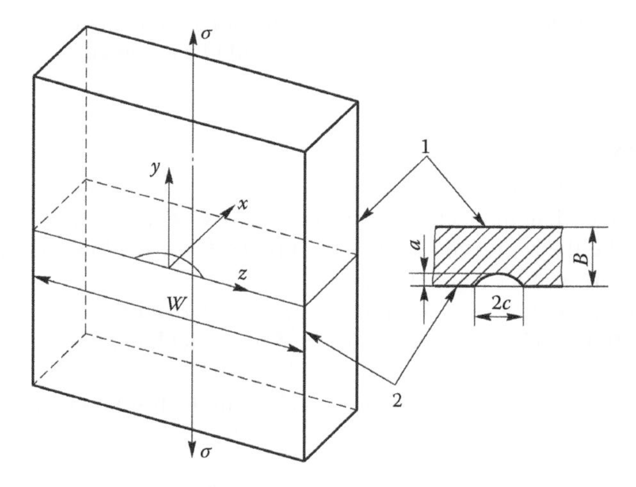

FIGURE 5.11 Plate with surface crack and a thickness of *B*. 1, Back surface; 2, front surface.

three-direction tension stress state; that is, the plane strain state. Therefore, even if the crack depth and ligament size do not meet the requirements, the critical value K_{IE} obtained is still within the K_{IC} range. It has a meaning absolutely different from the plane stress fracture toughness K_C obtained in the through crack test performed on this plate.

Because the surface crack specimen is essentially a type of plane strain state, the subcritical growth before fracture is so small that it can usually be ignored. Therefore, the maximum load when fracture occurs can be used to calculate K_{IE}. There is no need to draw the curve of load and opening displacement. However, when the surface crack specimen is used to measure the material's fracture toughness, there is no standard test method. It is only applicable to high-strength plates without obvious crack subcritical growth before fracture [5,6].

The principles and steps of testing K_{IE} are similar to those of testing K_{IC}. Next we only describe the principle of testing K_{IE}, which can be used to calculate the stress intensity factor and determine the specimen size and critical load.

5.2.1 Representation of Stress Intensity Factor K_I

The stress distribution around the semi-elliptical crack is an issue related to three-dimensional elastic mechanics. At present, there is no precise analytical solution, and approximate methods have to be used. Next we only describe the simplest Irwin approximate solution and the relatively more precise and easier Shah–Kobayashi solution.

5.2.1.1 Irwin Approximate Solution [7]

Irwin first calculated the stress intensity factor of a flat elliptical crack at both ends of the minor axis in an "infinitely large" body

$$K_I = \frac{\sigma\sqrt{\pi a}}{\Phi}, \tag{5.20}$$

where

σ is the tensile stress that is applied on both end planes of the "infinitely large" body and is perpendicular to the plane where the elliptical crack is on
a is the length of the ellipse's semi-minor axis
Φ is the complete elliptic integral of the second kind

$$\Phi = \int_0^{\pi/2} \left(1 - \frac{c^2 - a^2}{c^2} \sin^2\theta\right)^{1/2} d\theta, \tag{5.21}$$

where c is the length of the ellipse's semi-major axis. Then, the infinitely large body was cut into a semi-infinite body with a surface crack by a plane that contains the ellipse's major axis ($z - z$) and is perpendicular to the $x - z$ surface. The exposed surface is called the front surface. Then, this semi-infinite body was cut into a sheet by a plane that is parallel to the front surface. The surface obtained is called the back surface. As a result, the flat elliptical crack in the "infinitely large" body becomes the semi-elliptical crack of the finite plate (with a thickness of B), as shown in Figure 5.11. After these two free planes are used, the elastic constraint becomes small, the crack grows easily under tension, and the K_I value goes up. Irwin thought that free surface modification and plastic zone modification should be performed on Equation 5.20.

Irwin used the ratio between the crack tip's stress intensity factor K_I of the finite plate width double-edge through crack and the finite plate width central through crack (see Figure 5.12) to approximately estimate the total correction factor of the front surface and the back surface M_e:

$$M_e = \frac{(K_I)_{double\text{-}edge\,crack}}{(K_I)_{central\,crack}} = \frac{\sigma\sqrt{\pi a}\left[\dfrac{B}{\pi a}\left(\tan\dfrac{\pi a}{B} + 0.1\sin\dfrac{2\pi a}{B}\right)\right]^{1/2}}{\sigma\sqrt{\pi a}\left(\dfrac{B}{\pi a}\tan\dfrac{\pi a}{B}\right)^{1/2}} = \left[1 + \frac{0.1\sin(2\pi a/B)}{\tan(\pi a/B)}\right]^{1/2} \tag{5.22}$$

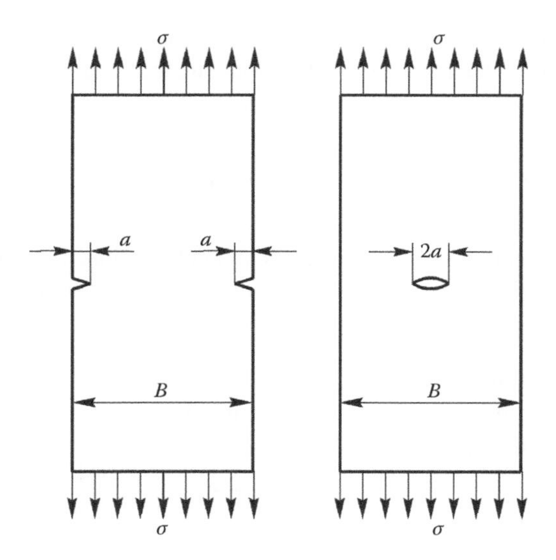

FIGURE 5.12 Plate with both edged through crack and central through crack.

When a/B is small, $\sin(2\pi a/B) \approx 2\pi a/B$, $\tan(\pi a/B) \approx \pi a/B$, and so

$$M_e = (1+0.2)^{1/2} \approx 1.1. \tag{5.23}$$

Considering the influence of the yielding zone existing near the crack tip, Irwin used the crack's effective depth $(a + r_y)$ to substitute for the crack's original depth a, when calculating K_I. Here, r_y should have this representation in the plane strain condition:

$$r_y = \frac{K_I^2}{4\sqrt{2}\pi\sigma_s^2}. \tag{5.24}$$

After multiplying by the free surface's correction factor $M_e = 1.1$ on the right side of Equation 5.20, and then substituting $(a + r_y)$ for a, we obtain

$$K_I = \frac{1.1\sigma\sqrt{\pi(a+r_y)}}{\Phi}.$$

After transposing and squaring, we have

$$(K_I\Phi)^2 = 1.21\sigma^2\pi\left(a + \frac{K_I^2}{4\sqrt{2}\pi\sigma_s^2}\right),$$

and subsequently,

$$K_I = \frac{1.1\sigma\sqrt{\pi a}}{\left(\Phi^2 - 0.212\dfrac{\sigma^2}{\sigma_s^2}\right)^{1/2}} = \frac{1.1\sigma\sqrt{\pi a}}{\sqrt{Q}}, \tag{5.25}$$

or

$$K_I = M_e M_P \frac{\sigma\sqrt{\pi a}}{\Phi}, \quad M_P = \frac{\Phi}{\sqrt{Q}}, \tag{5.26}$$

where
M_P is the plastic correction factor
Q is the crack shape factor

Because the crack exists and the cross section becomes smaller, sometimes the net section area is used to calculate the stress σ instead of the gross area. In this case, σ_n is used to substitute for σ in Equation 5.26. The averaged stress on the net section is

$$\sigma_n = \frac{P}{BW - \pi ac/2}, \tag{5.27}$$

where W is the plate width.

5.2.1.2 Shah–Kobayashi Solution [8]

Shah and Kobayashi calculated the front surface's correction factor M_1 using the following equation:

$$M_1 = 1 + 0.12\left(1 - \frac{a}{2c}\right)^2. \tag{5.28}$$

For the back surface's correction factor M_2, an alternating iterative method is used, gradually tending to the elastic solution of the elliptical crack inside the surface. If the interaction between M_1 and M_2 is not considered, the correction factor M_e is equal to the product of the correction factors of the front surface and the back surface, or $M_1 \cdot M_2$. Therefore, the equation for calculating the stress intensity factor on the ellipse's minor axis end is

$$K_I = \frac{M_e \sigma\sqrt{\pi a}}{\sqrt{Q}}. \tag{5.29}$$

This solution has sufficient theoretical basis to meet the free surface's boundary condition. a/B and $a/2c$ have widely applicable scope.

5.2.2 Requirements Related to Specimen Size

Strictly speaking, the specimen size—including the plate width, the crack depth, and the ligament size—should meet the plane strain condition and the linear elasticity small-scale yielding-zone requirements. Because the surface crack specimen's plate width W substitutes for the through-crack specimen's plate thickness B, the requirements related to the plate width can easily be met. However, it is difficult for the crack depth a and the ligament size $(B - a)$ to meet the requirement of $\geq 2.5(K_{IE}/\sigma_s)^2$. Until now, we could not in theory determine whether a stable K_{IE} value can be obtained by lowering it to one limit. Now, based on several test results and the opinion of a majority of institutes, we offer the following opinion.

1. *Thickness, crack depth, and ligament size*
To obtain the material's stable K_{IE} value, with the specimen thickness B maintaining a certain value, the crack depth a and the ligament size $(B - a)$ should also meet certain requirements. Obviously, only two items among B, a, and $(B - a)$ are independent.
In order to meet the plane strain condition for a shallow crack with $a/B < 0.5$, the thickness B, crack depth a, and ligament size $(B - a)$ are, respectively [4],

$$\left.\begin{aligned} B &\geq 1.0\left(\frac{K_{IE}}{\sigma_{0.2}}\right)^2 \\ a &\geq 0.5\left(\frac{K_{IE}}{\sigma_{0.2}}\right)^2 \\ B - a &\geq 0.5\left(\frac{K_{IE}}{\sigma_{0.2}}\right)^2 \end{aligned}\right\}. \tag{5.30}$$

In order to obtain a stable K_{IE} value for a deep crack with $a/B > 0.5$, the thickness and ligament size are, respectively [4],

$$\left.\begin{aligned} B &\geq 0.25(K_{IE}/\sigma_{0.2})^2 \\ B - a &\geq 0.10(K_{IE}/\sigma_{0.2})^2 \end{aligned}\right\}. \tag{5.31}$$

2. *Width and length*
The specimen width is an important issue in specimen design. From Irwin's original formula, the specimen width and length should meet the requirements of the "infinitely large" body. In testing practice, we also hope to minimize the influence of the size. Therefore, using references [4] from high-strength steel, we propose

$$\frac{W}{2c} \geq 3, \quad \frac{W}{B} \geq 6.$$

If the width is not large enough when the surface crack is deep for an aluminum alloy with high K_{IE}, bending will occur on the front surface where the crack is. Therefore, the aluminum alloy specimen needs a larger width [4]:

$$\frac{W}{2c} \geq 4,$$

or

$$\frac{W}{2c} \geq 5.$$

The specimen width requirement is also related to the crack depth. Generally, the deeper the crack is, the larger width it needs. Smith studied the influence of specimen width on the K_{IE} value [9] using an epoxy resin specimen and proposed that
When $a/2c < 0.3$ applies, $W/2c \geq 3$; and when $a/2c > 0.3$ applies, $W/2c \geq 5$.

For 30 CrMnSiAl medium-strength steel, testing and analyses [4] have studied the influence of specimen width on the fracture toughness value. If the plate width is too small compared with the plate thickness, the plane of the crack does not have sufficient constraint and transverse contraction occurs. Conversely, if the plate width is too large compared to the plate thickness, the crack imposes a relatively weak influence on the fracture. In both cases, we cannot obtain an effective fracture toughness value. Therefore, for the medium-strength steel specimen, its width is

$$\left.\begin{aligned} 4 \le \frac{W}{2c} \le 5 \\ 8 \le \frac{W}{B} \le 10 \end{aligned}\right\}.$$

For the specimen's working length l, we need to consider the loading method. Generally, it is better to use a dowel to transmit the force because it not only aligns the specimen under the force but also generates even internal force distribution. According to the Saint-Venant principle, to meet the condition of even internal force distribution, the working length should be [5]

$$L \ge 2W.$$

5.2.3 Determination of Critical Load

We have learned that the existing method of measuring a material's fracture toughness using a surface crack specimen is only applicable to a high-strength plate on which there is no obvious subcritical crack growth before fracture. In this condition, the maximum load (or fracture load) can be used to obtain the critical stress intensity factor K_{IE} value. Therefore, at present, K_{IE} testing is only performed on high-strength materials. The maximum load cannot be used to calculate the critical stress intensity factor for medium-strength materials if there is obvious subcritical crack growth before fracture. Therefore, the critical load of a medium-strength steel's surface crack specimen can be determined by using the relative crack growth [5]. If the crack growth $\Delta\alpha$ is equal to the size of the yielding zone on the crack tip $\gamma_y = (K_{IE}/\sigma_{0.2})^2/4\sqrt{2}\pi$, and if the requirement related to the initial crack depth α_0 is $\alpha_0 = 0.5(K_{IE}/\sigma_{0.2})^2$, then

$$\frac{\Delta\alpha}{\alpha_0} = \frac{\gamma_y}{\alpha_0} = \frac{1}{4\sqrt{2}\pi}\frac{(K_{IE}/\sigma_{0.2})^2}{0.5(K_{IE}/\sigma_{0.2})^2} \approx 10\%.$$

When calculating the crack growth $\Delta\alpha$ using the crack-opening displacement V, specimens with different crack depths should be used first. From the F–V curve based on the test, we can draw a calibration curve of the dimensionless $WEV/F = EV/\sigma B$ (where σ is the normal stress and E is the elastic modulus) and the relative crack size α/B. Using the calibration curve, we can correlate the crack's effective growth relative increment $\mathrm{d}\alpha/\alpha_0$ with the corresponding opening's displacement relative increment $\mathrm{d}V/V$ in the loading process:

$$\frac{\mathrm{d}V}{V} = H\frac{\mathrm{d}\alpha}{\alpha}, \tag{5.32}$$

where H is a factor when $\alpha = \alpha_0$ applies [4]. For specimens with a certain shape and a certain initial crack depth (when α_0/B is given), the factor H has a specific value. Please refer to the table [5]. Therefore, the opening's displacement relative increment corresponding to $\Delta\alpha/\alpha = 10\%$ is $\mathrm{d}V/V = 10\%H$.

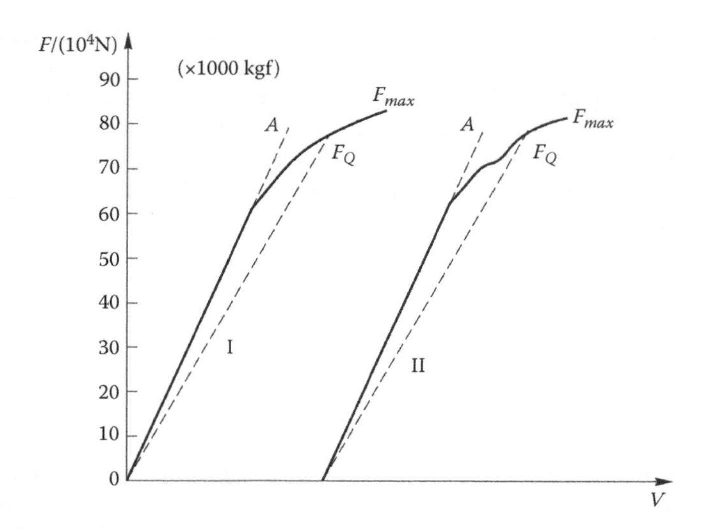

FIGURE 5.13 Medium-strength material.

In the F–V curve of a medium-strength steel specimen's surface crack (as shown in Figure 5.13), the gradient of the corresponding secant line $F/(V + dV)$ should be equal to $1/(1 + 10\%H)$ of the initial tangent line's gradient F/V. That is, the gradient of the secant line OF_Q is smaller than that of the initial tangent line OA by $10\%H$. The secant line OF_Q intersects the curve at the point F_Q, whose ordinate is the value in the critical load condition.

5.3 TESTING OF PLANE STRESS FRACTURE TOUGHNESS K_C

High-strength and thin-wall materials are used widely in modern industry, particularly as aviation and aerospace rapidly develop. To control fracture, we must know the material's plane stress fracture toughness K_C value. If high-strength and thin-wall components take the K_{IC} value as the design basis, the value is often conservative. The K_C value of thick materials (where the material's thickness does not meet the plane strain requirement) is related to the thickness B.

Although K_C testing methods have been studied for many years, they have only recently matured. Generally, the existing methods can be divided into two types: the direct measurement method and the indirect measurement method. The indirect measurement method system uses either a small specimen to measure the crack tip's critical opening displacement δ_C to calculate K_C (for an ideal plastic material and a low-hardening factor material, $K_C = \sqrt{E\sigma_s\delta_C}$ applies) or uses a cylinder blast [10,11] or a finite plate width strip tension test to measure some parameters for the calculation. However, these methods have not matured in theory and practice, and are seldom used. The direct measurement method requires the use of a large specimen. Although this method has not fully matured, it is used frequently. Therefore, this section describes the testing principles of the direct measurement method.

5.3.1 Representation of Stress Intensity Factor K_I

5.3.1.1 Representation of CCT Specimen's K_I

Center crack tension (CCT) specimens, compact tension (CT) specimens, and crack line wedge loaded (CLWL) specimens are often used to measure K_C. At present, the CCT specimen

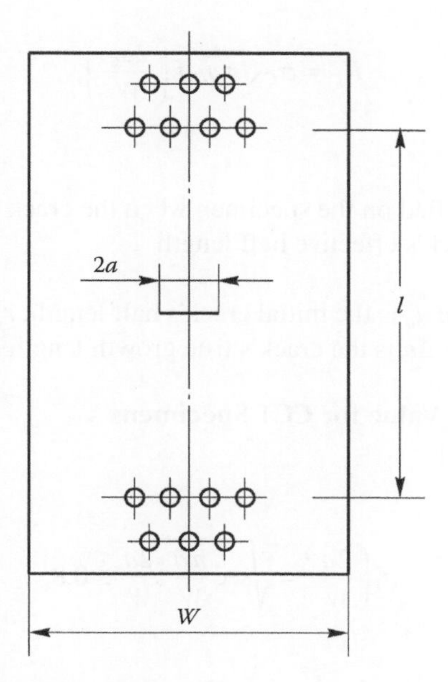

FIGURE 5.14 Central through crack specimen.

is used in most cases, as shown in Figure 5.14. The CCT specimen's stress intensity factor K_I is represented by

$$K_I = \sigma\sqrt{\pi a}\,f\left(\frac{2a}{W},\frac{l}{W}\right), \tag{5.33}$$

where

σ is the nominal stress applied on the specimen
a is the crack's half length
l is the specimen's working length
W is the specimen's width
f is a function of $2a/W$, l/W

During unstable crack growth (in the linear elastic condition), after substituting the critical value into Equation 5.33, we obtain

$$K_C = \sigma_C\sqrt{\pi a_C}\,f\left(\frac{2a_C}{W},\frac{l}{W}\right). \tag{5.34}$$

The analysis shows that when the specimen's end uses a single-dowel clip and $l/W \geq 2$, or when a multidowel clip is used and $l/W \geq 1.5$, its correction factor f is the value when $l/W \to \infty$, without significant influence. It is estimated that the error does not exceed 2%–3%. After that, the specimen's l/W meets this requirement. Equations 5.33 and 5.34 can be simplified thusly:

$$K_I = \sigma\sqrt{\pi a}\,f\left(\frac{2a}{W}\right), \tag{5.35}$$

$$K_C = \sigma_C \sqrt{\pi a_C} f\left(\frac{2a_C}{W}\right), \tag{5.36}$$

where

σ_C is the nominal stress applied on the specimen when the crack begins unstable growth
a_C is the corresponding crack's effective half length

That is, $a_C = a_0 + r_y + \Delta a$ where a_0 is the initial crack's half length, r_y is the plastic zone's equivalent growth length on one side, and Δa is the crack's true growth length on one side.

5.3.1.2 Correction Factor Value for CCT Specimens

The correction factor $f(2a/W)$ is

$$f_1\left(\frac{2a}{W}\right) = \sqrt{\sec\frac{\pi a}{W}}, \frac{2a}{W} \le 0.8, \tag{5.37}$$

or

$$f_2\left(\frac{2a}{W}\right) = 1 - 0.1\left(\frac{2a}{W}\right) + \left(\frac{2a}{W}\right)^2, \frac{2a}{W} \le 0.6. \tag{5.38}$$

At present, the most precise representation is

$$f_3\left(\frac{2a}{W}\right) = \left[1 - 0.025\left(\frac{2a}{W}\right)^2 + 0.06\left(\frac{2a}{W}\right)^4\right]\sqrt{\sec\frac{\pi a}{W}}. \tag{5.39}$$

5.3.2 Selection of Specimen Size

5.3.2.1 Specimen Width

Generally, large specimens are used to measure K_C. The key problem is how to select the minimum plate width (the requirement for the plate's ratio of length to width was described earlier) and to ensure that the measured K_C value is effective.

Most researchers accept Feddersen's semi-empirical result [4]. Fedderson proposed that the specimen's minimum width should be

$$W_{min} = 27r_y \approx 4.29\left(\frac{K_C}{\sigma_s}\right)^2. \tag{5.40}$$

The crack size is

$$2a = \frac{W_{min}}{3} \approx 1.43\left(\frac{K_C}{\sigma_s}\right)^2. \tag{5.41}$$

The corresponding critical stress (nominal stress) is

$$\sigma_C \leq \frac{2\sigma_s}{3}. \tag{5.42}$$

In fact, the averaged stress on the net section is $\sigma_n \leq \sigma_s$.

In many materials, the crack has obvious subcritical growth and the difference between $2a_0$ and $2a_C$ cannot be ignored. Most test results used are $2a_0$ (initial crack length). However, a calculated W_{min} based on Equation 5.41 is not the true minimum plate width. Cui Zhenyuan et al. proposed that [4]

$$W_{min} = 3.56\left(\frac{K_C}{\sigma_s}\right)^2. \tag{5.43}$$

This is lower than Feddersen's result by 17%. If the material's σ_s and plate width W are known, the highest effective K_C value of the specimen can be calculated based on the aforementioned equation. Values higher than that are invalid. The aforementioned equation is derived based on the $\sigma_n \leq \sigma_s$ condition. Generally, when $\sigma_n \approx \sigma_s$, the stress is too high, and it is difficult to say that the ligament part still dominates in the elastic stress field. In the common condition where $\sigma_n \leq 0.8\sigma_s$, the specimen plate width should be at least

$$W = 1.56W_{min} = 5.56\left(\frac{K_C}{\sigma_s}\right)^2 \approx 35r_y. \tag{5.44}$$

If the value of K_C/σ_s is estimated in advance, the required plate width can be calculated based on the aforementioned equation.

5.3.2.2 Initial Crack Length

To maximize the role of the plate width, according to the aforementioned analysis, $2a_C/W \approx 0.44$ should be maintained. Considering the great difference among various materials in subcritical crack growth, from crack initiation to unstable growth, there are different requirements on $2a_0/W$. Considering that the specimen ligament must also have a certain size, generally $2a_0/W \approx 1/3$ can be taken. For brittle materials with small subcritical growth, its value may be a little higher, for example, 0.33–0.40; for ductile materials with large subcritical growth, its value may be a little lower, for example, 0.25–0.30. For easy control, generally $2a_0/W \approx 0.3 \sim 0.35$ is used in the test.

5.3.3 Determination of K_C Value

There are many methods to determine the K_C value. Now we describe the crack growth resistance curve method (*R* curve method for short) and the load–displacement curve method (*F–V* curve method).

5.3.3.1 *R* Curve Method

The *R* curve is the graph of the relationship between the crack growth resistance *R* and the effective (or true) crack length *a* in the material, as shown in Figure 5.15. It shows the material's change with

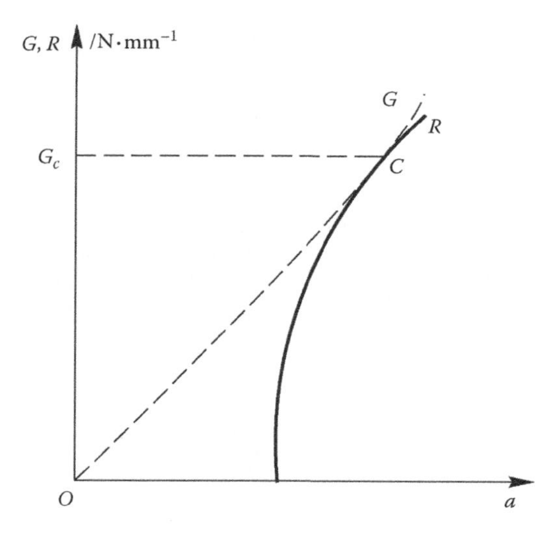

FIGURE 5.15 *R* curve.

fracture resistance in the crack's slow stable growth. In this process, the amount that the material resists the crack growth R is equal to the applied crack growth force G (under the plane stress, $G = K_I^2/E$). That is, $R = G$ until crack growth reaches the critical state. Afterwards, when applies, it changes to unstable growth.

On the F–V curve, as shown in Figure 5.16, the initial straight line is the elastic deformation stage. Then, the curve deflects to the right, which is caused by the crack tip plastic zone's equivalent growth and the crack's true growth. After the curve becomes horizontally oriented, unstable crack growth continues until fracture occurs. Therefore, the point of the critical state is

$$\frac{d\sigma}{dV} = 0. \tag{5.45}$$

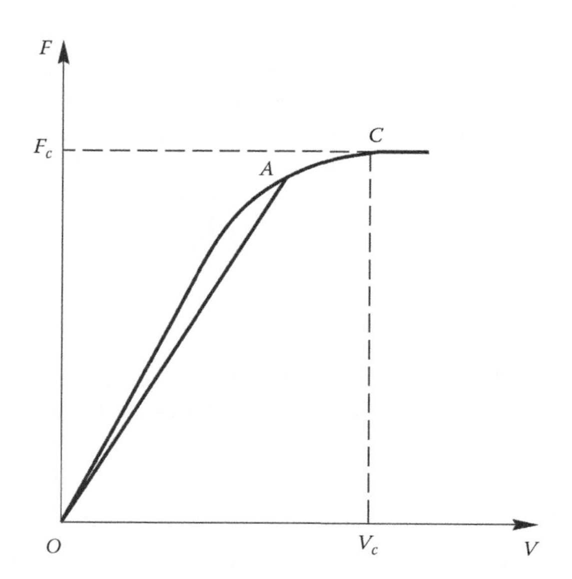

FIGURE 5.16 *F*–*V* curve.

$$G = f(\sigma, a) = R \quad \text{applies, so}$$

$$\frac{\partial R}{\partial \sigma}\cdot\frac{d\sigma}{dV}+\frac{\partial R}{\partial a}\cdot\frac{da}{dV}-\frac{\partial G}{\partial \sigma}\cdot\frac{d\sigma}{dV}-\frac{\partial G}{\partial a}\cdot\frac{da}{dV}=0. \tag{5.46}$$

After substituting Equation 5.45 into Equation 5.46, we obtain

$$\left.\frac{\partial R}{\partial a}\right|_{\sigma=\sigma_C}=\left.\frac{\partial G}{\partial a}\right|_{\sigma=\sigma_C}. \tag{5.47}$$

Therefore, the point of contact of the specimen's critical crack growth force G_C curve and R curve C (as shown in Figure 5.15) is the unstable crack growth point. The G value corresponding to this point is the critical crack growth force G_C, and the plane stress' fracture toughness K_C can be calculated.

To draw the R curve, we need to measure the specimen's crack effective length on any point A of the F–V curve, $2a_e = 2a_0 + 2r_y + 2\Delta a$ (as shown in Figure 5.16), where the initial crack length is $2a_0$; the plastic zone's equivalent growth length is $2r_y$; and the crack's true growth length is $2\Delta a$. When determining G_C (or K_C), we also need to measure the effective crack length $2a_C$ at the critical point C. There are many methods to measure the effective crack length. At present, the flexibility method is used in most cases.

5.3.3.2 *F–V* Curve Method

When calculating the K_C value with Equation 5.36, we need to determine the crack's effective length $2a_C$ at the critical point in the CCT specimen tension and the corresponding critical stress σ_C. It has been pointed out that the critical point is the initial point of unstable crack growth in the CCT specimen's slow tension process, or the point C, where the F–V curve (as shown in Figure 5.16) becomes oriented horizontally. After the critical point C is determined, the load corresponding to the critical point is the critical load F_C. The critical stress σ_C is obtained by dividing F_C by the specimen's section area BW. The flexibility method is used to calculate the critical crack length $2a_C$. Therefore, we can obtain the K_C value.

The aforementioned text describes two methods for determining a material's K_C value. Generally, we not only need to know the material's K_C value, but we also need to find out the resistance of the material to fracture in the entire process of slow crack growth. Therefore, in most cases, the R curve method is now used for determining K_C.

5.4 TESTING OF *J* INTEGRAL'S CRITICAL VALUE J_{IC}

It is easy to measure K_{IC} for high-strength and ultra-high-strength materials because such materials have high yielding point σ_s and low K_{IC}, so the specimen size $[B \geq 2.5(K_{IC}/\sigma_s)^2]$ is small. For medium- and low-strength materials, when the component size is large, or it is in the low-temperature condition, low-stress brittle fracture often occurs. However, such materials have low yielding points σ_s and high K_{IC}, so a very large specimen is required. This not only consumes materials but also requires large test equipment. At present, there are several methods for testing the fracture toughness of medium- and low-strength materials. The first method is to measure the J integral's critical value J_{IC} and convert it to K_{IC} based on the J integral principle. The second method is to represent fracture toughness using the critical value δ_C of the crack opening displacement (COD). The third method is to calculate K_{IC} from other material performance indicators, such as a Charpy V notch specimen's impact absorbed energy, based on test conclusions or semitheoretical analysis.

5.4.1 Testing Methods

The J deformation work change rate is defined as

$$J = -\frac{1}{B}\left(\frac{\partial U}{\partial a}\right)_{\Delta}. \tag{5.48}$$

where

- B is the specimen thickness
- U is the deformation work
- a is the crack length
- Δ is the displacement of the force position

J_{IC} can be measured in the testing method. At present, there are many methods to measure J_{IC}, such as the multispecimen method, the single-specimen method, and the resistance curve method. These are described next.

5.4.1.1 Multispecimen Method

A three-point bending test is performed on three to four specimens with the same size and different crack lengths. Draw a F–Δ curve as shown in Figure 5.17a. Each curve is divided into several sections to calculate the areas A_1, A_2, A_3, A_4,⋯ corresponding to the displacement Δ_1, Δ_2, Δ_3, Δ_4,⋯, as shown in Figure 5.17b. Then, this area is divided by the specimen's thickness B to obtain the corresponding U/B value, which is the ordinate when the crack length is a_1 and Δ is Δ_1,Δ_2,Δ_3,Δ_4,⋯ respectively, in Figure 5.17c. For other specimens with a crack length of a_2,a_3,a_4,⋯, the same method is used to draw the relationship curve of (U/B)–a in Figure 5.17c.

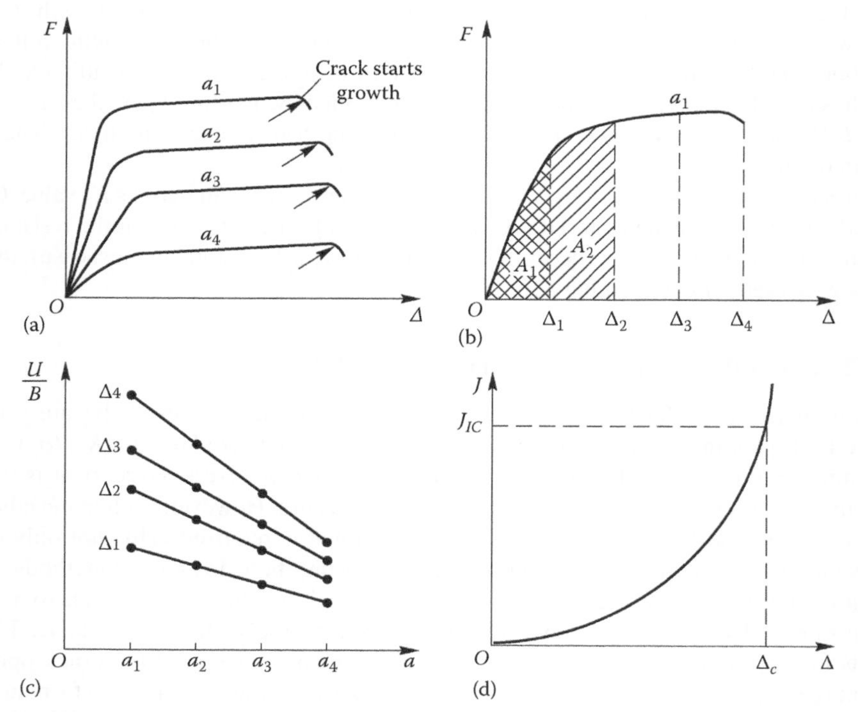

FIGURE 5.17 Measuring the J_{IC} value using the multispecimen method. (a) F–Δ curve. (b) Work with given displacement. (c) (U/B)–a curve with given displacement Δ. (d) J–Δ curve.

The (U/B)–a relationship is a set of approximate straight lines. The gradient of these straight lines with a negative sign, that is, $-(1/B)(\partial U/\partial a)$, is the J integral's value. Obviously, the gradient is related to Δ, so we can obtain the relationship between the J integral and the displacement Δ as shown in Figure 5.17d. As long as the critical displacement Δ_C value when the crack is initiated is determined, we can calculate the J_{IC} value.

5.4.1.2 Single-Specimen Method

Testing and calculation of the J_{IC} value using the multispecimen method is troublesome. However, references [12] concerning the plastic theory analysis of deep cracking and short span three-point bending specimens show that when it reaches the given displacement Δ or load F the following approximate relationship exists between the J integral and the deformation work absorbed by the specimen $U = \int_0^{\Delta} F d\Delta$ in the loading process and the crack length a or ligament size $(W - a)$:

$$J = \frac{2U}{B(W-a)}. \tag{5.49}$$

Based on this relationship, we only need to determine the deformation work of the critical displacement $U_C = \int_0^{\Delta_C} F d\Delta$ on one specimen's F–Δ load curve in order to calculate the J integral's critical value J_{IC}.

5.4.1.3 Resistance Curve Method

When medium-strength and low-strength material components initiate a crack, it does not mean that fracture will occur on the component. If the fracture toughness of the crack initiation point is used as the design basis, the value is too conservative. Therefore, in most cases, the resistance curve method is used to measure the fracture toughness of medium-strength and low-strength materials. We have already described how to determine the K_C value using the crack growth resistance curve method (or the R curve method). Similarly, in the slow and stable crack growth process, the fracture toughness parameter J is used to represent the relationship curve of the material's crack growth resistance J_R and the crack length growth Δa. This is called the J_R curve or the J_R–Δa resistance curve. The method used to determine the J integral value in the critical state on the J_R resistance curve (according to the component's working nature and actual needs) is called the resistance curve method.

The fracture toughness J_{IC} value is the J integral value in the critical state. Based on engineering safety analysis, fracture control, and toughness assessment, it generally has the following types:

1. Apparent crack initiation toughness J_i. This is the material's fracture resistance J_R when crack initiation begins. It is difficult to accurately determine the crack initiation point. Therefore, we can only determine it in a relative and conditional manner.
2. Conditional crack initiation toughness $J_{0.05}$. This is the material's fracture resistance J_R value when the apparent crack growth $\Delta a = 0.05$ mm applies.
3. Conditional fracture toughness $J_{0.2}$. This is the material fracture resistance J_R value when the apparent crack growth $\Delta a = 0.2$ mm applies. It represents the characteristic value of the resistance curve.
4. In addition, the maximum load's starting point, the unstable growth or load reduction point, and the resistance curve saturation point can also be used as the characteristic value of the resistance curve.

5.4.2 Determination of Critical Point

Because J_{IC} is determined based on the deformation work applied on the specimen at the critical point, J_{IC} accuracy depends directly on the selection of the critical point. The following can be used as the critical point: (1) the crack's initial growth point; that is, the crack initiation point; (2) the point where the crack growth is 2%a; (3) the point where the load just reaches the maximum load on the F–Δ curve; and (4) the point where the maximum load begins to go down on the F–Δ curve. Tests show that J_{IC} values calculated by taking the crack initiation point as the critical point are relatively concentrated. When certain conditions are met, the J_{IC} value is a material constant, independent of the specimen size. Therefore, it is reasonable to take the crack initiation point as the critical point.

It should also be noted that the crack initiation point varies to some degree with the testing instrument's sensitivity. The methods for determining the crack initiation point include the potential method, the resistance method, the oxidation method, the metallographic method, and the acoustic emission method. The resistance method is similar to the potential method. Next we outline the potential method, the metallographic method, and the acoustic emission method.

5.4.2.1 Potential Method

The potential method applies a constant value, stable current I on both ends of the specimen, then measures the potential U change on both sides of the crack. In the test, the displacement of the point where force is applied (Δ) is measured with a clamp-type electronic extensometer. The potential probes are welded on both sides of the crack. An X–Y recorder is used to automatically measure and draw the E–Δ curve. Due to crack growth, the potential difference goes up quickly. Therefore, based on the sudden change of the E–Δ curve, we can determine the crack initiation point. The measured E–Δ curve can basically be divided into the following three types:

1. An E–Δ curve is made up of straight lines AB, CD, and curve BC, as shown in Figure 5.18a. The AB straight-line section is caused by the elastic opening on the crack tip. The BC curve section is caused by plastic growth on the crack's front edge. The CD straight-line section is the increased potential linearity due to the crack growth. After scanning the specimen fracture unloaded near point C on one material, we find that the specimen crack starts crack initiation and grows forward. Therefore, point C is the crack initiation point where the crack begins to grow.
2. For high-strength materials or large specimen sizes, the BC curve section will disappear. The E–Δ curve is made up of straight lines AB, BD, as shown in Figure 5.18b. In this case, the intersection point B of these two straight lines can be taken as the crack initiation point.
3. For materials with good toughness values, or small specimens, the E–Δ curve is a smooth curve without a break, as shown in Figure 5.18c. In this case, other methods should be used to determine the crack initiation point.

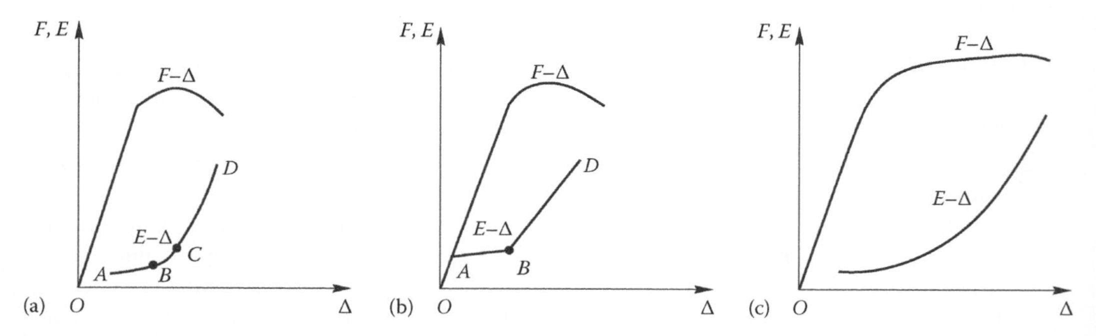

FIGURE 5.18 Three F–Δ and E–Δ curves for (a) typical elasto-plastic materials, (b) high-strength materials or large specimens, and (c) materials with good toughness or small specimens.

5.4.2.2 Metallographic Method

5.4.2.2.1 Multispecimen method

Four to six specimens with basically the same crack length are loaded on different points of the F–Δ curve (it is better to load one specimen on the crack growth that does not rarely occur and load other specimens on the other crack growths), as shown in Figure 5.19. After unloading, the specimen is split symmetrically along the direction perpendicular to the crack surface. The process of crack opening and growth is shown in Figure 5.20. The growth that we observe directly on the fracture is AC, which is larger by AB than the true growth BC. Therefore, the specimen is ground into metallurgical sections after being split. Under a metallurgical microscope, we can more accurately determine the crack initiation point (point B) based on the crack growth of the loaded specimens.

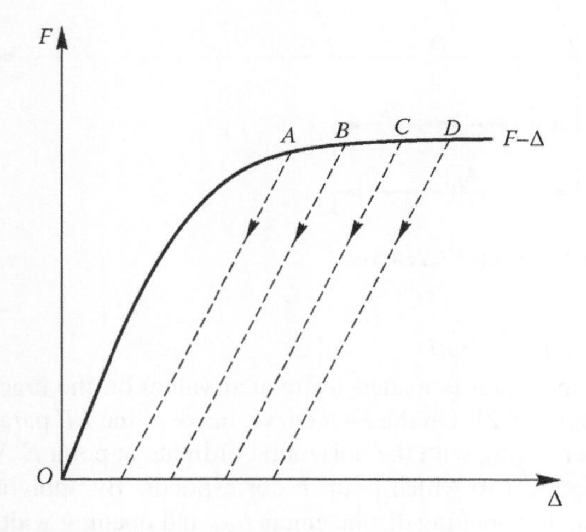

FIGURE 5.19 Unloading points of different specimens on F–Δ curve.

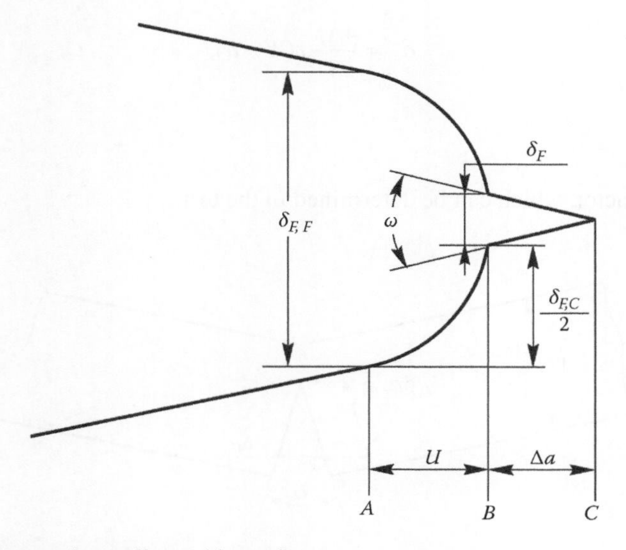

FIGURE 5.20 Crack growth metallographic section.

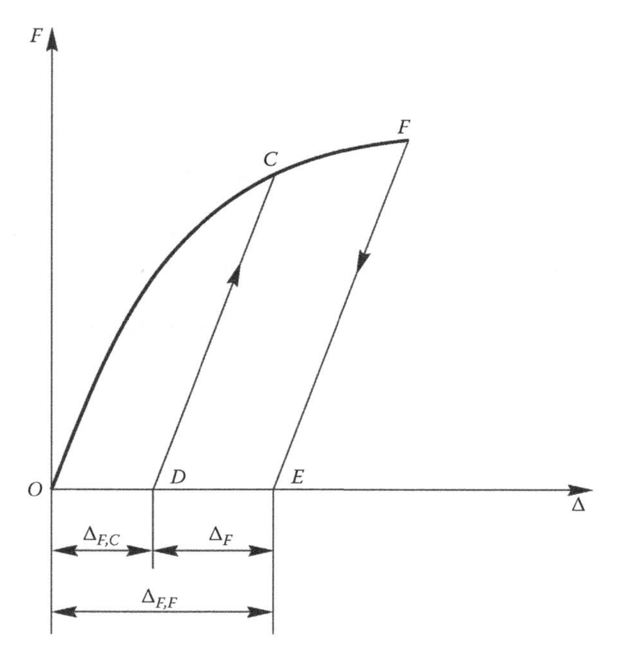

FIGURE 5.21 Crack initiation C on F–Δ curve.

5.4.2.2.2 Single-specimen method

Any point F—after the specimen is loaded (estimated value) on the crack initiation point C—is unloaded, as shown in Figure 5.21. On the F–Δ curve, make a line FE parallel to the initial tangent line crossing point F, intersecting with the horizontal ordinate at point E. When $OE = \Delta_{F,F}$ applies, this is the plastic displacement to which point F corresponds. By applying the concept of center rotation [4] and combining the opening displacement $\delta_{F,F}$ and opening width $\Delta\delta_F$ of the crack tip, as shown in Figure 5.20, we can accurately determine the crack initiation point.

The practice shows that when the yielding degree is deep, the following approximate relationship exists, as shown in Figure 5.22:

$$\delta_F = \frac{4\Delta_F}{S} r(W - a), \tag{5.50}$$

where

S is the span

r is the rotation factor, which can be determined in the test

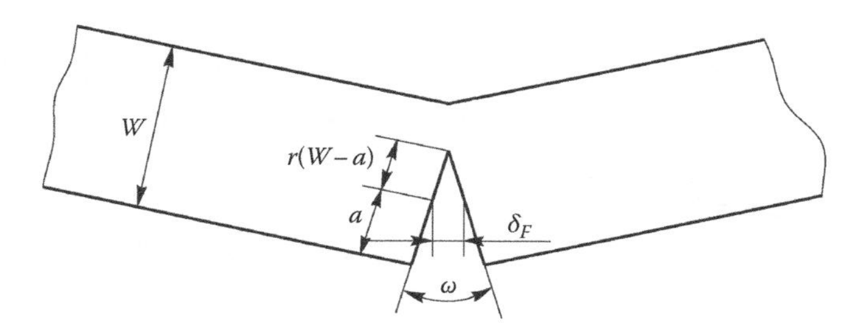

FIGURE 5.22 Three-point bending specimen after loading.

$$r = \frac{S\delta_{F,F}}{4\Delta_{F,F}\left[(W-a)+\Delta a\right]}, \tag{5.51}$$

where Δa is the crack growth length on the break point. From Equation 5.50, we can obtain the plastic displacement $\Delta_{F,C}$ of the point where force is applied during the crack initiation:

$$\Delta_{F,C} = \frac{S\delta_{F,C}}{4r(W-a)}. \tag{5.52}$$

When the ligament yielding degree is deep, in the case where the extension zone's width $\delta_{F,C}$ after crack initiation remains unchanged, and based on the metallographic section, we obtain $\delta_{F,C} = \delta_{F,F} - \Delta\delta_F$. After substituting it and Equation 5.51 into Equation 5.52, we obtain

$$\Delta_{F,C} = \left(1 - \frac{\Delta\delta_F}{\delta_{F,F}}\right)\left(1 + \frac{\Delta a}{W-a}\right)\Delta_{F,F}. \tag{5.53}$$

Calculate the $\Delta_{F,C}$ value. Then, from point D whose horizontal ordinate is $\Delta_{F,C}$ away from origin O in Figure 5.21, make a line parallel to the initial tangent line, intersecting the F–Δ curve at point C, which is the crack's initial point.

5.4.2.3 Acoustic Emission Method

In the loading process, first elastic deformation occurs on the specimen, then plastic deformation occurs on the crack end. When plastic deformation reaches the critical state, the crack begins to grow forward. Corresponding to the different stages given earlier, its acoustic emission characteristics are different. The acoustic emission rate S (or acoustic emission number N)–displacement Δ curve is recorded synchronously. Based on the characteristics of this curve, we can determine the crack initiation point. The principle of the acoustic emission test is that the specimen makes an acoustic emission signal, which is detected by an instrument probe. Then it is amplified by a preamplifier and an acoustic emission tester, from which the acoustic emission rate (or acoustic emission number) in the selected frequency range is input into an X–Y recorder. Meanwhile, the signal of the displacement Δ of the point where force is applied is also amplified and input in order to record the acoustic emission rate S-displacement Δ curve.

At present, there are differing opinions regarding which acoustic emission rate's peak value corresponds to the crack initiation point. Therefore, the potential method (or resistance method) and the acoustic emission method can be used together to determine the crack initiation point. Then the oxidation method or the metallographic method is used for validation in order to ensure the accuracy of the result.

5.5 TESTING OF COD'S CRITICAL VALUE δ_C

As described earlier, one of the methods for testing the fracture toughness of medium- and low-strength materials is to determine the COD (also represented as δ), whose critical value δ_C is used to represent the fracture toughness. In practice, δ_C can be measured with an indirect method using a small three-point bent specimen in the full yielding state. The measurement process is simple. Under certain conditions, the measured result is also stable. China uses the national standard GB2358-1980 Crack Opening Displacement (COD) Test Method.

Though the crack tip opening displacement is not a direct stress strain field parameter—and it is very difficult and uncertain for precise definition, analysis, and direct measurement—it can simply and effectively solve the actual problem because the indirect definition and measurement methods, as well some empirical relationships, are used. Therefore, it is still widely used in the engineering industry.

This section mainly describes the principle and method of testing the COD critical value δ_C, as well as the method used to test the δ_R–Δa curve.

5.5.1 δ_C REPRESENTATION

At present, different opinions exist regarding the COD definition. In the DD-19 Specification of the British Commission of COD Application and British Standard Institute, the opening displacement obtained by moving the crack surface's tangent line to the original crack tip is taken as the COD's practical definition (δ in Figure 5.23). However, Cai Qigong [13] proposed that the DD-19 definition should be described more precisely in this manner: after the elastic opening displacement is deducted, the opening displacement (δ' in Figure 5.23) is obtained by moving the straight-line part of the opening displacement curve, measured on the crack-free surface, to the original crack tip. Note that after elastic displacement is deducted, the elastic zone part should be a straight line. In the finite element calculation, it has been suggested that the plastic zone boundary or elastic–plastic boundary on the crack surface be taken as the measurement point to measure or calculate its opening displacement (δ'' in Figure 5.23). According to different COD definitions, the calculation of δ has different forms.

Because it is difficult to directly measure the crack tip opening displacement, the opening displacement on this end is determined indirectly, based on the deformation characteristics of a three-point bending specimen. On a rectangular section of the specimen, a mechanical notch is made. Then, a cannelure is cut on the symmetrical centerline of the mechanical notch by a molybdenum wire electric spark. Along the cannelure end, a fatigue crack with a certain length is made, as shown in Figure 5.24. In the three-point bending loading process, as in measuring K_{IC} of the three-point bending specimen, the load F and notch opening displacement V curve is drawn automatically. Therefore, the notch's opening displacement V, or force point displacement Δ, can be used calculate the crack tip opening displacement δ.

5.5.1.1 Representation Containing Rotation Factor *r*

Before crack initiation, assuming that the ligament has yielded, both arms of the specimen rotate around one point O, whose position depends on the rotation factor r. Assuming that the distance

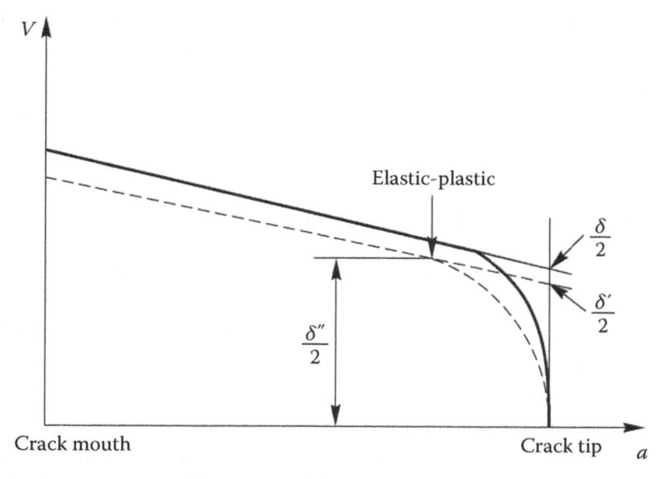

FIGURE 5.23 Crack tip opening displacement's different definitions.

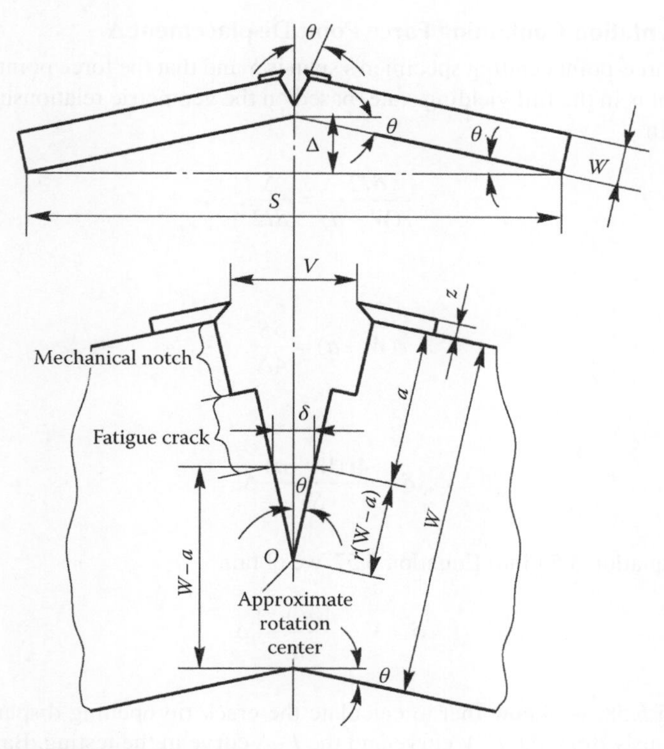

FIGURE 5.24 Fracture side view when bent.

from point O to the original crack end is $r(W - a)$, and that the blade thickness at both sides of the crack is z, the relationship between δ and V can be established using the triangle relationship.

$$\frac{\delta}{V} = \frac{r(W-a)}{r(W-a)+a+z}.$$

Thus,

$$\delta = \frac{r(W-a)}{r(W-a)+a+z}V. \tag{5.54}$$

From Equation 5.54, we know that if the rotation factor r is determined, we can use a clamp-type equation extensometer to measure the opening displacement between blades V_C in the critical state in order to calculate δ_C, the crack tip opening displacement value.

Determining the rotation factor r is the key to calculating δ_C and is rather difficult. The main methods include the test calibration method and the empirical formula method. Next we only describe the fixed rotation factor recommended by the DD-19 (British Standard Institution DD-19 standard):

$$r = \frac{1}{3}. \tag{5.55}$$

When $\delta = 0.0625$–0.625 mm, we can obtain a good approximate value of the crack tip opening displacement. When V is very small, namely, $\delta < 0.0625$, r should be less than 1/3. If $r = 1/3$ is used to calculate the δ value instead, a large error will occur.

5.5.1.2 Representation Containing Force Point Displacement Δ

Assume that the three-point bending specimen's span is S and that the force point displacement is Δ. When the ligament is in the full yielding state, based on the geometric relationship shown in Figure 5.24, we can obtain

$$\frac{\delta/2}{r(W-a)}=\frac{\Delta}{S/2},$$

so that

$$r(W-a)=\frac{\delta S}{4\Delta}, \tag{5.56}$$

or

$$\delta=\frac{4r(W-a)}{S}\Delta. \tag{5.57}$$

Substituting Equation 5.54 into Equation 5.57, we obtain

$$\delta=V-\frac{4(a+z)}{S}\Delta. \tag{5.58}$$

From Equation 5.58, we know that to calculate the crack tip opening displacement δ, we only need to synchronously draw the F–V curve and the F–Δ curve in the testing. Based on the opening displacement between blades V_C and the force point displacement Δ_C during the critical state in testing, we can calculate the crack tip opening displacement δ_C value.

5.5.1.3 Wells Representation

Wells thought that COD should be the crack tip opening displacement obtained after the crack surface's elastic displacement is deducted. His representation is

$$\left.\begin{aligned}\delta&=\frac{0.45(W-a)}{0.45(W-a)+a+z}(V-V'), \quad \text{when} \quad V\ge 2V'\\ \delta&=\frac{0.45(W-a)}{0.45(W-a)+a+z}\cdot\frac{V^2}{4V'}, \quad \text{when} \quad V<2V'\end{aligned}\right\}, \tag{5.59}$$

where $V'=\gamma\sigma_s W/E'$ is the elastic displacement of the crack surface in the plane stress state, $E'=E$; while in the plane strain state, and is a function of a/W.

From Equation 5.59, we know that we can calculate the crack tip opening's displacement δ_C value as long as we are measuring the opening displacement between blades V_C at the critical state in the test.

5.5.1.4 Representation Containing δ_e and δ_P

When calculating the J integral value [4],

$$J=J_e+J_P, \tag{5.60}$$

where

J_e is the J integral of the ideal linear elastic body specimen when the load is P

J_P is the J integral of the ligament plastic deformation

Similarly, δ may become

$$\delta = \delta_e + \delta_P, \tag{5.61}$$

where δ_e is the elastic opening displacement of the ideal linear elastic body crack tip when the load is P. Its calculation equation is

$$\left.\begin{aligned} \delta_e &= 0.5\frac{K_I^2}{E'\sigma_s}, \quad \text{Plane stress state} \\ \delta_e &= \frac{K_I^2}{E\sigma_s}, \quad \text{Plane strain state} \end{aligned}\right\}. \tag{5.62}$$

δ_P is the crack tip plastic opening displacement generated by the ligament plastic deformation. It can still be calculated from the plastic part V_P of the opening displacement between blades, based on the hinge model, so that

$$\delta_P = \frac{r(W-a)}{r(W-a)+a+z}V_P. \tag{5.63}$$

Finally, from Equations 5.62 and 5.63, and $r = 0.45$, we obtain the calculation equation in the plane strain state:

$$\delta = \delta_e + \delta_P = \frac{K_I^2(1-\nu^2)}{2E\sigma_s} + \frac{0.45(W-a)V_P}{0.45W+0.55a+z} \tag{5.64}$$

where K_I is the stress intensity factor.

For this definition, COD has clearer physical meaning and δ_e is calculated based on the load. Obviously δ_P does not include the influence of the crack surface's elastic deformation, and $\delta = \delta_e + \delta_P$ does not include the influence of the crack surface's elastic deformation. We also see that this definition relates COD to K_{IC} and the J integral. When the specimen becomes fractured in the linear elastic range, $\delta_P = 0$ applies. $\delta = \delta_e$ is directly related to K_{IC}. Generally, it is equivalent to the J integral, because in the rotation factor model the following equation applies:

$$\delta_P = \frac{4r(W-a)}{S}\Delta_P, \tag{5.65}$$

where Δ_P is the plastic part of the force point displacement. Therefore, an obvious relationship is established between COD and the J integral. Therefore, in COD testing, Equation 5.64 is used in most cases.

From Equations 5.64 and 5.65, we know that as long as the F–V curve or the F–Δ curve is drawn in the testing, we can determine the load F_C, the plastic opening displacement between the blades V_P, or the plastic force point displacement Δ_P in the critical state, and calculate the crack tip opening displacement value δ_C.

5.5.2 V_C Determination

The potential method can be used to determine the opening displacement's critical value V_C. In the testing process, two *X–Y* recorders are used to synchronously draw the *F–V* curve and the *E–t* (*t* is the time) curve, as shown in Figure 5.25. The *F–V* curve has four types.

5.5.2.1 First Type of *F–V* Curve

The load increases as the displacement goes up, until quick unstable fracture occurs. When it reaches the maximum load, we can hear a clear "burst" sound. In this case, a sharp break appears on the *E–t* curve, as shown in Figure 5.25a. The displacement at the maximum load F_{max} is taken as the critical displacement V_C.

5.5.2.2 Second Type of *F–V* Curve

Two similar curves have obvious "burst" platforms, as shown in Figure 5.25b and c. In the testing process, when the load reaches the "burst" load, we can also hear a "burst" sound, as shown in Figure 5.25c. After the second "burst" load, the specimen does not fracture quickly. The *F–V* curve extends downward stepwise, which is different from that in Figure 5.25b. Their *E–t* curve has two or more breaks, each of which corresponds to a "burst" sound and a "burst" platform on the *F–V* curve.

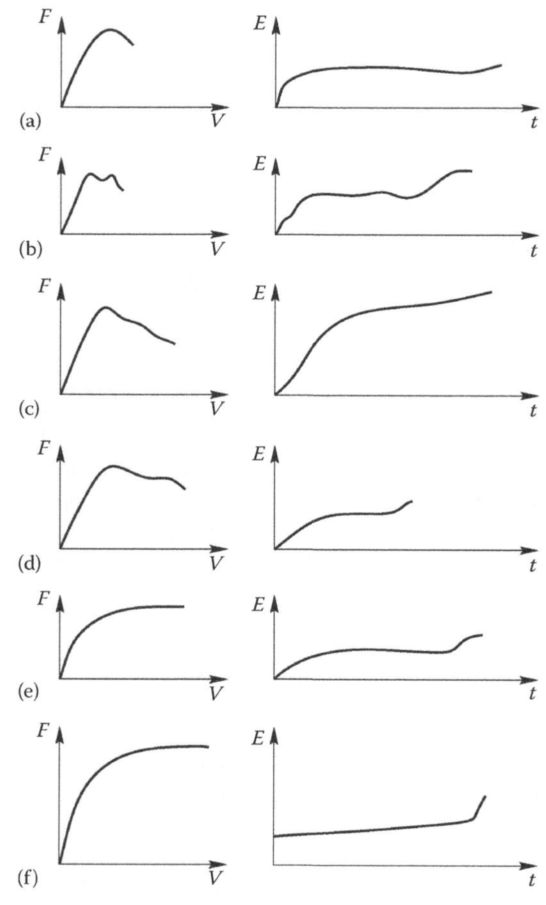

FIGURE 5.25 *F–V* curve and *E–t* curve in COD test. (a) The first type. (b,c) The second type. (d) The third type. (e,f) The fourth type .

Based on DD-19 rules, the crack tip crack initiation still corresponds to the first break, whose displacement is taken as the critical displacement V_C.

5.5.2.3 Third Type of *F–V* Curve

The load continues to decrease after passing the highest point, while the displacement continues to increase without a sudden gradient change or "burst" sound. We cannot directly determine the critical displacement value V_C from the *F–V* curve (as shown in Figure 5.25d). From the *E–t* curve we find that, after passing certain straight-line sections, the potential goes up quickly. According to the suggestion in DD-19 Annex A, a straight line is made along the curve's starting straight-line section. The time *t* corresponding to the contact point where the curve moves away from the straight line is taken as the starting point of the crack's slow growth. By selecting the corresponding value to it on the *F–V* curve, we can determine the critical displacement value V_C.

5.5.2.4 Fourth Type of *F–V* Curve

After the load reaches maximum value, it remains constant. In this case, the displacement increases as time goes by. For such a curve, we cannot directly determine the critical displacement value V_C on the *F–V* curve. The corresponding *E–t* curve has two cases. One is shown in Figure 5.25e, where the *E–t* curve has an obvious break. We can determine the critical displacement value V_C as shown in Figure 5.25d. The other one is shown in Figure 5.25f, where the *E–t* curve is smooth. It is hard to determine the break. According to DD-19's suggestion, the displacement when the maximum load is reached (i.e., the load platform's starting point) is taken as the critical displacement value V_C.

5.5.3 δ_R–Δa Curve

As pointed out earlier, crack growth resistance can be represented by the fracture toughness parameter *K* or *J*, or represented by the fracture toughness parameter δ. The δ_R curve is a curve that represents the relationship between the crack growth resistance δ_R and the instant crack length growth Δa, using the fracture toughness parameter δ (i.e., the δ_R–Δa curve). Figure 5.26 shows three typical δ_R curves of medium and lower strength steel.

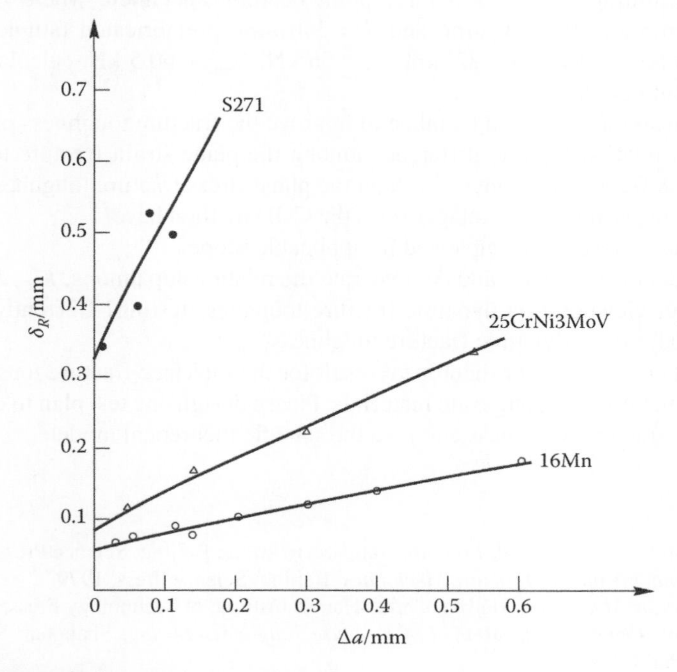

FIGURE 5.26 Three typical σ_R curves.

In the case where brittle unstable fracture or no stable crack growth exists before the burst point, we should determine the COD value δ_C of the unstable fracture point or burst point. In the case where brittle unstable fracture or stable crack growth exists before the burst, we should determine the COD value of the stable crack growth's starting point ($\Delta a = 0$ or $\Delta a = 0.05$ mm), and the unstable fracture point, respectively, represented by δ_i (or $\delta_{0.05}$) and δ_u.

A material's crack growth resistance curve obtained in testing not only provides the material's crack initiation resistance (δ_i) but also describes the crack's growth behavior after initiation. Therefore, when assessing materials and process quality, and analyzing for material safety, this method is more comprehensive than the elastic–plastic fracture toughness test method based on the crack initiation point.

The several fracture toughness testing methods described earlier are mainly used for bulk materials. However, due to the high-speed development of materials science, various new materials have appeared, such as polymers, thin-film materials, functional gradient materials, and nanowires. Therefore, testing the fracture performance of these new materials is challenging. Most traditional fracture toughness testing methods do not apply to these new materials; thus researchers have established some new fracture toughness testing methods. Miyata [14] used the indentation method to study the fracture toughness of glass materials in 1980. In the following decades, researchers widely applied the indentation method in fracture performance testing of various materials. They have also made great strides by designing various testing methods, such as the blister method and the buckling method, to study the fracture toughness of composite materials, thin-film materials, and intelligent materials [15–23].

EXERCISES

5.1 Discuss the physical meaning of the plane strain fracture toughness K_{IC} and its influencing factors from the viewpoint of combining macro and micro viewpoints.

5.2 Which basic requirements are there on the specimen when determining the plane strain fracture toughness K_{IC}? How can we measure an effective K_{IC} value?

5.3 When determining K_{IC} using a three-point bending specimen, where the specimen size is $B = 30$ mm and $W = 60$ mm, and $S = 240$ mm, prefabricated fatigue crack (including mechanical notch) depth $a = 32$ mm, $F_5 = 56$ kN, $F_{max} = 60.5$ kN, calculate the conditional fracture toughness K_Q.

5.4 What technical measures can be taken to improve the fracture toughness of a material?

5.5 What is the relationship and difference among the plane strain fracture toughness K_{IC}, the surface crack fracture toughness K_{IE}, and the plane stress fracture toughness K_C?

5.6 What advantages and disadvantages does the COD method have?

5.7 Describe the J_{IC} testing principle and its applicable scope.

5.8 Write a brief description of, and demonstrate the relationship among, K_{IC}, J_{IC}, and δ_C.

5.9 What is your viewpoint on dynamic fracture toughness testing? Currently, what is the best method to determine dynamic fracture toughness?

5.10 Currently there is no test or theoretical result for the interface fracture toughness of particle-reinforced metal-based composite materials. Please design one test plan to directly determine its interface fracture toughness and give the specific theoretical model.

REFERENCES

1. Chu W. Y., Lin S., Wang C. et al. *Fracture Toughness Testing*. Beijing: Science Press, 1979.
2. Chu W. Y. *Fundamentals of Fracture Mechanics*. Beijing: Science Press, 1979.
3. Wang D. *Fracture Mechanics*. Harbin, China: Harbin Institute of Technology Press, 1989.
4. Cui Z. Y. et al. *Principle and Method of Fracture Toughness Testing*. Shanghai: Shanghai Science & Technology Press, 1981.

5. Cui Z. Y. Exploring the fracture toughness K_{IE} and K_{IC} of a medium-strength steel plate using the surface-flaw test. *Eng Fract Mech*, 1980, 13(4): 775–789.
6. Cui Z. Y. Exploring the fracture toughness *K*IE and *K*IC of a medium-strength steel plate using the surface crack method. *Phys Test Chem Anal* (*Phys Test*), 1980, 16(3): 13–15.
7. Irwin G. R. Crack-extension force for a part-through crack in the plate. *J Appl Mech Trans ASME*, 1962, E29(4): 651–654.
8. Shah R. C. and Kobayashi A. S. On the surface flaw problem. The surface crack: Physical problems and computational solutions. *ASME*, 1972: 79–124.
9. Smith F. W. and Alavi M. J. Stress intensity factors for a penny shaped crack in a halfspace. *J Eng Fract Mech*, 1971, 3: 241–254.
10. Folias E. S. An axial crack in a pressurized cylindrical shell. *Int J Fract Mech*, 1965, 1(3): 104–113.
11. Hahn G. T., Sarrater M., and Rosenfeld A. R. Criteria for crack extension in cylindrical pressure vessels. *Int J Fract Mech*, 1969, 5(3): 187–210.
12. Chen C. Discussion of single sample JIC determination method. *New Metal Mater,* 1975, 11–12: 33–45.
13. Cai Q. G. Principle and application of elastic-plastic fracture mechanics. *New Metal Mater*, 1975, 11–12: 22.
14. Miyata N. and Jinno H. Use of Vickers indentation method for evaluation of fracture toughness of phase-separated glasses. *J Non-Cryst Solids*, 1980, 38–39: 391–396.
15. Lach R., Gyurova L. A., and Grellmann W. Application of indentation fracture mechanics approach for determination of fracture toughness of brittle polymer systems. *Polym Test*, 2007, 26(1): 51–59.
16. Li X. D. and Bhushan B. Measurement of fracture toughness of ultra-thin amorphous carbon films. *Thin Solid Films*, 1998, 315 (1–2): 214–221.
17. Roman A., Chicot D., and Lesage J. Indentation tests to determine the fracture toughness of nickel phosphorus coatings. *Surf Coat Technol*, 2002, 155 (2–3): 161–168.
18. Xia Z. H., Curtin W. A., and Sheldon B. W. A new method to evaluate the fracture toughness of thin films. *Acta Mater*, 2004, 52(12): 3507–3517.
19. Zhou Y. C., Hashida T., and Jian C. Y. Determination of interface fracture toughness in thermal barrier coating system by blister test. *J Eng Mater Tech ASME*, 2003, 125(2): 176–182.
20. Zheng X. J., Zhou Y. C., and Zhong H. Dependence of fracture toughness on annealing temperature in PZT thin films produced by metal organic decomposition. *J Mater Res*, 2003, 18(3): 578–584.
21. Zheng X. J. and Zhou Y. C. Nano-indentation fracture test of $Pb(Zr_{0.52}\ Ti_{0.48})O_3$ ferroelectric thin films. *Acta Mater*, 2003, 51(14): 3985–3997
22. Lee J. S. and Jang J. I. et al. An instrumented indentation technique for estimating fracture toughness of ductile materials: A critical indentation energy model based on continuum damage mechanics. *Acta Mater*, 2006, 54(4): 1101–1109.
23. Chen Z. and Gan Z. H. Fracture toughness measurement of thin films on compliant substrate using controlled buckling test. *Thin Solid Films*, 2007, 515(6): 3305–3309.

6 Residual Stresses in Materials

In practice, a material or component can fail without reason even though no external load is applied. It is very puzzling, at first, to observe that failure can occur without a load. After careful analysis, however, it has been discovered that stresses can be introduced inside a material or component during the manufacturing and machining process. Therefore, the internal stress state of a material or component has a significant influence on its reliability and service life. In order to eliminate these processing defects, it is necessary to measure and evaluate the residual stresses in a material or component and to take appropriate measures to reduce residual stresses or change their distributions.

Dozens of methods that measure residual stress have been developed since the 1930s, when these techniques were first introduced. In general, the traditional methods of measuring residual stresses can be broadly classified into two categories: destructive mechanical methods and nondestructive physical methods [1–4].

The extensive emergence and wide application of new materials introduced severe challenges to the traditional measuring methods and also provided new opportunities for the field of modern analysis and characterization techniques of materials [5–8]. The study of residual stresses in materials includes the following subjects: (1) taking appropriate measures to adjust residual stresses or changing their distributions; (2) reducing or eliminating the adverse effects of residual stresses on material properties such as static strength, brittle failure, resistance to stress corrosion cracking, and fatigue; and, (3) detrimental deformation such as dimensional deviation during or after the manufacturing process [9–12].

6.1 INTRODUCTION TO RESIDUAL STRESSES

6.1.1 Generation of Residual Stresses

6.1.1.1 Principle for the Generation of Residual Stresses

If there is no stress transfer from the surface of an object to its interior region under an external load, the stresses in equilibrium inside the object are called the inherent stresses or original stresses [9]. Without the external load, the stresses in equilibrium inside the object are called the residual stresses. Residual stress is one type of inherent stress, which is also called internal stress by some researchers.

In 1912, Martens and Heyn introduced a spring model, shown in Figure 6.1, to illustrate the generation of residual stresses. For the three springs in Figure 6.1, (a) is a free state and (b) is a state in which the upper and lower ends of the springs are connected to rigid plates. In (b), no external load is applied, but interaction occurs among the springs. Assume the initial lengths and elastic constants of the springs are l_1, l_2, l_3 and c_1, c_2, c_3, and the spacing between the rigid plates after connection is l. The forces P_1, P_2, P_3 generated by the springs are $P_1 = c_1(l - l_1)$, $P_2 = c_2(l - l_2)$, $P_3 = c_3(l - l_3)$, respectively. P_1, P_2, P_3 are residual stresses, and $P_1 + P_2 + P_3 = 0$.

The general procedure to generate residual stresses is illustrated in Figure 6.2. In the region R inside an object without the action of any stress, square A is cut out, as shown in (a). An arbitrary operation to cause volume change and shape change is applied to part A, which transforms part A into shape B, as shown in (b). Imagine that part A is put back into the original region R with a shape as in (c). Lateral forces must be applied to cause deformation, as shown in (d). If the applied force is removed, the result is illustrated in (e). The input part and its surrounding part must adjust to deformation, and the stress field is generated in this region. This is the state that generates the residual stresses.

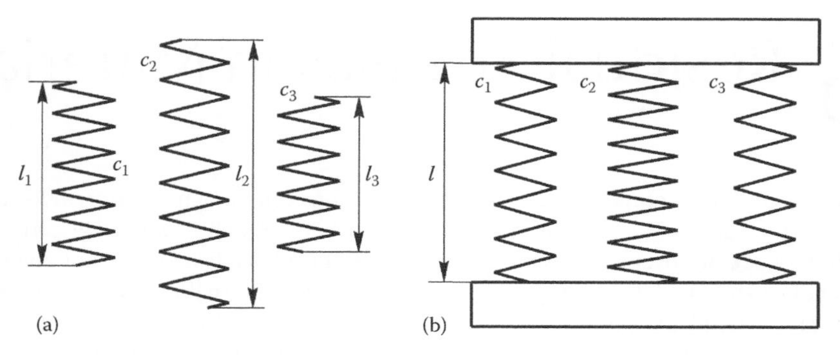

FIGURE 6.1 Schematic illustration of spring models. (a) Free state. (b) Constrained state.

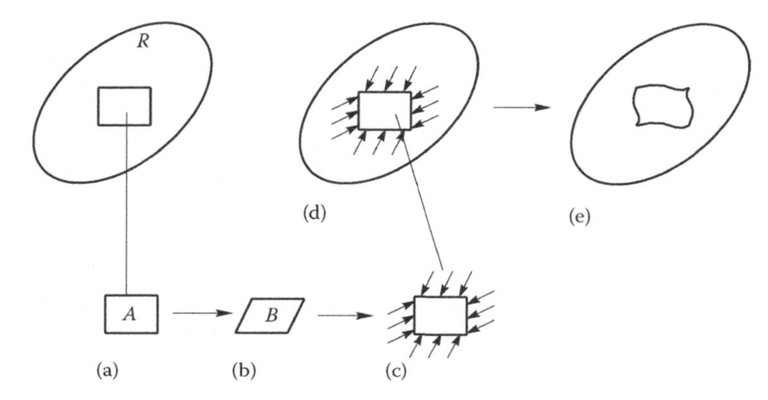

FIGURE 6.2 The generation of residual stresses. (a) Cutting out the square A. (b) Deforming part A. (c) Recovering to square A with force. (d) Putting square A back into its origin location. (e) Yielding the residual stress when the applied force is removed.

On the other hand, if the deformation of the cut part is measured, the residual stresses can be deduced. In fact, residual stresses are measured by groove-cutting or slice-cutting the object so that the residual stresses are partially or completely released. The deformation is measured by experimental methods, and the residual stresses are calculated.

6.1.1.2 Classification of Residual Stresses

Residual stresses occur for a variety of reasons, and they are very complex [2,6,7]. Residual stresses can be classified into two categories, according to the interaction ranges and the causes of residual stresses [13]:

1. Macroresidual and microresidual stresses in terms of the interaction range
   Take Figure 6.3 as an example. The residual stresses σ_α^{I} and σ_β^{I} in Figure 6.3a correspond to the macroscopic stresses in grains α and β, and are called residual stresses of the first order σ^{I}. Its magnitude, direction, and properties can be measured using common physical or mechanical methods. Microscopic residual stresses occur at the microscale, and can be further classified into two types, according to their range of action: (1) the stresses averaged over a single grain α or β, called residual stresses of the second order σ^{II}; and (2) the stresses acting inside the substructures of a grain, which are residual stresses of the third order σ^{III}. Mapping relationships exist between the three types of residual stresses and the crystalline microstructures (shown in Figure 6.3b).
2. The causes of residual stresses
   Macroscopic residual stresses are called volumetric stresses, while microscopic residual stresses are called textural stresses, or subtextural stresses.

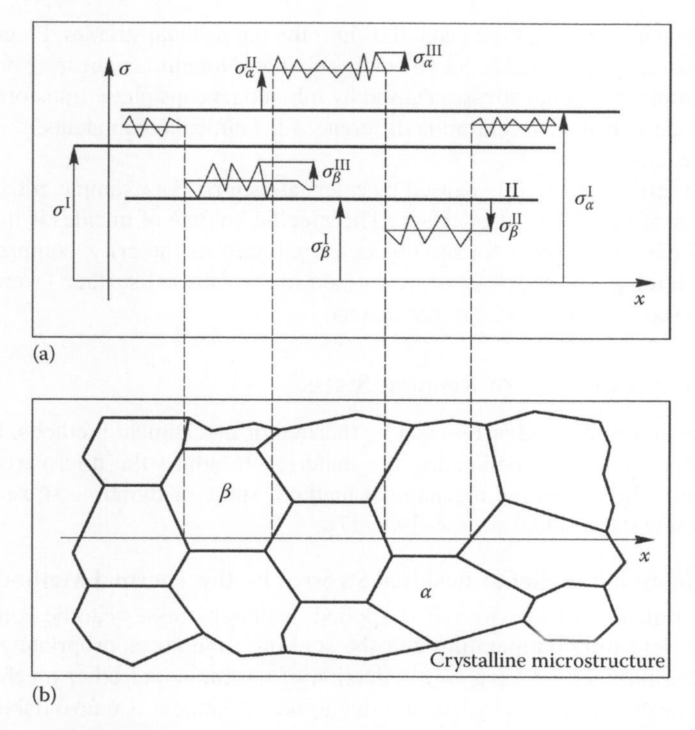

FIGURE 6.3 Schematic illustration of three types of residual stresses. (a) Three types of residual stresses. (b) The relative crystalline microstructures.

6.1.1.3 Origins of Residual Stresses

The origins of residual stresses include four types: (1) heterogeneous plastic deformation, (2) thermal stress, (3) phase transformation, and (4) chemical reaction. Each type has exterior and internal reasons [6,14–16].

1. Heterogeneous plastic deformation caused by mechanical load
   Under a mechanical load, each part of an object deforms nonuniformly. Some parts undergo large plastic deformation and some parts have small plastic deformation, while some parts only experience elastic deformation; but the whole object needs to keep integrity. As a result, when the externally applied mechanical load is removed, residual stresses are generated inside the object. The external cause is the nonuniform acting stress, while the internal cause can be ascribed to different concentrations of chemical compositions and different grain orientations, so that each part exhibits a different yield behavior.
2. Heterogeneous plastic deformation caused by thermal stress
   If the deformation under thermal stress introduced during the heating or cooling process is always in the elastic stage, then the thermal stress is reversible. As the temperature inside the body equalizes, stress disappears. In the heating or cooling process, the part of the object with high temperature has a much lower yield strength and first undergoes plastic deformation; the subsequent thermal stress is not reversible. Although the final temperature of each part tends to be identical, thermal stress will introduce residual stresses inside the body.
3. Inhomogeneous phase transformations
   During heating or cooling, a material may experience phase transformations, such as the martensitic transformation during quenching and precipitation during aging. Martensitic transformation incurs volume expansion. Nonuniform cooling processes cause nonuniform martensitic

transformation and volume expansion, thus introducing residual stresses. Precipitation introduces mismatching stress fields due to volume change. Nonuniform heating or cooling is the external reason for residual stresses caused by inhomogeneous phase transformations, while the internal cause is the concentration difference of chemical components.

4. Chemical reactions

   Chemical reactions are mainly caused by external factors. For example, nitriding of a steel forms a layer of nitrides on the surface. The specific volume of nitrides is higher than that of the steel matrix. In order for the object to maintain its integrity, compressive residual stresses with relatively high values are formed in the nitrided surface layer, while tensile residual stresses are introduced in the matrix.

6.1.2 Adjustment and Relief of Residual Stresses

Residual stresses can be adjusted or relieved by thermal or mechanical methods. Usually, the thermal method used is annealing, or heating the material, to adjust the microstructure to relax or relieve the residual stresses. In the mechanical method, static or dynamic stresses are applied to reduce or redistribute the residual stresses [9,16,17].

6.1.2.1 Adjustment and Relief of Residual Stresses by the Thermal Method

When the thermal method of stress relief is applied, residual stresses can be completely relieved, provided that the annealing temperature and the soaking time are appropriately selected. At the same time, the thermal method leads to a reduction of hardness and other mechanical properties of the material. The change of microstructure due to heat treatment is unavoidable. Thermal stress relieving is closely related to the phenomena of creep and stress relaxation. The most generally used annealing method is to heat the component at appropriately high temperatures for several hours to several days, and then cool slowly.

Mailender performed a pioneering study correlating the relief of residual stress by annealing, with the phenomenon of stress relaxation [9]. Mailender's experimental results revealed the stress relaxation state when the residual stress was relieved, and are introduced in detail here.

In the experiment, a steel specimen with a diameter of 14 mm and a length of 130 mm was subjected to uniform uniaxial tension, and then put into a high-temperature furnace. The strain was fixed and the stress relaxation state was investigated. The strain at temperature t is ε_e. At room temperature (20°C) and temperature t, the elastic moduli are E_{20} and $E_t (E_{20} > E_t)$, respectively; and the tensile stresses are σ_e and σ_{et}, respectively. When plastic deformation does not occur, the equations $\sigma_e = \varepsilon_e E_{20}$ and $\sigma_{et} = \varepsilon_e E_t$ hold true. Therefore, the stress at temperature t is

$$\sigma_{et} = \sigma_e \left(\frac{E_t}{E_{20}} \right). \tag{6.1}$$

When the temperature rises so that plastic deformation occurs, Equation 6.1 no longer holds true. After a soaking time z, if the permanent strain is ε_{bz}, the elastic strain ε_e is therefore reduced and the stress σ_{etz} can be expressed as

$$\sigma_{etz} = (\varepsilon_e - \varepsilon_{bz}) E_t = \varepsilon_{ez} E_t, \tag{6.2}$$

where ε_{ez} is the elastic strain after time z. Here, equation $\varepsilon_e = \varepsilon_{ez} + \varepsilon_{bz}$ is applied; that is, the total strain is the sum of the elastic strain and the plastic strain (see Chapter 1). The stress after cooling to 20°C can be expressed as

$$\sigma_{ez} = (\varepsilon_e - \varepsilon_{bz}) E_{20} = \varepsilon_{ez} E_{20}, \tag{6.3}$$

where σ_{ez} is the stress at room temperature after stress relaxation.

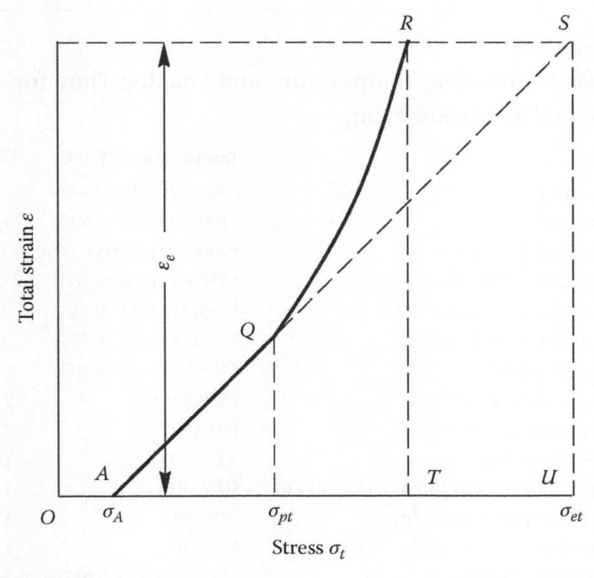

FIGURE 6.4 The stress–strain curve.

In practice, when studying stress relaxation with an initial elastic stress σ_{et}(= $\varepsilon_e E_t$), the relaxation stress σ_{etz} after time z can be obtained by the following method.

The specimen is first put into a furnace and heated to temperature t, then a small stress σ_A is applied, and subsequently a load is slowly applied. The stress and strain states can be illustrated in Figure 6.4. The strain is elastic before the stress reaches σ_{pt} (straight line AQ). After that, the strain increases inelastically (curve QR). The elastic stress OU at point S—found by extending the strain line AQ, and via the stress–strain relationship—is the initial stress σ_{et}. Figure 6.5 obtained in such a way shows the experimental results of stress relaxation for Ni-Cr (0.4%C; 1.5%Cr; 3%Ni) at 560°C. The values superimposed in the figure are the initial stresses with the units kg/mm^2. Regardless of the initial stress, stress can be significantly relaxed after 2 h or so. According to the

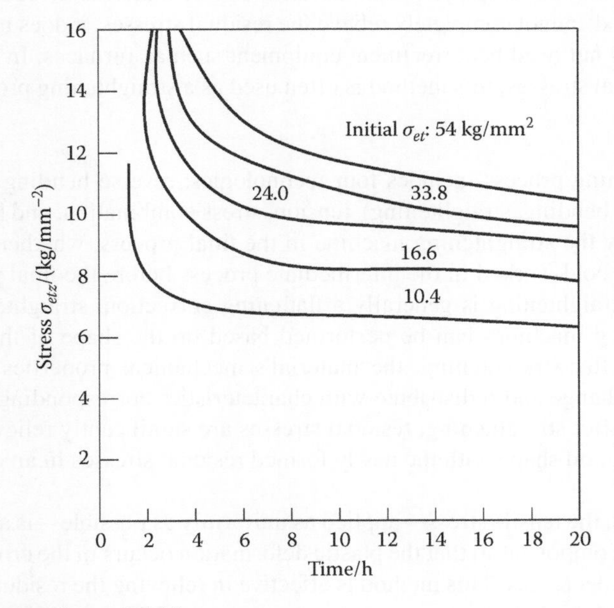

FIGURE 6.5 Stress relaxation at 560°C.

TABLE 6.1
Typical Annealing Temperature and Soaking Time for Residual Stress Relieving

Metal	Temperature/°F (°C)	Time/h
Gray cast iron	800(427)–1100(593)	5–1/2
Carbon steel	1100(593)–1250(677)	1
C-Mo steel (C < 0.2%)	1100(593)–1250(677)	2
C-Mo steel (0.2% < C < 0.35%)	1250(677)–1400(760)	3–2
Cr-Mo steel (2%Cr, 0.5%Mo)	1325(718)–1375(746)	2
Cr-Mo steel (9%Cr, 1%Mo)	1375(746)–1425(774)	3
Cr stainless steel	1425(774)–1475(802)	2
Cr-Ni stainless steel (316)	1500(816)	2
Cr-Ni stainless steel (310)	1600(871)	2
Copper alloy (Cu)	300(149)	1/2
Copper alloy (80 Cu-20 Zn, 70 Cu-30 Zn)	500(260)	1
Copper alloy (60 Cu-40 Zn)	375(191)	1
Copper alloy (64 Cu-18 Zn-18Ni)	475(246)	1
Nickel and molybdenum	525(274)–600(316)	3~1

According to the Handbook of Metals (1955): Appendix, the typical annealing temperatures and times of residual stress relieving for metals are listed in Table 6.1. If more thorough stress relief is expected, the materials should be annealed at the selected or higher temperatures for 5–10 h (for example, gray cast iron: 650°C, carbon steel: 600°C, alloyed steel: 800°C).

Appendix of the *Handbook of Metals* (1955), if the relaxation of residual stress is considered, the initial stress corresponds to the residual stress (Table 6.1).

6.1.2.2 Adjustment and Relief of Residual Stresses by the Mechanical Method

Mechanical stress relieving employs plastic deformation of materials to reduce residual stresses. Although this method cannot completely relieve the residual stresses, it does not impair mechanical properties, and does not need heat-treatment equipment such as furnaces. In addition to relief and adjustment of residual stresses, this method is often used as a straightening process for components.

1. Straightening
   The straightening process includes four technologies: reverse bending (roller straightening, rotational bending, straightening), tension, stress combination, and heating. The shape is corrected by the straightening machine in the final process, whether the component is hot-worked or cold-worked in the intermediate process before the final product. Although this type of straightening is generally a flattening correction, straightening by a variety of straightening machines can be performed based on the shape of the material and its application. After straightening, the material's mechanical properties change. Residual stresses also change and redistribute with characteristics corresponding to the straightening method. After straightening, residual stresses are significantly relieved, and the material obtains a good shape with the newly formed residual stresses in an equilibrium state.
2. Tension
   In this method, the tensile stress—applied as uniformly as possible—is applied to the cross section of the component so that the plastic deformation occurs in the cross section, and the residual stress decreases. This method is effective in relieving the residual stress in nonferrous metals with high ductility.

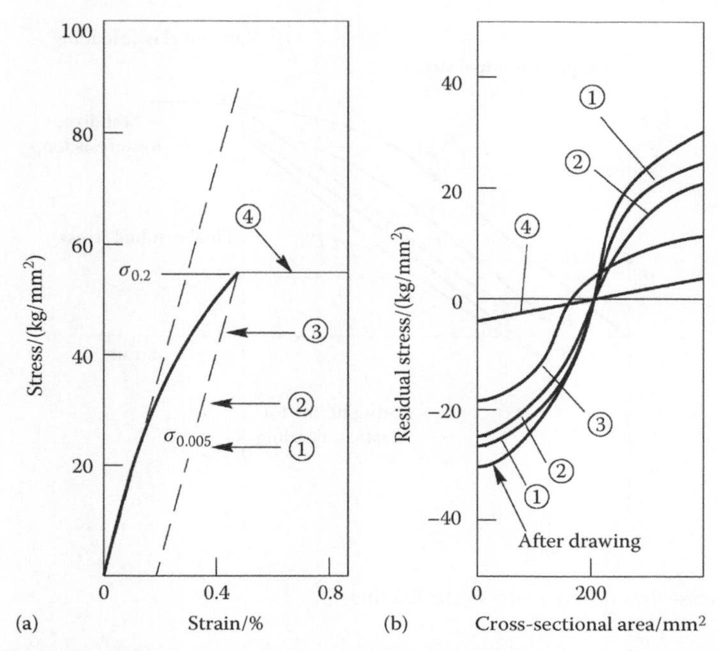

FIGURE 6.6 (a) The stress–strain curve during drawing and (b) change of residual stress by tension 9SMnPb2 ϕ 23 mm. (1) 20.8 kg/mm², (2) 36.6 kg/mm², (3) 46.4 kg/mm², and (4) 58.5 kg/mm².

For drawing specimens with ϕ23 mm of aluminum alloy 9SMnPb2, Figure 6.6 shows the stress–strain relationship (a) and residual stress–cross-sectional area relationship (b) during stress relieving by drawing. The stresses in the four stages shown in the figure denote the measured axial residual stresses after unloading, when drawing from the elastic limit ($\sigma_{0.005}$) to the yield stress ($\sigma_{0.2}$). Residual stress decreases even when tensile stress equal to the elastic limit is applied. When the applied external stress reaches the yield stress, residual stress is relieved to a large extent. The distribution of residual stress after drawing rotates toward the decreasing direction around the center at which the original residual stress is zero, as shown in Figure 6.6b.

3. Vibratory stress relieving

In vibratory stress relieving, the component is mounted on the appropriate stage and vibration is applied. This is a simple and economic method, and the principle is the same as that of fatigue, associated with the stability of residual stress under alternating stress. Under cyclic stress, residual stress relaxes when plastic deformation occurs. Here, residual stress can be taken as the mean stress.

Figure 6.7 shows the stress–strain curve under cyclic loading in Luban and Felger's study of residual stress. From this figure, the evolution of residual stress can be seen when the material is subjected to cyclic loading with a fixed strain range (the maximum strain is ε_C and the minimum strain is ε_B). Assume that the residual stress is *A* before the application of alternating stress. The hysteresis loop *ACDB* represents the evolution of the stress–strain relationship in the first loading cycle. When the stress is higher than *A*, the curve deviates from the straight line of elastic deformation, and plastic deformation continues until *C*. *CD* is parallel to the line of elastic loading while *DB* deviates again. This is caused by the so-called Bauschinger effect. After sufficient loading cycles, as long as the macroscopic crack does not propagate, the stress–strain curve forms a stabilized hysteresis loop, denoted by loop *C′EB″E′C*. The final residual stress is indicated by Point *E*. Observing the evolution of residual stress from *A* to *E* in Figure 6.7, we can learn the characteristics of the stress–strain curve and the influence of the applied load.

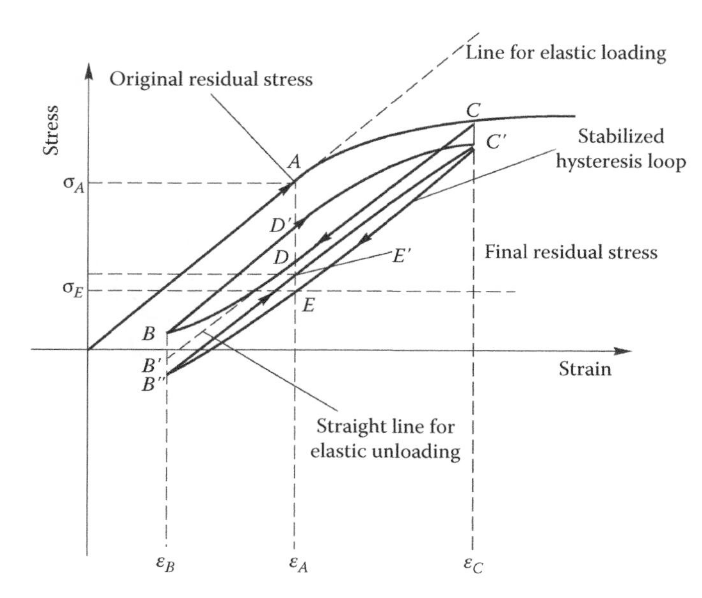

FIGURE 6.7 Stress–strain curve under cyclic loading.

4. Surface processing
   For a drawn or rolled rod or plate, the outer surface generally exhibits apparent tensile residual stress. The existence of surface tensile residual stress has adverse effects on fatigue, stress corrosion cracking, and other impairments. In order to eliminate this type of stress, and superimpose compressive residual stress, surface processing—such as extruding, surface rolling, shot peening, or secondary drawing—can be applied. These processes are more appropriately termed adjustment and redistribution of residual stress, rather than residual stress relieving.

6.2 MEASUREMENT OF RESIDUAL STRESSES

Technologies to measure residual stresses were introduced in the 1930s, and have since been developed into dozens of methods. These methods can be classified into two categories: destructive mechanical measurement and nondestructive physical measurement [1–4].

Mechanical measurement of residual stresses can be divided into three types: the calculation method, the experimental nondestructive method, and the experimental destructive method.

In the calculation method, the distributions of residual stresses are calculated theoretically in terms of the material's mechanical properties, the shape and dimensions of the component, and the change of external actions such as temperature and load.

Because residual stresses cause a change in component dimensions, measurement of residual stresses is called the experimental nondestructive method.

The measurement of residual stresses, in terms of the additional deformation when a component is cut from an object, is called the experimental destructive method. This method has two types: complete destruction of the component and partial destruction of the component. Physical methods to measure residual stresses, such as crystal x-ray diffraction and magnetism, are based on a material's physical properties. Residual stress can be measured directly without separating or cutting the material. Physical methods can measure not only macroscopic residual stresses, but also microscopic residual stresses. Mechanical methods measure macroscopic residual stresses, while the microscopic residual stresses are generally measured by the physical methods. In addition, application acoustics [18], optical scattering, Raman spectroscopy [19,20], and hardness [9] can also be used to measure residual stresses.

6.2.1 Mechanical Measurement Methods of Residual Stresses

When the mechanical method is used to measure residual stresses, an object with residual stresses is sectioned or cut so that the residual stresses are partially released. Then the deformation of the remaining part of the object is measured and the residual stresses can be calculated using elastic mechanics.

6.2.1.1 Sach's Method

6.2.1.1.1 Residual Stresses of a Circular Plate

Assume that the outer radius of a circular plate is R_2 and the inner radius is R_1 [9]. The situation is analyzed when peeling off from the inner layer. As shown in Figure 6.8a, when peeling off to a location at radius r, the residual stress $\sigma_{rr}(r)$ from R_1 to r is released. This is equivalent to applying $-\sigma_{rr}(r)$ in the inner surface of the remaining circular plate. Let $u_2(r)$ be the change of outer radius of the circular plate. According to the hollow cylinder theory of Lame, we have

$$E\frac{u_2(r)}{R_2} = \frac{2r^2}{R_2^2 - r^2}\sigma_{rr}(r), \tag{6.4}$$

where E is Young's modulus. The strain of the outer surface in the circumferential direction is $\varepsilon_{\theta\theta 2}(r)$, and $\varepsilon_{\theta\theta 2}(r) = u_2(r)/R_2$. From Equation 6.4, we obtain the radial residual stress $\sigma_{rr}(r)$:

$$\sigma_{rr} = E\varepsilon_{\theta\theta 2}(r)\frac{R_2^2 - r^2}{2r^2}. \tag{6.5}$$

At the same time, the residual stresses satisfy the following equilibrium equation:

$$\sigma_{\theta\theta} = \frac{\mathrm{d}(r\sigma_{rr})}{\mathrm{d}r}. \tag{6.6}$$

Substituting Equation 6.5 into Equation 6.6, the circumferential residual stress is

$$\sigma_{\theta\theta} = E\frac{\mathrm{d}\varepsilon_{\theta\theta 2}(r)}{\mathrm{d}r}\frac{R_2^2 - r^2}{2r} - E\varepsilon_{\theta\theta 2}(r)\frac{R_2^2 + r^2}{2r^2}. \tag{6.7}$$

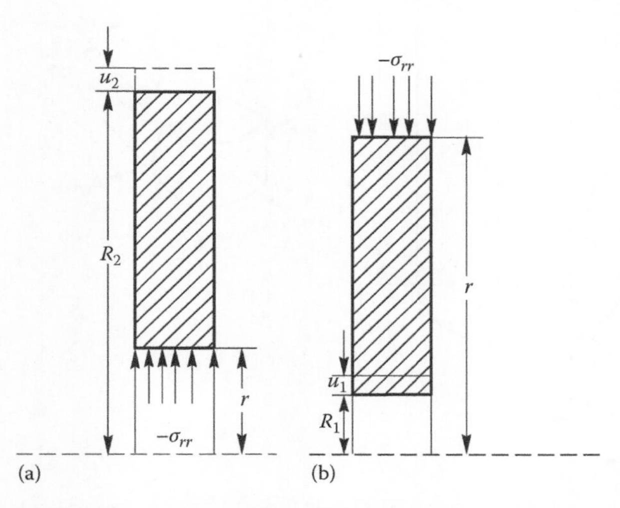

FIGURE 6.8 Inner layer removal (a) and outer layer removal (b) of a circular plate.

The circumferential strain $\varepsilon_{\theta\theta 2}(r)$ represents the change in the outer radius, and can be measured by electrical resistance strain gauges bonded to the outer surface. In the calculation of stress, the first term on the right side of Equation 6.7 requires a differential operation: the relationship curve of $\varepsilon_{\theta\theta 2}(r)$ versus radius r must be used.

Residual stresses in a circular plate can also be mechanically measured by the outer layer peeling method (Figure 6.8b); or, the overall distribution of residual stresses can be measured by combining the inner and outer layer peeling methods. The outer layer peeling method peels material layer by layer from the outer surface. After the outer layer is peeled, the change of inner radius is measured, which can then be used to calculate the residual stresses at the outer surface.

6.2.1.1.2 Residual Stresses in a Hollow Cylinder (The Inner Layer Removal Method)

As shown in Figure 6.9, the hollow cylinder has axisymmetric residual stress. Residual stresses in the axial, circumferential, and radial directions σ_{zz}, $\sigma_{\theta\theta}$, σ_{rr}, are only functions of r. Assume that the inner and outer radii are a and b, respectively, and that the length is sufficient in the axial direction.
Residual stresses should satisfy the equilibrium equation of an axisymmetric problem:

$$\frac{d\sigma_{rr}}{dr}+\frac{\sigma_{rr}-\sigma_{\theta\theta}}{r}=0. \tag{6.8}$$

Without external loads, the distribution of residual stress should satisfy the self-balanced condition in which the resultant forces in the axial and circumferential directions are zero:

$$\int_a^b 2\pi r\sigma_{zz}dr=0. \tag{6.9}$$

$$\int_a^b 2\pi r\sigma_{zz}dr=0. \tag{6.10}$$

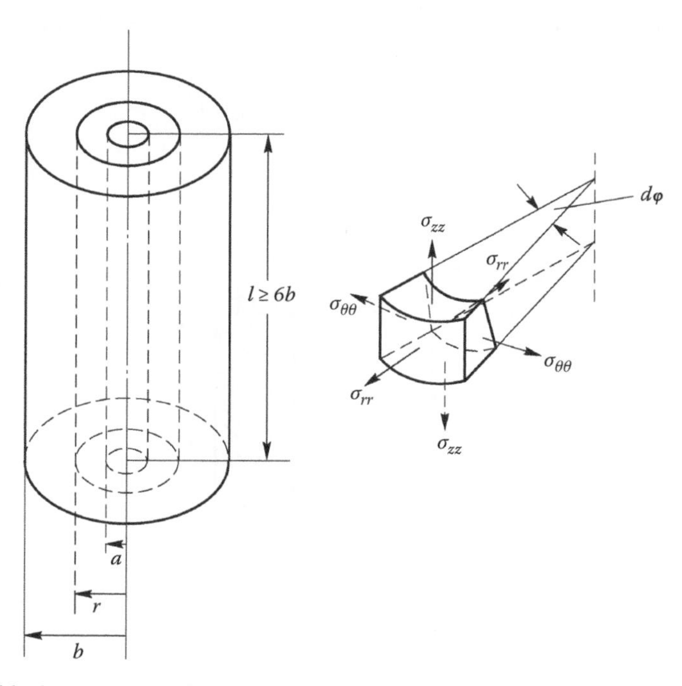

FIGURE 6.9 Residual stresses in hollow cylinder and inner layer removal.

Remove material layer by layer from the inner wall, and measure the changes of length and circumference of the outer surface of the cylinder when the inner radius changes from a to ρ. In terms of the measured axial strain and the circumferential strain, the three components of the original residual stresses σ_{zz}, $\sigma_{\theta\theta}$, and σ_{rr} are calculated according to elastic mechanics. The relationship between the original residual stresses and the removal depth ρ can be described by the following functions:

$$\sigma_{rr}(\rho)=\frac{E}{1-\nu^2}\left[\frac{b^2-\rho^2}{2\rho^2}\left(\varepsilon_{zz}^b+\nu\varepsilon_{\theta\theta}^b\right)\right], \tag{6.11}$$

$$\sigma_{zz}(\rho)=\frac{E}{1-\nu^2}\left[\frac{b^2-\rho^2}{2\rho^2}\frac{\mathrm{d}}{\mathrm{d}\rho}\left(\varepsilon_{zz}^b+\nu\varepsilon_{\theta\theta}^b\right)-\left(\varepsilon_{zz}^b+\nu\varepsilon_{\theta\theta}^b\right)\right], \tag{6.12}$$

$$\sigma_{\theta\theta}(\rho)=\frac{E}{1-\nu^2}\left[\frac{b^2-\rho^2}{2\rho}\frac{\mathrm{d}}{\mathrm{d}\rho}\left(\varepsilon_{\theta\theta}^b+\nu\varepsilon_{zz}^b\right)-\frac{b^2+\rho^2}{2\rho^2}\left(\varepsilon_{\theta\theta}^b+\nu\varepsilon_{zz}^b\right)\right], \tag{6.13}$$

where $\varepsilon_{\theta\theta}^b$ and ε_{zz}^b are the axial strain and the circumferential strain of the outer surface at the removal depth of $r = \rho$, respectively, and can be measured by the electrical resistance strain gauges bonded onto the outer surface. With the change of removal radius ρ, a series of strain data can be obtained: the relationships $\varepsilon_{\theta\theta}^b$, and ε_{zz}^b versus ρ. The distribution of the original residual stresses in the radial direction can be calculated in terms of the measured strain values of the outer surface. Figure 6.10 is an example showing the measurement of residual stresses of a hollow cylinder.

In the inner layer removal method, it is impossible to remove material to $\rho = b$. Therefore, the residual stresses of the outer surface cannot be directly obtained by this method, and can only be determined by extrapolation. The residual stresses using extrapolation should satisfy the self-balanced conditions represented by Equations 6.9 and 6.10. In addition, the residual stresses around the outer surface can be measured by the outer layer removal method.

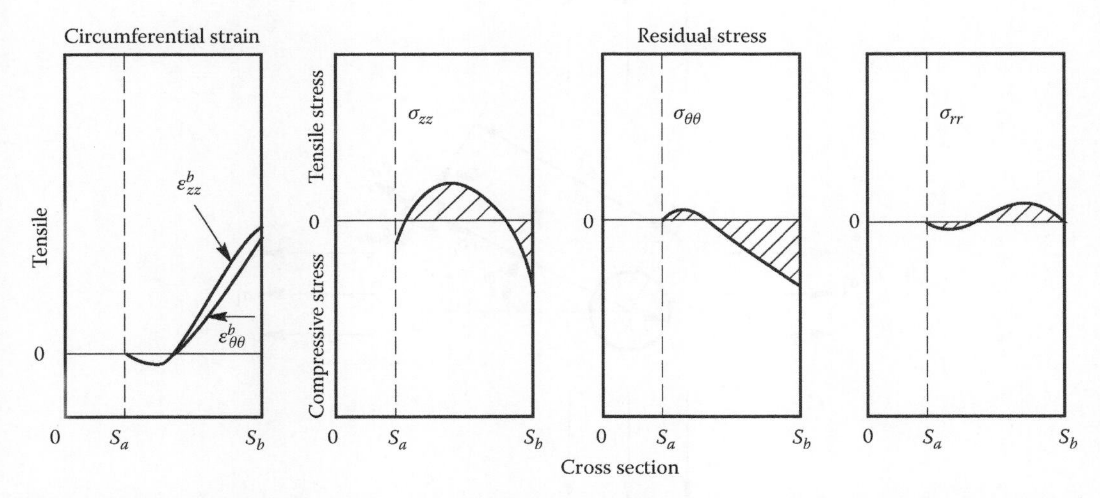

FIGURE 6.10 An example showing the measurement of residual stresses of a hollow cylinder by the inner layer removal method.

6.2.1.2 Hole Drilling Method

The basic idea of the hole drilling method is to measure the residual stresses in a plate by drilling a hole in the plate, and calculating the residual stresses in terms of the deformation around the hole [16,21]. As shown in Figure 6.11, we assume that residual stresses exist in the plate plane, and the residual stresses in the principal directions are σ_1 and σ_2, respectively. Drill a small hole with a diameter of $2R$ on the plate surface, and consider the residual stresses at distance r from the center of the hole, and at an angle of φ with respect to σ_1. The residual stresses in the radial and circumferential directions before drilling are $\sigma_{\theta\theta}$ and σ_{rr}, respectively, and they change due to drilling. Assuming that the additional stresses due to drilling are $\sigma'_{\theta\theta}$ and σ'_{rr}, we have

$$\sigma'_{\theta\theta} = \frac{1}{2}\frac{R^2}{r^2}\sigma_1 - \frac{3}{2}\frac{R^4}{r^4}\sigma_1 \cos 2\varphi + \frac{1}{2}\frac{R^2}{r^2}\sigma_2 + \frac{3}{2}\frac{R^4}{r^4}\sigma_2 \cos 2\varphi. \tag{6.14}$$

$$\begin{aligned}\sigma'_{rr} = &-\frac{1}{2}\frac{R^2}{r^2}\sigma_1 + \frac{3}{2}\frac{R^4}{r^4}\sigma_1 \cos 2\varphi - \frac{2R^2}{r^2}\sigma_2 \cos 2\varphi \\ &-\frac{1}{2}\frac{R^2}{r^2}\sigma_2 - \frac{3}{2}\frac{R^4}{r^4}\sigma_2 \cos 2\varphi + \frac{2R^2}{r^2}\sigma_2 \cos 2\varphi.\end{aligned} \tag{6.15}$$

Stresses $\sigma'_{\theta\theta}$ and σ'_{rr} are related to the deformation around the hole. If the associated strains can be measured, then the residual stresses σ_1 and σ_2 in the plate can be obtained. As shown in Figure 6.12, the strain gauge with length L is bonded in the radial direction at an angle φ. After drilling, the radial strain ε'_{rr} is introduced at the radius r in the region covered by the strain gauge, and the additional circumferential stress $\sigma'_{\theta\theta}$ and radial stress σ'_{rr} are introduced. The relationship between these stresses and the strain is

$$\varepsilon'_{rr} = \frac{\left(\sigma'_{rr} - \nu\sigma'_{\theta\theta}\right)}{E}. \tag{6.16}$$

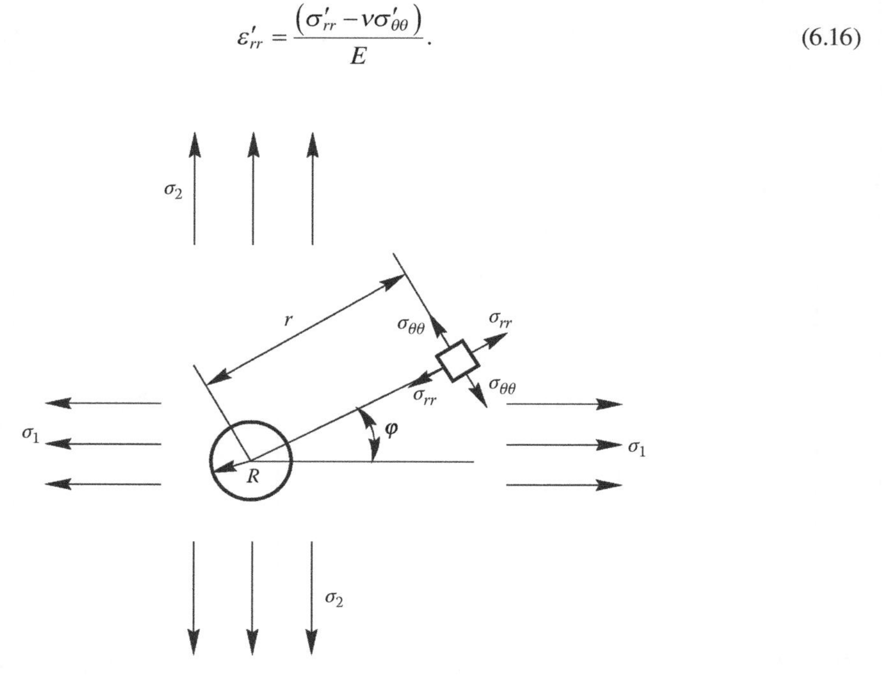

FIGURE 6.11 Stresses around a hole.

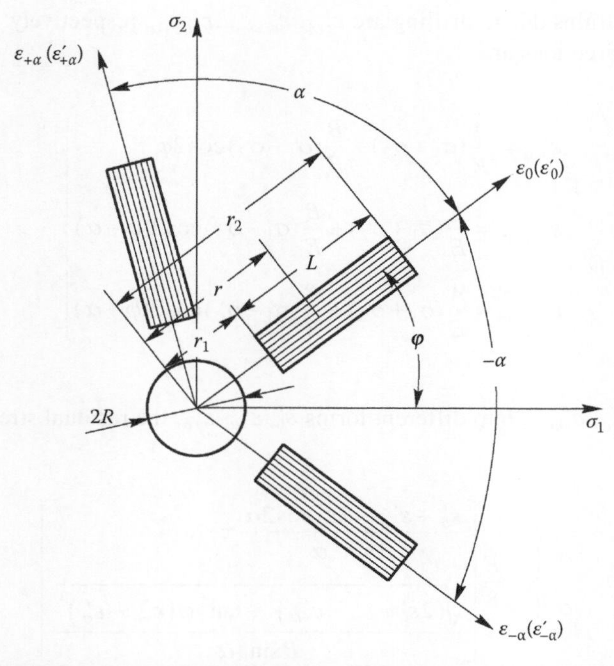

FIGURE 6.12 Strain gauges and their positions around a hole. ε_0, $\varepsilon_{-\alpha}$, $\varepsilon_{+\alpha}$ are strains under residual stress, ε'_0, $\varepsilon'_{-\alpha}$, $\varepsilon'_{+\alpha}$ are the strains in strain gauges.

The strain measured by the strain gauge is the average strain ε'_m. Assuming that the distances from the hole center to the two ends of the strain gauge are r_1 and r_2, we have

$$\varepsilon'_m = \frac{1}{r_2 - r_1}\int_{r_1}^{r_2} \varepsilon'_{rr}\mathrm{d}r. \tag{6.17}$$

Substituting Equations 6.14 and 6.15 into Equation 6.16, then substituting the result into Equation 6.17 and integrating, gives

$$\varepsilon'_m = \frac{A}{E}(\sigma_1 + \sigma_2) + \frac{B}{E}(\sigma_1 - \sigma_2)\cos 2\varphi, \tag{6.18}$$

where

$$\left.\begin{aligned} A &= -\frac{(1+\nu)}{2}\frac{R^2}{r_1 r_2} \\ B &= \frac{2R^2}{r_1 r_2}\left[-1 + \frac{(1+\nu)}{4}\frac{R^2\left(r_1^2 + r_1 r_2 + r_2^2\right)}{r_1^2 r_2^2}\right] \end{aligned}\right\}. \tag{6.19}$$

Thus, in the three directions which are each an equal distance to the hole center—that is, the three directions at angles φ, $\varphi - \alpha$, and $\varphi + \alpha$ with respect to σ_1—the same strain gauges are bonded.

The corresponding strains due to drilling are $\varepsilon'_{m,\varphi}$, $\varepsilon'_{m,\varphi-\alpha}$, $\varepsilon'_{m,\varphi+\alpha}$, respectively. From Equation 6.18, the strains in these directions are

$$\left.\begin{aligned} \varepsilon'_{m,\varphi} &= \frac{A}{E}(\sigma_1+\sigma_2)+\frac{B}{E}(\sigma_1-\sigma_2)\cos 2\varphi \\ \varepsilon'_{m,\varphi-\alpha} &= \frac{A}{E}(\sigma_1+\sigma_2)+\frac{B}{E}(\sigma_1-\sigma_2)\cos 2(\varphi-\alpha) \\ \varepsilon'_{m,\varphi+\alpha} &= \frac{A}{E}(\sigma_1+\sigma_2)+\frac{B}{E}(\sigma_1-\sigma_2)\cos 2(\varphi+\alpha) \end{aligned}\right\}. \tag{6.20}$$

Rewriting $\varepsilon'_{m,\varphi}$, $\varepsilon'_{m,\varphi-\alpha}$, $\varepsilon'_{m,\varphi+\alpha}$ into different forms ε'_0, $\varepsilon'_{-\alpha}$, $\varepsilon'_{+\alpha}$, the residual stresses σ_1, σ_2 and angle φ can be solved.

$$\left.\begin{matrix}\sigma_1\\ \sigma_2\end{matrix}\right\} = \frac{E}{8}\left[\frac{\varepsilon'_\alpha+\varepsilon'_{-\alpha}-2\varepsilon'_0\cos 2\alpha}{A\sin^2\alpha} \pm \frac{\sqrt{\left(2\varepsilon'_0-\varepsilon'_\alpha-\varepsilon'_{-\alpha}\right)^2+\tan^2\varphi\left(\varepsilon'_{-\alpha}-\varepsilon'_\alpha\right)^2}}{B\sin^2\alpha}\right] \tag{6.21a}$$

$$\tan 2\varphi = \left(\frac{\varepsilon'_{-\alpha}+\varepsilon'_\alpha}{2\varepsilon'_0-\varepsilon'_\alpha-\varepsilon'_{-\alpha}}\right)\tan\alpha. \tag{6.22a}$$

When $\alpha = 45°$, $\varepsilon'_\alpha = \varepsilon'_{45}$, $\varepsilon'_{-\alpha} = \varepsilon'_{-45}$, we have

$$\left.\begin{matrix}\sigma_1\\ \sigma_2\end{matrix}\right\} = \frac{E}{4}\left[\frac{\varepsilon'_{45}+\varepsilon'_{-45}}{A} \pm \frac{\sqrt{\left(2\varepsilon'_0-\varepsilon'_{45}-\varepsilon'_{-45}\right)^2+\left(\varepsilon'_{-45}-\varepsilon'_{45}\right)^2}}{B}\right] \tag{6.21b}$$

$$\tan 2\varphi = \left(\frac{\varepsilon'_{-45}-\varepsilon'_{45}}{2\varepsilon'_0-\varepsilon'_{45}-\varepsilon'_{-45}}\right). \tag{6.22b}$$

When $\alpha = 120°$, $\varepsilon'_\alpha = \varepsilon'_{120}$, $\varepsilon'_{-\alpha} = \varepsilon'_{-120}$, we have

$$\left.\begin{matrix}\sigma_1\\ \sigma_2\end{matrix}\right\} = \frac{E}{6}\left[\frac{\varepsilon'_{120}+\varepsilon'_{-120}+\varepsilon'_0}{A} \pm \frac{\sqrt{\left(2\varepsilon'_0-\varepsilon'_{120}-\varepsilon'_{-120}\right)^2+3\left(\varepsilon'_{-120}-\varepsilon'_{120}\right)^2}}{B}\right] \tag{6.21c}$$

$$\tan 2\varphi = -\sqrt{3}\left(\frac{\varepsilon'_{-120}-\varepsilon'_{120}}{2\varepsilon'_0-\varepsilon'_{120}-\varepsilon'_{-120}}\right). \tag{6.22c}$$

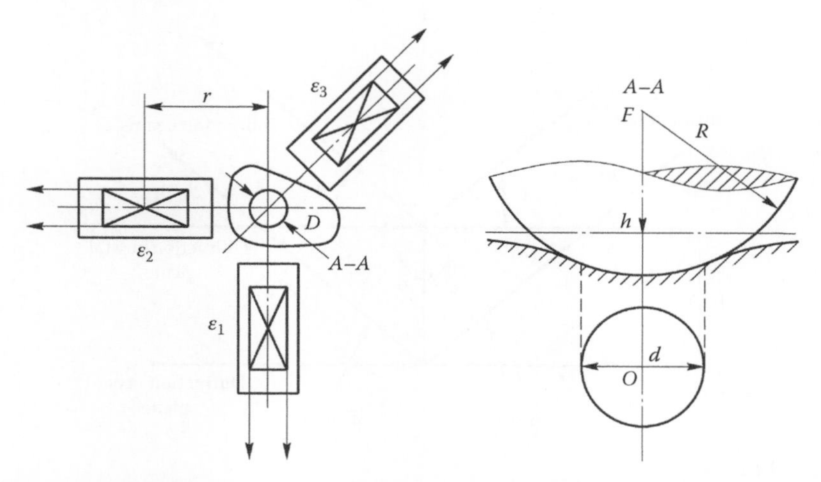

FIGURE 6.13 Schematic illustration of indentation method to measure residual stresses.

In addition, by eliminating the stresses in Equation 6.20, the corresponding strains ε_0, $\varepsilon_{-\alpha}$, ε_α of residual stresses in these directions can be solved. Residual stresses in the plate can be obtained using Mohr's circle of strains. This is called the graphical method.

6.2.1.3 Indentation Method

The indentation method is a new method used to measure residual stresses, and has emerged in recent years with the development of modern industries [3,4]. As illustrated in Figure 6.13, a bearing steel ball with diameter D is placed on the measuring point of the work piece. Impact power W, or static pressure P, is applied so that an indentation with a spherical crown shape of diameter d is formed on the surface of the work piece. The region around the indentation has superimposed stresses and strains, and the strain values can be measured by the strain gauge rosette around the indentation. The residual stresses at the measuring point influence the generated strains. Under fixed experimental conditions, there exists a good linear relationship between the strain increments at the measuring point, and the residual stresses. Therefore, for a given material, the calibration test can be conducted under certain experimental conditions to measure the material's stress–strain increment curve. The strain increments can also be measured by conducting the indentation test of the work piece under the same experimental conditions. The residual stresses in the work piece can be obtained in terms of the stress–strain increment curve in the calibration test.

6.2.2 Physical Measurement Methods of Residual Stresses

6.2.2.1 X-Ray Diffraction Method to Measure Residual Stresses

This method measures the residual stresses using the diffraction phenomenon of x-rays incident upon a material. Macroresidual stresses (type I and in some cases type II stresses) can be measured by the diffraction peak shift, while microstresses (type II and type III stresses) can be measured by diffraction peak broadening. Reference books that introduce the measurement of residual stresses by x-ray diffraction have already been published [1,10,11,13]. Zheng Xuejun et al. improved the traditional method of measuring macroresidual stresses, and applied it to the measurement of residual stresses in PZT thin films [6].

6.2.2.1.1 The Principle of Measurement

When the x-ray diffraction method is employed to measure residual stresses, the measured material must be a crystalline material [1,13]. In 1912, Bragg proved (as shown in Figure 6.14) that when

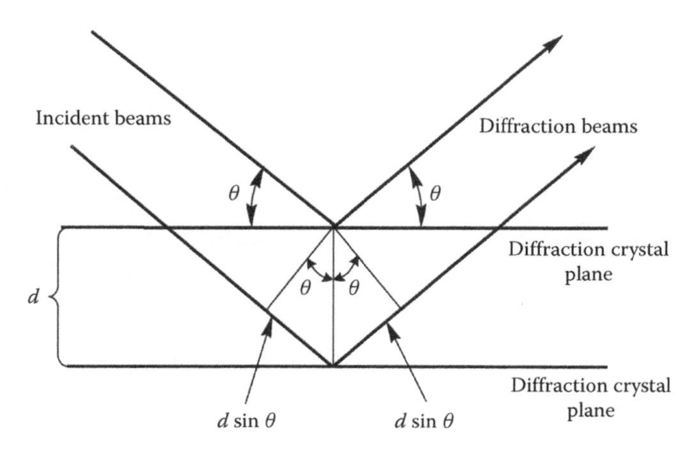

FIGURE 6.14 Bragg's diffraction.

x-ray beams are incident upon a crystalline material and diffraction occurs, the diffraction angle θ, the spacing of the lattice planes d, and the wavelength λ of the incident x-ray beams satisfy the following equation:

$$2d\sin\theta = n\lambda. \tag{6.23}$$

The scattered x-rays superpose and the intensity increases. This is the diffraction phenomenon, or Bragg's law. The diffraction angle θ is called Bragg's angle. Differentiating Bragg's equation gives

$$\frac{\Delta d}{d} = -\cot\theta\Delta\theta. \tag{6.24}$$

Equation 6.24 indicates that the diffraction angle 2θ changes when the interplanar spacing changes relatively by $\Delta d/d$ due to stress. Therefore, as long as the peak shift $\Delta\theta$ of a crystal plane in a certain diffraction direction on the specimen surface is measured, the change in spacing of the crystal planes can be calculated, and the stress in this direction can be calculated based on elastic mechanics.

In the x-ray diffraction method, the residual stress is obtained via the measurement of elastic strain. For an ideal polycrystalline material (where the grain size is small and uniform without preferred orientations), in the stress-free state, the spacing of the crystal planes of the same family in different orientations is equal. When a macroscopic stress σ_ϕ is applied, the spacing of the crystal planes of the same family in different orientations changes regularly with the grain orientation and the magnitude of stress, as shown in Figure 6.15. The change of interplanar spacing $d_{\phi\psi}$ in a certain orientation with respect to the stress-free state, $(d_{\phi\psi} - d_0)/d_0 = \Delta d/d_0$, reflects the elastic strain due to stress in the normal direction of the crystal plane, that is, $\varepsilon_{\phi\psi} = \Delta d/d_0$. Apparently, a functional relationship exists between the change of interplanar spacing in a certain orientation, and the applied stress. Therefore, the key to the measurement of stress is to establish the relationship between the undetermined residual stress σ_ϕ and the strain $\varepsilon_{\phi\psi}$ in a certain orientation.

Here, we discuss the measurement in the plane stress state (or the biaxial stress state). On the free surface of an object, the stress in the normal direction is zero. This plane stress assumption is reasonable when the gradient of stress inside the object along the direction normal to the specimen surface is very small, and the penetration depth of the x-ray is very shallow (on the order of 10 μm). As shown in Figure 6.16, the coordinate system is set up on the specimen surface, with the z-direction normal to the surface. By adjusting the orientation of the element, we can always find an appropriate orientation so that each surface of the element has only one of the three principal stresses σ_1, σ_2, σ_3 mutually perpendicular. The corresponding strains are ε_1, ε_2, ε_3. O-XYZ is the coordinate system of the principal stresses, and O-xyz represents the

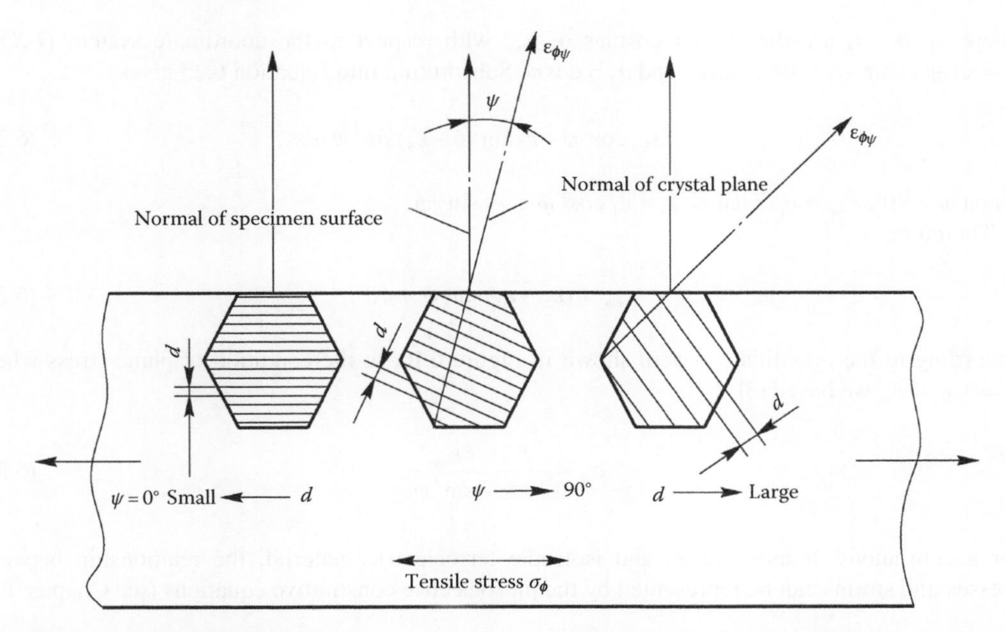

FIGURE 6.15 The relationship between stress and interplanar spacing of the same family of crystallographic planes in different orientations.

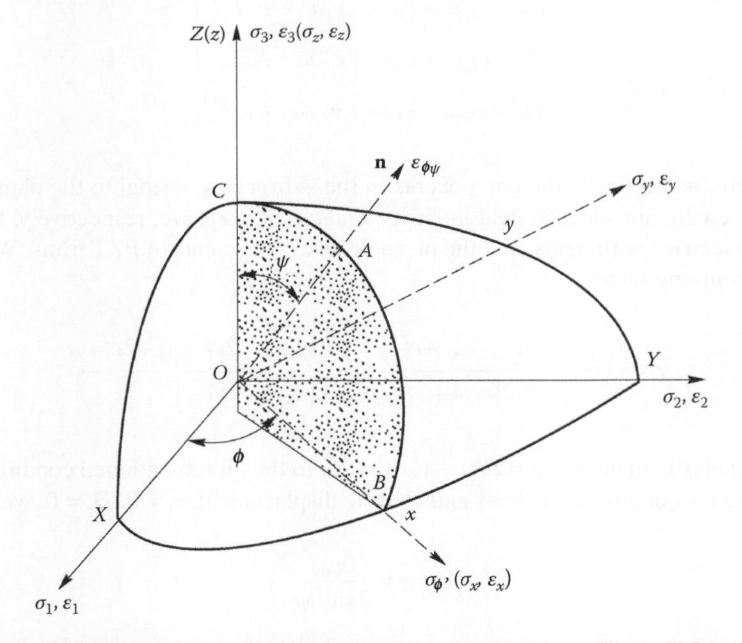

FIGURE 6.16 Stresses on the surface of a deformed object. (Adapted from Zheng, X.J. et al., *Acta. Mater.*, 52, 3313, Figure 2, 2004.)

direction of undetermined stress $\sigma_\phi(\sigma_x)$ and the directions of σ_y, σ_z normal to the undetermined stress, where ϕ is the angle between σ_ϕ and σ_1. The plane consisting of OZ and σ_ϕ is called "the plane of measuring direction." $\varepsilon_{\phi\psi}$, or $(\Delta d/d)_{\phi\psi}$, is the strain in an arbitrary direction **n** in this plane, and is the angle between it and OZ. The relationship between $\varepsilon_{\phi\psi}$ and the principal strains can be expressed as

$$\varepsilon_{\phi\psi} = a_2^2\varepsilon_1 + a_2^2\varepsilon_2 + a_3^2\varepsilon_3, \tag{6.25}$$

where a_1, a_2, a_3 are directional cosines of $\varepsilon_{\phi\psi}$ with respect to the coordinate system $O\text{-}XYZ$, $a_1 = \sin\psi\cos\phi$, $a_2 = \sin\psi\sin\phi$, and $a_3 = \cos\psi$. Substituting into Equation 6.25 gives

$$\varepsilon_{\phi\psi} = (\varepsilon_1\cos^2\phi + \varepsilon_2\sin^2\phi - \varepsilon_3)\sin^2\psi + \varepsilon_3. \tag{6.26}$$

When $\psi = 90°$, $\varepsilon_{\phi\psi} = \varepsilon_x$. That is, $\varepsilon_x = \varepsilon_1\cos^2\phi + \varepsilon_2\sin^2\phi$.

Therefore,

$$\varepsilon_{\phi\psi} = (\varepsilon_x - \varepsilon_3)\sin^2\psi + \varepsilon_3. \tag{6.27}$$

According to the coordinate system shown in Figure 6.16, in the condition of plane stress where $\sigma_z = 0$, $\varepsilon_z = \varepsilon_3$, we have [13]

$$\sigma_\phi = \frac{E}{1+\nu}\frac{\partial\varepsilon_{\phi\psi}}{\partial\sin^2\psi}. \tag{6.28}$$

For a continuous, homogeneous, and isotropic ferroelectric material, the relationship between stresses and strains can be represented by the piezoelectric constitutive equations (see Chapter 10):

$$\left.\begin{aligned}
\sigma_1 &= c_{11}\varepsilon_1 + c_{12}\varepsilon_2 + c_{13}\varepsilon_3 - e_{31}E_z \\
\sigma_2 &= c_{12}\varepsilon_1 + c_{11}\varepsilon_2 + c_{13}\varepsilon_3 - e_{31}E_z \\
\sigma_3 &= c_{13}\varepsilon_1 + c_{13}\varepsilon_2 + c_{33}\varepsilon_3 - e_{33}E_z \\
D_z &= e_{31}\varepsilon_1 + e_{31}\varepsilon_2 + e_{33}\varepsilon_3 + \in_{33} E_z
\end{aligned}\right\}, \tag{6.29}$$

where D_z, E_z are, respectively, the components in the z-direction normal to the plane of the film of electric displacement and electric field intensity, and c_{ijkl}, $\in_{ik}$, e_{kij} are, respectively, the elastic coefficients, the dielectric coefficients, and the piezoelectric coefficients of PZT films. We introduce the piezoelectric coupling factor

$$\chi = \frac{(c_{11}+c_{12})\left(c_{33}\in_{33} + e_{33}^2\right) - 4c_{13}e_{31}e_{33} - 2c_{13}^2\in_{33} + 2e_{31}^2c_{33}}{2(c_{13}\in_{33} + e_{31}e_{33}) + \left(c_{33}\in_{33} + e_{33}^2\right)}. \tag{6.30}$$

for an isotropic elastic material, $\chi = E/(1+\nu)$. Similar to the aforementioned condition, and considering the boundary conditions of stress and electric displacement $\sigma_3 = 0$, $D_z = 0$, we have [6]

$$\sigma_\phi = \chi\frac{\partial\varepsilon_{\phi\psi}}{\partial\sin^2\psi}. \tag{6.31}$$

This equation describes the relationship between the undetermined stress σ_ϕ and the changing rate of $\varepsilon_{\phi\psi}$ with respect to orientation ψ, and is the basic equation to calculate the undetermined stress. Under certain conditions of plane stress, $\varepsilon_{\phi\psi}$ is linear with $\sin^2\psi$. In order to derive a more practical formula to calculate the macroscopic stress in the method of x-ray diffraction, $\varepsilon_{\phi\psi}$ in Equation 6.31 should be converted to an expression with a diffraction angle. After a simple derivation, we obtain

$$\sigma_\phi = -\frac{\chi}{2}\cot\theta_0\frac{\pi}{180°}\frac{\partial 2\theta_{\phi\psi}}{\partial\sin^2\psi}, \tag{6.32}$$

where the unit of $2\theta_{\phi\psi}$ is "degree," not "radian." This is the basic equation to calculate the macroscopic stress in the plane stress state. In Equation 6.32, defining

$$K = -\frac{\chi}{2}\cot\theta_0 \frac{\pi}{180°}, \tag{6.33a}$$

$$M = \frac{\partial 2\theta_{\phi\psi}}{\partial \sin^2\psi}, \tag{6.33b}$$

we have

$$\sigma_\phi = KM. \tag{6.33c}$$

K is called the stress constant. It depends on the measured material and the diffraction angle of the selected diffraction plane (i.e., the spacing of the diffraction planes and the wavelength λ of the x-rays). For steels, the residual stress of a component is represented by the stress in primary ferrite. If CrK_α irradiation is taken as illumination, ($\lambda_{K\alpha} = 0.2291$ nm), and the {211} plane of ferrite is measured, the stress constant is $K = -318$ MPa/degree. Because the crystal is anisotropic, the values E, v of different {hkl} planes are different. Therefore, the value of K cannot be mechanically measured as the average elastic constant of a polycrystalline material. Rather, it is measured using a residual stress-free sample with a known applied stress. The stress constants of general materials are listed in Table 6.2. M is the slope of the straight line of $2\theta_{\phi\psi}-\sin^2\psi$. Since K is a negative value, when $M > 0$,

TABLE 6.2
Stress Testing Data Constants of General Materials

	Lattice Type	Lattice Constants	$E/10^3$ MPa	γ	Rad	(hkl)	$2\theta/(°)$	K/MPa/(°)
α-Fe (Ferrite, Martensite)	BCC	2.8664	206–216	0.28–0.3	CrK_α	(211)	156.8	−297.23
					CoK_α	(310)	161.35	−230.4
γ-Fe (Austenite)	FCC	3.656	192.1	0.28	CrK_β	(311)	149.6	−355.35
					MnK_α	(311)	154.8	−292.73
Al	FCC	4.0049	68.9	0.345	CrK_α	(222)	156.7	−92.12
					CoK_α	(420)	162.1	−70.36
					CoK_α	(331)	148.7	−125.24
					CuK_α	(333)	164.0	−62.85
Cu	FCC	3.6153	127.2	0.364	CrK_β	(311)	146.5	−245.0
					CoK_α	(400)	163.5	−118.0
					CuK_α	(420)	144.7	−258.92
Cu-Ni	FCC	3.595	129.9	0.333	CoK_α	(400)	158.4	−162.19
W C	HCP	a2.91	523.7	0.22	CoK_α	(121)	162.9	−466.0
		c2.84			CuK_α	(301)	146.76	−1118.18
Ti	HCP	a2.9504	113.4	0.321	CoK_α	(114)	154.2	−171.6
		c4.6831			CoK_α	(211)	142.2	−256.74
Ni	FCC	3.5238	207.8	0.31	CrK_β	(311)	157.7	−273.22
					CuK_α	(420)	155.6	−289.39
Ag	FCC	4.0856	81.1	0.367	CrK_α	(222)	152.1	−128.48
					CoK_α	(331)	145.1	−162.68
					CoK_α	(420)	156.4	−108.09
Cr	BCC	2.8845	—	—	CrK_α	(211)	153.0	—
					CoK_α	(310)	157.5	
Si	Diamond	5.4282	—	—	CoK_α	(531)	154.1	—

the stress is negative, indicating compressive stress. When $M < 0$, the stress is positive, indicating a tensile stress. If the relation $2\theta_{\phi\psi} - \sin^2\psi$ loses linearity, the state of the material deviates from the assumptions of the derived formula as listed in Table 6.2.

Considering the penetration depth of x-rays, the relationship $2\theta_{\phi\psi} - \sin^2\psi$ is influenced when an apparent stress gradient, nonplanar stress state or texture. In such cases, special methods are required to measure and calculate the residual stresses.

It can be seen from Equations 6.32 and 6.33 that at least two different diffraction angles ($2\theta_{\phi\psi}$), with different orientations ψ in the plane of the measuring direction, are required to solve the residual stresses on the specimen surface. Therefore, certain diffraction geometric conditions are employed to determine and change the orientation ψ of the diffraction plane. At present, the diffractometer is usually used to measure residual stresses. The general geometric conditions of diffraction have two modes: (1) iso-inclination mode and (2) side-inclination mode.

1. Iso-inclination mode

   As shown in Figure 6.17a, the geometric diffraction condition of the iso-inclination mode is the coincidence of the plane of the measuring direction and the scanning plane. The definition of the plane of measuring direction has been described before. The scanning plane consists of the incident direction, the normal of the diffraction plane (ON', the direction of $\varepsilon_{\phi\psi}$), and the diffraction direction. In this mode, there are two methods to determine the orientation of ψ.

   a. ψ is fixed

      In the conventional symmetric diffraction of a specimen using a diffractometer, the incident beam and the axis of the goniometer are arranged symmetrically on the two sides of the normal of the specimen surface. Goniometer and specimen rotate at angular velocities with the ratio of 2:1. In this condition, the diffraction plane of the recorded peak must be parallel to the specimen surface, as shown in Figure 6.18a, with $\psi = 0°$. From the position of $\psi = 0°$, the specimen is rotated independently by an angle ψ about the axis of the diffractometer, followed by $2\theta/\theta$ scanning and measurement. The angle between the normal of the diffraction plane and the normal of the specimen surface is equal to the rotation angle ψ, as shown in Figure 6.18b, with $\psi = 45°$. This method, in which the orientation ψ of the diffraction plane is directly determined and changed by setting the

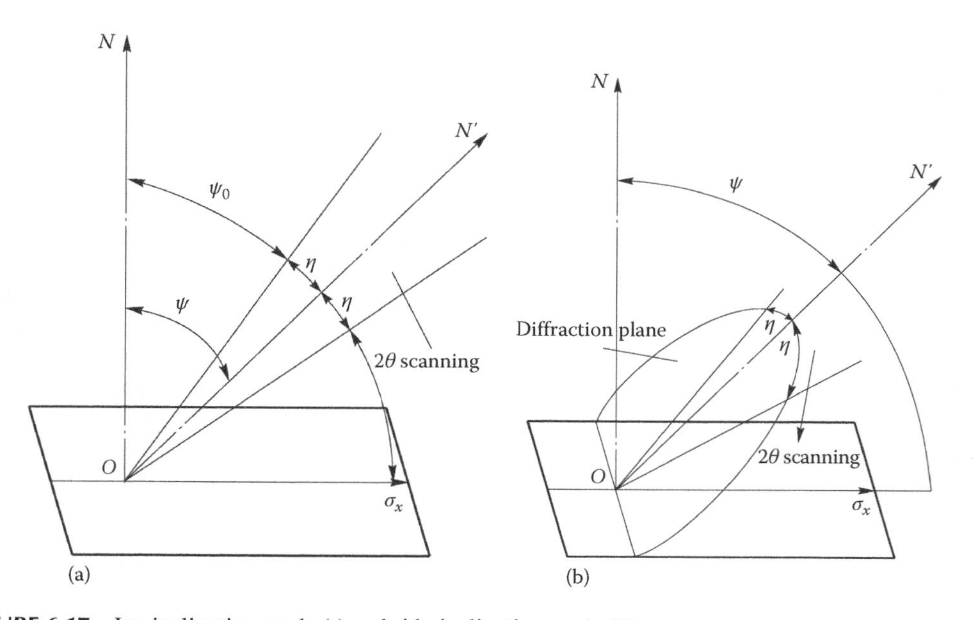

FIGURE 6.17 Iso-inclination mode (a) and side-inclination mode (b).

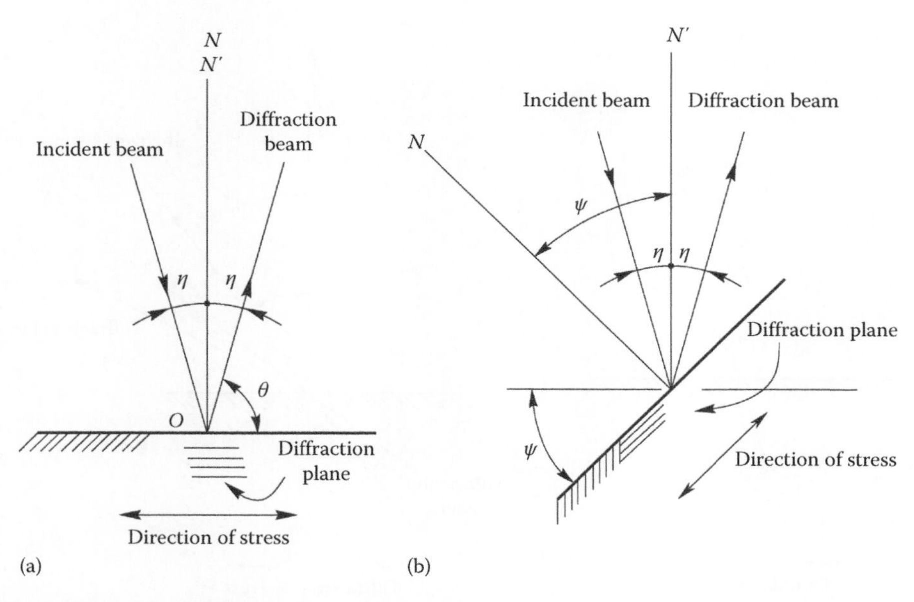

FIGURE 6.18 Fixed ψ method. (a) $\psi = 0°$. (b) $\psi = 45°$.

diffraction geometric condition, is called the fixed ψ method. This method is appropriate for measuring macroresidual stresses of small specimens using a diffractometer.

b. Fixed ψ_0 method

In engineering, it is often necessary to measure residual stresses in a mechanical part or a structural component. Its shape is complex, and the volume is large, and cannot be placed inside a diffractometer. In this case, the fixed ψ_0 method to measure stresses has been developed. In this method, the workpiece is fixed, the x-ray tube and the goniometer of the special instrument are mounted near the measuring site by column, crossbeam, or bracket; and the different ψ orientations are obtained by changing the incident direction of the x-ray. ψ_0 is the angle between the incident beam and the normal of the specimen surface. For a specified incident direction, the goniometer scans independently. According to the geometric condition in Figure 6.19, in terms of ψ_0 and the measured diffraction angle θ, we have $\psi = \psi_0 + (90° - \theta)$.

2. Side-inclination mode

Equation 6.33 shows that when the values of the residual stress and the stress constant K are fixed, the accuracy of the slope M is determined by $\Delta\sin^2\psi$. That is, the larger the difference in the measured crystal orientation ψ, the smaller is the relative error $\Delta 2\theta_{\varphi\psi}$. In the iso-inclination method, the change of ψ or ψ_0 is restricted by θ. In the fixed ψ method, the range of ψ is $0° \sim \theta$ (see Figure 6.18); in the fixed ψ_0 method, the range of ψ_0 is $0° \sim (2\theta–90°)$ (see Figure 6.19). When measuring the entire profile of the diffraction peak, it is necessary to determine the scanning range in terms of the broadening of the peak. The goniometer cannot accept the diffraction beam parallel to the specimen surface, thus the practical permission range of ψ should be narrower. When the workpiece shape is complex, the range of the orientation angle is restricted by the workpiece's shape, and the iso-inclination method cannot be applied to measure the residual stress. In these cases, the side-inclination method is adopted.

In side-inclination mode, the plane of the measuring direction is perpendicular to the scanning plane. Figure 6.17b compares this mode to the iso-inclination mode. In side-inclination mode, the goniometer scans on the plane perpendicular to the plane of the measuring direction. Moreover, the range of ψ is not constrained by the diffraction angle, but depends only on the shape of the measured sample. For samples with flat surfaces, the range of ψ in theory

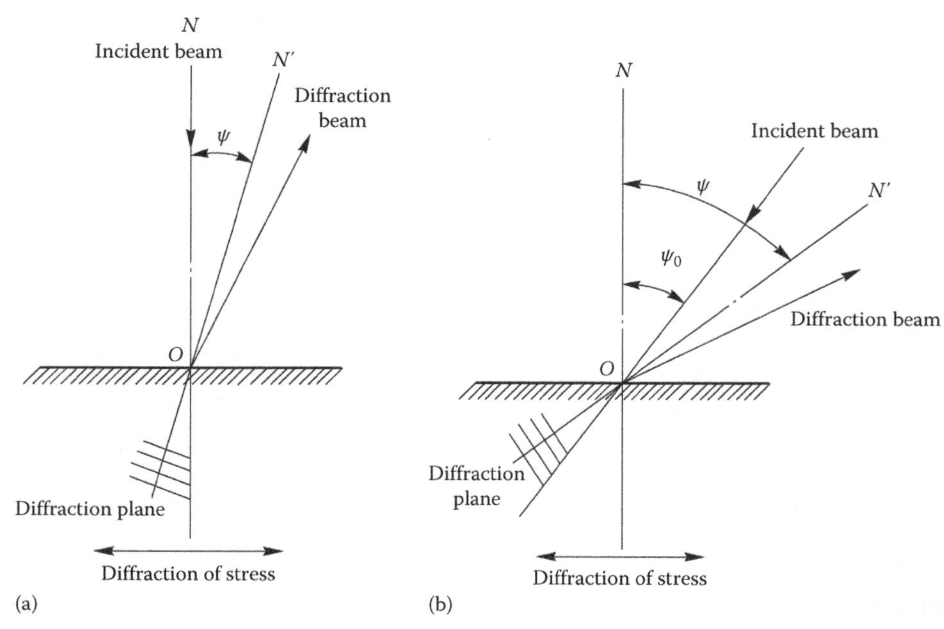

FIGURE 6.19 Fixed ψ_0 method. (a) $\psi_0 = 0°$. (b) $\psi_0 = 45°$.

TABLE 6.3
Measured Stress Data Using $\sin^2\psi$ Method (Iso-Inclination Mode)

ψ	0°	15°	30°	45°
$2\theta_\psi$	154.92	155.35	155.91	155.96
$\sin^2\psi$	0	0.067	0.25	0.707

can approach 90°. A fixed ψ method is the mode used to determine the orientation of ψ in side-inclination mode. Selection of the orientation angle can be a two-point method and a $\sin^2\psi$ method, and the formula to calculate stress is identical to that of the iso-inclination mode.

Take mild steel as an example. Table 6.3 lists the data for the measurement of surface stress using the $\sin^2\psi$ method of the iso-inclination mode. The illumination source is CrK_β radiation. For steel, the (211) peak is usually measured. From Bragg's equation, $2\theta_0 = 156.4°$. The $2\theta_\psi \sim \sin^2\psi$ curve is plotted using the data in Table 6.3. By the least square method, we obtain $M = 1.965°$. From Table 6.2, $K = -318.1$ MPa/degree. Therefore, $\sigma_0 = K_1M = -318.1 \times 196.5 = -625.1$ MPa.

6.2.2.2 Magnetic Method to Measure Residual Stresses

Under certain conditions, when the measured material is a ferromagnetic material (see Chapter 10) such as iron, steel, cast iron, and nickel–cobalt alloy, the magnetic method can be used to measure the residual stresses [9]. The magnetic method is a nondestructive method to measure the surface stress in materials, using the magneto-elastic effect of a ferromagnetic material. The principle of this method is simple, the apparatus is portable, application condition is not strict, and it is easy to operate. Therefore, the magnetic method has unmatched advantages in the measurement of residual stresses, especially for large components.

A ferromagnetic object consists of tiny domains which have been magnetized to a saturation state. In each domain, the magnetization direction is irregular. When it is placed in an external magnetic

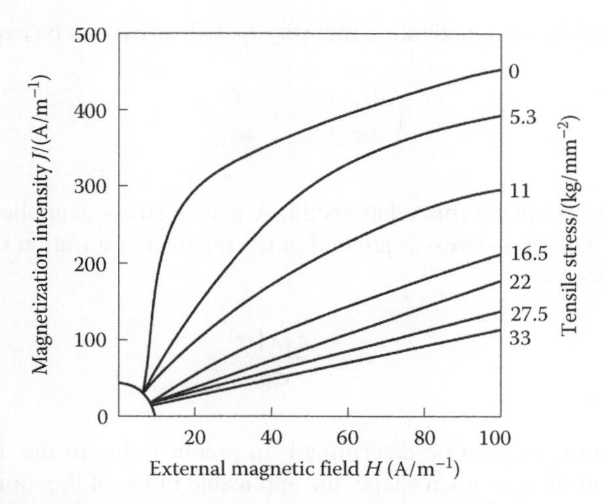

FIGURE 6.20 The influence of tensile stress on magnetization of nickel.

field, the magnetization direction of each domain is aligned to the direction of the magnetic field. We can imagine that the tiny domain has reached magnetic saturation. When it is placed into a magnetic field, its magnetization direction approaches the direction of the magnetic field. Magnetization intensity J increases with the increase of magnetic field strength H, and approaches a fixed value when the magnetic field strength reaches a certain value. This is called magnetic saturation. The tiny domain which has reached magnetic saturation—approximately 10^{-9} cm^3 in size—is called the magnetic domain. During the magnetization procedure, the magnetic domains rotate and displace. For the overall object, the length in the magnetization direction changes, which is called magnetostriction.

The magnetization of a ferromagnetic material is influenced by its crystallographic anisotropy, grain size, alloying elements, inclusions, and stresses. Internal stress generates additional drag force on the rotation and displacement of the magnetic domain. In this sense, the external stress and the internal residual stress are equivalent. The magnetization curve changes due to this type of stress. Figure 6.20 shows the influence of tensile stress on the magnetization of nickel.

Extensive experimental and theoretical studies concerning the influence of stress on the magnetism of ferromagnetic materials have been conducted. Becker, Kersten, and Preisach studied annealed Ni and Fe–Ni alloys and proposed the following equation [9]:

$$\psi_a = \frac{I_s^2}{3\lambda\sigma}, \tag{6.34}$$

where

ψ_a is the initial magnetic permeability
I_s is the saturation magnetization intensity
λ is the magnetostriction coefficient

Becker and Kersten also proposed the equation for quantitative evaluation of the internal stress. Assume that the saturation value of the magnetostriction coefficient is λ, that the magnetic anisotropy constant is K', and that the extent of the influence of stress on the magnetization direction depends on the value of $\lambda\sigma/K'$. When $\lambda\sigma < K'$, the easy magnetization direction is determined by the crystallographic direction; when $\lambda\sigma \gg K'$, the easy magnetization direction is determined by stress. When the internal stress σ_i is in an arbitrary direction and the magnetostriction is isotropic, and assuming that σ_{im} is the average value of the internal stress, we have

$$\psi_a = \frac{2I_s^2}{9|\lambda|\sigma_{im}}. \tag{6.35}$$

The variation of the residual magnetization intensity I_R with stress can be expressed as

$$\left(\frac{\mathrm{d}I_R}{\mathrm{d}\sigma}\right)_{\sigma\to 0} = \frac{I_s}{4\sigma_{im}}. \tag{6.36}$$

Förste measured the stress using this relationship. A tensile stress is applied to an unmagnetized ferromagnetic object. The same stress is applied in the magnetic saturation state. The difference in elongation is $\Delta\varepsilon$, and we have

$$\Delta\varepsilon = \frac{c\,|\lambda|\,\sigma}{\sigma_{im}}. \tag{6.37}$$

In terms of this equation, σ_{im} can be determined. In practice, due to the difference between the assumed properties and the specimen shape, the applicable range of the aforementioned equations is constrained.

The internal stress σ in Equations 6.34 through 6.37 can be caused either by an external load or residual stresses. Residual stresses can be classified into macroresidual stresses and microresidual stresses. The type I residual stresses are self-balanced on the entire cross section. The microscopic residual stresses existing at the grain level are Type II and Type III residual stresses. Because these residual stresses are in the same magnetic field, and the type I residual stresses are zero on the entire cross section, only type II and type III residual stresses have influence on the magnetism of the material.

6.2.2.3 Photoelastic Coating Method to Measure Residual Stresses

The photoelastic coating method to measure stress employs optical polarity [16]. When most transparent objects are applied stresses—even though they are isotropic in the stress free state—they exhibit anisotropy, and complex refraction occurs for the transmitted light. In order to perform the measurement, excellent adaptive materials must be used, such as phenol resins and alkyd resins.

In the photoelastic coating method, resins are coated on the specimen surface. The stress is calculated using the phase difference between the incident light and the polarized reflection light. Figure 6.21 illustrates the principle of this method. Consider the case in which the light ray (unpolarized light) is perpendicularly incident upon the specimen. In this case, the polarized light rays, after double refraction, exhibit a phase difference. Assume that the thickness of the coating is e. When the incident light ray is reflected at the bottom of the coating after traveling a distance $2e$ inside the coating, the polarized light rays with a phase difference of δ_1 are split. Assume that the principal stresses and strains in the coating are $\bar{\sigma}_1$, $\bar{\sigma}_2$ and $\bar{\varepsilon}_1$, $\bar{\varepsilon}_2$; and the stresses and strains in the specimen covered by the coating are σ_1, σ_2 and ε_1, ε_2. The relationship between phase difference and stresses is

$$\delta_1 = c(\bar{\sigma}_1 - \bar{\sigma}_2)2e, \tag{6.38}$$

where c is the experimentally measured photoelastic constant of the coating. The stresses and strains in the coating and in the specimen satisfy

$$\varepsilon_1 = \bar{\varepsilon}_1, \quad \varepsilon_2 = \bar{\varepsilon}_2, \tag{6.39}$$

$$\bar{E}\bar{\varepsilon}_1 = \bar{\sigma}_1 - \bar{v}\bar{\sigma}_2, \quad \bar{E}\bar{\varepsilon}_2 = \bar{\sigma}_2 - \bar{v}\bar{\sigma}_1, \tag{6.40}$$

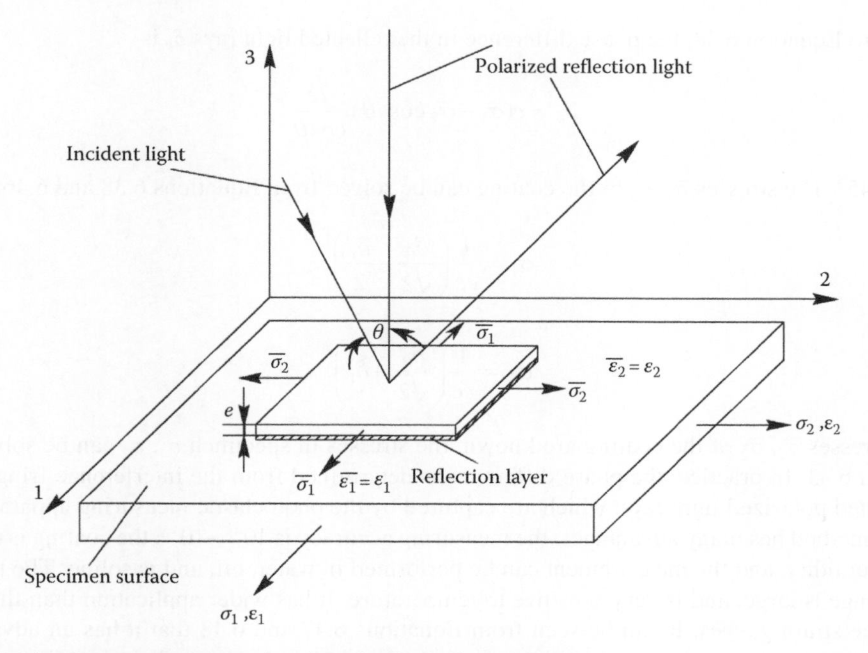

FIGURE 6.21 Stress measurement by photoelastic coating method.

where $\overline{E}$, $\overline{\nu}$ are Young's modulus and Poisson's ratio of the coating. Thus

$$\varepsilon_1 - \varepsilon_2 = \overline{\varepsilon}_1 - \overline{\varepsilon}_2 = \frac{1+\overline{\nu}}{\overline{E}}(\overline{\sigma}_1 - \overline{\sigma}_2). \tag{6.41}$$

Substituting Equation 6.38 gives

$$\varepsilon_1 - \varepsilon_2 = \frac{1+\overline{\nu}}{\overline{E}}\frac{\delta_1}{2ec}. \tag{6.42}$$

In terms of the stress–strain relationship of the specimen, we have

$$\sigma_1 - \sigma_2 = \frac{E}{1+\nu}\frac{1+\overline{\nu}}{\overline{E}}\frac{\delta_1}{2ec}. \tag{6.43}$$

In a uniaxial (one-dimensional) stress state where σ_1 or σ_2 is zero, the stress can be obtained directly from Equation 6.43. In the plane stress state, additional experimentation is required in order to solve σ_1 and σ_2. As shown in Figure 6.21, we consider the oblique incident light ray. The thickness of the transmission layer $2e_\theta$ is

$$2e_\theta = \frac{2e}{\cos\theta}. \tag{6.44}$$

At the same time, corresponding to the previous $\overline{\sigma}_2$, the stress $\overline{\sigma}_\theta$ in the plane perpendicular to the transmission light ray is

$$\overline{\sigma}_\theta = \overline{\sigma}_2 \cos^2\theta. \tag{6.45}$$

Similar to Equation 6.38, the phase difference in the reflected light rays δ_θ is

$$\delta_\theta = c(\bar{\sigma}_1 - \bar{\sigma}_2 \cos^2\theta)\frac{2e}{\cos\theta}. \tag{6.46}$$

Let $\theta = 45°$. The stresses $\bar{\sigma}_1$, $\bar{\sigma}_2$ of the coating can be solved from Equations 6.38 and 6.46:

$$\bar{\sigma}_1 = \frac{1}{ec}\left(\frac{\delta_\theta}{\sqrt{2}} - \frac{\delta_1}{2}\right), \tag{6.47}$$

$$\bar{\sigma}_2 = \frac{1}{ec}\left(\frac{\delta_\theta}{\sqrt{2}} - \delta_1\right). \tag{6.48}$$

If the stresses $\bar{\sigma}_1$, $\bar{\sigma}_2$ of the coating are known, the stresses in specimen σ_1, σ_2 can be solved using Equation 6.43. In practice, the phase difference is determined from the interference fringes of the transmitted polarized light rays, which are captured by the photoelastic measuring apparatus.

This method has many advantages: the measuring accuracy is 10^{-5} –10^{-6}, the coating is not sensitive to humidity, and the measurement can be performed in water, oil, and gasoline. The measured strain range is large, and is very sensitive to temperature. It has wider application than the electric resistance strain gauges. It can be seen from Equations 6.47 and 6.48 that it has an advantage in determining the stresses in two directions, from the incident light in one direction. Since it is very stable, photoelastic coatings can be used to measure residual stresses, strains in the groove cutting method and the hole drilling method, and welding stresses. Photoelastic coatings also have other applications, such as the measurement of residual stresses at the notch root of a component.

6.3 INFLUENCE OF RESIDUAL STRESSES ON THE MECHANICAL PROPERTIES OF MATERIALS

The influence of residual stresses can be approximately classified into the influence on static strength, brittle failure, stress corrosion cracking, fatigue, and the influence on the detrimental deformation due to dimensional deviation during or after the manufacturing and machining process.

6.3.1 Influence of Residual Stresses on Static Properties

As stated in Section 6.1.1, residual stresses are classified into macroresidual stresses and microresidual stresses [16]. It is the microstresses that have direct influence on a material's strength. Here, we discuss the influence of macroresidual stresses on the static strength and the static stability of structural components, and the hardness of materials.

6.3.1.1 Influence of Residual Stresses on Static Strength and Deformation

If an external load is applied to a component with residual stresses, the deformation of the entire component is influenced by the interaction of the applied stress and the residual stresses. Additionally, the residual stresses change after the removal of the applied load. Figure 6.22 shows a simple example to illustrate this procedure. On the cross section of the frame structural component in Figure 6.22a, residual stresses are present in a, b, and c as shown in Figure 6.22b. (The upward arrow and downward arrow represent tension and compression, respectively). At the two ends d of the component, a tensile load F is applied so that all a, b, c parts yield. After unloading, the tensile residual stress in the middle part a, and the compressive residual stress in the side parts b and c, are reduced, as shown in Figure 6.22c.

Next, the deformation of each part on the cross section under the tensile load is discussed. If the material is assumed to be ideal elastic-plastic, it exhibits stress–strain curves as shown in Figure 6.22d and e. The point 0 in the figure denotes the residual stress when the load is zero.

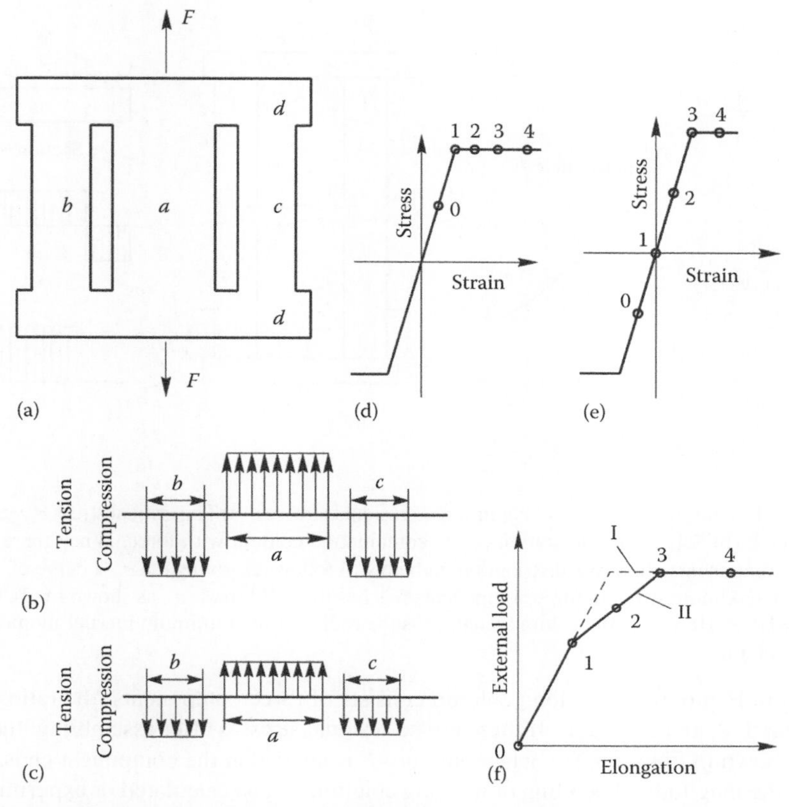

FIGURE 6.22 The change of residual stresses and deformation caused by the external load. (a) Component; (b) residual stresses before loading; (c) residual stresses after loading; (d) stress–strain curve of the middle part; (e) stress–strain curves of two side parts; (f) the load–elongation curve of all parts.

Figure 6.22f shows the relationship between the external load and the overall elongation. When the load reaches point l, the middle part a on the cross section yields. When the load reaches point 2, the middle part a undergoes plastic deformation while the side parts b and c on the cross section are still in the elastic state. When the load reaches point 3, both experience plastic deformation. Therefore, the curve between the load and elongation of the entire part is shown in Figure 6.22f. The states 1, 2, and 3 display the deformation behaviors shown in curve II. If the load is removed from these states, the residual stresses are reduced or released. Figure 6.22c shows the residual stresses after removal of the load from point 2.

For the plastic material with residual stresses as shown in this example, the entire cross section undergoes plastic deformation when the load goes beyond point 3. From this moment to the failure of the material, the deformation behavior is the same as the component without residual stresses. We can say that the residual stresses have no influence. That is, the influence of residual stress for a plastic material exists only before the entire cross section reaches plastic deformation. However, for a material hardened by heat treatment, this influence still exists in the plastic zone.

6.3.1.2 Influences of Residual Stresses on the Static Stability of Structural Components

For a composite component subjected to an applied compressive load, the influence of residual stress on its stability can be viewed as the influence of residual stress on longitudinal bending. After manufacturing or welding, residual stress in the component influences the bending strength. Therefore, it is important in practice to evaluate the residual stresses related to the longitudinal bending strength of a welded structural component. Here, we demonstrate a basic example.

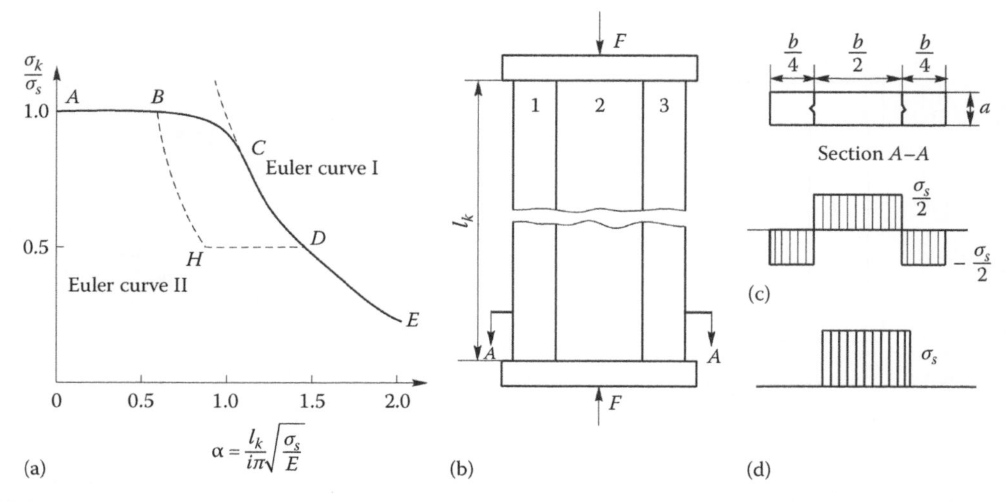

FIGURE 6.23 Longitudinal bending of column with residual stresses. (a) The critical stress σ_k versus the aspect ratio of column λ. (b) Schematic illustration of the combination column with force. When there is no residual stress in three components, the stress distribution in them will follow (c), giving the σ_k–λ curve of ABCDE in (a). However, if the maximum stress in three components reaches the yield stress σ_s, as shown in (d), the σ_k–λ curve will follow ABHDE. Here, l_k is the column length, i is the radius of the minimum inertial moment of the cross section of the column.

As shown in Figure 6.23, the long column consists of three components (the ratio of the cross-sectional areas 1, 2, and 3 is 1:2:1). If there are no residual stresses after assembling, the component stresses are shown in Figure 6.23c, when pressure F is applied at the component ends. The critical stress σ_k, for the longitudinal bending of the long column, can be calculated or experimentally measured. The relationship between σ_k and the aspect ratio of the column $\lambda = l_\kappa/i$ is represented by the curve $ABCDE$ in Figure 6.23a.

$$\sigma_k = \frac{F_k}{ab}, \tag{6.49}$$

where F_k is the critical load for longitudinal bending of the long column. And

$$F_k = \frac{\pi^2 EJ}{l_k^2}, \tag{6.50}$$

where

- E is the elastic modulus of the material
- J is the minimum inertial moment of the cross section of the long column.

Substituting Equation 6.50 into 6.49, we have

$$\sigma_k = \frac{\pi^2 E}{l_k^2} \times \frac{J}{F} = \frac{\pi^2 E}{\lambda^2}. \tag{6.51}$$

From Figure 6.23, for a compressive rod with large compliance, the design follows the Euler formula (6.51), or curve CDE. For a compressive rod with small compliance, the design follows the strength curve AB. For a rod with medium compliance, the critical stress can be approximately calculated by

$$\sigma_k = a_0 - b_0\lambda. \tag{6.52}$$

The designed stress σ of a compressive rod must satisfy $\sigma \leq \sigma_k$, where a_0 and b_0 are constants related to the properties of the material.

Assume that the three components after assembling have residual stresses as shown in Figure 6.23d. After a compressive load is applied, the relationship between the critical stress σ_k for longitudinal bending and the aspect ratio of the column follows the curve *ABHDE* in Figure 6.23a. When the compressive load increases so that the stress of the middle component (the sum of the residual stress and the applied stress) reaches the yield stress σ_s in Figure 6.23d—that is, when it reaches the elastic limit for longitudinal bending denoted by point *D* on Euler curve *I*—the stresses in the side components 1 and 3 are zero. For the entire section of the assembled components, $\sigma_k/\sigma_s = 0.5$. If the compressive stress continues to increase to $F + \Delta F$, the increased compressive load ΔF is completely borne by the two side components 1 and 3, and the relationship between the critical stress and the aspect ratio follows Euler curve II (curve *BHD*). Segment *HD* indicates that the practical aspect ratio of the assembled components decreases drastically after the middle component 2 enters the plastic state.

6.3.1.3 Influence of Residual Stress on Hardness

According to the measurement principle, hardness can be classified into indentation hardness and rebound hardness. Regardless of the type, hardness is influenced by the existence of residual stresses. For indentation hardness, residual stress influences the plastic deformation of the region surrounding the indent. For rebound hardness, residual stress influences the rebound energy, and thus the measured hardness changes. On the other hand, if this change is very large, hardness can be employed to measure the residual stress. However, the theoretical analysis of this influence is very difficult.

6.3.1.3.1 The Influence on Indentation Hardness

In order to analyze the influence of residual stresses on indentation hardness, the indentation method is simplified, as shown in Figure 6.24. Pressure F_0 is uniformly applied to the contact region. For the point below the contact region, the normal stresses in the *x*-direction and in the *y*-direction are σ_x, σ_y and the shear stresses τ_{xy} are

$$\sigma_x = \frac{-F_0}{2\pi}\left[2(\varphi_1 - \varphi_2) + \sin 2\varphi_1 - \sin 2\varphi_2\right], \tag{6.53}$$

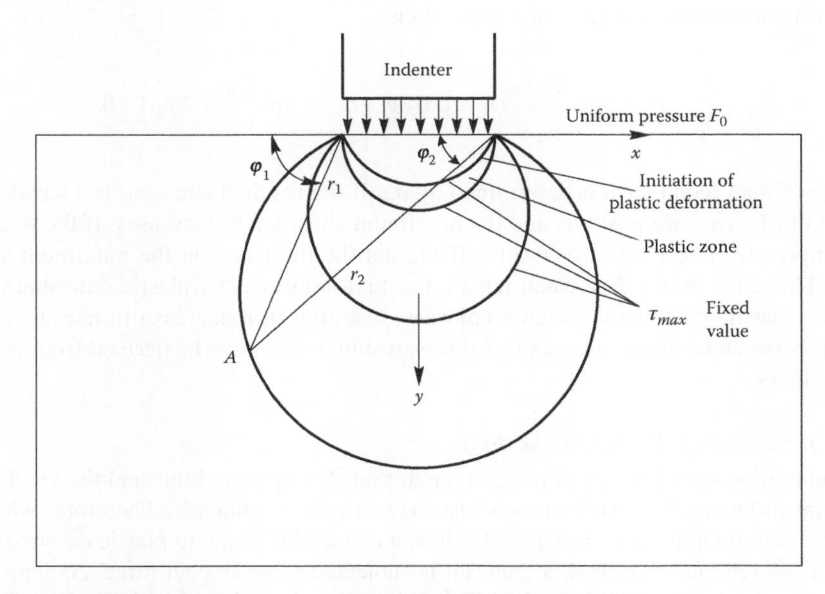

FIGURE 6.24 The maximum shear stress τ_{max} under a uniform contact pressure.

$$\sigma_y = \frac{-F_0}{2\pi}\left[2(\varphi_1 - \varphi_2) - \sin 2\varphi_1 - \sin 2\varphi_2\right]. \tag{6.54}$$

$$\tau_{xy} = \frac{F_0}{2\pi}(\cos 2\varphi_1 - \cos 2\varphi_2). \tag{6.55}$$

Thus, the maximum shear stress τ_{max} is

$$\tau_{max} = \pm\sqrt{\left(\frac{\sigma_x - \sigma_y}{2}\right)^2 + \tau_{xy}^2}, \tag{6.56}$$

$$\tau_{max} = \pm\frac{F_0}{\pi}\sin(\varphi_1 - \varphi_2). \tag{6.57}$$

The maximum shear stress τ_{max} occurs on the circle passing through the ends of the contact region. Therefore, plastic deformation initiates at the site when sin $(\varphi_1 - \varphi_2)$ reaches the maximum, that is when $(\varphi_1 - \varphi_2) = \pi/2$.

If residual stress σ_{xr} exists in the x-direction (assume that the residual stress in the y-direction $\sigma_{yr} = 0$), the stress in the x-direction in Equation 6.53 should be plus σ_{xr} on the right side. The maximum shear stress τ_{max} can be derived from Equations 6.53 through 6.56 taking σ_{xr} into consideration:

$$\tau_{\max} = \sqrt{\frac{F_0^2}{\pi}\sin^2(\varphi_1 - \varphi_2) - \frac{F_0}{\pi}\sin(\varphi_1 - \varphi_2)\cos(\varphi_1 + \varphi_2)\sigma_{xr} + \frac{\sigma_{xr}^2}{4}}. \tag{6.58}$$

In the aforementioned equation, the first term in the square root symbol is the square of the maximum shear stress when no residual stress exists. The second and third terms are related to the residual stress. Therefore, in terms of the magnitudes and signs of the second and third terms, the influence of residual stress on the initiation of plastic deformation can be determined. Assume that plastic deformation initiates at $(\varphi_1 - \varphi_2) = \pi/2$, then

$$\varphi_1 + \varphi_2 = \frac{\pi}{2} + 2\varphi_2, \quad \cos(\varphi_1 + \varphi_2) = \cos\left(\frac{\pi}{2} + 2\varphi_2\right) < 0,$$

and the sign of the second term is determined by σ_{xr}. If the residual stress σ_{xr} is a tensile stress, the second and third terms are positive, and the maximum shear stress increases. If the residual stress σ_{xr} is a compressive stress, the two terms offset, and the influence on the maximum shear stress decreases. This case shows that when tensile residual stress exists, plastic deformation initiates early and the plastic deformation region enlarges, resulting in a decrease in hardness. Using the relationship between residual stress and hardness, residual stress can be derived from the measurement of hardness.

6.3.1.3.2 Influence on Rebound Hardness

For rebound hardness, or Shore hardness, of a material, Young's modulus and the yield stress have the dominant influence. Residual stresses also have a minor influence. Therefore, when residual stress exists, even though the external load is low, a material is prone to plastic deformation. In the measurement of rebound hardness, a material is subjected to an impact force. As long as a slight plastic deformation is introduced inside the material, the rebound energy decreases, resulting in a

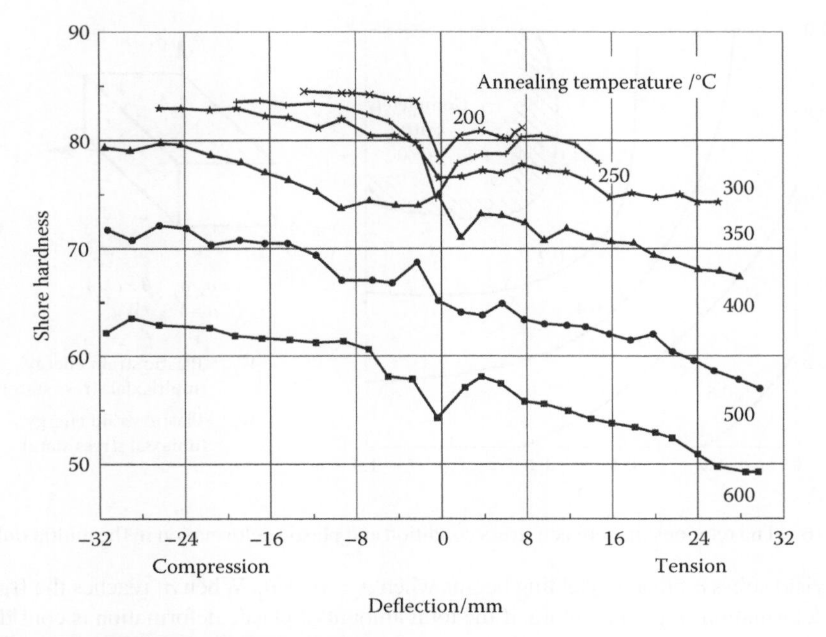

FIGURE 6.25 Shore hardness and bending stress.

decrease in rebound hardness. When the residual stress is tensile stress, the effect is more apparent. Figure 6.25 shows the relationship between Shore hardness and deflection. Here, the deflection of the horizontal axis corresponds to the bending stress. The result shows the same tendency as that of indentation hardness.

6.3.2 Influence of Residual Stresses on Brittle Failure and Stress Corrosion Cracking

6.3.2.1 Influence of Residual Stresses on Brittle Failure

As early as 1949, Green performed an experimental study on the influence of residual stresses on brittle failure [9]. He butt-welded 76 × 91 cm^2 mild steel plates with a thickness of 2 cm, and cut a Sawcut notch on each welding seam. The bending tests were conducted at a variety of temperatures for plate specimens with notches, with welded notches, and with the residual stresses removed by annealing. A relatively large portion of the residual stress along the joint direction of the welding seam around the notch is reserved. The general residual stress along the joint direction of the welding seam is the tensile stress close to the yield stress of the welds. At this moment, if the welds were cooled to −13°C, cracks could be introduced, even though no external load was applied. The reason for cracking is that the plasticity of the material decreases with the decrease in temperature. It can be seen from bending tests at different temperatures that the bending strength of welds with residual stresses was lower than that of welds with their residual stresses removed.

Besides temperature, the influence of residual stresses on brittle failure also depends on the fracture dimensions of the component. For a notched component, even in common dynamic bending testing, the dimensions of the fracture surface influence the absorbed energy and the ductile-brittle transition temperature. In addition, if the fracture dimensions are large, it is prone to the generation of a multiaxial stress state; that is, the distribution of residual stresses is influenced. Compared to the uniaxial stress state, the energy of plastic deformation in the multiaxial stress state is lower.

Now, we study the influence of residual stress on plastic strain energy under a multiaxial stress state. Figure 6.26 shows the relationship between the stress condition in the multiaxial stress state, and plastic deformation when three-dimensional stresses σ_1, σ_2, and σ_3 are applied. According to the Tresca yield criterion, plastic deformation begins when the difference between the principal stresses ($\sigma_1 - \sigma_3$) equals

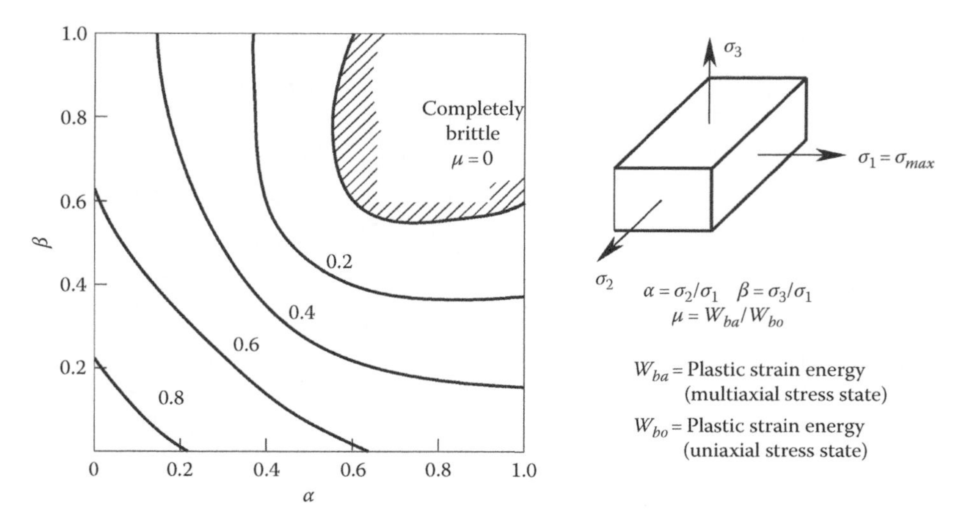

FIGURE 6.26 The relationship between stress condition and plastic deformation in the multiaxial stress state.

the tensile yield stress σ_s; that is, yielding begins when $\sigma_1 = \sigma_3 + \sigma_s$. When σ_1 reaches the fracture stress σ_R, plastic deformation stops. Therefore, if the total amount of plastic deformation is considered in this procedure, the range of plastic deformation corresponds to the stress range from σ_1 to σ_R. In the relationship $\sigma_1 = \sigma_3 + \sigma_s$, the larger is the value of σ_3, the larger is the value of σ_1, and thus the smaller is the amount of plastic deformation at stress σ_R when the crack initiates. The relationship of each stress with respect to the stress σ_1 is represented in Figure 6.26, as $\alpha = \sigma_2/\sigma_1$, $\beta = \sigma_3/\sigma_1$, $\mu = W_{ba}/W_{bo}$, and W_{ba}, W_{bo}, respectively, denote the plastic strain energy in the multiaxial stress state and the uniaxial stress state. When $\mu = 0$, the material is in a completely brittle state with zero deformation energy.

The relationship between brittle failure and temperature discussed previously indicates that the tensile residual stress can reduce the load bearing capacity of a component to zero. When a material's brittleness is determined by stress, a high value of the minimum principal stress results in embrittlement. If the residual stresses are in a multiaxial tensile stress state, they superimpose to the externally applied load, leading to a decrease in the value of μ. Once such a stress state is introduced, the material becomes vulnerable to embrittlement due to residual stresses. The residual stresses in general are in a multiaxial stress state. When the externally applied load is in a uniaxial stress state, the situation is opposite to the aforementioned case: the plastic deformation energy increases, and the material is prone to plastic deformation. When the externally applied external stresses are in a multiaxial stress state (although they superimpose to the residual stresses) the resultant stresses after superimposition are conducive to brittle failure.

6.3.2.2 Influence of Residual Stresses on Stress Corrosion Cracking

Both structural materials and functional materials are affected by environmental media [17]. The synergetic action of environmental media and residual stresses can mutually improve and accelerate the failure of a material, which is more serious than the solitary action of each one singly, or their simple superimposition (see Chapter 8). This is called stress corrosion cracking. In addition, corrosion is usually localized and selective, and the cracking is more often than not caused by pit corrosion. The crack propagates mainly along the direction perpendicular to the maximum principal stress, and at the microscopic scale, exhibits inter-granular or intra-granular cracking. The hazard of stress corrosion cracking is that it often occurs in a mildly corrosive media at low stress without warning, potentially resulting in a catastrophic accident. Therefore, the influence of residual stress on the failure of materials or components in corrosive environments has received much attention in academia and industry [22,23]. For metallic materials, corrosion is almost always electrochemical corrosion: when stress is applied, the corrosion potential changes.

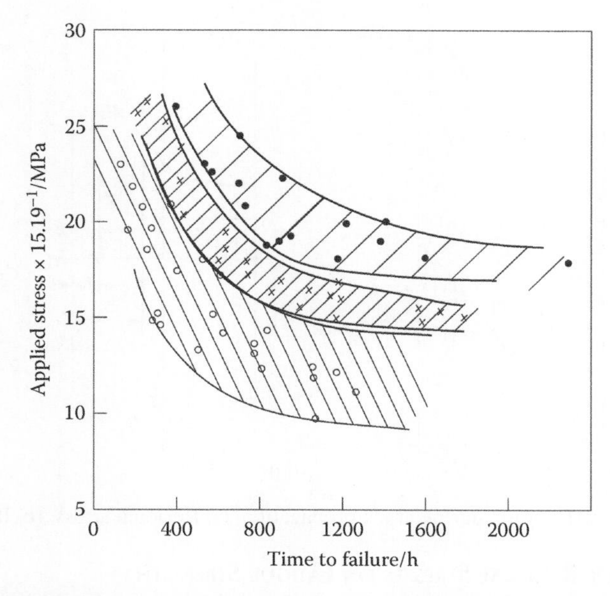

FIGURE 6.27 Influence of residual stress on stress corrosion cracking. Material: Aluminum alloy (0.4%Cu, 5.3%Zn, 2.8%Mg) (7.62 × 5.08 mm). The applied stress: bending stress by the external load; Corrosive media: 0.5 mol/L Sodium chloride + 0.0025 mol/L sodium carbonate. (•) A specimens (surface residual stress −4.557 to +4.557 MPa), (×) B specimens (surface residual stress −22.785 to −7.595 MPa), (o) C specimens (surface residual stress +136.71 to +7.595 MPa.)

Residual stress was introduced in an aluminum alloy by different processes, and stress corrosion cracking tests were performed with an applied external stress. Figure 6.27 shows the influence of residual stresses on the stress corrosion cracking of the material. Specimens A were cut after drawing, specimens B were quenched in hot water after drawing, and specimens C were surface-processed after drawing to introduce tensile residual stress. The surface residual stresses of specimens A and B were zero and compressive, respectively; while apparent tensile residual stresses existed in specimens C. Residual stress has significant influence on the performance of a material in stress corrosion cracking tests.

Both externally applied stress and residual stress have significant influence on crack propagation behavior. Here, the stresses are divided into applied stresses and residual stresses, and we further discuss their influences. For residual stress, it is self-balanced on the cross section. For externally applied stress, the stress on the cross section is balanced by the external load.

Next, we discuss the state of applied stress and residual stress during crack propagation. In this procedure, the apparent distribution of residual stress does not change. However, once the external load is applied, the stress on the cross section changes significantly. As shown in Figure 6.28, only residual stress exists in Figure 6.28a. When the tensile stress at the crack tip reaches 30% of the yield stress, the crack stops propagating. In Figure 6.28b, external tensile stress is applied. High stress is introduced at the crack tip, due to the notch effect, with the application of an external load. The deeper the crack, the higher the stress at the crack tip, and thus the faster the crack propagation rate.

As for crack propagation, the stress type, or the stress distribution, is very important. In practice, besides the applied stress, residual stress exists in almost all cases. In this situation, the residual stress has other important meanings: its sign, magnitude, and distribution. When it superimposes to the applied stress, it probably results in a stress state which is either suitable for, or adverse to, stress corrosion cracking. For residual stress, if compressive residual stress exists in the region contacting the corrosive medium, it is effective in preventing stress corrosion cracking. In practice, surface rolling, shot peening, and nitriding are recommended as on-site measures to prevent stress corrosion cracking, because these processes introduce compressive residual stresses on the material's surface.

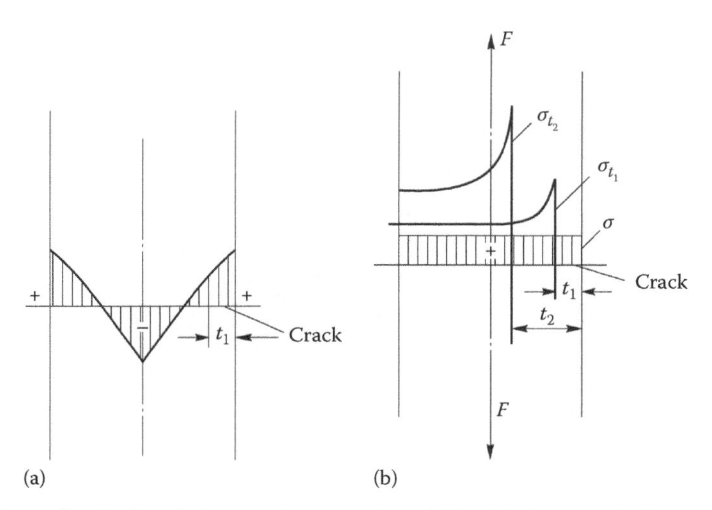

FIGURE 6.28 Stress distribution during crack propagation. (a) Residual stress. (b) External stress.

6.3.3 Influence of Residual Stresses on Fatigue Strength

The influence of residual stress on the fatigue strength of a material is of importance among the influences on the properties of materials under dynamic loading (see Chapter 7) [9]. In general, for components subjected to cyclic loading, compressive residual stresses improve fatigue strength while tensile residual stresses decrease fatigue strength.

6.3.3.1 Influence of Residual Stresses on Fatigue Strength Caused by Cold Working and Heat Treatment

Residual stresses caused by cold working and heat treatment influence the fatigue strength of materials. The influence of both macroresidual stress and microresidual stress need to be considered. The former can be viewed as superimposing temporarily on the applied stress, resulting in a change in magnitude of cyclic stress. The latter is caused by the heterogeneity of a material's microstructure. Under cyclic loading, plastic deformation in the microscopic region accumulates and leads to stress concentration in this region, and thus influences the generation of the crack. Compared to the influences on static strength, these influences are more important in practice.

1. Evolution of residual stress under cyclic loading
   In the study of the influence of residual stress on fatigue strength, residual stress is generally converted to mean stress. Many results of the influence of mean stress on fatigue strength have been obtained. When fatigue strength is correlated to applied cyclic stress and mean stress, a fatigue diagram is obtained. In practical simulations, if the resultant stress of mean stress and cyclic stress is sufficiently large, plastic deformation occurs and influences the mean stress.
   Here, we discuss a real case in which the residual stress changes under cyclic loading. Figure 6.29 shows the change of surface residual stress in a shot-peened circular rod under cyclic bending stress. In this case, the stress amplitude is above the fatigue limit. The larger the stress amplitude, the larger the decrease of residual stress. In addition, the residual stress decreases significantly after the first loading cycle. The aforementioned example demonstrates the change in residual stress under cyclic loading if residual stress exists. Even if residual stress does not exist in the original state, new residual stress can be generated under cyclic loading. This is very important in the detailed study of the change of residual stress.
2. Influence of residual stress caused by cold working
   Next, we introduce an example of the influence of residual stress on fatigue strength, caused by cold working. Figure 6.30 shows the relationship between the outer surface residual stress,

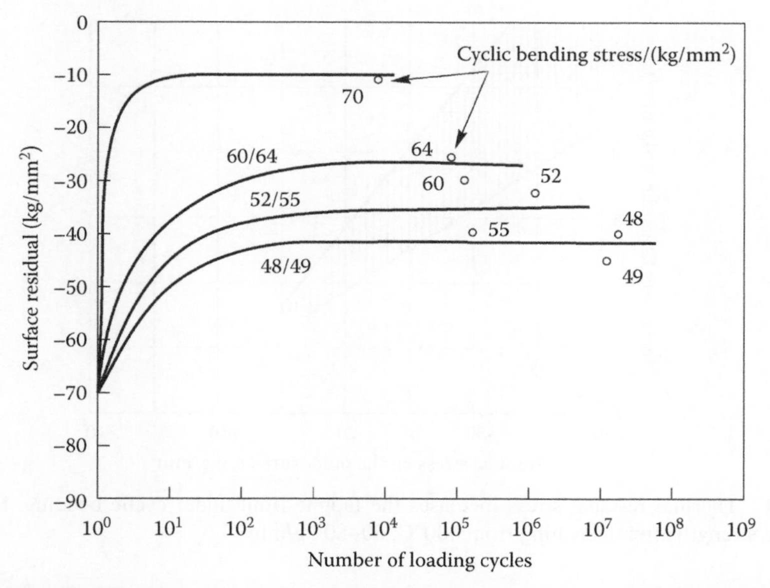

FIGURE 6.29 The change of residual stress in shot-peened steel due to cyclic loading. Material: Cr-Mo steel, ϕ7.52 mm; tensile strength, 90 kg/mm^2; rotating bending fatigue limit, 49 kg/mm^2.

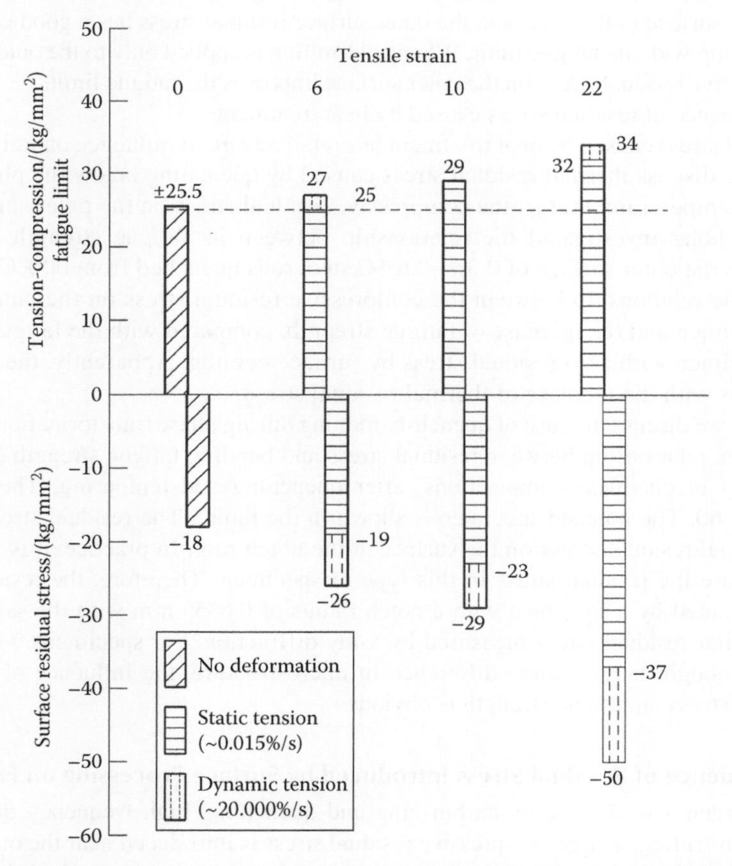

FIGURE 6.30 Residual stress and fatigue limit of steel after static or dynamic tension. Material: 0.4%C annealed, ϕ7.2 mm.

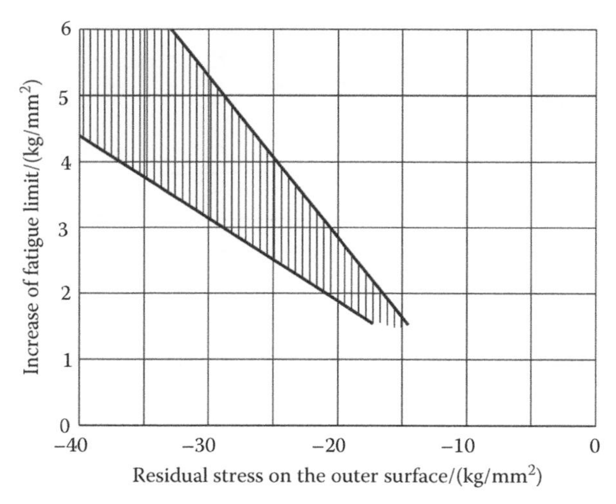

FIGURE 6.31 Thermal residual stress increases the fatigue limit under cyclic bending. Material: 0.3%–0.6%C; tensile strength after quenching from 600°C, 60–80 kg/mm^2.

and the tension-compression fatigue limit of specimens of 0.8%C annealed steel after static or dynamic plastic tension. Plastic tension causes a gradient distribution of residual stress from the surface to the core, and the outer surface residual stress has a good corresponding relationship with the fatigue limit. When cold-rolling is applied only to the outer surface, the compressive residual stress on the outer surface improves the fatigue limit.

3. The influence of residual stress caused by heat treatment

   Residual stress caused by heat treatment has relatively great influence on fatigue strength. First, we discuss thermal residual stress caused by quenching below the phase transformation temperature; that is, the compressive residual stress on the outer surface. Bühler and Bucholtz investigated the relationship between its fatigue strength and residual stress on the outer surface of 0.3%–0.6%C steel rods quenched from 600°C. Figure 6.31 shows the relationship between the compressive residual stress on the outer surface of the specimen and the increase of fatigue strength, compared with the fatigue strength of the specimen with zero residual stress by furnace cooling. Apparently, the fatigue limit increases with the increase of thermal residual stress.

   Next, we discuss the case of quench-hardening during phase transformation. Figure 6.32 shows the relationship between residual stress and bending fatigue strength of steels with a variety of chemical compositions, after quenching and tempering. The hardness is HRC50–60. The notched specimen is shown in the figure. The residual stress concerned is the axial residual stress on the surface at the notch root. In practice, it is very difficult to measure the residual stress in this type of specimen. Therefore, the residual stress is approximated by a specimen with a notch radius of 0.6355 mm with the same treatment and similar residual stress measured by x-ray diffraction. For specimens with this hardness, although there is some difference in microstructure, the influence of compressive residual stress on fatigue strength is obvious.

6.3.3.2 Influence of Residual Stress Introduced by Surface Processing on Fatigue Strength

To surface-harden a workpiece by carburizing and quenching, high frequency quenching, flame quenching, or nitriding, a large compressive residual stress is introduced near the outer surface. The distribution of residual stresses corresponds to the specific processing method, each of which has its own features. It is well known that the fatigue strength of a component increases significantly due

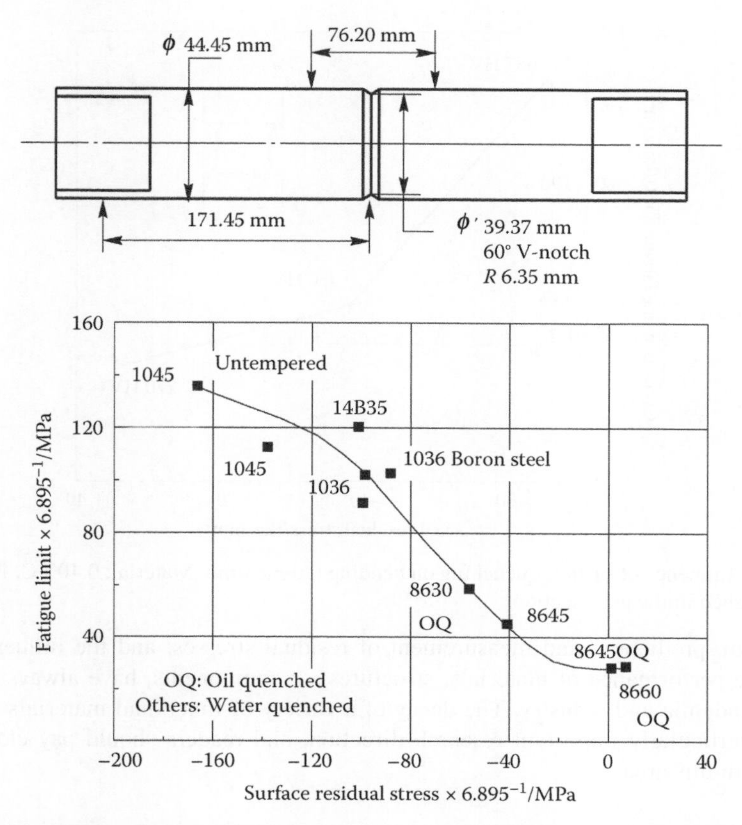

FIGURE 6.32 The relationship between residual stress and bending fatigue limit of steels after a variety of heat treatments.

to surface hardening processing. This is the superposition effect of the influence of microstructure after hardening, and the influence of compressive residual stress.

There are two typical distributions of residual stress after surface hardening: (1) the residual stress is small near the outer surface, increases drastically inward, and decreases drastically toward the core; (2) the residual stress is large near the outer surface, then decreases linearly and drastically toward the core. Regardless of the distribution, the same change of residual stress occurs under cyclic loading, just as in the previous example; and the change of residual stress near the outer surface caused by plastic deformation is observed. In the former case, a large compressive residual stress is introduced near the outer surface. On the contrary, in the latter case, residual stress reduces significantly. Figure 6.33 shows the experimental results of a rotating-bending-fatigue test using round specimens of 0.40%C, 1.0%Cr steel, high-frequency quenched to a short region in the gauge section. Presented here is the relationship of the fatigue limit versus the axial compressive residual stress on the outer surface of the damage region—the two ends of the quenched part—and the hardness. For specimens with surface hardness of HV270, when the surface residual stress in the axial direction is compressive and is increased in magnitude, the fatigue limit increases by 55%; when the hardness is further increased from HV380 to HV657, the fatigue limit increases by 190%.

Residual stresses exist everywhere, not only in traditional structural materials, but also in functional materials such as intelligent materials and information materials. Residual stresses also exist in large structural components, as well as in films and electronic components. These residual stresses not only impair the load-bearing capacity of structural materials, but also deprecate the proper functioning of materials. Examples are the decay of optical, magnetic, electric, ferroelectric, and ferromagnetic properties; shape memory capacity; and storage properties. Therefore, the origin

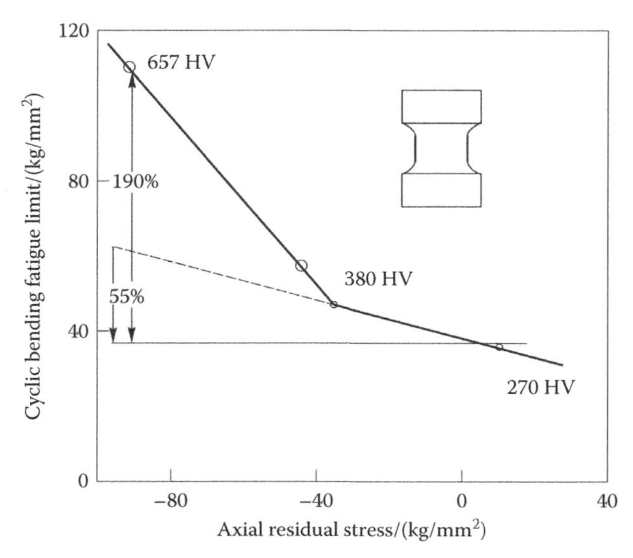

FIGURE 6.33 Influence of surface quenching on bending fatigue limit. Material: 0.40%C, 1.0%Cr steel high frequency quenched in the gage section.

and mechanism, prediction, and measurement of residual stresses, and the influence of residual stresses on the performance of materials, structures, or components, have always received much attention in academia and industry. The decay of a variety of functional materials due to residual stresses is a particularly important research direction, and readers should pay close attention to developments in this area.

EXERCISES

6.1 Describe the concept of residual stress.

6.2 The generation of residual stress is unavoidable. List several common residual stresses in components, explain the causes, and classify these stresses according to their influence ranges.

6.3 Enumerate three processes in daily life to remove and adjust the residual stresses, and indicate which methods they belong to.

6.4 Give several examples of measuring residual stresses and indicate which measuring methods they belong to.

6.5 The hole drilling method is used to measure the residual stress in the rear axle of a car. Set $\alpha = 45°$, when the residual stress is lower than 190 MPa, the experimental calibration gives $A/E = -0.3053$ Pa^{-1}, $B/E = -0.7152$ Pa^{-1}. Reading from the stress meter at the measuring point, the strains in the x-direction of the strain gauges at the weld seam due to hole drilling are $\varepsilon_0 = 5.67 \times 10^{-5}$, $\varepsilon_{-\alpha} = 1.25 \times 10^{-5}$, $\varepsilon_{+\alpha} = 8.35 \times 10^{-5}$. Calculate the stresses in this measuring point.

6.6 In the $\sin^2 \psi$ method, which situation and statement are correct?

1. The value of 2θ offsets to a lower angle with the increase of ψ; indicating the existence of residual tensile stress.
2. The value of 2θ offsets to a higher angle with the increase of ψ; indicating the existence of residual tensile stress.
3. The value of 2θ offsets to a lower angle with the increase of ψ; indicating the existence of residual compressive stress.
4. The value of 2θ offsets to a higher angle with the increase of ψ; indicating the existence of residual compressive stress.

6.7 To measure the stress in 7–3 brass, CoK is incidence upon the (400) plane. When $\psi = 0$, $2\theta = 150.4°$; when $\psi = 45°$, $2\theta = 150.9°$. Calculate the macroscopic stress on specimen surface ($a = 0.369\,5$ nm, $v = 0.35$, $E = 8.83 \times 10^4$ MPa).

6.8 Residual stresses have significant influence on the mechanical properties of materials. Summarize the influences of tensile residual stress and compressive residual stress on stress corrosion cracking.

6.9 Briefly describe the difference in the residual stresses caused by cold working and heat treatment.

6.10 Describe the distribution features of compressive residual stresses near the outer surface of a surface hardened workpiece by carburizing and quenching, high-frequency quenching, and flame quenching.

6.11 Derive Equation 6.32.

6.12 Experiment: The relationship of residual stress and hardness, and discuss in group of three to five students.

6.13 Experiment: The influence of residual stress on functional material properties, and discuss in group of three to five students.

6.14 Discussion: Can you bridge residual stress with the dimension effect in Chapter 4?

REFERENCES

1. Rowlands R. E. Residual stress. In: A. S. Kobayashi. Ed. *Handbook on Experimental Mechanics* (2nd revised edn.). New York: VCH Publication, 1993, pp. 785–828.
2. Sandeep K., Christopher D. R., and Peter K. D. X-ray characterization of annealed iridium films. *J Appl Phys*, 2002, 91: 1149–1154.
3. Gruninger M. F., Lawn B. R., Farabrugh E. N. et al. Measurement of residual stresses in coatings on brittle substrates by indentation fracture. *J Am Ceram Soc*, 1987, 70(5): 344–348.
4. Zhang T. Y., Chen L. Q., and Fu R. Measurements of residual stresses in thin films deposited on silicon wafers by indentation fracture. *Acta Mater*, 1999, 47(14): 3869–3878.
5. Zheng X. J., Yi W. M., Chen Y. Q. et al. Effects of annealing temperature on microstructure, ferroelectric and mechanical properties of $Bi_{3.15}Nd_{0.85}Ti_3O_{12}$ thin films. *Scr Mater*, 2007, 57(8): 675–678.
6. Zheng X. J., Li J. Y., and Zhou Y. C. X-ray diffraction measurement of residual stress in PZT thin films prepared by pulsed laser deposition. *Acta Mater*, 2004, 52(11): 3313–3332.
7. Zhou Y. C., Yang Z. Y., and Zheng X. J. Residual stress in PZT thin films prepared by pulsed laser deposition. *Surf Coat Technol,* 2003, 162(2–3): 202–211.
8. Zheng X. J., Yang Z. Y., and Zhou Y. C. Residual stresses in $Pb(Zr_{0.52}Ti_{0.48})O_3$ thin films deposited by metal organic decomposition. *Scr Mater*, 2003, 49: 71–76.
9. Shigeru Y. *The Generation and Countermeasures of Residual Stresses*. Beijing: China Machine Press, 1983.
10. Lu J. *Handbook of Measurement of Residual Stresses*. Bethel, CT: Society for Experimental Mechanics, 1996.
11. Macherauch E. and Hauk V. *Residual Stresses in Science and Technology*. Oberursel, Germany: DGM, 1986.
12. Lin L. H., Chen L. G., and Gu M. Y. Current status and development of residual stress measurement techniques. *Mechanisms*, 1998, 25(5): 46–49.
13. Zhou Y. *Material Analytical Methods*. Beijing: Machine Press, pp. 61–73, 2004
14. Sakashita Y. and Segawa H. Dependence of electrical properties on film thickness in $Pb(Zr_xTi_{1-x})O_3$ thin films produced by metalorganic chemical vapor deposition. *J Appl Phys*, 1993, 73(11): 7857–7863.
15. Kweon S. Y., Yi S. H., and Choi S. K. Intrinsic stress dependence of c-axis orientation ratio in $PbTiO_3$ thin films deposited by reactive sputtering. *J Vac Sci Tech A*, 1997, 15(1): 57–61.
16. Yuan F. R. and Wu S. L. *Measurements and Calculation of Residual Stresses*. Changsha, China: Hunan University Press, 1987.
17. Fang B. W. *Residual Stresses in Cold and Hot Worked Metals*. Beijing: Higher Education Press, 1991.
18. Wang Y. M. and Wang Y. L. Supersonic measurement system of residual stresses. *Instrum Technol*, 2004, 4: 23–24.

19. Narayanan S., Kalidindi S. R., and Schadler L. S. Determination of unknown stress states in silicon wafers using microlaser Raman spectroscopy. *J Appl Phys*, 1997, 82(5): 2595–2602.
20. Xu W. H., Lu D., and Zhang T. Y. Determination of residual stresses in $Pb(Zr_{0.53}Ti_{0.47})O_3$ thin films with Raman spectroscopy. *Appl Phys Lett*, 2001, 79(25): 4112–4116.
21. Díaz F. V., Kaufmann G. H., and Möller O. Residual stress determination using blind-hole drilling and digital speckle pattern interferometry with automated data processing. *Exp Mech*, 2001, 41(4):319–323.
22. Boven G. V., Chen W., and Rogge R. The role of residual stress in neutral pH stress corrosion cracking of pipeline steels. *Acta Mater*, 2007, 55(1): 29–42.
23. Ling X., Peng W. W., and Ma G. Influence of Laser peening parameters on residual stress field of 304 stainless steel. *J Press Vessel Technol*, 2008, 130: 021201.

7 Creep and Fatigue of Metals

In our daily lives, we discover that some tools fail only after repeated use, but others fail quickly. For example, cars and airplanes have limited lives. Why does this happen? According to "Mechanics of Materials" and the knowledge we have gained in Chapter 2, a material or component does not fail if the applied load is lower than the critical load, or if the stress intensity factor is lower than the fracture toughness of the material. Failure under cyclic loading—even though the maximum load is much less than the critical load under static load failure—is called fatigue failure.

In addition, we have observed a strange phenomenon in Figure 4.14a: in the nano-indentation test, the indentation depth increases continuously, although the load acting on the indentation is fixed. According to the constitutive relationship in Chapter 1, whether the material is elastic or elastic-plastic, the displacement or deformation is completely determined once the load is determined, and does not increase automatically with time. The observed phenomenon is called creep, which occurs often at high temperatures, particularly for metallic materials.

The failure of a material or a component due to creep and fatigue often happens in industry, and creep and fatigue usually take place together. Under the interaction of high temperature creep and fatigue, a material frequently fails at low stress and causes a catastrophic accident. Therefore, the most important property of this type of material is high temperature strength: the resistance of the material to deformation and fracture at high temperatures.

The deformation and fracture at high temperature is related to the hold time of the load. The description of mechanical behavior at high temperature must consider two mechanical and temporal parameters: creep rate and failure life. It is not necessary to consider the time factor for the case at room temperature. In addition, chemical or electrochemical reactions accelerate at high temperatures. Thus, the environment has a significant influence on the deformation and fracture of materials at high temperatures.

The theory of high-temperature deformation and fracture is based on diffusion theory and dislocation theory, and involves the disciplines of solid physics, metallography, elasto-plastic mechanics, fracture mechanics, and damage mechanics. It is an important branch of materials science. Internationally, high temperature deformation, fatigue failure, and fracture are among the hottest fields in materials science and engineering. In this chapter, we first introduce the fundamental knowledge on high temperature creep of metallic materials, the physics basis of creep, and the basic models of creep failure. We then introduce the fundamental knowledge of fatigue of metallic materials, mechanisms of fatigue failure, and models to predict fatigue life. Finally, we introduce the interaction mechanisms of high temperature creep and fatigue of metallic materials.

7.1 INTRODUCTION TO CREEP OF METALLIC MATERIALS

7.1.1 Concept of Creep

In Chapter 1, we discussed elasticity and plasticity in detail. We know that there is a determined corresponding relationship between stress and strain; that is, the strain is determined once the stress is known. In practice, as shown in Figure 4.14a, although the stress is fixed, the deformation or strain increases with time. How do we describe this phenomenon mathematically? Since time plays an important role, we introduce time t as an independent variable in the relationship of stress and strain, so that the degree of deformation is described by the strain rate $\dot{\varepsilon}$, which is the derivative of strain with respect to time. For simplicity, we assume that the stress σ is proportional to the strain rate $\dot{\varepsilon}$:

$$\sigma = \eta\dot{\varepsilon} = \eta\frac{d\varepsilon}{dt}. \quad (7.1)$$

This equation is the famous Newton's law for the description of a viscous fluid, where η is called the viscosity constant. This relationship is related to time. Assume stress σ_0 is applied to a one-dimensional rod, and that strain increases with time. From Equation 7.1, we have

$$d\varepsilon = \frac{\sigma_0}{\eta}dt.$$

Thus,

$$\varepsilon = \frac{\sigma_0}{\eta}t. \quad (7.2)$$

From the Equation 7.2, it is true that strain increases linearly with time when the stress is fixed. A very simple mathematical language, that is, Equation 7.1, can describe a natural phenomenon—creep. Creep is the relationship between deformation and time under a constant external load. The most typical feature of creep is that deformation of the material is closely related to time, as shown in Figure 7.1. Under an externally applied load, the material first undergoes elastic deformation ε_e. Although the externally applied stress is constant, deformation of the material increases slowly and gradually with time (creep deformation). When the external load is rapidly removed, the elastic deformation ε_e in the material recovers rapidly, while creep deformation recovers partially and slowly with time, and the remaining deformation

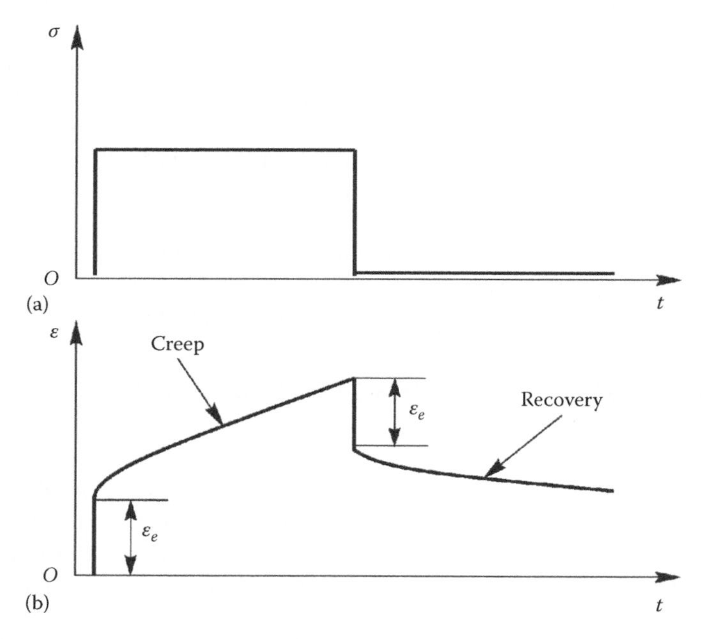

FIGURE 7.1 Under a constant stress, creep deformation varies and accumulates with time; after the removal of stress, creep deformation recovers tardily. (a) σ vs. t. (b) ε vs. t. (From Dowling, N.E., *Mechanical Behavior of Materials (Engineering Methods for Deformation, Fracture and Fatigue)*, Prentice Hall, Upper Saddle River, NJ, 1998; Taken from Figure 2.25.)

is permanent. It can be seen that during creep deformation of the material, the one-to-one mapping relationship between material deformation and stress, or external load, does not hold true. In addition, even though the applied stress is lower than the yield limit, the creep deformation is still not reversible.

7.1.2 Creep Curve

The most common creep test is the tensile or compressive test of a material under constant load, measuring the relationship between creep deformation and time, as shown in Figure 7.2. Using the illustrated apparatus, creep tests at different constant stresses and temperatures can be performed. The operation is simple, and it is easy to keep a constant stress. In the test, the specimen is heated to a certain temperature, and then the relationship of the strain of the specimen with time, at this temperature and constant stress, is recorded. The obtained strain–time curve as shown in Figure 7.3

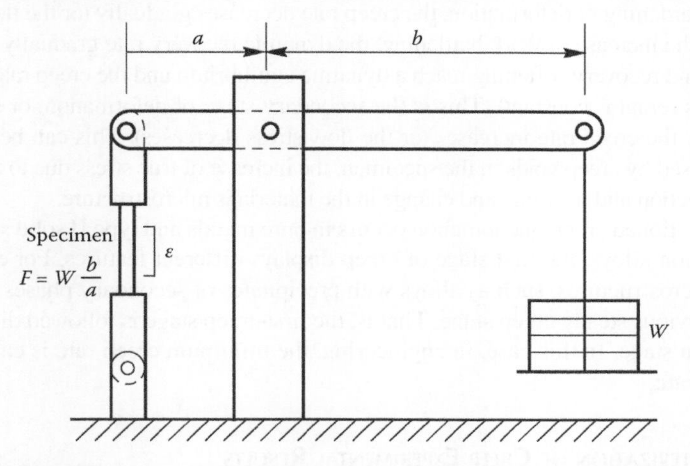

FIGURE 7.2 Schematic illustration of creep test setup. (From Dowling, N.E., *Mechanical Behavior of Materials (Engineering Methods for Deformation, Fracture and Fatigue)*, Prentice Hall, Upper Saddle River, NJ, 1998; Taken from Figure 15.3.)

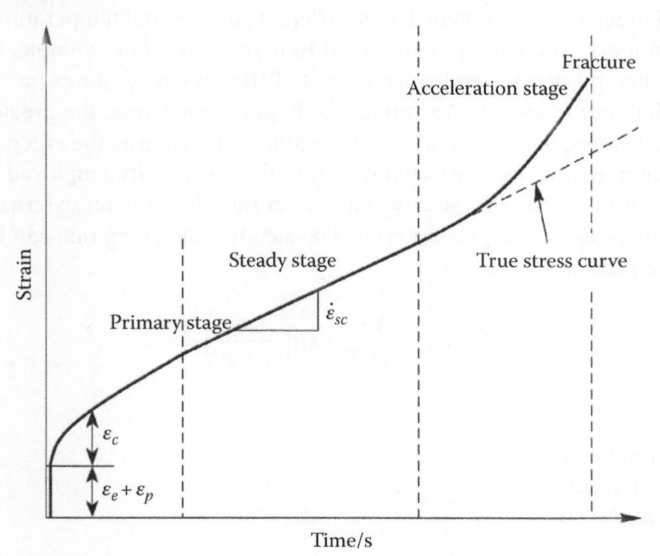

FIGURE 7.3 The evolution of strain with time in creep tests under constant load. (From Dowling, N.E., *Mechanical Behavior of Materials (Engineering Methods for Deformation, Fracture and Fatigue)*, Prentice Hall, Upper Saddle River, NJ, 1998; Taken from Figure 15.4.)

is the creep curve. Creep can be divided into three stages, according to the shape of the creep curve. In the first stage, when a constant load is applied, the material undergoes elastic strain ε_e, or even plastic strain ε_p (depending on the loading magnitude, the elastic modulus of the material, and the testing temperature), immediately followed by a gradual accumulation of creep deformation. In this stage, the strain rate $\dot{\varepsilon}$ initially increases with time, but decreases when approaching the end of the first stage, and gradually approaches a constant value. This stage is called the primary stage, or the transient stage. The second stage is the steady-state stage of creep deformation. In this stage, the creep rate is constant. In the third stage, the creep rate increases rapidly with time in an unsteady manner, until the material fractures. This stage is called the creep acceleration stage.

The shape of the creep curve reflects the work hardening and recovery softening procedures accompanying high temperature deformation. In the initial stage of creep, the deformation rate is very fast (or the flow stress is very low), indicating that the deformation resistance of the material is low. Subsequently, due to the work hardening of deformation, the creep rate decreases gradually (or the flow stress increases gradually). With the increase of work hardening, the dynamic recovery rate gradually increases. Finally, work hardening and recovery softening reach a dynamic equilibrium and the creep rate remains constant (or the flow stress remains constant). This is the secondary stage of deformation, or steady-state creep. In the third stage, the creep rate increases (or the flow stress decreases). This can be ascribed to stress concentration caused by creep voids in the specimen, the increase of true stress due to the decrease in the specimen cross section and necking, and change in the material's microstructure.

The above-mentioned creep phenomenon occurs in pure metals and type II solid solution alloys. For type I solid solution alloys, the first stage of creep displays different features. For engineering alloys with complex microstructures, such as alloys with precipitates of secondary phases, they probably do not exhibit the obvious steady creep stage. That is, the first creep stage is followed directly by the third creep acceleration stage. In this case, in engineering, the minimum creep rate is employed to replace the steady creep rate.

7.1.3 Characterization of Creep Experimental Results

The results of a single creep test are generally characterized by four parameters: (1) the loading stress σ, (2) the experimental temperature T, (3) the steady-state creep rate $\dot{\varepsilon}_{sc}$, and (4) the rupture time t_r. A series of creep tests are conducted at different stresses and temperatures. The creep resistance properties of a material can be represented in many ways. For example, creep experimental results are represented by stresses and creep rates at different temperatures, as shown in Figure 7.4 [2]. We discover that, at the same temperature, the larger is the stress, the greater is the creep rate. At the same applied stress, the higher is the temperature, the larger is the creep rate.

For metallic materials, the steady-state creep rate $\dot{\varepsilon}_{sc}$ is generally employed to characterize the creep properties of a material. The steady-state creep rate depends on external variables such as temperature and stress, as well as the material. The steady-state creep rate can be described by the following general equation:

$$\dot{\varepsilon}_{sc} = \frac{A_2\sigma^n}{d^qT}\exp\left(-\frac{Q}{RT}\right), \tag{7.3}$$

where

d denotes the average grain size
T is the absolute temperature
R is the universal gas constant

The coefficient A_2, the exponents n and q, and the activation energy Q all depend on the material and the corresponding creep mechanism.
The values of the creep exponents n and q for different creep mechanisms are summarized in Table 7.1.

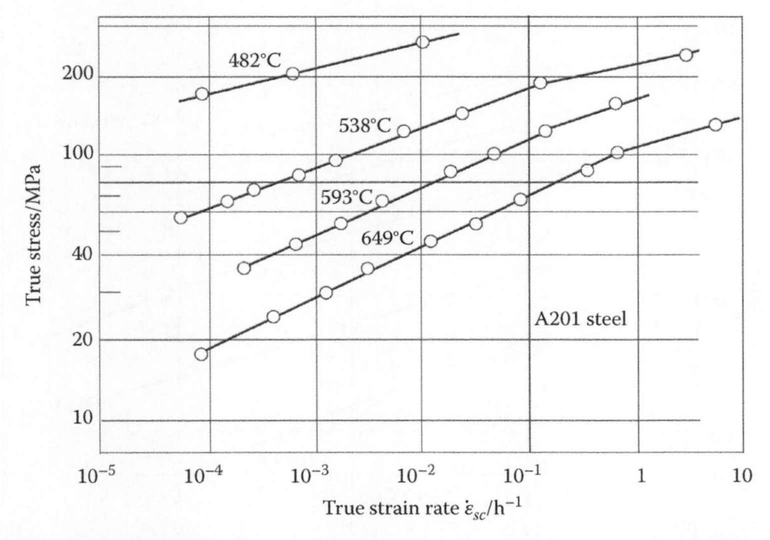

FIGURE 7.4 The relationship between true stress and steady-state creep rate at different temperatures of a mild steel. (Dowling, N.E., *Mechanical Behavior of Materials (Engineering Methods for Deformation, Fracture and Fatigue)*, Prentice Hall, Upper Saddle River, NJ, 1998; Askeland, D.R. and Phulé, P.P., *Essentials of Materials Science and Engineering*, Tsinghua University Press, Beijing, 2005; Taken from Figure 15.7.)

TABLE 7.1
Creep Exponentials for a Variety of Physical Mechanisms

Creep Mechanism	Exponent n	Exponent q	Description
Diffusion creep (Nabarro–Herring) creep)	1	2	Mainly caused by the vacancy diffusion via lattice
Diffusion creep (Cobleg creep)	1	3	Mainly caused by vacancy diffusion via grain boundary
Grain boundary sliding	2	2 or 3	Sliding caused by the vacancy diffusion via lattice ($q = 2$) or grain boundary ($q = 3$)
Dislocation creep (Power-law creep)	3 ~ 8	0	Dislocation movement (including climbing to bypass obstacles)

Source: Taken from Dowling, N.E., *Mechanical Behavior of Materials (Engineering Methods for Deformation, Fracture and Fatigue)*, Prentice Hall, Upper Saddle River, NJ, 1998.

In addition, as shown in Figure 7.5, creep experimental results are represented by the applied stresses σ and rupture times t_r at different temperatures [1]. The influence of applied stress and temperature on the creep rupture time follows Arrhenius' relationship:

$$t_r = K\sigma^n \exp\left(\frac{Q}{RT}\right), \tag{7.4}$$

where

K and n are constants related to material properties and stress
R is the gas constant
T is absolute temperature
Q is the activation energy for creep rupture

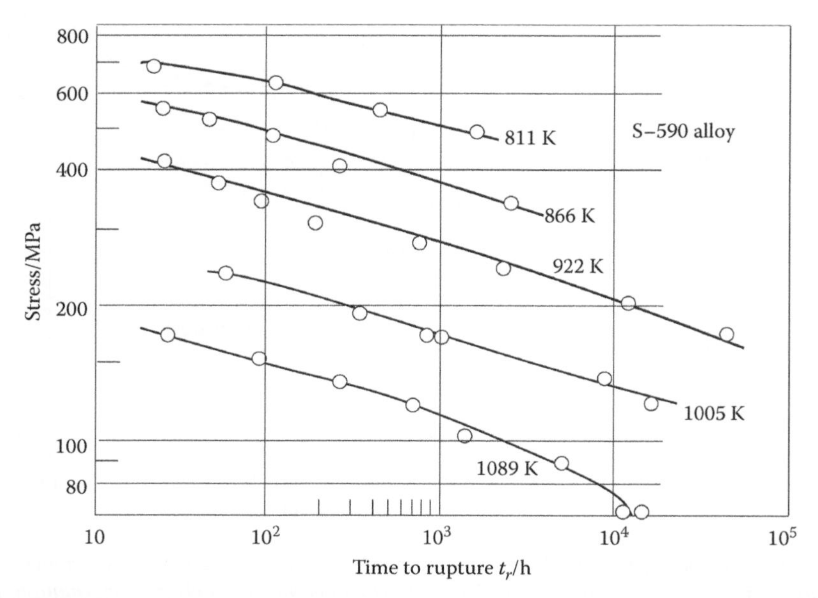

FIGURE 7.5 The relationship of the applied stress and rupture time of ferroalloy under different temperatures. (From Dowling, N.E., *Mechanical Behavior of Materials (Engineering Methods for Deformation, Fracture and Fatigue)*, Prentice Hall, Upper Saddle River, NJ, 1998; Taken from Figure 15.8.)

7.1.4 Relationship between Steady-State Creep Rate and Stress

The absolute majority of experimental results indicate that log $\dot{\varepsilon}_{sc}$ is linear with log σ for most materials at a relatively low stress at high temperature. Therefore, the relationship between $\dot{\varepsilon}_{sc}$ and σ is [3]

$$\dot{\varepsilon}_{sc} = A_1\sigma^n, \tag{7.5}$$

where

A_1 is a constant related to material properties and temperature

n is the stress exponent of the steady-state creep rate, whose value can be obtained from the slope of the line log $\dot{\varepsilon}_{sc} \sim$ log σ

Since the creep rate in Equation 7.5 is an exponential function of stress, the creep satisfying Equation 7.5 is called a power-law creep.

Figure 7.6 shows the experimental results of high purity aluminum [1]. It can be seen that log $\dot{\varepsilon}_{sc}$ is linear with logσ, and the slope of the line is $n = 5$. When the stress increases to a certain degree, the slope of log $\dot{\varepsilon}_{sc}$–log σ begins to increase and deviate from the linear relationship. This phenomenon is called the power law breakdown (PLB). In the PLB (low temperature, high stress) region, the relationship between creep rate and stress can be represented by an exponential function

$$\dot{\varepsilon}_{sc} = A_2 \exp(B\sigma), \tag{7.6}$$

where A_2 and B are constants related to material properties and temperatures. A unified equation is used to describe the relationship between $\dot{\varepsilon}_{sc}$ and σ from low stress to high stress [5]:

$$\dot{\varepsilon}_{sc} = A_3(\sinh \alpha\sigma)^n, \tag{7.7}$$

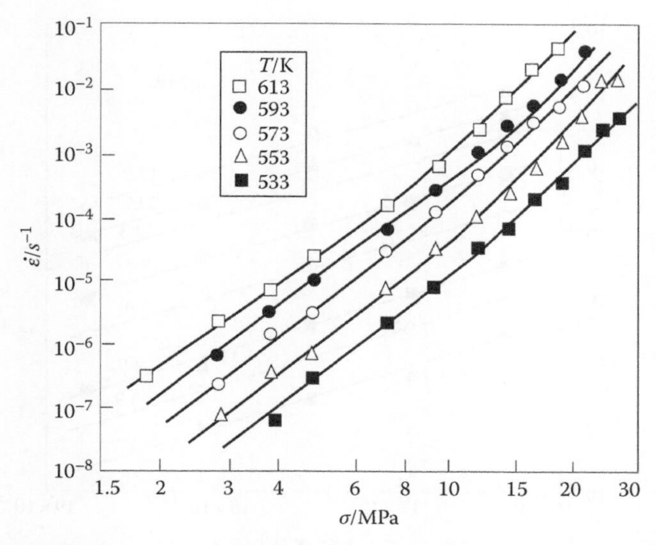

FIGURE 7.6 The relationship of steady-state creep rate and stress of high purity aluminum. (From Zhang, J., *Deformation and Fracture of Materials at High Temperature*, Science Press, Beijing, 2007; Taken from Figure 1.3.)

where A_3 and α are constants. At low stress, Equation 7.7 is reduced to Equation 7.5; while at high stress, Equation 7.7 approaches Equation 7.6. Although Equation 7.7 unifies the creep equation from low stress to high stress, we will see later in this chapter that Equations 7.5 and 7.6 represent completely different deformation mechanisms, and that the unified creep equation lacks a clear physical meaning. Therefore, we generally still use Equations 7.5 and 7.6 to represent creep at high stress and low stress, respectively.

7.1.5 Relationship between Steady-State Creep Rate and Temperature

Creep tests under a fixed stress at a series of temperatures reveal that the steady-state creep rate, $\log \dot{\varepsilon}_{sc}$, is linear with the reciprocal of the absolute temperature $1/T$, as shown in Figure 7.7. Therefore, the relationship between $\dot{\varepsilon}_{sc}$ and T can be described by Arrhenius' formula [1]

$$\dot{\varepsilon}_{sc} = C\sigma^n \exp\left(-\frac{Q}{RT}\right), \tag{7.8}$$

where

C is a constant related to material properties and stress
$R = 8.31$ J/(k·mol)
T is the absolute temperature
Q is the activation energy with the unit of kJ/mol

Its value can be obtained from the slope of the straight line.

Since dislocation substructure in the steady-state creep stage is primarily determined by stress normalized by the elastic modulus (Equation 7.9), and is not related to the creep testing temperature [4], creep activation energy thus denotes the variation of the creep rate with temperature at a certain stress with a certain dislocation substructure:

$$Q = \left[\frac{R\ln(\dot{\varepsilon}_{sc1}/\dot{\varepsilon}_{sc2})}{(1/T_2 - 1/T_1)}\right]_{\sigma/E,S}, \tag{7.9}$$

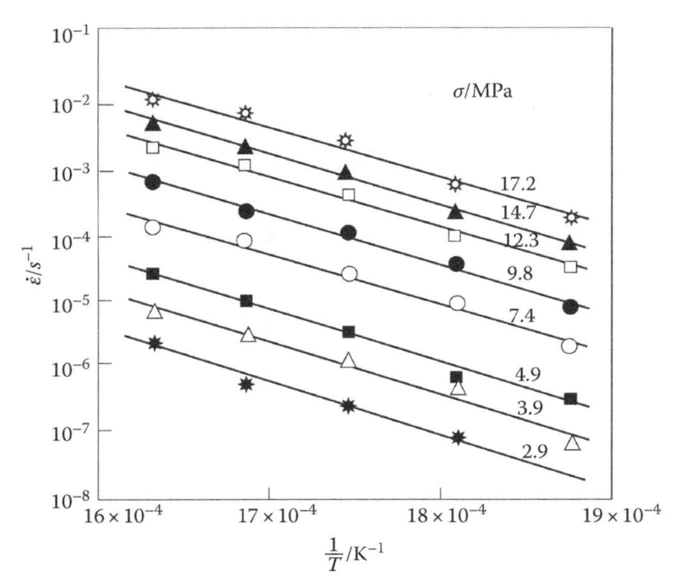

FIGURE 7.7 The relationship between steady-state creep rate and temperature of high purity aluminum. (From Zhang, J., *Deformation and Fracture of Materials at High Temperature*, Science Press, Beijing, 2007; Taken from Figure 1.4.)

where the subscript σ/E denotes the normalized stress, and S represents the dislocation substructure. In addition, the elastic modulus, the stacking default energy, and the grain size of metallic materials have a significant influence on the creep rate. Many scholars over the world have performed extensive and detailed studies. Interested readers can refer to the associated literature [4].

7.1.6 Application Examples

Researchers have conducted experimental studies on the improvement of creep resistance properties and tensile mechanical properties of Ti–22Al–27Nb alloys [6]. By adding 1at% Mo and Fe elements, stabilizing the β phase to optimize the microstructure of Ti–22Al–27Nb alloys, a new alloy material (Ti–22Al–11Nb–2Mo–1Fe) is manufactured. The creep properties and microstructure of two alloys at room temperature and high temperature are tested, analyzed, and compared. Experimental investigations are performed in the ambient atmosphere at room temperature, and in a vacuum (10^{-3} Pa) at a temperature of 650°C, with a tensile strain rate of $3 \times 10^{-4}\ s^{-1}$. The microstructure of the Ti–22Al–11Nb–2Mo–1Fe alloy after heat treatment is shown in Figure 7.8. The grain boundaries and phases can be clearly seen in the microstructure.

The curves of creep deformation versus time of two alloys are shown in Figure 7.9. It can be seen that the creep curve consists of three stages: the primary transient creep stage, the steady-state creep stage, and the accelerated creep stage. It is apparent that the creep resistance properties of the new alloy, Ti–22Al–11Nb–2Mo–1Fe with added trace elements, are superior to that of the original alloy Ti–22Al–27Nb. By quantitative analysis, it was discovered that—at 650°C and at an applied stress of 350 MPa—the time to reach the required creep deformation strain of 1% for the alloy Ti–22Al–11Nb–2Mo–1Fe is approximately 5.6 times longer than that for the Ti–22Al–27Nb alloy. The corresponding steady-state creep rate is 3 times lower than that of the Ti–22Al–27Nb alloy.

Figure 7.10 shows the relationship between the steady-state creep rate and the applied stress of a Ti–22Al–11Nb–2Mo–1Fe alloy at 700°C and 650°C. According to the power law of the

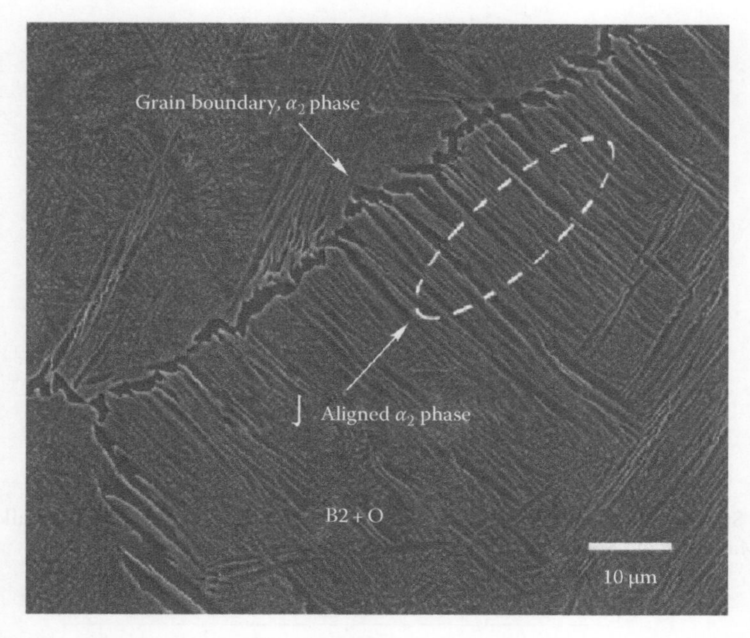

FIGURE 7.8 SEM micrograph of microstructure of Ti–22Al–11Nb–2Mo–1Fe alloy. (From Mao, Y. et al., *Scr. Mater.*, 57, 261, 2007; Taken from Figure 1.)

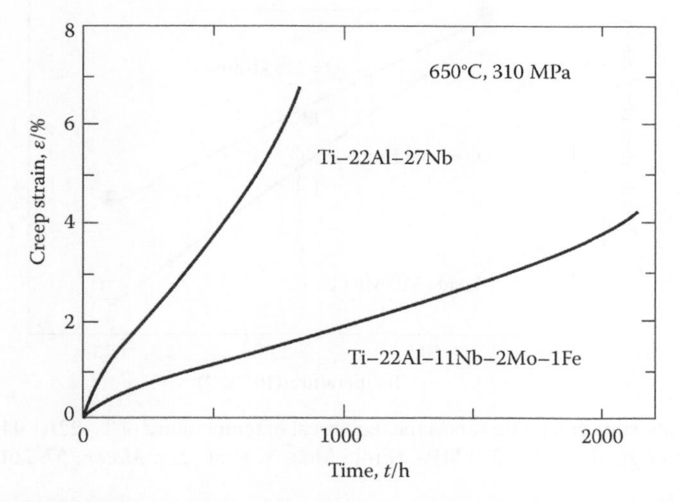

FIGURE 7.9 Creep strain–time curves of Ti–22Al–11Nb–2Mo–1Fe and Ti–22Al–27Nb alloys. (From Mao, Y. et al., *Scr. Mater.*, 57, 261, 2007; Taken from Figure 2.)

creep Equation 7.5, the creep exponent of the material is the slope of the straight line in the figure. Under different experimental conditions, the creep features are different, so that the exponents of n are different. Figure 7.11 represents the relationship between the steady-state creep rate, and the reciprocal of the loading temperature of a Ti–22Al–11Nb–2Mo–1Fe alloy at 310 MPa. The activation energy is obtained in terms of the creep equation power law (Equation 7.8) and the least square method; that is, the slope of the straight line in the figure. They discovered that the activation energy of the improved Ti–22Al–11Nb–2Mo–1Fe alloy, which is approximately $360 \pm 40\ kJ \cdot mol^{-1}$, is higher than that of the original Ti–22Al–27Nb alloy, which is approximately $246 \pm 20\ kJ \cdot mol^{-1}$.

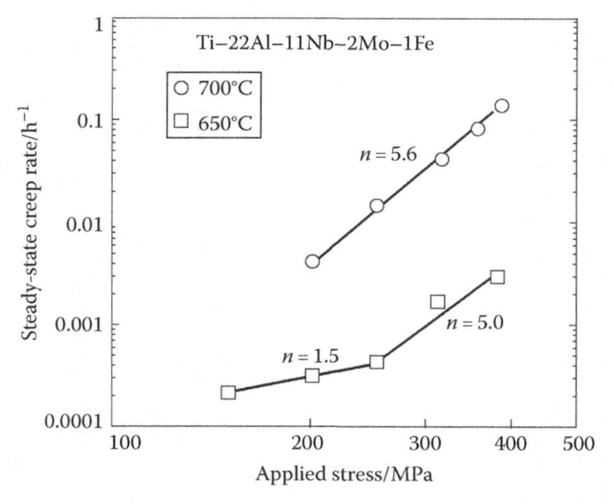

FIGURE 7.10 Steady-state creep rate versus applied stress of Ti–22Al–11Nb–2Mo–1Fe alloy at 700°C and 650°C. (From Mao, Y. et al., *Scr. Mater.*, 57, 261, 2007; Taken from Figure 3.)

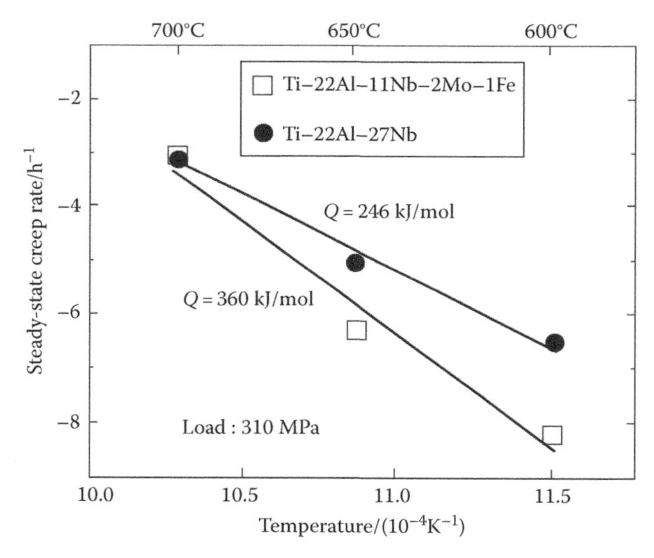

FIGURE 7.11 Steady-state creep rate versus the reciprocal of temperature of Ti–22Al–11Nb–2Mo–1Fe alloy and Ti–22Al–27Nb original alloy at 310 MPa. (From Mao, Y. et al., *Scr. Mater.*, 57, 261, 2007; Taken from Figure 4.)

Boehlert et al. analyzed the creep mechanisms of grain boundary sliding and dislocation climbing of the original alloy Ti_2AlNb [7]. The creep mechanism can be divided into two regions in terms of the experimental conditions. In each region, the creep mechanism is obtained in terms of the values of n and Q. When $n \approx 2$ and $Q = 256 \sim 311$ kJ·mol^{-1}, material creep is dominated by grain boundary sliding; when $n > 3.5$ and $Q > 320$ kJ·mol^{-1}, material creep is controlled by dislocation climbing. In general, at high temperatures, the dislocation climbing mechanism is accompanied by lattice self-diffusion. It can be seen that the creep of the original alloy Ti–22Al–27Nb is dominated by grain boundary sliding, while the creep of the new alloy Ti–22Al–11Nb–2Mo–1Fe is dominated by dislocation climbing. The microstructure of the steady-state creep stage at 650°C and 310 MPa is shown in Figure 7.12. Grain boundary sliding and cracking is not found, and this proves the correctness of the above-mentioned experimental results.

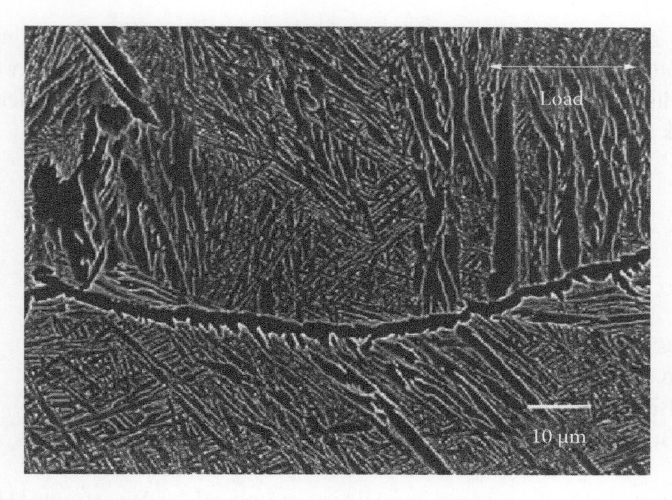

FIGURE 7.12 SEM micrograph of alloy Ti–22Al–11Nb–2Mo–1Fe in the steady-state creep state at 650°C and 310 MPa. (From Mao, Y. et al., *Scr. Mater.*, 57, 261, 2007; Boehlert, C.J., *Mater. Sci. Eng. A*, 267, 1, 82, 1999; Taken from Figure 5.)

7.2 CREEP MECHANISMS AND CREEP MECHANISM DIAGRAMS OF METALLIC MATERIALS

7.2.1 Creep Mechanisms of Metallic Materials

Creep deformation of a metallic material is closely related to the stress level, temperature, loading, and the crystallographic type (i.e., single crystal or polycrystalline material). Compared to polycrystalline materials, the single crystal has a relatively large elongation in the creep initial stage, and the duration for the steady-state creep stage is relatively long. Extensive studies indicate that metallic materials have the following primary creep mechanisms, and that several mechanisms usually coexist and interact [1,5].

1. Dislocation glide, which includes dislocation motion under thermal activation on slip planes, or by passing the obstacles. This mechanism appears at high stress, when $\sigma/G > 10^{-2}$. When dislocations are obstructed by obstacles (such as precipitates, solute atoms, and other dislocations), the creep rate is relatively easy to determine [10].
2. Dislocation creep, which includes dislocation motion that bypasses obstacles by thermal activation, and the action of vacancy or interstitial atoms. This mechanism generally appears in the range of $10^{-4} < \sigma/G < 10^{-2}$. Dislocation creep is mainly caused by dislocation slip and vacancy diffusion, and it plays a major role in a large number of engineering structures [8–10]. Among many theoretical frameworks, the theory proposed by Orowan and Bailey is relatively reasonable. According to this theory, the steady-state creep rate denotes the dynamic equilibrium between the strain hardening rate and the thermal recovery rate. The strain hardening rate is denoted by $h = \partial\sigma/\partial\varepsilon$, and the thermal recovery rate due to dislocation rearrangement and annihilation is represented by $r = -\partial\sigma/\partial t$. A steady creep condition is established if the thermal recovery rate is fast enough, and the strain hardening rate is sufficiently slow, so that they reach an equilibrium state:

$$\dot{\varepsilon}_{sc} = \frac{r}{h} = \frac{-\dfrac{\partial\sigma}{\partial t}}{\dfrac{\partial\sigma}{\partial\varepsilon}} \tag{7.10}$$

The physics model of dislocation creep must predict the values of h and r. This work has been finished by Gittus [1], whose mechanism-based model agreed very well with experimental results. This model is primarily based on stress in the three-dimensional dislocation network and dislocation diffusion motion:

$$\dot{\varepsilon}_{sc} = \frac{16\pi^3 c_j D_{sd} G b}{kT}\left(\frac{\sigma}{G}\right)^3, \tag{7.11}$$

where

c_j denotes the density of dislocation jogs (steps)
D_{sd} is the diffusion coefficient
b is the Burgers vector
k is the Boltzmann constant

The earliest dislocation creep model was proposed by Weertman [11–13], which is mainly based on dislocation climb. When the temperature rises, gliding dislocations accumulate gradually when their motion is obstructed by obstacles. After accumulation to a certain degree, dislocations surmount the obstacles and continue to slip, and the above process is repeated. Although dislocation slip determines the degree of creep deformation, the speed of the process is controlled by dislocation climb. Because dislocation climb requires the diffusion of vacancy or impurity atoms, atom diffusion dominates the rate of dislocation jogs. In this model, the creep rate is proportional to the third power of stress. However, in creep tests of many metallic materials, the creep rate is proportional to the 3–8 power of stress. For most metallic materials, the creep rate is proportional to the fifth power of stress. Therefore, for creep testing at a high stress level, when the temperature reaches half of the melting point, the steady-state creep rate is generally represented by a power law:

$$\dot{\varepsilon}_{sc} = \frac{A D_{sd} G b}{kT}\left(\frac{\sigma}{G}\right)^n, \tag{7.12}$$

where A and n are material constants. The diffusion coefficient D_{sd} can be written as

$$D_{sd} = D_0 \exp\left(-\frac{Q}{kT}\right). \tag{7.13}$$

Substituting Equation 7.13 into Equation 7.12 gives

$$\dot{\varepsilon}_{sc} = B\sigma^n \exp\left(-\frac{Q}{kT}\right). \tag{7.14}$$

After appropriate modification to Equation 7.12 [14], it can be applied to high-temperature creep dominated by grain boundary diffusion, or to low temperature creep dominated by vacancies.

Figure 7.13 shows the variation of the dimensionless steady-state creep rate ($\dot{\varepsilon}_{sc}/D_{sd}$) with the dimensionless applied stress (σ/G). The central region of stress follows the power-law creep. When the stress is relatively low ($\sigma/G < 5 \times 10^{-6}$), it exhibits a linear

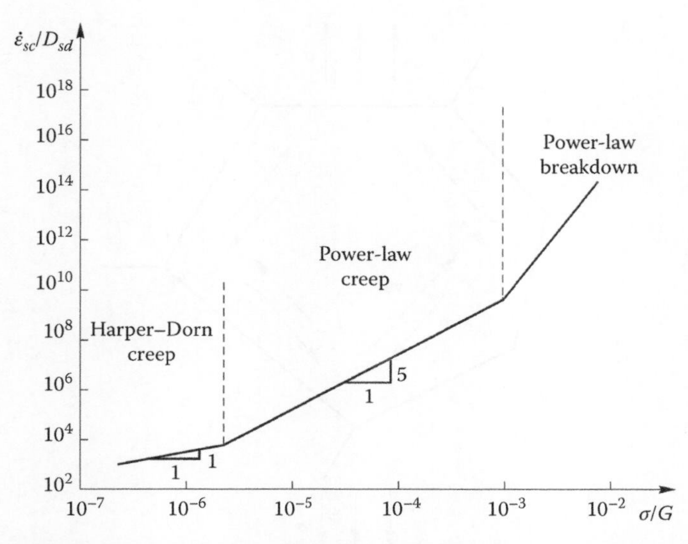

FIGURE 7.13 Influence of dimensionless stress on dimensionless steady-state creep rate. (From Dieter, G.E., *Mechanical Metallurgy*, 3rd edn., Tsinghua University Press, Beijing, 2006; Taken from Figure 13.10.)

relationship (slope $n = 1$). This is the well-known Harper–Dorn creep [15]. The reason is that the dislocation density no longer changes with stress, and dislocation climb dominates the creep process. In regions with relatively high stress ($\sigma/G > 10^{-3}$), power-law creep breaks down and the measured creep rate is larger than the predicted value by Equation 7.12. This region can be described by the Sellars–Tegart equation or the Wu–Sherby equation [16]:

$$\dot{\varepsilon}_{sc} = \frac{AD}{\alpha^n b^2}\left(\sinh \alpha \frac{\sigma}{E}\right)^n. \tag{7.15}$$

At the point where the power-law creep stage terminates, $\alpha = (\sigma/E)^{-1}$.

3. Diffusion creep. At high temperature and low stress ($\sigma/G < 10^{-4}$), the vacancies and interstitial atoms of the material move and diffuse inside the crystal. According to the studies of Nabarro–Herring and Coble, the creep process under this condition is dominated by stress-introduced diffusion of atoms. Stress changes the chemical potentials of atoms at the grain boundaries in polycrystalline materials. This results in the diffusion of vacancies from tensile regions to compressive regions, and the diffusion of atoms in the reverse direction, leading to the elongation of crystals in a certain direction (see Figure 7.14). The corresponding Nabarro–Herring creep equation is [17]

$$\dot{\varepsilon}_{sc} \approx \frac{14\sigma b^3 D_{sd}}{kTd^2}, \tag{7.16}$$

where

d is the diameter of the grain

D_{sd} is the diffusion coefficient inside the grain

It can be seen that creep rate decreases with the increase in grain size. This equation is mainly appropriate for the creep of metallic materials at high temperature with low stress.

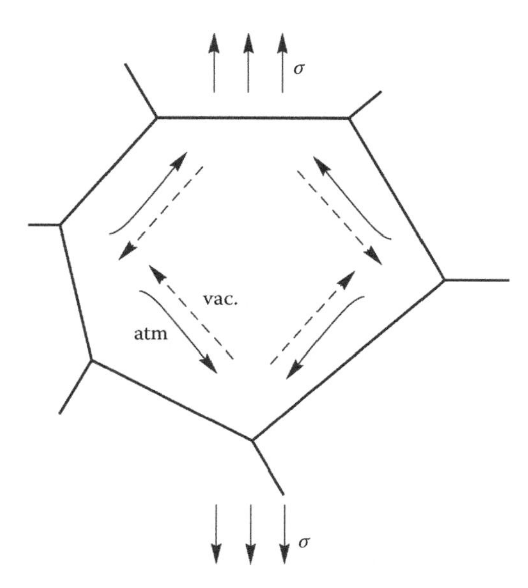

FIGURE 7.14 Creep mechanism dominated by vacancy diffusion inside a grain. (From Dowling, N.E., *Mechanical Behavior of Materials (Engineering Methods for Deformation, Fracture and Fatigue)*, Prentice Hall, Upper Saddle River, NJ, 1998; Taken from Figure 2.26.)

At low temperature and low stress, grain boundary diffusion plays a major role. Coble-type creep can be described by the following equation [18]:

$$\dot{\varepsilon}_{sc} \approx \frac{50\sigma b^4 D_{gb}}{kTd^3}, \tag{7.17}$$

where D_{gb} is the diffusion coefficient at the grain boundaries. We should note that in Nabarro–Herring creep equation, the creep rate is proportional to D_{sd}/d^2; but in the Coble creep equation, the creep rate is proportional to D_{gb}/d^3.

4. Grain boundary sliding mainly includes intergranular shear slip. Although grain boundary sliding has no great influence on the steady-state creep rate, it plays an important role in intergranular cracking. In creep mechanisms, grain boundary sliding plays a supporting role in maintaining the continuity of grains [19].

7.2.2 Diagram of Creep Mechanisms

A very practical way to explain a variety of creep deformation mechanisms is to use constitutive equations, and diagrams of creep deformation mechanisms. These mechanisms are plotted in the stress–temperature space, as shown in Figure 7.15. Each region in the diagram indicates the creep mechanism which dominates, and which plays a major role under the combined condition of stress and temperature. Using Equations 7.12 through 7.17, stress can be viewed as a function of temperature, and thus the boundary conditions of different regions can be calculated. Two different creep mechanisms at the boundaries can have the same creep deformation rate. For example, for an identical temperature ($T/T_m = 0.8$), when the stress is relatively low, creep deformation is mainly Nabarro–Herring creep. When the stress increases gradually, creep deformation enters the region of power-law creep (dislocation creep, T_m is the melting point temperature of the material). And when the stress is further increased, the creep deformation of metallic materials is mainly dominated by dislocation sliding caused by thermal activation. The top region in the diagram denotes the region for slip in ideal crystalline materials.

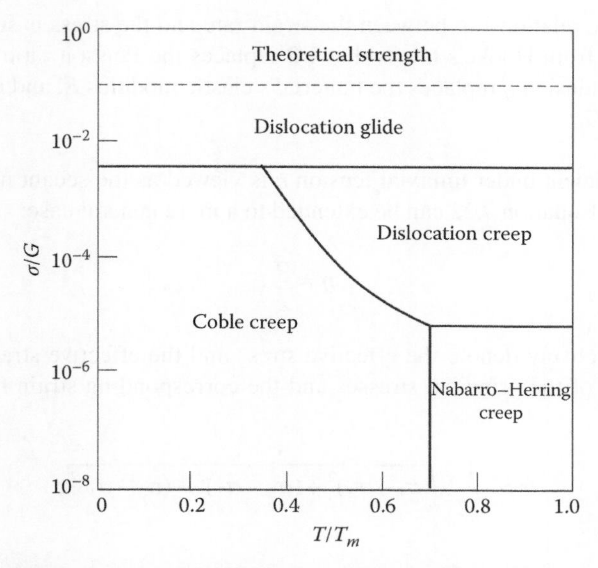

FIGURE 7.15 Creep deformation mechanisms of metallic materials. (From Dieter, G.E., *Mechanical Metallurgy*, 3rd edn., Tsinghua University Press, Beijing, 2006; Zeng, C.H. and Zou, S.J., *Fatigue Analysis Method and Application*, National Defense Industry Press, Beijing, 1991; Taken from Figure 13.11.)

7.2.3 Creep Deformation under Complex Stress State

In practical applications, metallic materials frequently experience creep deformation at high temperatures under complex (or three dimensional) stress states [1]. The processing method is similar to that of uniaxial tensile experimental data. Assume that the material is uncompressible, and thus the volume change should be zero. Correspondingly, the volumetric deformation rate is also zero:

$$\dot{\varepsilon}_x + \dot{\varepsilon}_y + \dot{\varepsilon}_z = 0 \tag{7.18}$$

Now consider a state with a uniaxial tensile stress σ_x. The deformation rate is

$$\dot{\varepsilon}_x = \frac{\sigma_x}{\eta}(\sigma_y = \sigma_z = 0), \tag{7.19}$$

where η is the viscosity coefficient under uniaxial tension in Equation 7.1, with the unit MPa·s. At the same time, the strain rates in the other two directions are:

$$\dot{\varepsilon}_y = \dot{\varepsilon}_z = -\frac{\dot{\varepsilon}_x}{2} = -\frac{1}{2}\left(\frac{\sigma_x}{\eta}\right)(\sigma_y = \sigma_z = 0) \tag{7.20}$$

If stresses in the directions y and z exist, similar deformations can be obtained. Therefore, the strain rates in different directions under a complex stress state are:

$$\dot{\varepsilon}_x = \frac{1}{\eta}\left[\sigma_x - \frac{1}{2}(\sigma_y + \sigma_z)\right], \quad \dot{\varepsilon}_y = \frac{1}{\eta}\left[\sigma_y - \frac{1}{2}(\sigma_x + \sigma_z)\right], \quad \dot{\varepsilon}_z = \frac{1}{\eta}\left[\sigma_z - \frac{1}{2}(\sigma_y + \sigma_x)\right]. \tag{7.21}$$

Similarly, the corresponding shear strain rates are:

$$\dot{\gamma}_{xy} = \frac{3\tau_{xy}}{\eta}, \quad \dot{\gamma}_{yz} = \frac{3\tau_{yz}}{\eta}, \quad \dot{\gamma}_{xz} = \frac{3\tau_{xz}}{\eta}. \tag{7.22}$$

It can be seen that the relationship between the strain rate and the stress is similar to Hooke's law. The main difference from Hooke's law is that 1/2 replaces the Poisson ratio ν, the viscosity coefficient under unaxial tension η replaces the material's elastic modulus E, and $\eta/3$ replaces the material's shear modulus G.

If the viscosity coefficient under uniaxial tension η is viewed as the secant modulus of the stress–strain rate curve, the Equation 7.22 can be extended to a more general case:

$$\eta = \frac{\bar{\sigma}}{\bar{\dot{\varepsilon}}}, \tag{7.23}$$

where $\bar{\sigma}$ and $\bar{\dot{\varepsilon}}$ respectively denote the effective stress and the effective strain rate, which can be represented in terms of the principal stresses and the corresponding strain rates by the following equations:

$$\bar{\sigma} = \frac{1}{\sqrt{2}}\sqrt{(\sigma_1-\sigma_2)^2+(\sigma_2-\sigma_3)^2+(\sigma_3-\sigma_1)^2}, \tag{7.24}$$

$$\bar{\dot{\varepsilon}} = \frac{\sqrt{2}}{3}\sqrt{(\dot{\varepsilon}_1-\dot{\varepsilon}_2)^2+(\dot{\varepsilon}_2-\dot{\varepsilon}_3)^2+(\dot{\varepsilon}_3-\dot{\varepsilon}_1)^2}. \tag{7.25}$$

Equation 7.25 is similar to the effective plastic strain in plastic deformation theory, in which plastic strains instead of strain rates are used.

The relationship between strain and time is generally expressed by

$$\varepsilon = \varepsilon_e + \varepsilon_p + B\sigma^m t + D\sigma^\alpha \tag{7.26}$$

where B, m, D, and σ are empirical constants from creep experiments at given temperature, ε_e and ε_p are respective elastic and plastic strain. If the strain rate and stress follow the uniaxial relation as $\dot{\varepsilon} = \mathrm{f}(\sigma)$, the following equation can be deduced by Equation 7.26,

$$\dot{\varepsilon}_{sc} = B\sigma^m. \tag{7.27}$$

The creep strain rate $\dot{\varepsilon}_{sc}$ and the stress σ under a complex stress state are generally replaced by the equivalent strain rate $\bar{\dot{\varepsilon}}$ and the equivalent stress $\bar{\sigma}$:

$$\bar{\dot{\varepsilon}} = B\bar{\sigma}^m \tag{7.28}$$

7.3 INTRODUCTION TO FATIGUE OF METALLIC MATERIALS

Strength, stiffness, and fatigue life are three basic requirements of engineering structures and mechanisms. Fatigue failure is one of the main reasons for the failure of engineering structures and mechanisms. The peak value of the cyclic load which leads to fatigue failure is generally much lower than the "safe" load estimated in terms of static fracture. Therefore, the study of the fatigue of structural components is very important.

7.3.1 Definition of Fatigue

Fatigue is a branch of solid mechanics. It primarily studies the strength of materials or structures under cyclic loading, and the relationship between the life of materials or structures and their stress states. Failure may occur when a material or a structure is repeatedly subjected to an alternating load, even if the stress never goes beyond the material's strength limit, or is even lower than the elastic limit. The failure of materials or structures under cyclic loading is called fatigue failure.

There are many apparent differences in nature between fatigue failure and traditional failure under static load.

1. Static failure is a one-time failure under the maximum load; fatigue failure is a failure after many cycles of cyclic loading. Fatigue failure does not occur in a short time period, but only after some time has elapsed, or even after a very long time.
2. When static stress is lower than the yield limit or strength limit, static failure does not occur. When cyclic loading is much less than the static strength limit, or even less than the yield limit, fatigue failure may occur.
3. Static failure is generally accompanied by obvious plastic deformation, while fatigue failure does not have obvious macroscopic plastic deformation, even for metals with good ductility. Similar to the failure of a brittle material, it is difficult to detect fatigue failure in advance, indicating that fatigue failure is more dangerous.
4. The fracture surface under static load generally exhibits a coarse granular or fibrous feature; the fracture surface of fatigue failure always has two regions: one is flat and smooth, the other exhibits a coarse granular or fibrous feature.
5. The resistance to static failure depends on the material itself, while the resistance to fatigue failure depends on the constituents of the material: the shape, size, and surface conditions of its components; the service condition; and the environment.

7.3.2 Classification of Fatigue Failure

Fatigue failure of an engineering component is mainly influenced by the type and magnitude of externally applied loads, the frequency and the number of cycles of cyclic loading, the service temperature, and the environmental medium. Fatigue failure can be classified according to the following factors:

1. At the microscopic level, the initiation of fatigue cracking is related to the local microscopic plastic deformation. However, at the macroscopic level, when the magnitude of cyclic loading is relatively low, elastic strain dominates fatigue. Fatigue life is relatively long, and is called stress fatigue or high-cycle fatigue ($N_f > 10^5$, where N_f denotes the number of cycles when fatigue failure occurs).
2. When the magnitude of cyclic stress is relatively high, plastic strain dominates. Fatigue life is relatively short, and is called strain fatigue or low-cycle fatigue ($N_f < 10^5$).
3. Mechanical fatigue is fatigue failure that occurs solely under applied cyclic stress or strain.
4. Creep–fatigue is fatigue failure under the synthetic action of cyclic loading and high temperature.
5. Thermo-mechanical fatigue is fatigue under the synthetic effect of cyclic loading and cyclic temperature.
6. Corrosion fatigue is fatigue under cyclic loading in a corrosive chemical medium or in an embrittled medium.
7. Sliding contact fatigue and rolling contact fatigue is fatigue under the synthetic action of cyclic loading and sliding contact, or rolling contact between materials.
8. Fretting fatigue is fatigue caused by the synthetic action of cyclic stress and the minute relative motion and friction sliding at contact surfaces.

Most failures of mechanisms or structural components are caused by one of the above-mentioned fatigue processes.

7.3.3 Fatigue Load

Fatigue load is repeated alternating loading, leading to fatigue failure. Fatigue load is applied repeatedly, in many cycles, and the magnitude and direction of the load changes in each cycle. Fatigue load is generally classified into determined load or random load. A determined load has two types. In type 1, the amplitude of the load never changes, and is called the constant amplitude load, as shown in Figure 7.16a. In type 2, the load magnitude changes regularly, and is called programmed load or block load, as shown in Figure 7.16b. For a random load, the load magnitude changes randomly, as shown in Figure 7.16c. Most loads are random loads.

There are two different loading modes in fatigue testing: load-controlled or flexible loading, and deformation-controlled or stiff loading. In flexible loading, the load magnitude is kept constant during the test, the displacement of the specimen is not constrained, and fatigue damage changes with the stiffness of the loading frame. In stiff loading, the specimen displacement is kept constant during the test, and the load changes with the stiffness of the loading frame. In engineering, the usual types of fatigue loading include bending fatigue, rotating fatigue, torsion fatigue, impact fatigue, and axial fatigue.

7.3.4 Cyclic Stress

In fatigue loading, cyclic stress denotes stress that alternately varies with time [5]. It is also defined as cyclically changed stress with time between two peak stresses. In order to clearly show the change of stress with time, we plot the variation of stress σ with time t as a sine wave, shown in Figure 7.17. Each periodic change of stress is called a stress cycle. In a stress cycle, the maximum value is called the maximum stress σ_{max}, and the minimum value is called the minimum stress σ_{min}. The algebraic mean value of the maximum stress and the minimum stress is called the mean stress σ_m. The difference between the maximum stress and the mean stress, or between the mean stress and the minimum stress, is called the stress amplitude σ_a. The ratio of the minimum stress to the maximum stress is called the stress ratio R, a representation of the change of stress.

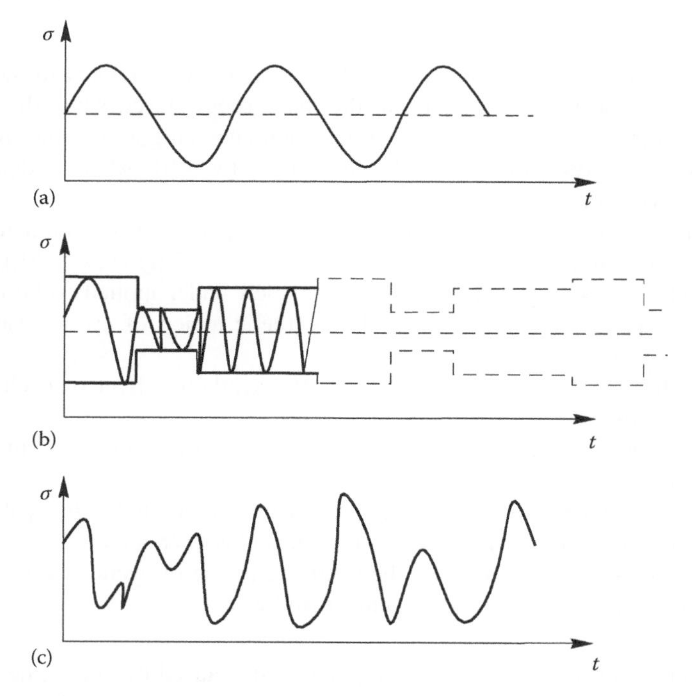

FIGURE 7.16 Types of fatigue loading. (a) Constant amplitude load, (b) block load, and (c) random load.

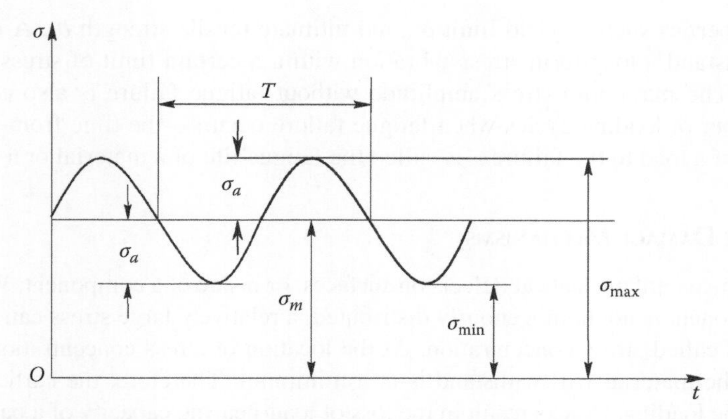

FIGURE 7.17 Schematic of typical constant-amplitude cyclic stress. (From Dieter, G.E., *Mechanical Metallurgy*, 3rd edn., Tsinghua University Press, Beijing, 2006; Taken from Figure 12.2.)

The stress range $\Delta\sigma$ is two times the stress amplitude σ_a. The ratio of stress amplitude σ_a to mean stress σ_m is called the stress amplitude ratio A. The above stresses exist in the following relationships:

$$\Delta\sigma = \sigma_{\max} - \sigma_{\min} = 2\sigma_a, \quad \sigma_{\mathrm{m}} = \frac{\sigma_{\max} + \sigma_{\min}}{2} \tag{7.29}$$

$$R = \frac{\sigma_{\min}}{\sigma_{\max}}, \quad A = \frac{\sigma_a}{\sigma_m} = \frac{1-R}{1+R}. \tag{7.30}$$

From Equations 7.29 and 7.30, we can derive the following equations:

$$\Delta\sigma = \sigma_{\max}(1-R), \quad \sigma_m = \frac{\sigma_{\max}}{2}(1+R), \tag{7.31}$$

$$R = \frac{1-A}{1+A}, \quad A = \frac{1-R}{1+R}. \tag{7.32}$$

If the mean stress is equal to zero, this type of stress is called the fully reversed cyclic stress, denoted by $R = -1$. If the minimum stress is equal to zero, it is denoted by $R = 0$.

7.4 FATIGUE FAILURE AND FATIGUE MECHANISMS OF METALLIC MATERIALS

7.4.1 Fatigue Strength and Fatigue Limit

The fatigue properties of a material or structural component are measured by fatigue strength. Fatigue strength is the strength of a material or structural component under cyclic loading. The magnitude of fatigue strength is measured by the fatigue limit. Fatigue limit is defined theoretically as the upper limit of cyclic stress under which the specimen does not fail after an indefinite large number of loading cycles (i.e., $N_f \to \infty$). It is impossible to experimentally measure the theoretical fatigue limit. In engineering practice, fatigue limit is defined as the cyclic stress the specimen can bear with a given fatigue life. When the stress is $R = -1$, fatigue limit is denoted by σ_{-1}. In other words, fatigue limit is the maximum stress $\sigma_{\max}$ at a given stress ratio R, that can be applied to a material without causing fatigue failure. High-cycle fatigue limit is called the endurance limit. Fatigue limit is mainly measured by a fatigue test, and can sometimes be estimated from other mechanical properties such as yield limit σ_y, and ultimate tensile strength σ_b. A material or component can withstand a long-term stress vibration within a certain limit of stress without causing fatigue failure. The maximum stress amplitude without fatigue failure is also called the fatigue limit. The number of loading cycles when fatigue failure occurs—the time from the beginning of the application of a load to the failure—is called the fatigue life of a material or a component.

7.4.2 Fatigue Damage Mechanisms

Fatigue cracks in general nucleate at defects on surfaces, or inside of a component. When the internal stress of a component is not homogenously distributed, a relatively large stress can appear in a local region, which is called stress concentration. At the location of stress concentration, the number of loading cycles the material will withstand is at a minimum. Therefore, the earliest cracks appear there under cyclic loading. Cracks result in the loss of load-bearing capacity of a part of the material of a component, and the increase of average stress in the remaining material. Since a component generally has residual strength, cracking does not lead to immediate failure, but it introduces a sharp notch at the tip of the crack, and forms a new region of stress concentration. As the component sees more and more use, cracking occurs at this location. In this way, the crack becomes larger and larger, and the material that can bear load becomes less and less. Eventually, the component fails when the remaining material is not sufficient to bear the load. Therefore, the procedure of fatigue failure is: stress concentration → nucleation of crack → new stress concentration → crack propagation → final failure. Alternatively, we can say that the procedure of fatigue failure consists of: nucleation of fatigue crack → growth of microcracks → appearance of macrocracks → final failure.

The fatigue process consists of three stages: fatigue crack nucleation, crack subcritical propagation, and final unsteady propagation. The fatigue life N_f consists of the period for fatigue crack nucleation N_0, and the period for fatigue subcritical propagation N_p. Understanding the physics processes of individual stages is very important in recognizing the nature of fatigue, analyzing the causes of fatigue, taking measures to increase strength and toughness, and increasing fatigue life [4].

7.4.2.1 Nucleation of Fatigue Crack

Macroscopic fatigue cracks develop from the formation, growth, and coalescence of microscopic cracks [4]. There is still no united standard for crack length during the nucleation period of fatigue crack. Usually, a crack of 0.05–1.0 mm is taken as the nucleus of a fatigue crack, and the period of fatigue crack nucleation is thus determined. Extensive investigation shows that microscopic fatigue cracking is caused by non-uniform local slip and microscopic cracking, which primarily includes: (1) cracking by surface slip; (2) cracking at the secondary phase, precipitates, or other interface; and (3) cracking at grain boundaries or sub-grain boundaries.

1. Cracking in slip bands
   Experimentation shows that cyclic slip occurs and cyclic slip bands form when metals are subjected to cyclic stress ($\sigma > \sigma_{-1}$), even though the stress is lower than the yield stress. Compared to the formation of homogeneous slip bands under static loading, cyclic slip bands are severely heterogeneous, and are concentrated in some local weak regions. It is difficult to remove the cyclic slip bands on specimen surfaces with electro-polishing. Even if they are removed, they reappear in the original regions when a cyclic load is applied. These permanent or reappearing slip bands are called persistent slip bands. Experiment strongly indicates that persistent slip bands are generated in weak regions of the material. Persistent slip bands generally form near the specimen surface, and their depth is shallow. With the increase of loading cycles, the persistent slip band becomes wider and wider. When its width increases to a certain degree, due to the dislocation pileup and intersection, microcracks can form in persistent slip bands. During the broadening process of persistent slip bands, extrusions and intrusions may appear. Thus, stress concentration and voids

can be generated there, resulting in microcracks after a certain number of loading cycles. Extrusions and intrusions have been widely observed in experimentation, and cracks form there. Therefore, improving the slip resistance of materials (by solid-solution strengthening and grain refinement strengthening, for example) prevents the nucleation of fatigue cracks, and enhances fatigue strength.

2. Cracking at phase boundaries
   In fatigue analysis, it is frequently observed that many fatigue sources are caused by secondary phase or inclusions. Therefore, decreasing the brittleness of secondary phase or inclusions, increasing the strength of the phase boundary, and controlling the amount, morphology, size, and distribution of secondary phase or inclusions, can suppress or delay the nucleation of fatigue cracks around secondary phase or inclusions, and improve fatigue strength.
3. Cracking at grain boundaries
   Because polycrystalline materials have grain boundaries and different orientations of neighboring grains, the motion of dislocations in a grain is obstructed at the grain boundaries. Dislocations pile up at the grain boundary and introduce stress concentration there. Under cyclic stress, stress concentration at the grain boundaries cannot be relaxed, and the stress peak becomes higher and higher. When it surpasses the grain boundary strength, cracking occurs at the grain boundaries. From the viewpoint of grain boundary cracking, all factors that weaken grain boundaries and coarsen grains are prone to result in cracking at the grain boundaries, and the impairment of fatigue strength. These factors include low melting-point inclusions or detrimental elements, composition segregation, temper embrittlement, hydrogen absorption at grain boundaries, and grain coarsening. On the contrary, these factors can also be measures to strengthen grain boundaries and purify grains, to suppress cracking at the grain boundary, and to improve fatigue strength.

7.4.2.2 Propagation of Fatigue Crack

After a fatigue crack nucleates, it propagates. Depending on the propagation direction, crack propagation can be divided into two stages. In the first stage, microcracks form in the individual intrusions on the surface, and then propagate inward along the direction of the primary slip (the direction of the maximum shear stress) as in pure shear. In the propagation stage, most microcracks stop propagation, and only a few microcracks propagate through two to three grains. In this stage, the crack propagation rate is very low, at the order of 0.1 μm per cycle. The first stage of crack propagation can be observed in many ferrous alloys, aluminum alloys, and titanium alloys. However, for notched specimens, the first stage may disappear. Since the crack propagation rate is very low in the first stage, the total amount of propagation is very small. It is difficult to analyze the fracture surface of this stage, which is usually without morphologic features except for some traces of scratches. However, in some hardened materials, periodic cleavage or quasi-cleavage patterns—even crystal sugar-like patterns due to intergranular cracking—can sometimes be observed.

During the first stage of crack propagation, due to the continuous obstruction of grain boundaries, crack propagation turns gradually toward the direction perpendicular to the tensile stress, and enters the second stage. At room temperature and in a corrosion-free environment, fatigue crack propagation is transgranular. During most cycles in this stage, the crack propagation rate is approximately 10^{-5}–10^{-2} mm per cycle. Therefore, the second stage consists of the main part of the subcritical propagation of fatigue cracking. The fracture surface of the second stage exhibits a pattern with slightly curved and parallel trenches, called fatigue striations. These are microscopic traces when the fatigue crack propagates forward. Each striation can be viewed as the propagation trace of each loading cycle. The crack propagation direction is perpendicular to the striations.

7.4.3 General Behavior of Fatigue Crack Propagation

7.4.3.1 Different Regions of the Fatigue Crack Propagation

For a given material and its corresponding experimental conditions, fatigue crack propagation behavior can generally be described by the relationship between the fatigue crack propagation rate $\mathrm{d}a/\mathrm{d}N$, and the range of the stress intensity factor ΔK, as shown in Figure 7.18. The curve of $\mathrm{d}a/\mathrm{d}N$ versus ΔK in log-log coordinates exhibit three regions with an antisigmoidal shape. Region A is close to the threshold value. The feature of this stage is the fatigue crack propagation threshold value ΔK_{th}.

When $\Delta K < \Delta K_{th}$, fatigue cracking does not propagate, or propagates minimally; but when the stress intensity factor goes beyond the threshold value, the crack propagation rate increases rapidly. In region B, the crack propagation rate follows Paris' power law

$$\frac{\mathrm{d}a}{\mathrm{d}N} = A_1(\Delta K)^m, \tag{7.33}$$

where A_1 and m are constants that depend on the material, the frequency and waveform of loading, and the environment. For most ductile metals and alloys, the typical slope of this linear region is between 2 and 4. In region C, the maximum stress intensity factor of the cyclic loading $K_{\max}$ approaches the material's fracture toughness K_{IC}, and the crack propagation rate increases drastically, followed by unsteady propagation, until fracture.

7.4.3.2 Microscopic Process of Fatigue Crack Propagation [4]

Extensive experimental investigations show that fatigue crack propagation can generally be divided into two stages, as shown in Figure 7.19. It has been observed in many ductile materials that, when the crack length is less than the dimension of a few grains, deformation is constrained in a single slip system, and fatigue cracking advances along the primary slip plane. This stage is called the first stage of crack propagation, which approximately corresponds to the region near the threshold value (Region A) in Figure 7.18. In this stage, the stress intensity factor is relatively small and it

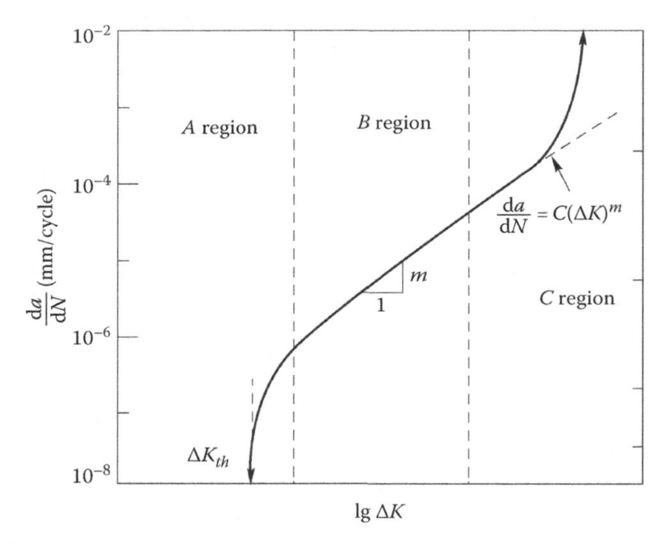

FIGURE 7.18 Different stages of fatigue crack propagation. (From Zhang, J., *Deformation and Fracture of Materials at High Temperature*, Science Press, Beijing, 2007; Taken from Dowling, N.E., *Mechanical Behavior of Materials (Engineering Methods for Deformation, Fracture and Fatigue)*, Prentice Hall, Upper Saddle River, NJ, 1998; Taken from Figure 22.21.)

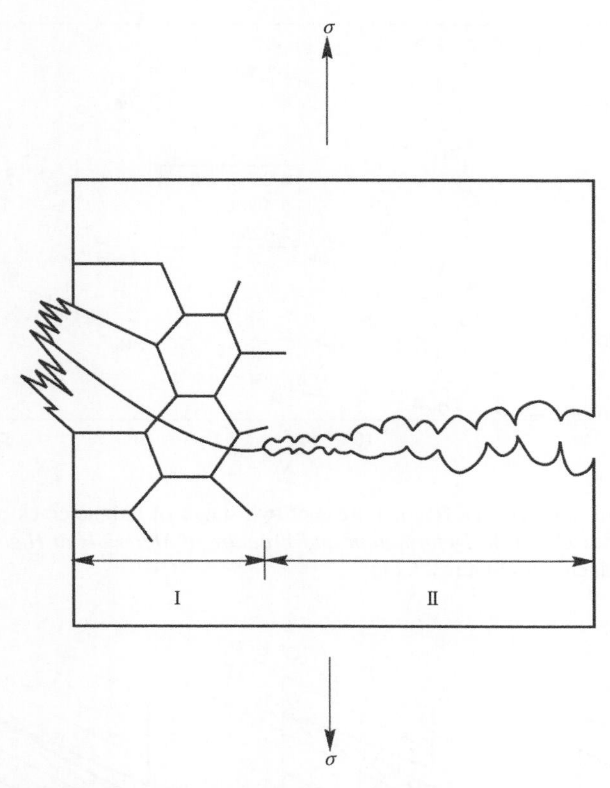

FIGURE 7.19 Two stages of fatigue crack propagation. (From Zhang, J., *Deformation and Fracture of Materials at High Temperature*, Science Press, Beijing, 2007; Taken from Figure 22.22.)

is not sufficient to cause apparent plastic deformation ahead of the crack tip, and the dimensions of the plastic zone are generally less than the grain size. Because cracking propagates along slip planes, the fracture surface exhibits cleavage facets. When the dimensions of the plastic zone are much larger than the diameter of the grain, the material exhibits another fatigue fracture mode: in this case, the macroscopic path of fatigue crack propagation is approximately perpendicular to the tensile load. This stage is called the second stage of fatigue crack propagation, and normally corresponds to region B in Figure 7.18.

The fraction areas of the two stages on the fracture surface depend on the strain amplitude. When the strain amplitude is very large, the second stage occupies most of the fracture surface area; with a decrease in strain amplitude, the area fraction of the first stage increases gradually. Figure 7.20 shows the experimental results of pure Ni at room temperature, and of 316 stainless steel at 400°C and 625°C, under fully reversed cyclic loading. With an increase of the number of cycles to failure, the fraction area of the first stage increases gradually, while the fraction area of the second stage decreases gradually. Two different materials, and identical materials at different temperatures, display the same tendency, indicating that fatigue fracture mechanisms are almost irrelevant to material and temperature.

Many mechanisms have been proposed for the first stage of crack propagation, one of which is illustrated in Figure 7.21 [4]. From the compressive half cycle to the tensile half cycle, the crack length increases from C_1 to C_2 due to slip on the slip plane, as shown in Figure 7.21a and b. When the load changes to compression again, due to slip in the reverse direction, the upper and lower sides of the crack are equidistant to the crack tip, and this distance is equal to $(C_1 + C_2)/2$. The crack propagates at $\Delta C = (C_2 - C_1)/2$, as shown in Figure 7.21c.

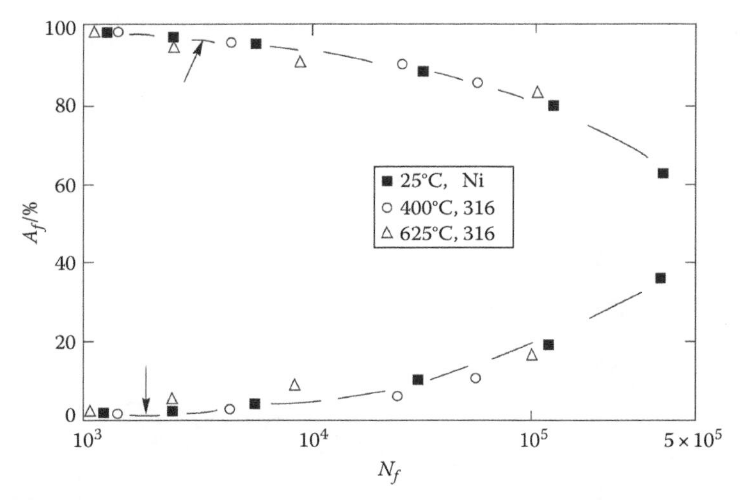

FIGURE 7.20 Relationship between fraction areas of two stages of fatigue crack propagation on fracture surface A_f and N_f. (From Zhang, J., *Deformation and Fracture of Materials at High Temperature*, Science Press, Beijing, 2007; Taken from Figure 22.23.)

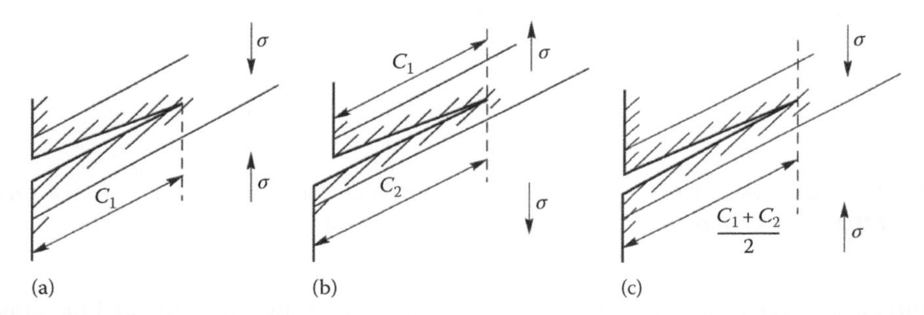

FIGURE 7.21 Model for fatigue crack propagation in Stage 1. (a) Under tension. (b) Under compression. (c) Under tension again. (From Zhang, J., *Deformation and Fracture of Materials at High Temperature*, Science Press, Beijing, 2007; Taken from Figure 22.24.)

The most obvious feature of the second stage of fatigue crack propagation is the pattern of fatigue striations, as shown in Figure 7.22. If the externally applied load remains unchanged, the spacing of striations can be mapped to the propagation distance in each loading cycle. This is widely used in fatigue failure analysis to identify the history of crack propagation. Many models have been proposed to interpret the formation of fatigue striations. The model of continuous blunting followed by re-sharpening, proposed by Laird [5], is widely accepted. Figure 7.23 illustrates this model.

In the tensile stage of cyclic loading, the crack tip blunts due to double slip, and the crack propagates at a distance, at the same order of crack tip opening displacement, as shown in Figure 7.23a through c. When loaded in the reverse direction, the crack tip re-sharpens into a double notch, as shown in Figure 7.23d and e, and the crack tip blunts again in the next tensile stage, as shown in Figure 7.23f. In each loading cycle, the crack propagates some distance, and forms fatigue striations. Fatigue striations in quasi-crystalline or amorphous solids can be explained in terms of the propagation mechanisms of crack tip blunting and re-sharpening, but fatigue striations are not formed in all cases.

In addition, formation of fatigue striations is significantly influenced by the environment. Many alloys do not exhibit fatigue striations in a vacuum, and the crack propagation rate in a vacuum is

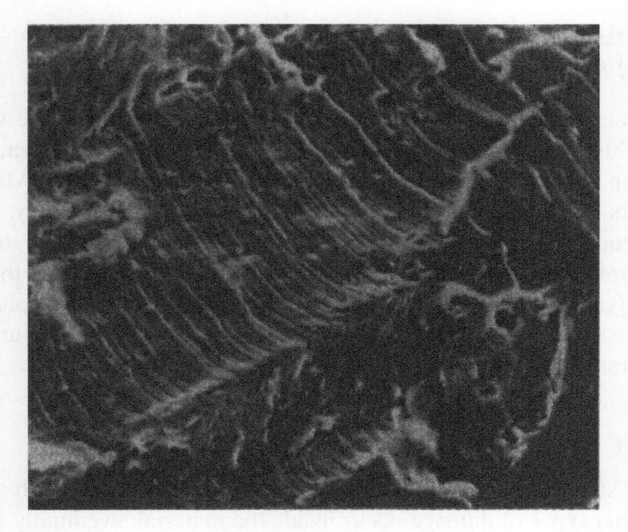

FIGURE 7.22 Fatigue striations on the fracture surface. (From Zhang, J., *Deformation and Fracture of Materials at High Temperature*, Science Press, Beijing, 2007; Taken from Figure 22.25.)

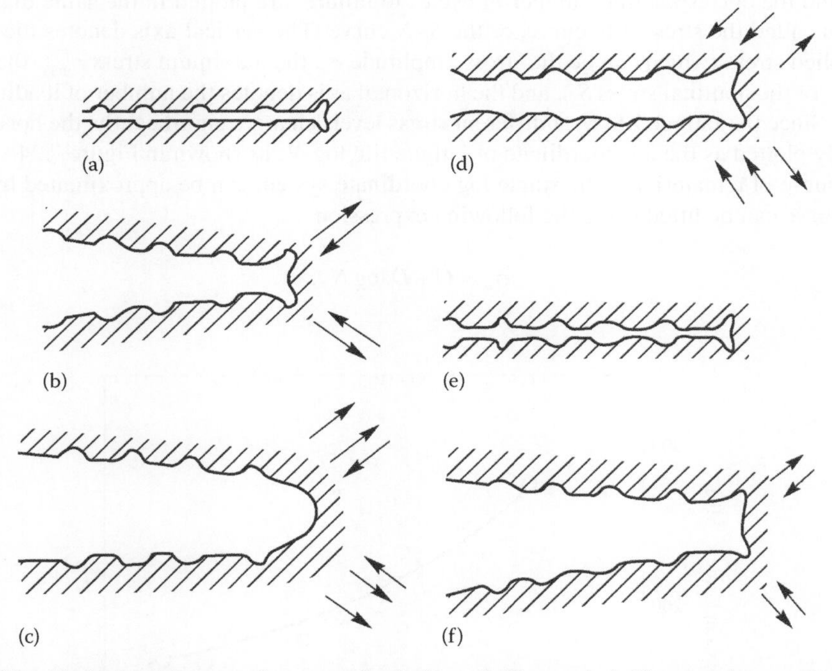

FIGURE 7.23 Model for fatigue crack propagation in Stage 2. The crack tip blunts and propagates a distance equivalent to the opening displacement during the tensile stage of cyclic loading (a–c), while during the compress stage, the crack tip re-sharpens (d and e), until in the next tensile stage, the crack tip blunts again (f). (From Zhang, J., *Deformation and Fracture of Materials at High Temperature*, Science Press, Beijing, 2007; Taken from Figure 22.26.)

apparently lower than it is in humid air. The reason is that slip steps in a vacuum do not oxidize, and the slip during tension and compression is reversible. If the newly formed fresh slip steps in the tensile stage of cyclic loading are oxidized or chemically corroded, the reversibility of slip motion when the load is reversed is suppressed. Therefore, the striation spacing, or the propagation distance of each cycle (propagation rate) is influenced by the environment.

7.5 METHODOLOGY OF STUDY OF FATIGUE FAILURE IN METALLIC MATERIALS

It has been 150 years since the advent of the study of mechanical failure caused by fatigue in engineering [20]. This was first done by Albert, a Germany mining engineer, in 1829. He applied cyclic load to an iron chain in a mining elevator, to examine its reliability. After Albert, many scientists and engineers made outstanding contributions in this field. Currently, there are three main approaches to the study of fatigue failure in metallic materials [1,5]: (1) the stress-based approach, or the study of the relationship between nominal stress or mean stress and the number of loading cycles in the investigated region; (2) the strain-based approach, which emphasizes local yielding problems with high stress; and (3) the fracture mechanics approach, which analyzes fatigue crack propagation using fracture mechanics.

7.5.1 *S–N* Curve

When a material or structural component is subjected to a cyclic stress with sufficient magnitude, fatigue cracks or other types of damage occur inside the material, eventually leading to the failure and fracture of the entire material. Experiments have shown that the larger is the applied stress, the smaller is the number of cycles to failure. In high-cycle fatigue tests, the different applied cyclic stresses, and the corresponding number of cycles to failure, are plotted in the same diagram. This diagram is called the stress–life curve, or the *S–N* curve. The vertical axis denotes the magnitude of the applied stress (which can be the stress amplitude σ_a, the maximum stress $\sigma_{\max}$, the minimum stress $\sigma_{\min}$, or the nominal stress S_a), and the horizontal axis denotes the number of loading cycles to failure N_f. Since the fatigue life N_f at different stress levels changes significantly, the horizontal axis is generally plotted as the log coordinate of fatigue life log N_f, as shown in Figure 7.24.

If the curve of a material in the single log coordinate system can be approximated by a straight line, the curve can be fitted using the following expression:

$$\sigma_a = C + D\log N_f, \tag{7.34}$$

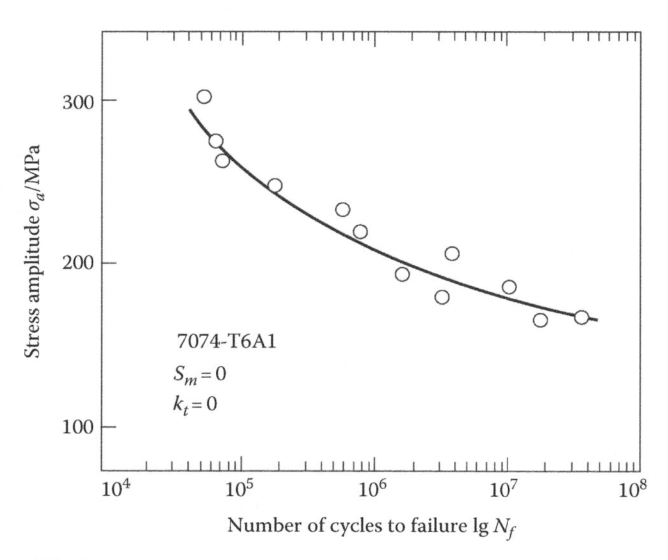

FIGURE 7.24 Stress–life (*S–N*) curve of an aluminum alloy in rotating-bending fatigue test. (From Dowling, N.E., *Mechanical Behavior of Materials (Engineering Methods for Deformation, Fracture and Fatigue)*, Prentice Hall, Upper Saddle River, NJ, 1998; Taken from Figure 9.4.)

where C and D are fitting constants. If the S–N curve of a material in log-log coordinates can be approximated by a straight line, the curve can be fitted by:

$$\sigma_a = AN_f{}^B. \tag{7.35}$$

Equation 7.35 is generally transformed into another form:

$$\sigma_a = \sigma'_f(2N_f)^b. \tag{7.36}$$

The fitting coefficients in Equations 7.35 and 7.36 satisfy

$$A = 2^b\sigma'_f \quad (B = b) \tag{7.37}$$

These fitting coefficients are generally obtained by fitting the experimental data. It is found that the coefficient σ'_f is approximately equal to the material's true fracture strength $\tilde{\sigma}_f$ in the uniaxial tensile test.

Figure 7.25 plots two basic types of S–N curves. The fatigue life of all materials increases with a decrease in stress amplitude. For a ferrous metal or alloy, the S–N curve approaches a horizontal asymptote, and the longitudinal coordinate of the horizontal asymptote is the fatigue limit σ_e. When σ_{max} is larger than σ_e, fatigue failure occurs in the specimen after a certain number of cycles. When σ_{max} is less than σ_e, the specimen can withstand an indefinite number of cycles without failure. In general, if a ferrous metal or alloy does not fail after 2×10^6–2×10^7 cycles, it is regarded as being able to withstand an indefinite number of cycles. For a nonferrous metal or alloy (such as pure aluminum and aluminum alloys), the fatigue curve does not have such an asymptote. The maximum stress corresponding to 10^7–10^8 cycles is artificially taken as the conditional fatigue limit.

Figure 7.26 shows the relationship between the mean stress σ_m and the stress amplitude σ_a under conditions with the same fatigue life.

It is apparent that, in order to keep the same fatigue life, the applied stress amplitude decreases gradually with the increase of mean stress. They exhibit an inverse proportional relationship.

In addition, the shape of the S–N curve is mainly influenced by material type, mean stress, the geometry of the component, the chemical environment, testing temperature, frequency, and residual stress [1]. Under some special experimental conditions with special microstructures,

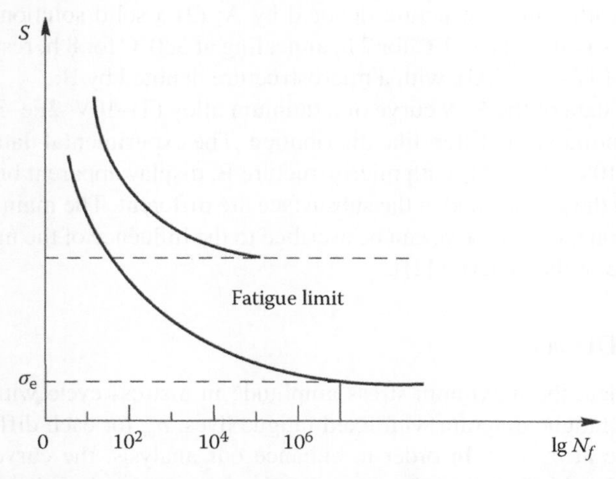

FIGURE 7.25 The relationship of applied stress and lg N_f. (From Zeng, C.H. and Zou, S.J., *Fatigue Analysis Method and Application*, National Defense Industry Press, Beijing, 1991; Taken from Figure 2.9.)

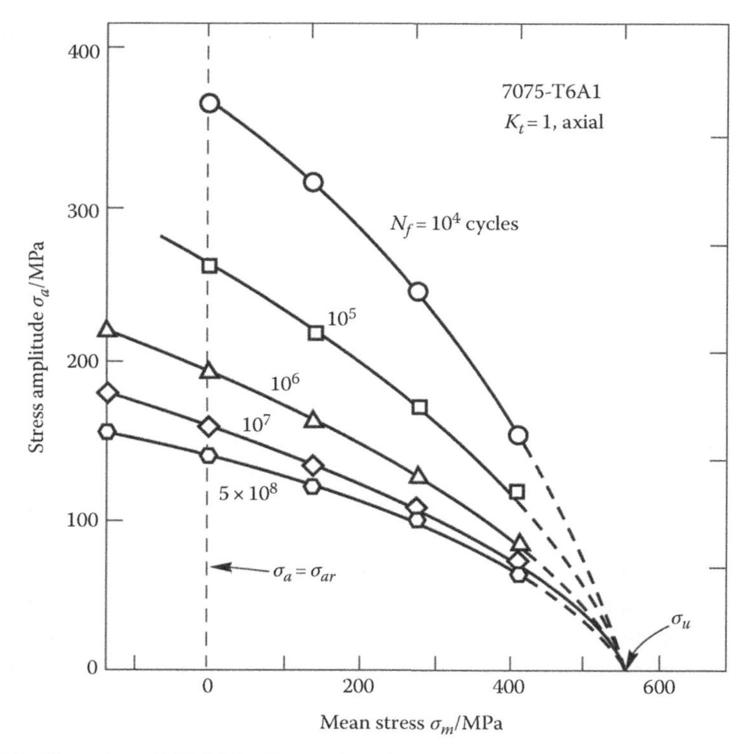

FIGURE 7.26 *S–N* curves of 7075-T6 alloy under given fatigue life. (From Dowling, N.E., *Mechanical Behavior of Materials (Engineering Methods for Deformation, Fracture and Fatigue)*, Prentice Hall, Upper Saddle River, NJ, 1998; Taken from Figure 9.34.)

S–N curves display a binary or step attribute. That is, the *S–N* curve data at different locations are completely different or discontinuous [21–24]. For example, Chandran et al. discovered the binary attribute of *S–N* curves when they studied the fatigue properties of a metastable β titanium alloy (Ti–10V–2Fe–3Al), as shown in Figure 7.27 [21]. The experimental conditions used to obtain the two microstructures A and B are (1) a solid solution of powders (volume fraction of α_p particles is 45%) at 780°C for 2 h, annealing at 525°C for 8 h, resulting in a metastable β titanium alloy (Ti–10V–2Fe–3Al), with a microstructure denoted by A; (2) a solid solution of powders (volume fraction of α_p particles is 10%) at 780°C for 2 h, annealing at 580°C for 8 h, resulting in a metastable β titanium alloy (Ti–10V–2Fe–3Al), with a microstructure denoted by B.

The experimental data of the *S–N* curve of a titanium alloy (Ti–10V–2Fe–3Al) with microstructure A, exhibit discontinuous and step-like distribution. The experimental data of the *S–N* curve of a titanium alloy (Ti–10V–2Fe–3Al) with microstructure B, display apparent binary distribution: the experimental data on the surface and at the subsurface are different. The main reason for the binary or step-like attribute on the *S–N* curve can be ascribed to the influence of the manufacturing process on the microstructure of the material [21].

7.5.2 Goodman Diagram

For a smooth specimen, the maximum stress amplitude in a stress cycle with zero mean stress is defined as σ_{ar}. In the fatigue diagram with fixed fatigue lives, σ_{ar} for each different fatigue life corresponds to the value at $\sigma_m = 0$. In order to enhance our analysis, the curves in Figure 7.26 are replotted as the relationship between the normalized stress amplitude σ_a/σ_{ar}, and σ_m, as shown in Figure 7.28. In this way, the complex relationship between mean stress and different fatigue lives can

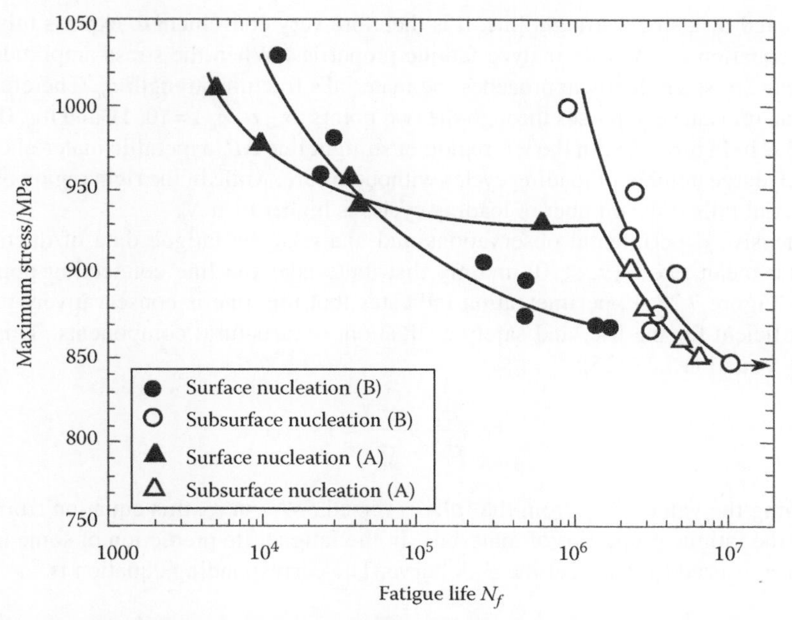

FIGURE 7.27 *S–N* curves of β titanium alloys (Ti–10V–2Fe–3Al) with two different metastable microstructures. (From Chandran, K.S.R. and Jha, S.K., *Acta Mater.*, 53, 1867, 2005; Taken from Figure 9.)

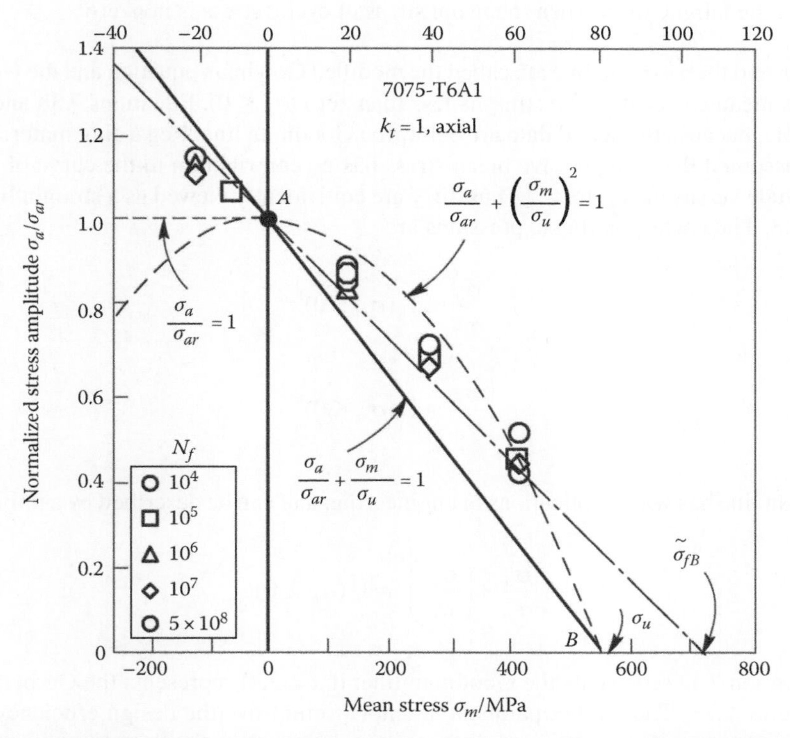

FIGURE 7.28 Relationship between normalized stress amplitude and mean stress of 7075-T6 aluminum alloy. (From Dowling, N.E., *Mechanical Behavior of Materials (Engineering Methods for Deformation, Fracture and Fatigue)*, Prentice Hall, Upper Saddle River, NJ, 1998; Taken from Figure 9.36.)

be approximated by a single straight line. It is therefore very convenient to express this line with an appropriate equation in order to analyze fatigue properties. When the stress amplitude approaches zero, the mean stress gradually approaches the material's fracture strength σ_u. Therefore, no matter whether a line or a curve, it passes through the two points $(\sigma_m, \sigma_a/\sigma_{ar}) = (0, 1)$, and $(\sigma_u, 0)$, denoted by points *A* and *B* in Figure 7.28. In the left region of straight line *AB*, a metallic material can withstand an indefinitely large number of loading cycles without failure; while in the right region of straight line *AB*, the material fails if the number of loading cycles is higher than N_f.

From extensive experimental observations and analysis, the fatigue data of ductile materials, with a tensile mean stress ($\sigma_m \geq 0$), mainly distribute near the line connecting points *A* and *B*, as shown in Figure 7.28. Experimentation indicates that this line is conservatively placed, and is therefore sufficient for the life, and safety evaluation, of structural components. This line can be described by equation 7.38 [25]:

$$\frac{\sigma_a}{\sigma_{ar}} + \frac{\sigma_m}{\sigma_u} = 1. \tag{7.38}$$

After obtaining the value of σ_{ar} from the fully reversed *S–N* curve, this equation can be employed to represent the fatigue properties of materials. In the fatigue life prediction of some materials, the fatigue limit σ_e is used to represent the *S–N* curve. The corresponding equation is

$$\frac{\sigma_e}{\sigma_{er}} + \frac{\sigma_m}{\sigma_u} = 1, \tag{7.39}$$

where

σ_{er} denotes the fatigue limit when the mean stress of cyclic stress is zero
σ_e denotes the fatigue limit when the mean stress of cyclic stress is nonzero

This equation and the associate line are called the modified Goodman equation and the Goodman line.

When the mean stress of cyclic stress is less than zero ($\sigma_m \leq 0$), Equations 7.38 and 7.39 are no longer suitable, because the actual data are below the Goodman line. Regarding material safety, it is sometimes assumed that compressive mean stress has no contribution to the curve of "normalized stress amplitude versus mean stress." Thus, they are equivalently viewed as a straight line, as shown in Figure 7.28. The corresponding expressions are

$$\frac{\sigma_a}{\sigma_{ar}} = 1 \quad (\sigma_m \leq 0). \tag{7.40}$$

$$\frac{\sigma_e}{\sigma_{er}} = 1 \quad (\sigma_m \leq 0). \tag{7.41}$$

The Goodman line has wide applications in engineering, and can be described by a unified equation:

$$\frac{\sigma_a}{\sigma_{ar}} + \left(\frac{\sigma_m}{\sigma_u}\right)^x = 1 \quad (\sigma_m \geq 0). \tag{7.42}$$

If $x = 1$, Equation 7.42 represents the Goodman line; if $x = 2$, it represents the Gerber parabola, as shown in Figure 7.28. The Gerber parabola attempts to improve the design efficiency of this rule when the mean stress is low, using a mean stress permission value that is not too conservative.

For low ductility metallic materials (like high-strength steel), the "stress amplitude versus mean stress curve" is very close to the Goodman line; for some ductile materials, the curve is close to the

Gerber parabola; for some brittle metallic materials (like cast iron), the experimental data are below the Goodman line. Therefore, for a specific material, a special equation is needed to describe its fatigue properties. Currently, the main approaches include: (1) replacing the fracture strength σ_u in Equation 7.38 by the rectified true fracture strength $\tilde{\sigma}_{fB}$, and (2) replacing the fracture strength σ_u in Equation 7.38 by σ'_f, obtained at $\sigma_m = 0$. The equations are, respectively,

$$\frac{\sigma_a}{\sigma_{ar}} + \frac{\sigma_m}{\tilde{\sigma}_{fB}} = 1. \tag{7.43}$$

$$\frac{\sigma_a}{\sigma_{ar}} + \frac{\sigma_m}{\sigma'_f} = 1. \tag{7.44}$$

This approach was proposed by Morrow in 1968 to modify Goodman line. For most ductile metallic materials, the constant σ'_f is approximately equal to $\tilde{\sigma}_{fB}$, and both are larger than the fracture strength σ_u. The two equations are denoted by dotted lines in Figure 7.28, which better agree with experimental data.

Another important equation to describe the fatigue properties of metallic materials is the SWT (Smith, Watson, Topper) equation:

$$\sigma_{ar} = \sqrt{\sigma_{\max} \cdot \sigma_a} \qquad (\sigma_{\max} > 0), \tag{7.45}$$

where $\sigma_{\max} = \sigma_m + \sigma_a$. This equation includes the same variables (σ_a, σ_m, σ_{ar}), and is not dependent on the influence of the material constant σ'_f. Another form of the SWT equation is

$$\sigma_{ar} = \sigma_{\max}\sqrt{\frac{1-R}{2}} \qquad (\sigma_{\max} > 0). \tag{7.46}$$

The SWT equation can describe very well the fatigue of most metallic materials, and it is particularly suitable for aluminum alloys. In general, the selection of equations regarding mean stress should compare favorably with experimentally measured data. That is, in terms of a specific material, the appropriate equation is selected to fit and describe the experimental data.

7.5.3 Fatigue Failure of Materials under Complex Stress States

Goodman equation (Equation 7.38) is proposed for uniaxial stress. In many engineering applications, the applied stresses are two dimensional or three dimensional [5]. As we select the single stress of σ_u or σ_y, we need to use a combined stress to determine the corresponding fatigue failure condition of the material [26–28]. For example, the equivalent deformation energy is a reasonable engineering method. The strain energy Q when applying a uniaxial stress to σ is

$$Q = \left(\frac{1+v}{3E}\right)\sigma^2. \tag{7.47}$$

When the principal stresses σ_1, σ_2, σ_3 are not zero, the equivalent strain energy is

$$Q = \left(\frac{1+v}{3E}\right)\left(\sigma_1^2 + \sigma_2^2 + \sigma_3^2 - \sigma_1\sigma_2 - \sigma_2\sigma_3 - \sigma_3\sigma_1\right). \tag{7.48}$$

Assume that (a) the main features of cyclic behavior are equivalent to the cyclic deformation between two given strain energies; and also assume that (b) the maximum value of the mean stress is determined by the von Mises yield criterion. We can derive the combined value of σ_1, σ_2, and σ_3, corresponding to the fatigue life of N_f. The endurance limit σ_R can be expressed as

$$\sigma_R(\sigma_1,\sigma_2,\sigma_3)=\left(\sigma_{1\max}^2+\sigma_{2\max}^2+\sigma_{3\max}^2-\sigma_{1\max}\sigma_{2\max}-\sigma_{2\max}\sigma_{3\max}-\sigma_{3\max}\sigma_{3\max}\right)^{1/2}$$
$$-\left(1-\frac{\sigma_a}{\sigma_y}\right)\left(\sigma_{1m}^2+\sigma_{2m}^2+\sigma_{3m}^2-\sigma_{1m}\sigma_{2m}-\sigma_{2m}\sigma_{3m}-\sigma_{3m}\sigma_{1m}\right)^{1/2} \quad (7.49)$$

where σ_{1m}, σ_{2m}, and σ_{3m} respectively are the mean value of the principal stresses σ_1, σ_2, and σ_3.

7.6 CYCLIC STRESS–STRAIN CURVES OF METALLIC MATERIALS

7.6.1 Cyclic Deformation Behavior of Single Crystals

Similar to the previously analyzed stress-controlled fatigue, cyclic strain can also control the fatigue process, and fatigue failure of materials, when the strain amplitude in a loading cycle remains constant [4]. Strain-controlled cyclic loading generally appears during cyclic thermal loading, when two ends of a component are fixed and a change of temperature causes thermal expansion and contraction. Figure 7.29 shows the stress–strain hysteresis loop under strain-controlled cyclic loading. The specimen is loaded to the maximum strain, then unloaded, and the load is applied in the reverse direction. Due to the Bauschinger effect, the material yields at a relatively lower stress in the compressive direction than in the tensile direction. After reaching the minimum strain, the material is loaded again in the positive direction until tensile yielding. Thus a stress–strain hysteresis loop is formed, which is generally represented by strain amplitude, total strain range, stress amplitude, and total stress range.

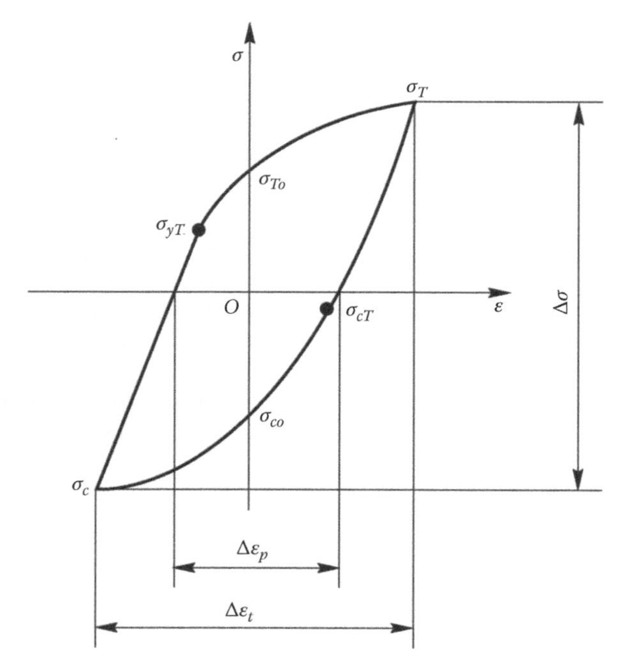

FIGURE 7.29 Stress–strain hysteresis loop under strain-controlled cyclic loading. (From Zhang, J., *Deformation and Fracture of Materials at High Temperature*, Science Press, Beijing, 2007; Taken from Figure 22.1.)

Here, the total strain range $\Delta\varepsilon_t$ includes the elastic strain range ($\Delta\varepsilon_e = \Delta\sigma/E$), and the plastic strain range $\Delta\varepsilon_p$. The width of the hysteresis loop depends on the magnitude of the cyclic strain.

Experimentation indicates that well-annealed pure metals exhibit cyclic hardening in the initial stage of cyclic deformation, as shown in Figure 7.30. Under strain-controlled (constant strain amplitude) cyclic loading, with the increase of loading cycles, stress amplitude increases rapidly, then the hardening rate decreases gradually and eventually reaches a saturation state. For cyclic hardening under stress-controlled (constant stress amplitude) cyclic loading, the total strain amplitude decreases gradually and reaches a saturation state. A stabilized stress–strain hysteresis loop is called the saturated hysteresis loop. Some materials may exhibit cyclic softening, in which the stress amplitude decreases gradually to a saturation value with the increase of loading cycles.

Because plastic deformation is irreversible, it changes the material during cyclic loading; and also changes the material's stress–strain hysteresis loop. Depending on the original state, metallic materials may display cyclic hardening, cyclic softening, or cyclic saturation. Not all metallic materials display the aforementioned three cyclic deformation behaviors, being influenced by the original state of the material, and experimental conditions. In general, after 100 cycles, the material's hysteresis loop essentially reaches a saturation state. A new saturation state can be reached when the strain amplitude is changed to a new value. The cyclically stabilized stress–strain curve is different from the monotonic stress–strain curve under static load. To determine the cyclic stress–strain curve, connect the tips of the stabilized hysteresis loops at a series of strain amplitudes obtained in strain-controlled constant amplitude loading, as shown in Figure 7.31.

The cyclic stress–strain curve can generally be described by a power law:

$$\Delta\sigma = K'(\Delta\varepsilon_p)^{n'}, \tag{7.50}$$

where

n' is the exponent for cyclic strain hardening

K' is the coefficient for cyclic strength

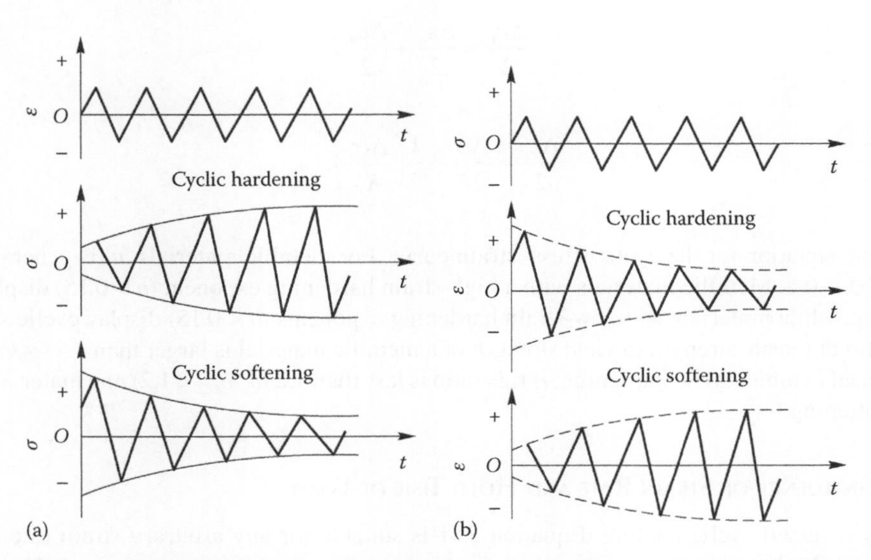

FIGURE 7.30 Cyclic hardening and softening. (a) Strain-controlled mode, and (b) stress-controlled mode. (From Zhang, J., *Deformation and Fracture of Materials at High Temperature*, Science Press, Beijing, 2007; Taken from Figure 22.2.)

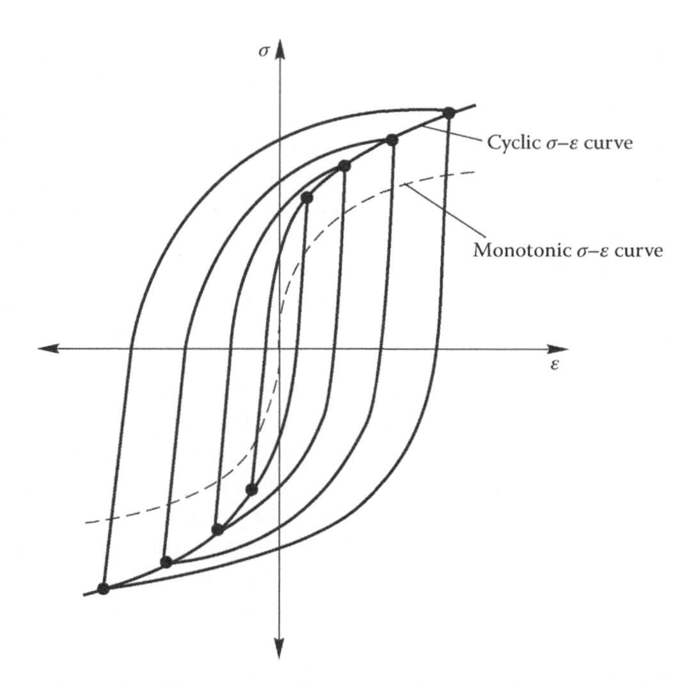

FIGURE 7.31 Comparison of cyclic stress–strain curve and monotonic stress–strain curve for cyclic-hardening material. (From Dieter, G.E., *Mechanical Metallurgy*, 3rd edn., Tsinghua University Press, Beijing, 2006; Zhang, J., *Deformation and Fracture of Materials at High Temperature*, Science Press, Beijing, 2007; Taken from Figure 12.12.)

This equation indicates that the cyclic stress–strain curve can be represented by a single straight line in a log-log coordinate system. However, cyclic deformation may change with the variation of strain amplitude. Usually, the slopes in the high-cycle fatigue region and the low-cycle fatigue region are different. This can be explained by

$$\frac{\Delta\varepsilon_t}{2} = \frac{\Delta\varepsilon_e}{2} + \frac{\Delta\varepsilon_p}{2}, \tag{7.51}$$

$$\frac{\Delta\varepsilon_t}{2} = \frac{\Delta\sigma}{2E} + \frac{1}{2}\left(\frac{\Delta\sigma}{K'}\right)^{1/n'}. \tag{7.52}$$

This is the equation for the cyclic stress–strain curve. For metallic materials, n' is generally in the range of 0.1–0.2. Metallic materials with a high-strain hardening exponent ($n > 0.15$) display cyclic hardening, while materials with low-strain hardening exponents ($n < 0.15$) display cyclic softening. If the ratio of tensile strength to yield strength of a metallic material is larger than 1.4 ($\sigma_b/\sigma_{0.2} > 1.4$), the material exhibits cyclic hardening; if this ratio is less than 1.2 ($\sigma_b/\sigma_{0.2} < 1.2$), the material exhibits cyclic softening.

7.6.2 Influence of Strain Rate and Hold Time of Load

For fully reversed cyclic loading, Equation 7.51 is suitable for any arbitrary strain rate $\dot{\varepsilon}_t$. When $\dot{\varepsilon}_t < 10^{-6}\text{s}^{-1}$, high temperature creep changes the unloading line into a curve, and the measured plastic strain amplitude $\Delta\varepsilon_p$ includes a creep strain component. If the strain at the maximum tensile strain is held for a period of time, the hysteresis loop changes into that shown in Figure 7.32.

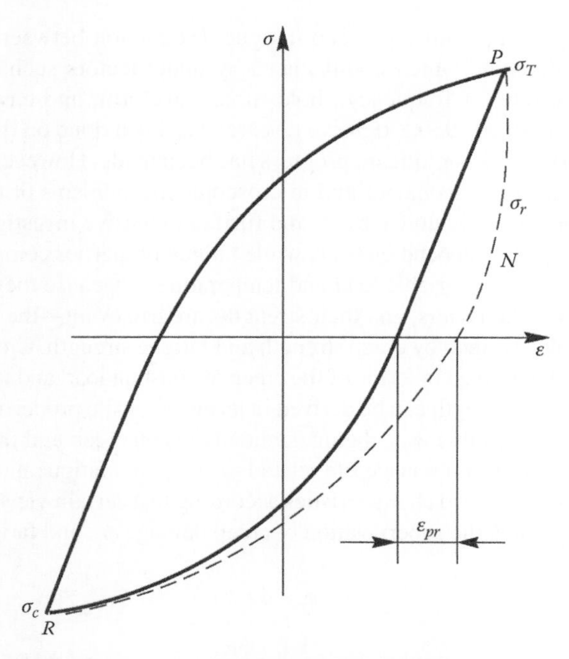

FIGURE 7.32 Influence of holding load on stress–strain hysteresis loop. (From Zhang, J., *Deformation and Fracture of Materials at High Temperature*, Science Press, Beijing, 2007; Taken from Figure 23.4.)

If the strain remains unchanged, the stress relaxes from σ_T to σ_r along *PN*. At the end of the hold time, unloading from σ_r and compressing in the reverse direction to σ_C, results in a creep strain ε_{pr}:

$$\varepsilon_{pr} = \frac{1}{E}(\sigma_T - \sigma_r). \tag{7.53}$$

In order to let the hysteresis loop close—due to the existence of extra tensile strain ε_{pr}—a larger stress in the reverse direction must be applied ($\sigma_C > \sigma_T$), indicating that the hold of tensile strain shifts the hysteresis loop in the compression direction. If asymmetric strain (or stress) is applied to offset the shift of the hysteresis loop, the hysteresis loop after hold of load regains symmetry.

To some extent, the discussed cyclic deformation features of ductile single crystals are also applicable to polycrystalline metallic materials. However, in engineering alloys, the heterogeneity of microstructures such as grain boundaries, inclusions, and secondary phases, complicate the cyclic deformation of materials. The method used to obtain the cyclic stress–strain curve is similar to that of single crystal materials. Single crystal materials exhibit cyclic hardening during the initial stage of cyclic deformation. Engineering materials may display cyclic hardening, or cyclic softening, depending on their microstructures. After an initial cyclic hardening or softening stage, the hysteresis loop reaches a stabilized (or saturation) state. The trace connecting the tips of saturated stress–strain hysteresis loops gives the cyclic stress–strain curve. The cyclic stress–strain curve of a polycrystalline material consists of different regions, and the formation of persistent slip bands has apparent influence on the deformation [29,30].

7.7 INTERACTION OF CREEP AND FATIGUE

Creep can occur even at elevated temperatures with a relatively low cyclic frequency, even if the applied load changes continuously with time (no hold time of load). In the analysis of low-cycle fatigue at high temperature, creep must be taken into consideration. In a general sense, low-cycle fatigue at high temperature is creep–fatigue. Interaction between creep and fatigue is a very complicated phenomenon, which is influenced by many factors such as strain amplitude (or stress amplitude), strain rate (or frequency), hold time, waveform, and environment (temperature and atmosphere) [31–36]. Since 1950, extensive research has been done on the creep–fatigue behavior of a variety of materials, and significant progress has been made. However, due to the complexity of the problem, the macroscopic behavior and microscopic mechanisms of the interaction between creep and fatigue are not yet fully understood, and further extensive investigation is required.

Creep properties completely depend on time, while fatigue properties completely depend on cyclic loading. Under the condition of variable load and temperature—because they have different mechanisms and display different behaviors, and their strengths are irrelevant—the strength of the material should be synthetically determined by creep strength and fatigue strength. Creep strength at changing load or temperature can be derived in terms of the creep at constant load and temperature, and its rupture characteristics. Fatigue strength can be derived in terms of tensile properties or fatigue properties, which are irrelevant to time. In this way, the interaction between creep and fatigue can be essentially solved. Therefore, the strength and service life related to creep and fatigue must be derived according to a theory which combines the two characteristics according to a certain viewpoint. For this problem, the representative viewpoint is the superposition of creep damage ϕ_c, and fatigue damage ϕ_f:

$$\phi = \phi_c + \phi_f \le 1. \tag{7.54}$$

Fracture or failure occurs when ϕ reaches 1. The creep damage ϕ_c is the sum of the ratio of time Δt, accounting for alternating stress or alternating temperature over the static creep rupture life:

$$\phi_c = \sum_{l=1}^{l=k} \left(\frac{t}{t_d} \right)_l , \tag{7.55}$$

where

t is the time for the application of creep load

t_d is the permitted rupture time under loading condition l

Fatigue damage ϕ_f is the ratio of the number of cycles N to the number of cycles to failure, under pure fatigue loading N_d:

$$\phi_f = \sum_{j=1}^{j=k} \left(\frac{N}{N_d} \right)_j , \tag{7.56}$$

where

N is the number of cycles under loading condition j

N_d is the fatigue life under loading condition j

The above viewpoint on damage is called the linear damage rule. The creep damage expressed by Equation 7.55 is called the accumulative damage rule, or the life-fraction rule.

7.7.1 Creep–Fatigue Waveform

The service conditions of structural components are complex and different. Creep–fatigue tests in the laboratory generally adopt several waveforms, as shown in Figure 7.33. The first column represents strain-controlled strain waveforms. The second column represents the corresponding stress

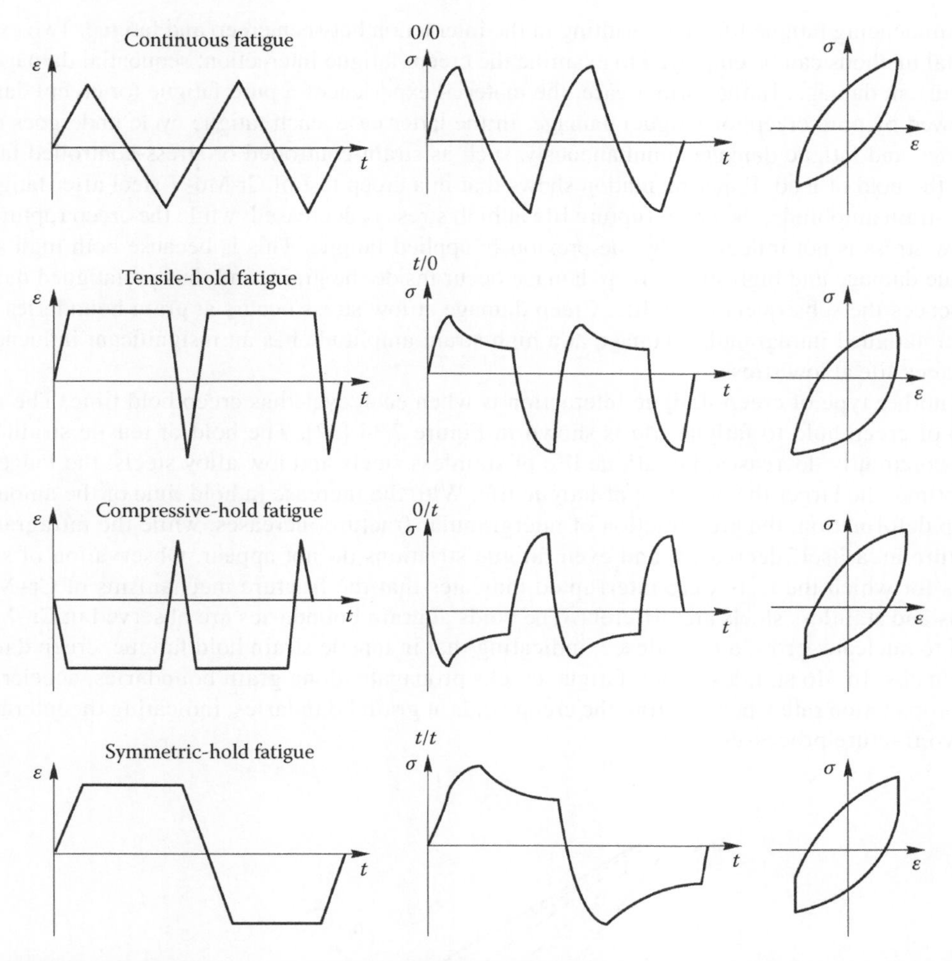

FIGURE 7.33 Waveforms of creep–fatigue. (From Zhang, J., *Deformation and Fracture of Materials at High Temperature*, Science Press, Beijing, 2007; Taken from Figure 23.1.).

waveforms, in which stress relaxation occurs during the hold of constant strain. The third column denotes the corresponding stress–strain hysteresis loops. The first waveform represents symmetric fatigue without hold of load, called continuous fatigue, and denoted by 0/0. If necessary, asymmetric fatigue—fast tension–slow compression (F/S) or slow tension–fast compression (S/F)—can be adopted. The second waveform is the tensile-hold creep–fatigue; denoted by *t*/0, where *t* is the hold time. The third waveform is the compressive-hold creep–fatigue, denoted by 0/*t*. The fourth waveform is the symmetric-hold creep–fatigue, denoted by *t*/*t*. The mean strain or mean stress of all waveforms does not have to be zero. The hold times for tension and compression do not have to be equal. The load with unequal tensile and compressive hold times is denoted by t_1/t_2.

7.7.2 Nature of Creep–Fatigue Interaction

The nature of creep–fatigue interaction is the relationship between creep damage and fatigue damage. The main mode of fatigue damage is the transgranular propagation of cracks. The main mode of creep damage is the nucleation and growth of voids at grain boundaries, but damage inside the grain is possible at high stress. When creep and fatigue occur sequentially or concurrently, one damage influences the developments of other damages, thus accelerating or retarding total damage and influencing fatigue life, and resulting in the interaction between creep and fatigue. Two experimental methods can be employed to examine the creep–fatigue interaction: sequential damage and concurrent damage. In the former case, the material experiences a pure fatigue (or creep) damage followed by pure creep (or fatigue) damage. In the latter case, each fatigue cycle undergoes creep damage and fatigue damage simultaneously, such as strain-controlled or stress-controlled fatigue with the hold of load. Experimentation shows that in a creep test of Cr-Mo-V steel after fatigue at high strain amplitude, the creep rupture life at high stress is decreased; while the creep rupture life at low stress is not influenced by the previously applied fatigue. This is because both high strain fatigue damage and high stress creep damage occur inside the grain, and thus prefatigued damage influences the subsequent creep life. Creep damage at low stress occurs at grain boundaries, thus the prefatigued intragranular damage at a high-strain amplitude has an insignificant influence on the creep life at low stress.

Another type of creep–fatigue interaction is when each cycle has creep hold time. The influence of creep hold to fatigue life is shown in Figure 7.34 [37]. The hold of tensile strain peak ($t/0$) apparently decreases the fatigue life of stainless steels and low alloy steels: the longer the hold time, the larger the decrease of fatigue life. With the increase in hold time or the amount of creep deformation, the area fraction of intergranular fracture increases, while the intragranular fracture area itself decreases, and even fatigue striations do not appear. Observation of specimens for which the tests were interrupted indicates that the fracture mechanisms of Cr–Mo–V steels and stainless steels are different. The voids at grain boundaries are observed in Cr–Mo–V steel to nucleate, grow, and coalesce, indicating that in tensile strain hold fatigue, creep damage dominates. In 316 stainless steel, fatigue cracks propagate along grain boundaries, accelerating the propagation rate when meeting the creep voids at grain boundaries, indicating the interaction of two fracture processes.

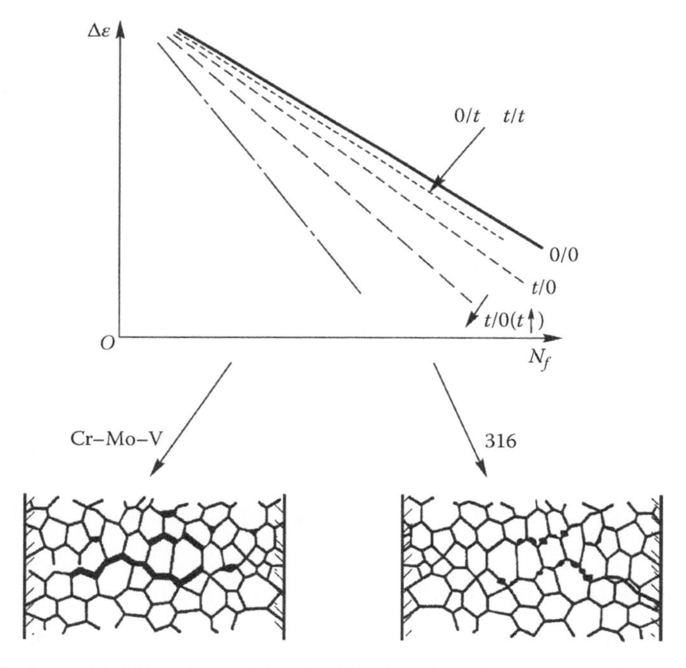

FIGURE 7.34 Influence of holding time on fatigue life N_f. (From Zhang, J., *Deformation and Fracture of Materials at High Temperature*, Science Press, Beijing, 2007; Plumbridge, W.J., Metallography of high temperature fatigue, in Skelton, R.P., ed., *High Temperature Fatigue-Properties and Prediction*, Elsevier Applied Science, London, U.K., 1987, pp. 177–228; Taken from Figure 23.2.)

Creep–fatigue behavior is strongly influenced by the material's environment. In an oxidative or corrosive environment, surface damage accelerates the nucleation of surface fatigue cracks, and the oxidation and corrosion of the crack tip and the crack surface influence crack tip morphology and crack closure behavior, thus influencing the crack propagation rate. During the hold time, creep damage and environmental damage occur concurrently, and are joined to time-dependent damage. Time-dependent damage relies on material properties. Creep damage dominates in alloys with good oxidization resistance but bad creep resistance, while environment damage dominates for materials with bad oxidization resistance but good creep resistance. Time-dependent damage is also related to waveform. Environment damage is similarly sensitive to tensile stress hold and compressive stress hold, while creep damage is sensitive to tensile stress hold, but not sensitive to compressive stress hold.

7.7.3 Creep–Fatigue Fracture Mechanism Diagram

Creep–fatigue interaction is complex, not only because fatigue life is influenced by the interaction of many factors but also because the results from different materials, and the same material from different heat numbers, as well as from different researchers, are inconsistent or even contradictory [3,4]. Even so, attempts are being made to establish criteria to demarcate the range of a variety of creep–fatigue interactions, one of which is the creep–fatigue fracture mechanism diagram proposed by Hales.

Figure 7.35 schematically illustrates the creep–fatigue fracture mechanism diagram, which is the binary section of a three-dimensional diagram. The third axis is the time axis perpendicular to the paper. The binary diagram shows the relationship between the total strain range $\Delta\varepsilon_t$ and the fatigue life N_f, when the tensile strain hold time is a constant. There are four curves in the diagram. *aed* and *abc* are the fatigue damage starting curve and the fatigue fracture curve, while *be* and *cdf* are the creep damage starting curve and the creep rupture curve.

Extensive experimentation indicates that the relationship between fatigue life and the total strain range for many materials can be described by the Coffin–Manson relationship:

$$N_f \cdot \Delta\varepsilon_t^{\alpha} = C, \tag{7.57}$$

where C and α are constants. Thus, the fatigue fracture curve can be determined in terms of Equation 7.57.

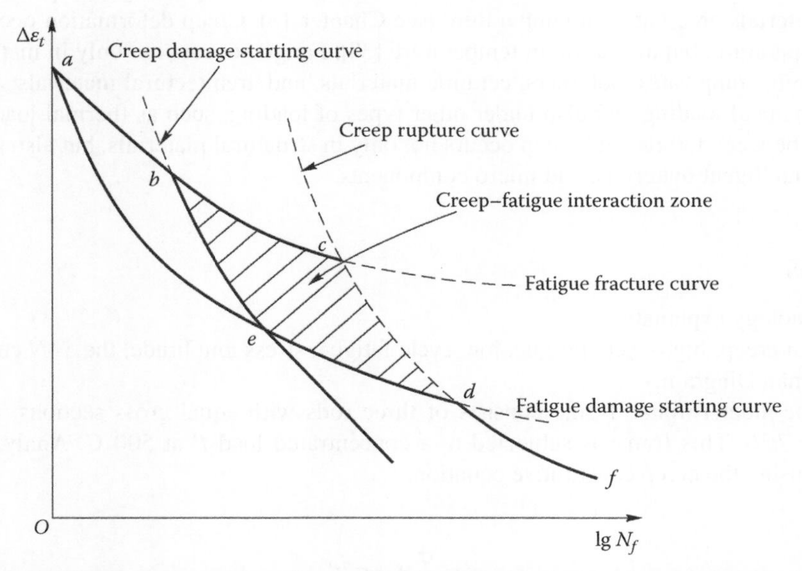

FIGURE 7.35 Diagram of creep–fatigue fracture mechanisms (schematic). (From Zhang, J., *Deformation and Fracture of Materials at High Temperature*, Science Press, Beijing, 2007; Taken from Figure 23.3.)

Previous discussion has pointed out that fatigue crack propagation consists of two stages. In the first stage, the crack propagates along the primary slip plane, and the crack length can be one to two grains. In the second stage, the crack propagates in the direction perpendicular to the stress axis until fracture, and is characterized by fatigue striations on the fracture surface; each striation corresponds to one cycle. The proportion of each stage in the total fatigue life depends on the total strain amplitude. When the strain amplitude is large, the second stage dominates; when the strain amplitude is small, the first stage dominates. Maiya proposed that the number of cycles to the beginning of the second stage can be established in terms of the number of striations on the fracture surface, and the total fatigue life; and that the starting of the second stage can be taken as the starting of fatigue damage. This method can be used to determine the *aed* curve. The creep rupture curve *cdf* can be established in terms of a few short-time creep data, and the relationship between rupture time and creep strain proposed by Hales. It is much more difficult to determine the creep damage starting curve. It is generally agreed that the nucleation of creep damage at the microscopic level (such as creep voids) occurs in the later part of creep life, and corresponds to the third stage of creep. But this is not always correct. For example, in 316 stainless steel, voids at grain boundaries nucleate at the early stage of creep, and are irrelevant to the third creep stage. Because different materials exhibit different behaviors, Plumbridge suggested taking 20% of rupture life as the creep damage starting curve, which is parallel to the creep rupture curve (the horizon axis denotes the log value of N_f) [37].

A diagram of the creep–fatigue fracture mechanism can be used to explain the experimental phenomenon, and to predict the fracture mode. The diagram shows that (1) when the strain amplitude is large, fatigue damage dominates, (2) fatigue fracture occurs before the starting of creep damage, and (3) fatigue life is determined by the *ab* curve. When the strain amplitude is very small, creep damage dominates, creep rupture occurs before the starting of fatigue damage, and fatigue life is determined by the *df* curve. When the strain amplitude goes between these two cases (the region filled with the slant lines), creep–fatigue interaction may occur. The smaller the strain amplitude, the larger the creep damage and the shorter the fatigue life. If the hold time increases, creep damage and rupture curves shift toward lower fatigue life. Damage caused by neutron irradiation also shifts the creep curves toward lower fatigue life.

In this chapter, we have only discussed creep behavior at high temperature, and fatigue failure of metallic materials. In fact, not only metallic materials creep at high temperature, but also ceramic materials creep at high temperature (see Chapter 13). Creep deformation occurs not only at high temperature, but also at room temperature [38]. Fatigue occurs not only in metallic materials, but also in composites, polymers, ceramic materials, and architectural materials; and not only under mechanical loading, but also under other types of loading, such as thermal loading [39,40]. Interaction between fatigue and creep occurs not only in structural materials, but also in functional materials, intelligent materials, and micro components.

EXERCISES

7.1 Terminology explanation:
Explain creep, high-cycle fatigue, low-cycle fatigue, stress amplitude, the *S–N* curve, and the Goodman Diagram.

7.2 A static-indeterminate frame consists of three rods with equal cross sections, as shown in Figure 7.36. This frame is subjected to a concentrated load *P* at 500°C. Analyze the creep stress using the creep constitutive equation.

$$\varepsilon = \frac{\sigma}{E} + A\sigma^n t^m$$

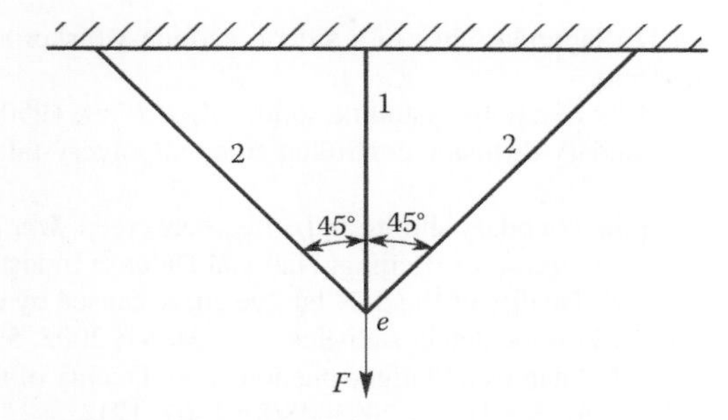

FIGURE 7.36 Schematic illustration for Exercise 7.2.

The material parameters are $E = 2^*10^5$ MPa, $A = 5^*10^{-13}$, $n = 2$, $m = 0.7$, and $F = 20$ kN. The cross-sectional area is $S = 1$ cm^2, and the rod length is $l_1 = 10$ cm.

7.3 In the previous problem, the angle between rods 1 and 2 is θ. Analyze the creep stress using the time-hardening theory $\dot{\varepsilon}_c = \sigma^n B(t)$.

7.4 Illustrate the essential difference between fatigue failure and static failure.

7.5 What are the empirical laws for fatigue behavior?

7.6 A 2024-T4 aluminum alloy is subjected to cyclic stress. The maximum stress is $\sigma_{max} = 430$ MPa, the minimum stress is $\sigma_{min} = 172$ MPa. Estimate its fatigue life.

7.7 Illustrate the meaning of the fatigue life–load curve in low-cycle fatigue.

7.8 Interpret the meaning of creep accumulation damage.

7.9 Discussion: Why does a metallic material at the microscale, creep at room temperature?

REFERENCES

1. Dowling N. E. *Mechanical Behavior of Materials* (*Engineering Methods for Deformation, Fracture and Fatigue*). Upper Saddle River, NJ: Prentice Hall, 1998.
2. Askeland D. R. and Phulé P. P. *Essentials of Materials Science and Engineering*. Beijing: Tsinghua University Press, 2005.
3. Nabarro F. R. N. Do we have an acceptable model of power-law creep? *Mater Sci Eng A*, 2004, 387–389(15): 659–664.
4. Zhang J. *Deformation and Fracture of Materials at High Temperature*. Beijing: Science Press, 2007.
5. Dieter G. E. *Mechanical Metallurgy*. 3rd ed. Beijing: Tsinghua University Press, 2006.
6. Mao Y., Hagiwara M., and Emura S. Creep behavior and tensile properties of Mo- and Fe-added orthorhombic Ti-22Al-11Nb-2Mo-1Fe alloy. *Scr Mater*, 2007, 57: 261–264.
7. Boehlert C. J. Microstructure, creep, and tensile behavior of a Ti–12Al–38Nb (at.%) beta + orthorhombic alloy. *Mater Sci Eng A*, 1999, 267(1): 82–98.
8. Poirier J. P. *Plastic Deformation at High Temperature of Crystals*. Translated by Guan Deling. Dalian, China: Dalian University of Technology Press, 1989, p. 70.
9. Mu X. *Creep Mechanics*. Shanxi, China: Xian Jiaotong University Press, 1990.
10. Somekawa H., Hirai K., and Watanabe H. Dislocation creep behavior in Mg-Al-Zn alloys. *Mater Sci Eng A*, 2005, 407: 53–61.
11. Weertman J. Creep of polycrystalline aluminium as determined from strain rate tests. *J Mech Phys Solid*, 1956, 4: 230–235.
12. Weertman J. Creep of indium, lead and some of their alloys with various metals. *Trans AIME*, 1960, 218: 207–210.
13. Garofalo F. An empirical relation defining the stress dependence of minimum creep rate in metals. *Trans Met Soc AIME*, 1963, 227(2): 351–355.
14. Robinson S. L. and Sherby O. D. Mechanical behavior of polycrystalline tungsten at elevated temperature. *Acta Met*, 1969, 17: 109–113.
15. Harper J. G. and Dorn J. E. Creep of aluminum under extremely small stresses. *Acta Met,* 1957, 5: 654–658.

16. Wu M. Y. and Sherby O. D. Superplasticity in a silicon carbide whisker reinforced aluminum alloy. *Acta Met*, 1984, 32: 1561–1566.
17. Herring C. Diffusional viscosity of a polycrystalline solid. *J Appl Phys*, 1950, 21: 437–440.
18. Coble R. L. A. model for boundary-diffusion controlled creep in polycrystalline materials. *J Appl Phys*, 1963, 34: 1679–1672.
19. Raj R. and Ashby M. F. On grain boundary sliding and diffusional creep. *Met Trans*, 1971, 2: 1113–1116.
20. Yao W. *Fatigue Life Analyses of Structures*. Beijing: National Defense Industry Press, 2004.
21. Chandran K. S. R. and Jha S. K. Duality of the S–N fatigue curve caused by competing failure modes in a titanium alloy and the role of Poisson defect statistics. *Acta Mater*, 2005, 53: 1867–1881.
22. Jha S. K. and Chandran K. S. R. An unusual fatigue phenomenon: Duality of the S–N fatigue curve in the β-titanium alloy Ti–10V–2Fe–3Al. *Scr Mater*, 2003, 48(8): 1207–1212.
23. Close F. E. and Isgur N. The origins of quark–hadron duality: How does the square of the sum become the sum of the squares? *Phys Lett B*, 2001, 509(1): 81–86.
24. Tokaji K. High cycle fatigue behaviour of Ti–6Al–4V alloy at elevated temperatures. *Scr Mater*, 2006, 54(12): 2143–2148.
25. Shen M. H. H. Reliability assessment of high cycle fatigue design of gas turbine blades using the probabilistic Goodman diagram. *Int J Fatigue*, 1999, 21(7): 699–708.
26. Bernasconi A., Foletti S., and Papadopoulos I. V. A study on combined torsion and axial load fatigue limit tests with stresses of different frequencies. *Int J Fatigue*, 2008, 30(8): 1430–1440.
27. Karolczuk A. and Macha E. Selection of the critical plane orientation in two-parameter multiaxial fatigue failure criterion under combined bending and torsion. *Eng Frac Mech*, 2008, 75(3): 389–403.
28. Sendeckyj G. P. Constant life diagrams— A historical review. *Int J Fatigue*, 2001, 23: 347–353.
29. Li H., Nishimura A., Nagasaka T., and Muroga T. Fatigue life and cyclic softening behavior of JLF-1 steel. *Fusion Eng Des*, 2007, 82(15): 2595–2600.
30. Gloanec A. L., Jouiad M., Bertheau D., Grange M., and Hénaff G. Low-cycle fatigue and deformation substructures in an engineering TiAl alloy. *Intermetallics*, 2007, 15(4): 520–531.
31. Kim J. B., Lee H. Y., Park C. G., and Lee J. H. Creep-fatigue test of a SA 316SS structure and comparative damage evaluations based upon elastic and inelastic approaches. *Int J Press Vessels Pip*, 2008, 85(8): 550–556.
32. Fournier B., Sauzay M., Caës C., Noblecourt M. et al. Creep–fatigue–oxidation interactions in a 9Cr-1Mo martensitic steel. Part III: lifetime prediction. *Int J Fatigue*, 2008, 30(10): 1797–1812.
33. Sih G. C. and Tang X. S. Micro/macro-crack growth due to creep–fatigue dependency on time–temperature material behavior. *Theor Appl Fract Mech*, 2008, 50(1): 9–22.
34. Fournier B., Sauzay M., Caës C., Noblecourt M. et al. Creep–fatigue–oxidation interactions in a 9Cr–1Mo martensitic steel. Part I: Effect of tensile holding period on fatigue lifetime. *Int J Fatigue*, 2008, 30(4): 649–662.
35. Fournier B., Sauzay M., Caës C., Noblecourt M. et al. Creep-fatigue-oxidation interactions in a 9Cr–1Mo martensitic steel. Part II: Effect of compressive holding period on fatigue lifetime. *Int J Fatigue*, 2008, 30(4): 663–676.
36. Kwofie S. and Chandler H. D. Fatigue life prediction under conditions where cyclic creep–fatigue interaction occurs. *Int J Fatigue*, 2007, 29(12): 2117–2124.
37. Plumbridge W. J. Metallography of high temperature fatigue. In: R. P. Skelton, Ed. *High Temperature Fatigue-Properties and Prediction*. London: Elsevier Applied Science, 1987, pp. 177–228.
38. Ma Z. S., Long S. G., Zhou Y. C., and Pan Y. Indentation scale dependence of tip-in creep behavior in Ni thin films. *Scr Mater*, 2008, 59: 195–198.
39. Zhou Y. C. and Hashida T. Thermal fatigue failure induced by delamination in thermal barrier ceramic coating. *Int J Fatigue*, 2002, 24(2–4): 407–417.
40. Zhou Y. C. and Hashida T. Thermal fatigue in thermal barrier coating. *JSME Int J*, 2002, A45(1): 57–64.

8 Mechanical Properties of Materials in Environmental Media

The previous chapters primarily discussed the mechanical behavior of materials under external mechanical loads. But, in fact, materials and components are always used in certain ambient media. Environmental media refers to the media that surrounds the materials (including gas, liquid, and solid media) and allows certain interface reactions with the materials. For example, bridges and steel rails are surrounded by a damp atmosphere; oil platforms and vessel hulls are surrounded by sea water; underground oil and gas pipelines, and cables, are surrounded by soil. Synergic action and mutual promotion between the environmental media and stress will cause faster material damage, crack initiation, and growth. Therefore, material damage under the combined action of environmental media and stress is more severe than under the independent action or simple summation of the two factors.

In short, materials suffer from mechanical property reduction, and even early cracking, under the combined action of environmental media and stress. This phenomenon is called environmentally induced cracking (EIC) or environmentally assisted cracking (EAC). The stress conditions of materials or structures vary. Synergic action between the media and stress under different conditions results in different forms of EAC. Based on stress conditions, EAC phenomena can be divided into stress corrosion cracking (SCC), corrosion fatigue cracking (CFC), corrosive wear (CW), and fretting corrosion (FC). Based on the fracture mechanism, EAC can be divided into stress corrosion due to anodic dissolution on the crack tip, hydrogen embrittlement (HE), or hydrogen induced cracking (HIC) due to cathodic hydrogen evolution. In addition, EAC also includes radiation damage (RD) due to radiation by high speed electron, neutron, or ion currents; and liquid or solid metal embrittlement (LME or SME) due to contact with liquid or solid metals at high temperature, and other factors.

With the rapid development of the aviation, aerospace, marine, atomic energy, petroleum, chemical, and other industries, requirements for the mechanical properties of the materials become more stringent, while the conditions of environmental media in contact with the components are harsher. Therefore, EAC attracts more attention from engineering designers and materials researchers. The present chapter primarily describes EAC for materials under the combined actions of stress and corrosive media; including SCC, HE, CFC and CW; and then briefly describes RD, and LME or SME.

8.1 STRESS CORROSION CRACKING

8.1.1 Stress Corrosion Cracking and Its Cracking Characteristics

Stress corrosion cracking refers to embrittlement cracking that occurs on a material under the combined action of specific corrosive media and static stress. SCC is not a simple summation of stress and corrosive media, but cracking that occurs according to a special mechanism under their combined action. Generally, a small stress is required when SCC occurs. If the material is not in a specific corrosive media, however, cracking will not occur under a small stress. But even if no stress exists, the material will still suffer from a small amount of corrosion in this environmental media. The reason why SCC is dangerous is that it often occurs in relatively mild media, and in a relatively low stress state where no obvious potential signs of cracking appear. As a result, SCC often leads to a catastrophic accident.

Generally, SCC has the following characteristics. It is static stress, usually in the form of tensile stress that results in stress corrosion fracture. It is much lower than the material's yield strength. The higher the tensile stress, the shorter the cracking time becomes. From recent research, we find that [1–4], under pressure stress, SCC also occurs in stainless steel, aluminum alloys, and copper alloys. However, it is longer than tensile stress-induced corrosion by 1–2 orders of magnitude with respect to the initiation time and is also different from tensile stress-induced correction in the fracture shape. It should be noted that the stress may be external stress or residual stress [5,6].

The fracture resulting from stress corrosion is embrittlement cracking, without obvious plastic deformation.

Stress corrosion in heavy metals or alloys only occurs in specific media. For example, corrosion fracture occurs on α brass in ammonia solution, while it occurs on β brass in water. Similarly, austenitic stainless steel has high SCC susceptibility (generally called chloride embrittlement) in chloride, while ferrite stainless steel is susceptible to chloride solution. For the combination of SCC susceptible materials and media, see Table 8.1 [7–10].

The crack growth of stress corrosion is slow, at a rate of 10^{-9} to 10^{-6} m/s. It is much larger than the corrosion rate when there is no stress, but much less than the rate of cracking, due simply to the mechanical factor.

SCC mostly initiates from the corrosion pit on the surface, while its growth path is usually perpendicular to the direction of the tensile stress.

When stress corrosion cracks grow, there is often branching, as shown in Figure 8.1.

Fracture surfaces are dull grey, and often have "mud cracks" (see Figure 8.2a) and corrosion pits (Figure 8.2b).

TABLE 8.1
Common Metal Materials and Their Susceptible Media

Alloy Materials	Environmental Media
Low-carbon steel	Hot nitrate solution, carbonate solution, peroxide
Carbon steel and low-alloy steel	Sodium hydroxide, iron trichloride solution, hydrocyanic acid, boiling magnesium chloride ($w(MgCl_2) = 42\%$) solution, sea water
High-strength steel	Distilled water, damp atmosphere, chloride solution, hydrogen sulfide
Austenitic stainless steel	Chloride solution, high-temperature high-pressure oxygenated ultrapure water, sea water, F^-, Br^-, $NaOH–H_2S$ solution, $NaCl–H_2O_2$ solution
Copper alloy	Ammonia vapor, mercuric salt solution, SO_2 contained atmosphere, diluted ammonia solution, iron trichloride, nitric acid solution
Nickel alloy	Sodium hydroxide solution, ultrapure water vapor
Aluminum alloy	Sodium chloride solution, sea water, water vapor, SO_2 contained atmosphere, molten sodium chloride, Br^-, I^- solution
Magnesium alloy	Nitric acid, sodium hydroxide, cyhalofop acid solution, distilled water, $NaCl–H_2O_2$ solution, $NaCl–K_2CrO_4$ solution, marine atmosphere, $SO_2–CO_2$ damp air
Titanium alloy	Cl^-, Br^-, I^- contained solution, N_2O_4, carbinol, trichloroethylene, organic acid

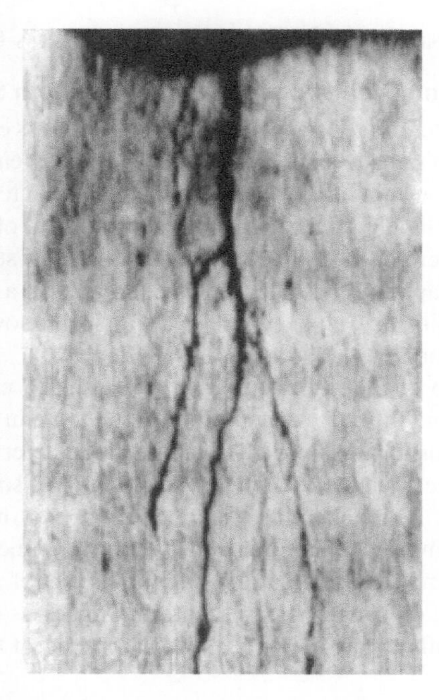

FIGURE 8.1 SCC branching. (From Shu, D.L., *Mechanical Properties of Engineering Materials*, China Machine Press, Beijing, 2003; Taken from Figure 6.2.)

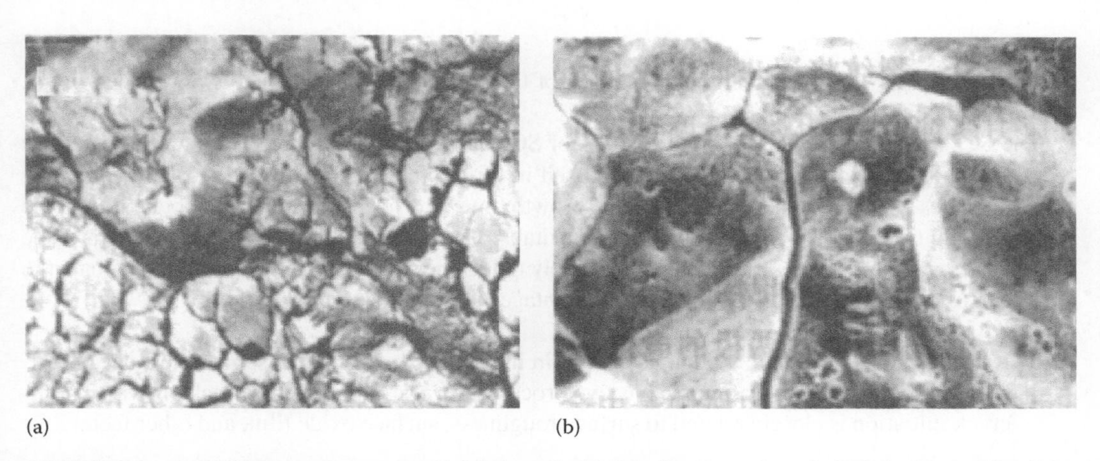

(a) (b)

FIGURE 8.2 SCC fracture microgram. (a) Mud crack (TEM). (b) Corrosion pit (SEM). (From Shu, D.L., *Mechanical Properties of Engineering Materials*, China Machine Press, Beijing, 2003; Taken from Figure 6.3.)

Stress corrosion cracks can be transgranular or intergranular, or even a mixture of the two. Transgranular cracking involves cleavage or quasi-cleavage, and V-shaped or feather markings can be seen.

As we have seen, stress, environmental, and material conditions must combine for SCC to occur. Sometimes, the necessary corrodant or alloy combination and strain can develop during service, even though the three conditions were initially not all present. For example, under normal conditions, boiler water does not cause SCC in carbon steel. However, repeated boiling and condensation in the course of boiler operation can increase the alkali concentration in boiler water and in structural cracks, ultimately leading to SCC [10].

8.1.2 Testing Methods and Evaluating Indicators of Stress Corrosion

8.1.2.1 Testing Methods and Evaluating Indicators of Smooth Samples

In early research on SCC, a material's resistance to stress corrosion was evaluated according to the duration of the cracking of a smooth sample under the combined action of tensile stress and chemical media.

The test is divided into a constant load test and a constant strain test. The simplest constant load test is to apply a tensile stress by hanging weights on the lower end of a corroded smooth flat plate, or a round bar tensile specimen; or by using a lever, a hydraulic system, or a spring. This method can use different specimen shapes, such as a ring shape, a U shape, a Y shape, a camber beam, and so on. Because of its practicality, as well as its use of a simple and low-cost sample and clamp, this method is mostly used in factory and laboratories.

During the test, one group of the same specimens are used to measure their cracking time t_f under different stress levels, and the σ–t_f curve is plotted, as shown in Figure 8.3. The cracking time t_f increases as the external tensile stress goes down. When the external stress is lower than a particular value, the cracking time t_f tends to be indefinitely long. This stress is called the critical stress without stress corrosion σ_{SCC}. If the cracking time t_f continues to increase slowly as the external stress goes down, the minimum stress resulting in SCC in a given time is called the conditional critical stress σ_{SCC}. This critical stress is used for studying the impact of alloy elements, organizational structure, and chemical media on a material's susceptibility to stress corrosion.

Occasionally, the media influencing factor β is used to represent the susceptibility of the stress corrosion:

$$\beta = \frac{\psi_{air} - \psi_{media}}{\psi_{air}} \times 100\%, \tag{8.1}$$

where ψ_{air} and ψ_{media} are the section shrinkages of the specimen when the test is performed in air, and in the media, respectively.

The SCC curve $\sigma - t_f$, and the critical stress of SCC σ_{SCC}, is a good representation of a material's SCC susceptibility. However, as the specimen used in this method is smooth, the measured cracking time t_f includes the crack initiation time and the growth time. The former accounts for about 90% of the total cracking time, but the actual component inevitably has cracks, or similar defects. Therefore, the critical stress indictor of SCC σ_{SCC} cannot objectively reflect the resistance of the cracked component to stress corrosion. In addition, the following disadvantages exist in this testing method:

1. It has dispersed test data, which may result in incorrect conclusions in some cases, because the fracture of the smooth sample has two processes: crack initiation and crack growth. The crack initiation is closely related to surface roughness, surface oxide film, and other factors.

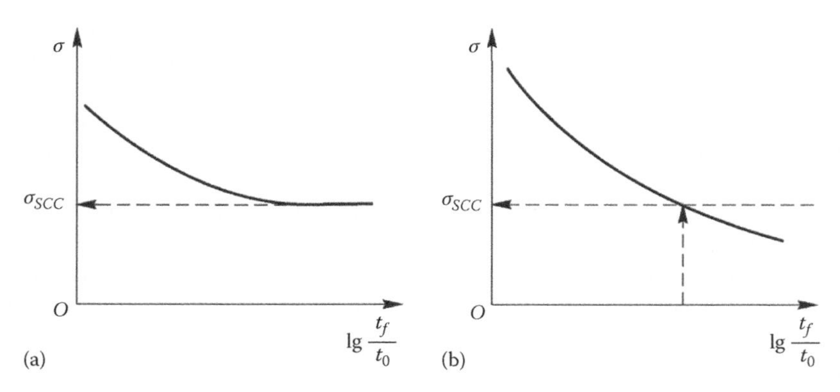

FIGURE 8.3 SCC curve of smooth specimen. (a) With stress limit. (b) Without stress limit.

The test data obtained are too dispersed and are sometimes even illusory. For example, the US Naval Research Laboratory has done stress corrosive property research on the high-strength titanium alloy Ti-8Al-1Mo-1V [11]. Because there was compact oxide film on the surface, when the stress corrosion test was performed on a smooth sample in 3.5% (by mass) NaCl solution, it was hard to generate a crack. The cracking time was so long, it was thought that this alloy was a new generation of materials for building submarine hulls. However, when testing was performed on a sample with a crack, cracking occurred in a short time. Therefore, this material is actually very susceptible to a simple 3.5% (by mass) NaCl solution.

2. The test is unable to correctly establish the change pattern of the crack growth rate. As this traditional method uses nominal stress as the driving force of crack growth, it cannot reflect the stress state on the crack tip. This problem cannot be solved unless fracture mechanics is used in research on SCC.
3. The test is time consuming and not applicable to engineering design.

8.1.2.2 Evaluating Indicator of Cracked Sample

8.1.2.2.1 SCC's Critical Stress Intensity Factor K_{ISCC}

The sample, with crack, is placed in a specific medium. Under constant load or constant displacement, determine the change of the stress intensity factor K_I with the cracking time t_f due to crack growth, based on which material's stress corrosion resistance is obtained. The cracking time t_f increases as the stress intensity factor K_I goes down. When the K_I value goes down to a specific value, SCC ceases. The K_I value at this time is called the critical stress intensity factor, or threshold value, of the stress corrosion, and is represented by K_{ISCC}. See Figure 8.4.

For the stress corrosion test of a Ti-8Al-1Mo-1V alloy in 3.5% (mass fraction) NaCl solution, $K_{IC} = 100\ \text{MPa}\cdot\text{m}^{1/2}$ applies. When the initial K_I value is 40 MPa · m$^{1/2}$, the sample becomes fractured in several minutes. However, if the K_I value is slightly lower, the fracture time is much longer. In this alloy/media setup, K_{ISCC} = 38 MPa · m$^{1/2}$ [11]. Therefore, it can be found through comparison that

For $K_I < K_{ISCC}$, under stress, the materials or parts can be used in the corrosive media safely without fracture for a long time.

For $K_{ISCC} < K_I < K_{IC}$, under the combined action of corrosive environment and stress, the crack shows subcritical growth. As the crack gradually grows, the crack tip's K_I value goes up continuously. When it is equal to K_{IC}, cracking occurs.

For $K_I > K_{IC}$, after the initial load is applied, cracking occurs immediately.

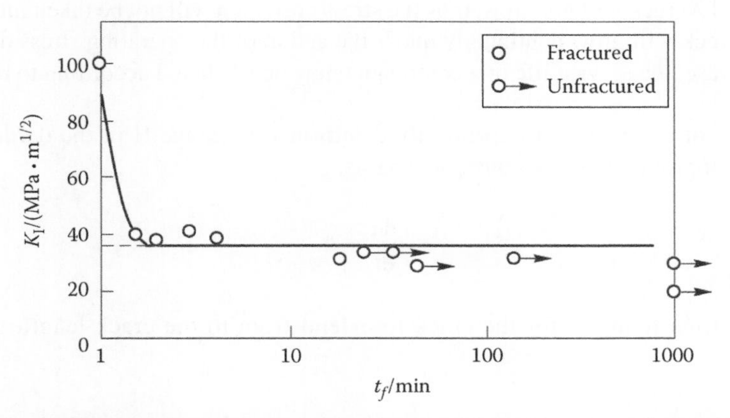

FIGURE 8.4 Relationship curve between fracture time t_f and K_I. (From Liu, R.T. et al., *Mechanical Properties of Engineering Materials,* Harbin Institute of technology Press, Harbin, China, 2001; Taken from Figure 8.1.)

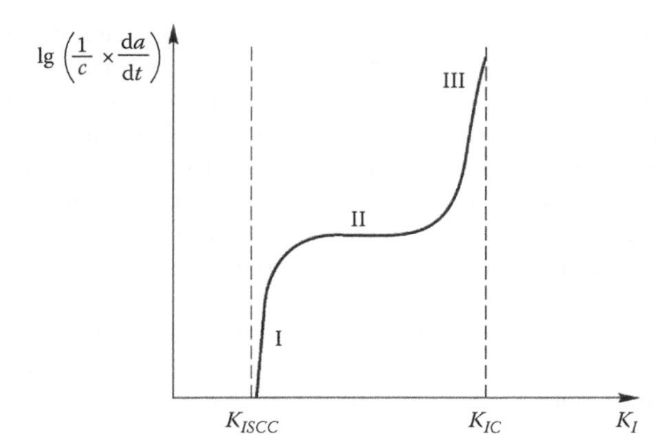

FIGURE 8.5 SCC's relationship curve.

For most metal materials, the K_{ISCC} value is constant in the specific chemical media. Therefore, K_{ISCC} can be used as the mechanical property indicator of metal materials. It represents the cracking toughness of materials with macrocracks in the stress corrosion condition.

8.1.2.2.2 SCC's Crack Growth Rate d*a*/d*t*

As seen earlier, when $K_I > K_{ISCC}$ applies on the top of the stress corrosion crack, the crack will keep growing. The crack growth in a unit time is called the crack growth rate, typically represented by d*a*/d*t*.

The test shows that d*a*/d*t* is related to K_I. On a $1g(\mathrm{d}a/\mathrm{d}t) - K_I$ coordinate diagram, its relationship curve is shown in Figure 8.5. The curve is divided into three stages: In Stage I, when K_I is just larger than K_{ISCC}, the crack accelerates the growth accidently after an initiation period, and the $\mathrm{d}a/\mathrm{d}t - K_I$ curve is nearly parallel to the vertical axis. In Stage II, the curve has a horizontal section and d*a*/d*t* is nearly unrelated to K_I; because, at this time, branching occurs on the crack tip as shown in Figure 8.1. The crack growth is mainly controlled by the electrochemical process, so it is closely related to the material and environment. In Stage III, the crack length is close to the critical size and d*a*/d*t* depends on K_I again, increasing sharply as K_I goes up. In this case, the material goes into the transition stage of unstable growth. When K_I is equal to K_{IC}, unstable growth occurs, finally leading to cracking.

8.1.2.2.3 Service Life of SCC Components

The safety and lifespan of the component can be assessed based on K_{ISCC} and the crack growth rate d*a*/d*t*. Since the component is safe when $K_I < K_{ISCC}$, the critical crack length K_{ISCC} may be calculated with. If $a < a_0$, the crack will not grow, thus the stress corrosion will not be taken into consideration. If $a > a_0$, the crack will grow continually under the action of the operating stress due to stress corrosion. In this case, the service life of a component may be predicted according to the crack growth rate.

The lifespan of a component is primarily determined by Stage II in the $\mathrm{d}a/\mathrm{d}t - K_I$ curve. In Stage II, d*a*/d*t* approximates to a constant. That is,

$$\frac{\mathrm{d}a}{\mathrm{d}t} = A. \tag{8.2}$$

The period of time required for the crack to extend from to the crack length a_2 at the end of Stage II is

$$t_f = \frac{a_2 - a_0}{A}, \tag{8.3}$$

where a_2 can be calculated with a stress intensity factor $K_{I2} = Y\sigma\sqrt{a_2}$ at the end of Stage II (see Equation 2.11). That is,

$$a_2 = \left(\frac{K_{I2}}{Y\sigma}\right)^2. \tag{8.4}$$

When substituting the calculated a_2 into Equation 8.3, the lifespan of the component can be calculated. Because Stage I and Stage III, with cracks growing, are not taken into account, the lifespan thus calculated is a conservative value.

8.1.2.3 Testing Method of Cracked Sample

8.1.2.3.1 Constant Load Method

The most common constant load method is a hanging beam bending device with constant load, as shown in Figure 8.6. The sample used is similar to the three-point bending sample with pre-cracking. One end of the sample is fixed on the frame, while the other end is connected to the arm. Weights are added on the end of the arm by weight. The sample goes into the solution tank, leaving the pre-crack immersed in the chemical media. In the entire test process, the load is a constant. Therefore, as the crack grows, the crack tip's stress intensity factor K_I increases. K_I can be calculated with the following equation :

$$K_I = \frac{4.12M(\alpha^{-3} - \alpha^3)^{1/2}}{BW^{3/2}}, \tag{8.5}$$

where

M is the bending moment on the crack interface, $M = F \times L$
B is the sample thickness
W is the sample width
$\alpha = 1 - a/W$
a is the crack length

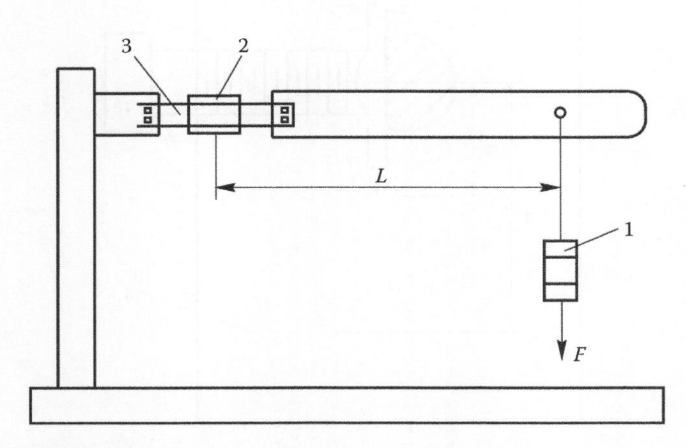

FIGURE 8.6 Hanging beam bending test device. (1) Weight, (2) solution tank, and (3) sample.

In the K_{ISCC} test, like K_{IC}, the sample size meets the plane strain requirement (see Chapters 2 and 5):

$$a_{\min} = B_{\min} = (W - a)_{\min} = \frac{W_{\min}}{2} = 2.5\left(\frac{K_{IC}}{\sigma_S}\right)^2, \tag{8.6}$$

where

$a_{\min}$ is the minimum crack length
$B_{\min}$ is the minimum sample thickness
$W_{\min}$ is the minimum sample width
σ_S is the material's yield strength

During the test, one constant load F is maintained until the sample becomes fractured. The cracking time t_f is recorded and Equation 8.5 is used to calculate the initial stress intensity factor K_I. The aforementioned test is repeated on one group of samples with the same size, under different constant loads, to obtain a series of t_f and corresponding K_I; based on which a $K_I - t_f$ curve is plotted as shown in Figure 8.4. K_I, corresponding to the infinite cracking time, is K_{ISCC}. In actual testing, a long deadline (typically 100–300 h) can be used as the basis for K_{ISCC}. In addition, the hanging beam bending test can also be used to obtain the da/dt – K_I curve.

8.1.2.3.2 Constant Displacement Method

A bolt is used to load the compact tensile specimen. As shown in Figure 8.7, one screw bar engages with the upper part of the sample, standing against the lower surface of the crack, so that there is one crack opening displacement corresponding to one initial load. The sample itself is loaded using this method. Actually, this is the compact tensile sample in Figure 5.2. When the crack grows, while the displacement is a constant, the load goes down. Therefore, the K_I value also goes down. When the K_I value is equal to K_{ISCC}, the crack stops growing.

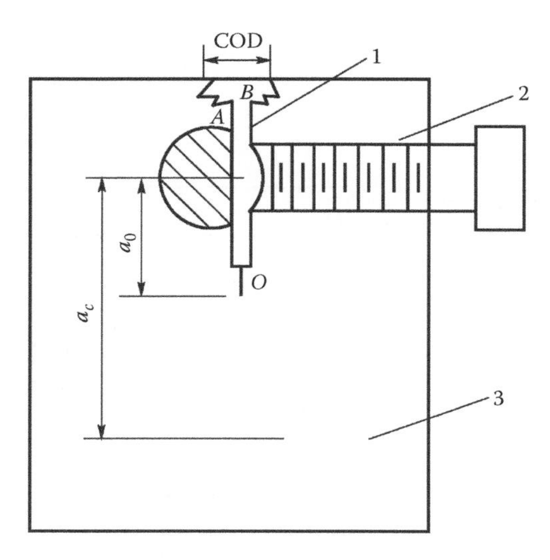

FIGURE 8.7 Constant displacement test method. (1) Pre-crack, (2) bolt, and (3) sample.

For the constant displacement specimen, the crack tip's stress intensity factor is

$$K_I = \frac{Ff(a/W)}{B\sqrt{W}}, \tag{8.7}$$

where

F is the applied load
a is the crack length, that is, the distance from the screw's centerline to the crack tip
B and W are respectively the sample's thickness and width
$f(a/W)$ is the shape factor function, also called the correction factor

Equation 5.8 is $f(a/W)$'s representation.

During the test, a screw is used to load to the required displacement δ; then, it is placed in the specific media. Then the crack length on the sample surface is measured regularly, obtaining the a–t curve, and $\mathrm{d}a/\mathrm{d}t$, by making a tangent line. Meanwhile, a is substituted into Equation 8.7 to obtain the corresponding K_I value. As a result, a $\mathrm{d}a/\mathrm{d}t - K_I$ curve can be plotted. After the crack completely stops growing, the sample is taken out. The crack length after the cracking stops (a_c) is measured precisely. The result is substituted into Equation 8.7 to obtain the critical stress intensity factor K_{ISCC}.

K_{ISCC} is an important mechanical property indicator. When K_I is less than the critical stress intensity factor, the crack does not grow. Therefore, it can be used for an infinite life design. Meanwhile, the K_{ISCC}/K_{IC} value is an indicator for evaluating the susceptibility of the material to stress corrosion. For high-strength materials, K_{ISCC}/K_{IC} is in the range of 1/4 to 1/6. Generally, when $K_{ISCC}/K_{IC} > 0.6$, the material is not susceptible to stress corrosion.

8.1.3 SCC Mechanisms

Although scientists have come up with many SCC mechanisms, until now there has been no mechanism that can satisfactorily explain all SCC phenomena. However, from the viewpoint of interaction between a material and the environmental media, the following models are now widely accepted.

8.1.3.1 Anode Rapid Dissolution Theory

Hoar and Hines [12] first came up with the anode rapid dissolution theory. This theory indicates that once the crack is initiated, stress concentration on the crack tip will result in quick yield occurring on its front edge. Dislocation inside the crystal continuously moves along the glide plane and reaches the front edge surface of the crack tip, generating a large quantity of instant active dissolved particles, which leads to rapid dissolution of the crack tip (anode). Figure 8.8 shows the crack tip, anodic dissolution, and the crack growth model. When the crack growth is $\mathrm{d}a$; the dissolved metal weight is $\rho A\mathrm{d}a$ (ρ is the density and A is the crack side's area); the number of dissolved ions is $\rho A\mathrm{d}aN_A/M$ (M is the amount of substance, N_A is Avogadro's constant); and the quantity of carried electricity is $\Delta Q = \mathrm{d}a\rho AN_AZe/M$ (Z is the ionic valency), or the anodic dissolution current density is $i_a = \mathrm{d}I/\mathrm{d}t$, $I = \Delta Q/A$ is the current intensity, then

$$i_a = \frac{ZF\rho}{M}\frac{\mathrm{d}a}{\mathrm{d}t}, \quad \frac{\mathrm{d}a}{\mathrm{d}t} = \frac{Mi_a}{ZF\rho}, \tag{8.8}$$

where F is the Faraday constant, which is the product of N_A multiplied by the unit charge e; that is, $F = N_Ae = 96{,}500$ C/mol.

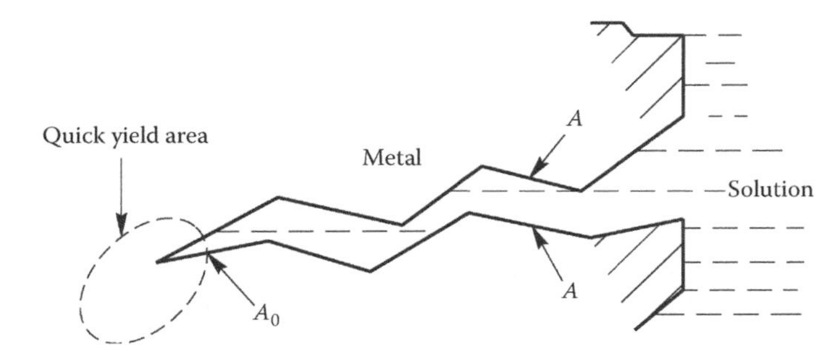

FIGURE 8.8 Model of crack growth caused by metal anode dissolution on the crack tip.

In a boiling $MgCl_2$ solution, when #18-8 stainless steel is not under tensile stress, the anodic dissolution current density is only 10^{-5} A/cm^2. However, in the stress corrosion condition, anodic current density on the crack tip is up to 0.4–2.0 A/cm^2, which means that the crack growth rate is 0.5–5 mm/h [10]. This result shows that the measured anodic current density complies with the rapid dissolution theory.

8.1.3.2 Occluded Cell Theory

The occluded cell theory has the following key points: (1) Under the combined action of stress and corrosive media, micropits or crack sources appear on the defective points of the metal surface. (2) The micropit and crack source have a channel so narrow that it is hard for the solution inside and outside the pore to convect and diffuse, forming a so-called occluded area. (3) Because the anode reaction and the cathode reaction co-exist, the metal atoms become ions and go into solution: Me → Me^{2+} + 2e. On the other hand, the electrons bond with oxygen in the solution to form hydroxyl radical ions: $1/2\,O_2 + H_2O + 2e \rightarrow 2OH^-$. However, in the occluded area, oxygen is consumed in a short time and cannot be replenished, and so anode reaction occurs eventually. (4) Hydrolysis of metal ions inside the crack generates H^+ ions, leading to a reduced PH value: $Me^{2+} + 2H_2O \rightarrow Me(OH)_2 + 2H^+$. As the metal ions and hydrogen ions inside the crack increase, Cl^- ions outside the crack move into the crack to maintain electric neutrality, forming corrosive hydrochloric acid, which accelerates corrosion inside the crack due to auto-catalysis. The freezing method is used to cool down the sample. After the solution inside the crack is clotted, the sample is broken to take the solution out of the crack and measure its pH value. The result shows that the solution inside the crack indeed becomes acidic. The occluded cell theory gives a good explanation for why corrosion does not occur on some corrosion resistant alloys (like stainless steel, aluminum alloys, and titanium alloys) in sea water, and the fact that the chloride can easily cause pits and stress corrosion on the metal.

8.1.3.3 Passive Membrane Theory

The passive membrane theory is also called the slip-dissolution theory. This theory indicates that, under the action of corrosive media, a passive membrane layer appears on the metal surface, which prevents the metal from further corrosion; that is, it is in a passive state. Under the action of stress or active ions (Cl^-), the passive membrane may easily rupture, exposing the active metal surface. The corrosive media immerses and dissolves the active metal along one path, finally leading to SCC. Figure 8.9 shows the mechanism of slip, resulting in membrane rupture. Figure 8.9a shows that when the membrane is not ruptured, and the stress is small, the oxide film remains undamaged and maintains good plasticity. Figure 8.9b shows that if the membrane is undamaged and has high bonding strength, even as the external stress goes up, dislocation only builds up on the glide plane, so the base metal is not exposed. When the stress reaches a certain level, the membrane becomes ruptured after dislocation begins. In addition, the membrane thickness h and the slip step height b are important. When $b \geq h$, new base metal may be exposed

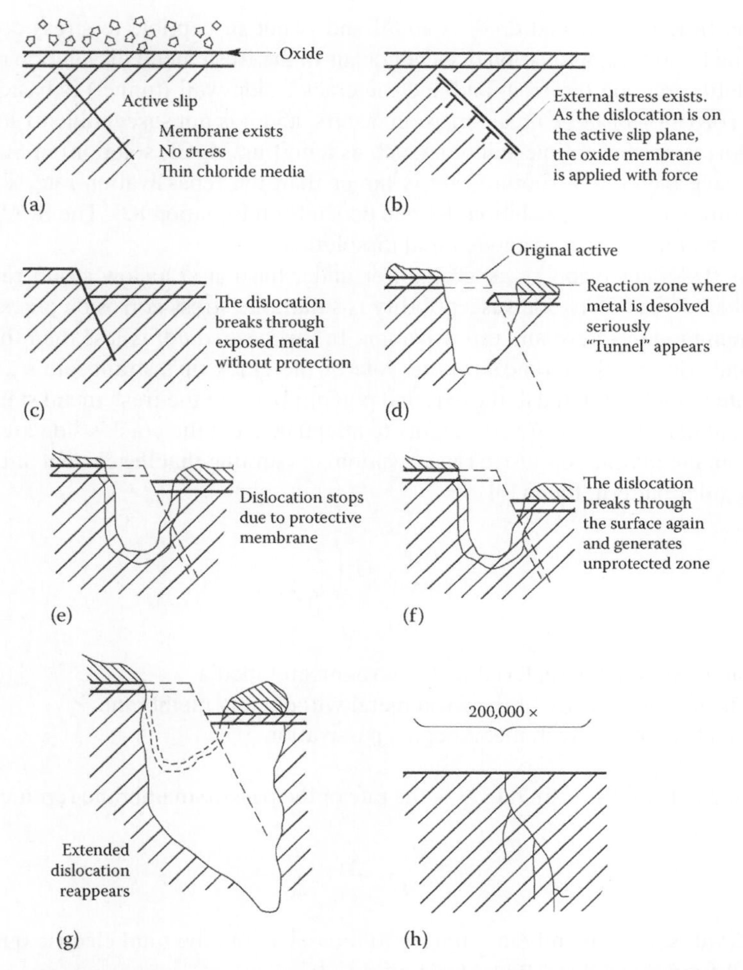

FIGURE 8.9 Metal SCC slip-dissolution mechanism. (a) The oxide film is not ruptured with small stress. As the external stress going up, (b) the dislocation may only build up on the glide plane without exposure of the base metal if the oxide film is unbroken and bonding tightly to the substrate. When the stress reaches a certain level, (c) the oxide film breaks and the fresh base metal may be exposed. If the base metal contacts the corrosive media, anode rapid dissolution occurs, (d) forming a "tunnel." Meanwhile, a passive membrane appears again on the metal surface resulting in a passive dissolved zone once again. Thus, (e) the dislocation builds up again. Under the action of stress or active ions, (f) the dislocation starts again, leading to surface passive membrane rupture. (g) Rapid dissolution occurs on the exposed metal. The membrane is ruptured, then repaired, and then ruptured again, repeatedly. (h) Finally, stress corrosion rupture occurs.

easily, as shown in Figure 8.9c. When the base metal contacts the corrosive media, anode rapid dissolution occurs, forming a "tunnel," as shown in Figure 8.9d. Meanwhile, passive membranes also appear on the metal surface on the tunnel in the corrosive media, so the dissolved zone goes into the passive state again. In this case, the dislocation stops moving, so that the dislocation builds up again, as shown in Figure 8.9e. Under the action of stress or active ions, dislocation starts again, leading to surface passive membrane rupture.

Rapid dissolution occurs on the exposed metal. The aforementioned procedure is repeated. The membrane is repaired repeatedly (repassivation), then is ruptured and dissolved repeatedly, finally leading to stress corrosion rupture, as shown in Figure 8.9f through h.

From the slip-dissolution mechanism, we know that stress corrosion is the result of the membrane rupture rate matching the re-membranization rate. If the repassivation rate is fast, the metal's local dissolution time is short and da/dt is small and is not susceptible to stress corrosion. If the repassivation rate is slow, repassivation does not occur in the fresh metal dissolution process. In this case, the dissolution rate on the front end and the crack's side wall (tunnel) is basically the same, so disc-shaped corrosion or large area corrosion occurs, and it is not susceptible to stress corrosion. For the stress corrosion susceptible solution (such as a boiling $MgCl_2$ solution) in Stage II of crack growth, the passive membrane rupture rate is larger than the repassivation rate, so the crack tip grows forward through metal dissolution. Its rate depends on Equation 8.8. The SCC mechanism at this time is also the mechanism of anode rapid dissolution.

In Stage I of stress corrosion crack growth, or under tension at a slow strain rate in a neutral environment where stress corrosion susceptibility is small, the stress corrosion crack is realized by matching the membrane rupture and repassivation. In this case, da/dt is less than that in Equation 8.8. da/dt depends on the membrane's rupture rate or the crack tip's strain rate $\dot{\varepsilon}_{ct}$, as well as the repassivation rate (which is related to the external potential). Once the fresh metal starts passivation, the anodic current density $i = \Delta Q/(A\Delta t)$ begins to attenuate (A is the crack's side area and Δt is the time interval from membrane rupture to repassivation). Assuming that the current attenuation meets the following regular pattern, then [13]

$$i = \frac{\Delta Q}{(A\Delta t)} = i_0\left(\frac{t}{t_0}\right)^{-n}, \tag{8.9}$$

where

n is a constant related to the material and environmental media
i_0 is the anodic current density of the fresh metal without any membrane
t_0 is the moment when the fresh metal begins passivation

Assuming that the crack tip's strain rate $\dot{\varepsilon}_{ct}$ is the rate of the passive membrane rupture strain ε_f, then

$$\dot{\varepsilon}_{ct} = \frac{\varepsilon_f}{\Delta t}, \quad \Delta t = \frac{\varepsilon_f}{\dot{\varepsilon}_{ct}}. \tag{8.10}$$

In the time interval Δt from membrane rupture to repassivation, the total electric quantity is equal to the integral of Equation 8.9 ($t = 0$ to $\Delta t = \varepsilon_f/\dot{\varepsilon}_{ct}$):

$$\Delta Q = Ai_0\int_0^{\Delta t}\left(\frac{t}{t_0}\right)^{-n} \mathrm{d}t = \frac{Ai_0t_0^n(\varepsilon_f/\dot{\varepsilon}_{ct})^{1-n}}{(1-n)}. \tag{8.11}$$

After substituting into Equation 8.8, we obtain

$$\frac{\mathrm{d}a}{\mathrm{d}t} = \frac{M}{ZF\rho}\frac{\Delta Q}{A\Delta t} = \frac{M}{ZF\rho}\frac{\Delta Q}{A(\varepsilon_f/\dot{\varepsilon}_{ct})} = \frac{i_0Mt_0^n\dot{\varepsilon}_{ct}^n}{Z\rho F(1-n)\varepsilon_f^n}. \tag{8.12}$$

Mao [14] thought that Equation 8.12 does not take the impact of potential into consideration, so he added it in the current's attenuation equation:

$$i = i_0\exp(-\beta t), \tag{8.13}$$

$$i_0 = 4 + 6E, \quad \beta = 0.03 - E,$$

where both i_0 and β are empirical formulas [14] (i_0 is in units of A/cm^2 and β is in unit of s^{-1}); E is V_{SCE}, V_{SCE} is the potential of the saturated calomel electrode (SCE). Assuming that the rate of the slip step formation on the crack tip is v, the membrane rupture time interval is $\Delta t = 1/v$ and the total integral current within Δt time is

$$i_a = \frac{i_0}{(1/v)} \int_0^{1/v} \exp(-\beta t)\mathrm{d}t = i_0 v \frac{\left[1-\exp(-\beta/v)\right]}{\beta}. \tag{8.14}$$

The rate of slip step formation v is related to $\dot{\varepsilon}_{ct}$. Assuming that the number of active slipbands in a unit length is N, the number of dislocations in each slipband is n, the height of step generated by each dislocation along the glide plane is b, and that the height along the loading direction is $b \cos \varphi$ (φ is the angle between the slip direction and the tensile axis), we obtain

$$\dot{\varepsilon}_{ct} = vN\, nb \cos \varphi. \tag{8.15}$$

After substituting this into Equation 8.8, we get

$$\frac{\mathrm{d}a}{\mathrm{d}t} = \frac{M}{ZF\rho} \frac{\dot{\varepsilon}_{ct}(4+6E)}{Nnb\cos\varphi(0.03-E)} \left\{1 - \frac{\exp\left[-Nnb\cos\varphi(0.03-E)\right]}{\dot{\varepsilon}_{ct}}\right\}, \tag{8.16}$$

where $\mathrm{d}a/\mathrm{d}t$ is in cm/s. Remember that the empirical formula $(4+6E)$ is in A/cm^2, while $(0.03-E)$ is in s^{-1}. After substituting the data of 304 steel in $MgCl_2$, we find that $\mathrm{d}a/\mathrm{d}t$ increases in a linear manner as $\dot{\varepsilon}_{ct}$ goes up (equivalent to Stage I). However, when $\dot{\varepsilon}_{ct} \geq 10^{-2}\ \mathrm{s}^{-1}$ applies, it tends to a constant value; that is, it goes into Stage II. Furthermore, when the external potential is less than the open-circuit potential, $\mathrm{d}a/\mathrm{d}t$ increases as the potential goes up. However, when it exceeds the open-circuit potential E_c, it tends to a stable value: the anode polarization does not have too much influence on $\mathrm{d}a/\mathrm{d}t$ [14].

8.1.4 SCC Countermeasures

Because stress corrosion is a result of synergistic action between the material, the environmental media, and stress, measures should be taken to prevent and reduce the material's tendency toward SCC. These measures are described in the following paragraphs.

1. Improve the material.
   a. Try to select materials on which no SCC occurs in the given environment or test and screen the currently available materials. For example, copper has high stress corrosion susceptibility to ammonia, so copper alloys should not be used in manufacturing components that have contact with ammonia. In high-concentration chloride media, we can use low-carbon high-chrome ferrite stainless steel that does not contain nickel or copper, or that only contain a little nickel or copper; or we can use nickel chromium stainless steel with a high silicon content; or nickel-based and ferrum-nickel-based corrosion-resistant alloys can be used.
   b. Develop alloys that resist stress corrosion or have high K_{ISCC}.
   c. Reduce impurities in the materials and improve the purity by using new metallurgical processes. Changing the structure by heat treatment, eliminating the segregation of harmful substances, and refining the grain can all reduce the material's stress corrosion susceptibility.

2. Eliminate or reduce the stress.
   a. Improve the structural design to avoid or reduce local stress concentration.
   b. Reduce and eliminate residual stress, which is a main reason for stress corrosion, mainly caused by design and processing of the metal component. It can be reduced by heat treatment annealing and shot blasting (see Chapter 6).
   c. Perform the structure design according to *Fracture Mechanics*. Because it is inevitable that components have macroscopic cracks and defects, designing using *Fracture Mechanics* is more reliable than using the traditional mechanics method. In a corrosive environment, the material's K_{ISCC} and da/dt are determined in advance. It is important to identify the component's permissible critical crack size a_c based on the conditions of application.
3. Improve the environmental media.
   a. Improve the material's application condition by reducing and eliminating harmful chemical ions that promote SCC. For example, reducing chloride ions in the cooling water and in vapor by water purification is effective in preventing chlorine embrittlement of austenitic stainless steel.
   b. Add an inhibiter to the media. Each material–environment system has some substances that can inhibit or retard stress corrosion. These substances can change the potential, accelerate membrane formation, avoid adsorbing harmful substances, and influence the electrochemical reaction kinetics that prevent or retard stress corrosion.
   c. Use a protective coating. A material's susceptibility to stress corrosion can be reduced by using an organic coating to isolate the material surface from the environment or by using metal that is not as susceptible to the environment as the membrane of the material.
   d. Adopt electrochemical protection. The stress corrosion phenomenon appears on the material only when it is in a certain electrode potential range in the media. Therefore, the external potential method can be used to bring the material's potential in the media far away from the susceptible potential zone. It is also a measure to prevent stress corrosion.

8.2 HYDROGEN EMBRITTLEMENT

Because the hydrogen atom is small, it can move easily in the lattice of various metals and alloys. Hydrogen penetration retards the performance of metals and alloys. Therefore, hydrogen in metals is a harmful element. Even a small amount of hydrogen, e.g., 0.0001% (by weight), can result in metal embrittlement (ME). Embrittled cracking of a metal material under the combined action of hydrogen and stress is called hydrogen-induced cracking or hydrogen embrittlement (HE) [6–10,15]. Stress that results in HE may be external stress or residual stress. Hydrogen in the metal may inherently exist inside the material or be adsorbed from the surface. The former means that hydrogen is adsorbed by the metal in the melting process, and in subsequent processing such as welding, pickling, or plating. The latter means that hydrogen is adsorbed by the metal component in service from the hydrogen-contained environmental media, such as solutions, damp air, or other environments that contain H_2 and H_2S.

8.2.1 Types of Hydrogen Embrittlement

Generally, hydrogen is solid-dissolved in the metal in an interstitial atomic state. For most industrial alloys, hydrogen's dissolubility decreases as the temperature goes down. Hydrogen in the metal can be aggregated by diffusion on large defects such as cavities, bubbles, and cracks, in the form of hydrogen molecules. In addition, hydrogen can also react with some transitional, rare earth, or alkaline earth metal elements to generate hydride, or react with the second phase in the metal to generate gas products. For example, hydrogen in steel can react with carbon atoms in cementite to generate methane. Because hydrogen exists in the metal in different forms, and reacts with the metal in different ways, there are different types of HE. The main types include

1. Hydrogen attack
   As the hydrogen reacts with the second phase (e.g., carbon) in the metal to generate high-pressure gas CH_4, the base metal's bonding force on the grain boundary decreases, leading to ME. CH_4 bubbles must be formed by adsorbing inclusions or second phase particles in the metal. These second-phase particles usually exist on the grain boundary. Therefore, a hydrogen attack embrittlement crack usually grows along the grain boundary, forming a grain-shaped fracture. It takes a certain amount time to form CH_4 and accumulate to a certain level. Therefore, the hydrogen attack process has an incubation period. The higher the temperature is, the shorter the incubation period becomes. The temperature when hydrogen attack occurs on steel is generally 300°C–500°C. When it is lower than 200°C, hydrogen attack will not occur.
2. White spot
   Sometimes, after pickling on the steel section, we can find long hair-like cracks, with a width of approximate 1 μm. Hence, the white spot is also called "cracking." If the steel is broken, a white elliptical spot can be seen on the fracture, that is, a so-called white spot. It is generally thought that the white spot is caused by excessive hydrogen in the steel. The hydrogen in the forge piece (sosoloid), during fast cooling after forging, is over-saturated due to reduced dissolubility and separated out from the sosoloid. If the separated hydrogen atoms do not escape, they are aggregated at the defect and become hydrogen molecules in the steel. In this case, the hydrogen's volume expands sharply and the local pressure goes up gradually, leading to local tearing and microcracking.
3. Hydride embrittlement
   Because the IVB or VB group of metals (such as titanium, α-iron alloy, nickel, vanadium, aluminum, and their alloys) have high affinity for hydrogen, they can easily generate hydride embrittlement, leading to ME. The HE susceptibility of metal materials to hydride increases as the temperature goes down and the notch on the component becomes sharp. The crack often grows along the interface of hydride and base. Therefore, hydride can be seen on the fracture. The hydride's shape and distribution have obvious influences on ME. If the grain is thick, the hydride shows a slice shape on the grain boundary, easily generating great stress concentration and danger. If the grain is fine, the hydride shows block discontinuous distribution, which is less dangerous to the metal.
4. Hydrogen-induced delayed cracking
   High-strength steel or $\alpha + \beta$ titanium alloys contain the proper quantity of hydrogen in a solid-dissolved state (existing inherently or adsorbed from environmental media). Under the continuous action of stress lower than the yield strength, after a certain incubation period, cracking is initiated inside the metal, particularly in the three-directional tensile stress zone. The crack grows gradually and finally results in embrittlement cracking. This delayed cracking phenomenon, due to the action of hydrogen, is called hydrogen-induced delayed cracking. In the case of HE in engineering, most refer to hydrogen-induced delayed cracking. This HE has the following characteristics: (1) It occurs only within a certain temperature range. For example, high-strength steel is most susceptible at room temperature. (2) It improves the strain rate. The susceptibility of the material to HE becomes lower. (3) It obviously reduces elongation of the metal material. However, when the hydrogen content exceeds a certain value, the elongation does not change anymore. The section shrinkage decreases as the hydrogen content goes up. The higher is the material strength, the more sharply it goes down. (4) Its crack path is related to the stress level. Testing on 40 CrNiMo steel shows that [10], when the stress intensity factor K is high, cracking is transgranular; when K is intermediate, cracking is mixed cracking (quasi-cleavage and micropore); when K is low, cracking is intergranular. In addition, the cracking type is also related to the impurities content. When the content of impurities is high, more impurities are aggregated on the grain boundary, so more hydrogen can be adsorbed, leading to intergranular cracking. Improving purity can cause intergranular cracking to transition to transgranular cracking.

8.2.2 HIDC's Resistance Indicator and Testing Method

8.2.2.1 Threshold Stress σ_{HC}

From the first and second laws of thermodynamics, we know that, at constant temperature, a substance's free enthalpy can be represented as

$$\Phi = \Phi^0 + nRT\ln\left(\frac{p}{p_0}\right), \tag{8.17}$$

where Φ^0 is the standard free enthalpy variable in J. For a gas, its standard state is pure gas under one atmospheric pressure. For a solution, its standard state is pure substance or 1% of solution. p is the gas pressure at temperature T, p_0 is the standard atmospheric pressure, n is the substance's mole number, $R = 8.314$ J/(mol · K) is the gas constant, and T is the Kelvin temperature.

For any reaction $aA_1 + bA_2 \rightarrow cA_3 + dA_4$ (a, b, c, d are mole numbers), the free enthalpy is changed to

$$\Delta\Phi = (c\Phi_3 + d\Phi_4) - (a\Phi_1 + b\Phi_2) = \Delta\Phi^0 + RT\ \ln\frac{(p_3)^c\ (p_4)^d}{(p_1)^a\ (p_2)^b}. \tag{8.18}$$

It can be proven that [15] the relationship between the free enthalpy of 1 mol of pure substance and its concentration C is $\mu = \mu_0 + RT\ln C$ (the reader should deduce this using thermodynamic theory). Hence, the pressure p_i in the aforementioned equation may also be the concentration C_i. For example, when A_1 is gas, it is pressure p_1, otherwise liquid is concentration C_1.

H_2 is broken down into H and goes into the metal. In the balanced state ($1/2\,H_2 \rightleftharpoons H$), $\Delta\Phi = 0$ applies. From Equation 8.18, we know that

$$\Delta\Phi^0 = -RT\ln\left(\frac{C_H}{\sqrt{p}}\right) \Rightarrow C_H = e^{-\Delta\Phi^0/RT}\sqrt{p} = S\sqrt{p}, \tag{8.19}$$

where p is the hydrogen pressure and $S = e^{-\Delta\Phi^0/RT}$. Therefore, the dissolubility of hydrogen in the metal C_H is directly proportional to the square root of the hydrogen pressure. This is Sievert's Law.

If the concentration gradient $\partial C/\partial x$ exists in the crystal, the hydrogen will diffuse and move from the areas of high concentration to the areas of low concentration. The diffusion process is described by Fick's Law:

$$J = \frac{i}{F} = -D\frac{\partial C}{\partial x},\quad \frac{\partial C}{\partial t} = D\frac{\partial^2 C}{\partial^2 x}, \tag{8.20}$$

where

J is the diffusion flux (mol/(cm^2 · s))
i is the current density
F is the Faraday constant
D is the diffusion factor (cm^2/s), which can be measured in the test

Hydrogen generates a strain field in the interstitial sites, represented by the three main strains $\varepsilon_i (i = 1, 2, 3)$. It can interact with the external stress, represented by the three main stresses $\sigma_i (i = 1, 2, 3)$. Its interaction energy U is equal to the difference between the chemical potential μ (free enthalpy of 1 mol of pure substance) when there is stress, and when there is no stress, $\mu_\sigma - \mu_0$. Based on thermodynamic theory and Equation 1.69, it is easily demonstrated that (the reader is encouraged to derive this)

$$U = \mu_\sigma - \mu_0 = -V(\varepsilon_1 \sigma_1 + \varepsilon_2 \sigma_2 + \varepsilon_3 \sigma_3), \tag{8.21}$$

where V is the metal molar volume, $V = a^3 N$. If the strain is in spherical symmetry, that is, if $\varepsilon_1 = \varepsilon_2 = \varepsilon_3 = (1/3)(\Delta V/V)(1/\Delta n)$ (where Δn is hydrogen's mole number), then the aforementioned equation is changed to $U = \mu_\sigma - \mu_0 = -V\varepsilon_1(\sigma_1 + \sigma_2 + \sigma_3) = -3V\varepsilon_1\sigma_H$, where σ_H is the averaged stress. From the partial molar volume of hydrogen in the metal (change of the metal volume due to one mole of hydrogen added at constant temperature and pressure),

$$\overline{V}_H = \frac{\partial V}{\partial n} = \frac{\Delta V}{V}\frac{V}{\Delta n} = 3\varepsilon_1 V,$$

and we obtain

$$U = \mu_\sigma - \mu_0 = -\overline{V}_H \sigma_H. \tag{8.22}$$

From $\mu = \mu_0 + RT \ln C$ and Equation 8.20, we obtain $J = -\frac{DC}{RT}\frac{\partial \mu}{\partial x}$, which means that a chemical potential gradient can also cause hydrogen diffusion. Furthermore, the diffusion flux as a result of the stress gradient is $J_\sigma = -\frac{DC}{RT}\frac{\partial \mu}{\partial x} = \frac{DC\overline{V}_H}{RT}\frac{\partial \sigma_h}{\partial x}$. When $J_C = -D\frac{\partial C}{\partial x}$ as a result of the concentration gradient, Fick's first law is changed to

$$J = J_C + J_\sigma = -D\frac{\partial C}{\partial x} + \frac{DC\overline{V}_H}{RT}\frac{\partial \sigma_H}{\partial x}. \tag{8.23}$$

When diffusion reaches the stable state, $J = 0$ applies, so we obtain [15]

$$C_\sigma = C_0 \exp\left(\frac{\sigma_H \overline{V}_H}{RT}\right). \tag{8.24}$$

If the hydrogen's strain field is not spherically symmetric ($\varepsilon_1 \neq \varepsilon_2 = \varepsilon_3$), hydrogen concentration in the balanced state can be calculated by using the interaction energy in Equation 8.21:

$$C_\sigma = C_0 \exp\left(\frac{\sigma_H \overline{V}_H}{RT}\right), \tag{8.25}$$

where ε_i is the hydrogen's mole strain field.

When constant stress is applied, hydrogen is aggregated at maximum stress via stress-induced diffusion. Although the external unidirectional stress is σ, local stress concentration occurs on the inclusion or second phase's tip. Hence,

$$\sigma_H = \alpha\sigma, \tag{8.26}$$

where α is the stress concentration factor related to the inclusion shape. As the external stress σ goes up, $C_\sigma = C_0 \exp(\alpha \overline{V}_H \sigma / RT)$ increases. When C_σ is equal to the critical value C_{th}, HIC initiates and grows, finally resulting in delayed cracking. The corresponding external stress is the HIDC's threshold value σ_{Hc}, or the minimum external stress when HIDC occurs. In this case, from Equation 8.24 we obtain

$$\sigma_{Hc} = \frac{RT}{\alpha \overline{V}_H}(\ln C_{th} - \ln C_0). \tag{8.27}$$

When $\sigma < \sigma_{Hc}$, even after an infinite time (stress-induced hydrogen diffusion has reached the balanced state), C_σ does not equal the critical value C_{th}, so HIC will not occur.

8.2.2.2 Threshold Stress Intensity Factor K_{IHC}

Stress concentration exists on the crack tip. From Equation 2.13, for the stress field of the crack tip on the crack extension line ($\theta = 0$) in the plane strain condition, the averaged stress is σ_H: (the reader is encouraged to derive this)

$$\sigma_H = \frac{2(1+\nu)K_I}{3\sqrt{\pi r}}. \tag{8.28}$$

As the external stress intensity factor K_I goes up, the concentration of hydrogen aggregated through stress-induced diffusion C_σ increases continuously. When $K_I = K_{IHC}$, after a sufficient time, the maximum hydrogen concentration $C_\sigma = C_{th}$, and HIC initiates and then grows. We obtain

$$K_{IHC} = \frac{3RT\sqrt{\pi r}}{2(1+\nu)\overline{V}_H}(\ln C_{th} - \ln C_0). \tag{8.29}$$

K_{IHC} is the HIDC's threshold stress intensity factor.

8.2.2.3 Crack Growth Rate d*a*/d*t*

Because K_{IHC} is much less than the cracking toughness K_{IC}, when $K_{IH} < K_I < K_{IC}$, after an incubation period, the hydrogen aggregated through stress-induced diffusion is equal to the critical concentration C_{th}, so HIC occurs. Based on test results, we can obtain the *a*–*t* curve, and calculate the HIC growth rate d*a*/d*t*. Similar to the relationship curve of the crack's instant stress intensity factor K_I, and the d*a*/d*t*–K_I curve during stress corrosion, it also has three stages. Stage I occurs when K_I is a little larger than K_{IHC}, and after an incubation period, the crack grows and d*a*/d*t* goes up sharply. When the crack goes into the stable growth stage, it is called Stage II, during which d*a*/d*t* is unrelated to K_I. When K_I is close to K_{IC}, the crack growth speeds up, and is called Stage III.

The test method of studying HE is basically the same as for stress corrosion. By applying a certain stress on a smooth sample or a notched sample, and charging the hydrogen dynamically with the same current density and then measuring the HIDC time, we can establish a K_I–t_f curve similar to Figure 8.4, and identify the HIDC's critical stress σ_{HC}. For a premade cracked sample,

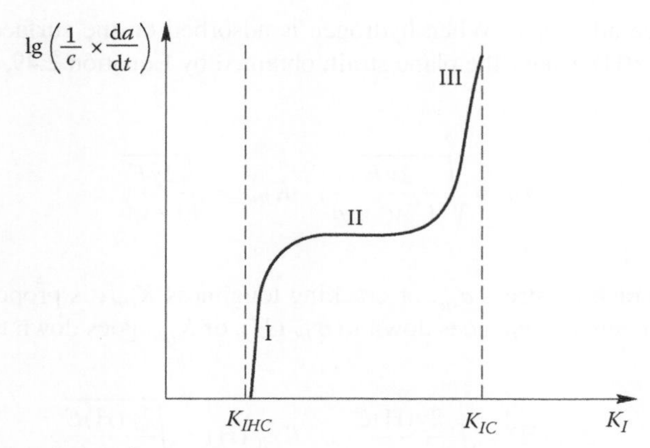

FIGURE 8.10 HIDC's $\lg\left(\frac{1}{c}\times\frac{da}{dt}\right)$ relationship curve.

in the electrolytic cathode hydrogen charging or gas hydrogen charging condition, the relationship curve between the crack growth rate da/dt and the stress intensity factor K_I can be measured, as shown in Figure 8.10.

In addition, the material's susceptibility to HE can also be represented by the section shrinkage change in a tensile test before and after charging hydrogen in the smooth sample:

$$I = \frac{\psi_0 - \psi_H}{\psi_0} \times 100\%, \tag{8.30}$$

where ψ_H and ψ_0 are, respectively, the section shrinkage of the contained hydrogen and the free hydrogen samples. In addition, there is also the idea of reflecting the material's susceptibility to HE using the sample's cracking specific work change (i.e., the area under the true stress–strain curve) before and after charging hydrogen.

8.2.3 Hydrogen Embrittlement Mechanism

8.2.3.1 Hydrogen Pressure Theory

If the metal contains supersaturated hydrogen, hydrogen molecules can appear at various defects, resulting in huge pressure. This is the so-called hydrogen blistering. When supersaturated hydrogen inside the metal diffuse to hydrogen blistering, the pressure inside the hydrogen blister goes up, and the size of the hydrogen blister becomes larger. When hydrogen pressure inside the hydrogen blister exceeds a critical value, embrittlement cracking occurs. When the hydrogen blister size becomes larger, the pressure inside the hydrogen blister goes down, so the supersaturated hydrogen diffuses to this hydrogen blister, fully supplementing its pressure change and maintaining the metal's HE condition. Meanwhile, hydrogen atoms diffused inside the hydrogen blister, under high pressure, soon become hydrogen molecules. This eliminates the possibility of high-pressure hydrogen back-diffusion inside the hydrogen blister.

This theory is often used to explain the HE mechanism inside a metal material. It is a persuasive way to explain HIC and white spotting.

8.2.3.2 Theory of Surface Energy Decrease after Hydrogen Adsorption

The changing value of the surface free energy in a unit area is called the surface energy; that is, surface tension. Surface energy is related not only to the crystal type and crystal orientation, but also to the surface adsorbate. When hydrogen is adsorbed on the surface, the surface energy goes down from γ to γ(H). Under the plane strain obtained by Equation 2.49, a Mode I crack in an infinite body is

$$\sigma_{Hc} = \sqrt{\frac{2\gamma E}{(1-v^2)\pi a}}, \quad K_{IHC} = \sqrt{\frac{2\gamma E}{1-v^2}}. \tag{8.31}$$

We can see that the cracking stress σ_{Hc}, or cracking toughness K_{IHC}, is proportional to $\sqrt{\gamma}$. When hydrogen causes γ to decrease, σ_{Hc} goes down to $\sigma_{HC}(H)$, or K_{IHC} goes down to $K_{IHC}(\mathrm{H})$:

$$\sigma_{HC}(\mathrm{H}) = \sqrt{\frac{2\gamma(\mathrm{H})E}{(1-v^2)\pi a}}, \quad K_{IHC}(\mathrm{H}) = \sqrt{\frac{2\gamma(\mathrm{H})E}{1-v^2}}. \tag{8.32}$$

The theory of surface energy decrease by hydrogen adsorption states that a metal's HE is the phenomenon where cracking of the metal's internal or external surface is due to the decrease of surface energy after the adsorption of active substance (hydrogen), causing the crack to grow and the material's mechanical property to drop.

8.2.3.3 Weak Bond Theory

The weak bond theory says that when the local stress concentration σ_y is equal to the atomic bonding force σ_{th}, the atomic bond is broken, nucleating the microcrack. Solid-dissolved hydrogen atoms make the atomic bonding force decrease from σ_{th} to $\sigma_{th}(\mathrm{H})$. As a result, the critical external stress required for local stress concentration, decreases from σ_{Hc} to $\sigma_{HC}(\mathrm{H})$; or the critical stress intensity factor K_{IHC} decreases to $K_{IHC}(\mathrm{H})$.

If the local stress concentration is σ_y and the number of atoms in a unit area is n, the concentration force is σ_y/n. When it is equal to the atomic bonding force σ_{th}, the crack nucleation basis is

$$\sigma_y \geq n\sigma_{th}. \tag{8.33}$$

Oriani [15] assumed that $\sigma_{th}(\mathrm{H}) = \sigma_{th} - \alpha C_H$ (where α is the comparison factor and C_H is the hydrogen concentration) and thought that $\sigma_y = K_I/k\sqrt{\rho}$ (where k is the value factor). When Equation 8.33 is met, the external stress intensity factor is equal to the cracking toughness K_{IC}:

$$\frac{K_{IHC}}{k\sqrt{\rho}} = \frac{K_{IC}}{k\sqrt{\rho}} - \alpha n C_H, \tag{8.34}$$

where C_H is the hydrogen concentration after stress-induced diffusion.

For a metal material (before the microcrack nucleates in any condition) when the stress in the crack tip zone is equal to the yield stress σ_s, local plastic deformation occurs. Hence, the stress near the crack tip is equal to σ_s, but $n\sigma_{th}$ is much larger than $5\sigma_s$ [15]. Obviously, Equation 8.33 is not established. However, when the external stress is high, it is possible that stress concentration on the crack tip or in the dislocation-free zone is equal to the bonding force, so the crack nucleation criterion is possibly established. Another stress concentration comes from the dislocation group. When the dislocation pile-up group is large enough, local stress on the front of the pile-up group is equal to σ_{th}. In other words, only when local plastic deformation occurs first, it is possible that the local stress concentration is equal to the atomic bonding force reduced by the hydrogen. However, the assumption that the bonding force goes down linearly as the hydrogen concentration increases is not experimentally proven.

The material's modulus of elasticity is a performance parameter that represents the force among atoms, so we can reflect the variation of the atomic bonding force based on its variation. If this point can be demonstrated through testing, the weak bond theory can be indirectly proven to be correct.

8.2.3.4 Dislocation Theory

Dislocation theory states that HE is a result of the interaction of hydrogen atoms and dislocations. Hydrogen atoms in the sosoloid are aggregated under blade dislocation, forming a Conttrell air mass. At a certain strain rate, the dislocations move together with the Conttrell air mass. When they meet defects such as grain boundaries and second phase particles, the interaction between them and the hydrogen atoms causes the dislocation to move with difficulty, so hydrogen atoms are aggregated at these locations. This phenomenon is also called the Conttrell air mass's pinning effect on the dislocation. When the cracked body is applied with force, a plastic deformation zone appears on the crack tip, resulting in a local high dislocation density that aggregates more hydrogen atoms. Hydrogen atoms are aggregated toward the crack tip, leading to a reduced atomic bonding force and local material embrittlement.

In this theory, the HE crack growth has a jumping feature. Figure 8.11 shows the HE crack growth process of high strength steel. When hydrogen atoms in the steel become aggregated on the crack tip, the crack does not grow. Only when the hydrogen concentration in this local zone is equal to a critical value does cracking occur at the junction of the plastic zone before the crack and the elastic zone. When the new small crack is connected to the old crack, the old crack grows for a certain distance and then stops. Afterward, it develops and grows again. Finally, when the crack is equal to the critical crack size through subcritical growth, unstable crack growth occurs. The resistance variation in this growth method and process is shown in Figure 8.11.

This theory explains why HE is related to temperature and strain rate. When the deformation rate is constant and the temperature is low, hydrogen atoms diffuse slowly and cannot catch up with the dislocation movement, so they cannot form an air mass. Therefore, few hydrogen atoms are aggregated on the crack tip, and hydrogen embrittlement will not occur. When the temperature is high, the hydrogen atoms diffuse quickly. However, although they can form an air mass, the thermal effect causes aggregated hydrogen atoms to leave the air mass, and diffuse all around.

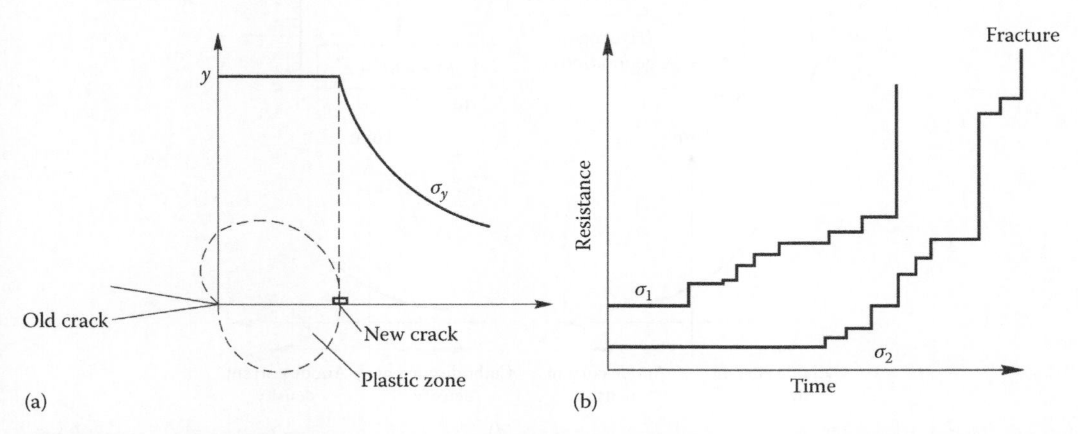

FIGURE 8.11 HIC's growth process and method. (a) Crack growth process. (b) Resistance variation in the crack growth process.

Therefore, the concentration of hydrogen atoms around the dislocation begins to drop. The material's plasticity goes up, while its embrittlement goes down. When the temperature continues to rise, the hydrogen mass is damaged thoroughly by thermal diffusion and HE is thoroughly eliminated. For steel in the general deformation condition, the temperature range to which it is most susceptible to HE is near room temperature.

8.2.4 Relationship between Hydrogen Embrittlement and Stress Corrosion

Both stress corrosion and HE are delayed cracking phenomenon due to the combined action of static stress and chemical media. They have a closed relationship. Figure 8.12 is the electrochemical schematic of stress corrosion and HE of steel in specific chemical media. From the schematic we discover that, when stress corrosion occurs, there is always hydrogen separation. The separated hydrogen can easily lead to HE. The difference is that stress corrosion is an anodic dissolution process (Figure 8.12a), forming a so-called anode active channel that results in metal cracking, while HE is a cathode hydrogen adsorption process (Figure 8.12b). When discussing the delayed cracking of one specific alloy-chemical media system, we generally use the polarization test method; that is, we use the influence of the external current on the crack time or the crack growth rate under a static load. When the crack time is shortened by applying a small anodic current, it is due to stress corrosion (Figure 8.12c); when the crack time is shortened by applying a small cathodic current, it is hydrogen-induced delayed cracking (HIDC) (Figure 8.12d).

For a particular cracked component, we can also differentiate it from the fracture appearance. Table 8.2 [9] is the comparison between stress corrosion and HE of steel in the fracture appearance.

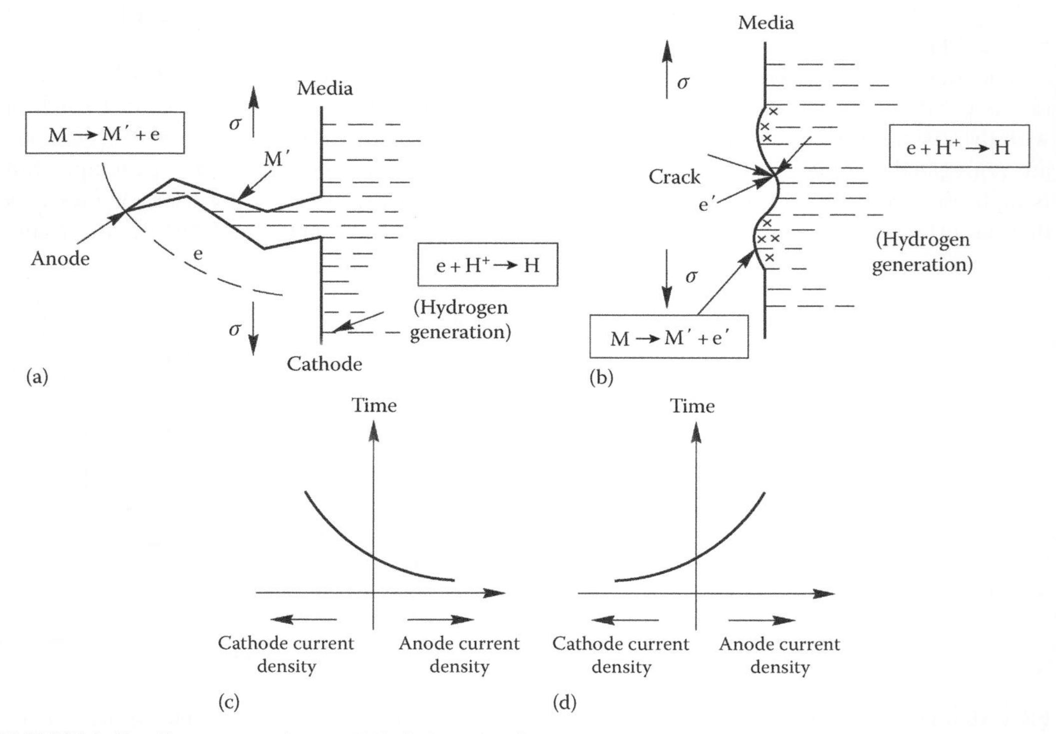

FIGURE 8.12 Stress corrosion vs. HIDC. (a and c) Stress corrosion. (b and d) HIDC.

TABLE 8.2
Comparison between Stress Corrosion and HIDC of Steel in Fracture Appearance

Type	Cracking Source Location	Fracture Macroscopic Characteristics	Fracture Microcosmic Characteristics	Secondary Crack
Stress corrosion	Surely on the surface, no exception, usually at the tensile stress concentration like sharp corner, slide mark, pitting, pit, etc.	Brittle, dark color, even black, obvious boundary with the final static fracture zone, darkest color in the cracking source zone	Generally intergranular cracking, occasionally transgranular cleavage cracking, many corrosion products, special ions like chlorine, sulfur, etc., most corrosion products on the cracking source	Many
HIDC	Mostly under the surface, occasionally at the surface stress concentration, as the external stress increases, the cracking location moves toward the surface	Brittle, bright, no corrosion when the steel is fractured, even corrosion when placed in the corrosive environment	Mostly intergranular cracking, occasionally transgranular cleavage or quasi-cleavage cracking, usually many torn edges on the grain boundary, dimple in some places, if not placed in the corrosive environment, generally no corrosion products	None or seldom

8.2.5 Measures of Preventing Hydrogen Embrittlement

Hydrogen embrittlement is related to factors like environment, load, and the type of material. Therefore, we can prevent it in three ways.

1. Environment: Try to cut the path through which hydrogen goes into the metal, or control one key link on this path to slow down the reaction speed on this link, so that the hydrogen does not enter, or seldom enters, the metal. For example, surface coating is used to isolate the component surface from hydrogen in the environmental media. An inhibiter can also be added into the hydrogen contained media.
2. Load: During component design ad processing, eliminate all factors that generate residual tensile stress. On the contrary, surface treatment can generate a residual pressure stress layer, which can effectively prevent HIDC. During design, try to make the K_I value (when the part is in service) be less than K_{IHC}.
3. Material: Reasonable material selection and the correct hold and cold tooling process are also important to prevent HE of the component.

8.3 CORROSION FATIGUE CRACKING

8.3.1 Definition and Features of Corrosion Fatigue

Corrosion fatigue refers to brittle cracking occurring on the material or component under the combined action of alternating stress and a corrosive environment. This fracture is much more serious than that caused by alternating stress or corrosion individually. In engineering technology, corrosion fatigue is one of the most important reasons why cracking accidentally occurs on a safely designed metal component. For example, an oil rod steel has a short life, so millions of dollars are invested to change the oil rod every year in America's petroleum industry. CFC often occurs on marine propellers, towing wire ropes for mines, automobile springs and axles, and turbine blades, among others. It also occurs in other industries such as in the chemical industry, the atomic energy industry, and in aerospace.

Corrosion fatigue has the following features:

1. The corrosive environment is not specific. As long as the environmental media is corrosive to the metal, together with alternating stress, corrosion fatigue will occur. This is much different from stress corrosion. Corrosion fatigue does not need a specific combination between metal and environmental media. Therefore, corrosion fatigue is more ubiquitous.
2. Corrosion fatigue's *S–N* curve is different from mechanical fatigue's *S–N* curve in its shape. From Figure 8.13, we can see that there is no fatigue limit with infinite life. Generally, the stress value in the specified cycles (typically 10^7 cycles) is taken as the corrosion fatigue limit and represents the resistance of the material to corrosion fatigue.
3. There is no specific relationship between the corrosion fatigue limit and static strength. From Figure 8.14 we see that there is almost no difference in the fatigue limit among steels with different tensile strengths in sea water. This means that in a corrosive media, improving the material's static strength makes no contribution to its resistance to fatigue.
4. CFC primarily comes from surface corrosion pits or surface defects, usually from several sources. Most of them are transgranular cracks.

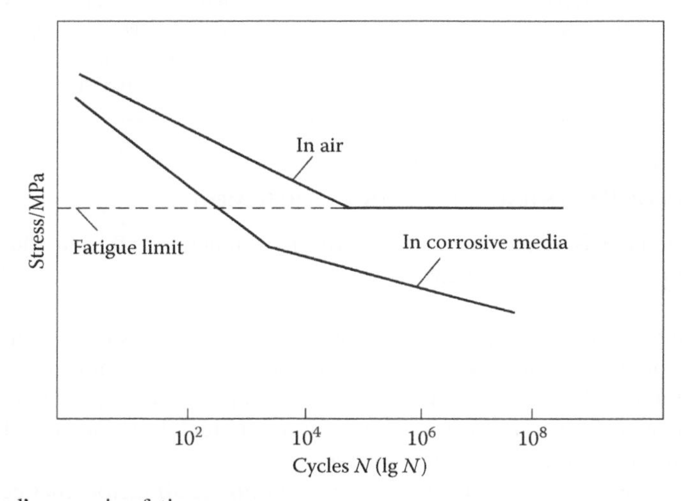

FIGURE 8.13 Steel's corrosion fatigue curve.

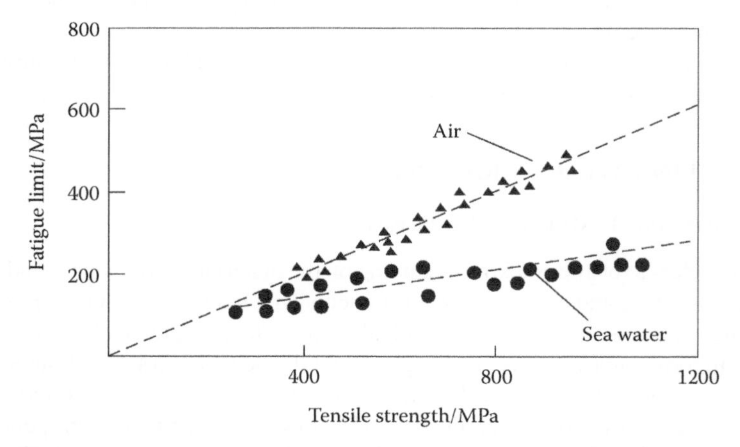

FIGURE 8.14 Media influence on steel's corrosion fatigue strength. (From Sun, Q.X., *Corrosion and Protection of Materials*, Metallurgical Industry Press, Beijing, 2001; Taken from Figure 3.22.)

8.3.2 Corrosion Fatigue Mechanism

In the corrosion fatigue process, two basic damage forms exist. The first is micro metal repeated slip caused by cyclic stress, which forms a slip band, and is the basic reason for fatigue damage. The second is corrosion damage caused by electrochemical reaction between corrosive media and metal. When these two damages co-exist, their action is not a simple summation. Obvious mutual promotion and interaction exists between these two basic damages.

There are three common corrosion fatigue mechanism models:

1. Slip-dissolution mechanism
   In this theory, it is thought that cyclic alternating stress results in uneven metal deformation. In the deformation zone, violent slip occurs and a slip step appears. When the alternating stress goes up, a fresh surface is exposed on the slip step. The atoms on this fresh surface have higher activity than those inside the step, so they are dissolved first in the corrosive media. When the alternating stress goes down, reverse slipping occurs on the metal. However, because the slip step that was exposed in the corrosive media when the stress went up has been dissolved, it cannot be closed. As a result, under the action of repeated alternating load, unceasing dissolution of the slip step promotes the initiation and growth of the corrosion fatigue crack (see Figure 8.15).
2. Pitting-stress concentration mechanism
   This mechanism emphasizes the important role of pitting for initiation of the corrosion fatigue crack. The pit's notch effect causes the bottom to become the originating point for corrosion fatigue. Repeated alternating stress promotes pitting formation, and the corrosion fatigue crack initiates more quickly. This mechanism gives a good explanation for corrosion fatigue having several sources. However, it has been discovered that in media without pitting, corrosion fatigue still occurs. Meanwhile, in some cases where pitting occurs, there is no obvious influence on corrosion fatigue life. Therefore, the current common viewpoint is that although pitting can promote the development of corrosion fatigue, it is not the sole reason for corrosion fatigue.
3. Surface membrane fracture mechanism
   In environmental media, an oxide membrane appears on the surface of many alloys. Generally, this oxide membrane can protect the alloy to a certain extent. However, due to the difference between specific volume and structure, internal stress exists between the membrane and the base alloy (see Chapter 11). Under an external cyclic alternating load,

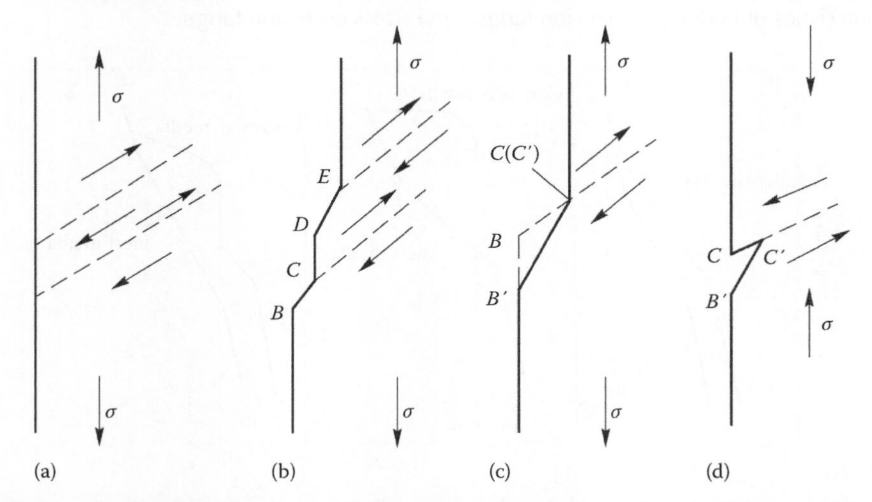

FIGURE 8.15 Corrosion fatigue's slip-dissolution mechanism. (a) Local strain area. (b) Slip step formation. (c) New surface generated by slip step dissolution. (d) Crack initiation.

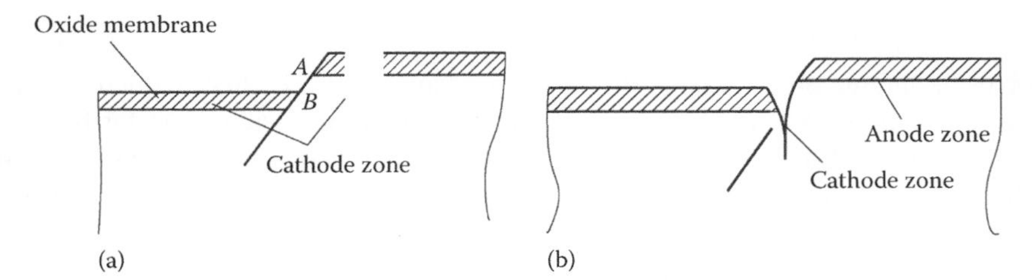

FIGURE 8.16 Corrosion fatigue crack formed by surface membrane fracture. (a) Fresh metal surface AB was exposed by the fracture of the slip oxide membrane. (b) Anode zone formed where the membrane fracture.

fracture may occur on the surface membrane and a fresh surface is exposed in Figure 8.16(a). The location where the membrane is fractured becomes the micro anode, while its surrounding area becomes a cathode covered with an oxide membrane. Therefore, the combined action of alternating load and media leads to initiation and growth of the corrosion fatigue crack, as shown in Figure 8.16b.

8.3.3 Corrosion Fatigue Crack Growth

Corrosion fatigue crack growth is usually related to the external stress strength. The relationship between the fatigue crack growth rate da/dN and the stress intensity factor range ΔK has three stages, where the second stage complies with the Paris Equation: $da/dN = c(\Delta K)^m$. However, corrosion fatigue growth behavior in environmental media is obviously different. The relationship between its crack growth rate da/dN and the crack tip stress intensity factor ΔK can be divided into three types, as shown in Figure 8.17. Figure 8.17a is the true corrosion fatigue (Mode A). The corrosion fatigue of an aluminum alloy in water is of this type. The media's influence causes the threshold value ΔK_{th} to decrease and the crack growth rate $(da/dN)_{CF}$ to increase. When K_I is close to K_{IC}, the media's influence goes down. Figure 8.17b is stress corrosion fatigue (Mode B). It has the characteristics of the summation of alternating load fatigue and stress corrosion. When $K_I < K_{ISCC}$, the media's influence can be ignored. When $K_I > K_{ISCC}$, the media has great influence on $(da/dN)_{CF}$. $(da/dN)_{CF}$ goes up sharply and has a horizontal step. Steel's corrosion fatigue in hydrogen media is of this type. The third case is shown in Figure 8.17c, and is a mixture of Mode A and Mode B (Mode C). Corrosion fatigue of most engineering alloys and environmental media is of this type. From the figure we find that Mode C has the characteristics of both true corrosion fatigue and stress corrosion fatigue.

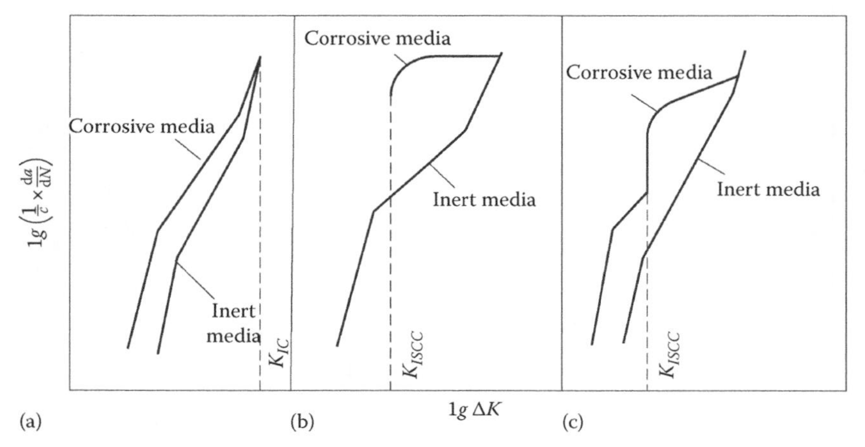

FIGURE 8.17 Basic types of corrosion fatigue crack growth curves. (a) True corrosion fatigue. (b) Stress corrosion fatigue. (c) Mixed corrosion fatigue.

Regarding the quantitative study of corrosion fatigue crack growth, there are two primary models: the summation model and the competition model.

8.3.3.1 Summation Model

The linear summation model states that the corrosion fatigue crack's growth rate $(\mathrm{d}a/\mathrm{d}N)_{CF}$ is the sum of the stress corrosion crack's growth rate in the same environment $(\mathrm{d}a/\mathrm{d}N)_{SCC}$, and the sole mechanical fatigue crack's growth rate in an inert environment $(\mathrm{d}a/\mathrm{d}N)_{CF}$:

$$\left(\frac{\mathrm{d}a}{\mathrm{d}N}\right)_{CF}=\left(\frac{\mathrm{d}a}{\mathrm{d}N}\right)_{SCC}+\left(\frac{\mathrm{d}a}{\mathrm{d}N}\right)_{F}, \tag{8.35}$$

where $(\mathrm{d}a/\mathrm{d}N)_{SCC}$ is the stress corrosion crack growth generated in one stress cycle. If the time period in one cycle is $\tau(\tau = 1/f)$, the following equation applies:

$$\left(\frac{\mathrm{d}a}{\mathrm{d}N}\right)_{SCC}=\int_{\tau}\left(\frac{\mathrm{d}a}{\mathrm{d}t}\right)_{SCC}\mathrm{d}t, \tag{8.36}$$

where $(\mathrm{d}a/\mathrm{d}N)_{SCC}$ is the stress corrosion crack's growth rate. Wei [16] has used this model to estimate the fatigue crack growth of high strength steel in dry hydrogen, distilled water, and vapor, as well as a titanium alloy in a salt solution. The result shows that when $K_I > K_{ISCC}$, it is identical with test results. However, the premise of this model is that corrosion fatigue and SCC are controlled by the same mechanism, and the interaction between fatigue and stress corrosion is not considered. Therefore, on this basis, Wei [17] considered the alloy's $\mathrm{d}a/\mathrm{d}N$–ΔK relationship in an inert environment and in a corrosive environment, and modified the linear model:

$$\left(\frac{\mathrm{d}a}{\mathrm{d}N}\right)_{CF}=\left(\frac{\mathrm{d}a}{\mathrm{d}N}\right)_{m}+\left(\frac{\mathrm{d}a}{\mathrm{d}N}\right)_{cf}, \tag{8.37}$$

where

- $(\mathrm{d}a/\mathrm{d}N)_m$ is the sole mechanical fatigue crack's growth rate in an inert environment
- $(\mathrm{d}a/\mathrm{d}N)_{cf}$ is the difference between the crack's growth rate in a corrosive environment and in an inert environment, representing the interaction between cyclic plastic deformation and chemical reaction

One more reasonable description in physics is that fatigue crack growth in the media is caused by simultaneous occurrence of two parallel processes: mechanical fracture and chemical–mechanical fracture. Therefore, the summation model can be represented as

$$\left(\frac{\mathrm{d}a}{\mathrm{d}N}\right)_{CF}=\left(\frac{\mathrm{d}a}{\mathrm{d}N}\right)_{m}(1-\theta)+\left(\frac{\mathrm{d}a}{\mathrm{d}N}\right)_{C}\theta, \tag{8.38}$$

where

- $(\mathrm{d}a/\mathrm{d}N)_C$ is the sole corrosion fatigue crack's growth rate
- θ is the sole corrosion fatigue's component in the crack growth
- $(1-\theta)$ is the mechanical fatigue's component
- θ can be obtained by fracture analysis

8.3.3.2 Competition Model

The competition model states that corrosion fatigue crack growth is an independent competition process between mechanics and chemistry. The corrosion fatigue crack's growth rate is equal to the higher rate, not their summation. The competition model can also be deemed as one special case of Equation 8.38. If $(da/dN)_m << (da/dN)_C$, the competition model states that $\theta = 1$. In this case, the competition model is identical to the summation model.

8.3.4 Measures of Preventing Corrosion Fatigue

1. Correct material selection and control is still the basic factor. Generally, materials with good pitting resistance have high corrosion fatigue strength. Their inclusion in steel, particularly manganese sulfide (MnS), has great influence on corrosion fatigue, because it is usually the source of pitting. The influence of an alloy's composition and structure on corrosion fatigue is nothing like as obvious as it is on stress corrosion and HE.
2. Reasonable design and improvement of the manufacturing process is important to control corrosion fatigue. Geometric configurations that may result in high stress concentration and crevice corrosion should be avoided. Heat treatment is used to eliminate internal stress, or shot blasting is used to make the part's surface in the pressure stress state, which can effectively inhibit corrosion fatigue fracture.
3. Treatment is made on the working media, and electrochemical protection is provided on the structure. The working media treatment is mainly used in a closed system. For example, removing the oxygen in a solution and adding chromate or emulsified oil can extend steel's corrosion fatigue life. Cathode protection is usually used for corrosion fatigue control on marine environment metal structures, with good effect. However, this is not applicable to acidic media and applications where HE fracture occurs.

8.4 CORROSIVE WEAR

8.4.1 Definition and Features of Corrosive Wear

Corrosive wear (CW), also called wear corrosion, refers to surface fracture phenomena that occurs on a solid under other solids' mechanical friction action and the action of corrosive media; or under the action of a corrosive media's self-friction. It generally occurs in corrosive media, and is a result of mechanical, chemical, or electrochemical factors and their interaction. It widely exists in mechanical equipment in various industries including the petroleum, chemical fiber, coal mining, and power industries. It is the main reason why devices and components fail in equipment used in these fields (such as water turbines, pumps, valves, pipes, and sprayers).

In the CW process, both corrosion and wear occur on the component. However, it is much different from corrosion or wear separately, because in the CW condition, due to the media's corrosion, the material's surface performance worsens and the material's mechanical wear increases. On the other hand, in the mechanical wear condition, the corrosion rate goes up sharply. Hence, the CW process has both a mechanical factor and an electrochemical factor, as well as their interaction, which has much higher fracture influence on the material than the sum of corrosion and wear individually, leading to accelerated material fracture and loss.

8.4.2 Corrosive Wear Mechanism

Material loss due to CW is not the sum of weight loss from corrosion and dry wear separately, but is much larger than their sum. In other words, interaction exists between corrosion and wear.

Corrosion can accelerate wear, and in turn wear can accelerate corrosion. To quantify the relationship between CW and corrosion and wear, material loss due to CW is generally represented as

$$W = W_C + W_W + \Delta W, \tag{8.39}$$

$$\Delta W = \Delta W_C + \Delta W_W, \tag{8.40}$$

where W is the material's total loss due to CW, generally measured by the weighing method or the surface topography method. W_C is the sole corrosion weight loss (weight loss in static status), generally calculated by using the corrosion current density in the corrosion weight loss method (soaking) or the electrochemical method. W_W is the sole wear weight loss, generally calculated by using dry wear in the air or wear under the cathode protection potential in corrosive media. ΔW is the interaction between corrosion and wear, calculated by using $W - W_C - W_W$. ΔW_C is the value of wear accelerating the corrosion, generally calculated by using the corrosion current density measured by the electrochemical method in the wear condition. And finally, ΔW_W is the value of corrosion accelerating the wear, generally calculated by using $W - W_C - W_W - \Delta W_C$.

CW interaction is usually represented as mutual acceleration. Madsen [18] studied the erosion corrosion of several metal materials and gave the data of interaction: the interaction of common low alloy steel accounts for 25%–33% of the total CW loss; the interaction of 316 stainless steel accounts for 55%–62%, or 1/3–2/3 of the total weight loss in the CW interaction. Yue et al. [19] studied the erosion corrosion of high chrome cast iron in a dilute sulphuric acid mortar media. The interaction of CW is related to the media's pH value. When the pH value is low, total weight loss is mainly caused by wear. In this case, the interaction is high and corrosion makes the primary contribution to the wear increment; that is, corrosion accelerates the wear. When the pH value increases, the percentage of the sole wear weight loss (dry wear) goes up, but the interaction becomes obviously low.

In testing we also find that, in some cases, the total loss of a material's CW is less than the dry wear in air. For example, the CW of a Ti-6Al-4V alloy in 0.5 mol/L H_2SO_4 solution and 1Cr18Ni9Ti and 1Cr28Ni32 stainless steel in 0.5 mol/L H_2SO_4 solution is less than the dry wear in air [20,21]. The reason why this phenomenon occurs is primarily because the media's corrosivity is weak, so most of the material loss is wear. Compared with the weight loss of the wear in air, the media works as a lubricant, so the friction factor and wear weight loss are reduced.

From Equations 8.39 and 8.40 we find that the interaction of CW includes wear accelerating the corrosion, and corrosion accelerating the wear. Let us consider the first case.

1. In a corrosive media, friction force damages the passive membrane on the material's surface, so the corrosion potential becomes a negative shift and the corrosion trend becomes higher. If the media's repassivation is not able to repair the damaged passive membrane, a fresh active metal surface is exposed, and the inside and outside of the grinding crack constitute a corrosion electrolytic cell, which accelerates corrosion of the material.
2. In a system without a surface membrane, friction removes the corrosion product, so the new metal surface is exposed. Furthermore, plastic deformation, dislocation aggregation, or induced microcracking occurs on the surface, so it is in the high-energy zone, which becomes a cathode zone in the corrosion primary battery, accelerating the material's corrosion.

For the case where corrosion accelerates wear, there are three aspects to consider.

1. Corrosion can increase the metal surface's roughness.
2. Because the metal structure is uneven, corrosion will damage the material's grain boundary or other structures, reducing the material's bonding strength. When the grinding head slides, the material peels off easily, and wear increases.

3. In a system with a passive membrane, because a surface shearing force exists, the passive membrane becomes cracked and torn in large pieces, and brittle peel-off occurs, accelerating material loss. In this case, the material is not ground off gradually, but is cracked and torn in large pieces due to embrittlement of the material itself, or surface membranes in the corrosive media. Therefore, the material's wear increases multiply.

From the aforementioned analysis we know that, similar to low stress brittle fracture in stress corrosion, premature failure caused by an multiplied increase of material wear due to the corrosion factor also exists in the CW process. Therefore, this phenomenon of corrosion accelerating wear is called a material's environmental embrittlement in CW.

8.4.3 Relationship between Corrosive Wear and Stress Corrosion, Hydrogen Embrittlement, and Corrosion Fatigue

Cracking, corrosion, and wear are three main forms of material failure. Similar to a material's stress corrosion, HE, and corrosion fatigue, CW also leads to fracture caused by the synergistic action of mechanics and electrochemistry. Compared with SCC, CW has environmental embrittlement with the following features:

1. *Load form.* Both fracture forms are applied with external stress. The CW environment-induced external load is shearing stress, while SCC is tensile stress. Shearing stress results in microcracking on the material's surface, and the surface material peels off in subsequent wear, increasing material loss. Tensile stress causes cracking on the metal surface, which grows toward it from the inside, and finally leads to cracking and discarding of the material. The tensile stress that results from the material's SCC can be eliminated by taking measures, but the shearing force in CW cannot be eliminated, because it exists when the material is in service.
2. *Specific system between the material and the media.* Generally, the material environment system in CW can lead to SCC of the material. For example, austenitic stainless steel in a contained chloride ion acidic solution, high-strength low-alloy steel in a dilute acid solution, copper alloy in ammonia or an ammonium solution. Furthermore, media concentration influences both the SCC system and the CW system. However, SCC closely depends on whether the material is passivated in the media, that is, it is susceptible to the potential zone (particularly stainless steel's SCC behavior). In CW, the external shearing force is often large enough to damage the surface membrane, and some systems do not have a membrane at all, so it is unrelated to the potential zone where the material is located.
3. *Failure form.* The fracture form of SCC is brittle cracking. Once cracking occurs, accidents may result, leading to the discarding of the workpiece. CW is material surface fracture under the combined action of strain fatigue and chemical media. Wear causes cracking to occur on the material's surface, while corrosion accelerates crack growth. Material loss is primarily surface embrittlement peel-off. The workpiece's life is shortened. Its failure rate is lower than that of SCC.
4. *Threshold value.* The CW system has a critical load and a media concentration threshold value. When it is larger than the threshold value, material loss rate increases as the load and concentration go up. This is similar to the critical stress σ_{SCC} in SCC.

8.4.4 Protection Measures of Corrosive Wear

Because CW is not a material's intrinsic performance, but a system characteristic of its condition of use, it can be controlled and prevented in the material, the environmental media, and electrochemical protection.

Material. An appropriate material is selected according to the extent of the influence of the mechanical factor and the electrochemical factor in the conditions of use. For a low-load (weak wear–weak corrosion) environment, common materials are acceptable and there is no special requirement. For a high-load (strong wear–weak corrosion) environment, mechanical wear is the primary reason for material loss, so the material must have good resistance to wear and resistance to corrosion. For a low-load (weak wear–strong corrosion) environment, the material must have good resistance to corrosion in order to resist the environment's violent corrosion. For a high-load (strong wear–strong corrosion) environment, the material's resistance to corrosion can be improved by high alloying, proper heat treatment to realize structural homogenization, no strong cathodic second phase, grain refining, and so on. Deformation strengthening is the most effective method to improve resistance to wear. This can also be applied to the cathodic second phase.

Environmental media. In a CW environment, the material's CW loss can be reduced greatly by adding cathodic membrane-forming inhibiters, such as chromate, nitrite, orthophosphate, silicate, and benzoate. For example, pure iron's CW can be reduced by adding 5% $NaNO_2$, 0.1% $CaCl_2$, or 3% Na_2HPO_4 (mass fraction) in a 1% NaCl solution.

Electrochemical protection. For a material–environment combination not susceptible to hydrogen (common carbon steel, or cast iron in a sea water system), cathodic protection measures can be taken to improve the material's resistance to CW.

Surface treatment. The material's resistance to CW can be improved by surface treatment, such as Ni-P chemical plating, Ni-P/SiC plating, carbonitriding, and TiN ultrahard membrane vapor deposition.

8.5 OTHER ENVIRONMENTALLY ASSISTED CRACKING OR EMBRITTLEMENT ISSUES

Since the 1960s, the nuclear energy industry has experienced rapid development. Meanwhile, the problems of material embrittlement in a radiation environment, and the occurrence of liquid metal (used as the heat carrier) embrittlement in the reactor exist. To ensure the safety of the atomic reactor and prevent brittle cracks occurring on the reactor's pressure container, we need to study the mechanism, influence factors, and assessment indicators of a metal's material radiation embrittlement, and LME, in order to provide a basis for selecting materials and preventing brittle cracking of the atomic energy reactor's pressure container.

8.5.1 Radiation Embrittlement

8.5.1.1 Radiation Effect

There are many types of rays with high strength in the reactor. However, for metal materials, the primary influence comes from fast neutrons. After the metal material is impacted by these energy-carrying particles, defects like micropores, depleted atom zones, faults, dislocation loops, and phase changes occur, and radiation damage results. Material performance changes due to radiation damage defects are called the radiation effect. Many tests show that, after being irradiated by fast neutrons, the material's radiation behavior mainly includes increased strength, reduced plasticity (radiation hardening), reduced toughness, increased fracture appearance transition temperature (radiation embrittlement), increased creep rate (radiation creep), geometric size changes, reduced density (radiation growth, radiation swelling), and so on [22–27]. In this section, we only describe radiation hardening and the radiation embrittlement effect related to mechanical property changes.

8.5.1.1.1 Radiation Hardening

After impact by fast neutrons, low-carbon steel has a tensile stress–strain curve as shown in Figure 8.18. The austenitic stainless steel stress–strain curve before and after irradiation is shown in Figure 8.19. These curves show that, after a certain dosage of neutron radiation, metal strength

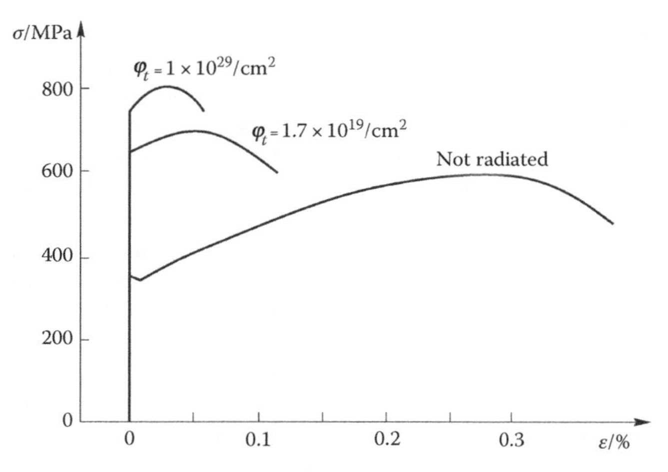

FIGURE 8.18 Low carbon steel's stress–strain curve before and after neutron radiation (φ_t is the dosage of irradiation). (From Wang, J.H. et al., *Mechanical Behavior of Materials*, Tianjin University Press, Tianjin, China, 2006; Taken from Figure 6.38.)

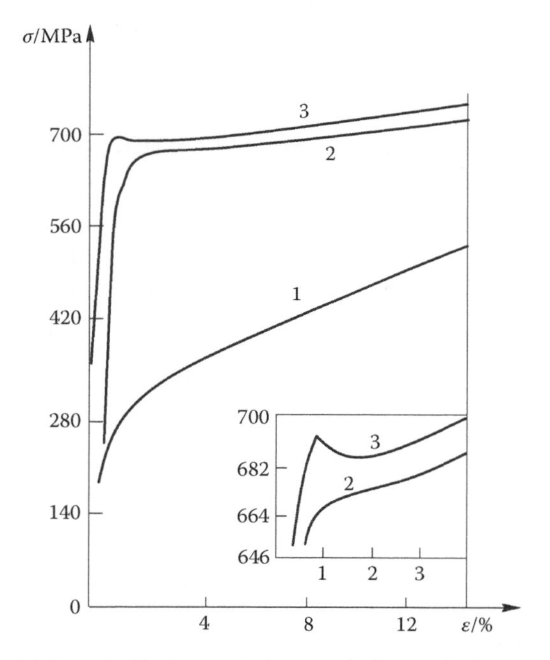

FIGURE 8.19 Austenitic stainless steel's stress–strain curve before and after neutron radiation. 1, Not radiated; 2, 95% radiation, deformation rate is 0.01/min; 3, 95% radiation, deformation rate is 0.05/min. (From Wang, J.H. et al., *Mechanical Behavior of Materials*, Tianjin University Press, Tianjin, China, 2006; Taken from Figure 6.39.)

clearly increases and the yield strength doubles, while the processing hardening rate goes down. Although tensile strength increases, it is not as susceptible as the yield strength. After a high dosage of radiation (e.g., $10^{20}/cm^2$), constriction occurs immediately after yield, without the even hardening stage. In this case, the yield strength is the maximum strength. Meanwhile, the even elongation clearly goes down. In the worse case, it even becomes zero.

After studying different materials, we find that it is a universal law of nearly all materials after irradiation that their strength goes up and their plasticity goes down, as shown in Table 8.3. In addition, we also find that some metal materials have obvious yielding points during the original

TABLE 8.3
Fast Neutron Radiation's Influence on a Material's Mechanical Property

Materials	Radiation Dosage (cm^{-2})	Yield Strength (MPa) Before Radiation	Yield Strength (MPa) After Radiation	Tensile Strength (MPa) Before Radiation	Tensile Strength (MPa) After Radiation	Elongation (%) Before Radiation	Elongation (%) After Radiation
Mo	5×10^{19}	643	682	587	716	24	22
Cu	5×10^{19}	58	208	186	233	42	27
Ni	5×10^{19}	247	426	405	432	34	23
Hastelly-x	2.5×10^{20}	339	730	769	899	52	42
Inconel-x	1.6×10^{20}	708	1089	1107	1168	28	14

Source: Wang J.H. et al., *Mechanical Properties of Materials*, Tianjin University Press, Tianjin, China, 2006.

deformation (such as body-centered cubic structure iron, molybdenum, niobium, and low-carbon steel); but after irradiation, the yield phenomenon is usually not obvious, and may even disappear. However, for face-centered cubic metals that do not have the yield effect (such as polycrystal aluminum, nickel, copper and their alloys, and austenitic stainless steel), we find the yielding point after irradiation (Figure 8.19). This means that in the case of a point defect's dislocation pinning in different metals, different changes occur after the radiation.

8.5.1.1.2 Radiation Embrittlement

Embrittlement occurring on a structural material under the action of high-speed electrons, neutrons, or ion currents is called radiation embrittlement. Most of this influence comes from the neutrons. Figure 8.20 shows the cracking toughness of container steel before and after fast neutron radiation. After fast neutron radiation, the steel's toughness goes down and embrittlement occurs.

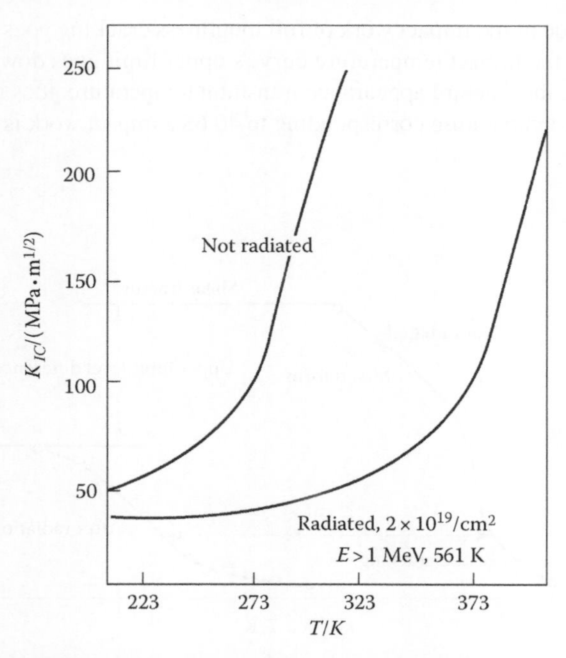

FIGURE 8.20 Container steel's relationship between fracture toughness and test temperature before and after irradiation. (From Wang, J.H. et al., *Mechanical Behavior of Materials*, Tianjin University Press, Tianjin, China, 2006; Taken from Figure 6.40.)

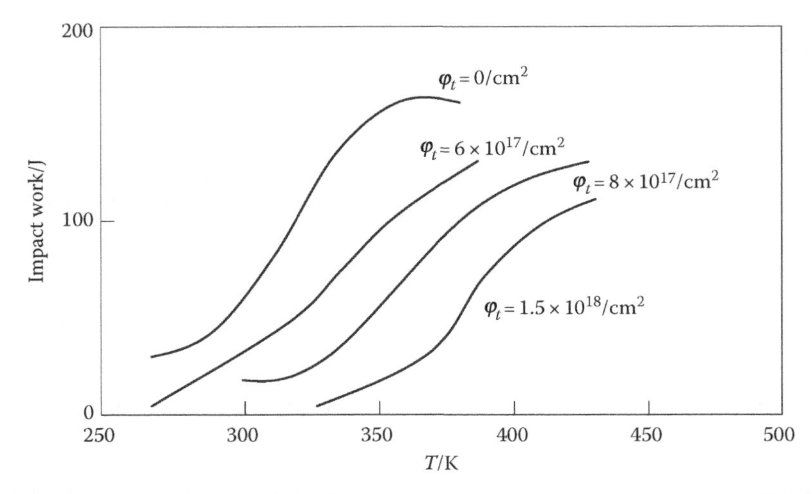

FIGURE 8.21 Fast neutron radiation dosage's influence on steel's impact work–test temperature curve. (From Wang, J.H. et al., *Mechanical Behavior of Materials*, Tianjin University Press, Tianjin, China, 2006; Taken from Figure 6.41.)

The radiation embrittlement of ferrite steel with a body-centered cubic lattice is usually measured by its fracture appearance transition temperature behavior. Figure 8.21 shows that as the neutron radiation dosage goes up, the steel's impact work–test temperature curve moves rightward, and the fracture appearance transition temperature goes up. Radiation embrittlement of austenitic steel with a face-centered cubic lattice is usually measured by the steel's plasticity decrease.

Generally, radiation's influence on body-centered cubic lattice ferrite steel impacts toughness, and its test temperature curve is shown in Figure 8.22. Radiation embrittlement has the following characteristics:

1. The absolute value of the impact work of full toughness cracking goes down. After irradiation, the level of the impact temperature curve's upper limit goes down by ΔE.
2. After irradiation, the fracture appearance transition temperature goes up by ΔT. It is widely accepted that the temperature corresponding to 40.68 J impact work is taken as the fracture

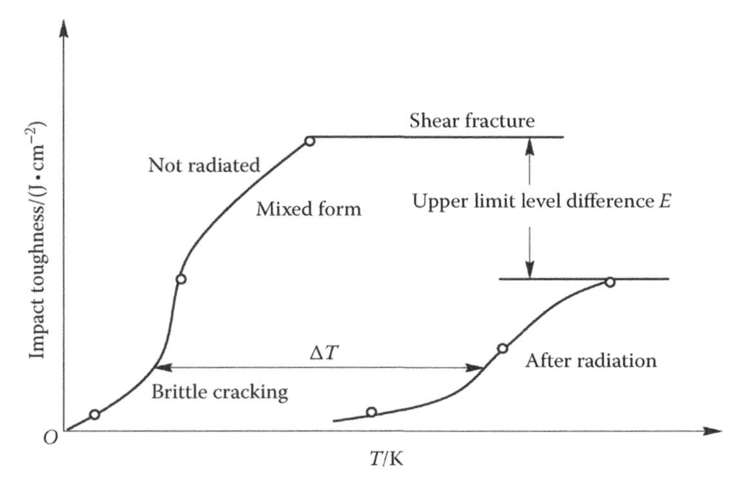

FIGURE 8.22 Radiation's influence on ferrite steel Charpy-V impact toughness and test temperature curve. (From Wang, J.H. et al., *Mechanical Behavior of Materials*, Tianjin University Press, Tianjin, China, 2006; Taken from Figure 6.42.)

appearance transition temperature [10]. ΔE and ΔT are two important performance indicators in the assessment of radiation embrittlement. Many sources [28,29] measured ΔE and ΔT value of the steel after irradiation and established the relationship between it and the radiation dosage in order to deduce an empirical formula with ΔE and ΔT. For example, Carpenter [30] studied the change of the fracture appearance transition temperature of A302B and A212B steel after irradiation and came up with the following empirical formula:

$$\Delta T = a + b\,\lg(\phi t) + c\,\lg(\phi t)^2, \tag{8.41}$$

where

a, *b*, and *c* are a constant

ϕt is the radiation dosage, cm^{-2}

The test result of A302B steel in different heat treatment states without annealing after irradiation is shown in Figure 8.23. The empirical formulas of ΔT and ϕt of the upper limit and lower limit are

$$\Delta T = 11097 + 1333.9\lg(\phi t) + 40.06[\lg(\phi t)]^2. \tag{8.42}$$

$$\Delta T = 9228 + 1065.3\lg(\phi t) + 30.58[\lg(\phi t)]^2. \tag{8.43}$$

It should be noted that the empirical formula is only applicable to the tested steel type. In fact, the material's embrittlement increase after irradiation does not solely depend on the radiation dosage. It also depends on the radiation temperature, steel composition, phase structure, and other factors. Because radiation causes low-carbon steel's embrittlement temperature to rise, it often happens that the steel is in the embrittlement state at room temperature. The radiation

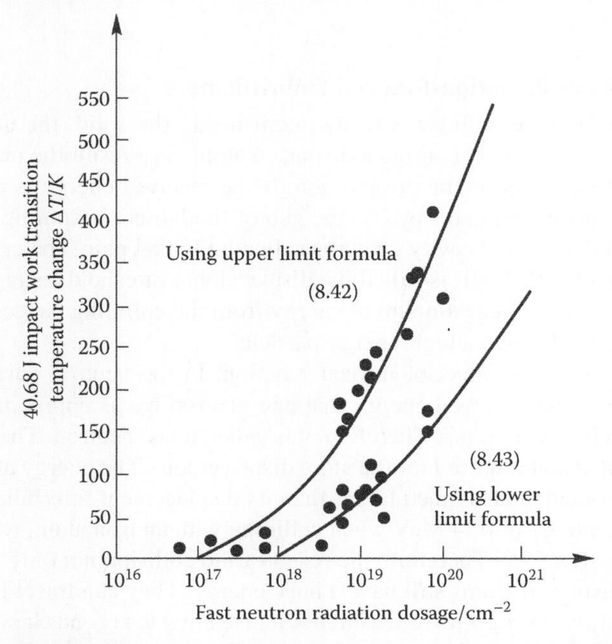

FIGURE 8.23 Radiation's influence on A302 steel ΔT. (From Wang, J.H. et al., *Mechanical Behavior of Materials*, Tianjin University Press, Tianjin, China, 2006; Taken from Figure 6.43.)

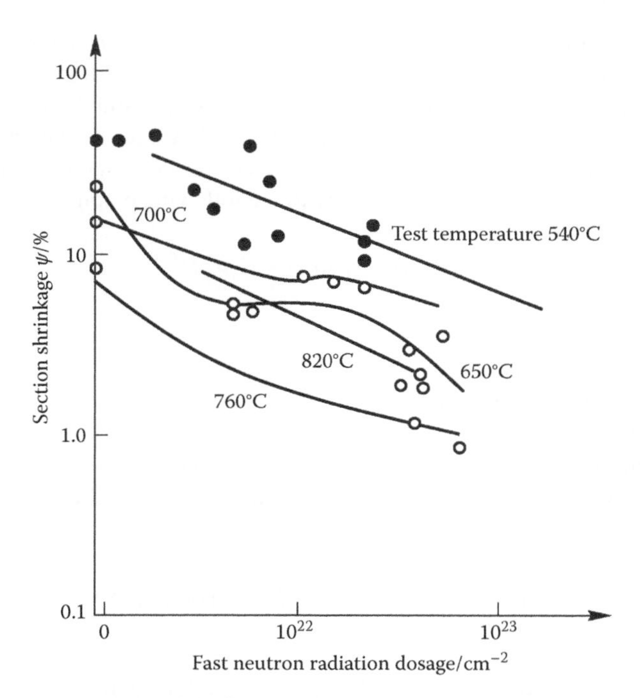

FIGURE 8.24 Fast neutron radiation's influence on AISI316 austenitic stainless steel's plasticity. (From Wang, J.H. et al., *Mechanical Behavior of Materials*, Tianjin University Press, Tianjin, China, 2006; Taken from Figure 6.44.)

embrittlement of austenitic steel with a face-centered cubic lattice is usually measured by its plasticity decrease (see Figure 8.24). Regardless of the test temperature, austenitic steel's radiation embrittlement increases as the fast neutron dosage goes up and is susceptible to the temperature.

8.5.1.2 Mechanism of Radiation-Induced Embrittlement

When a high-energy particle collides with the atom inside the solid, the atom is impacted and leaves the equilibrium position, becoming a displaced atom; as a result the original place becomes a vacancy. The displaced atom is the positive ion. If the received energy is not large enough, the ion quickly stops its movement and stays on the gap of the lattice, becoming the interstitial atom. This pair of interstitial atom and cavity defect is called a Frenkel pair. The critical value of energy required for forming a Frenkel pair is called the displacement threshold energy. It is about 25 eV. If the displaced atom absorbs a huge amount of energy from the colliding particle and has a high rate of motion, it can also be deemed a high-energy particle.

The neutron current is a product of nuclear reaction. In the neutron current generated in the nuclear fission process, the averaged energy that one neutron has is approximately 2 MeV, with a speed of approximately 2×10^9 cm/s. Therefore, it is called a fast neutron. The energy of a fast neutron is much larger than that required for the atom displacement. The energy absorbed by the lattice atom after being impacted is also much larger than its displacement threshold energy. One 2 MeV neutron can generate energy of 0.14 MeV when colliding with an iron atom, while the displacement threshold energy is only 25 eV. Therefore, the result of the collision not only forms a Frenkel pair, but most first class displaced atoms still have a huge energy. They can travel in the lattice at a very high rate, continuously colliding with lattice atoms, and leading to second class, third class, and even higher class displaced atoms. The higher is the class of the displaced atom, the more limited energy it has, and the shorter is its travel distance. Therefore, one displaced atom and cavity intensive zone

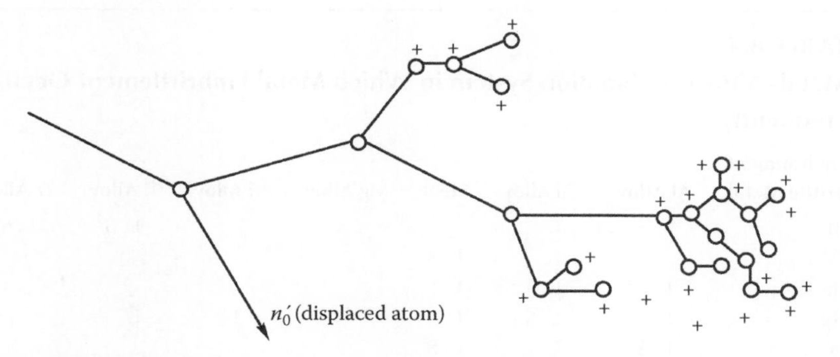

FIGURE 8.25 Displacement effect. (o) vacancy, (+) interstitial atom. (From Wang, J.H. et al., *Mechanical Behavior of Materials*, Tianjin University Press, Tianjin, China, 2006; Taken from Figure 6.45.)

with very high density is formed in the last part of the displaced atom collision. It is the aggregate of one group of Frenkel defects, called the displacement spike, as shown in Figure 8.25.

The thermal spike corresponds to the displacement spike. When the neutron, displaced atom, or excited electron moves inside the substance, side impact also occurs between it and the atom. This does not necessarily result in lattice atom displacement, but intensifies the thermal vibration of the lattice atom. In the displacement spike zone, the displaced atom finally converts its own energy into thermal energy. There are so many displaced atoms in this zone that the converted thermal energy can heat this zone to a very high temperature.

The vacancy cluster in the displacement spike can be aggregated into a cavity, while the interstitial atom can form a dislocation loop. Therefore, when steel is under high neutron dosage radiation, defects like cavities and dislocation loops may appear. However, due to their lattice distortion, these defects pin the dislocation so it is difficult for the dislocation to start on the glide plane, increasing the stress required for relief of the dislocation pinning. In addition, after the dislocation begins, like an obstacle, the radiation defect on the glide plane blocks the dislocation movement, increasing the stress required for the dislocation movement, so that the material's hardening, plasticity, and toughness decreases. The impact work decreases, while the ductile-brittle transition temperature goes up.

Another mechanism of metal radiation embrittlement is that radiation can generate gaseous fission products inside the metal. These reaction products are often diffused to the crystal defect, where they are aggregated into a bubble. When the bubble pressure is large enough, internal cracking occurs. For example, an air bubble appears on austenitic steel after irradiation. It is easily formed in the cavity or the vacancy goes into the air bubble and grows. They are connected together to form a large grain boundary crack.

In addition, during irradiation, local heating results in local crystal expansion. Some crystals may have serious thermal expansion anisotropy, so stress appears in the polycrystal with disordered orientation.

8.5.2 Phenomena and Features of Fluid (Solid) Metal Embrittlement

When a low-melting point metal contacts a high-melting point metal, at a certain temperature and tensile stress, brittle cracking phenomenon occurring in the high-melting point metal is called ME. The low-melting point metal may be solid or liquid. LME has a much higher hazard and crack growth rate than does SME.

Although the ME phenomenon is not as ubiquitous as stress corrosion, hydrogen-induced cracking, corrosion fatigue, and CW, it often occurs in actual engineering. For example, in the industrial galvanization process, brittle cracking occurs on plating tanks and equipment. In the aerospace industry, ME occurs on cadmium plated titanium alloys, high-strength steel, and aluminum alloys. The material's combination on which ME may easily occur is shown in Table 8.4 [10].

TABLE 8.4
Metal–Alloy Combination System in Which Metal Embrittlement Occurs Frequently

Environment Brittle Metal	Al Alloy	Cu Alloy	Steel	Mg Alloy	Ni Alloy	Ti Alloy	Zr Alloy
Bi		L				L, S	L, S
Cd			L, S				
Ga	L	L	L				
Hg	L	L, S	L		L	L	
In	L, S	L	L, S				
Li		L	L		L		
Na	L, S	L		L	L		
Pb		L	L, S				
Sn	L	L	L				
Zn	L		L, S	L			

Note: L and S, respectively, indicate liquid and solid.

From the aforementioned text, we know that ME is a comprehensive process covering material combination, temperature, stress (strain) rate, metallurgy, and chemical factors. Fracturing caused by ME often has the following characteristics:

1. Tensile stress must exist. It may be working stress or residual stress.
2. ME occurs only in a specific metal–metal combination system (Table 8.4); for example, in ME on cadmium plated steel plates.
3. ME occurs only at a specific temperature and stress (strain) rate. When it is lower or higher than this temperature range, ME becomes weak, or even disappears. If the stress (strain) rate is changed, the temperature range will change.
4. Low-melting point metal must be able to dampen high-melting point metal. However, generally they are mutually insoluble and cannot form an intermetallic compound. Certainly, there are exceptions. For example, Zn and Sn can cause embrittlement to occur on steel.
5. ME cracking is generally a branch crack or a mesh crack connecting to the main crack. The fracture is mostly an intergranular type or a transgranular cleavage type. The fracture surface often contains low melting point metal. This is the key to identifying ME cracking.

8.5.3 Mechanism of Metal Embrittlement

Like stress corrosion, hydrogen-induced cracking, and corrosion fatigue, ME fracture also involves the process of crack initiation, growth, and cracking. The initiation of ME cracking, according to the coherent view, is that the environment embrittlement atoms are diffused into the material's surface, and form a membrane layer, which improves the material's rheological strength. Hence, it is helpful to crack initiation and increases the metal's susceptibility to embrittlement.

ME crack growth involves two mechanisms: Stoloff, Johnson, Kamdar, and Westwood came up with the model of adsorbing an environment brittle particle to reduce cohesive energy (SJKW model) [31]. This model states that the environment adsorption particles are adsorbed by the crack tip (Figure 8.26), so the material's atomic binding energy (Figure 8.27a) and key bonding force decreases (Figure 8.27b), leading to reduced cracking strength. This model is identical with the cleavage cracking characteristics on the fracture.

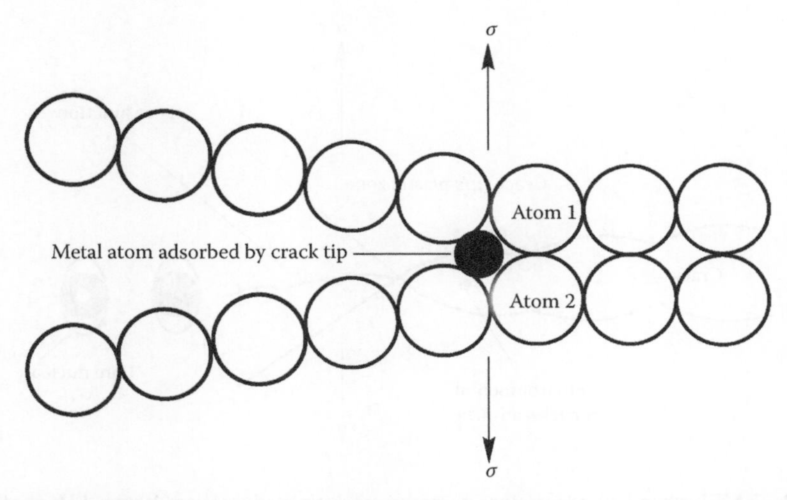

FIGURE 8.26 Mode of crack tip adsorbing brittle particle to reduce cohesive energy. (From Wang, J.H. et al., *Mechanical Behavior of Materials*, Tianjin University Press, Tianjin, China, 2006; Taken from Figure 6.55.)

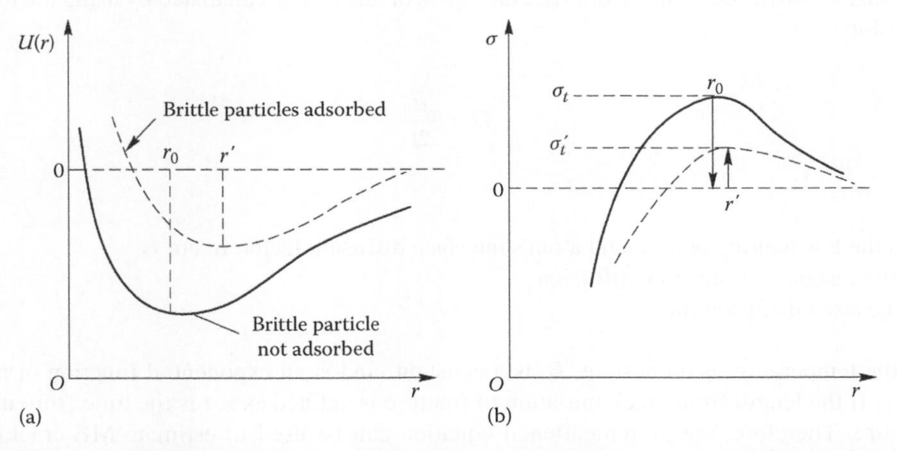

FIGURE 8.27 Binding energy, binding strength, and interatomic distance curve before and after the crack tip adsorbing brittle particle. (a) Binding energy. (b) Binding strength. (From Wang, J.H. et al., *Mechanical Behavior of Materials*, Tianjin University Press, Tianjin, China, 2006; Taken from Figure 6.56.)

To explain the toughness characteristics of some ME fractures, Lynch [32] came up with the model of environmental brittle particle adsorption reducing plastic rheological resistance (the Lynch model). This model states that the particle adsorbed on the crack tip (Figure 8.28) reduces the material's plastic rheological resistance, so that the crack tip's shear stress goes up. The crack is easily connected to the hole at the back (the cracking between the second phase and the base interface), accelerating the crack growth.

Regardless of the SJKW model or the Lynch model, the necessary condition for ME fracture to occur is that environment embrittlement particles reach and are adsorbed at the crack tip. If environment embrittlement particles, because liquid metal has low viscosity and good fluidity, so the particles can reach the crack tip in a flowing method. Therefore, LME has a high crack growth rate, sometimes up to 0.1 m/s. When embrittled particles are solid, they can reach to the crack tip by the solid substance's sublimation or surface diffusion, so that the SME crack's growth rate is related to the substance's heat of sublimation, and the activation energy of surface diffusion.

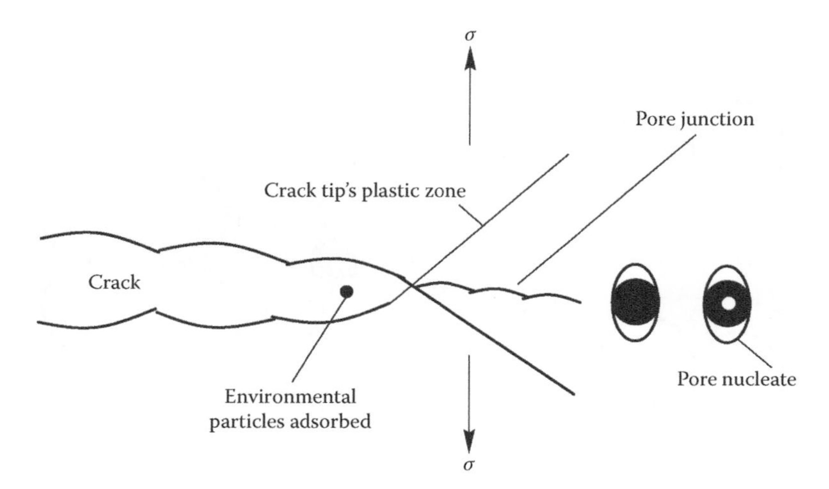

FIGURE 8.28 LME's chemical adsorption increases crack tip's slip. (From Wang, J.H. et al., *Mechanical Behavior of Materials*, Tianjin University Press, Tianjin, China, 2006; Taken from Figure 6.57.)

The surface diffusion controlled SME crack growth rate can be calculated by using the following relationship:

$$D_s \approx \frac{x^2}{2t}, \tag{8.44}$$

where

D_s is the low melting point metal atom's interface diffusion factor, in cm²/s
x is the distance of the atom diffusion
t is the atom diffusion time

When the temperature is a constant, D_s is a constant, and is an exponential function of the temperature. If the length from crack initiation to fracture is defined as x, t is the time from initiation to fracture. Therefore, the aforementioned equation can be used to estimate ME cracking life. Meanwhile, from the aforementioned equation we find that the crack length maintains a parabolic curve relationship with the time, and maintains an exponential function relationship with the temperature.

In addition, the ME crack's susceptibility is also related to the combination of strain rate and temperature. When the strain rate goes up, it is necessary to increase the ambient temperature to make the embrittlement particles quickly move to the crack tip, so that ME fracture occurs on the material in the tensile stress condition.

EXERCISES

8.1 Explain the following terms: (1) stress corrosion; (2) hydrogen attack; (3) white spot; (4) hydride embrittlement; (5) hydrogen-induced delayed cracking; (6) corrosion fatigue; (7) corrosive wear; (8) radiation embrittlement; and (9) radiation damage.

8.2 Explain the meaning of the following mechanical property indicators: (1) K_{ISCC}; (2) da/dt; (3) K_{IHC}; (4) σ_{SCC}; and (5) K_{IC}.

8.3 Describe the condition of stress corrosion occurring on a metal. What characteristics does the material's stress corrosion have?

8.4 For a large sample with a notch on one side, the crack growth rate under long-time load action is observed. It is found that the material in a corrosive media goes into Stage I and Stage II in an accelerated manner, but does not go into Stage III. When the premade crack's depth a is 3 mm, under 50 MPa load action, the crack just grows. When the crack grows to 5 mm, it goes into Stage II, where $da/dt = 2 \times 10^{-6}$ mm/s applies. What is the material's K_{ISCC}? How long does the crack stay in Stage II ($K_{IC} = 20\ \text{MPa} \cdot \sqrt{\text{m}}$)?

8.5 For a high-strength steel, $\sigma_{0.2}$ = 1400 MPa, $K_{IC} = 77.5\ \text{MPa} \cdot \sqrt{\text{m}}$; when it is in water $K_{ISCC} = 21.3\ \text{MPa} \cdot \sqrt{\text{m}}$; when the crack grows to Stage II, $da/dt = 2 \times 10^{-5}$ mm/s; when Stage II ends, $K_I = 62\ \text{MPa} \cdot \sqrt{\text{m}}$. A component made of this material is in service in water, and the working tensile stress is $\sigma = 400$ MPa. After the defect detection, it is found that there is a semicircle crack with a radius of $\alpha_0 = 4$ mm on this component's surface. Try to estimate its remaining life.

8.6 What is hydrogen-induced delayed cracking? Why does high strength steel's hydrogen-induced delayed cracking occur at a certain strain rate and temperature range?

8.7 Describe the method of differentiating high-strength steel's stress corrosion and hydrogen-induced delayed cracking.

8.8 There is a M24 bolt welding bridge made of high-strength bolt and 40B steel. Its tensile strength is 1200 MPa and its tensile stress is 650 MPa. During use, due to the influence of damp air and rain, a cracking accident occurs. It is found from the fracture that the crack has obvious intergranular cracking characteristics from the screw thread root, followed by the fast brittle cracking part. There are many corrosion products on the fracture and many secondary cracks. Try to analyze the reason why cracking occurs on this bolt and consider measures to prevent cracking.

8.9 There is a missile fastener made of ultra-high-strength steel. After passing testing, it is sealed and stored in a cave. After several years, an obvious crack is found on the fastening part. Try to explain why.

8.10 Explain the influence of environmental media on a metal material's fatigue performance, and why.

8.11 Which forms does a metal material's corrosion fatigue crack growth curve have?

8.12 An iron bar is embedded in a damp zone. Explain why the part with the most serious corrosion is on the section that is just exposed from the ground.

8.13 For a vessel in sea water, try to analyze the cracking forms that may appear on its propeller. How can you simulate its working condition in the lab, and evaluate its damage?

8.14 Discussion: How can you study corrosive wear by using fracture mechanics theory, electrochemical theory, and thermodynamic theory?

REFERENCES

1. Sang C., Wang Y. B., Chu W. Y., and Hsiao C. M. Stress corrosion cracking of alpha+ beta brass in ammonia solution under compressive stress. *Scr Metall Mater*, 1991, 25(12): 2751–2756.
2. Chen H. N., Chen L. S., Lin Q. H., and Long X. Stress corrosion test for clad plate weldments with compressive stress treatment using the anti-welding-heating method. *Corrosion*, 1999, 55: 626–630.
3. Liu X. D. and Frankel G. S. Effects of compressive stress on localized corrosion in AA2024-T3. *Corros Sci*, 2006, 48(10): 3309–3329.
4. Miura K., Ishigami I., and Usui T. Effects of compressive stress on corrosion-protective quality and its maintenance under a corrosive environment for TiN membranes deposited by reactive HCD ion plating. *Mater Trans*, 2004, 45: 102–111.
5. Boven G. V., Chen W., and Rogge R. The role of residual stress in neutral pH stress corrosion cracking of pipeline steels. *Acta Mater*, 2007, 55(1): 29–42.
6. Ling X., Peng W. W., and Ma G. Influence of Laser peening parameters on residual stress field of 304 stainless steel. *J Press Vessel Technol*, 2008, 130: 021201.
7. Zheng X. L. *Mechanical Properties of Materials* (Ver. 2). Xi'an, China: Northwestern Polytechnical University, 2000.

8. Sun Q. X. *Corrosion and Protection of Materials*. Beijing: Metallurgical Industry Press, 2001.
9. Liu R. T. *Mechanical Properties of Engineering Materials*. Harbin, China: Harbin Institute of Technology Press, 2001.
10. Wang J. H., Zheng J. P, Liu J. C. et al. *Mechanical Properties of Materials*. Tianjin, China: Tianjin University Press, 2006.
11. Gray H. R. Effects of hot-salt stress corrosion on titanium alloys. *Metal Eng Quart*, 1972, 12: 10–17.
12. Hoar T. P. and Hines J. G. The stress-corrosion cracking of austenitic stainless steels. Part 1 mechanism of the process in hot magnesium-chloride solutions. *J Iron Steel Inst*, 1956, 182: 124–143.
13. Ford F. P. Quantitative prediction of environmentally assisted cracking. *Corrosion*, 1996, 52(5): 375–395.
14. Mao X. and Li D. The crack propagation rate in stress corrosion cracking. *Metall Mater Trans A*, 1995, 26: 641–646.
15. Chu W. Y., Qiao L. J, Chen Q. Z. et al. *Cracking and Environmental Cracking*. Beijing: Science Press, 2000.
16. Wei R. P. and Landes J. D. The effect of D_2O on fatigue-crack propagation in a high-strength aluminum alloy. *Int J Fract,* 1969, 5(1): 69–71.
17. Wei R. P. and Gao M. Reconsideration of the superposition model for environmentally assisted fatigue crack growth. *Scripta Metall*, 1983, 17(7): 959–962.
18. Madsen B. W. Measurement of erosion-corrosion synergism with a slurry wear test apparatus. *Wear*, 1988, 123(2): 127–142.
19. Yue Z. Y., Zhou P., and Shi J. Some factors influencing corrosion-erosion performance of Materials. In: K. C. Ludema Ed. *Wear of Materials: International Conference on Wear of Materials*, Houston, TX, New York: ASME, 1987, 763–770.
20. Jiang X. X., Li S. Z., Duan C. T. et al. Study of the corrosive wear of Ti-6Al-4V in acidic medium. *Wear*, 1989, 129(2): 293–301.
21. Jiang X. X., Li S. Z., Lin X. P. et al. Corrosive wear behavior of stainless steel in dilute sulfuric acid solution. *Acta Metall Sin*, 1991, 27(4): B21–B29.
22. Klueh R. L., Shiba K., and Sokolov M. A. Embrittlement of irradiated ferritic/martensitic steels in the absence of radiation hardening. *J Nucl Mater*, 2008, 377(3): 427–437.
23. Simos N., Kirk H. G., Thieberger P. et al. Radiation damage studies of high power accelerator materials. *J Nucl Mater*, 2008, 377(1): 41–51.
24. Henry J., Averty X., Dai Y. et al. Tensile behaviour of 9Cr–1Mo tempered martensitic steels irradiated up to 20 dpa in a spallation environment. *J Nucl Mater*, 2008, 377(1): 80–93.
25. Nishiyama Y., Onizawa K., Suzuki M. et al. Effects of neutron-radiation-induced intergranular phosphorus segregation and hardening on embrittlement in reactor pressure vessel steels. *Acta Mater*, 2008, 56(16): 4510–4521.
26. Tsang D. K. L., Marsden B. J., Vreeling J. A. et al. Analyses of a restrained growth graphite radiation creep experiment. *Nucl Engg Des*, 2008, 238(11): 3026–3030.
27. Akiyoshi M. and Yano T. Neutron-radiation effect in ceramics evaluated from macroscopic property changes in as-irradiated and annealed specimens. *Prog Nucl Energ*, 2008, 50(2–6): 567–574.
28. Gaganidze E., Schneider H. C., Dafferner B. et al. High-dose neutron radiation embrittlement of RAFM steels. *J Nucl Mater*, 2006, 355(1–3): 83–88.
29. Song S. H. and Weng L. Q. Embrittlement of a Cr-Mo low-alloy steel due to low-temperature neutron radiation. *Acta Metall Sin (Eng Lett)*, 2006, 19(1): 20–26.
30. Gilbert R. W., Griffiths M., and Carpenter G. J. C. Amorphous intermetallics in neutron irradiated zircaloys after high fluences. *J Nucl Mater*, 1985, 135(2–3): 265–268.
31. Kamdar M. H. Embrittlement by liquid metals. *Prog Mater Sci*, 1973, 15(4): 289–374.
32. Lynch S. P. Environmentally assisted cracking: Overview of evidence for an adsorption-induced localised-slip process. *Acta Metall*, 1988, 36(10): 2639–2661.
33. Shu, D. L. *Mechanical Properties of Engineering Materials*. Beijing: China Machine Press, 2003.

9 Macro- and Microcomputational Materials Mechanics

Materials scientists and engineers use computers to investigate and solve problems in materials. If experimental investigation and theoretical analysis are called the first and the secondary method, the emerging of numerical simulations can be viewed as the third method, which is causing a structural change in scientific research and development. It promotes the cross-penetration of multiple disciplines and accelerates the transition from fundamental research to technological applications. Computational materials mechanics is a discipline that solves material-dependent mechanics problems based on the theories of materials science and mechanics, by employing modern electronic computers and numerical methods. It intersects each branch in materials science and mechanics, expands the research and applications of each branch in materials science and mechanics, and simultaneously develops its own theories and approaches.

This chapter first introduces the structural hierarchy of materials and computational materials science, then introduces the primary simulation approaches for mechanical properties of materials at the macro-, micro-, and nanoscales. The finite element method (FEM) is introduced at the macroscale, the methodology for the simulation of elasto-plastic deformation of polycrystals is introduced at the mesoscale, and molecular dynamics simulation is introduced at the nanoscale. Finally, multiscale analysis of deformation and failure of materials is introduced.

9.1 STRUCTURAL HIERARCHY OF MATERIALS AND COMPUTATIONAL MATERIALS SCIENCE

Historically, human beings knew materials very early on. However, studies on materials have been primarily performed via experimentation. A material with a predetermined property is obtained mainly or completely through experimentation. However, material properties not only depend on chemical composition, but also are closely related to microstructure, which itself depends on chemical composition and manufacturing processes. Therefore, relying completely on experimentation is impractical. Since the 1950s—with the invention and wide application of computers and the continuous emerging of new analytical and measurement facilities and technologies—theories in materials science have been developing rapidly, resulting in a new branch called computational materials science. At the same time, because of a deeper understanding of length scales and the hierarchy of microstructure, breakthrough progress has been made at new scales. Due to the development of computational materials science, the dream of materials scientists and engineers—the design of materials—is gradually coming true [1]. Therefore, the formation and development of computational materials science is an important mark for material disciplines that are evolving from "engineering" to "science."

9.1.1 Material System and the Structural Hierarchy of Materials

The so-called material system consists of the following elements: material properties, prices, constituents, microstructure, processing conditions, spatial coordinates (positions), and time. Properties and prices form one group of variables, while constituents, microstructure, processing conditions, spatial coordinates, and time form another group of variables. The purpose of studying the material system is to understand the correlation between two groups of variables, and among the variables in each group, so that we can (1) explain material behavior in terms of processing conditions and further the design of materials and manufacture; and (2) produce materials with predetermined microstructure and properties for a relatively low cost and with relatively easy processing conditions of implementation and control [1].

In the field of materials science and engineering, the classification of the structural hierarchy of materials is not yet united, but we can distinguish some broad categories. For example, according to the spatial dimensions of the investigated object, some material design scientists classify material structures into three scales [1].

1. Scale at engineering design. This dimension corresponds to macroscopic materials, and the concern is the design for manufacturing and service performance of bulk materials.
2. Scale at the continuum model. The typical dimension is at the order of 1 μm, and a material is taken as a continuous medium regardless of the behavior of individual atoms or molecules.
3. Scale at the microscopic design. This spatial dimension is at the order of 1 nm, and the concern is the design at the atomic or molecular scale.

There are four scales in terms of spatial dimension.

1. Macroscale: The scale at which the human being's daily life takes place. These temporal and spatial domains can be reached directly by a human being, or through the aid of devices or mechanisms. The basic necessities of life and production take place at this scale. This spatial dimension ranges from 1 mm to tens of thousands of kilometers.
2. Mesoscale: This scale comes between "macroscale" and "microscale." Its dimension is at the order of the millimeter. In materials science, the representative structure at this scale is the grain. At this scale, microscopic structures such as texture, composition segregation, the grain boundary effect, capillary absorption, percolation, and catalysis, all play roles.
3. Microscale: This dimension is at the order of the micron. For many years, with the aid of optical microscopy, electronic microscopy, x-ray diffraction, and the electronic microprobe (EMPA), extensive studies have been conducted for crystalline and amorphous materials at this scale. Many of these approaches have become the conventional analytical methods in materials science.
4. Nanoscale: This dimension ranges from the nanometer to the micron; that is, $10^{-6} \sim 10^{-9}$ m, approximately corresponding to the dimension of an aggregate of dozens or several hundred atoms. At this scale, quantum properties appear and the studied object cannot be treated as a "continuum," and simply represented by statistically averaged values. Defects and dopants in the microstructure play significant roles.

From the viewpoint of physics, microscopic models consider complicated electron-quantum effects, and the basic properties of a material at this scale are generally quantum-mechanical. At the nanoscale, matter is simulated by an aggregate of particles united under the potentials of atomic interaction. The basic properties of a material at this scale are studied by statistical mechanics and molecular dynamics. At the mesoscale, continuum elements with intrinsic microstructure are adopted, and the mechanics of a continuous medium based on Newton mechanics is employed to study their basic properties [2].

9.1.2 Generation and Main Methodologies of Computational Materials Science

Computational materials science stems directly from the idea of "material design." Material design predicts the constituents, structures, and properties of a new material via theories and computations, or develops a new material with specified properties via theoretical design. The idea of "material design"

TABLE 9.1
Mapping Relationship between Approaches in Material Simulations and Spatial Dimensions (Nano- to Microscale)

Spatial Dimension (m)	Simulation Method	Typical Applications
$10^{-10} \sim 10^{-6}$	Metropolis Monte Carlo	Thermodynamics, diffusion, ordered system
$10^{-10} \sim 10^{-6}$	Cluster variation method	Thermodynamics system
$10^{-10} \sim 10^{-6}$	Ising model	Magnetic system
$10^{-10} \sim 10^{-6}$	Bragg–Williams–Gorsky model	Thermodynamics system
$10^{-10} \sim 10^{-6}$	Molecular field model	Thermodynamics system
$10^{-10} \sim 10^{-6}$	Molecular dynamics models (embedded atom method, shell potentials, empirical pair potentials, bond-order potentials, effective medium theory, secondary polar moment potentials, etc.)	Structures and dynamic characteristics of lattice defects
$10^{-12} \sim 10^{-8}$	Ab initio (i.e., the first-principles) molecular dynamics method (includes tight-binding potential and local density functional theory)	Structures and dynamic characteristics of simple lattice defects, material constants

started in the 1950s, and the field of "material design science" was proposed in the 1980s. With the progress of numerical simulation methods in materials science and engineering, an emerging and exciting inter-disciplinary branch had developed, called "computational materials science." In computational materials science, several scales are classified according to the dimension of structure. At each scale, according to the problem, the appropriate model is selected, then numeric simulation is carried out. Tables 9.1 through 9.3 list the mapping relationships between models and approaches in the material simulation at each scale and spatial dimension [3]. This gives us a general view of the approaches used to solve these types of questions. It can be seen from the tables that some approaches (such as molecular dynamics, dislocation dynamics, large-scale finite element analysis) are probably only suitable for a certain scale, which other approaches (such as the percolation model and the cell automata model) do not have an "intrinsic" dimension, and are almost suitable for any dimension and structural scale.

9.1.3 Trend of Computational Materials Science

Although computational materials science stems from the idea of material design, its complete description includes two parts—computational analysis and modeling. Currently, computational materials science has developed sufficiently to establish its own independent and complete theoretical framework. It is promising and has greatly promoted the development of materials science. Based on its current state, it promotes the following [1]:

1. The unity of theories of three types of materials. Computational materials science is suitable for metals, inorganic nonmetallic materials, and polymers; and its theoretical achievements further promote the development of a united materials science. It should be noted that computational materials science has made more and more outstanding achievements in the field of metallic materials, which can serve as enlightenment and reference for scientists and engineers in nonmetallic material fields.
2. The unity of studies of material structures at different scales. In computational materials science, although different specific approaches are employed in the study of structures at

TABLE 9.2
Mapping Relationship between Approaches in Material Simulations and Spatial Dimensions (Micro- to Mesoscale)

Spatial Dimension (m)	Simulation Approach	Applications
$10^{-10} \sim 10^{0}$	Cell automata	Recrystallization, grain growth, phase transformation, hydrodynamics, recrystallization texture, crystal plasticity
$10^{-7} \sim 10^{-2}$	Spring model	Fracture mechanics
$10^{-7} \sim 10^{-2}$	Vertex model, topological mesh model, grain boundary dynamics	Grain coarsening, recrystallization, secondary recrystallization, nucleation, regenerative recovery, grain growth, fatigue
$10^{-7} \sim 10^{-2}$	Geometric model, topology model, compositional model	Recrystallization, grain growth, secondary recrystallization, recrystallization texture, solidification, crystal configuration
$10^{-9} \sim 10^{-4}$	Dislocation dynamics	Crystal plasticity, regenerative recovery, microstructure, dislocation distributions, thermal activation energy
$10^{-9} \sim 10^{-5}$	Dynamic Ginzburg–Landau phase field model	Diffusion, interface motion, formation and coarsening of precipitations, grain coarsening in polycrystals and multiphase materials, conversion between isomorphism and nonisomorphism, Type II superconductor
$10^{-9} \sim 10^{-5}$	Multistate dynamic Potts model	Recrystallization, grain growth, phase transformation, recrystallization texture

Source: Raabe, D., *Computational Materials Science*, Chemical Industry Press, Beijing, 2002; Taken from Table 1.2.

different scales, the modeling processes and simulation steps are primarily the same. This unified form provides a good basis for the construction of the theoretical framework and the operation framework of computational materials science. This harmonic, unified, and stylized operation makes it easier for people to master and apply it, and paves the way for its extension and applications.

3. The unity of three research methods in applications. In the past, the field of materials science relied more on experimentation. Currently, the generation and development of computational materials science complements research methods, and promotes the advances of theories in materials science. This has moved materials science to the forefront of modern science.
4. The unity of materials science and materials engineering. In the past, it was difficult for some subjects and fields to communicate and collaborate. For example, there is not much connection between materials engineers in furnace and mechanical equipment, and materials scientists. Recently, materials scientists have classified the problems and approaches related to furnace, mechanics in mechanisms, heat and mass transfer, and fluid mechanics. These methodologies include the finite difference method (FDM), the finite element

TABLE 9.3
Mapping Relationship between Approaches in Material Simulations and Spatial Dimensions (Meso- to Macroscale)

Spatial Dimension (m)	Simulation Approach	Typical Applications
$10^{-5} \sim 10^{0}$	Large-scale FEM finite difference method, linear iteration, boundary element method	Solution of difference equations at macroscale (mechanics, electromagnetic field, hydrodynamics, temperature field)
$10^{-6} \sim 10^{0}$	Crystal plasticity finite element model, microstructure-based finite element model	Microstructure-sensitive mechanical properties of multicomponent alloys, fracture mechanics, structure, crystal slip, solidification
$10^{-6} \sim 10^{0}$	Taylor–Bishop–Hill model, relaxation and constraint model, Voigt mode, Sachs model, Reuss model, Hashin–Shtrikman model, self-consistent model	Elasticity and plasticity of multiphase or polycrystalline material, homogeneity of heterogeneous microstructure, recrystalline texture, Taylor factor, crystal slip, homogenization of microstructures
$10^{-8} \sim 10^{0}$	Cluster model	Elasticity of polycrystals
$10^{-10} \sim 10^{0}$	Percolation model	Nucleation, fracture mechanics, phase transformation, plasticity, current transport, superconductor

Source: Raabe, D., *Computational Materials Science*, Chemical Industry Press, Beijing, 2002; Taken from Table 1.3.

method (FEM), and the boundary element method (BEM), which delve into the scope of computational materials science at the macroscale. In addition, bridges have been established among the different industries in engineering: between traditional ceramics (daily ceramics, construction ceramics, electric ceramics, and chemical ceramics) and special ceramics (functional ceramics and structural ceramics). This unity and fusion will promote the full development of materials science and engineering, and motivate great progress in science and technology.

9.2 COMPUTATIONAL MATERIAL MECHANICS AT THE MACROSCALE

Computational material mechanics at the macroscale is based on the mechanics of a continuous medium, and solves differential equations of continuous media using computational mathematical approaches. We studied the equilibrium equations in Chapter 1. For simple cases, analytical solutions can be obtained in terms of equilibrium equations and boundary conditions. For complicated cases, analytical solutions are not available. What should we do? For the time being, general approaches include the BEM, the FDM, and the FEM. Here, we introduce the FEM. The other two methods can be found in the Reference [3].

The idea of "finite elements," proposed by mechanicians and mathematicians, is one of the greatest achievements of the twentieth century. It enables airplanes to aviate freely in the sky, and allows the safe operation of many large structures such as buildings, bridges, and nuclear power plants. Finite element analysis employs mathematical approximation to simulate real physical systems. Using simple and interactive elements, a limited number of unknowns can be used to approximate the unlimited number of unknowns of real systems [4].

9.2.1 Generation of the Finite Element Method

In order to overcome the difficulty of dealing with continuity problems, engineers and mathematicians put forward a variety of discrete methods. These methods are approximations: when the number of discrete variables increases, they approach the expected, real, continuous solutions. Mathematicians and engineers implemented discreteness as solutions to continuous problems using different approaches, and established the general method applicable to the governing differential equations of the problems. These methods include finite difference approximation, weighted residual methods, and the extremum of functions with appropriate definitions. On the other hand, engineers often deal with this type of problem more directly. They establish discrete elements to simulate a limited part of a continuous region [5].

The term "finite element" stems from the viewpoint of "direct simulation" in engineering. Clough first adopted this terminology. This is extremely important. Conceptually, it is easier to understand this approach; and in computation, a unified method can be applied to a variety of problems, and standard computational programs can be developed.

The first successful attempt of the modern FEM was to extend the rigid frame displacement method to the plane problem in elastic mechanics. This was the achievement obtained by Turner and Clough when they analyzed the structure of the airplane in 1956. They first gave the correct solution of a plane problem using triangle elements. Their research work initiated new approaches to solving complex elastic plane problems with electronic computers. In 1960, Clough further investigated elastic plane problems, and for the first time, proposed the terminology "finite element method." The efficiency of the FEM was recognized. In the past 40 years, the applications of the FEM have been extended from plane elastic problems to three-dimensional problems and shell problems; from static equilibrium problems to stability problems, dynamic problems, and vibration problems. The analyzed objects have been extended from elastic materials to plastic, viscoelastic, viscoplastic, and composite materials; from solid mechanics to fluid mechanics and heat transfer. In engineering, it has been extended from analysis and verification to optimized design, and has been combined with auxiliary computer design technologies [6].

With the development of computer technologies, FEMs have exhibited general applicability to a variety of mechanics problems. Under the strong push of requirements in industrial applications, mathematicians and mechanicians have gradually established the mathematical theorems of FEMs, general solutions, and programs. In traditional mechanics, a variety of problems in structural mechanics and solid mechanics—such as rods, plates, shells, bulk materials, their governing equations and solution methods—are apparently different, and belong to different research fields. But currently, for the first time in history, the field of mechanics provides a unified approach, and tools, to industries in the solution of a variety of problems: finite element analysis software packages. The universality of this approach makes it rapidly and extensively applicable in the civil, aerospace, and mechanical industries. Now, engineers can accurately predict the mechanical characteristics of skyscrapers, oversea bridges, vehicles, and rockets, and simulate many high speed collisions, explosions, and complex flow phenomena. Due to its vigorous power, this method expands rapidly to other fields in mechanics and in physics [7].

9.2.2 Matrix Representation of Elastic Mechanics and Variational Principles

Finite element analysis frequently uses the matrix forms of the governing equations of elastic mechanics, and its equivalent variational principles. The detailed derivation is described in Chapter 1 and in other references [4]. Here, a plane problem (plane stress problem or plane strain problem) is taken as an example.

9.2.2.1 Matrix Representation of Governing Equations

The fundamental equations in elastic mechanics include equilibrium equations, geometric equations, physics equations, and additional boundary conditions. Here, we introduce the matrix representation of governing equations, and the boundary conditions of plane problems.

1. Equilibrium equation

$$\left.\begin{aligned}\frac{\partial \sigma_x}{\partial x}+\frac{\partial \tau_{xy}}{\partial y}+f_x=0\\ \frac{\partial \sigma_y}{\partial y}+\frac{\partial \tau_{yx}}{\partial x}+f_y=0\end{aligned}\right\}. \tag{9.1}$$

Expressed in matrix form as

$$\begin{bmatrix}\frac{\partial}{\partial x} & 0 & \frac{\partial}{\partial y}\\ 0 & \frac{\partial}{\partial y} & \frac{\partial}{\partial x}\end{bmatrix}\begin{Bmatrix}\sigma_x\\ \sigma_y\\ \tau_{xy}\end{Bmatrix}+\begin{Bmatrix}f_x\\ f_y\end{Bmatrix}=0. \tag{9.2}$$

The abbreviated formula is

$$[A]^T[\sigma]+[f]=0, \tag{9.3}$$

where $[\sigma]=\begin{Bmatrix}\sigma_x\\ \sigma_y\\ \sigma_{xy}\end{Bmatrix}, [f]=\begin{Bmatrix}f_x\\ f_y\end{Bmatrix}, [A]=\begin{bmatrix}\partial/\partial x & 0\\ 0 & \partial/\partial y\\ \partial/\partial y & \partial/\partial x\end{bmatrix}$, and where $[\sigma]$, $[f]$, and $[A]$ are called, respectively, the stress matrix, the body force matrix, and the differential operator matrix.

2. Geometric equations

$$\varepsilon_x=\frac{\partial u}{\partial x},\quad \varepsilon_y=\frac{\partial v}{\partial y},\quad \varepsilon_{xy}=\frac{1}{2}\left(\frac{\partial u}{\partial y}+\frac{\partial v}{\partial x}\right). \tag{9.4}$$

Represented in matrix form as

$$\begin{Bmatrix}\varepsilon_x\\ \varepsilon_y\\ \varepsilon_{xy}\end{Bmatrix}=\begin{bmatrix}\frac{\partial}{\partial x} & 0\\ 0 & \frac{\partial}{\partial y}\\ \frac{\partial}{\partial y} & \frac{\partial}{\partial x}\end{bmatrix}\begin{Bmatrix}u\\ v\end{Bmatrix}. \tag{9.5}$$

Abbreviated as

$$[\varepsilon]=[A][u], \tag{9.6}$$

where $[\varepsilon]=\begin{Bmatrix}\varepsilon_x\\ \varepsilon_y\\ \varepsilon_{xy}\end{Bmatrix}, [u]=\begin{Bmatrix}u\\ v\end{Bmatrix}$ are called, respectively, the strain matrix and the displacement matrix.

3. Physics equations
   The physics equations of the plane stress problems are

$$\sigma_x=\frac{E}{1-v^2}(\varepsilon_x+v\varepsilon_y),\quad \sigma_y=\frac{E}{1-v^2}(\varepsilon_x+v\varepsilon_y),$$

$$\tau_{xy}=\frac{E}{2(1+v)}\varepsilon_{xy}t. \tag{9.7}$$

In the matrix forms,

$$\begin{Bmatrix}\sigma_x\\ \sigma_y\\ \tau_{xy}\end{Bmatrix}=\frac{E}{1-v^2}\begin{bmatrix}1 & v & 0\\ v & 1 & 0\\ 0 & 0 & \dfrac{1-v}{2}\end{bmatrix}\begin{Bmatrix}\varepsilon_x\\ \varepsilon_y\\ \varepsilon_{xy}\end{Bmatrix}t. \tag{9.8}$$

Abbreviated as

$$[\sigma]=[D][\varepsilon], \tag{9.9}$$

where $[D]=E/1-v^2\begin{bmatrix}1 & v & 0\\ v & 1 & 0\\ 0 & 0 & 1-v/2\end{bmatrix}$ is called the elastic matrix of plane stress problems, and E, ν are the elastic modulus and the Poisson ratio of the material. For the plane strain problem, the physics equations can also be written in the aforementioned forms, but the E in $[D]$ is replaced by $E/(1-v^2)$, and ν is replaced by $v/(1-v)$.

4. Boundary conditions
   Stress boundary conditions:

$$l(\sigma_x)_S+m(\tau_{xy})_S=T_x,\quad m(\sigma_y)_S+l(\tau_{xy})_S=T_y. \tag{9.10}$$

Represented in matrix form as

$$\begin{bmatrix}l & 0 & m\\ 0 & m & l\end{bmatrix}\begin{Bmatrix}\sigma_x\\ \sigma_y\\ \tau_{xy}\end{Bmatrix}=\begin{Bmatrix}T_x\\ T_y\end{Bmatrix}. \tag{9.11}$$

Abbreviated as

$$[n][\sigma] = [T], \tag{9.12}$$

where $[n] = \begin{bmatrix} l & 0 & m \\ 0 & m & l \end{bmatrix}, [T] = \begin{Bmatrix} T_x \\ T_y \end{Bmatrix}$ are called, respectively, the direction cosines of the outward normal of the boundary, and the face force matrix of the given boundary.

The displacement boundary conditions are

$$u = \overline{u}, \quad v = \overline{v}, \tag{9.13}$$

and are rewritten in matrix form as

$$[u_S] = [\overline{u}], \tag{9.14}$$

where $[u_S] = [u \quad v]^T$ is the displacement matrix on the boundary, and $[u] = [\overline{u} \quad \overline{v}]^T$ is the displacement matrix of the given boundary.

5. For the given boundary conditions, solving Equations 9.3, 9.6, and 9.9 simultaneously gives the displacements, strains, and stresses.

9.2.2.2 Variation Principles

The mechanics analysis of structures based on the fundamental equations of elastic mechanics is called the static method. Another analytical method used in elastic mechanics is called the energy method, and the variation method. The variational equations, the expression for the energy principle, and the expression for the virtual work principle in elastic mechanics are energy expressions for the fundamental equations of elastic mechanics. From the variations of the displacements, the virtual displacement principle and the minimum potential energy principle can be derived, which are equivalent to the equilibrium equations and stress boundary conditions represented by the displacements. From the variation of stresses, the virtual work principle and the minimum complementary energy principle can be derived, which are equivalent to the displacement compatibility conditions and the displacement boundary conditions. Next, we use plane problems as examples to introduce the virtual displacement principle and the minimum potential energy principle, which are the bases for the displacement method in finite element analysis.

1. Virtual displacement principle
   Assume an elastic material undergoes virtual displacement, denoted by u^*, v^*, and the introduced virtual strains are $\varepsilon_x^*, \varepsilon_y^*, \varepsilon_{xy}^*$, represented by matrixes

$$\left[u\right] = \begin{Bmatrix} u^* \\ v^* \end{Bmatrix} \text{ and } \left[\varepsilon\right] = \begin{Bmatrix} \varepsilon_x^* \\ \varepsilon_y^* \\ \varepsilon_{xy}^* \end{Bmatrix}. \tag{9.15}$$

In terms of the virtual displacement principle of an elastic object, the necessary and sufficient condition for the equilibrium state of an elastic object is that the total virtual work done by the external force is equal to the total virtual deformation work of the internal stress in the elastic object, for any virtual displacements. The virtual work equation for the virtual displacement principle of the plane problem of an elastic object is

$$\iint_{\Omega}\left(f_x u^* + f_y v^*\right)\mathrm{d}x\mathrm{d}y + \int_{S}\left(T_x u^* + T_y v^*\right)\mathrm{d}S$$

$$= \int_{\Omega}(\sigma_x \varepsilon_x^* + \sigma_y \varepsilon_y^* + \tau\varepsilon_{xy}^*)\,\mathrm{d}x\mathrm{d}y, \quad (9.16)$$

where

Ω is the interior region of the elastic object
S is the boundary for the known surface stress T

The virtual work equation can be expressed compactly in a matrix form as

$$\iint_{\Omega}[u]^T[f]\mathrm{d}x\mathrm{d}y + \int_{S}[u]^T[T]dS = \iint_{\Omega}[\varepsilon]^T[\sigma]\mathrm{d}x\mathrm{d}y. \quad (9.17)$$

2. The minimum potential energy principle
According to the minimum potential energy principle, the necessary and sufficient condition for a virtual displacement to be the true displacement is that the total potential energy of the elastic object reaches the minimum values with the virtual displacement. This principle can be represented by the following equation

$$\delta V = \iint_{\Omega}\left[\left(\frac{\partial\sigma_x}{\partial x} + \frac{\partial\tau_{xy}}{\partial y} + f_x\right)\delta u + \left(\frac{\partial\sigma_y}{\partial y} + \frac{\partial\tau_{xy}}{\partial x} + f_y\right)\delta v\right]\mathrm{d}x\mathrm{d}y$$

$$-\int_{S_0}[(l\sigma_x + m\tau_{xy} + f_x)\,\delta u + (m\sigma_y + l\tau_{xy} + f_y)\,\delta v]\mathrm{d}S = 0. \quad (9.18)$$

The aforementioned equation can be written in matrix form:

$$\delta V = \iint_{\Omega}\left\{[A]^T[\sigma]+[f]\right\}^T[\delta u]\mathrm{d}x\mathrm{d}y + \int_{S_0}\left\{[T]^T - [n]^T[\sigma]\right\}[\delta u]\mathrm{d}S = 0, \quad (9.19)$$

where $[\delta u] = \begin{Bmatrix}\delta u\\ \delta v\end{Bmatrix}$. If the variation of virtual displacement is taken as the variation of displacement, the minimum potential energy principle can be derived from the virtual displacement principle.

9.2.3 Analytical Procedures in the Finite Element Method

No matter whether a rod or an elastic continuous body, finite element analysis usually consists of three principal procedures: (1) discretization of structure, (2) element analysis, (3) assembly of elements, and (4) solving of the whole system [6].

9.2.3.1 Discretization of Structure

During the discretization of a structure, the first thing to do is to select the appropriate element. Figure 9.1 shows the shapes of a variety of elements. The intersection point of edges (end point) is called a node.

Discretization of structure means to divide the original structure into an assembly of discrete elements. This procedure is also called meshing by elements. The attributes of selected elements replace or simulate the attributes of the corresponding part in the original structure. Figure 9.2 enumerates the meshes of some structures.

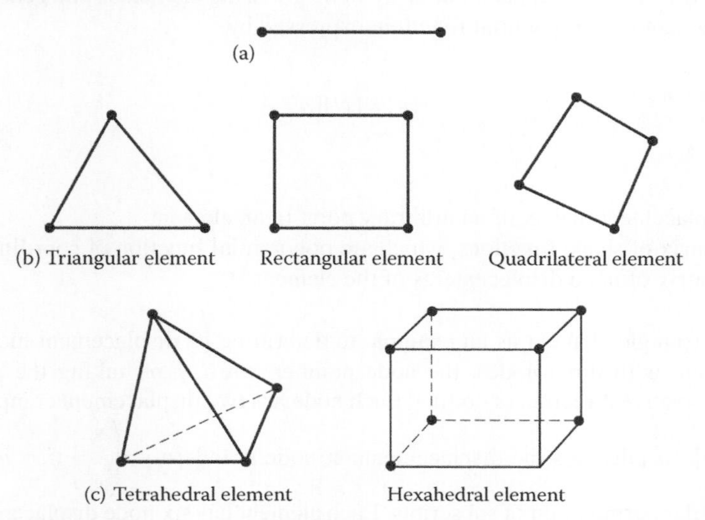

FIGURE 9.1 The element shapes. (a) One-dimensional element. (b) Two-dimensional element. (c) Three-dimensional element.

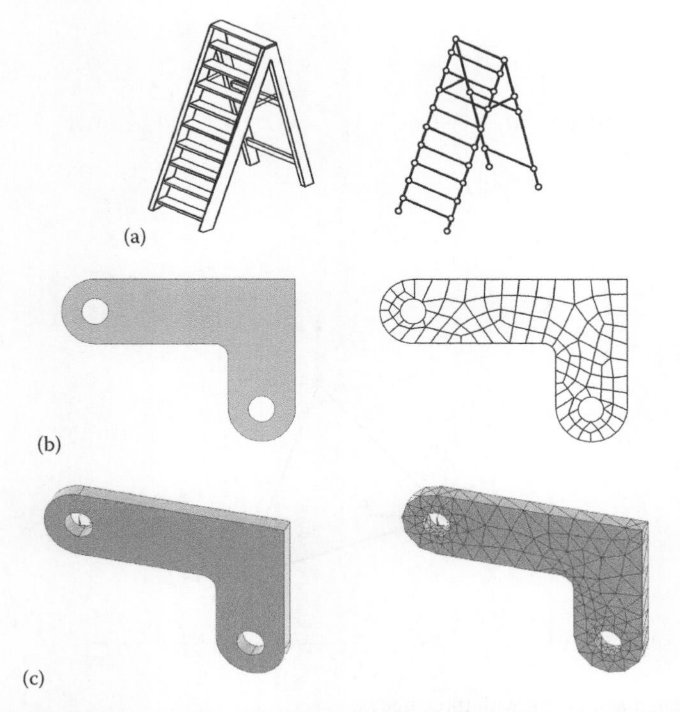

FIGURE 9.2 Physical systems and finite element models. (a) Meshing with one-dimensional elements. (b) Meshing with two-dimensional elements. (c) Meshing with three-dimensional elements.

9.2.3.2 Element Analysis

9.2.3.2.1 Determination of Displacement Mode

The reasonable representation of the true continuous structure using the discretized finite elements is implemented by the reasonable assumption of a displacement distribution in each element. This assumption directly determines the accuracy of the final solutions. Therefore, the determination of the displacement mode is the key to the FEM. The important rule in determining the displacement mode is to keep the displacement compatibility of neighboring elements. The general assumption is that the displacement is a polynomial function, expressed by

$$[u]=[N][a^e], \tag{9.20}$$

where

$[u]$ is the displacement matrix of an arbitrary point in an element
$[N]$ is the matrix of shape functions, which are polynomial function of coordinates of nodes
$[a^e]$ is the matrix of node displacements of the element

We take the triangle element as an example to determine its displacement mode. For a typical triangular element with three nodes, the node numbers are i, j, m, taking the counterclockwise direction as the positive direction of coding. Each node has two displacement components, as shown in Figure 9.3 [5]. $[a_i]$ denotes the displacements at node i, and $[a_i]=\begin{pmatrix} u_i \\ v_i \end{pmatrix}$ (i, j, m), where (i, j, m) denotes the circular permutation of subscripts. Each element has six node displacements or six nodal degrees of freedom:

$$[a^e]=\begin{pmatrix} [a_i] \\ [a_j] \\ [a_m] \end{pmatrix}=[u_i \quad v_i \quad u_j \quad v_j \quad u_m \quad v_m]^T. \tag{9.21}$$

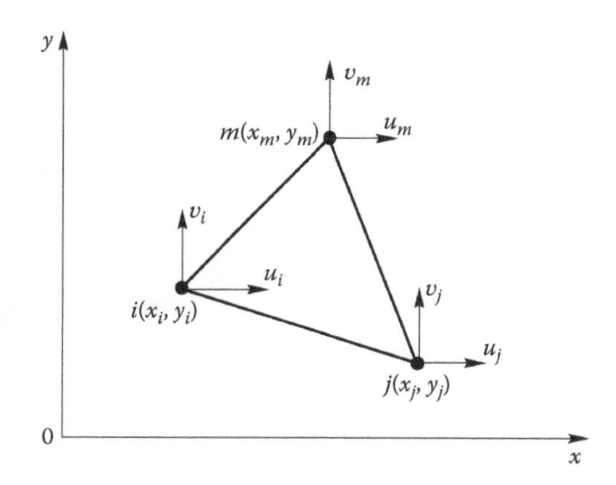

FIGURE 9.3 Triangular element with three nodes.

In the FEM, the displacement mode or displacement function of an element is approximated by a polynomial, because the polynomial computation is simple and convenient, and it can approximate the segment of any smooth curve. The displacement functions of a triangular element with three nodes take first-order polynomials:

$$u = \beta_1 + \beta_2 x + \beta_3 y \quad v = \beta_4 + \beta_5 x + \beta_6 y, \tag{9.22}$$

where
(u, v) are displacements at an arbitrary point of the triangular element
$\beta_1 \sim \beta_6$ are undermined coefficients

In the first expression of the aforementioned equation, substituting the coordinates (x_i, y_i) of node i gives the displacement u_i in the x-direction of node i. Similarly, u_j and u_m can be obtained. They are represented by

$$u_i = \beta_1 + \beta_2 x_i + \beta_3 y_i \quad u_j = \beta_1 + \beta_2 x_j + \beta_3 y_j \quad u_m = \beta_1 + \beta_2 x_m + \beta_3 x_m. \tag{9.23}$$

From Equation 9.23, we have

$$\begin{aligned} \beta_1 &= \frac{1}{2A}(a_i u_i + a_j u_j + a_m u_m) \\ \beta_2 &= \frac{1}{2A}(b_i u_i + b_j u_j + b_m u_m) \\ \beta_3 &= \frac{1}{2A}(c_i u_i + c_j u_j + c_m u_m) \end{aligned} \tag{9.24}$$

where $A = 1/2\begin{vmatrix} 1 & x_i & y_i \\ 1 & x_j & y_j \\ 1 & x_m & y_m \end{vmatrix}$, and a_i, b_i, and c_i are

$$a_i = x_j y_m - x_m y_j, \quad b_i = y_j - y_m, \quad c_i = -x_j + x_m \quad (i, j, m); \tag{9.25}$$

and where (i, j, m) denotes the circular permutation of subscripts. Similarly,

$$\begin{aligned} \beta_4 &= \frac{1}{2A}(a_i v_i + a_j v_j + a_m v_m) \\ \beta_5 &= \frac{1}{2A}(b_i v_i + b_j v_j + b_m v_m), \\ \beta_6 &= \frac{1}{2A}(c_i v_i + c_j v_j + c_m v_m) \end{aligned} \tag{9.26}$$

Substituting $\beta_1 - \beta_6$ into Equation 9.22, displacement functions can be expressed as the functions of nodal displacements:

$$u = N_i u_i + N_j u_j + N_m u_m \quad v = N_i v_i + N_j v_j + N_m v_m, \tag{9.27}$$

where

$$N_i = \frac{1}{2A}(a_i + b_i x + c_i y) \quad (i, j, m). \tag{9.28}$$

N_i, N_j, N_m are called interpolation functions or shape functions of the element. For the example here, they are the first-order functions of coordinates x, y. a_i, b_i, c_i, …, c_m are constants, determined by three nodal coordinates of three nodes. If written as

$$[N] = \begin{bmatrix} N_i & 0 & N_j & 0 & N_m & 0 \\ 0 & N_i & 0 & N_j & 0 & N_m \end{bmatrix}, \tag{9.29}$$

then $[N]$ is the matrix of interpolation functions or shape functions.

9.2.3.2.2 *Representation of Element Strain by Nodal Displacements*

In terms of geometric equations, the following matrix equations can be obtained,

$$[\varepsilon] = [B][a^e], \tag{9.30}$$

$$[B] = [A][N], \tag{9.31}$$

where

- $[\varepsilon]$ is the strain matrix of an arbitrary point in an element
- $[B]$ is the strain matrix, taking the partial differential derivative of shape functions with respect to coordinates as its elements
- $[A]$ is the matrix of differential operator, as seen in Equation 9.3.

9.2.3.2.3 *Element Stress Represented by Nodal Displacements*

According to the physical equations, the following matrix equations can be established,

$$[\sigma] = [D][\varepsilon] = [D][B][a^e] = [S][a^e], \tag{9.32}$$

$$[S] = [D][B], \tag{9.33}$$

where

- $[\sigma]$ is the stress matrix for an arbitrary point in the element
- $[S]$ is the stress matrix of the element
- $[D]$ is the elastic matrix of the element related to its material properties, seen in Equation 9.9

9.2.3.2.4 Construction of Element Stiffness Matrix Using Virtual Displacement Principle or the Minimum Potential Energy Principle

In terms of the virtual work Equation 9.16, the virtual work equation of an element can be expressed as

$$[F^e]=[K^e][a^e], \tag{9.34}$$

$$[K^e]=\iint_{\Omega}[B]^{\mathrm{T}}[D][B]t\mathrm{d}x\mathrm{d}y, \tag{9.35}$$

where

- t is the thickness of the element
- $[F^e]$ is the matrix of the nodal forces of the element
- $[K^e]$ is the stiffness matrix of the element. For an element with three nodes (with vertex as i, j, m),

we have

$$[F^e]=[F_{ix} \quad F_{iy} \quad F_{jx} \quad F_{jy} \quad F_{mx} \quad F_{my}]^T,$$

$$[K^e]=[B]^T[D][B]tA=\begin{bmatrix}[K_{ii}] & [K_{ij}] & [K_{im}]\\ [K_{ji}] & [K_{jj}] & [K_{jm}]\\ [K_{mi}] & [K_{mj}] & [K_{mm}]\end{bmatrix},$$

where A is the area of the triangular element [5]. Substituting the elastic matrix $[D]$ and the strain matrix $[B]$, any block matrix can be represented as

$$[K_{rs}]=[B_r]^T[D][B_s]tA=\frac{Et}{4(1-\nu^2)A}\begin{bmatrix}K_1 & K_3\\ K_2 & K_4\end{bmatrix} \quad (r,s=i,j,m),$$

where

$$K_1=b_rb_s+\frac{1-\nu}{2}c_rc_s,\quad K_2=\nu c_rb_s+\frac{1-\nu}{2}b_rc_s,$$

$$K_3=\nu b_rc_s+\frac{1-\nu}{2}c_rb_s,\quad K_4=c_rc_s+\frac{1-\nu}{2}b_rb_s,$$

where $a_i, b_i, c_i, \ldots, c_m$ are constants, determined by the nodal coordinates of the element, as seen in Equation 9.25.

9.2.3.3 Assembly of Equilibrium Equations of the Whole System

In this step, the equilibrium equations at node i can be established as

$$\sum_e \left[F_i^e\right] = \sum_e \left[F_E^e\right], \quad (9.36)$$

where

$\sum_e$ denotes the sum for all elements around node i

$\left[F_E^e\right]$ is the matrix for equivalent loads at node i of the element loads

The equilibrium equations for all nodes can be established, and expressed in a matrix form:

$$[K][a] = [F_E], \quad (9.37)$$

where

$[K]$ is the stiffness matrix of the assembly, and its element is the assembly of the stiffness of each element $[K^e]$

$[a]$ is the matrix of assembly displacement

$[F_E]$ is the matrix of equivalent loads of nodes

The assembly of $[K]$ and $[F_E]$ and the imposition of boundary conditions are discussed in detail in Reference [6].

9.2.3.4 Matrix for Solving Nodal Displacements and Calculation of Stresses

For linear problems, the equilibrium equations of the assembly (9.37) are linear algebra equations. Solving these equations gives nodal displacements, and the stresses in each element can be calculated using Equation 9.32. For most complicated structures, Equation 9.37 is a high-order linear equation. Linear algebra equations can be solved by direct methods and iteration methods. Direct methods include the Gaussian elimination method and the Gauss–Jordan elimination method. In the iteration method, an initial solution is assumed, and iteration is carried out according to specific algorithms. The error of the solution is examined in each iteration, and it decreases constantly with the increase of the iteration number, until it satisfies the accuracy of the solution [6].

The computation work of the FEM is mainly in the generation and solution of the system of equations. Due to the rapid development of computers and the emergence of a variety of numerical computational methods, FEMs proposed in the 1950s developed rapidly in the 1960s.

In order for users to use finite element software simply, conveniently, and rapidly, preprocess initial data automatically and use a postprocessor to display the results in contours, figures, and tables.

9.2.4 Brief Introduction to the Nonlinear Finite Element Method

The basic features of linear elastic mechanics is that the equilibrium equations are linear equations of the deformation state; the relationship between strain and displacement in the geometric equations is linear; the relationship between stresses and strains in the physical equations is linear; and the external forces on the force boundaries, or the displacements on the displacement boundaries, are independent or linearly dependent on the deformation state.

In practice, if any of the aforementioned equations or boundaries does not meet these features, then the problem is nonlinear. In terms of the specific features of the equations and the boundary conditions, nonlinear problems can be divided into three types.

1. *Material nonlinear problems.* In this type of problem, the relationship between stresses and strains in physical equations is no longer linear. For example, stress concentration takes place at the location where the geometry of the structure changes discontinuously (such as at notches and cracks). When the external load reaches a certain value, plastic deformation take place at this locality first. Although most of the regions in the structure are elastic, the linear elastic relationship between stresses and strains at this locality no longer holds true.

   In another example, creep deformation occurs in structures serving at high temperatures for extended time periods, which cannot be described by the linear elastic constitutive equations. The relationship between stresses and strains is nonlinear, regardless whether plastic deformation, or creep deformation, is inelastic deformation that cannot be recovered.
2. *Geometric nonlinear problems.* The characteristic of this type of problem is that large displacement and rotation occur in the structure under load. For example, the large deformation of a fishing pole in Figure 9.4; the large flexibility of shell structures; and yielding and over-yielding. At this time, the material may still be in the linear elastic state. However, the equilibrium equations of structure must be established based on the configuration after deformation, in order to consider the influence of deformation on equilibrium. At the same time, due to the large deformation and large rotation, the geometric equation cannot be simplified by a linear form; that is, the expressions for strains should include the second-order terms.
3. *Boundary nonlinear problems.* The typical example of this type of problem is the contact and collision of two objects. The position and range of the contact boundary and the force distribution and magnitude on the contact surface cannot be predetermined, and depend on the solution to the whole problem. Another example is that the magnitude and direction of the external force on the force boundary linearly depend on deformation. For example, the acting direction of the distributed pressure on the surface of a thin-walled structure changes when the structure deforms. Thus, in geometric nonlinear problems, the change of load needs to be considered at the same time.

Many practical engineering problems involve one of the three types of nonlinearity. There are also many practical problems in which three types of nonlinearity occur simultaneously, such as the collision of cars and the forging of metallic materials. During the collision or forging process, structures or materials experience huge deformation; materials are in a plastic flow state, and the contact regions and their interaction vary rapidly, and are accompanied by interaction with nonstructural factors, such as heat.

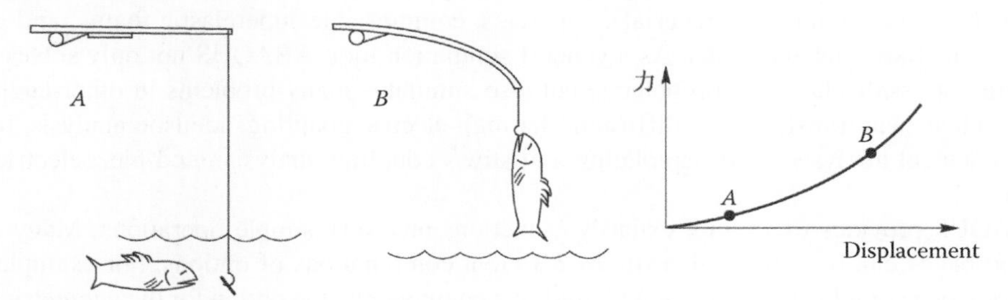

FIGURE 9.4 Geometric nonlinear bend of a fishing pole induced by a fat fish, resulting in a nonlinear force-displacement curve (A–B).

Due to the complexity of nonlinear problems, the cases in which analytical solutions are available are very limited. With the successful application of the FEM in linear analyses, great progress has been made in nonlinear analyses. Numerical solutions for many different types of engineering problems have been obtained, and a lot of general software packages for nonlinear analysis have been developed, such as LS-DYNA finite element software. The theories and algorithms for the nonlinear FEM can be found in References [6,7].

9.2.5 Finite Element Analysis Software Packages

Currently, the widely used finite element software packages include ANSYS structural software, ABAQUS mechanical finite element analysis software, MSC.NASTRAN structural software, ADINA Systems structural analysis software, COSMOS finite element analysis software, and LMS/FALANCS fatigue life analysis software, among others. Next, we briefly introduce ANSYS structural analysis software and ABAQUS mechanical finite element analysis software.

9.2.5.1 ANSYS Structural Analysis Software

ANSYS software is a general finite element analysis software at large scale, integrating the analysis of structures, fluids, electric fields, magnetic fields, and acoustic fields. It was developed by ANSYS in the United States, one of the biggest finite element analysis corporations in the world. It interfaces with most CAD software packages, with the implementation of data sharing and data exchanging, such as Pro/Engineer, NASTRAN, and AutoCAD. It is one of the advanced CAD tools in the design of modern products.

The software mainly includes three modules: the preprocessing module, the computation and analysis module, and the postprocessing module. The preprocessing module provides a powerful tool for the modeling and meshing of objects. Users can conveniently construct finite element models. The analytical computation module covers structural analysis (linear analysis, nonlinear analysis, and highly nonlinear analysis), fluid dynamic analysis, electric–magnetic field analysis, acoustic field analysis, piezoelectric analysis, the coupling of multiple physics fields, the interaction among multiple physical media, sensitivity analysis, and optimization analysis. In order to show the internal structure the postprocessing module represents simulation results using color contour displays, gradient displays, vector displays, flownet of particles, stereographic slicing, transparent displays, and semi-transparent displays. The simulation results can also be output or printed in figures, charts, and curves. The software provides more than 100 element types that simulate different structures and materials in engineering.

9.2.5.2 ABAQUS Mechanical Finite Element Analysis Software

ABAQUS is a powerful finite element analysis software package for engineering simulations, covering problems from relatively simple linear analyses to many very complicated nonlinear analyses. ABAQUS has a rich element library which can simulate an arbitrary geometric shape. It has a materials library that can simulate the properties of typical engineering materials, including metals, rubbers, polymers, composite materials, concretes, compressible superelastic foams, and geological materials like soils and rocks. As a general simulation tool, ABAQUS not only solves a lot of structural (stress/displacement) problems, but also simulates many problems in other engineering fields, such as heat transfer, mass diffusion, thermal–electric coupling, acoustic analysis, rock and soil mechanical analysis (fluid percolating and stress coupling analysis), and piezoelectric media analysis.

ABAQUS provides users with extensive functions and very simple operations. Many complicated problems can be simulated easily by different combinations of options. For example, in the simulation of a complicated problem with multiple components, the option for the geometric dimensions of each component is combined with the option for its associated material properties. In most simulations, even in highly nonlinear problems, users only need to provide some engineering data such as the geometric shape, material properties, boundary conditions, and loading conditions. In a nonlinear analysis, ABAQUS can automatically select the loading increment and the convergence limit. It not only selects appropriate parameters, but also continuously adjusts parameters, which guarantees that an accurate solution will be reached. By defining parameters accurately, users can also effectively control numerical simulation results.

9.3 COMPUTATIONAL MICROMECHANICS OF MATERIALS

The terminology "micro," frequently encountered in the literature, has several scopes with different mechanical analysis characteristics. Roughly speaking, "micro" in mechanics can be classified into "meso" and "nano." The terminology "micromechanics" was advocated by Dr. Xuesheng Qiang, which investigates the deformation and failure of solids accounting for microstructures, and using the continuum mechanics approach. The length scale of micromechanics is generally at the order of one micron. In this case, the English word "micro" correlates with micromechanics very well. Nanomechanics studies the problems at a finer length scale at the nanoscopic level. Nanomechanics studies nanocrystals and nanomaterials, but more generally, the mechanical behavior of general solid materials at the nanoscale [2].

Most metallic materials in engineering are polycrystalline materials, consisting of an aggregate of single crystals with different sizes, shapes, and orientations at the microscale. The physical processes of plastic deformation of polycrystals mainly include dislocation slip, twinning, and sliding at grain boundaries. When grains are randomly oriented, the macroscopic mechanical properties of polycrystals are isotropic. However, under certain conditions, preferred orientations are formed, and the macroscopic mechanical properties of polycrystals are anisotropic. For example, large plastic deformation results in deformation texture in polycrystals [8].

The next section first introduces the homogenization processing of polycrystals, then illustrates the simulation approach for elastic–plastic deformation of polycrystals [9]. Other approaches in computational micromechanics and applications can be found in References [10–13].

9.3.1 Homogenization of Polycrystals

Single-phase polycrystals generally contain a large number of grains with different orientations. The distribution of grain orientations in polycrystals is called crystallographic texture. Each individual grain exhibits high anisotropic responses to external mechanical loads or magnetic fields. Therefore, when all grains have different orientations, predicting the overall response of the whole specimen is complicated and difficult. For example, elastic anisotropy obviously depends on the components of the elastic tensor, while plastic anisotropy is caused by geometric factors such as dislocation slip, crack propagation, and nonthermal mechanisms (twinning and martensitic transformation, etc.). When multiphase polycrystalline materials are involved, the heterogeneous problem is very complex. Not only does the orientation of each constituent vary with location, but the inherent elastic and plastic responses also varies with location.

Due to the inhomogeneous microstructure at the microscale or grain level, both stresses and strains are nonuniform. In order to overcome the complexity of the problem, the isostrain and isostress approaches are proposed, as shown in Figure 9.5 [3].

In Figure 9.5, the symbol "⊥" denotes dislocation. Since the direction of each dislocation is different, the stresses and strains in individual grains under the external load are inhomogeneous. However, in the numerical simulation for polycrystals at the meso-macro scale, approximate homogenization processing for stress or strain is necessary in order to clearly describe the detailed information of the microstructure, and to obtain the solutions of material response with sufficient accuracy. Figure 9.5a is an isostrain model, in which the strain of each grain is assumed to be uniform, but the stress is not uniform. Figure 9.5b is an isostress model, in which the stress in each grain is uniform, but the strain is not uniform [3].

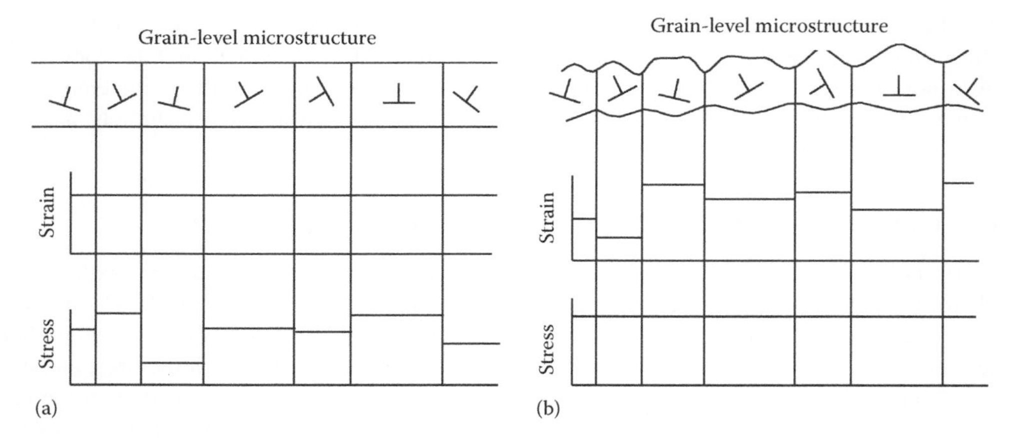

FIGURE 9.5 Inhomogeneous microstructure at mesoscale. (a) Isostrain model. (b) Isostress model.

9.3.2 Simulation Approaches for the Deformation of Polycrystals

A polycrystalline material is an aggregate of single crystals with different sizes, shapes, and orientations. Assuming that neighboring grains contact each other and do not crack after deformation, the equilibrium conditions and compatibility conditions of continuum mechanics are satisfied. The polycrystalline model predicts, on the basis of statistical averaging, the deformation and properties of polycrystals using the properties of single crystals.

We can simulate the microscopic deformation of a face-centered cubic (FCC) polycrystalline material under tension using the isostrain model of polycrystalline materials, combined with the FEM. Considering the influence of grain orientations on deformation, the deformations of grains at different tensile stresses are obtained.

Figure 9.6 shows the FCC polycrystalline model with $10 \times 10 \times 10 = 1000$ grains, subjected to a uniform tension in the z-direction [9,14]. Each element represents an FCC single crystal. Figure 9.7 illustrates the slip planes of the FCC single crystal. There are 4 slip planes and 12 slip directions, listed in Table 9.4.

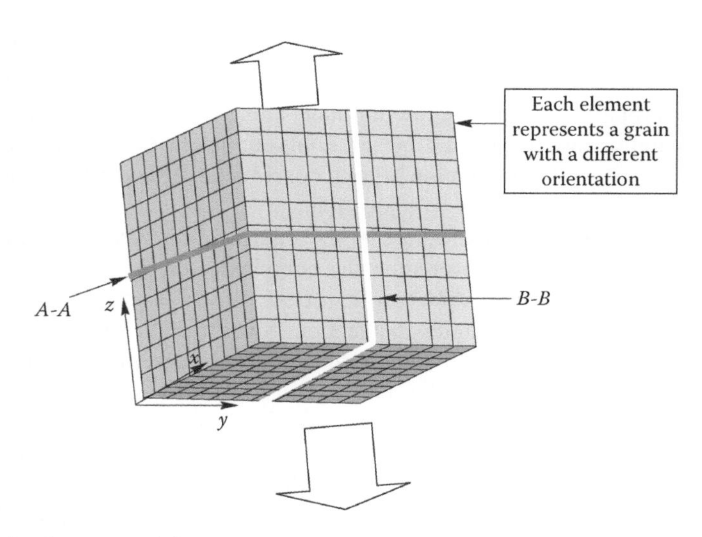

FIGURE 9.6 Finite element model.

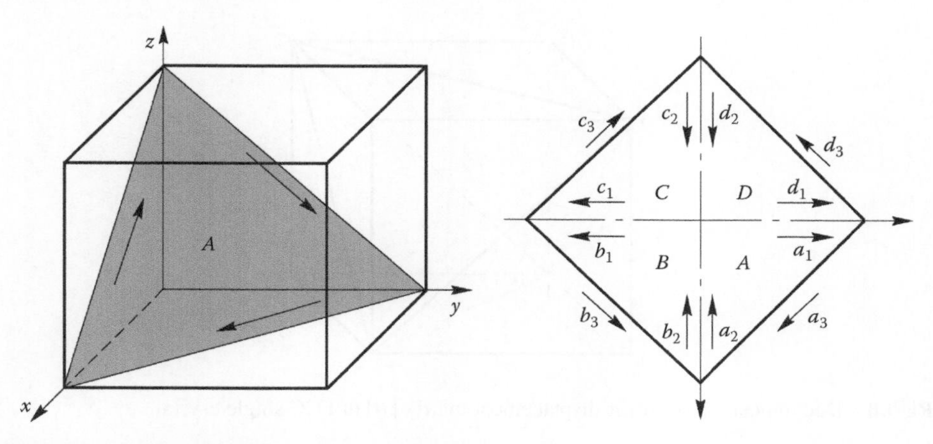

FIGURE 9.7 Slip planes of FCC single crystal.

TABLE 9.4
Slip Planes and Slip Directions of FCC Single Crystal

Slip Plane	Slip Direction		
$A(1\ 1\ 1)$	$a_1[0\ 1\ \bar{1}]$	$a_2[\bar{1}\ 0\ 1]$	$a_3[1\ \bar{1}\ 0]$
$B(1\ \bar{1}\ 1)$	$b_1[0\ \bar{1}\ \bar{1}]$	$b_2[\bar{1}\ 0\ 1]$	$b_3[1\ 1\ 0]$
$C(\bar{1}\ \bar{1}\ 1)$	$c_1[0\ \bar{1}\ \bar{1}]$	$c_2[1\ 0\ 1]$	$c_3[\bar{1}\ 1\ 0]$
$D(\bar{1}\ 1\ 1)$	$d_1[0\ 1\ \bar{1}]$	$d_2[1\ 0\ 1]$	$d_3[\bar{1}\ \bar{1}\ 0]$

9.3.2.1 Fundamental Equations for the Simulation

According to the virtual work principle, we have the following equation:

$$\int_V [\sigma]^T [\delta\Delta\varepsilon]\,\mathrm{d}V = [F]^T [\delta\Delta u], \tag{9.38}$$

where

$[\sigma]$ is the matrix for microscopic stresses in the current step
$[\delta\Delta\varepsilon]$ is the matrix for the virtual strain increments
$[\delta\Delta u]$ is the matrix for the virtual displacement increments
$[F]$ is the matrix for the nodal forces
$\mathrm{d}V$ is the volume increment
$[\sigma]$ denotes six-dimensional vectors

$$[\sigma]^T = [\sigma_x,\ \ \sigma_y,\ \ \sigma_z,\ \ \tau_{yz},\ \ \tau_{zx},\ \ \tau_{xy}], \tag{9.39}$$

where the superscript T denotes transpose.

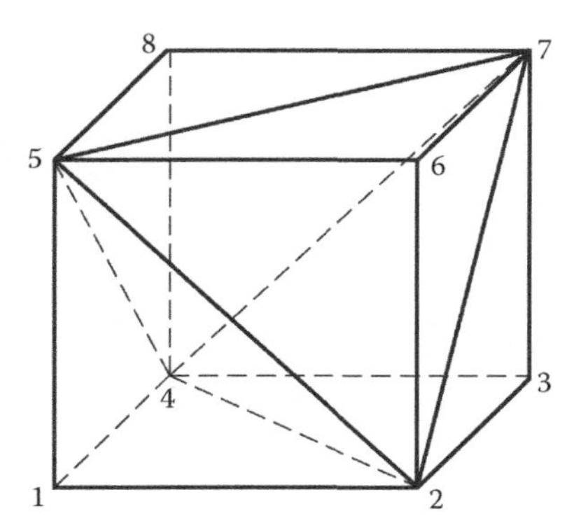

FIGURE 9.8 Decomposition of strain-displacement matrix $[B]$ of FCC single crystal.

Microscopic strain increments can be calculated as

$$[\Delta\varepsilon] = [B][\Delta u], \tag{9.40}$$

where

$[\Delta u]$ represents the nodal displacement increments

$[B]$ is the 6 × 6 strain-displacement matrix for hexahedral elements

A hexahedral element consists of five tetrahedral elements, as shown in Figure 9.8. The specific expressions can be found in Reference [14].

Substituting Equation 9.40 into Equation 9.38, we have

$$[F] = \int [B]^T [\sigma] \mathrm{d}V. \tag{9.41}$$

Assuming that the strain increment and grain rotation in each time step is small, we obtain the following constitutive relationship:

$$[\Delta\sigma] = [D]([\Delta\varepsilon] - [\Delta\varepsilon^p]). \tag{9.42}$$

Here, $[\Delta\varepsilon^p]$ denotes the matrix for plastic strain increments, and $[D]$ is a 6 × 6 elastic matrix

$$[D] = \frac{2G}{1-2\nu}\begin{bmatrix} 1-\nu & \nu & \nu & 0 & 0 & 0 \\ \nu & 1-\nu & \nu & 0 & 0 & 0 \\ \nu & \nu & 1-\nu & \frac{1-2\nu}{2} & 0 & 0 \\ 0 & 0 & 0 & \frac{1-2\nu}{2} & 0 & 0 \\ 0 & 0 & 0 & 0 & \frac{1-2\nu}{2} & 0 \\ 0 & 0 & 0 & 0 & 0 & \frac{1-2\nu}{2} \end{bmatrix}. \tag{9.43}$$

The stress increments are

$$[\Delta\sigma] = [\sigma] - [\sigma_l], \tag{9.44}$$

$[\sigma_I]$ represents the stress matrix of the previous step, and the current stress matrix $[\sigma]$ is replaced by

$$[\sigma]=[D]([\Delta\varepsilon]-[\Delta\varepsilon^P])+[\sigma_I]. \tag{9.45}$$

Substituting Equation 9.45 into Equation 9.41, we have

$$[K][\Delta u]=[F]+[F_P]. \tag{9.46}$$

where

$$[K]=\int[B]^T[D][B]dV,\quad [F_P]=\int[B]^T([D][\Delta\varepsilon^P]-[\sigma_I])\,dV. \tag{9.47}$$

9.3.2.2 Determination of Plastic Strain

It can be seen from the aforementioned analysis that $[\Delta u]$ can be obtained from Equation 9.46, then $[\Delta\varepsilon]$ can be obtained from Equation 9.40. $[\sigma]$ can be obtained in terms of $[\Delta\varepsilon^P]$, and $[\sigma_I]$ via Equation 9.45.

If a_i represents the unit vector normal to the slip plane, b_i the unit vector parallel to slip direction, and $\Delta\gamma^{(r)}$ the shear strain increment of the rth slip system, then the plastic strain increment of each grain can be calculated as

$$\Delta\varepsilon_{ij}^p=\sum_r L_{ij}^{(r)}\Delta\gamma^{(r)}, \tag{9.48}$$

where

$$L_{ij}=\frac{1}{2}(a_ib_j+a_jb_i), \tag{9.49}$$

$$[L]=[L_{xx},\quad L_{yy},\quad L_{zz},\quad 2L_{yz},\quad 2L_{zx},\quad 2L_{xy}]^T. \tag{9.50}$$

9.3.2.3 Crystallographic Orientation [9,14]

A macroscopic coordinate system (x, y, z) and a microscopic coordinate system x', y', z' are correlated by Euler angles (φ, θ, ψ), as shown in Figure 9.9. For large deformations, we must consider

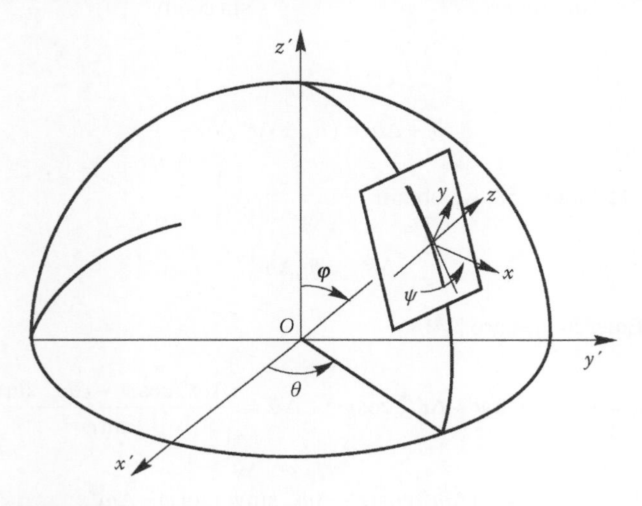

FIGURE 9.9 Euler angle (φ, θ, ψ).

the lattice rotation. Generally, the total rotational increment $\Delta\omega_{ij}$ consists of the elastic rotational increment $\Delta\omega_{ij}^{e}$, the plastic slip increment $\Delta\omega_{ij}^{p}$, and the lattice rotational increment $\Delta\omega_{ij}^{*}$, such that

$$\Delta\omega_{ij} = \frac{1}{2}\left(\frac{\partial \Delta u_j}{\partial x_i} - \frac{\partial \Delta u_i}{\partial x_j}\right) = \Delta\omega_{ij}^{e} + \Delta\omega_{ij}^{p} + \Delta\omega_{ij}^{*}. \tag{9.51}$$

The rotation caused by plastic slip is given by

$$\Delta\omega_{ij}^{p} = \sum_{r} W_{ij}^{(r)} \Delta\gamma^{(r)}. \tag{9.52}$$

Here,

$$W_{ij} = \frac{1}{2}\left(a_i b_j - a_j b_i\right), \quad \left[\Delta\omega^{p}\right] = \left[\Delta\omega_{yz}^{p}, \Delta\omega_{zx}^{p}, \Delta\omega_{xy}^{p}\right]^{T}, \quad \left[W\right] = \left[W_{yz}, W_{zx}, W_{xy}\right]^{T}. \tag{9.53}$$

Assuming that elastic rotation can be ignored, the lattice rotation can be calculated as

$$\Delta\omega_{ij}^{*} = \Delta\omega_{ij} - \Delta\omega_{ij}^{p}. \tag{9.54}$$

The change of slip direction can be calculated by

$$\Delta b_i = b_j \Delta\omega_{ji}^{*}. \tag{9.55}$$

The relationship between macroscopic basis vectors b_i and microscopic basis vectors b_i' is

$$b_i = R_{ji} b_j'. \tag{9.56}$$

The coordinate transformation matrix R_{ji} is

$$R_{ji} = \begin{bmatrix} \cos\varphi\cos\theta\cos\psi - \sin\theta\sin\psi & -\cos\varphi\cos\theta\sin\psi - \sin\theta\cos\psi & \sin\varphi\cos\theta \\ \cos\varphi\sin\theta\cos\psi + \cos\theta\sin\psi & -\cos\varphi\sin\theta\sin\psi + \cos\theta\cos\psi & \sin\varphi\sin\theta \\ -\sin\varphi\cos\psi & \sin\varphi\sin\psi & \cos\varphi \end{bmatrix}. \tag{9.57}$$

The lattice rotation is

$$b_i + \Delta b_i = (R_{ji} + \Delta R_{ji}) b_j'. \tag{9.58}$$

From Equations 9.57 and 9.58, we obtain

$$\Delta R_{ij} = R_{ik} \Delta\omega_{kj}^{*}. \tag{9.59}$$

The changes in Euler angles are [14]

$$\Delta\varphi = -(\Delta\omega_{yz}^{*}\sin\psi + \Delta\omega_{zx}^{*}\cos\psi) \quad \Delta\theta = \frac{(\Delta\omega_{yz}^{*}\cos\psi - \Delta\omega_{zx}^{*}\sin\psi)}{\sin\varphi}$$

$$\Delta\psi = -(\Delta\omega_{yz}^{*}\cos\psi - \Delta\omega_{zx}^{*}\sin\psi)\cot\varphi - \Delta\omega_{xy}^{*}. \tag{9.60}$$

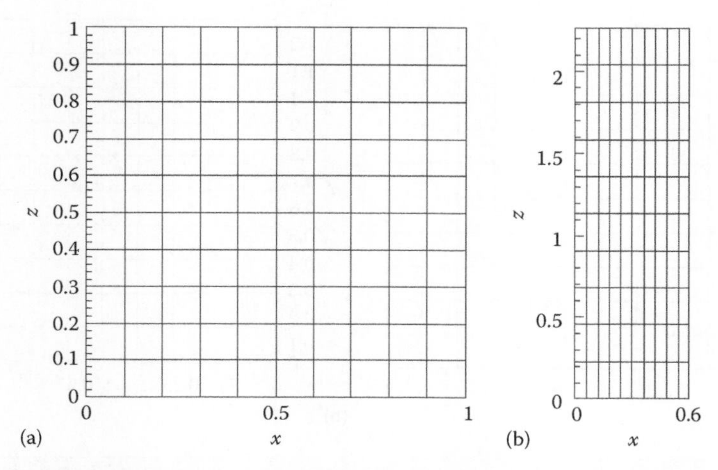

FIGURE 9.10 Microscopic deformations of *B-B* section in Figure 9.6 at different stages when the grain orientations are the same. (a) Tensile strain 0%. (b) Tensile strain 130%.

9.3.2.4 Simulation Results and Discussion [15]

Figure 9.10 shows microscopic deformation of the *B-B* section in Figure 9.6 at different tensile strains, when all grains have the same crystallographic orientation (the Euler angles are $\varphi = 0$, $\theta = 0$, $\psi = 0$). Figure 9.10a represents the initial configuration and Figure 9.10b shows the case where the tensile strain is 130%.

It can be seen from Figure 9.10 that the grains in each deformation stage are uniform. This is because each grain has the same orientation; that is, its Euler angles are $\varphi = 0$, $\theta = 0$, $\psi = 0$. Thus the material is homogeneous at the microscopic level. When it is subjected to uniform tension, the deformation is also uniform.

Figure 9.11 shows microscopic deformations of the *B-B* section in Figure 9.6, when grains are differently oriented (i.e., φ, θ and ψ are random angles), in which Figure 9.11a represents the initial configuration, Figure 9.11b for a tensile strain of 25%, Figure 9.11c for a tensile strain of 61%, and Figure 9.11d for a tensile strain of 150%.

It can be seen from Figure 9.11 that the grains at different deformation stages are nonuniform. In particular, lattice distortion increases with the increase of deformation. Although the shape of each grain at microscale is identical, the orientation in each grain is different, and randomly orientated. Their properties are anisotropic. Therefore, the microdeformation is nonuniform, although it is subjected to uniform tension.

Figure 9.12 shows the microscopic deformation of the *A-A* section in Figure 9.6 when grains are randomly oriented. (a) is for tensile strain of 50% and (b) for the tensile strain of 150%.

We can see from Figure 9.12 that, due to the random orientation of each grain and its anisotropic properties, the grains are nonuniform in each deformation stage. The lattice distortion increases with the increase of tensile strain.

9.4 COMPUTATIONAL NANOMECHANICS OF MATERIALS

Since the 1990s, nanotechnology has developed into an emerging technology. It is regarded as one of the key technologies in the twenty-first century, and is being extensively studied. It has been realized that we are advancing from the microelectronic age to the nanotechnology age. Mechanics workers are being brought into a scientific research field that is neither traditionally macroscopic nor traditionally microscopic. Nanomechanics is the most important part of nano-science and nanotechnology. The mechanical behavior of solids at nanoscale is an important part of the science of matter.

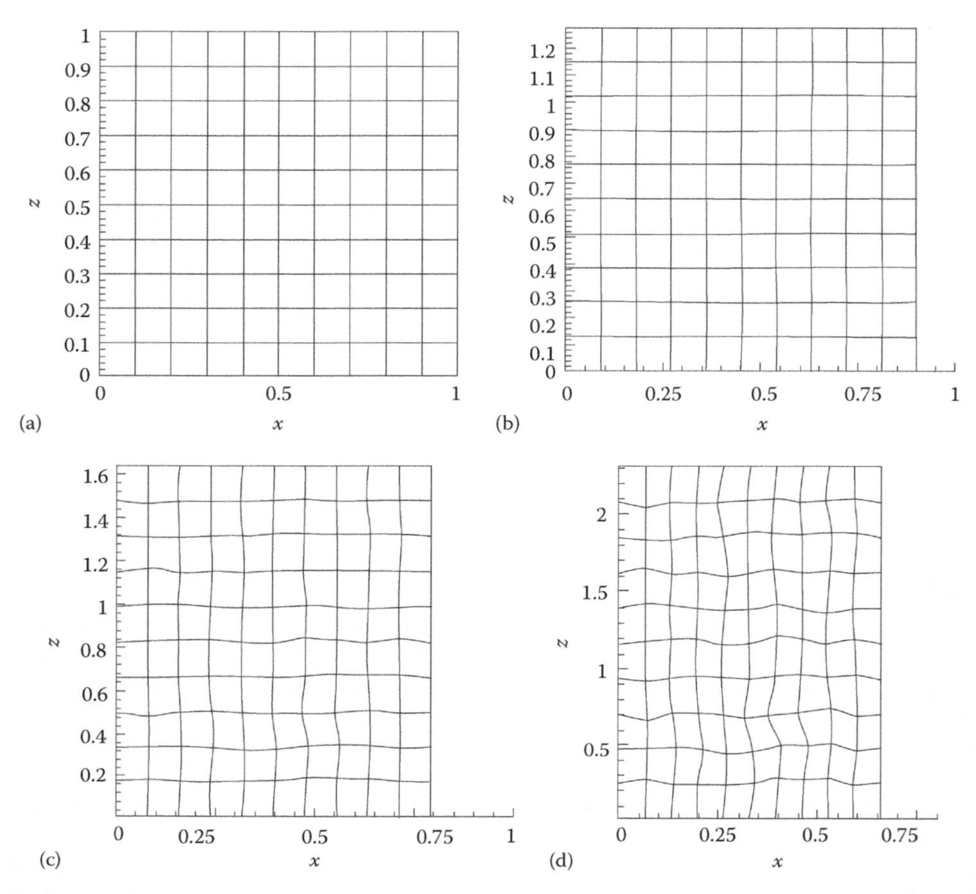

FIGURE 9.11 Deformation of *B-B* section in Figure 9.6. (a) Initial configuration, 0. (b) Tensile strain of 25%. (c) Tensile strain of 61%. (d) Tensile strain of 150%.

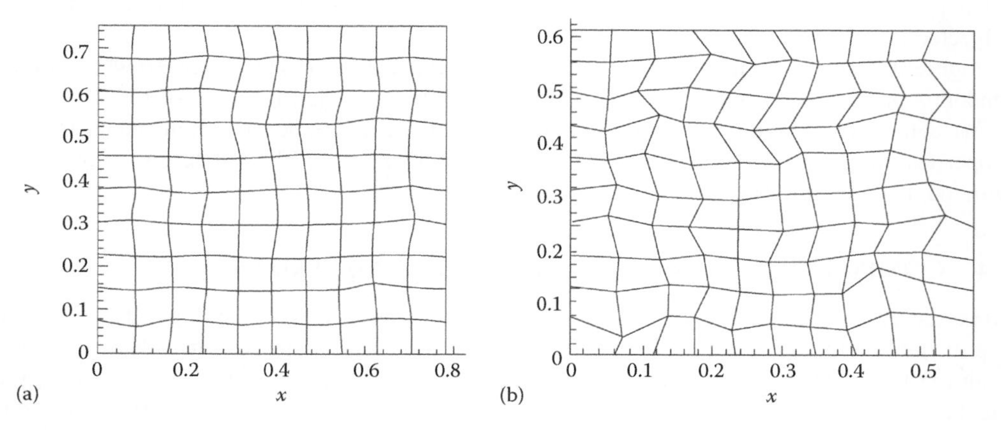

FIGURE 9.12 Deformation of *A-A* section in Figure 9.6. (a) Tensile strain of 50%. (b) Tensile strain of 150%.

According to research methodology, nanomechanics can be divided into computational nanomechanics, experimental nanomechanics, and nanomechanics theories [16]. In this section, we only introduce computational nanomechanics. The general methods for computational mechanics at nanoscale mainly include molecular dynamics, Monte Carlo simulations, and the quasi-continuum method. The molecular dynamics method is the most extensively employed computational nanomechanics method, and is classified into classical molecular dynamics (including multibody molecular dynamics and variable charge molecular dynamics) and modern molecular dynamics (tight-binding molecular dynamics, density functional molecular dynamics, and first principle molecular dynamics).

In 1957, Alder and Wainwright first adopted the molecular dynamics method to solve the hard-sphere state equations of gases and liquids [17], which is the original work on the simulation of the macroscopic behavior of a mass using the molecular dynamics method. After the 1980s, with the rapid development of computer technologies, molecular dynamics has been widely applied in fields such as materials, mechanics, chemical reactions, pharmacy, physics, microelectronics, MEMS, and in biomacromolecules. Together with the Monte Carlo method and the first principle method, it becomes the primary technology in microscopic simulation. Particularly within the last 10 years, with the emerging and development of nanotechnology, molecular dynamics simulation is the most widely used method in computational mechanics at nanoscale. A detailed description is beyond the scope of this book, but in the next sections we briefly introduce classic molecular dynamics.

9.4.1 Fundamental Principle of Molecular Dynamics

9.4.1.1 Solution of Motion Equations

From the viewpoint of computational mathematics, the molecular dynamics (MD) method is an initial value problem, and a lot of algorithms have been developed for this type of problem. We consider simple point particles in the microcanonical ensemble (i.e., the number of atoms N, volume V, and energy E are unchanged). Assuming that the ensemble consists of N particles, the Hamiltonian of the ensemble can be described as

$$\xi = \frac{1}{2}\sum_i \frac{p_i^2}{m_i} + \sum_{i<j} u(r_{ij}), \tag{9.61}$$

where

r_{ij} is the distance between particles i and j

m_i is the mass of particle i

The first term of the aforementioned equation is the kinetic energy of the particles, and the second term is the interaction energy among particles or potential energy.

In classical mechanics, the Hamiltonian can derive the motion equations with different forms, and the solving algorithms of the various equations are different. Although the motion equations in different forms are equivalent in mathematics, they are not equivalent in numerical calculation. We first consider the evolution of a system in phase space (the concept of phase space can refer to statistic physics) along the orbit of constant energy, that is, the microcanonical ensemble. From Newton's equation, we have

$$\frac{\mathrm{d}^2 r_i(t)}{\mathrm{d}t^2} = \frac{1}{m_i}\sum_{i<j} F_i(r_{ij}). \tag{9.62}$$

The analytical solution of the second-order differential equation can be obtained by integrating twice from 0 to t. The first integral obtains the velocity, and the second integral gives position. It is necessary to know not only the initial positions but also the initial velocities. The initial positions determine the contribution of potential energy to the total energy, and the initial velocities determine the contribution of kinetic energy. After the initial conditions are determined, the system moves along the path in phase space with constant energy.

In order to solve this differential equation using numerical methods, the second-order differential at the left side of Equation 9.62 is discretized, resulting in the explicit central difference method:

$$\frac{d^2 r_i}{dt^2} = \frac{1}{h^2}[r_i(t+h) - 2r_i(t) + r_i(t-h)] = \frac{1}{m_i}F_i(t), \tag{9.63}$$

where h is the step size of time. Thus, Equation 9.62 can be written as

$$r_i(t+h) = \frac{2r_i(t) - r_i(t-h) + F_i(t)h^2}{m_i}. \tag{9.64}$$

Let $t_n = nh, r_i^n = r_i(t_n)$, $F_i^n = F_i(t_n)$. Then Equation 9.64 becomes

$$r_i^{n+1} = \frac{2r_i^n - r_i^{n-1} + F_i^n h^2}{m_i}. \tag{9.65}$$

Starting from the initial positions, all of the subsequent positions can be determined from the aforementioned recurrence relation. That is, the particle position at the moment $n + 1$ can be derived from the positions at the two moments just before the current moment.

The aforementioned concurrence relation formula only gives positions. But we still need to know velocity to calculate the kinetic energy. The velocity can be expressed as

$$v_i^n = \frac{(r_i^{n+1} - r_i^{n-1})}{2h}. \tag{9.66}$$

Equations 9.65 and 9.66 together with the initial conditions are called the Verlet algorithm.

The Verlet algorithm is concise in operation and moderate in its memory requirement. One shortcoming is that the position $r(t + h)$ is obtained by summation of a small term and the difference between two big terms, $2r(t)$ and $r(t - h)$. This is vulnerable to losing accuracy. The Verlet algorithm has other shortcomings. For example, there is no explicit velocity term in the equation. It is difficult to obtain the velocity before the position of the next step is obtained. In addition, it is not a self-starting algorithm: the position must be calculated from the positions at the moments t and $t - h$.

The Verlet algorithm can be re-expressed, so that it gives a more stable numerical algorithm. Let

$$z_i^n = \frac{(r_i^{n+1} - r_i^n)}{h}. \tag{9.67}$$

Equations 9.65 and 9.66 become

$$\begin{cases} r_i^n = r_i^{n-1} + hz_i^{n-1} \\ z_i^n = z_i^{n-1} + m_i^{-1}hF_i^n \end{cases}. \tag{9.68}$$

Equation 9.67 gives $2v_i^n = z_i^n + z_i^{n-1}$. Thus,

$$z_i^n = v_i^n + \frac{hF_i^n}{2m_i}. \tag{9.69}$$

Substituting Equation 9.69 into 9.68, we obtain the velocity form of molecular dynamics of NVE (the microcanonical ensemble):

$$r_i^{n+1} = r_i^n + hv_i^n + \frac{1}{2m_i}h^2F_i^n \quad v_i^{n+1} = v_i^n + \frac{h(F_i^{n+1} + F_i^n)}{2m_i}. \tag{9.70}$$

The aforementioned velocity form is superior to the original Verlet algorithm in many aspects; in particular, the position and velocity at the same moment can be obtained simultaneously. In addition, numerical stability is enhanced. In the practical molecular dynamic simulation, another algorithm, the leap-frog algorithm, is also frequently used [18]. This algorithm is expressed as

$$v\left(t+\frac{h}{2}\right) = v\left(t-\frac{h}{2}\right) + m_i^{-1}hF(t), \quad r(t+h) = r(t) + v\left(t+\frac{h}{2}\right)h. \tag{9.71}$$

Compared with the Verlet algorithm, the leap-frog algorithm includes an explicit velocity term, and the computation is slightly smaller. It also has an obvious shortcoming: position and velocity are not synchronous. This means that when the position is determined, the contribution of kinetic energy to the total energy cannot be calculated simultaneously. In the molecular dynamic simulation, the accuracy of the algorithm is an extremely important issue, because both the Verlet algorithm and the leap-frog algorithm are derived from the central difference method, and the errors of position and velocity are at the order of $0(h^3)$. Extensive investigations indicate that such accuracy is sufficient for most molecular dynamics simulations. Since the leap-frog algorithm has a smaller and faster computation and better accuracy, it is frequently adopted in simulations. In addition, there are many other high-order algorithms, such as the Geer algorithm [19], the Beeman algorithm [20], and the Toxvaerd algorithm [21], which are not discussed in detail here.

9.4.1.2 Interatomic Potential

The key step to accurately describing nanomechanical behavior at the atomic scale is the determination of the interaction among atoms, that is, the interatomic potential among atoms. In theoretical mechanics, the interaction of atoms is equal to the negative value of the gradient of the total potential of the atoms. If the total potential of the system is U_{tot}, the force acting on an arbitrary atom is

$$\boldsymbol{F} = -\nabla U_{tot}. \tag{9.72}$$

If the adopted interatomic potential cannot accurately predict the particle traces, the simulation results are not accurate. Therefore, the interatomic potential has the most direct and essential influence on the simulation results. Here, we briefly introduce the pair potential and the multibody potential, both of which are frequently used in molecular dynamics simulations.

In the early stage of molecular dynamics simulation, the pair potential is often adopted. In the pair potential, atomic interaction is constricted to two atoms, and is not related to other particles. Therefore, when calculating the interaction between two atoms, the influence of other particles is not considered. This is a relatively good approximation in some cases. The potential of particle i in the system is

$$U_i = \frac{1}{2}\sum_j u(r_{ij}). \tag{9.73}$$

The total potential of the system is

$$U_{tot} = \frac{1}{2}\sum_i \sum_j u(r_{ij}), \tag{9.74}$$

where

- $u(r_{ij})$ is the interaction potential between particles (i, j), called the pair potential or the two-body potential
- r_{ij} is their distance

The Lennard–Jones (L–J) potential [22] is the most general pair potential:

$$\phi(r_{ij}) = 4\varepsilon\left[\left(\frac{\sigma}{r_{ij}}\right)^{12} - \left(\frac{\sigma}{r_{ij}}\right)^{6}\right], \tag{9.75}$$

where ε and σ are parameters. For most materials, these parameters have been determined. This type of potential is relatively appropriate to describe the interaction between inert atoms. In addition, the Morse potential [23] is also a popular pair potential:

$$\phi_{ij}(r_{ij}) = A\left[e^{-2\alpha(r_{ij}-r_0)} - 2e^{-\alpha(r_{ij}-r_0)}\right], \tag{9.76}$$

where A, r_0, α are empirical parameters.

In practice, the object is often a system of multiple particles with relatively strong interactions. The change of state of one atom influences the change of other particles. In this case, the interaction is not simple between two particles, but is a multibody interaction. Due to inherent shortcomings of the pair potential, new potentials are explored to overcome them. The multibody potential was developed in the early 1980s when Daw and Baskes first proposed the embedded atom method (EAM) [24]. The basic idea of the EAM potential is to divide the total potential of crystals into two parts: one part for the pair potential of the interaction of atoms on the crystal lattice, and the other part for the embedded energy of atoms embedded in the electron cloud, which represents the multibody interaction. The functions of pair potential and embedded potential in the EAM potential can be selected in terms of experience.

In the EAM, the total potential of crystals can be expressed as

$$U = \sum_i F_i(\rho_i) + \frac{1}{2}\sum_{j\neq i} \phi_{ij}(r_{ij}). \tag{9.77}$$

The first term F denotes the embedded energy; the second term represents the pair potentials, taking different forms in terms of needs. ρ_i is the sum of electronic cloud densities at atom i of all atoms except atom i, expressed as

$$\rho_i = \sum_{j \neq i} \rho_j(r_{ij}) \tag{9.78}$$

where
$\rho_j(r_{ij})$ is the charge density at atom i contributed by the electrons of atom j
r_{ij} is the distance between atom i and atom j

For each metal, the parameters in the embedded energy function and in the pair potential function can be determined by best fitting the macroscopic parameters of the metal.

In terms of the tight-binding theory of electronic energy bands in metals, Finnis and Sinclair [25] developed a potential function equivalent to EAM in mathematics and proposed the functional expression of the multibody interaction potential: the embedded energy function is in the form of a square root. Based on this, Ackland and Vitek. [26] developed the multibody potential functions of Cu, Al, Ni, and Ag by fitting the elastic constants, lattice constants, vacancy formation energy, coalescence energy, and the relationship between the pressure and volume of metals.

9.4.2 Isothermal Molecular Dynamics

In some cases, we need to hold or control temperature in molecular dynamics simulation. Even in the simulation of equilibrium processes of NVE, temperature is frequently adjusted to the expected value. For example, we need to simulate the quasi-static tension process of a rod. Since the tension rate is sufficiently slow, the temperature of the rod remains constant during the tension process. Thus we need to control the temperature of the entire system to a constant. If the temperature of the system at a moment is not equal to the expected value, how is the temperature of the system adjusted? We can adjust the temperature of the system by changing the velocities of atoms in the system. Since the temperature of the system is related to the kinetic energy, the relationship between the instantaneous temperature of the system and the kinetic energy of system can be described by

$$E_k = \sum_{i=1}^{N} \frac{1}{2} m_i v_i^2 = \frac{k_B T}{2}(3N - N_c), \tag{9.79}$$

where
k_B is the Boltzman constant
N_c is the constraint number of the system (e.g., the constraint number for a system with constant volume is 3)
$3N - N_c$ is the total number of degrees of freedom of the system

From Equation 9.79, we obtain the relationship between temperature and the velocities of atoms in the system:

$$T = \frac{1}{k_B T(3N - N_c)} \sum_{i=1}^{N} m_i v_i^2. \tag{9.80}$$

If the temperature of system $T(t)$ at moment t is not equal to the required temperature T_{req}, we can multiply the velocity of each atom by a scale factor λ. The temperature of the system becomes

$$T_{new} = \frac{1}{k_B T(3N - N_c)} \sum_{i=1}^{N} m_i (\lambda v_i)^2 = \lambda^2 T(t). \tag{9.81}$$

Letting the rescaled temperature be equal to the required temperature $T_{new} = T_{req}$, the scalar factor can be easily calculated from Equation 9.81: $\lambda = \sqrt{T_{req}/T(t)}$. Therefore, the simplest method to control temperature is to multiply the velocities of all atoms by a scalar factor $\lambda = \sqrt{T_{req}/T(t)}$. Here, $T(t)$ is the instantaneous temperature of the current system calculated by Equation 9.80, and T_{req} is the expected reference temperature.

Another way to keep a constant temperature is the Berendsen method [27]. In this method, the simulated system is contacted with a heat bath that has an expected temperature. Acting as a heat source, the heat bath absorbs heat from, or releases heat to, the system. For each simulation step, the velocities are rescaled so that the rate of temperature change is proportional to the temperature difference between the heat bath and the system

$$\frac{\mathrm{d}T(t)}{\mathrm{d}t} = \frac{1}{\tau}(T_{bath} - T(t)), \tag{9.82}$$

where τ is a coupling parameter, and its magnitude determines the coupling degree between the heat bath and the system. In this method, the system temperature exponentially approaches the expected temperature value. From Equation 9.82, the temperature change of each step is

$$\Delta T = \frac{h}{\tau}[T_{bath} - T(t)]. \tag{9.83}$$

From Equation 9.81, the difference between the temperature of the system before scaling and after multiplying the velocity of each atom by the scalar factor, can be calculated as

$$\Delta T = (\lambda^2 - 1)T(t). \tag{9.84}$$

Comparing Equations 9.83 and 9.84, we obtain the scalar factor λ

$$\lambda = \sqrt{1 + \frac{h}{\tau}\left[\frac{T_{bath}}{T(t)} - 1\right]}. \tag{9.85}$$

The larger is the value of τ, the weaker is the coupling. On the contrary, the smaller the value of τ, the stronger is the coupling. When the coupling parameter is equal to step size h from Equation 9.85, we have $\lambda = \sqrt{(T_{bath}/T(t))}$, and this algorithm is equivalent to the simple velocity rescaling method. The advantage of this method is that it allows the fluctuation of system temperature around the expected temperature value.

In the simulation of the microcanonical ensemble, the Andersen method [28] is also often used to adjust temperature. In the method proposed by Andersen, the simulation system is coupled with a heat bath with an imposed constant temperature. The coupling is represented by random impulse forces occasionally acting on the randomly selected atoms. In the isothermal molecular dynamics simulation using the Andersen method, a constant temperature is achieved by random collision with the strict heat bath. Nose introduced another method to implement isothermal MD simulation [29]. Hoover further extended Nose's algorithm, and developed the most frequently used Nose-Hoover thermostat [30]. Limited by the length, the specific algorithms for the Andersen thermostat and the Nose-Hoover thermostat are not introduced in detail here, and can be found in Reference [30].

9.4.3 Applications of Molecular Dynamics in the Fracture Behavior of Materials

Because molecular dynamics can trace the motion trajectory of each atom, it depicts the evolution of material at the atomic scale in different mechanics behaviors. This evolutionary process at the atomic scale is difficult to implement in experiments and in other simulation methods. Therefore, molecular dynamics is widely used to investigate the deformation behaviors of materials under tension, compression, bending, torsion, and nanoindentation [31–34]. Here, we take the uniaxial tension of gold nanowire as an example to briefly introduce the applications of molecular dynamics in nanomechanics.

Gold nanowire was first fabricated in 1998, and the process of formation of the monoatomic chain until fracture under tension was observed using a transmission electron microscope (TEM) [35]. However, this process cannot be observed in macroscopic experiments. Using molecular dynamics, Edison et al. [34] simulated the unaxial tension process of gold nanowire. The Verlet algorithm was adopted, the step size was 1 fs, and the system temperature was controlled to be a constant.

The simulation is carried out as follows:

1. The initial configuration is the FCC single crystal of gold nanowire with a length of 24 Å in the (111) direction, and each layer (or atomic plane) has seven atoms. The system is relaxed (i.e., thermal motion of atoms is allowed with the given initial condition and no external influence) until the system reaches a steady state.
2. The entire nanowire is elongated to 0.5 Å.
3. The system temperature is raised to 400 K.
4. The system is annealed to approximately 30 K. Repeat steps 2–4 until the nanowire fractures.

Figure 9.13a through d shows the atomic configurations of gold nanowire at four different lengths during tension until fracture. It can be seen from Figure 9.13a that, when the nanowire extends

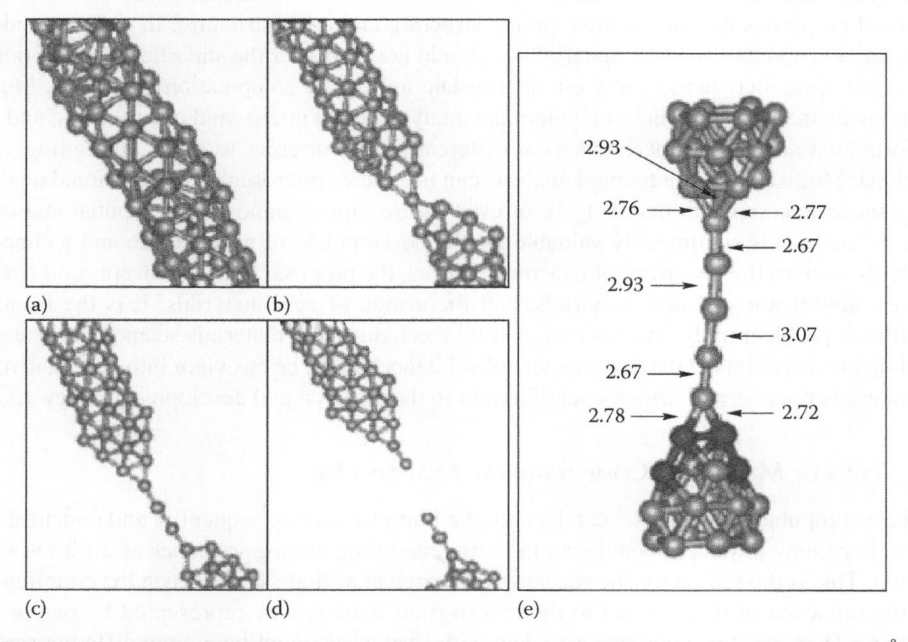

FIGURE 9.13 The atomic configurations of gold nanowire at different lengths during tension. (a) 25.5 Å. (b) 33.0 Å. (c) 38.0 Å. (d) 43.0 Å. (e) is the instantaneous configuration of nanowire before fracture, the numerical values are atomic bond lengths (unit: Å). (From Edison, Z. et al., *Comput. Mater. Sci.*, 30(1), 73, 2004; Taken from Figure 1.)

to 25.5 Å, the original plane with seven atoms becomes a ring with six atoms, and thus the entire nanowire becomes a hollow nanotube. Like carbon nanotubes, gold nanotubes can be viewed as the wrap up of gold atoms on the (111) crystal plane. When the length of the nanowire reaches 33 Å, due to the "necking" of the nanowire, a "neck" structure consisting of a single atom is formed, as shown in Figure 9.13b. With continuous elongation, more atoms enter the neck and form a monoatomic chain, as shown in Figure 9.13c, until the formation of an atomic chain with a length of five atoms, as shown in Figure 9.13e. When the nanowire reaches 41 Å, one bond length of the monoatomic chain increases abruptly from 3.1 to 4.3 Å, indicating the final fracture of the nanowire.

With further tension, the upper and lower parts of the nanowire simply separate, as shown in Figure 9.13d, and the structure does not change. Figure 9.13e gives the final configuration before fracture, and gives the bond lengths of the monoatomic chain. Before fracture, the longest bond length in the monoatomic chain is approximately 3.07 Å, and the other bond lengths are in the range of 2.67–2.93 Å, and the total length of the atomic chain is 11.3 Å. All of these features agree very well with experimental results [35]. We can see from this example that molecular dynamics can give the deformation process at the nanoscale of materials, which is not available in many macroscopic experiments. This is one of the most successful examples in computational mechanics at nanoscale.

9.5 MULTISCALE COMPUTATIONAL ANALYSIS

9.5.1 Necessity of Multiscale Computational Analysis [36]

We know from the previous several sections that the traditional FEM is developed based on continuum mechanics. It can simulate deformation at the macroscopic scale, but it cannot simulate the influence of microstructure on deformation, due to its limitations. Molecular dynamics can simulate the deformation behavior of materials at the atomic scale. Nowadays, for a general computer condition, molecular dynamics can consider up to one million atoms, occupying a volume with a one-dimensional length of 0.1 μm. However, the grain size of a traditional metal is at the order of the micron. Thus, molecular dynamics cannot simulate deformation at the macroscopic scale.

Material properties depend on their atomic structure and microstructure. In order to understand the deformation and damage of a material, we should not stop after the superficial description at the macroscopic scale, but should carry out appropriate multiscale computational analysis. Multiscale computational analysis combines computational analyses at the micro- and macroscales, and reveals the deformation and failure mechanisms at different scales, in order to avoid misleading via false phenomena. Multiscale computational analysis can be macro–microscale computational analysis, or micro–nanoscale computational analysis, or even macro–micro–nanoscale computational analysis. Multiscale analysis is significantly valuable for the development of new science and technology in many fields, such as the discovery of new mechanisms, the proposal of new concepts and new methodologies, modification of new materials, and the design of new materials. It is the frontier and most difficult problem in the intersection of solid mechanics and materials science. It incubates the jumped-up growth points in the development of solid mechanics, brings vigor into quantitative analysis in materials science, and brings scientific rigor to the research and development of new materials.

9.5.2 Types of Multiscale Computational Analysis [36]

Multiscale computational analysis can be classified into two types: sequential and concurrent computation. For many physical phenomena in nature, coupling among variables at different scales is not strong. This is the reason for the success of research at a single scale. When the coupling is very weak, the influence of other scales to the investigated scale can be represented by one or several parameters. There are two approaches to deal with very weak coupling among different scales.

One is the phenomenological approach, which determines these parameters in terms of empirical methods or data fitting. It should be pointed out that this approach at a single scale has limitations.

Its accuracy is constrained to intervals of parameter fitting, and it is difficult to reveal the physical mechanisms of the observed phenomena.

The other is the multiscale sequential approach. The parameters are determined using a physics-based approach. The parameters at large scales can be determined from the computational analysis at small scales. Concurrent computational analysis is simultaneous computation at multiple scales. It is appropriate when strong coupling exists between the variables at one scale and the variables at other scales, such as turbulence and crack propagation in elastoplastic media.

Although material properties depend on atomic structure and microstructure of materials, the influences of the two structures on material properties and failure are different. Based on this difference, multiscale analysis can be divided into two correlated categories.

The first category is multiscale analysis from the atomic scale (or ion scale) to microscale of a continuous medium. The influence of atomic structure becomes very important, and the quantum effect and microscale of continuous medium must be considered. This is called the Type I problem in multiscale analysis. Its length scale ranges from the angstrom to the micron, covering the nanoscale. Due to the importance of nano science and technology, it is the hot spot of current research. It is very important to explore new phenomena, clarify new mechanisms for deformation and failure, and develop new materials. The core task of this type of multiscale analysis is to develop efficient approaches that smoothly link atomic scale analysis and continuum analysis, and to improve their computational accuracy and efficiency.

The second category is micro–meso–macro–multiscale analysis. In many practical problems, the chemical compositions of materials are known, and the influence of electronic and atomic structures is clear. Material properties mainly depend on the microstructure of materials. The main purpose of multiscale analysis is to obtain the influence of microstructure on macroscopic properties. This requires micro-, meso-, and macroscale analysis. The core task of this type of problem is to develop effective approaches to study the relationship among the sets of variables at three scales of the continuum medium, in addition to the submicro and the submeso scales. This type of problem covers a lot of practical engineering problems, including modeling and solving, when extending mesoscale analysis to a corresponding analysis of the microscale.

9.5.3 Multiscale Simulation Combining the Finite Element Method and Molecular Dynamics [2]

The failure of a material begins with the separation of atoms or molecules at the crack tip, and then gradually forms macroscopic cracks until the material fractures. Molecular dynamics can simulate the mechanical behavior of a certain number of atoms or molecules, but cannot simulate the mechanical behavior of a macroscopic material under loading. Finite element analysis can simulate the mechanical behavior of continuous media under loading, but cannot simulate the mechanical behavior of atoms or molecules. How is the fracture of a material completely simulated from the atomic or molecular scale to macroscopic crack propagation? This requires a multiscale simulation that combines the FEM and molecular dynamics.

Figure 9.14 is the overlapping model of a continuous medium–discrete particles generally used in the multiscale simulation of finite elements and molecular dynamics. Figure 9.14a is a straight interface, and Figure 9.14b is a serrated interface [2]. The outer portion of this figure is the radial mesh of finite elements for the outer mesoscale region. The mesh has 20 layers in the radial direction (only the two innermost layers are displayed in the figure). Each layer has 24 elements in the circumferential direction. There are 480 elements in total. The finite element mesh has an inner radius of 10 nm and an outer radius of 30 μm. The entire model implements the transition from mesoscale to nanoscale. The model of atoms for the inner nanoscale region has 2310 atoms, with a serrated interface structure. The radius of the inner nanoscale region is 12.5 nm, and the motion of atoms under force is simulated by molecular dynamics. For the outer mesoscale region, the FEM is used to calculate the motion of discrete dislocations around the crack tip.

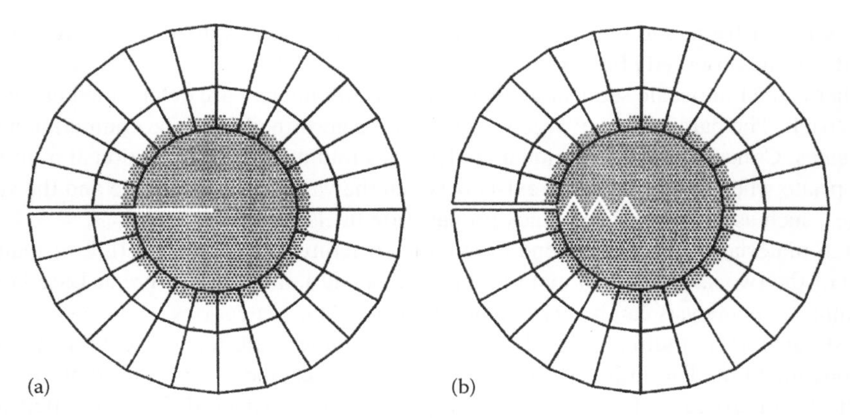

FIGURE 9.14 Initial configuration for combined atoms–continuous medium simulation for interface fracture. (a) Straight interface. (b) Serrated interface. (From Yang, W., *Macro and Microscopical Fracture Mechanics*, National Defense Industry Press, Beijing, 1995; Taken from Figure 11.1.)

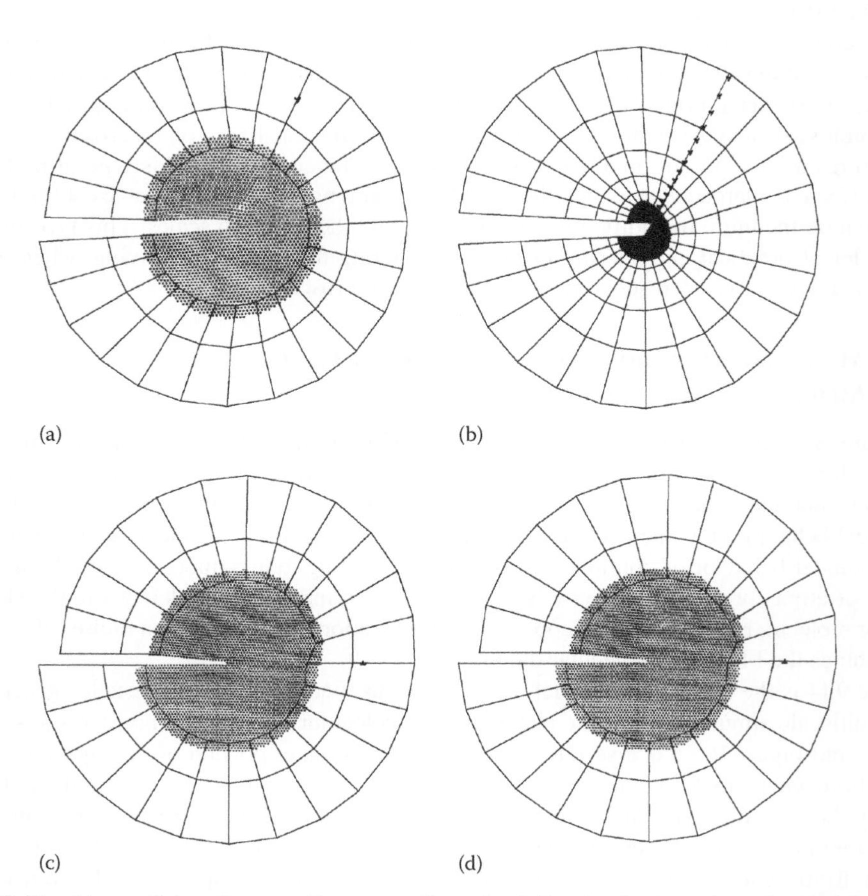

FIGURE 9.15 Atoms–dislocations–continuous medium simulation for fracture along the straight interface $\left(|K| = 0.8\ \text{MPa}\sqrt{\text{m}}\right)$. (a) $\psi = 15°$, $t = 4$ ps. (b) $\psi = 15°$, $t = 30$ ps. (c) $\psi = 120°$, $t = 4$ ps. (d) $\psi = 120°$, $t = 30$ ps. (From Yang, W., *Macro and Microscopical Fracture Mechanics*, National Defense Industry Press, Beijing, 1995; Taken from Figure 11.4.)

At the interface of atoms and a continuous medium, the automatically designed particles-continuous medium overlapping model is adopted. The overlapping region has about four to five atoms. In the overlapping region, the particle and its residing element has the same displacement and stress. The advantage of this type of model is that there is a natural transition of mechanical parameters between the mesoscale and the nanoscale, and it is convenient to describe the stride hierarchy motion of dislocations.

Now, consider an interface crack where the crack lies between two different materials. This case is frequently met in films or composites (see Chapters 11 and 14). Assume the material above the interface is aluminum, and that below the interface is an ion bonding ceramic material. The lattice constants of this material are compatible with those of aluminum, but the macroscopic elastic modulus is 10 times higher than that of aluminum. Both materials have a Poisson ratio of 0.3. Let ψ denote the mixing degree of Type I and Type II fracture modes at a location one atom ahead of the crack tip, where $\psi = \arctan(K_{II}/K_I)$. Figure 9.15 shows the dislocation emission at the straight interface in the macro-microscale simulation. Figure 9.16 shows the macro–microfracture behavior of the serrated interface. In both figures, (a) and (b) represent the mixing degree at the crack tip of $\psi = 15°$, while (c) and (d) represent $\psi = 120°$. The former is close to a type I crack, and the latter is close to a Type II crack.

In Figure 9.15a and c, the first dislocation emitted from the atomic region at the crack tip just passes through the continuous medium-discrete atoms overlapping layer, and the corresponding time is $t = 4$ ps. In Figure 9.15b and d, the array of dislocations is displayed due to the continuous

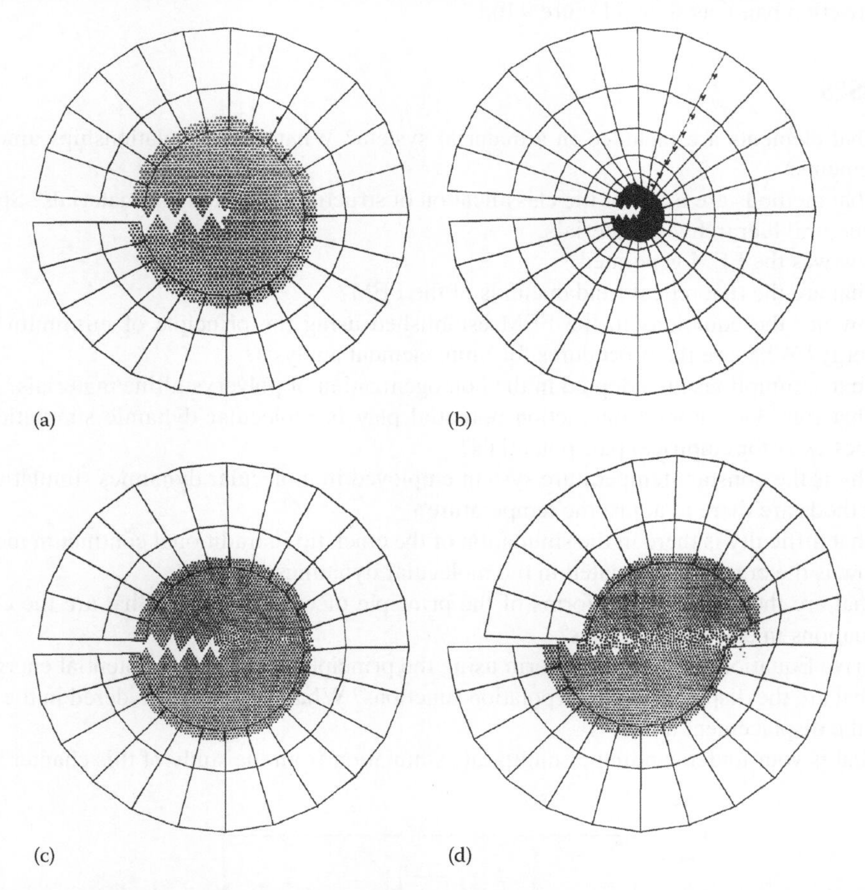

FIGURE 9.16 Atoms–dislocations–continuous medium simulation for fracture along the serrated interface, $\left(|K| = 0.8\,\text{MPa}\sqrt{\text{m}}\right)$. (a) $\psi = 15°$, $t = 4$ ps. (b) $\psi = 15°, t = 30$ ps. (c) $\psi = 120°, t = 4$ ps. (d) $\psi = 120°, t = 30$ ps. (From Yang, W., *Macro and Microscopical Fracture Mechanics*, National Defense Industry Press, Beijing, 1995; Taken from Figure 11.5.)

emission of dislocations, and the corresponding time is $t = 30$ ps, or 6000 time steps. When $\psi = 15°$, dislocations are emitted from the slip planes which are at 60° with respect to the interface. From Figure 9.16b, at $t = 30$ ps, eight dislocations have passed through the continuous medium-discrete particle overlapping layer. In the near-Type I interface fracture, the serrated interface lowers the rate of dislocation emission from the crack tip, but does not change the emission mode.

As shown in Figures 9.15b and 9.16b, at the same moment $t = 30$ ps and the same loading amplitude $|K| = 0.8\,\text{MPa}\sqrt{\text{m}}$, the number of dislocations passing through the overlapping layer emitted from the serrated interface (eight dislocations) is lower than that emitted from the straight interface (13 dislocations). The ceramic phase below the interface is much harder than the metallic phase above the interface, and the morphology of the serrated interface disturbs the dislocation nucleation condition. Thus, the serrated hard constraint in the dislocation nucleation region at the crack tip retards the nucleation of dislocations.

When $\psi = 120°$, the situations are completely different. For the straight interface, dislocations can be emitted continuously from the crack tip, and move along the interface, as shown in Figure 9.15c and d. For the serrated interface, under the same far-field loading conditions and loading time step size, the evolution at the crack tip changes, as shown in Figure 9.16c and d. The morphology of interlocked serrated interfaces disturbs the emission of dislocations, as seen in Figure 9.16c. Further continuous loading leads to the collapse of atomic bonding at the interface, and forms a separated slip and friction band, as shown Figure 9.16d.

EXERCISES

9.1 What elements are included in a material system? What are the relationships among these elements?

9.2 What methods are used for the classification of structural hierarchy of materials? Specify the structural hierarchy of materials.

9.3 How was the FEM generated?

9.4 What are the theoretical fundamentals of the FEM?

9.5 How are the equations in the FEM established using the principle of minimum potential energy? What are the procedures for finite element analysis?

9.6 What assumptions are adopted in the homogenization of polycrystalline materials?

9.7 What role does atomic interaction potential play in molecular dynamic simulation? What types exist for empirical pair potentials?

9.8 Why is the constant temperature system employed in molecular dynamics simulation? What methods are there to adjust the temperature?

9.9 What difficulty is there in the simulation of the crack tip in traditional continuum mechanics? How is the crack tip simulated in the molecular dynamics method?

9.10 What are the two different forms of the principle of virtual work? What are the equivalent equations in elastic mechanics?

9.11 Derive Equation 9.19 in matrix form using the principle of minimum potential energy.

9.12 What are the displacement interpolation functions? What rules are considered in the selection of the displacement mode?

9.13 What is your understanding of multiscale simulation from the study of this chapter?

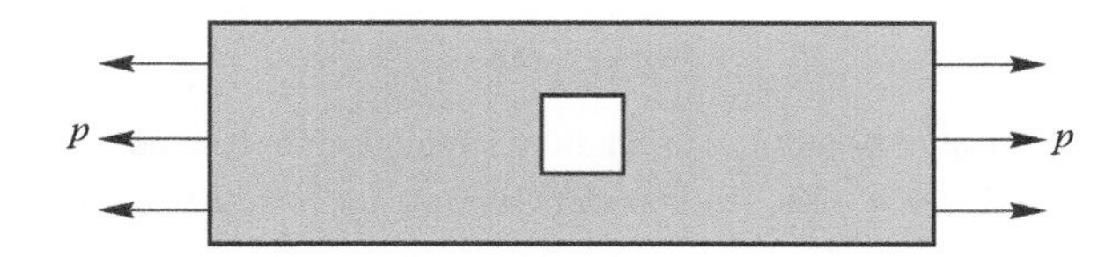

FIGURE 9.17 Schematic illustration of Exercise 9.14.

9.14 As shown in Figure 9.17, a thin rectangular plate has a central square hole with an edge length of 2 cm. The uniform tensile stress p is applied at two ends. The plate has a length of 12 cm, a width of 4 cm and a thickness of 1 cm. $E = 2.0 \times 10^5$ MPa, $\nu = 0.3$, and $p = 5$ MPa. Solve the internal stresses in the plate using the FEM.

9.15 What potential function is most generally used to describe the interaction among metallic atoms? What is its basic idea?

9.16 Discussion: the advantages of multiscale analysis and its long-term and important influences on the development of science and technology.

REFERENCES

1. Liming G. Computational materials science and gradation of materials structure. *Chin J Ceram*, 2002, 25(2): 69–74.
2. Wei Y. *Macro and Micro Fracture Mechanics*. Beijing: National Defense Industry Press, 1995.
3. Robert D. *Computational Materials Science*. Beijing: Chemical Industry Press, 2002.
4. Haixia Y., Qing Z., and Jianguo S. *Fundamentals of Computational Mechanics*. Nanjing, China: Hehai University Press, 2001.
5. Siyuan C. The methodology in finite element method. *J Chongqing Univ*, 2002, 7(4): 61–63.
6. Maocheng W. *Finite Element Method*. Beijing: Tsinghua University Press, 2003.
7. Belytschko T., Liu W. K., and Moran B. *Nonlinear Finite Elements for Continua and Structures*. Translated by Zhuang Zhuo. Beijing: Tsinghua University Press, 2002.
8. Guang Z., Keshi Z., and Lu F. Algorithm and applications of polycrystal plasticity modeling under finite deformation. *Chin J. Appl Mech*, 2004, 21(1): 96–100.
9. Takahashi H., Motohashi H., Tokuda M., and Abe T. Elastic-plastic finite element polycrystal model. *Int J Plasticity*, 1994, 10: 63–80.
10. Zhao Z., Mao W., Roters F. et al. A texture optimization study for minimum earing in aluminium by use of a texture component crystal plasticity finite element method. *Acta Materialia*, 2004, 52: 1003–1012.
11. Zaefferer S., Kuo J. C., and Zhao Z. On the influence of the grain boundary misorientation on the plastic deformation of aluminum bicrystals. *Acta Mater*, 2003, 51: 4719–4735.
12. Raabe D. Grain-scale micromechanics of polycrystal surfaces during plastic straining. *Acta Mater*, 2003, 51: 1539–1560.
13. Raabe D. and Zhao Z. Study on the orientational stability of cube-oriented FCC crystals under plane strain by use of a texture component crystal plasticity finite element method. *Scripta Mater*, 2004, 50: 1085–1090.
14. Shizhong S., Mingzhe L., and Dongping L. Prediction of plastic anisotropy and texture in FCC metal by finite element polycrystal model. *Acta Metall Sinica*, 2001, 37: 531–536.
15. Long S. G., Zhou Y. C., and Pan Y. Computation of deformation-induced textures in electrodeposited nickel coating. *Trans Nonferr Metal Soc China*, 2006, s16: 232–238.
16. Wei Y., Xinling M., Hongtao W. et al. Advances in nanomechanics. *Adv Mech*, 2002, 32: 161–174.
17. Alder B. J. and Wainwright T. E. Phase transition for a hard-sphere system. *J Chem Phys*, 1957, 27: 1208–1209.
18. Honeycutt R. W. The potential calculation and some applications. *Methods Comput Phys*, 1970, 9: 136–211.
19. Gear C. W. The numerical integration of ordinary differential equations of various orders. Report ANL 7126, Argonne National Laboratory, 1966.
20. Beeman D. Some multistep methods for use in molecular dynamics calculations. *J Comp Phys*, 1976, 20: 130–139.
21. Toxvaerd S. Algorithms for canonical molecular dynamics simulations. *Mol Phys*, 1991, 72: 159–168.
22. Heermann D. W. *Computer Simulation Methods in Theoretical Physics*. Translated by Qin Kecheng. Beijing: Beijing University Press, 1996.
23. Cotterill R. M. J. and Doyama M. In: R. Hasiguti, Ed. *Lattice Defects and Their Interactions*. New York: Gordon and Breach Science Publishers Inc., 1967, 62–75.
24. Daw M. S. and Baskes M. I. Embedded atom method-derivation and application to impurities, surfaces, and other defects in metals. *Phys Rev B*, 1984, 29(12): 8486–8495.
25. Finnis M. W. and Sinclair J. E. A simple empirical n-body potential for transition-metals. *Philos Mag A*, 1984, 50(1): 45–55.

26. Ackland G. J. and Vitek V. Many-body potentials and atom-scale relaxations in noble-metal alloys. *Phys Rev B*, 1990, 41(15): 10324–10333.
27. Berendsen H. J. C., Postma J. P. M., and Gunsteren W. F. V. Molecular dynamics with coupling to an external bath. *J Chem Phys*, 1984, 81: 3684–3690.
28. Anderson H. C. Molecular dynamics simulations at constant press and/or temperature. *J Chem Phys*, 1980, 72: 2384–2393.
29. Nose S. A unified formulation of the constant temperature molecular dynamics method. *J Chem Phys*, 1984, 81: 511–519.
30. Hoover W. G. Canonical dynamics: Equilibrium phase-space distributions. *Phys Rev A*, 1985, 31: 1695–1697.
31. Zhu T., Li J., and Yip S. Atomistic study of dislocation loop emission from a crack tip. *Phys Rev Lett*, 2004, 93: 25503.
32. Chen S. D., Ke F. J., Bai Y. L. et al. Atomistic investigation of the effects of temperature and surface roughness on diffusion bonding between Cu and Al. *Acta Mater*, 2007, 55: 3169–3175.
33. Denis S. and Miller R. E. Atomic-scale simulations of nanoindentation-induced plasticity in copper crystals with nanometer-sized nickel coatings. *Acta Mater*, 2006, 54: 33–45.
34. da Silva E. Z., da Silva A. J. R., Fazzio, A. et al. Breaking of gold nanowires. *Comput Mater Sci*, 2004, 30: 73–76.
35. Rodrigues V. and Ugarte D. Real-time imaging of atomistic process in one-atom-thick metal junctions. *Phys Rev B*, 2001, 63: 073405.
36. Jinghong F. *Multiscale Analysis of Deformation and Failure of Materials*. Beijing: Science Press, 2008.

10 Mechanical Properties of Smart Materials

Human understanding of natural phenomena such as sound, light, electricity, and magnetism, and the use of the material effects of these phenomena enable us to understand the world and use the knowledge acquired to serve mankind. Smart materials or structures are gradually emerging that can sense the external environment and internal state changes. Smart structures and devices have been widely used in information technology, in new materials technology, and in high-tech fields such as aerospace; and increasingly show their great advantages. Smart devices often use functional materials such as piezoelectric/ferroelectric materials, ferromagnetic materials, and shape memory alloy (SMA) materials; and commonly work in the electricity–force–magnetism–thermal coupling load environment. The design of smart devices must consider the deformation, vibration, and instability of structural components and systems occurring in external magnetic fields, electric fields, and large temperature gradient environments. Their reliability is of great concern. The primary concerns of the mechanical properties of smart materials include the determination of the constitutive relations of anisotropic smart materials, the field analysis of multifield coupling, and the determination of basic mechanical parameters. This chapter introduces the basic concept of smart materials and the most widely used SMAs, magnetostrictive and ferromagnetic SMA materials, ferroelectric and piezoelectric materials, and their related mechanical properties.

10.1 INTRODUCTION TO SMART MATERIALS

10.1.1 Concepts and Characteristics of Smart Materials

Smart structures are a new concept, developed in the 1990s. They have the ability to perceive the external environment and internal state changes, and can respond accordingly [1–4]. To achieve these functions, smart materials in smart structures should have the following characteristics: (1) sensing capability: detecting the external environment and internal state changes, such as electric and magnetic fields, temperature, and changes in stress and strain; (2) actuating capability: making appropriate responses to external or internal changes by adjusting and changing their states, particularly their posture; and (3) information processing capability: making a judgment from the external environment and internal state changes, and selecting the appropriate actuating mode for an appropriate response. These capabilities are represented in Figure 10.1.

According to their function, smart materials can be divided into actuating materials and sensing materials. Their simple diagrams are shown in Figure 10.2. Actuating materials under external fields such as an electric field will have deformation, resulting in mechanical displacement. Sensing materials are capable of perceiving changes in the external fields, such as mechanical deformation, and generating electrical signals or other signals. Many smart materials have both the ability to sense and actuate and can be used as both a sensing material and an actuating material.

To achieve the function of actuating and sensing, the microstructure of smart materials under external fields often evolves. Therefore, they are also known as active materials, as opposed to passive materials, whose microstructure cannot be changed. Moreover, because of microstructure evolution, the performance of smart materials often shows a marked nonlinearity, which presents great difficulties in theoretical analysis.

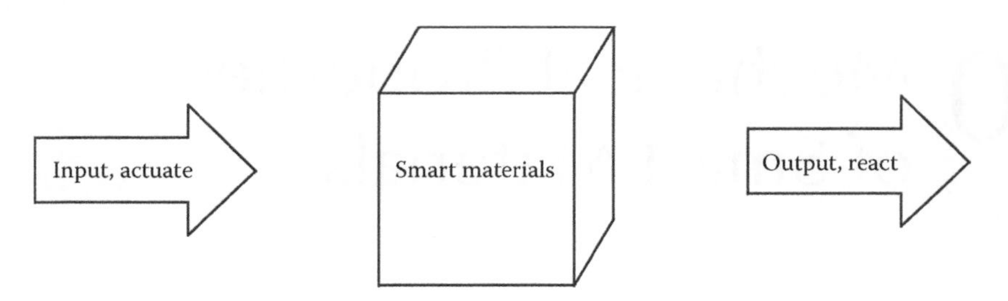

FIGURE 10.1 Diagram of smart materials.

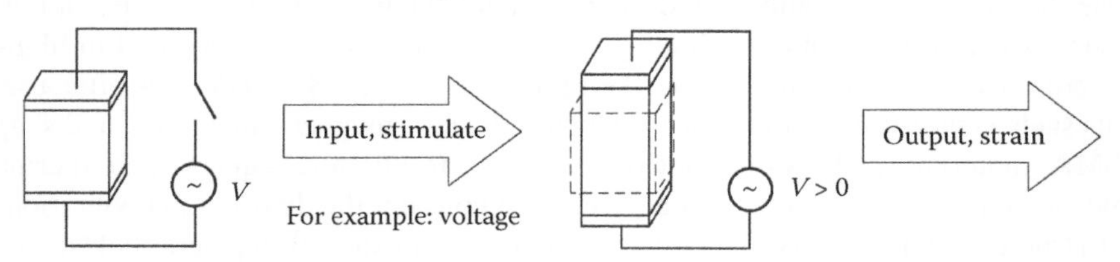

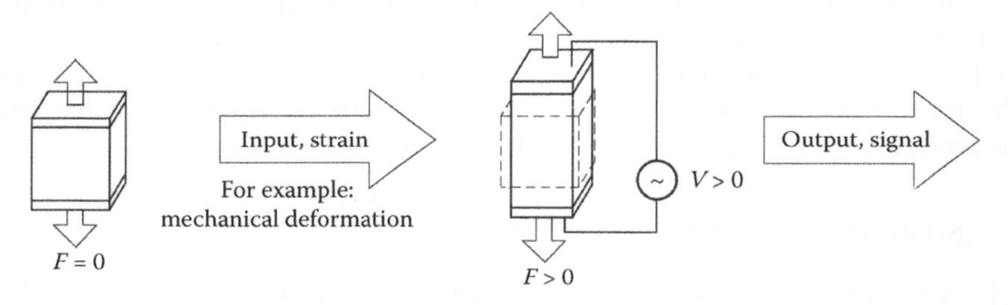

FIGURE 10.2 Diagram of actuating materials and sensing materials.

10.1.2 Applications of Smart Materials

Smart materials can not only bear loads like structural materials but can also perceive internal and external environmental changes. Smart materials, through appropriate responses, can achieve self-diagnosis as well as adaptive and self-repair functions by altering their physical properties. Therefore, smart structural materials usually possess a wide range of potential applications. Figure 10.3 shows the VCR magnetic head positioner, which is another type of smart structure. It is actually made with a piezoelectric ceramic and a sensor electrode, capable of sensing the position and orientation of the head. It also has a positioning electrode actuating the head, enabling it to reach the designated location [3].

Generally speaking, the application of smart materials can be divided into the following categories.

10.1.2.1 Structural Inspection

Smart structures can be used for the real-time measurement of strain, temperature, and cracking within a structure, as well as the detection of fatigue and damage, and are therefore able to monitor the structure and predict its life expectancy. For example, smart structures such as fiber-optic sensor arrays and polyvinylidene fluoride (PVDF) sensors on aircraft wings and frames, and on reusable space launch vehicles, are used for lifecycle real-time monitoring, damage assessment, and prediction of life expectancy. Space stations and other large in-orbit systems using fiber-optic smart structures can execute damage assessment and implement self-diagnosis by real-time damage detection of rendezvous and docking collisions, meteorite impacts, or other causes.

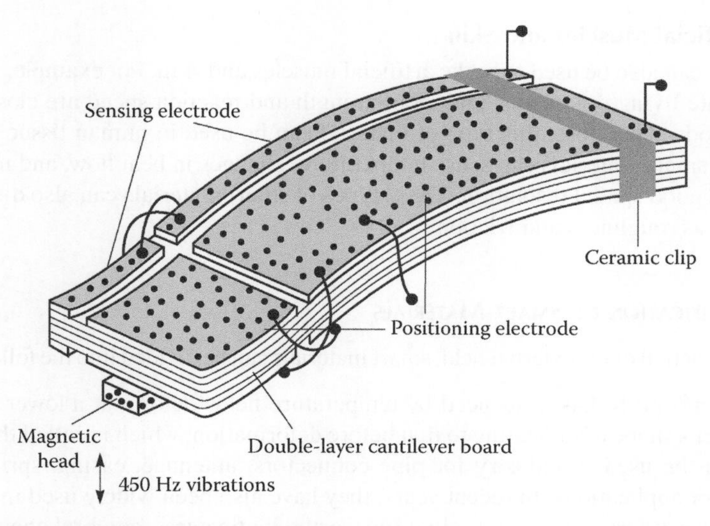

FIGURE 10.3 Diagram of magnetic head positioner.

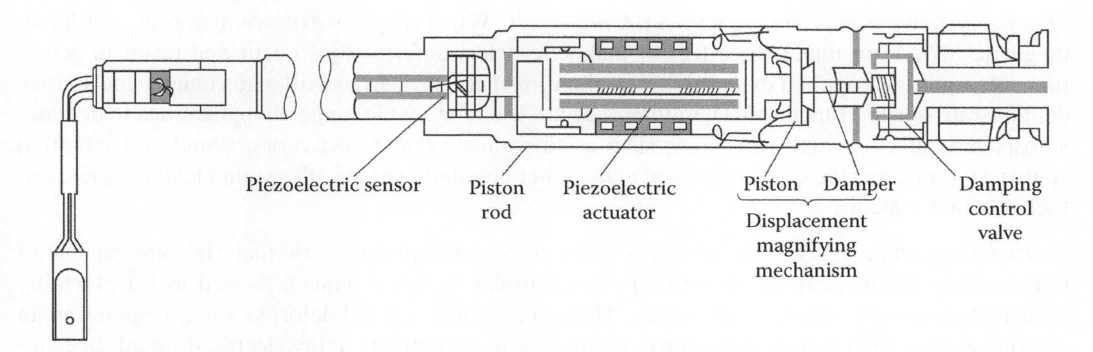

FIGURE 10.4 Schematic diagram of smart structure for automotive vibration reduction.

10.1.2.2 Vibration Control

Smart structures can also be used in aviation and aerospace systems to eliminate harmful vibrations, reduce the interference of electronic systems, and improve system reliability. Smart structures for ships can suppress noise transmission and improve the acoustic stealth performance of submarines and warships. Smart structures for ground vehicles can enhance the performance of military vehicles and ride comfort. Figure 10.4 is a schematic diagram of such structures, including sensors made by piezoelectric materials for automotive vibrations. It also has actuators made with piezoelectric materials that actuate water systems to offset automotive vibrations.

10.1.2.3 Adaptive Structures

Adaptive aircraft wings made with smart structures can sense, in real time, changes in the external environment and actuate the wing's bending and torsion, and adjust the airfoil and angle of attack to obtain the best aerodynamic performance, thus reducing the wing drag coefficient and extending wing fatigue life. For example, when an aircraft during flight encounters vortices or fierce headwinds, smart materials in the aircraft wing can quickly deform, causing the wing to change its shape and thereby enabling the aircraft to fly stably.

10.1.2.4 Artificial Muscles and Skin

Smart materials can also be used to make artificial muscles and skin. For example, bio-elastic materials can simulate living organisms, and their strength and reaction speed are close to the muscles of the human body. Therefore, this type of material can be used in human tissue repair. Artificial skin made with smart materials can sense temperature, changes in heat flow, and measure a variety of stresses, with good spatial resolution. These types of smart materials can also distinguish surface conditions such as roughness and friction.

10.1.3 Classification of Smart Materials

According to the action of the external field, smart materials can be divided into the following categories:

Shape memory alloys: SMAs influenced by temperature fields deform at a lower temperature and restore the object's shape after heating to that before deformation, which is called the shape memory effect. They can be used in industry for pipe connectors, antennae, casing, springs, posts, glass frames, and other applications. In recent years, they have also been widely used in medical devices such as thrombosis filters, orthopedic spinal rods, orthodontic wires, cerebral aneurysm clips, bone plates, intramedullary nails, artificial joints, contraceptive devices, cardiac repair components, artificial kidneys using a micropump, and so on.

Magnetostrictive and ferromagnetic SMA materials: When magnetostrictive materials and ferromagnetic SMAs are under the effect of magnetic fields, deformations occur and result in actuation. In response to applied external forces, their magnetic properties will also change, and so they demonstrate sensing functions. Therefore, they are used for a wide range of applications in defense, aerospace, and other high-tech areas such as microdisplacement actuators, sound and vibration control systems, marine exploration and underwater communications, ultrasound technologies, and fuel injection systems.

Ferroelectric and piezoelectric materials: Ferroelectric and piezoelectric materials are capable of transforming mechanical signals (stress) into electrical signals (voltage), as well as transforming electrical signals into mechanical signals. They can produce a small deformation in response to an external electric field and also transform the small deformation into a tiny electrical signal. Because of this unique feature, piezoelectric materials have a wide range of potential applications in smart systems. For example, an actuator made with multilayer piezoelectric ceramics can generate tens of microns of displacement within a few milliseconds. The responding speed is unmatched by other materials, making piezoelectric ceramics necessary for high-precision and high-speed actuators. Because piezoelectric materials produce a measurable electrical signal in response to an applied external stress, they can be used as sensors in smart materials systems.

In the following sections, we present a detailed introduction on these three types of materials.

10.2 SHAPE MEMORY ALLOYS

SMAs exhibit shape-memory effects when responding to temperature changes. Their mechanical deformation, occurring at a lower temperature, can be restored when heated above a certain transition temperature. At the same time, they are able to withstand larger reversible deformation at higher temperatures and thus have superelasticity. As described later, these effects originate from the SMA's unique microstructure [5–9].

10.2.1 Shape Memory Effect and Superelasticity

The shape memory effects and superelasticity of SMAs can be described by their stress–strain curves at different temperatures, as shown in Figure 10.5. At lower temperatures, the SMA shows

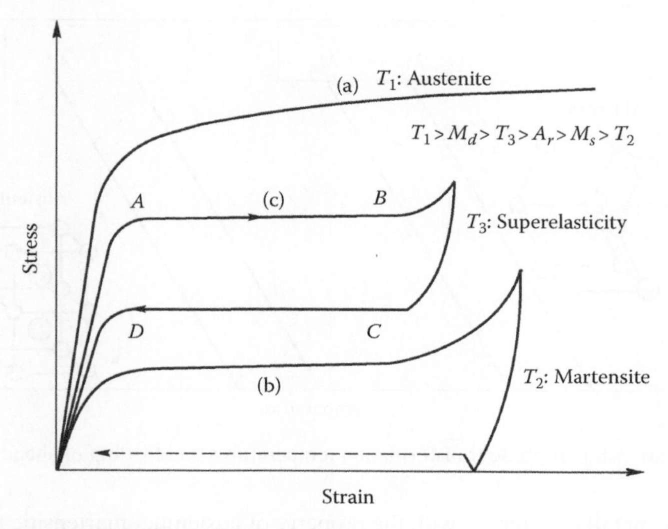

FIGURE 10.5 Shape memory alloy stress–strain curves.

smaller elastic deformation in the initial loading period, followed by larger plastic deformation. Deformation is retained after unloading, but it can be fully restored by heating. This is what we usually refer to as the shape memory effect. At moderate temperatures, the SMA, when loaded, can withstand larger deformation. This deformation is reversible and can be fully recovered after unloading without heating. This is the so-called superelasticity. At higher temperatures, the SMA's stress–strain curve is no different from other ordinary metals. After elastic deformation, irreversible plastic deformation occurs, which cannot be restored even by heating.

10.2.2 Microstructure and Memory Mechanisms of Shape Memory Alloys

The unique properties of the SMA come from its microstructure. At high temperatures, SMAs are usually a cubic structure of austenite; at low temperatures, a phase change occurs and the material transitions into a monoclinic structure of martensite. These crystal structures are shown in Figure 10.6. SMAs have four phase transformation temperatures: (1) martensite start (M_S), (2) martensite finish (M_F), (3) austenite start (A_S), and (4) austenite finish (A_F) temperatures. These phase transformation temperatures are affected by stress and will increase under loading. The relationship of phase transformation temperature and stress is shown in Figure 10.7.

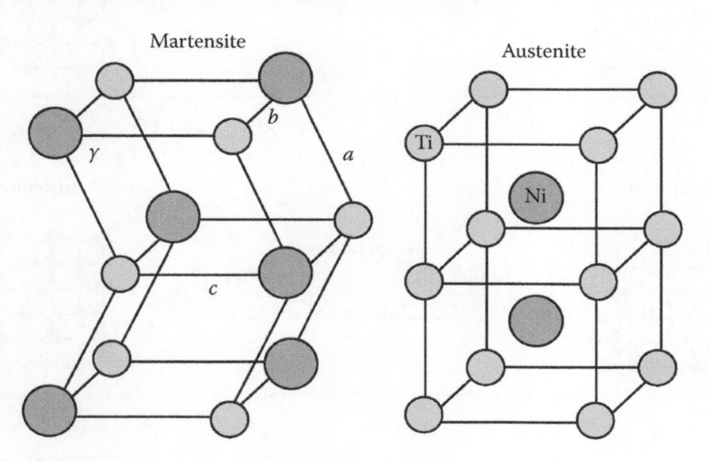

FIGURE 10.6 Phase transformation of shape memory alloys.

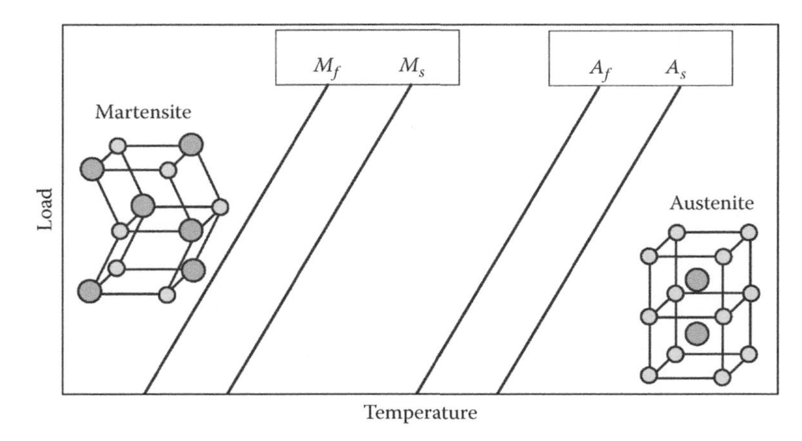

FIGURE 10.7 Relationship of phase transformation temperatures and loading of shape memory alloys.

There are many metallic materials with the property of austenitic–martensitic transformation, but materials with shape memory effects are rare. This is because the shape memory effect has very strict requirements for the crystal lattice constants, which require that the material's volume remain the same after phase transformation. Upon cooling, in the absence of an applied load, the SMA transforms from austenite into twinned martensite. Although its microstructure changes, no observable macroscopic shape and size changes occur. This is called self-accommodation, as shown in Figure 10.8. Self-accommodation is the key to shape memory effects. The crystal lattice constants of martensite and austenite must meet a very stringent relationship: the volume of the material is unchanged after the martensitic transformation. This is why the austenitic–martensitic transformation is very common, but SMAs are rare.

As mentioned earlier, the SMA transforms from austenite into self-accommodated twinned martensite, which, from a macroscopic view, is the same shape and size as austenite. At this point, upon loading, the twinned martensite will transform into a single variant of martensite, showing large macrodeformation. This deformation cannot be restored after unloading. The variant is a state: for instance, the cubical structure and the tetragonal structure are all variants. Upon heating, the martensite transforms into the cubic structure of austenite. The macroscopic

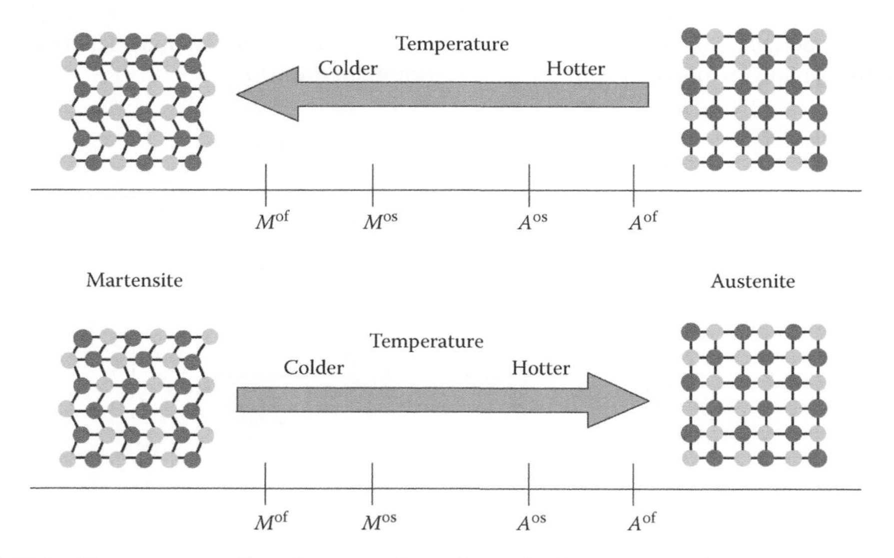

FIGURE 10.8 Shape memory alloy phase transformation and self-accommodation.

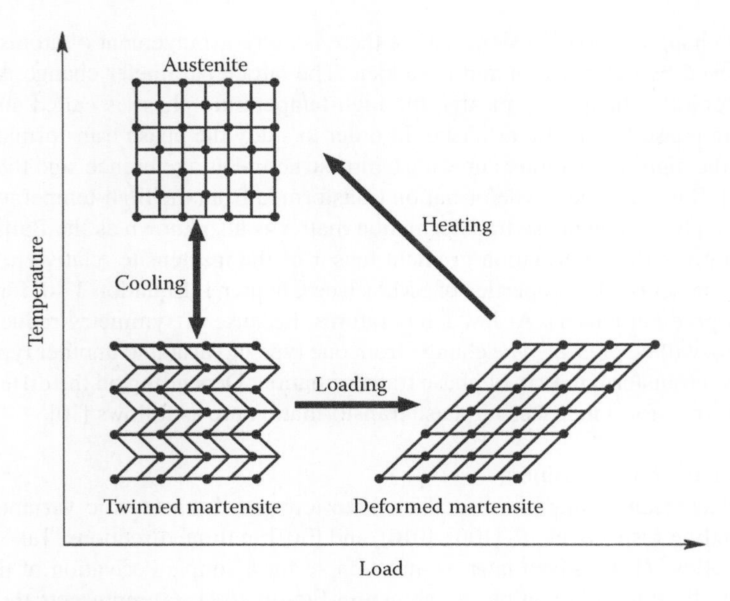

FIGURE 10.9 Micromechanism of shape memory effect.

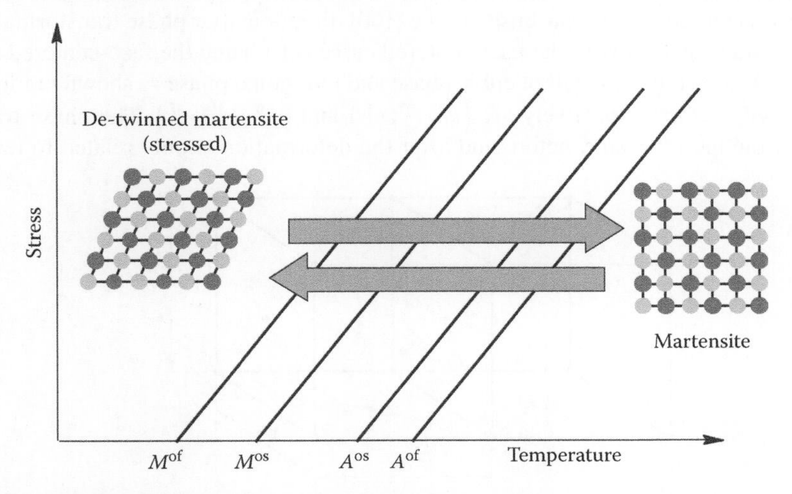

FIGURE 10.10 Micromechanism of shape memory alloy superelasticity.

shape and size are no different from the self-accommodated twinned martensite. Therefore, the material can be restored to the shape and size before loading, showing the shape memory effect. This process is shown in Figure 10.9.

As discussed earlier, the phase transformation temperatures are related to the applied load. At higher temperatures, applying a load to the austenitic SMA will increase the phase transformation temperature, and thus cause stress-induced martensitic transformation. At the time of impact of the applied load, the introduced martensite is a single variant, showing a greater macrodeformation. After unloading, the martensite transforms into austenite, and thus deformation is completely restored. This is the micromechanism [5–9] of superelasticity, and the process is shown in Figure 10.10.

10.2.3 Mathematical Models of Shape Memory Alloys

Martensitic transformation is a first-order, nondiffusion, and solid-to-solid phase transformation. Its unit cell structure at high temperatures is different from its structure at low temperatures.

This structural change is nondiffusion; that is, there is no re-arrangement of atoms. The structural change is achieved by deformation and is sudden. The lattice parameter change is noncontinuous when the temperature changes. Typically, the high-temperature phase is called austenite, and the low-temperature phase is called martensite. In order to study the phase transformation behavior of SMAs, we use the high-temperature cubic austenitic structure as a reference, and the phase transformation matrix $[U]$ to describe the deformation transformed from the high-temperature phase to the low-temperature phase. The phase transformation matrix is also known as the Bain transformation matrix—which gives the deformation gradient tensor of the martensite relative to austenite—and has significant impact on the properties of SMAs (see Chapter 1, Equation 1.40, for the component of deformation gradient tensor). At low temperatures, because of symmetry reduction, equivalent multiple variants will be present. The change from one type of variant to another type of variant can be described by a transform matrix of phase transformation. Depending on the different martensitic structures, their transform matrices of phase transformation are as follows [10].

1. Cubic-to-tetragonal transition
   In this phase transformation, a total of three tetragonal martensitic variants exist, each corresponding respectively to [100], [010], and [001] in three directions. Take the indium–thallium alloys (InTl) given later as an example for a simple derivation of the transform matrix of phase transformation. As shown in Figure 10.11, (a) represents the lattice cells of the cubic austenite before phase transformation and (b) represents the lattice cells of the single-variant tetragonal martensite in the [100] direction after phase transformation. They correspond respectively to the face-centered cubic (FCC) and the face-centered tetragonal (FCT). To select the unit cell of cubic phase and tetragonal phase as shown in Figure 10.11, the lattice vectors respectively are $\left(e_1^A, e_2^A, e_3^A\right)$ and $\left(e_1^M, e_2^M, e_3^M\right)$. The phase transformation, or the lattice vector, before and after the deformation is only related to temperature

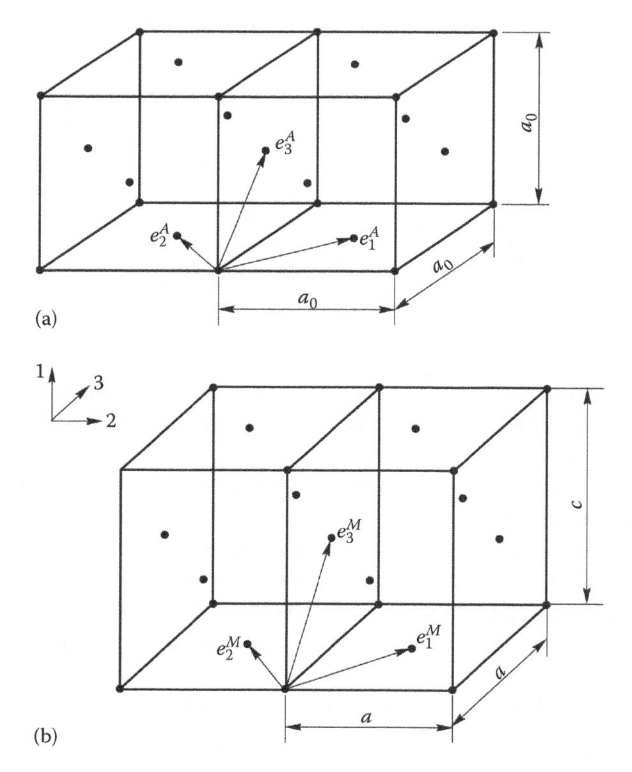

FIGURE 10.11 Lattice vectors of unit lattice cell. (a) Cubic austenite. (b) Tetragonal martensite.

and is barely influenced by thermal expansion. Referring to Figure 10.11a and b, and the coordinate system on the left of the figure, we can obtain a_0 and a, c, which are the lattice constants corresponding to the cubic phase and the tetragonal phase, respectively:

$$\left[\boldsymbol{e}_1^A\right]=\frac{1}{2}\begin{pmatrix}0\\a_0\\a_0\end{pmatrix}\quad\left[\boldsymbol{e}_2^A\right]=\frac{1}{2}\begin{pmatrix}0\\-a_0\\a_0\end{pmatrix}\quad\left[\boldsymbol{e}_3^A\right]=\frac{1}{2}\begin{pmatrix}a_0\\0\\a_0\end{pmatrix}$$
$$\left[\boldsymbol{e}_1^M\right]=\frac{1}{2}\begin{pmatrix}0\\a\\a\end{pmatrix}\quad\left[\boldsymbol{e}_2^M\right]=\frac{1}{2}\begin{pmatrix}0\\-a\\a\end{pmatrix}\quad\left[\boldsymbol{e}_3^M\right]=\frac{1}{2}\begin{pmatrix}c\\0\\a\end{pmatrix}\tag{10.1}$$

In the phase transformation process, considering the deformation caused by the phase transformation is significant. Ignoring the smaller deformation caused by the thermal expansion and contraction effects, we can obtain a matrix $\boldsymbol{U}_1$, which satisfies the following equation:

$$\boldsymbol{e}_i^M=\boldsymbol{U}_1\boldsymbol{e}_i^A\tag{10.2}$$

Or,

$$\begin{pmatrix}0&0&\dfrac{c}{2}\\\dfrac{a}{2}&-\dfrac{a}{2}&0\\\dfrac{a}{2}&\dfrac{a}{2}&\dfrac{a}{2}\end{pmatrix}=\boldsymbol{U}_1\begin{pmatrix}0&0&\dfrac{a_0}{2}\\\dfrac{a_0}{2}&-\dfrac{a_0}{2}&0\\\dfrac{a_0}{2}&\dfrac{a_0}{2}&\dfrac{a_0}{2}\end{pmatrix}.\tag{10.3}$$

From Equation 10.3, in the direction of the cubic phase to the [100] single variant of the tetragonal phase, we obtain the transform matrix:

$$\boldsymbol{U}_1=\begin{pmatrix}\dfrac{c}{a_0}&0&0\\0&\dfrac{a}{a_0}&0\\0&0&\dfrac{a}{a_0}\end{pmatrix}=\begin{pmatrix}\beta&0&0\\0&\alpha&0\\0&0&\alpha\end{pmatrix}.\tag{10.4}$$

Here, $\alpha = a/a_0$ and $\beta = c/a_0$. Employing a similar operation to solve $\boldsymbol{U}_1$, other transform matrices can be obtained. The transform matrices from the cubic phase to the tetragonal phase can be generally described as

$$\boldsymbol{U}_1=\begin{bmatrix}\beta&0&0\\0&\alpha&0\\0&0&\alpha\end{bmatrix},\quad\boldsymbol{U}_2=\begin{bmatrix}\alpha&0&0\\0&\beta&0\\0&0&\alpha\end{bmatrix},\quad\boldsymbol{U}_3=\begin{bmatrix}\alpha&0&0\\0&\alpha&0\\0&0&\beta\end{bmatrix}.\tag{10.5}$$

TABLE 10.1
Parameters of Phase Transformation Matrix of Some Materials from Cubic Phase to Tetragonal Phase

Number of Martensitic Variants $N = 3$		
Materials	α	β
In-23at.%Tl	0.9889	1.0212
In-32at.%Pb	1.0208	0.9688
Ni-36at.%Al	0.9392	1.1302
Ni-49.4at.%Al	0.912	1.194
Fe-24at.%Pt	1.0868	0.8503
Fe-7.9at.%Cr-1.1at.%C	1.1176	0.8243
Fe-22at.%Ni-0.8at.%C	1.1083	0.819
Fe-31at.%Ni-0.3at.%C	1.1241	0.8059
Fe-7at.%Al-1.5at.%C	1.0946	0.8546
Fe-7at.%Al-2at.%C	1.0833	0.8727

For indium–thallium alloys (InTl), $a_0 = 4.7445$ Å, $a = 4.691\,9$ Å, $c = 4.845\,1$ Å, therefore, $\alpha = 0.9889$, and $\beta = 1.0212$.

Many other alloys are the result of phase transformation from the cubic phase to the tetragonal phase, and their phase transformation matrices are described in Equation 10.4. This matrix is determined by α and β; α and β are determined by the lattice constants of the materials. Table 10.1 lists some of the specific values of the alloys.

2. Cubic-to-orthorhombic transition

In this phase transformation, a total of six orthorhombic–martensitic variants exist. The lattice constants of the cubic phase are a_0, a_0, and a_0; and the lattice constants of the orthorhombic phase are a, b, and c, as shown in Figure 10.12. Thus, the transform matrix of phase transformation can be determined by

$$\alpha = \frac{\sqrt{2}a}{a_0}, \quad \beta = \frac{b}{a_0}, \quad \gamma = \frac{\sqrt{2}c}{a_0}.$$

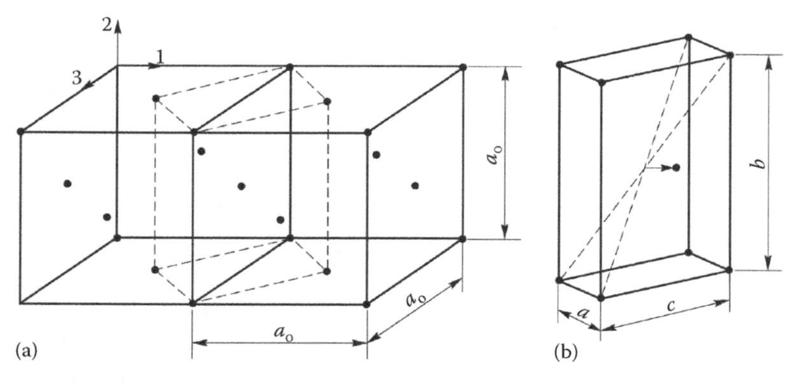

FIGURE 10.12 Unit lattice cell. (a) cubic phase, (b) orthorhombic phase.

The six transformation matrices are, respectively,

$$
U_1 = \begin{bmatrix} \frac{\alpha+\gamma}{2} & 0 & \frac{\alpha-\gamma}{2} \\ 0 & \beta & 0 \\ \frac{\alpha-\gamma}{2} & 0 & \frac{\alpha+\gamma}{2} \end{bmatrix}, \quad U_2 = \begin{bmatrix} \frac{\alpha+\gamma}{2} & 0 & -\frac{\alpha-\gamma}{2} \\ 0 & \beta & 0 \\ -\frac{\alpha-\gamma}{2} & 0 & \frac{\alpha+\gamma}{2} \end{bmatrix},
$$

$$
U_3 = \begin{bmatrix} \frac{\alpha+\gamma}{2} & \frac{\alpha-\gamma}{2} & 0 \\ \frac{\alpha-\gamma}{2} & \frac{\alpha+\gamma}{2} & 0 \\ 0 & 0 & \beta \end{bmatrix}, \quad U_4 = \begin{bmatrix} \frac{\alpha+\gamma}{2} & -\frac{\alpha-\gamma}{2} & 0 \\ -\frac{\alpha-\gamma}{2} & \frac{\alpha+\gamma}{2} & 0 \\ 0 & 0 & \beta \end{bmatrix}, \tag{10.6}
$$

$$
U_5 = \begin{bmatrix} \beta & 0 & 0 \\ 0 & \frac{\alpha+\gamma}{2} & \frac{\alpha-\gamma}{2} \\ 0 & \frac{\alpha-\gamma}{2} & \frac{\alpha+\gamma}{2} \end{bmatrix}, \quad U_6 = \begin{bmatrix} \beta & 0 & 0 \\ 0 & \frac{\alpha+\gamma}{2} & -\frac{\alpha-\gamma}{2} \\ 0 & -\frac{\alpha-\gamma}{2} & \frac{\alpha+\gamma}{2} \end{bmatrix}
$$

α, β, and γ are determined by the lattice constants of the crystals. For example, CuAlNi alloys have $a_0 = 5.836$ Å, $a = 4.3823$ Å, $b = 5.3563$ Å, $c = 4.223$ Å. In this way, α, β, and γ are 1.0619, 0.9178, and 1.0231, respectively.

3. Cubic–monoclinic transition
   There are two types of cube–monoclinic phase transformation, each with 12 variants. For the first type of phase transformation, the cubic phase lattice constant is a_0; the monoclinic phase lattice constants are a, b, and c, and the angle is β, as shown in Figure 10.13.

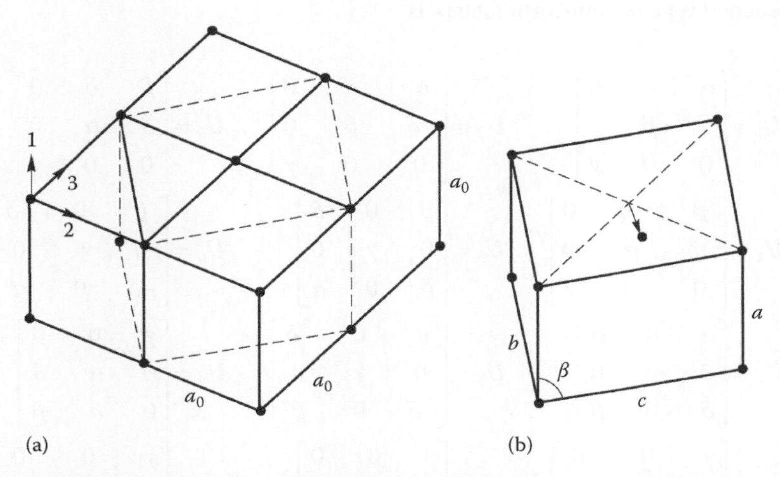

FIGURE 10.13 The unit cell of cube–monoclinic phase transformation. (a) Cubic phase. (b) Monoclinic phase.

The transform matrix of phase transformation is determined using the following γ, ε, α, and δ:

$$\gamma = \frac{a\left(\sqrt{2}a + c\sin\beta\right)}{a_0\sqrt{2a^2 + c^2 + 2\sqrt{2}ac\sin\beta}} \qquad \varepsilon = \frac{ac\cos\beta}{\sqrt{2}a_0\sqrt{2a^2 + c^2 + 2\sqrt{2}ac\sin\beta}},$$

$$\alpha = \frac{1}{2\sqrt{2}a_0}\left[\frac{c\left(c + \sqrt{2}a\sin\beta\right)}{\sqrt{2a^2 + c^2 + 2\sqrt{2}ac\sin\beta}} + b\right] \qquad \delta = \frac{1}{2\sqrt{2}a_0}\left[\frac{c\left(c + \sqrt{2}a\sin\beta\right)}{\sqrt{2a^2 + c^2 + 2\sqrt{2}ac\sin\beta}} - b\right]$$

In this way, the 12 transform matrices are

$$\begin{gathered}
\boldsymbol{U}_1 = \begin{bmatrix} \gamma & \varepsilon & \varepsilon \\ \varepsilon & \alpha & \delta \\ \varepsilon & \delta & \alpha \end{bmatrix} \quad
\boldsymbol{U}_2 = \begin{bmatrix} \gamma & -\varepsilon & -\varepsilon \\ -\varepsilon & \alpha & \delta \\ -\varepsilon & \delta & \alpha \end{bmatrix} \quad
\boldsymbol{U}_3 = \begin{bmatrix} \gamma & -\varepsilon & \varepsilon \\ -\varepsilon & \alpha & -\delta \\ \varepsilon & -\delta & \alpha \end{bmatrix} \\
\boldsymbol{U}_3 = \begin{bmatrix} \gamma & -\varepsilon & \varepsilon \\ -\varepsilon & \alpha & -\delta \\ \varepsilon & -\delta & \alpha \end{bmatrix} \quad
\boldsymbol{U}_5 = \begin{bmatrix} \alpha & \varepsilon & \delta \\ \varepsilon & \gamma & \varepsilon \\ \delta & \varepsilon & \alpha \end{bmatrix} \quad
\boldsymbol{U}_6 = \begin{bmatrix} \alpha & -\varepsilon & \delta \\ -\varepsilon & \gamma & -\varepsilon \\ \delta & -\varepsilon & \alpha \end{bmatrix} \\
\boldsymbol{U}_7 = \begin{bmatrix} \alpha & -\varepsilon & -\delta \\ -\varepsilon & \gamma & \varepsilon \\ -\delta & \varepsilon & \alpha \end{bmatrix} \quad
\boldsymbol{U}_8 = \begin{bmatrix} \alpha & \varepsilon & -\delta \\ \varepsilon & \gamma & -\varepsilon \\ -\delta & -\varepsilon & \alpha \end{bmatrix} \quad
\boldsymbol{U}_9 = \begin{bmatrix} \alpha & \delta & \varepsilon \\ \delta & \alpha & \varepsilon \\ \varepsilon & \varepsilon & \gamma \end{bmatrix} \\
\boldsymbol{U}_{10} = \begin{bmatrix} \alpha & \delta & -\varepsilon \\ \delta & \alpha & -\varepsilon \\ -\varepsilon & -\varepsilon & \gamma \end{bmatrix} \quad
\boldsymbol{U}_{11} = \begin{bmatrix} \alpha & -\delta & \varepsilon \\ -\delta & \alpha & -\varepsilon \\ \varepsilon & -\varepsilon & \gamma \end{bmatrix} \quad
\boldsymbol{U}_{12} = \begin{bmatrix} \alpha & -\delta & -\varepsilon \\ -\delta & \alpha & \varepsilon \\ -\varepsilon & \varepsilon & \gamma \end{bmatrix}
\end{gathered} \tag{10.7}$$

For the NiTi alloy (49.75at.% Ni), $a_0 = 3.015$ Å, $a = 2.889$ Å, $b = 412$ Å, $c = 4.622$ Å, $\beta = 96.8°$. Therefore, $\alpha = 1.0243$, $\gamma = 0.9563$, $\varepsilon = -0.04266$, $\delta = 0.05803$.

In addition, the CuZr alloy also belongs to this type of phase transformation: $\alpha = 1.0348$, $\gamma = 1.0229$, $\delta = 0.1067$, $\varepsilon = -0.0929$.

The second type of transform matrix is

$$\begin{gathered}
\boldsymbol{U}_1 = \begin{bmatrix} \alpha & \delta & 0 \\ \delta & \beta & 0 \\ 0 & 0 & \gamma \end{bmatrix} \quad
\boldsymbol{U}_2 = \begin{bmatrix} \alpha & -\delta & 0 \\ -\delta & \beta & 0 \\ 0 & 0 & \gamma \end{bmatrix} \quad
\boldsymbol{U}_3 = \begin{bmatrix} \beta & \delta & 0 \\ \delta & \alpha & 0 \\ 0 & 0 & \gamma \end{bmatrix} \\
\boldsymbol{U}_4 = \begin{bmatrix} \beta & -\delta & 0 \\ -\delta & \alpha & 0 \\ 0 & 0 & \gamma \end{bmatrix} \quad
\boldsymbol{U}_5 = \begin{bmatrix} \beta & 0 & \delta \\ 0 & \gamma & 0 \\ \delta & 0 & \alpha \end{bmatrix} \quad
\boldsymbol{U}_6 = \begin{bmatrix} \beta & 0 & -\delta \\ 0 & \gamma & 0 \\ -\delta & 0 & \alpha \end{bmatrix} \\
\boldsymbol{U}_7 = \begin{bmatrix} \alpha & 0 & \delta \\ 0 & \gamma & 0 \\ \delta & 0 & \beta \end{bmatrix} \quad
\boldsymbol{U}_8 = \begin{bmatrix} \alpha & 0 & -\delta \\ 0 & \gamma & 0 \\ -\delta & 0 & \beta \end{bmatrix} \quad
\boldsymbol{U}_9 = \begin{bmatrix} \gamma & 0 & 0 \\ 0 & \alpha & \delta \\ 0 & \delta & \beta \end{bmatrix} \\
\boldsymbol{U}_{10} = \begin{bmatrix} \gamma & 0 & 0 \\ 0 & \alpha & -\delta \\ 0 & -\delta & \beta \end{bmatrix} \quad
\boldsymbol{U}_{11} = \begin{bmatrix} \gamma & 0 & 0 \\ 0 & \beta & \delta \\ 0 & \delta & \alpha \end{bmatrix} \quad
\boldsymbol{U}_{12} = \begin{bmatrix} \gamma & 0 & 0 \\ 0 & \beta & -\delta \\ 0 & -\delta & \alpha \end{bmatrix}
\end{gathered} \tag{10.8}$$

TABLE 10.2
Parameters of the Second Type of Phase Transformation Matrix in Some Materials from Cubic Phase to Monoclinic Phase Transformation

	Number of Martensitic Variants $N = 12$			
Materials	α	β	γ	δ
Cu-39.3at.%Zn	0.9955	1.0935	0.9109	0.015
Cu-15at.%Zn-17at.%Al	1.0101	1.0866	0.9093	0.0249
Cu-20at.%Zn-12at.%Ga	0.9998	1.0888	0.9096	0.0167
Cu-14wt.%Al-4wt.%Al	1.0004	1.1038	0.9133	0.0202

Here α, β, γ, δ, and ε all are determined by the crystal lattice constants. Table 10.2 lists the parameters of the second type of phase transformation matrix in some materials from cubic to monoclinic phase transformation.

Usually, after the completion of austenite–martensite phase transformation, multiple martensitic variants coexist. The average transform matrix, or the average component of the deformation gradient tensor, is

$$\boldsymbol{U} = \sum_{i=1}^{N} \lambda_i \boldsymbol{U}_i, \tag{10.9}$$

where λ_I is the volume ratio of each variant. The average transform matrix describes the macroscopic deformation of the martensite relative to the austenite. As mentioned earlier, the key to the shape memory effect is the formation of a self-accommodated martensitic structure; that is, the macroscopic volume of the martensite is the same as austenite. The necessary and sufficient condition for the formation of self-accommodated structures is that the transform matrix of each variant satisfies

$$\det \boldsymbol{U}_i = 1. \tag{10.10}$$

This is the volume conservation in the process of martensite phase transformation. Upon application of the load, the volume ratio of each variant of the self-accommodated martensitic structure will change, expressing as macroscopic deformation. The macroscopic deformation can be restored at high temperatures after the austenitic transformation: this is what we call the shape memory effect. However, the recoverable macroscopic deformation of martensite is limited: only those macroscopic deformations through the change of variant volume ratio are recoverable. The plastic deformation of single-variant martensite at high temperatures cannot be restored. Figure 10.14 shows the recovery curves of the experimental test for a spring made of an NiTi SMA after heat treatment [11]. Here, the recovery ratio is in fact the residual strain. The experimental results show that different heat-treatment processes can improve the recovery properties of NiTi SMAs in varying degrees and that the optimized heat treatment process is at 550°C in air cooling.

10.3 MAGNETOSTRICTIVE MATERIALS AND FERROMAGNETIC SHAPE MEMORY ALLOYS

10.3.1 Magnetocrystalline Anisotropy

Magnetocrystalline anisotropy refers to the different performance of the increased free energy of a single ferromagnetic crystal to align itself along different crystal axes in the process of magnetization. Magnetocrystalline anisotropy is smaller when a ferromagnetic crystal magnetizes along the

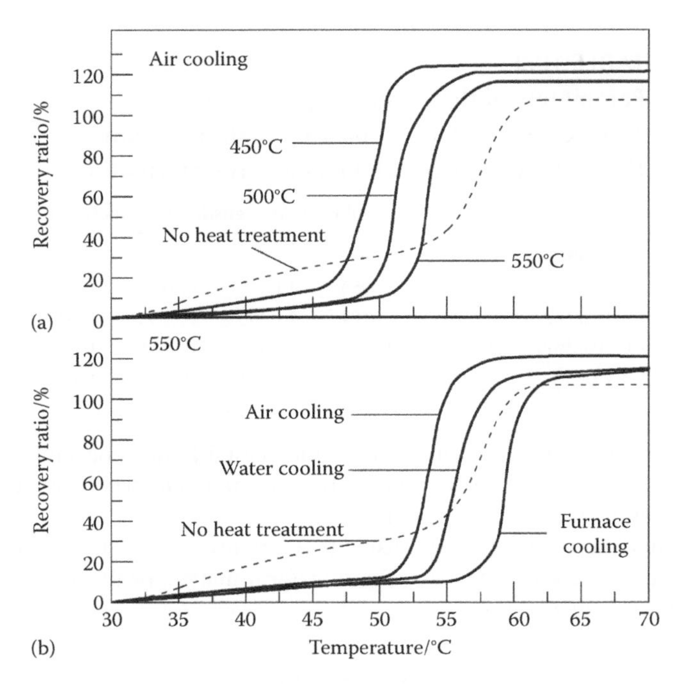

FIGURE 10.14 Recovery curves of NiTi shape memory alloy after heat treatment. (a) Air cooling. (b) 550°C. (From Wang, G. et al., *Adv. Eng. Mater.*, 8, 107, 2006; Taken from Figure 4.)

direction of an easy magnetization axis, while the magnetocrystalline anisotropy is greater when a ferromagnetic crystal magnetizes along the direction of a hard magnetization axis. The energy difference of magnetization along different directions of crystal axes represents the magnetocrystalline anisotropy difference along different directions of the crystal axes. The easy magnetization directions of various materials are different; for example, the easy magnetization axis of an iron crystal is [100] and for a nickel crystal it is [111].

The generation of magnetocrystalline anisotropy is complicated. The influence of the crystal field on the electron orbit, and the combined effects of electron spin–orbit coupling, should be taken into account. Magnetocrystalline anisotropy energy is generated by magnetic interaction between the electron spin, as well as the mutual coupling effect between the spin magnetic moment and the orbital magnetic moment. Atoms or ions in the solid crystal lattice—due to the effects of the static crystalline field on each of the nearby atoms [12]—cause the electron orbital moment to lose its isotropic symmetry in space. Thus, the distribution of the electron cloud becomes the anisotropic shape. The anisotropic interaction energy between the electron spin is generated through mutual coupling between the electron spin moment and the orbital moment.

For the simple cubic crystal system, take [100], [010], and [001] for the three x-, y-, and z-coordinate axes. Set the magnetization vector as M_s and the direction cosine of the x-, y-, and z-axes as m_1, m_2, and m_3, respectively. Then the magnetocrystalline anisotropy energy of cubic crystals—the Hemholtz free energy F_k, according to the direction cosine m_i ($i = 1, 2, 3$)—has the following expanded form:

$$\begin{aligned} F_K &= A_0 + A_1(m_1 + m_2 + m_3) + A_2(m_1m_2 + m_2m_3 + m_3m_1) \\ &\quad + A_3\left(m_1^2 + m_2^2 + m_3^2\right) + A_4\left(m_1m_2^2 + m_1m_3^2 + m_2m_1^2 + m_2m_3^2\right. \\ &\quad \left. + m_3m_1^2 + m_3m_2^2\right) + A_5\left(m_1^4 + m_2^4 + m_3^4\right) \\ &\quad + A_6\left(m_1^2m_2^2 + m_2^2m_3^2 + m_3^2m_1^2\right) + \cdots, \end{aligned} \tag{10.11}$$

where A_i ($i = 0, 1, 2\ldots$) are the expansion coefficients. Each easy magnetization axis includes both positive and negative directions, and in the cubic crystal, the macroscopic effect of the interchange of the three directions [100], [010], and [001] is the same. Therefore, Equation 10.11 can only include the even order of m_1, m_2, and m_3.

Using the trigonometric relationship

$$m_1^2 + m_2^2 + m_3^2 = 1 = \left(m_1^2 + m_2^2 + m_3^2\right)^2$$
$$= m_1^4 + m_2^4 + m_3^4 + 2\left(m_1^2 m_2^2 + m_2^2 m_3^2 + m_3^2 m_1^2\right) \tag{10.12}$$

it can be seen that constants such as A_3 and A_5 can be included in A_0. Therefore, the general mathematical expression of F_k is [12–16]

$$F_K = K_0 + K_1\left(m_1^2 m_2^2 + m_2^2 m_3^2 + m_3^2 m_1^2\right) + K_2\left(m_1^2 m_2^2 m_3^2\right) + \cdots \tag{10.13}$$

where

K_0 is a constant

K_1 and K_2 are magnetocrystalline anisotropy constants

Usually, the magnetocrystalline anisotropy constant of materials K_1 and K_2 can be tested through experimental methods, including the single-crystal magnetization curve method, the magnetic torque method, the ferromagnetic resonance method, and the saturation approach law for polycrystalline magnetic materials [12–17].

10.3.2 Magnetostrictive Effect

The magnetostrictive effect refers to the deformation of materials in the presence of an external magnetic field, which exists in almost all ferromagnetic materials. Usually, the magnetostrictive effect in ferromagnetic materials is very small. For example, the strain of nickel, iron, and cobalt materials is about ten-millionths. However in Terfenol-D materials, the magnetostrictive strain may reach one-thousandth. This strain is also related to the applied stress on the material. Moderate compressive stress can help to improve the magnetostrictive strain of the materials as shown in Figure 10.15, which shows the relationship of the magnetostrictive strain under different pressures and magnetic fields [18].

The magnetostrictive effect originates from the microstructure of materials. Ferromagnetic materials possess self-magnetization vectors, and the magnetization vector will rotate under the applied magnetic field, causing magnetostriction, as shown in Figure 10.16. The applied compressive stress is able to make the magnetization vector in a zero magnetic field deflect to the direction of the vertical stress, thus causing greater magnetostrictive strain in the magnetic field, as shown in Figure 10.17. Therefore, for $K_1 > 0$ cubic crystals, if the 90° and 180° domain wall displacements occur simultaneously, the elongation should be [12,16]

$$\frac{\delta l}{l} = \lambda, \tag{10.14}$$

where λ is the magnetostrictive strain. The magnetostrictive strain changes with an increase of the external magnetic field and eventually reaches a saturation value [19] when the magnetostriction reaches a certain saturation value, called the saturation magnetostriction λ_s. The reason for this behavior is due to the spontaneous deformation of the crystal lattice within each

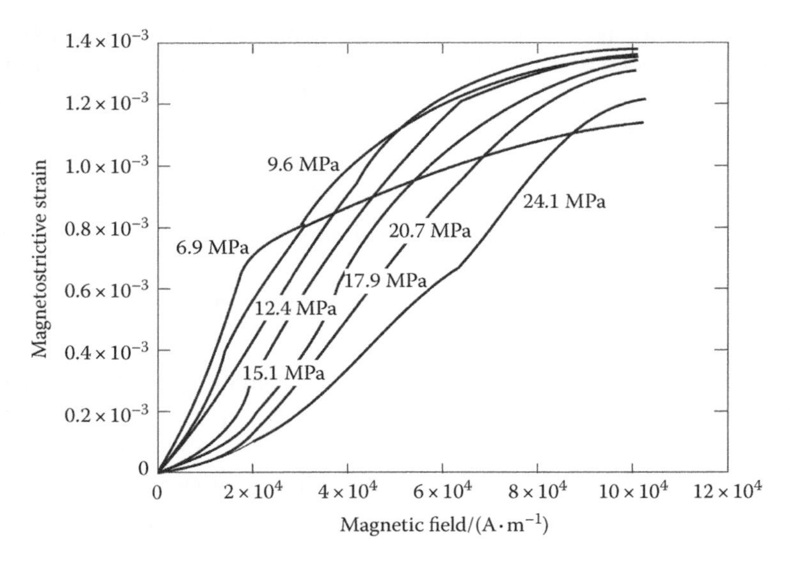

FIGURE 10.15 Magnetostrictive strain curves of Terfenol-D.

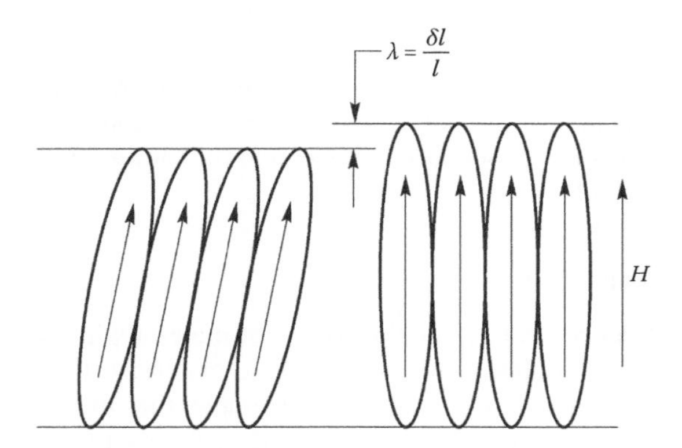

FIGURE 10.16 Micromechanism of the magnetostrictive effect.

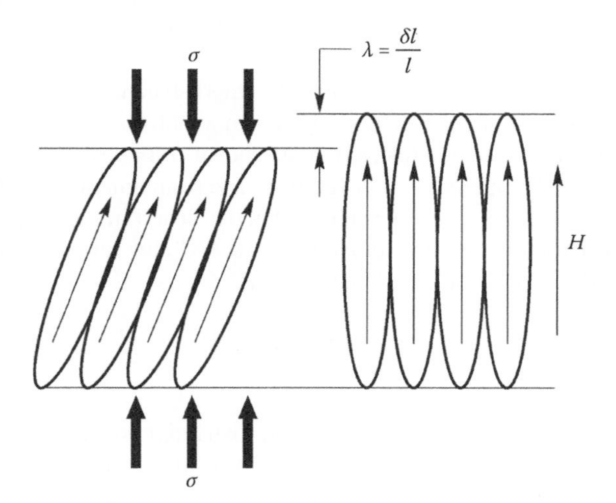

FIGURE 10.17 Influence of compressive stress on magnetostrictive effect.

domain along the magnetization direction of the magnetic domain. The strain axis rotates with the rotation of the magnetization of the magnetic domain, resulting in a deformation in the sample as a whole.

Magnetostriction is one of the basic properties of ferromagnetic materials, originating from the combined effects of the crystal field, spin–orbit coupling, and elastic deformation. For different magnetic materials, the mechanism is not identical. Linear magnetostriction results from the decrease of the magnetocrystalline anisotropy energy and the magnetoelastic energy: thus in single crystals, the linear magnetostriction is anisotropic. Volume magnetostriction is caused by the effect of the exchange energy between electron spins; thus it is isotropic. Magnetostrictive deformations, however, are the result of the total free energy minimization of the ferromagnetic crystal after deformation. This section mainly discusses the linear magnetostrictive effect.

The magnetoelastic energy caused by the interaction of magnetization and deformation, and the elastic energy caused by pure deformation, plays an important role in the magnetostrictive effect. Magnetostriction arises from the balance of decreases in magnetoelastic energy and increases in elastic energy. Thus, if the elastic modulus of a material decreases, the magnetostriction will certainly increase. The structural phase transformation is one of the mechanisms that reduces the elastic modulus. When the high-temperature cubic phase at a transition temperature transforms into the low-symmetry phase, with the decreasing of the temperature to the transition point, the elastic modulus corresponding to the deformation is also reduced. A typical example is the Jahn–Teller distortion. Magnetostriction to some extent reduces the elastic modulus of ferromagnetic materials. Under the appropriate stress, the magnetizing intensity can rotate freely. Magnetostriction produces an additional deformation that has nothing to do with the symbol λ, and thus reduces Young's modulus E. This phenomenon is called the ΔE effect. In ferromagnetic metals having the ΔE effect, mechanical vibrations produce a vortex through the rotational vibration of magnetizing intensity, which leads to additional internal friction.

Various features of the magnetostriction phenomenon indicate that when ferromagnets are affected by an external magnetic field, that magnetization can lead to magnetostriction and cause the change of object's geometric dimensions. In turn, applying tensile or compressive stress on materials will cause the magnetic properties of materials to change, which is the piezomagnetic effect [12–16]. Such an effect is called the inverse effect of magnetostriction. Because of this effect, magnetostriction plays an important role in determining the magnetic domain structure. Magnetostriction also impacts cubic anisotropy, because when the magnetizing intensity rotates, lattice domain switching occurs, resulting in a change in the magnetoelastic and elastic energy.

10.3.3 Ferromagnetic Shape Memory Alloys

Ferromagnetic SMAs [11] are new actuator materials that can produce a significant strain under the action of magnetic fields, up to 6% in the Ni-Mn-Ga alloy. Such materials not only have a ferromagnetic phase transformation but also have an austenitic–martensitic transformation, and thus have the features of conventional SMAs. However, they only work in the martensitic structure. Because the actuating mechanism is not temperature but the external magnetic field, they have a high frequency response. Like other SMAs, ferromagnetic SMAs at low temperatures usually transform into twinned martensite. At the same time, they have a strong magnetic anisotropy, and their twinned structures will produce a significant strain under external magnetic field changes, or stress changes. The actuating mechanism is shown in Figure 10.18.

Under nonloading conditions, ferromagnetic SMAs are twinned martensite. The applied magnetic field will make the magnetization vector of one of the variants rotate to the direction of the magnetic field. At the same time, due to the strong anisotropy, the twinned variants will produce shear deformation resulting in a significant strain. Their microstructures and macroscopic size changes under the effect of a magnetic field are shown in Figure 10.19.

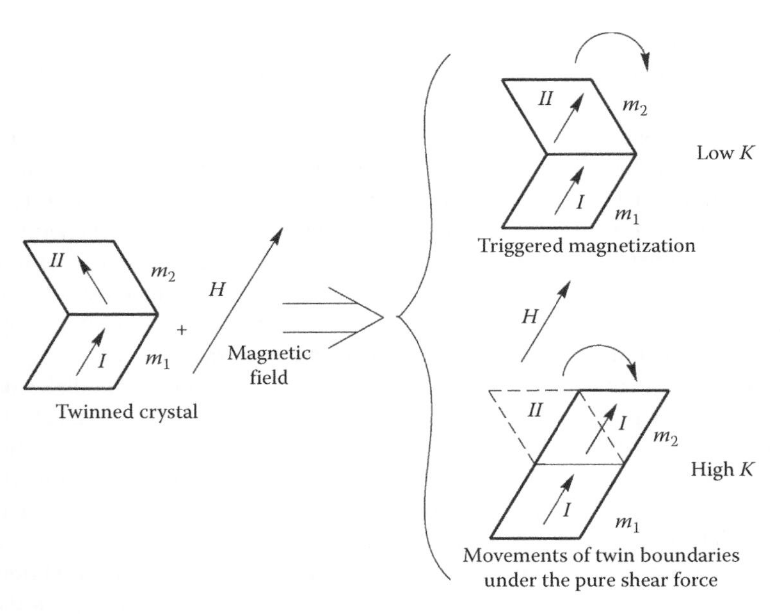

FIGURE 10.18 Mechanisms of ferromagnetic shape memory alloy.

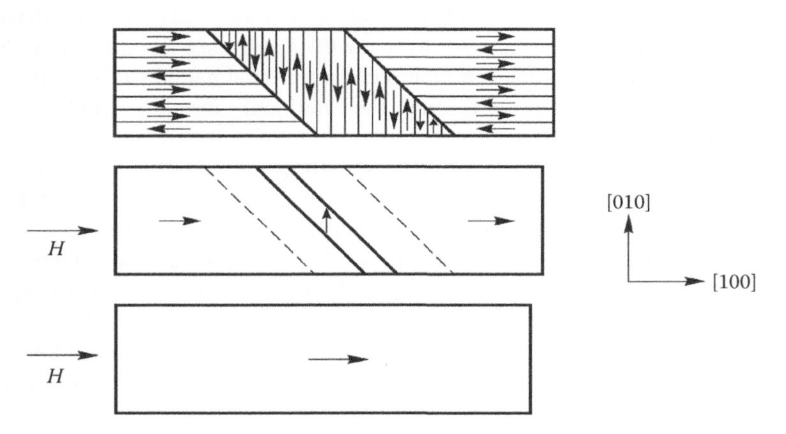

FIGURE 10.19 Microstructural and macrodimensional changes of ferromagnetic shape memory alloy in the magnetic field.

10.3.4 Mathematical Models of Magnetic Couplings

The emergence of the magnetostrictive effect is due to the coupling of magnetic quantities and mechanical quantities in the system energy. This magnetic coupling energy is generated by the interaction of the magnetization and the deformation, known as the magnetoelastic energy. The magnetoelastic energy can be derived from a method similar to the magnetic anisotropy energy. To this end, the magnetoelastic energy can be expanded according to the component of deformation:

$$F_{magnetoelasticity} = F^0_{magnetoelasticity} + \sum_{i \geq j} \left(\frac{\partial F_{magnetoelasticity}}{\partial \varepsilon_{ij}} \right) \varepsilon_{ij} \tag{10.15}$$

where ε_{ij} is the component of the strain tensor. After omitting the constant term in (10.15), the magnetoelastic energy is anisotropic.

Therefore, the coefficient $(\partial F_{magnetoelasticity} / \partial \varepsilon_{ij})$ should be related to the direction cosine $m_i(i = 1, 2, 3)$ of the magnetization vector. From cubic crystal symmetry considerations, similar to the derivation of the magnetocrystalline anisotropy in Section 10.3.1, we can obtain

$$\begin{aligned} \frac{\partial F_{magnetoelasticity}}{\partial \varepsilon_{11}} &= B_1 m_1^2 \qquad \frac{\partial F_{magnetoelasticity}}{\partial \varepsilon_{21}} = B_2 m_2 m_1 \\ \frac{\partial F_{magnetoelasticity}}{\partial \varepsilon_{21}} &= B_2 m_2 m_1 \qquad \frac{\partial F_{magnetoelasticity}}{\partial \varepsilon_{32}} = B_2 m_3 m_2 \\ \frac{\partial F_{magnetoelasticity}}{\partial \varepsilon_{33}} &= B_1 m_3^2 \qquad \frac{\partial F_{magnetoelasticity}}{\partial \varepsilon_{31}} = B_2 m_3 m_1. \end{aligned} \tag{10.16}$$

Here, B_1 and B_2 are the coefficients representing the interaction of magnetization and deformation. Therefore, the magnetoelastic energy (omit the constant term) is

$$\begin{aligned} F_{magnetoelasticity} &= B_1 \left(m_1^2 \varepsilon_{11} + m_2^2 \varepsilon_{22} + m_3^2 \varepsilon_{33} \right) \\ &+ B_2 (m_2 m_1 \varepsilon_{21} + m_3 m_2 \varepsilon_{32} + m_3 m_1 \varepsilon_{31}) \end{aligned} \tag{10.17}$$

In order to consider the effect of elastic energy caused by a pure deformation, magnetic crystals can be treated as nonmagnetic pure elastomers. From the elasticity analysis, the elastic energy can be obtained:

$$\begin{aligned} F_{elasticity} &= \frac{1}{2} c_{11} \left(\varepsilon_{11}^2 + \varepsilon_{22}^2 + \varepsilon_{33}^2 \right) + \frac{1}{2} c_{44} \left(\varepsilon_{21}^2 + \varepsilon_{32}^2 + \varepsilon_{31}^2 \right) \\ &+ c_{12} (\varepsilon_{11} \varepsilon_{22} + \varepsilon_{22} \varepsilon_{33} + \varepsilon_{33} \varepsilon_{11}). \end{aligned} \tag{10.18}$$

Therefore, the total energy of magnetic crystals without applied stress is

$$F = F_K + F_{elasticity} + F_{magnetoelasticity}. \tag{10.19}$$

When the crystal reaches a steady state, according to the energy minimization principle,

$$\begin{aligned} \frac{\partial F}{\partial \varepsilon_{11}} &= B_1 m_1^2 + c_{11} \varepsilon_{11} + c_{12} (\varepsilon_{33} + \varepsilon_{22}) = 0 \\ \frac{\partial F}{\partial \varepsilon_{22}} &= B_1 m_2^2 + c_{11} \varepsilon_{22} + c_{12} (\varepsilon_{11} + \varepsilon_{33}) = 0 \\ \frac{\partial F}{\partial \varepsilon_{33}} &= B_1 m_3^2 + c_{11} \varepsilon_{33} + c_{12} (\varepsilon_{11} + \varepsilon_{22}) = 0 \\ \frac{\partial F}{\partial \varepsilon_{21}} &= B_2 m_2 m_1 + c_{44} \varepsilon_{21} = 0 \\ \frac{\partial F}{\partial \varepsilon_{32}} &= B_2 m_3 m_2 + c_{44} \varepsilon_{32} = 0 \\ \frac{\partial F}{\partial \varepsilon_{31}} &= B_2 m_3 m_1 + c_{44} \varepsilon_{31} = 0. \end{aligned} \tag{10.20}$$

From the linear algebraic Equation 10.20, the three components of the normal strain and the three components of the shear strain are respectively

$$\begin{aligned} \varepsilon_{ij} &= \frac{B_1\left[c_{12} - m_i^2(c_{11} + 2c_{12})\right]}{(c_{11} - c_{12})(c_{11} + 2c_{12})} \qquad i, j = 1, 2, 3; \quad i = j \\ \varepsilon_{ij} &= -\frac{B_2 m_i m_j}{c_{44}} \qquad i, j = 1, 2, 3; \quad i \neq j \end{aligned} \tag{10.21}$$

Next we calculate and analyze the linear magnetostriction in any direction. We assume that the direction cosine in any direction is β_1, β_2, β_3 and that the linear magnetostriction $(\delta l/l)_{\beta_i}$ can be expressed by the strain tensor ε_{ij}. Set the coordinate of any point (x_0, y_0, z_0) in the cubic crystal as (x, y, z) after deformation. The length of the radius vector can be represented respectively as r_0 and r.

By the definition of ε_{ij},

$$\begin{aligned} x &= (1+\varepsilon_{11})x_0 + \frac{1}{2}\varepsilon_{21}y_0 + \frac{1}{2}\varepsilon_{31}z_0 \\ y &= \frac{1}{2}\varepsilon_{21}x_0 + (1+\varepsilon_{22})y_0 + \frac{1}{2}\varepsilon_{32}z_0 \\ z &= \frac{1}{2}\varepsilon_{31}x_0 + \frac{1}{2}\varepsilon_{32}y_0 + (1+\varepsilon_{33})z_0 \end{aligned} \tag{10.22}$$

From the definition of the direction cosine,

$$x_0 = r_0\beta_1, \quad y_0 = r_0\beta_2, \quad z_0 = r_0\beta_3. \tag{10.23}$$

Equation 10.22 then becomes

$$\begin{aligned} x &= r_0\left(\beta_1 + \varepsilon_{11}\beta_1 + \frac{1}{2}\varepsilon_{21}\beta_2 + \frac{1}{2}\varepsilon_{31}\beta_3\right) \\ y &= r_0\left(\beta_2 + \varepsilon_{22}\beta_2 + \frac{1}{2}\varepsilon_{21}\beta_1 + \frac{1}{2}\varepsilon_{32}\beta_3\right) \\ z &= r_0\left(\beta_3 + \varepsilon_{33}\beta_3 + \frac{1}{2}\varepsilon_{31}\beta_1 + \frac{1}{2}\varepsilon_{32}\beta_2\right) \end{aligned} \tag{10.24}$$

Taking into account the small deformation, $\varepsilon_{ij} \ll 1$, the term $\varepsilon_{ij}\varepsilon_{kl}(i \geq j, k \geq l)$ can be omitted, and thus from Equation 10.24 we can obtain the length of the radius vector after spontaneous deformation r as

$$r = r_0\left(1 + 2\sum_{i \geq j}\varepsilon_{ij}\beta_i\beta_j\right)^{1/2}. \tag{10.25}$$

Magnetostrictive strain is defined as

$$\lambda = \frac{r - r_0}{r_0} = \left(\frac{\delta l}{l}\right)_{\beta_i}. \tag{10.26}$$

Expand Equation 10.25 using a Taylor series, and take the linear term of ε_{ij}, substituted into (10.26). Thus we can get an approximation:

$$\lambda = \left(\frac{\delta l}{l}\right)_{\beta_i} = \sum_{i\geq j} \varepsilon_{ij}\beta_i\beta_j = \sum_{i=j} \varepsilon_{ij}\beta_i^2 + \sum_{i>j} \varepsilon_{ij}\beta_i\beta_j. \quad (10.27)$$

Take the strain in Equation 10.21, substitute into Equation 10.27, and the magnetostrictive strain can be obtained:

$$\left(\frac{\delta l}{l}\right)_{\beta_i} = -\frac{B_1}{c_{11}-c_{12}}\left(m_1^2\beta_1^2 + m_2^2\beta_2^2 + m_3^2\beta_3^2 - \frac{1}{3}\right) - \frac{B_2}{c_{44}}\left(m_2m_1\beta_2\beta_1 + m_3m_2\beta_3\beta_2 + m_3m_1\beta_3\beta_1\right) - \frac{B_1}{3\left(c_{11}+2c_{12}\right)} \quad (10.28)$$

The third term of Equation 10.28 is the constant term independent of direction. The coefficients of the first two terms are called the magnetostriction coefficient.

If the measured direction of linear magnetostriction is the same as that of the magnetization vector, then $m_i = \beta_i$. Omitting the constant term independent of direction, Equation 10.28 can be simplified to

$$\left(\frac{\delta l}{l}\right)_{m_1=\beta_i} = -\frac{B_1}{c_{11}-c_{12}}\left(\beta_1^4 + \beta_2^4 + \beta_3^4 - \frac{1}{3}\right) - \frac{B_2}{c_{44}}\left(\beta_2^2\beta_1^2 + \beta_3^2\beta_2^2 + \beta_3^2\beta_1^2\right) \quad (10.29)$$

When the measuring direction is [100], $\beta_1 = 1, \beta_2 = \beta_3 = 0,$ and we obtain

$$\left(\frac{\delta l}{l}\right)_{[100]} = \lambda_{[100]} = -\frac{2B_1}{3(c_{11}-c_{12})}. \quad (10.30)$$

When the measuring direction is [111], $\beta_1 = \beta_2 = \beta_3 = 1/\sqrt{3}$, and we obtain

$$\left(\frac{\delta l}{l}\right)_{[111]} = \lambda_{[111]} = -\frac{B_2}{3c_{44}}. \quad (10.31)$$

Therefore, the magnetostrictive Equation 10.29 can be expressed as

$$\left(\frac{\delta l}{l}\right)_{\beta_i} = \frac{3}{2}\lambda_{[100]}\left(m_1^2\beta_1^2 + m_2^2\beta_2^2 + m_3^2\beta_3^2 - \frac{1}{3}\right) + 3\lambda_{[111]}(m_2m_1\beta_2\beta_1 + m_3m_2\beta_3\beta_2 + m_3m_1\beta_3\beta_1). \quad (10.32)$$

Equation 10.32 is the magnetostrictive formula in the direction of any $(\beta_1, \beta_2, \beta_3)$, where λ_{100}, λ_{111} are, respectively, the magnetostriction coefficients of the crystal along the two [100] [111] directions.

Compare Equation 10.32 with Equation 10.27. We can obtain the strain of materials under no applied stress as

$$\varepsilon_{ij} = \frac{3}{2}\lambda_{[100]}\left(m_i^2 - \frac{1}{3}\right), \quad i, j = 1, 2, 3; \quad i = j. \tag{10.33}$$

$$\varepsilon_{ij} = 3\lambda_{[111]}m_i m_j, \quad i, j = 1, 2, 3; \quad i > j. \tag{10.34}$$

It can be found that the strain of the material is related to its magnetization vector. This is the magnetostrictive effect, where λ_{100} corresponds to the strain of the magnetization vector along the [100] direction and λ_{111} corresponds to the strain of the magnetization vector along the [111] direction. For isotropic magnetostriction, the magnetostrictive strains are $\lambda_{100} = \lambda_{111} = \lambda_{110} = \lambda_s$. The magnetostriction of polycrystalline materials is isotropic because the total magnetostriction is the average of the deformation of each grain even if $\lambda_{100} \neq \lambda_{111}$ for each grain. Averaging the orientation of different grains, we can obtain the average longitudinal magnetostriction as

$$\bar{\lambda}_0 = \frac{2}{5}\lambda_{[100]} + \frac{3}{5}\lambda_{[111]}. \tag{10.35}$$

In cubic crystal isotropic materials, $\lambda_{100} = \lambda_{111} = \lambda_s$. Using trigonometric relations to simplify Equation 10.32, the magnetostriction in cubic crystal isotropic materials is

$$\lambda = \frac{3}{2}\lambda_s\left(\cos^2\theta - \frac{1}{3}\right). \tag{10.36}$$

In cubic crystals, the magnetostriction of anisotropy in polycrystalline materials with the grains randomly oriented is

$$\bar{\lambda} = \frac{3}{2}\bar{\lambda}_0\left(\cos^2\theta - \frac{1}{3}\right). \tag{10.37}$$

where

θ represents the angle between the direction of magnetizing intensity and the measuring direction of magnetostriction

λ_s represents the saturation magnetostriction coefficient along the direction of the magnetic field

$\bar{\lambda}_0$ represents the polycrystalline magnetostriction coefficient [12]

Using approaches similar to the those given earlier in solving the magnetostriction, we can substitute the strain in Equation 10.21 into Equation 10.17, which is the equation of magnetoelastic energy. Omitting the constant term, and using (10.30), (10.31), (10.33), and (10.34) to derive the general expression of the magnetoelastic energy in cubic crystals, we obtain

$$U = -\frac{3}{2}(c_{11} - c_{12})\lambda_{100}\sum_{i=1}^{3}\varepsilon_i\left(m_i^2 - \frac{1}{3}\right) - 3c_{44}\lambda_{111}\sum_{i>j}\varepsilon_{ij}m_i m_j, \tag{10.38}$$

where

c_{11}, c_{12}, and c_{44} are elastic constants

m_i is the magnetization vector

ε_{ij} is the component of the strain tensor

We know that the magnetostrictive effect exists in all ferromagnetic materials. However, if the ferromagnetic material has a martensitic transformation, there may be a ferromagnetic memory effect. In this case, in order to describe the ferromagnetic martensite, we not only need the transfer matrix of phase transformation but also the magnetization vector. The derivation process for the transfer matrix of phase transformation is similar to that of SMAs described in Section 10.2.3, and thus is not repeated here. Because the magnetization vector has six directions in the tetragonal ferromagnetic martensite, there are six variants for the tetragonal ferromagnetic martensite:

$$\begin{aligned}
&\boldsymbol{U}_1=\begin{bmatrix}\beta&0&0\\0&\alpha&0\\0&0&\alpha\end{bmatrix},\quad \boldsymbol{m}_1=\begin{bmatrix}m\\0\\0\end{bmatrix},\quad \boldsymbol{U}_2=\begin{bmatrix}\beta&0&0\\0&\alpha&0\\0&0&\alpha\end{bmatrix},\quad \boldsymbol{m}_2=\begin{bmatrix}-m\\0\\0\end{bmatrix}\\
&\boldsymbol{U}_3=\begin{bmatrix}\alpha&0&0\\0&\beta&0\\0&0&\alpha\end{bmatrix},\quad \boldsymbol{m}_3=\begin{bmatrix}0\\m\\0\end{bmatrix},\quad \boldsymbol{U}_4=\begin{bmatrix}\alpha&0&0\\0&\beta&0\\0&0&\alpha\end{bmatrix},\quad \boldsymbol{m}_4=\begin{bmatrix}0\\-m\\0\end{bmatrix}\\
&\boldsymbol{U}_5=\begin{bmatrix}\alpha&0&0\\0&\alpha&0\\0&0&\beta\end{bmatrix},\quad \boldsymbol{m}_5=\begin{bmatrix}0\\0\\m\end{bmatrix},\quad \boldsymbol{U}_6=\begin{bmatrix}\alpha&0&0\\0&\alpha&0\\0&0&\beta\end{bmatrix},\quad \boldsymbol{m}_6=\begin{bmatrix}0\\0\\-m\end{bmatrix}
\end{aligned} \tag{10.39}$$

For orthorhombic ferromagnetic martensite, there are a total of 12 variants:

$$\begin{aligned}
&\boldsymbol{U}_1=\begin{bmatrix}\alpha&\delta&0\\\delta&\alpha&0\\0&0&\beta\end{bmatrix},\quad \boldsymbol{m}_1=\begin{bmatrix}m\\m\\0\end{bmatrix}\quad \boldsymbol{U}_2=\begin{bmatrix}\alpha&\delta&0\\\delta&\alpha&0\\0&0&\beta\end{bmatrix},\quad \boldsymbol{m}_2=-\begin{bmatrix}m\\m\\0\end{bmatrix}\\
&\boldsymbol{U}_3=\begin{bmatrix}\alpha&-\delta&0\\-\delta&\alpha&0\\0&0&\beta\end{bmatrix},\quad \boldsymbol{m}_3=\begin{bmatrix}-m\\m\\0\end{bmatrix}\quad \boldsymbol{U}_4=\begin{bmatrix}\alpha&-\delta&0\\-\delta&\alpha&0\\0&0&\beta\end{bmatrix},\quad \boldsymbol{m}_4=-\begin{bmatrix}-m\\m\\0\end{bmatrix}\\
&\boldsymbol{U}_5=\begin{bmatrix}\alpha&0&\delta\\0&\beta&0\\\delta&0&\alpha\end{bmatrix},\quad \boldsymbol{m}_5=\begin{bmatrix}m\\0\\m\end{bmatrix}\quad \boldsymbol{U}_6=\begin{bmatrix}\alpha&0&\delta\\0&\beta&0\\\delta&0&\alpha\end{bmatrix},\quad \boldsymbol{m}_6=-\begin{bmatrix}m\\0\\m\end{bmatrix}\\
&\boldsymbol{U}_7=\begin{bmatrix}\alpha&0&-\delta\\0&\beta&0\\-\delta&0&\alpha\end{bmatrix},\quad \boldsymbol{m}_7=\begin{bmatrix}-m\\0\\m\end{bmatrix}\quad \boldsymbol{U}_8=\begin{bmatrix}\alpha&0&-\delta\\0&\beta&0\\-\delta&0&\alpha\end{bmatrix},\quad \boldsymbol{m}=-\begin{bmatrix}-m\\0\\m\end{bmatrix}\\
&\boldsymbol{U}_9=\begin{bmatrix}\beta&0&0\\0&\alpha&\delta\\0&\delta&\alpha\end{bmatrix},\quad \boldsymbol{m}_9=\begin{bmatrix}0\\m\\m\end{bmatrix}\quad \boldsymbol{U}_{10}=\begin{bmatrix}\beta&0&0\\0&\alpha&\delta\\0&\delta&\alpha\end{bmatrix},\quad \boldsymbol{m}_{10}=-\begin{bmatrix}0\\m\\m\end{bmatrix}\\
&\boldsymbol{U}_{11}=\begin{bmatrix}\beta&0&0\\0&\alpha&-\delta\\0&-\delta&\alpha\end{bmatrix},\quad \boldsymbol{m}_{11}=\begin{bmatrix}0\\-m\\m\end{bmatrix}\quad \boldsymbol{U}_{12}=\begin{bmatrix}\beta&0&0\\0&\alpha&-\delta\\0&-\delta&\alpha\end{bmatrix},\quad \boldsymbol{m}_{12}=-\begin{bmatrix}0\\-m\\m\end{bmatrix}
\end{aligned} \tag{10.40}$$

When these variants coexist, the components of the average gradient tensor of phase transformation, and the magnetization vector, are, respectively,

$$\boldsymbol{U}=\sum_{i=1}^{N}\lambda_i\boldsymbol{U}_i,\quad \boldsymbol{m}=\sum_{i=1}^{N}\lambda_i\boldsymbol{m}_i. \tag{10.41}$$

Under the effect of an external magnetic field, the shape–volume ratio of the variant alters, causing the change $[U]$. Thus it results in the macroscopic deformation of the material, which is the mechanism of ferromagnetic SMAs. The size of the deformation is determined by the lattice constants, which are usually much larger than the magnetostrictive strain.

When ferromagnetic crystals are subject to an external force or the originally existing internal stress—for example, in the preparation process of ferromagnetics, which are cooled down from high temperatures, there is always internal stress present—deformation will occur in ferromagnetic crystals due to stress. At this time, in addition to considering the magnetic anisotropy energy and the magnetostrictive energy, we must also take into account the strain energy caused by the internal stress. In general, the form of the stress is more complicated, and here we only discuss a simple but important case, in which the stress along a certain direction is a simple tensile stress.

Suppose the crystal is a cubic crystal, and the direction of stress (the three crystal axes as coordinate system) is $(\gamma_1, \gamma_2, \gamma_3)$, and its size is σ. Then, from the elasticity, the component of the stress tensor is

$$\sigma_{ij} = \sigma\gamma_i\gamma_j. \tag{10.42}$$

Note that, as defined in Chapter 1, the normal stress σ is positive in stretch and negative in compression. If the strain tensor caused by the stress is $\varepsilon_{ij}^{\sigma}$, then the total component of the strain tensor is $\varepsilon_{ij} = \varepsilon_{ij}^{0} + \varepsilon_{ij}^{\sigma}$, where ε_{ij}^{0} is the component of the strain tensor caused by the magnetostriction. At this time the ferromagnetic crystal's Hemholtz free energy F should increase the strain energy $\sum_{i\geq j}\sigma_{ij}\varepsilon_{ij}$, that is,

$$F = F_K + F_{elasticity} + F_{magnetoelasticity} + F_{stress}. \tag{10.43}$$

Based on steady-state conditions, we have

$$\begin{aligned}
\frac{\partial F}{\partial \varepsilon_{11}} &= B_1 m_1^2 + c_{11}\varepsilon_{11} + c_{12}(\varepsilon_{33} + \varepsilon_{22}) + \sigma\gamma_1^2 = 0 \\
\frac{\partial F}{\partial \varepsilon_{22}} &= B_1 m_2^2 + c_{11}\varepsilon_{22} + c_{12}(\varepsilon_{11} + \varepsilon_{33}) + \sigma\gamma_2^2 = 0 \\
\frac{\partial F}{\partial \varepsilon_{33}} &= B_1 m_3^2 + c_{11}\varepsilon_{33} + c_{12}(\varepsilon_{11} + \varepsilon_{22}) + \sigma\gamma_3^2 = 0 \\
\frac{\partial F}{\partial \varepsilon_{21}} &= B_2 m_2 m_1 + c_{44}\varepsilon_{21} + \sigma\gamma_2\gamma_1 = 0 \\
\frac{\partial F}{\partial \varepsilon_{32}} &= B_2 m_3 m_2 + c_{44}\varepsilon_{32} + \sigma\gamma_3\gamma_2 = 0 \\
\frac{\partial F}{\partial \varepsilon_{31}} &= B_2 m_3 m_1 + c_{44}\varepsilon_{31} + \sigma\gamma_1\gamma_3 = 0
\end{aligned} \tag{10.44}$$

where

c_{11}, c_{12}, and c_{44} are the elastic constants
m_i is the component of the magnetization vector
ε_{ij} is the component of the strain tensor
B_1, B_2 represent the coefficients of the interaction of the magnetization and the deformation

By solving the linear algebraic equations in (10.44), we can obtain three normal strain components of the strain $\varepsilon_i (i = 1, 2, 3)$ and three shear strain components $\varepsilon_{ij} (i, j = 1, 2, 3; i \neq j)$:

$$\varepsilon_{ij} = \frac{B_1\left[c_{12} - m_i^2(c_{11} + 2c_{12})\right]}{(c_{11} - c_{12})(c_{11} + 2c_{12})} + \frac{\sigma\left[c_{12} - \gamma_i^2(c_{11} + 2c_{12})\right]}{(c_{11} - c_{12})(c_{11} + 2c_{12})} \quad i, j = 1, 2, 3; \quad i = j \tag{10.45}$$

$$\varepsilon_{ij} = -\frac{B_2 m_i m_j}{c_{44}} - \frac{\sigma\gamma_i\gamma_j}{c_{44}} \quad i, j = 1, 2, 3; \quad i \neq j \tag{10.46}$$

Substituting ε_{ij} into the expression $\sum_{i \geq j} \sigma_{ij}\varepsilon_{ij}$ of the strain energy caused by the internal stress, we can obtain the strain energy:

$$F_\sigma = \frac{3}{2}\lambda_{[100]}\sigma\left(m_1^2\gamma_1^2 + m_2^2\gamma_2^2 + m_3^2\gamma_3^2\right) + 3\lambda_{[111]}\sigma(m_1 m_2 \gamma_1 \gamma_2 + m_2 m_3 \gamma_2 \gamma_3 + m_3 m_1 \gamma_3 \gamma_1). \tag{10.47}$$

When $\lambda_{111} = \lambda_{100} = \lambda$, using the trigonometric relations, we can simplify Equation 10.47 to

$$F_\sigma = \frac{3}{2}\lambda\sigma\cos^2\theta, \tag{10.48}$$

where θ is the angle between the direction of stress $(\gamma_1, \gamma_2, \gamma_3)$ and the direction of the magnetization vector (m_1, m_2, m_3). The strain energy caused by the internal stress is much greater than the elastic energy. It should be noted that for tension (pull), σ should be positive and for compression, σ should be negative.

Strain gauge technology is commonly used for measuring magnetostriction. When the sample bonded with a strain gauge elongates, the electrical resistance wire of the strain gauge will elongate, and its electrical resistance changes accordingly.

Here is how to measure λ_{100} and λ_{111} in the cubic materials: first of all, cut a disc from the single crystal along the $[0\bar{1}0]$ surface, and bond a strain gauge whose axis is along the [100] or the [111] direction; then, have the external magnetic field rotate in the disc surface and measure the strain in the [100] or the [111] direction. It is noteworthy that the strain gauge will change the stiffness of the sample if the amount of glue used to secure the strain gauge on the sample is too heavy, or the sample is too thin, which will result in lower sensitivity of the strain gauges. On the other hand, if the sample is too thick, the demagnetization factor will be too large, so that the sample is difficult to magnetize to reach saturation. Additionally, capacitor technology can be used to measure the magnetostriction, which can also be calculated to obtain the magnetoelastic-coupling coefficient.

Based on the mathematical model of magnetic coupling, Figures 10.20 and 10.21 display the curves—obtained through binding and nonbinding theoretical calculations—showing the relationship between magnetic fields and the parameters of related properties of the ferromagnetic SMA NiMnGa (martensite bars) [19,20]. Figures 10.20 and 10.21 show the relationship between magnetic fields and the axial strain, magnetic susceptibility, and volume fraction (the magnetization angle of a variant) of NiMnGa corresponding to a compressive stress $\sigma_0 = 0.6$ MPa and 3.0 MPa. It can be seen that at a higher compressive stress, the nonbinding model is truer to actual test results. The reason is

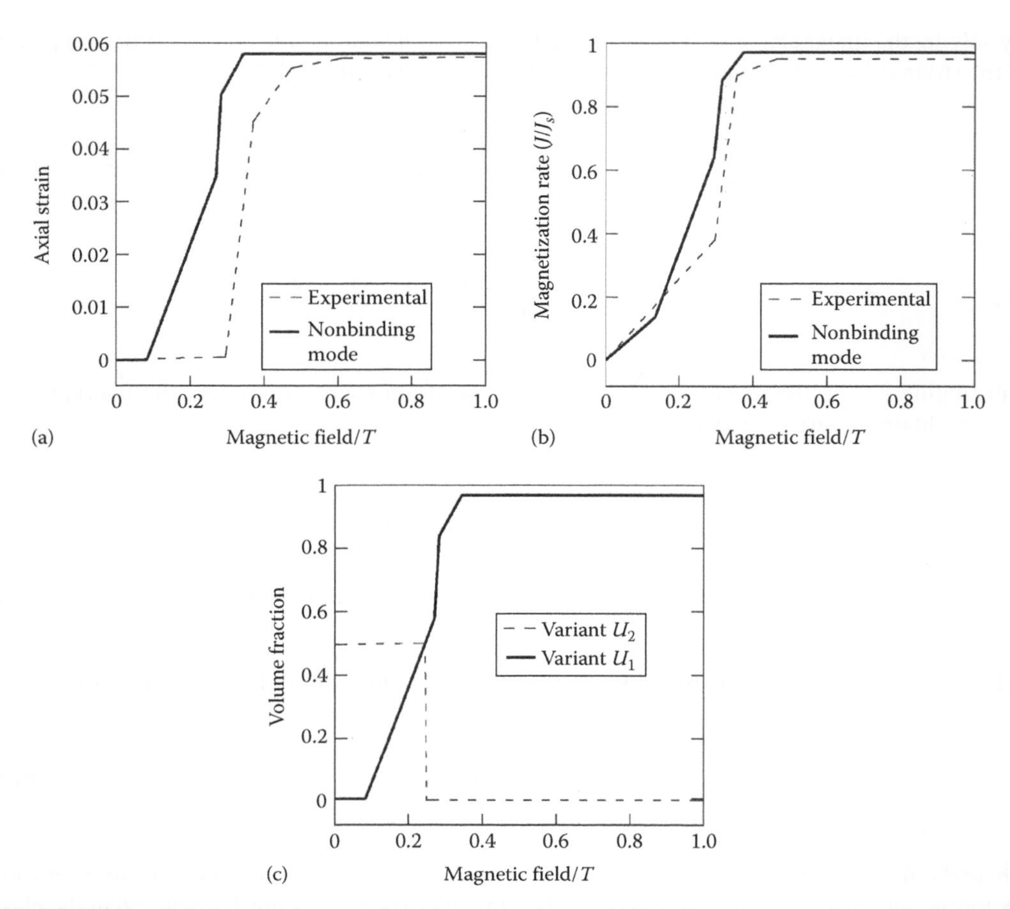

FIGURE 10.20 Relative parameters of martensite variants of NiMnGa rod versus the applied magnetic field under a compressive axial stress of 0.6 MPa. (a) Axial strength, (b) magnetization rate, and (c) change of partial volume fraction with the magnetic field. (From Ma, Y.F. and Li, J.Y., *Appl. Phys. Lett.*, 90, 172504, 2007; Taken from Figure 1.)

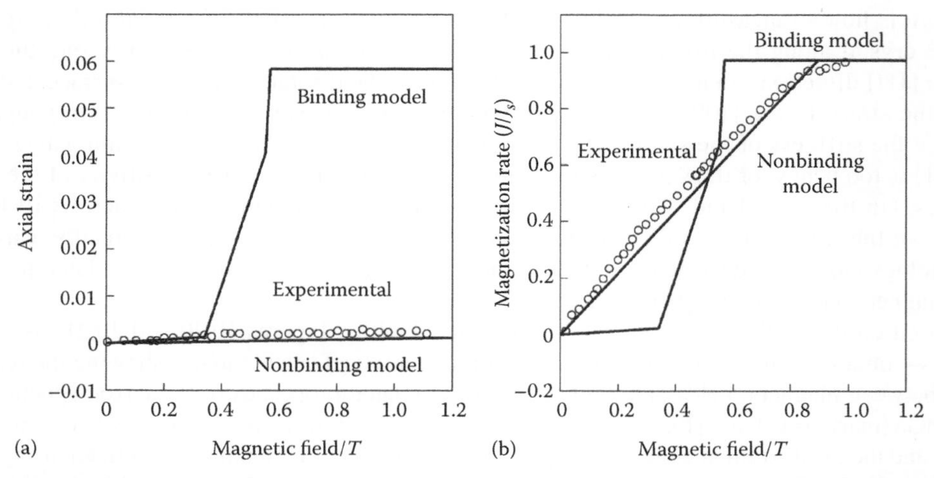

FIGURE 10.21 Relative parameters of martensite variants of NiMnGa rod versus the applied magnetic field under a compressive axial stress of 3 MPa. (a) Axial strain. (b) Magnetization rate.

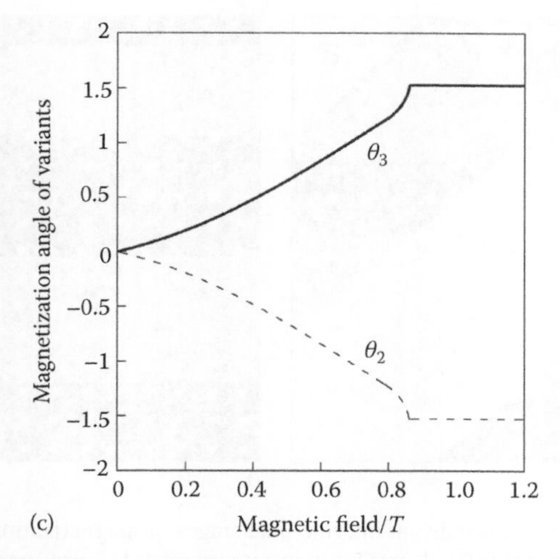

FIGURE 10.21 (continued) Relative parameters of martensite variants of NiMnGa rod versus the applied magnetic field under a compressive axial stress of 3 MPa. (c) Change of magnetization angle of variants with the magnetic field. (From Li, L.J. et al., *Appl. Phys. Lett.*, 92, 172504, 2008; Adapted from Figure 2.)

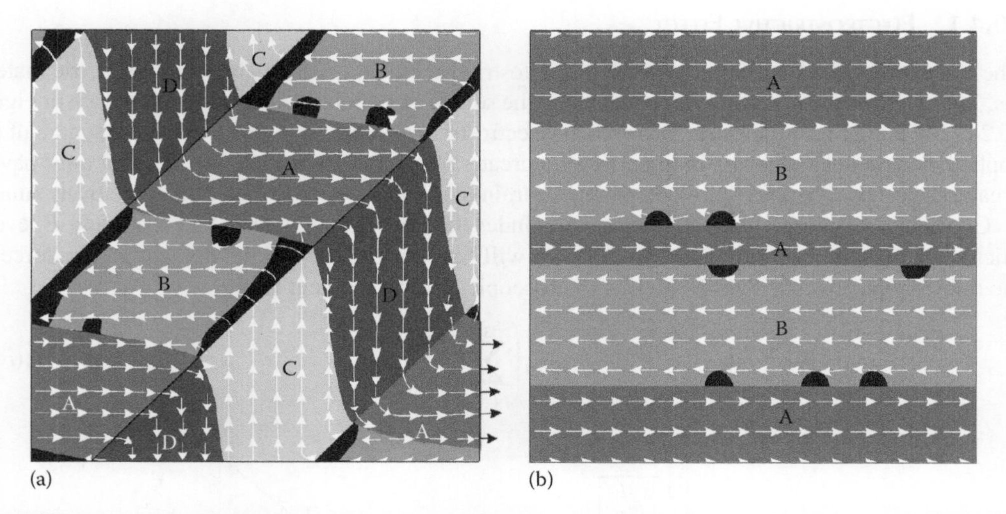

FIGURE 10.22 The formation of magnetoelastic domain. (A and B represent the variant 1, C and D represent variant 3, and arrows represent the magnetization direction.) (a) The second-order domain morphology of the ferromagnetic shape memory alloy under the condition of clamped periodic boundary. (b) The first-order domain morphology of the ferromagnetic shape memory alloys under stress conditions.

that at higher compressive stresses, it is difficult for the rotation of the magnetic domains (occurring mainly in the magnetic crystal) to cause domain wall displacement. We use the phase-field theoretical calculations to verify this conclusion. Figures 10.22 and 10.23 show the magnetic domain evolution of the SMAs NiMnGa under clamped periodic boundary conditions and stress conditions. This calculation, and the experimental results [21,22], are in good agreement. It confirms that at higher compressive stresses, the magnetic domain rotation plays a major role. Moreover, at a lower compressive stress, the domain wall displacement of ferromagnetic materials will occur, forming a new magnetic domain structure.

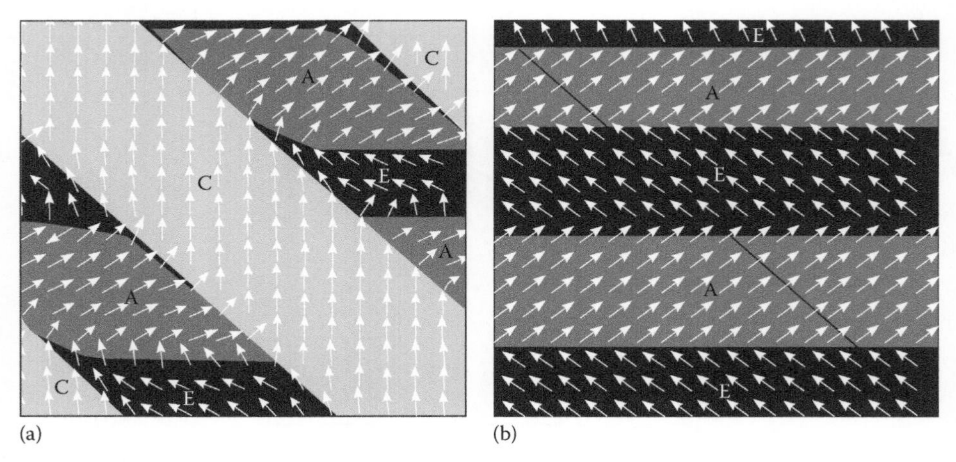

FIGURE 10.23 Magnetoelastic domains at intermediate stages of magnetization under different compressive stresses of (a) 0.6 MPa and (b) 3.0 MPa. A and E represent variant 1, C represent variant 3; the arrow is used to indicate the magnetization direction. (From Li, L. et al., *Appl. Phys. Lett.* 92(17), 172504; Taken from Figure 3.)

10.4 FERROELECTRIC AND PIEZOELECTRIC MATERIALS

10.4.1 Electrostrictive Effect

The electrostrictive effect, similar to the magnetostrictive effect, exists in almost all dielectric materials. The electrostrictive strain is proportional to the square of its polarization vector, as shown in Figure 10.24 [23]. For a linear dielectric material, the electrostrictive strain is proportional to the square of the applied electric field intensity. In order to get a greater electrostrictive effect, the material must have a greater dielectric constant. The electrostrictive strain in normal dielectric materials is relatively small.

Consider an isotropic dielectric material. Under the effect of an external electric field $\boldsymbol{E}$, every microstructural unit of the dielectric material will have an electric dipole moment $\boldsymbol{p}$. The force of the microscopic electric field $\boldsymbol{E}'$ on this microscopic electric moment is

$$F_i = p_i\left(\sum_{j=1}^{3}\frac{\partial E'_j}{\partial x_j}\right). \tag{10.49}$$

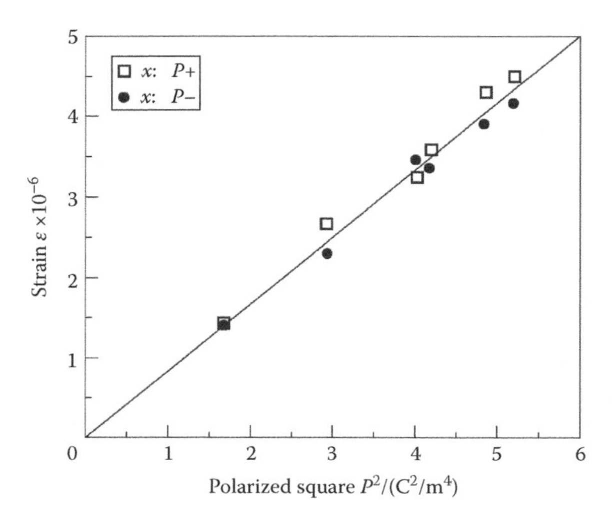

FIGURE 10.24 Electrostrictive effect.

Note that F_i and p_i represent, respectively, the component of $\boldsymbol{F}$ and $\boldsymbol{p}$. Summing the unit volume of a dielectric material in the aforementioned equation, we can obtain the macroscopic volume (or body) force density:

$$f_i = \frac{1}{\Delta V}\sum_{\Delta V} p_i\left(\sum_{j=1}^{3}\frac{\partial E'_j}{\partial x_j}\right), \quad f_i = N_{p_i}\left(\sum_{j=1}^{3}\frac{\partial E'_j}{\partial x_j}\right), \tag{10.50}$$

where N is the number of structural units in the unit volume of the dielectric material. Because we want the microscopic electric field gradient $\sum_{j=1}^{3}\partial E'_j/\partial x_j$ to be placed outside the summation symbol, it should be averaged. From the physics viewpoint, the average of the microscopic electric field gradient is equal to the macroscopic quantity $\sum_{j=1}^{3}\partial E_j/\partial x_j$, and N_p is equal to the macroscopic polarization intensity $\boldsymbol{P}$. Hence, we have

$$f_i = P_i\left(\sum_{j=1}^{3}\frac{\partial E_j}{\partial x_j}\right). \tag{10.51}$$

$\boldsymbol{F}$ and f are force and force density, respectively.

Using $\boldsymbol{P} = \chi\varepsilon_0\boldsymbol{E}$ and $\varepsilon = 1 + \chi$, where χ is the polarizability and ε_0 is the vacuum dielectric constant, Equation 10.51 can be rewritten as

$$f_i = (\varepsilon - 1)\varepsilon_0 E_i\left(\sum_{j=1}^{3}\frac{\partial E_j}{\partial x_j}\right). \tag{10.52}$$

Because the curl of the electrostatic field is equal to zero, the vector formula (reader: prove this) is

$$\boldsymbol{E}\nabla\cdot\boldsymbol{E} = \frac{1}{2}\nabla(\boldsymbol{E}\cdot\boldsymbol{E}) - \boldsymbol{E}\times\nabla\times\boldsymbol{E} = \frac{1}{2}\nabla(\boldsymbol{E}\cdot\boldsymbol{E}). \tag{10.53}$$

Thus,

$$f_i = \frac{1}{2}(\varepsilon - 1)\varepsilon_0\frac{\partial}{\partial x_i}\left(\sum_{j=1}^{3}E_jE_j\right). \tag{10.54}$$

We assume that the dielectric material is incompressible, so the dielectric constant ε is a constant. In practice, due to the effect of the volume force f, the density ρ of the dielectric material will change. Knowing the volume force, we can link the volume force to the stress by the equilibrium equations in Chapter 1.

The aforementioned discussion suggests that dielectric materials under the effect of an electric field $\boldsymbol{E}$ will exhibit stress. The relationship between the size of this stress and the quadratic term of $\boldsymbol{E}$ is linear. This effect is called electrostriction. The stress proportional to the quadratic term of $\boldsymbol{E}$ will cause the dielectric to produce a corresponding strain. Note that when $\varepsilon > 1$, the electrostrictive effect will cause the dielectric to absorb in the direction of the greatest density of the power lines and has nothing to do with the direction of the electric field. This is because the external field $\boldsymbol{E}$ causes the polarization electric moment $\boldsymbol{p}$, so that when $\boldsymbol{E}$ reverses, $\boldsymbol{p}$ also reverses. Because the relationship of the electrostrictive effect with both $\boldsymbol{p}$ and $\boldsymbol{E}$ is linear, the stress and the quadratic term of $\boldsymbol{E}$ also have a linear relationship.

Generally speaking, the electrostrictive effect is relatively weak, and for a long time its practical significance was unrecognized. After 1979, significant electrostrictive strain has been found in materials with a high dielectric constant, and in ferroelectric materials with temperatures slightly higher than the Curie point. Thereafter, some solid electrostrictive materials [24,25] useful in technical applications have also been found. Because of their good reproducibility, fast response time, good temperature stability, and aging stability, they are particularly suitable for sophisticated microdisplacement modulation devices.

The previous discussion on electrostriction was only qualitative for solid dielectric materials. Due to the nature of the parameters of the tensor of solids, as well as the strong interaction between the various structural units in the solid, the problem becomes very complicated. The phenomenological theory of thermodynamics does yield experimental quantitative methods that describe the electrostrictive effect, which will not be repeated here.

10.4.2 Ferroelectric Effect

Ferroelectric materials possess a spontaneous polarization vector when the applied electric field is zero. Flipping of the spontaneous polarization vector will occur under the effect of an applied electric field or a stress field. These materials are known as ferroelectrics. Consider the unit cell of barium titanate as an example to describe the spontaneous polarization of ferroelectrics [26]. The Curie temperature of the barium titanate unit cell is 120°C, as shown in Figure 10.25. When above the Curie temperature, the crystal is in the centrosymmetric cubic phase, and the cell edge length is a_0, as shown in Figure 10.25a. Titanium ions occupy the body center of the unit cell, the oxygen ions occupy the surface center of the unit cell, and barium ions occupy the vertices of the unit cell. When the temperature drops to between 0°C and 120°C, the centrosymmetric cubic phase transforms into a noncentrosymmetric tetragonal crystal structure, producing Landau's symmetrical incompleteness. Titanium ions in the body center of the unit cell will undergo relative displacement to the surrounding oxygen ions. These exist all around, for a total of six balanced addresses. Titanium ions will chose one, as shown in Figure 10.26 [26,27].

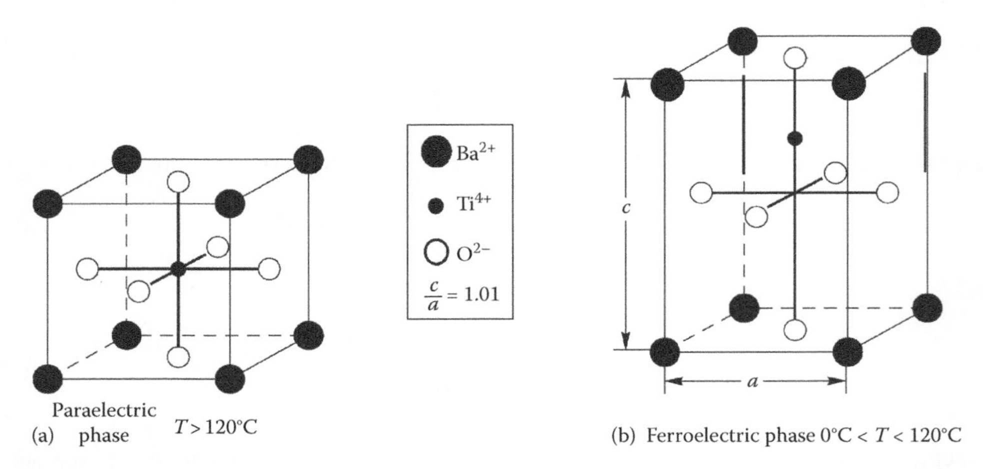

FIGURE 10.25 Schematic diagram of barium titanate unit cell. (a) The cubic phase. (b) Tetragonal phase.

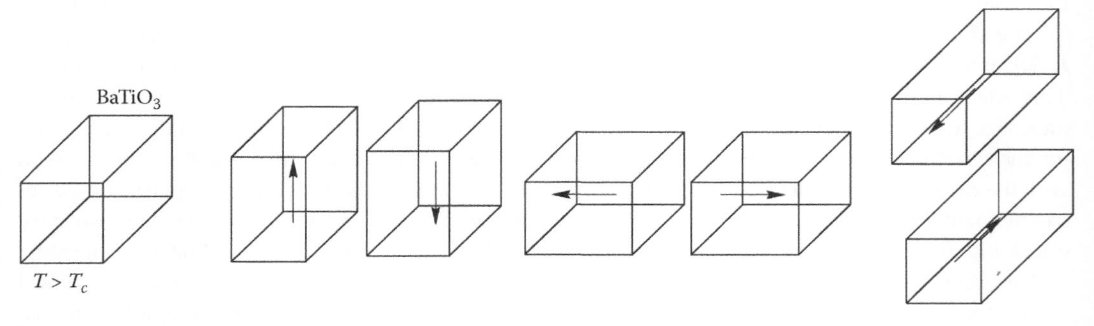

FIGURE 10.26 Ferroelectric phase transition and variants.

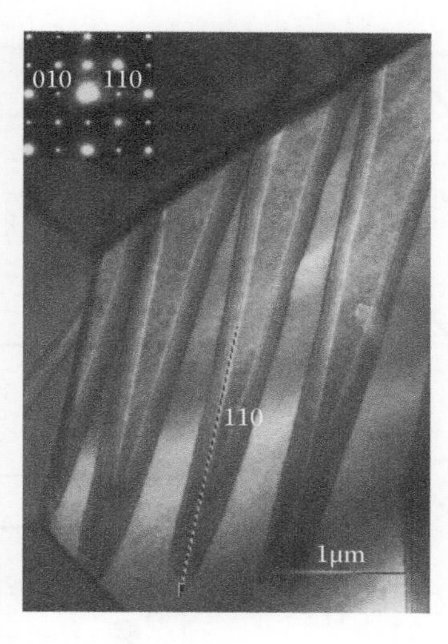

FIGURE 10.27 Ferroelectric domain structures.

After spontaneous polarization, the scale of the unit cell along the displacement direction of the central ion becomes *c*, and the side length of the unit cell perpendicular to the displacement direction becomes *a*, as shown in Figure 10.25b.

At room temperature, the lattice constants of barium titanate in the tetragonal phase are $a = 0.3992$ nm and $c = 0.4032$ nm. At temperatures above the Curie temperature, barium titanate is in the cubic phase; at this point it is a nonpolar crystal, known as the paraelectric phase. Below the Curie temperature, the barium titanate is in the tetragonal phase; it is a polar crystal known as the ferroelectric phase. Devonshire extended Landau's second-order phase transformation theory to a first-order phase transformation, and established a more systematic thermodynamic theory of paraelectric–ferroelectric phase transformation, called the Landau–Devonshire thermodynamic theory of ferroelectric phase transformation. The interested reader can refer to the monographs or textbooks [26,28].

In ferroelectrics, the microzone composed of adjacent unit cells with the same spontaneous polarization orientation is called the "electric domain" [26,29,30]. The different regions have different polarization vectors, and a typical photo is shown in Figure 10.27. The flip of domains is called domain switching. Now we will use tetragonal barium titanate after spontaneous polarization as an example to discuss the problem of domain switching in the presence of an applied electric field and stress field. First, consider only the applied electric field [26]. If the angle between the direction of the applied electric field and the direction of polarization is greater than 45°, it is possible to displace the central titanium ion. When the applied electric field exceeds the critical electric field (defined as the coercive field E_c), the central titanium ion will migrate to the other five eccentric tetragonal phases. Thus, the polarization intensity points to the closest direction of the applied electric field. Typical examples are shown in Figure 10.28.

Because there are, all around, a total of six potential wells in the perovskite structure, the direction of polarization could only have 180° switching, or 90°switching. Figure 10.28a and b show 180° switching and 90° switching. The 180° or 90° switching of the polarization direction in the electric domain is referred to as the distortion of the 180° or 90°. Domain switching causes significant changes in spontaneous strain and polarization intensity, and thus determines the macroscopic response of ferroelectric ceramics. Ferroelectric domain switching is the microscopic physical mechanism resulting in macroscopic hysteresis and ferroelectric nonlinear effects. The flipping of the electric domains in ferroelectric materials due to the effect of the electric field performs a macroscopic ferroelectric hysteresis loop, as shown in Figure 10.29.

Figure 10.29 indicates that domain switching of ferroelectrics near the coercive electric field results in the mutation of the electric displacement (polarization vector). See the up-arrow and

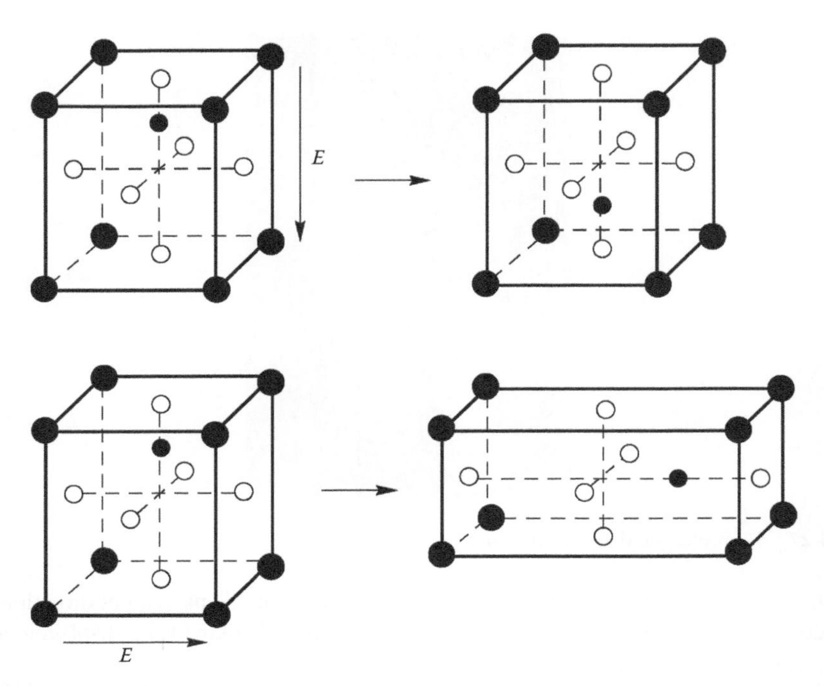

FIGURE 10.28 Tetragonal ferroelectric crystal cells under the electric field have 180° or 90° shift along the polarization direction.

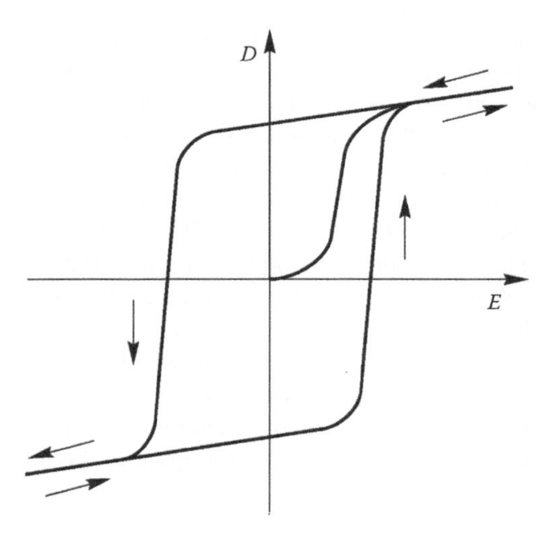

FIGURE 10.29 Electric field–electric displacement hysteresis loop of ferroelectrics.

down-arrow in the figure corresponding to the mutation curve, resulting in an electric hysteresis loop. Linear dielectric relations can describe the slash at the top and bottom side of the electric hysteresis loop.

Now consider the corresponding electric field–strain curves, as shown in Figure 10.30. The electric domain flipping at the coercive field may cause mutation of the ferroelectric strains, as seen in the figure on both sides of the upward arrows, thus resulting in the butterfly curve (or butterfly loop). Linear dielectric relations can still describe the slash far away from the coercive field.

Now we will discuss domain switching caused by the stress field [26]. Stress can cause the direction of ferroelectric polarization to flip by 90°. When applying a large enough compressive stress

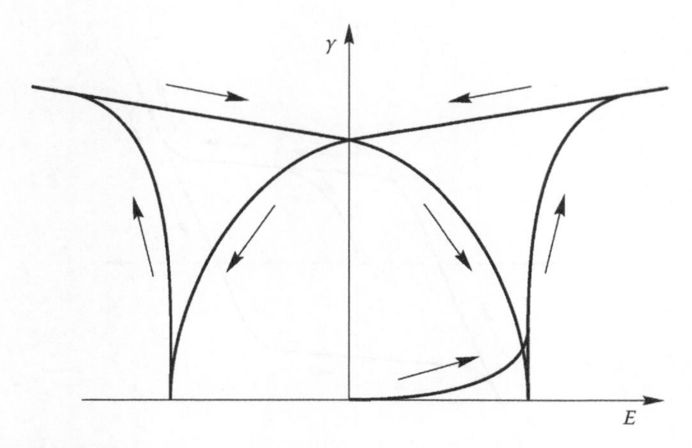

FIGURE 10.30 Electric field–strain curves.

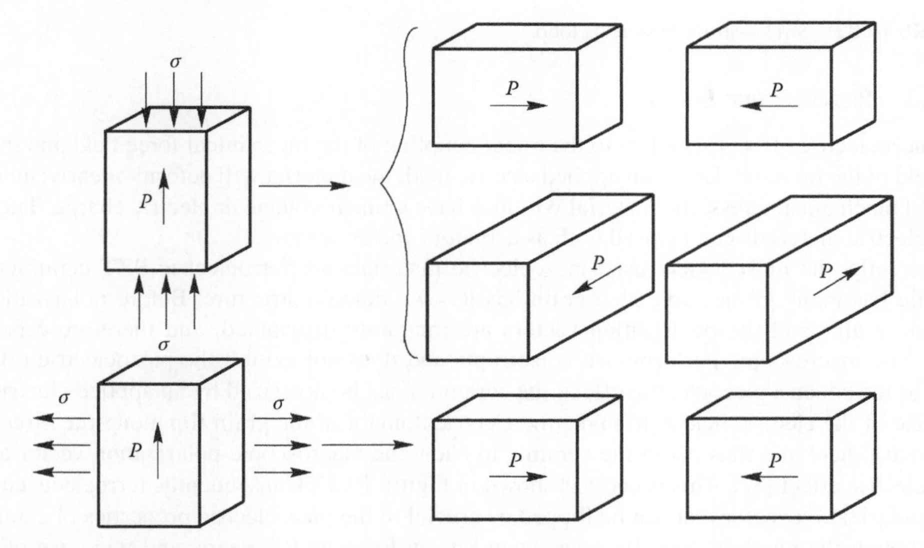

FIGURE 10.31 Tetragonal ferroelectric unit cells have 90° change along with the polarization direction under the action of stresses.

along the direction of the polarization vector, the central ion will migrate by a difference of 90° to the eccentric tetragonal phase (see Figure 10.31). The polarization direction then has a 90° switching. When applying a large enough tensile stress perpendicular to the polarization vector direction (see Figure 10.31), the central ion will migrate to the position of the tensile stress direction. The polarization direction then has a 90° switching.

Under compressive stress, there are four possible variants; while under the effect of tensile stress, there are two possible variants. The distortion caused by the stress field is called ferroelastic behavior. Now consider the corresponding stress–strain curve, as shown in Figure 10.32. The ferroelectric strain mutates at the critical stress, resulting in domain flipping, as shown by the almost parallel right and left arrows in the figure, and thus forms the stress–strain hysteresis loop.

The formation of electric domains is the result of taking the minimum of the system free energy. In ferroelectric materials, the evolution and flip of the electric domain is the genesis of polarization flips. The basic process of domain movement with corresponding polarization flips is as follows: (1) new domain nucleation, (2) the vertical extension of domains, (3) the horizontal extension of domains, and (4) the merging of domains.

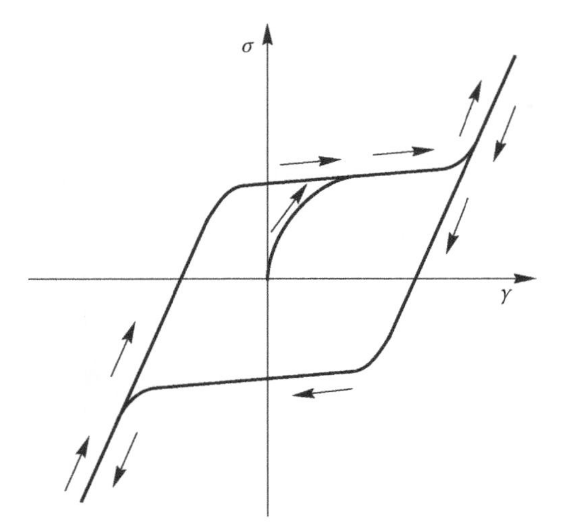

FIGURE 10.32 Stress–strain hysteresis loop.

10.4.3 Piezoelectric Effect

The piezoelectric effect [31] refers to the linear coupling of the mechanical force field and the electric field of the material. Under an applied electric field, the material will deform linearly; and under applied mechanical stress, the material will also have a linear voltage or electric charge. Therefore, piezoelectric materials can be used both as actuators and as sensors.

Currently, the most widely used piezoelectric materials are ferroelectric PZT ceramics. This ceramic has many grains, and each grain has its own domain structure. Before polarization, the ceramic grains and the polarization vectors are randomly distributed, and therefore offset each other. The macroscopic performance is isotropic and does not exhibit the piezoelectric effect. In order to introduce a piezoelectric effect, the ceramic must be polarized by an applied electric field. The role of the electric field is to make the electric domain in the grain flip along the direction of the electric field and thus cause the ceramic to show the macroscopic polarization vector and the piezoelectric effect [32]. This process is shown in Figure 10.33. Consequently, ferroelectricity, having a polarization vector that can be flipped, is crucial to the piezoelectric properties of ceramics.

Although the piezoelectric effect has been known for over 100 years, and wide use of piezoelectrics in technology has been in use for nearly half a century, there are still many difficulties in the microscopic theoretical research of piezoelectricity. As early as 1920, Born tried to calculate

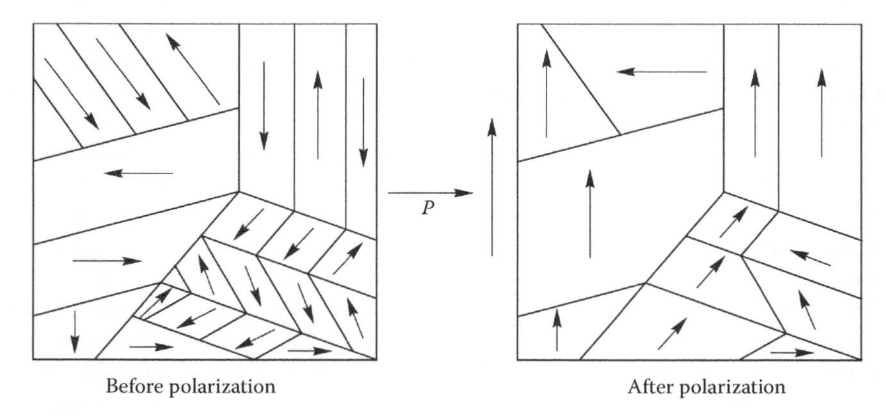

FIGURE 10.33 Polarization of the piezoelectric ceramics.

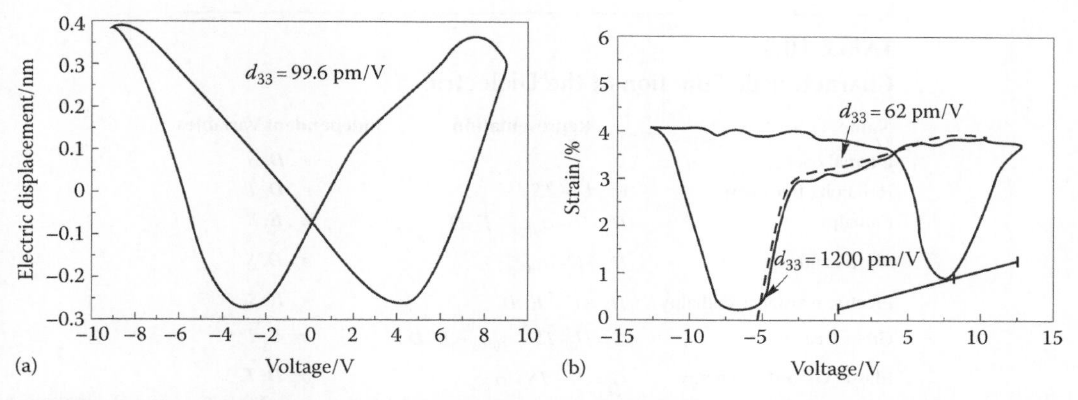

FIGURE 10.34 Butterfly curves of (a) Pb $(Zr_{0.2}Ti_{0.8})O_3$ nano thin-film and (b) Pb $(Zr_{0.2}Ti_{0.8})O_3$ nano fibers. (From Zhou, Z. et al., *Appl. Phys. Lett.*, 90, 052902, 2007; Taken from Figure 2a.)

the piezoelectric constant of cubic ZnS. Many other scholars have also made attempts. However, although ZnS has the simplest structure in piezoelectrics—with only one nonzero independent piezoelectric constant—the theoretical results, and even the sign, are often difficult to match with experimental results. Most of the research on the piezoelectric effect is limited to the development of materials research, the measurement of parameters, and the application of piezoelectric technologies.

Because of the macroscopic morphology of the material, the measuring method for the piezoelectric coefficient, and the parameters describing the piezoelectric properties, vary. For the measurement of the piezoelectric coefficient of bulk materials, a quasi-static d_{33} measuring instrument is commonly used. For the piezoelectric coefficient of thin films and one-dimensional nanofibers, it is necessary to use atomic force microscopy (AFM) with the piezo force module. Figure 10.34a and b, show, respectively, the butterfly curves obtained from testing Pb $(Zr_{0.52}Ti_{0.48})$ O_3 (PZT), nano-thin films [33] and Pb $(Zr_{0.2}Ti_{0.8})$ O_3, nano-fibers [34]. Because nanofibers are not constrained by the substrate, it can be seen from Figure 10.34 that the largest piezoelectric coefficient is 1,200 pm/V, far greater than the piezoelectric coefficient 99.6 pm/V of the thin-film structure.

Based on the analysis of electromechanic functions, the difference between the electrostrictive effect [35] and the piezoelectric effect is that the former is a second-order effect, existing in any dielectric material; while the latter is a first-order effect that may occur only in noncentrosymmetric dielectric materials. Under an external electric field, the first-order piezoelectric effect and the second-order electrostrictive effect appear simultaneously. Generally speaking, the first-order effect is more significant than the second-order effect, but both can sometimes have the same order of magnitude.

10.4.4 Mathematical Models of Electromechanical Couplings

According to thermodynamic theory and after the appropriate selection of independent variables, only a thermodynamic function can fully determine the equilibrium properties of a uniform system. This function is called the characteristic function. The state of a homogeneous elastic dielectric can be characterized by temperature T, entropy S, stress σ_{ij}, strain ε_{ij}, electric field E, and the electric displacement vector D (or polarization vector P). In order to compose the characteristic function of the dielectric, one independent variable can be selected from any of the three pairs of variables (the thermodynamic variables T and S, the mechanical variables σ_{ij} and ε_{ij}, and electrical variables E_i and D_i (or P_i). Such a choice constitutes eight different characteristic functions. Their names and representations are shown in Table 10.3 [28].

The ferroelectrics of different structures correspond to different deformation gradient tensors and polarization vectors [36–38]. The derivation process of the deformation gradient matrix is exactly

TABLE 10.3
Characteristic Function of the Dielectric

Name	Representation	Independent Variables
Internal energy	U	$\varepsilon_{ij}, \boldsymbol{D}, S$
Hemholtz free energy	$\psi = U - TS$	$\varepsilon_{ij}, \boldsymbol{D}, T$
Enthalpy	$H = U - \sigma_{ij}\varepsilon_{ij} - \boldsymbol{E} \cdot \boldsymbol{D}$	$\sigma_{ij}, \boldsymbol{E}, S$
Elastic enthalpy	$H_1 = U - \sigma_{ij}\varepsilon_{ij}$	$\sigma_{ij}, \boldsymbol{D}, S$
Electric resistance enthalpy	$H_2 = U - \boldsymbol{E} \cdot \boldsymbol{D}$	$\varepsilon_{ij}, \boldsymbol{E}, S$
Gibbs free energy	$G = U - TS - \sigma_{ij}\varepsilon_{ij} - \boldsymbol{E} \cdot \boldsymbol{D}$	$\sigma_{ij}, \boldsymbol{E}, T$
Elastic Gibbs free energy	$G_1 = U - TS - \sigma_{ij}\varepsilon_{ij}$	$\sigma_{ij}, \boldsymbol{D}, T$
Electric Gibbs free energy	$G_2 = U - TS - \boldsymbol{E} \cdot \boldsymbol{D}$	$\varepsilon_{ij}, \boldsymbol{E}, T$

analogous to that of the ferromagnetic material. For example, in tetragonal ferroelectrics, there are a total of six variants corresponding to the deformation gradient and the polarization vector:

$$
\begin{aligned}
&\boldsymbol{U}_1 = \begin{bmatrix} \beta & 0 & 0 \\ 0 & \alpha & 0 \\ 0 & 0 & \alpha \end{bmatrix}, \quad \boldsymbol{p}_1 = \begin{bmatrix} p \\ 0 \\ 0 \end{bmatrix} \quad \boldsymbol{U}_2 = \begin{bmatrix} \beta & 0 & 0 \\ 0 & \alpha & 0 \\ 0 & 0 & \alpha \end{bmatrix}, \quad \boldsymbol{p}_2 = \begin{bmatrix} -p \\ 0 \\ 0 \end{bmatrix} \\
&\boldsymbol{U}_3 = \begin{bmatrix} \alpha & 0 & 0 \\ 0 & \beta & 0 \\ 0 & 0 & \alpha \end{bmatrix}, \quad \boldsymbol{p}_3 = \begin{bmatrix} 0 \\ p \\ 0 \end{bmatrix} \quad \boldsymbol{U}_4 = \begin{bmatrix} \alpha & 0 & 0 \\ 0 & \beta & 0 \\ 0 & 0 & \alpha \end{bmatrix}, \quad \boldsymbol{p}_4 = \begin{bmatrix} 0 \\ -p \\ 0 \end{bmatrix} \\
&\boldsymbol{U}_5 = \begin{bmatrix} \alpha & 0 & 0 \\ 0 & \alpha & 0 \\ 0 & 0 & \beta \end{bmatrix}, \quad \boldsymbol{p}_5 = \begin{bmatrix} 0 \\ 0 \\ p \end{bmatrix} \quad \boldsymbol{U}_6 = \begin{bmatrix} \alpha & 0 & 0 \\ 0 & \alpha & 0 \\ 0 & 0 & \beta \end{bmatrix}, \quad \boldsymbol{p}_6 = \begin{bmatrix} 0 \\ 0 \\ -p \end{bmatrix}.
\end{aligned}
\tag{10.55}
$$

For hexagonal ferroelectrics, there are a total of eight variants corresponding to the deformation gradient and the polarization vector [37,39]:

$$
\begin{aligned}
&\boldsymbol{U}_1 = \begin{bmatrix} \eta & \delta & \delta \\ \delta & \eta & \delta \\ \delta & \delta & \eta \end{bmatrix}, \quad \boldsymbol{p}_1 = \begin{bmatrix} p \\ p \\ p \end{bmatrix} \quad \boldsymbol{U}_2 = \begin{bmatrix} \eta & \delta & \delta \\ \delta & \eta & \delta \\ \delta & \delta & \eta \end{bmatrix}, \quad \boldsymbol{p}_2 = -\begin{bmatrix} p \\ p \\ p \end{bmatrix} \\
&\boldsymbol{U}_3 = \begin{bmatrix} \eta & -\delta & -\delta \\ -\delta & \eta & \delta \\ -\delta & \delta & \eta \end{bmatrix}, \quad \boldsymbol{p}_3 = \begin{bmatrix} -p \\ p \\ p \end{bmatrix} \quad \boldsymbol{U}_4 = \begin{bmatrix} \eta & -\delta & -\delta \\ -\delta & \eta & \delta \\ -\delta & \delta & \eta \end{bmatrix}, \quad \boldsymbol{p}_4 = -\begin{bmatrix} -p \\ p \\ p \end{bmatrix} \\
&\boldsymbol{U}_5 = \begin{bmatrix} \eta & \delta & -\delta \\ \delta & \eta & -\delta \\ -\delta & -\delta & \eta \end{bmatrix}, \quad \boldsymbol{p}_5 = \begin{bmatrix} p \\ p \\ -p \end{bmatrix} \quad \boldsymbol{U}_6 = \begin{bmatrix} \eta & \delta & -\delta \\ \delta & \eta & -\delta \\ -\delta & -\delta & \eta \end{bmatrix}, \quad \boldsymbol{p}_5 = -\begin{bmatrix} p \\ p \\ -p \end{bmatrix} \\
&\boldsymbol{U}_7 = \begin{bmatrix} \eta & -\delta & \delta \\ -\delta & \eta & -\delta \\ \delta & -\delta & \eta \end{bmatrix}, \quad \boldsymbol{p}_7 = \begin{bmatrix} p \\ -p \\ p \end{bmatrix} \quad \boldsymbol{U}_8 = \begin{bmatrix} \eta & -\delta & \delta \\ -\delta & \eta & -\delta \\ \delta & -\delta & \eta \end{bmatrix}, \quad \boldsymbol{p}_8 = -\begin{bmatrix} p \\ -p \\ p \end{bmatrix}.
\end{aligned}
\tag{10.56}
$$

When these variants coexist, the average deformation gradient and the polarization vector are

$$U = \sum_{i=1}^{N} \lambda_i U_i, \quad p = \sum_{i=1}^{N} \lambda_i p_i. \tag{10.57}$$

Under the effect of an applied electric field or a force field, the shape–volume ratio of the variant changes, causing a change in $[U]$, and thus resulting in macroscopic deformation of the material. The lattice constant determines the size of the deformation, which is usually much larger than the linear piezoelectric strain [40].

The piezoelectric effect is the linear coupling of the electric field and the force field, and exists when the electric field is not large. The constitutive equations of piezoelectric materials in the three-dimensional condition depend on a variety of independent variables. There are a total of four constitutive equations, Equations 10.58 through 10.61 [41].

Proceeding from Gibbs free energy, according to Table 10.3, use the stress and the electric field as independent variables (assuming temperature is constant) to obtain the first category of piezoelectric equations:

$$\varepsilon_{ij} = S_{ijkl}^{E} \sigma_{kl} + d_{ijk}^{T} E_k \quad D_i = d_j \sigma_{ij} + a_{ij}^{\sigma} E_j. \tag{10.58}$$

Proceeding from electric Gibbs free energy, according to Table 10.3, use the strain and electric field as independent variables to obtain the second category of piezoelectric equations:

$$\sigma_{ij} = C_{ijkl}^{E} \varepsilon_{kl} - e_{ijk}^{T} E_k \quad D_i = e_{ijk} \varepsilon_{jk} + \alpha_{ij}^{\varepsilon} E_j. \tag{10.59}$$

Proceeding from elastic Gibbs free energy, according to Table 10.3, use the stress and electric displacement as independent variables to obtain the third category of piezoelectric equations:

$$\varepsilon_{ij} = S_{ijkl}^{D} \sigma_{kl} + g_{ijk}^{T} D_k \quad E_i = -g_{ikl} \sigma_{kl} + \beta_{ik}^{\sigma} D_k. \tag{10.60}$$

Proceeding from Helmholtz free energy, according to Table 10.3, use the strain and electric displacement as independent variables to obtain the fourth category of piezoelectric equations:

$$\sigma_{ij} = C_{ijkl}^{D} \varepsilon_{kl} - h_{ijk}^{T} D_k \quad E_i = -h_{ikl} \varepsilon_{kl} + \beta_{ik}^{\varepsilon} D_k. \tag{10.61}$$

where

C_{ijkl}^{D} is the component of the elastic stiffness fourth-order matrix in the constant electric displacement, called the open circuit elastic stiffness matrix

h_{ikl} is the piezoelectric stiffness matrix in units of V/m

β_{ik}^{ε} is the dielectric isolation rate matrix in the constant strain—referred to as the clamping dielectric isolation rate matrix—and the unit is m/F

S_{ijkl}^{D} is the elastic compliance matrix in the constant electric displacement, known as the open circuit compliance matrix, and the unit is m^2/N

g_{ikl} is the piezoelectric voltage constant matrix, and the unit is $V \cdot m/N$

β_{ik}^{σ} is the dielectric isolation rate matrix in the constant stress, known as the free dielectric isolation rate matrix

S_{ijkl}^{E} is the elastic compliance constant matrix in the constant electric field, called the short-circuit compliance matrix

d_{ikl} is the piezoelectric strain constant matrix, and the unit is m/V

α_{ik}^{σ} is the dielectric constant matrix in the constant stress, called the free dielectric constant matrix

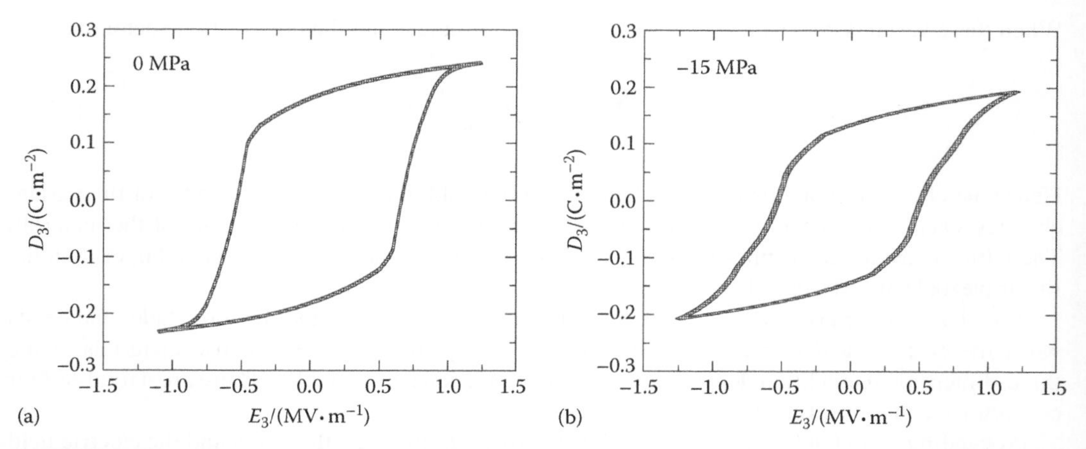

FIGURE 10.35 Electric hysteresis loops of PZT-51 under compressive stress. (a) 0 MPa. (b) −15 MPa. (From Li, C.Q. and Fang, D.N., *Acta Mech. Sinica.*, 32(1), 34, 2000; Taken from Figure 8.)

The aforementioned four types of piezoelectric equations reflect the laws that govern piezoelectric materials. They are interrelated; and these relationships are reflected in the relationship between the coefficients of the various types of piezoelectric equations.

Experiments that have tested ferroelectric materials with the characteristics of ferroelectric hysteresis loops and butterfly curves verify the aforementioned mathematical model of electromechanical coupling [42].

For example, Figures 10.35 and 10.36 give experimental test results of PZT-51 ferroelectric materials under the effect of electromechanical coupling. These are the ferroelectric hysteresis loops of the electric displacement D_3 and electric field E_3, and the butterfly curves of strain ε_{33} and the electric field E_3. Figure 10.35a and b show these results under the effect of nonstress and the effect of stress. In Figures 10.35 and 10.36, it is obvious that the shape of the hysteresis loop and the butterfly curve change with compressive stress. When compressive stress increases, the remnant polarization of ferroelectric material PZT-51 was significantly reduced, the coercive electric field decreased, and the residual strain was significantly reduced. These experimental test results are in good agreement [42] with calculations of the mathematical model of electromechanical coupling.

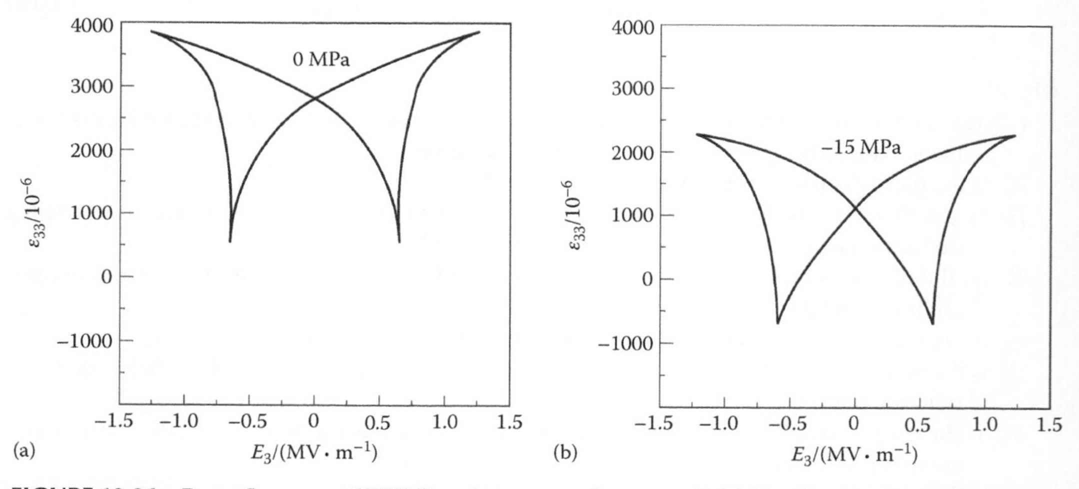

FIGURE 10.36 Butterfly curves of PZT-51 under compressive stress. (a) 0 MPa. (b) −15 MPa. (From Li, C.Q. and Fang, D.N., *Acta Mech. Sinica.*, 32(1), 34, 2000; Taken from Figure 9.)

EXERCISES

10.1 What is the difference between the electrostrictive effect and the piezoelectric effect?

10.2 Is it necessary for piezoelectric materials to have ferroelectricity? Will ferroelectric materials always have ferroelectricity?

10.3 What is the difference between SMAs and ferromagnetic SMAs?

10.4 What is the impact of stress on the properties of ferromagnetic materials?

10.5 Consider a cubic ferromagnetic crystal with large positive magnetocrystalline anisotropy ($K_1 > 0$) in the demagnetization state, containing only [100] and [111] domains, and now make it magnetize along the [010] direction. Calculate the functional relationship between the elongations in the [010] direction and the magnetization intensity in the same direction.

10.6 Prove: The magnetization power that magnetizes a ferromagnetic cubic crystal to its saturation along different crystal axis is

$$W_{[110]} - W_{[100]} = \frac{K_1}{4}.$$

$$W_{[111]} - W_{[100]} = \frac{K_1}{3} + \frac{K_2}{27}.$$

10.7 What types of electric domain structures help to improve the piezoelectricity of a $BaTiO_3$ single crystal?

10.8 Derive the transfer matrix $[U_1] = \begin{bmatrix} \frac{\alpha+\gamma}{2} & 0 & \frac{\alpha-\gamma}{2} \\ 0 & \beta & 0 \\ \frac{\alpha-\gamma}{2} & 0 & \frac{\alpha+\gamma}{2} \end{bmatrix}$ of the SMA having cubic–orthorhombic transformation. α, β, and γ are respectively determined by the lattice constant of crystals.

10.9 A ferroelectric single crystal is in a single-domain state, and the polarization vector is along the [001] direction. Under the effect of a strong electric field along the [100] direction, the domain flips. Find the change of the deformation gradient and the polarization vector before and after the domain flip.

10.10 Derive Equation 10.61 from Equation 10.59.

REFERENCES

1. Jiang D. S. and Claus R. O. *Device Structure and Application of Smart Materials*. Wuhan, China: Wuhan University Press, 2000, pp. 1–5.
2. Yao K. D. and Cheng G. X. *Smart Materials*. Beijing: Chemical Industry Press, 2002, pp. 1–7.
3. Tao B. Q. *Smart Materials and Structures* (2nd edn.). Beijing: National Defence Industry Press, 1999, pp. 1–22.
4. Gong C. S. and Zhang K. L. *New Functional Materials*. Beijing: Chemical Industry Press, 2001, pp. 59–79.
5. Yang D. Z. *Smart Materials and Intelligent Systems*. Tianjin: Tianjin University Press, 2000: 104–140.
6. Wayman C. M. and Shimizu K. The shape memory ('marmen') effect in alloys. *J Metal Sci*, 1972, 6(9): 175–183.
7. Xu Z. Y., Jiang B. H., Yang D. Z. et al. *Shape Memory Materials*. Shanghai: Shanghai Jiaotong University Press, 2000.
8. Zhou K., Kang H. et al. Fan Qiandong translation. *Shape Memory Alloys* [M] Beijing: Mechanical Industry Press, 1992, pp. 23–32.
9. Wayman C. M. Some applications of shape-memory alloys. *J Metals*, 1980, 32(6): 129–137.
10. Bhattacharya K. *Microstructure of Martensite: Why It Forms and How It Gives Rise to the Shape-Memory Effect*. New York: Oxford University Press, 2003, pp. 55–58.

11. Wang G. C., Yang G., Huang Y. B. et al. Effect of heat treatment and thermochemical treatment on linear recovery property of TiNi shape memory alloy. *Adv Eng Mater*, 2006, 8: 107–111.
12. Cheng G. Y. *Ferromagnetic School*. Beijing: Higher Education Press, 1961, pp. 168–184.
13. Chikazumi S., Ge Shihui translation. *Ferromagnetic Physics*. Lanzhou, China: Lanzhou University Press, 2002, pp. 283–316.
14. Jiang S. and Li W. *Condensed Matter Magnetic Physics*. Beijing: Science Press, 2003, pp. 215–237.
15. Wan D. F. and Ma X. L. *Magnetic Physics*. Beijing: Electronic Industry Press, 1994, pp. 165–218.
16. Jahn H. A. and Teller E. Stability of polyatomic molecules in degenerate electronic state. I. Orbital degeneracy. *Proc Roy Soc Lond A*, 1937, A161: 220.
17. BecKe R. and Doring W. *Ferromaegnetismus*. Berlin: Springer, 1939, p. S284.
18. Jiles D. C. and Thoelke J. B. Magnetization and magnetostriction in Terbium-dysprosium-iron alloys. *Phys Stat Sol*, 1995, 147: 535–551.
19. Ma Y. F. and Li J. Y. Magnetization rotation and rearrangement of martensite variants in ferromagnetic shape memory alloys. *Appl Phys Lett*, 2007, 90: 172504.
20. Li L. J., Li J. Y., Shu Y. C. et al. Magnetoelastic domains and magnetic field-induced strains in ferromagnetic shape memory alloys by phase-field simulation. *Appl Phys Lett*, 2008, 92: 172504.
21. Heczko O. Magnetic shape memory effect and magnetization reversal. *J Magn Magn Mater*, 2005, 290: 787–794.
22. Armstrong J. N., Sullivan M. R., and Romancer M. L. Role of magnetostatic interactions in micromagnetic structure of multiferroics. *J Appl Phys*, 2008, 103: 023905.
23. Sundar V. and Newnham R. E. Electrostriction and polarization. *Ferroelectrics*, 1992, http://www.informaworld.com/smpp/title~content=t713617887~db=all~tab=issueslist~branches=135-v135135: 431–446.
24. Jang S. J., Uchino K., and Cross L. E. Electrostrictive behavior of lead magnesium niobate based ceramic dielectrics. *Ferroelectrics*, 1980, 27: 31–35.
25. Cross L. E., Jang S. J., and Newnham R. E. Large electrostrictive effects in relaxor ferroelectrics. *Ferroelectrics*, 1980, 23: 187–192.
26. Yang W. *Mechatronic Reliability*. Beijing: Tsinghua University Press, 2001, pp. 87–106.
27. Zhu Z. X., Li J. F., Lai F. P. et al. Phase structure of epitaxial Pb(Zr,Ti)O3 thin films on Nb-doped SrTiO3 substrates. *Appl Phys Lett*, 2007, 91: 222910.
28. Zhong W. L. *Physics of Ferroelectri*cs. Beijing: Science Press, 1998: 271–274.
29. Shu Y. C. and Bhattacharya K. Domain patterns and macroscopic behaviour of ferroelectric materials. *Philos Mag B*, 2001, 81: 2021–2054.
30. Shu Y. C., Yen J. H., Chen H. Z. et al. Constrained modeling of domain patterns in rhombohedral ferroelectrics. *Appl Phys Lett*, 2008, 92: 052909.
31. Li Y. and Qin Z. K. *Measurement of Piezoelectric and Ferroelectric Materials*. Beijing: Science Press, 1984, pp. 15–17.
32. Li J. Y., Rogan R. C., Üstündag E. et al. Domain swithcing in polycrystalline ferroelectric ceramics. *Nat Mater*, 2005, 4: 776–781.
33. Xie S. H., Li J. Y., Liu Y. Y. et al. Electrospinning and multiferroic properties of NiFe2O4-Pb(Zr0.52Ti0.48) O3 composite nanofibers. *J Appl Phys*, 2008, 104: 024115.
34. Zhou Z. H., Gao X. S., Wang J. et al. Giant strain in PbZr0.2Ti0.8O3 nanowires. *Appl Phys Lett*, 2007, 90: 052902.
35. Li J. Y. Exchange coupling in P(VDF-TrFE) copolymer based all-organic composites with giant electrostriction. *Phys Rev Letts*, 2003, 90: 217601.
36. Liu D. and Li J. Y. The enhanced and optimal piezoelectric coefficients in single crystalline barium titanate with engineered domain configurations. *Appl Phys Lett*, 2003, 83: 1193–1195.
37. Li J. Y. and Liu D. On ferroelectric crystals with engineered domain configurations. *J Mech Phys Solids*, 2004, 52: 1719–1742.
38. Liu J. J., Zhou Y. C., Soh A. K. et al. Engineering domain configurations for enhanced piezoelectricity in barium titanate single crystals. *Appl Phys Lett*, 2006, 88: 032904.
39. Liu D. and Li J. Y. Domian-engineered Pb(Mg1/3Nb2/3)O3-PbTiO3 crystals: Enhanced piezoelectricity and optimal domain configurations. *Appl Phys Lett*, 2004, 84: 3930–3932.
40. Damjanovic D. Ferroelectric, dielectric and piezoelectric properties of ferroelectric thin films and ceramics. *Rep Prog Phys*, 1998, 61: 1267–1324.
41. Du S. Y. and Wang B. *The Micromechanics of Composites*. Beijing: Science Press, 1998: 274–276.
42. Li C. Q. and Fang D. N. Ferroelectric ceramic PZT experimental structure research. *Adv Mech*, 2000, 32: 34–41.

11 Mechanical Properties of Thin Films

Why does a coat of whitewash on a wall surface often shed or peel off? A vast majority of the metallic surfaces of primary electronic components are coated with a man-made, or naturally occurring, protective coatings layer. Such layers also frequently peel off. Why is that so? Because almost all primary components of electronic devices are made of thin films, the shedding or peeling off of such thin films is one of the important reasons why these devices fail. Why do such thin films peel off? Thin films come in many different types and varieties. When they go "bad," what common and regular characteristics do they share? This chapter begins with an overview of thin films, and then proceeds to discuss their mechanical properties in terms of stress–strain analysis, residual stress, and their fracture toughness; as well as the polarization of ferroelectric thin films and the bending of ductile thin films.

11.1 AN OVERVIEW

The history of thin films can be traced back over a thousand years. However, the emergence of thin films as a new field in science and technology is a recent development, occurring about 30 years ago. At the present time, thin film materials have become an important branch in the field of materials science. It involves other branches of learning such as physics, chemistry, electronics, and metallurgy; and has a wide scope of application, particularly in national defense, aviation, aeronautics, in the electronics industry, and in the chemical industry. It has become the most active area in materials science, and is also gradually evolving into an independent field of study—thin film science.

Thin films can be manufactured from a single chemical element, or from inorganic as well as organic compounds. They can also be made of solid, liquid, or gaseous matter. Thin films, like lumpy bodies, can be formed from single crystals, polycrystals, microcrystals, nanocrystals, and multilayer and super-crystalline structures. Prior to a discussion of thin film materials, we must define what a thin film is. How "thin" should a film be in order to be considered a thin film? As we know, the term "thin film" came on the scene as a result of scientific and technological progresses. Sometimes the word film is synonymous with the words "coating," "layer," or "foil." However, there are some differences in these terms. Thin films are often characterized in terms of their thickness. Generally speaking, a film is taken with a thickness that is capable of existing independently, in the absence of a substrate, and which is approximately 1 μm. Following constant scientific and technological development and advancement, the thin film field has also been constantly expanded and diversified. Each diversified application imposes different requirements on film thickness. A thin film is defined as a layer of a certain thickness deposited by physical, chemical, or other processes on the surface of a metallic or nonmetallic substrate. It is made of a different substrate and serves as a covering layer for reinforcement, protection, or other purposes [1–3]. At the present time, many methods for preparing thin films exist, such as the gas phase generation method (gas phase epitaxy), the vapor phase generation method (vapor phase epitaxy), oxidation, diffusion, and electroplating. Each of these methods involves a number of different approaches.

Thin films can be classified into many different types [1]. In terms of their nature, thin films can be classified as natural and composite. In accordance with their usage, thin films can be classified as wear-resistant or wear-tolerating, decorative, conducting, magnetic, or piezoelectric. In accordance with

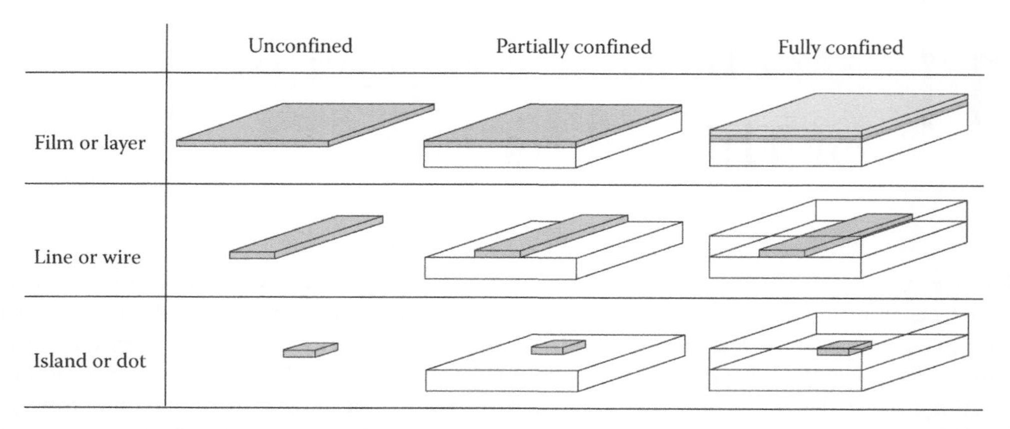

FIGURE 11.1 Geometric shapes and constraints of thin films. (From Freund, L.B. and Suresh, S., *Thin Film Materials: Stress, Defect Formation, and Surface Evolution*, Cambridge University Press, Cambridge, 2003.)

their mechanical properties, thin films can be classified as brittle and ductile. In accordance with their geometric shapes and constraints, they can be classified as free of constraint, partially constrained, and completely constrained.

This data are summarized in Figure 11.1.

The following is a brief explanation of the geometric shapes of thin films [1]. It should be noted that it is their geometric shapes that determine the orientation of their reference coordinate systems. Furthermore, the geometric shapes are characterized by the relative dimensions in a three-dimensional orthogonal coordinate system. The extent of constraint is based upon the structure of the film, and the interaction between the film and the deformable solid material with which the film is combined, or in contact with. In the former case, compatibility of deformations is called for, while in the latter case, restriction in relative movement is required. Geometric shapes can be divided into film (or layer), line (or thread), and island (or dot). Constraint can be classified as free, partial, and total. Figure 11.1 is a summary of the hierarchy of thin films based upon relative physical sizes, not referring to any length unit that reflects the basic structure of the material. The aforementioned illustration should be helpful in understanding the basic concepts pertaining to various practical applications in the field.

According to Figure 11.1, when the dimension of a film in one direction is smaller than its dimensions in the other two directions, it is referred to as a thin film. The word "smaller" here means that the stated dimension is at least 20 times less than the dimensions in the other two directions, and could be several hundred times less. When the dimensions in two directions are both smaller than the third, the film is referred to as line or thread. When its dimensions in all three directions are relatively smaller than those it surrounds, it is referred to as an island or a dot. As far as the degree of constraint in inhibiting deformation or displacement is concerned, if the boundaries associated with the small dimension can displace freely without any restriction, such a configuration is referred to as free. On the other hand, if all boundaries associated with the small dimension are constrained against deformation, such a configuration is referred to as totally constrained. Realistically, constraint occurs along the interface or the common boundaries shared with another material. If displacements in certain (but not all) directions are unconstrained, the configuration is referred to as partially constrained.

Our main focus in this chapter is on partially constrained thin films. This type of film can be used in a variety of ways. They all have a fixed shape, size, and flatness; and have a thickness of less than 10 μm.

Because it is not possible to rely on the mechanical strength of the film itself to maintain its status quo, it has to be attached to a variety of substrates. Consequently, the strength of the bond between film and substrate will unavoidably affect the various film properties. A poor bond at any boundary surface may render the film useless. Moreover, during the process of making the film, its structure may be very much affected by the manufacturing process, because some intrinsic stress may develop within the film. The difference between the expansion coefficients of the substrate, and the film, may also give rise to stress in the film. Any excessive stress may cause the film to curl or crack up, and render it useless. Therefore, in all fields of application, the film bond strength and stress are of primary concern. Under certain circumstances, other thin film mechanical properties must be considered. In the case of super-hard thin films used in enhancing substrate hardness and the ability to tolerate friction, the film hardness and its ability to withstand friction, as well as its capacity to tolerate wear and tear, must be investigated.

The microstructure of thin films and their physical and chemical properties are entirely different from those of bulk materials. Similarly, the mechanical properties of thin films are different from the corresponding properties of bulk materials. What are the mechanical properties of thin films that warrant our attention? How can we acquire knowledge of those properties? In their studies of the properties of thin films, many scholars have made outstanding contributions. In this author's opinion, the most authoritative work on thin films up to the present time is the monograph entitled *Thin Film Materials – Stress, Defect Formation and Surface Evolution* by L. B. Freund and S. Suresh, published by Cambridge University Press in 2003. I strongly recommend this monograph to readers who are interested in thin films. This chapter is concerned mainly with the characterization of the mechanical properties of thin films.

11.2 ELASTIC MODULUS AND STRESS–STRAIN RELATIONSHIP OF THIN FILMS

11.2.1 Elastic Modulus of Thin Films [4]

Elastic modulus is one of the basic parameters in materials science. Because of the differences in certain basic properties, the elastic modulus of a thin film can be completely different from those of a bulk material of the same composition. Presently, the three main approaches for measuring elastic modulus are dynamic techniques, static techniques, and indentation methods [5].

1. The dynamic techniques—including the ultrasound method and the vibratory reed method—utilize the resonance frequency of a material to determine its elastic modulus. These two methods have their limitations, however. The ultrasound method works when the thin film is made of a single pure material, while the vibratory reed method is useful only when the density of the thin film is known.
2. The static techniques make use of the Hooke's law to determine elastic modulus, as in the axial tensile method and the bubbling method. These methods require an accurate determination of the stress and strain of the test sample, and the stripping of the sample from its substrate. They call for a high degree of precision in instrumentation and sample preparation. Hence, the success rate of these methods is low.
3. Indentation methods involve the use of a diamond indenter to press into the test sample. The elastic modulus can be determined from the load vs. indenter displacement curve. Nano-indentation is one of the most typical indentation methods used for this purpose, but the elastic modulus thus determined is often subject to the influence of the substrate. Next, we discuss in particular the three-point bending method and indentation methods.

11.2.1.1 Three-Point Bending Method [5]

As shown in Figure 11.2, a load F is applied at the midpoint on the surface of a film-substrate system, between two symmetrically located supports. The resulting deflection is measured at the load point. Let the distance between the two supports be L, and the film thickness and the substrate thickness be t_f and t_s, respectively.

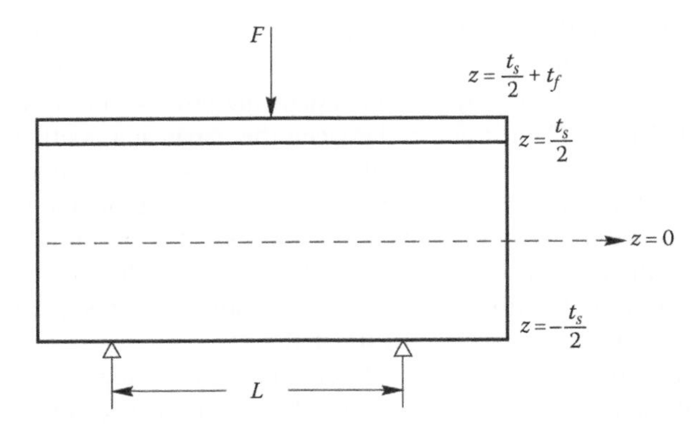

FIGURE 11.2 Schematic diagram of the three-point bending method.

For small deflections, ΔF (the load increment) and $\Delta\delta$ (the corresponding deflection increment) can be related as follows:

$$\Delta F = \frac{48\Delta\delta}{L^3}EI, \tag{11.1}$$

in which EI is the flexural rigidity of the beam. The flexural rigidity of the composite beam consisting of the substrate plated by a thin film is given by

$$EI = E_s I_s + E_f I_f. \tag{11.2}$$

In this equation, E_s and E_f are, respectively, the elastic moduli of the substrate and the film, and I_s and I_f are, respectively, their moments of inertia with respect to the z-axis

$$I_s = \int_{-t_s/2}^{t_s/2} y^2 b\,dy, \tag{11.3a}$$

and

$$I_f = \int_{t_s/2}^{t_s/2+t_f} y^2 b\,dy, \tag{11.3b}$$

where b is the width of the film's cross-sectional area.

When experimental results show that the load F at the midpoint, and the corresponding deflection δ at the same point, closely follow a linear relationship, Equation 11.1 can be used to determine the flexural rigidity EI. If the elastic modulus E_s of the substrate is known, then Equation 11.2 can immediately be used to determine the elastic modulus E_f of the film.

Experimental errors originate mainly from deflection measurements themselves, and can also result when the load point and the measurement point are not exactly at the midpoint. Therefore, in small deflection studies, stringent requirements are called for in the resolution power of sensors used to measure deflections, as well as in positioning the load, and ensuring that deflection measurements are made exactly at the midpoint of the thin-film system.

11.2.1.2 Indentation Method [6,7]

The use of the indentation method as a means for measuring the elastic moduli of thin films began in the 1970s. In recent years, nano-indentation methods have been widely used in measuring the mechanical properties of thin films. The special features of this technology are its high resolution with respect to load and deflection measurements, and its ability to continuously record the load and displacement data during loading and unloading. Therefore, it is most suitable for measuring the mechanical properties of thin films. The use of these methods to determine the hardness H, elastic modulus E, and the creep deformation of thin films, is based upon Snedden's theoretical analysis on the relationship between symmetrical indenter loading and penetration depth. A loading–unloading curve in connection with nanoindentation is shown in Figure 4.10. According to the method proposed by Oliver and Pharr [6], the hardness can be obtained from the slope at the top of the unloading curve using Equation 4.22 [6] as follows:

$$S = \frac{\mathrm{d}F}{\mathrm{d}h} = \frac{2}{\sqrt{\pi}} E^* \sqrt{A}, \tag{11.4}$$

where

F and h are, respectively, the indenter load and indentation depth

A is the indenter's projected area of contact, and the reduced elastic modulus E^* is defined as follows:

$$\frac{1}{E^*} = \frac{1-\nu^2}{E} + \frac{1-\nu_i^2}{E_i}. \tag{11.5}$$

where

E and E_i in the aforementioned equation are, respectively, the elastic moduli of the film being tested and the indenter

ν and ν_i are, respectively, their Poisson's ratios

The hardness of the material H being tested is defined as

$$H = \frac{F_{max}}{A}. \tag{11.6}$$

Once A, $\mathrm{d}F/\mathrm{d}h$, and F_{max} have been determined, Equations 11.4 through 11.6 can be used to obtain the elastic modulus E of the film and its hardness H.

11.2.2 Stress–Strain Relationship of Thin Films

The stress–strain relationship is a basic mechanical property of materials. In the course of preparing and using materials, one must first have an in-depth understanding of the stress–strain relationship. Similarly, in the case of ductile thin films, we also need a detailed understanding of their stress–strain relationship. In Chapter 3, we saw that the basic methods used for investigating the stress–strain relationship of materials involves the use of unidirectional tensile force, or unidirectional compressive force. Are these methods still applicable to thin films? Obviously, they are no longer very suitable because it is not easy to get test thin-film samples suitable for unidirectional tension and compression analyses. However, it is possible to extract data for thin films from experimental results obtained from a film-substrate composite system. In the next section, we briefly describe the basic idea on obtaining the stress–strain relationship for thin films, using the tensile method and the indentation method.

11.2.2.1 Tensile Method [8]

Assume that the stress–strain relationship $\sigma_s(\varepsilon)$ for the substrate is known. Then, using the stress–strain relationship $\sigma_c(\varepsilon)$ for a composite system—consisting of a substrate plated with a film on both of its surfaces—the stress–strain relationship $\sigma_f(\varepsilon)$ for the film can be obtained as follows:

$$\sigma_f(\varepsilon) = \frac{\left[\sigma_c(\varepsilon)(t_s + 2t_f) - \sigma_s(\varepsilon)t_s\right]}{2t_f}. \tag{11.7}$$

where

s and f denote, respectively, substrate and film
t_s and t_f, their thicknesses

It is assumed in Equation 11.7 that, in the tensile process, the film and substrate do not detach from each other and that they undergo an identical deformation ε. That is,

$$\varepsilon = \varepsilon_s = \varepsilon_f. \tag{11.8}$$

11.2.2.2 Indentation Method [9,10]

A great deal of effort has been made on the analysis of the stress–strain relationship pertaining to thin films, by using the indentation method. However, satisfactory results have not as yet been reached. Regarding brittle materials, only measurements of their elastic modulus and hardness suffice; and so the task is easier. But in the case of ductile materials, determining their stress–strain relationship is more complicated. If we assume that the film is comparatively thick, and if effects of the substrate can be ignored in indentation experiments, readers may point out that shallower indentations would perhaps do the job. But then the question would be: how shallow should the indentations be? On the other hand, according to Chapter 4, if they are too shallow, scale effects will possibly arise. What could then be done? Currently, experience tells us that when the indentation exceeds 4–5 μm, the scale effect can be neglected; and when the indentation is less than 1/3 of the film thickness, the effect of the substrate can be ignored. Based upon these restrictive conditions, we now introduce the basic idea of Dao et al. [9] on how the indentations by a sharp indenter can be used to determine the elasto-plastic stress–strain relationship.

Assume that the uni-axial stress–strain relationship of a ductile material is shown in Figure 11.3. The material obeys Hooke's law before yield, and obeys a power law after yield. That is,

$$\sigma = E\varepsilon, \quad \sigma \le \sigma_y, \tag{11.9a}$$

$$\sigma = R\varepsilon^n, \quad \sigma \ge \sigma_y, \tag{11.9b}$$

where σ_y, n, E, and R are, respectively, the yield strength, the strain-hardening exponent, the elastic modulus, and a strengthen coefficient. When the stress reached the yield point, we have

$$\sigma_y = E\varepsilon_y = R\varepsilon^n, \tag{11.10}$$

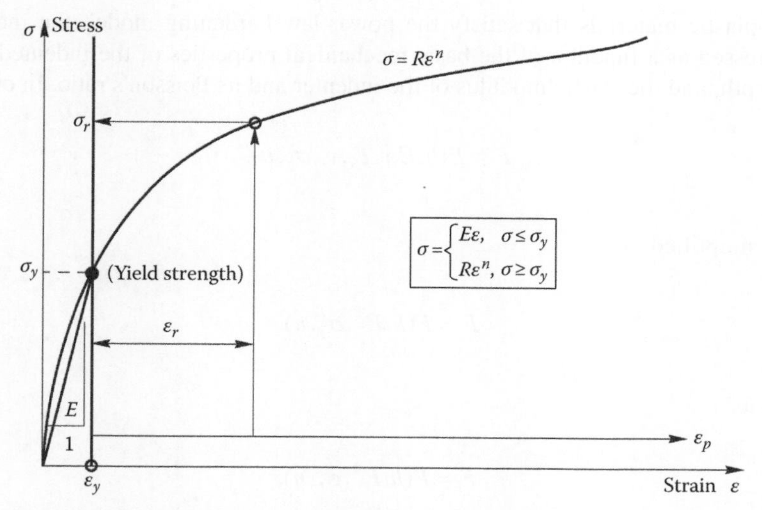

FIGURE 11.3 Stress–strain curve. (ε_y, σ_y) is the yield strain and yield stress. ($\varepsilon = \varepsilon_y + \varepsilon_p$; σ and ε are the coordinates of any arbitrary point on the nonlinear portion of the curve, so no subscripts are needed. For a point before yield, ε_p is of course zero.) (From Dao, M. et al., *Acta Mater.*, 49(19), 3899, 2001; Taken from Figure 2.)

where ε_y is the strain corresponding to σ_y. The stress–strain relationship can be shown to be as follows:

$$\sigma = E\varepsilon, \quad \sigma \le \sigma_y, \tag{11.11a}$$

$$\sigma = \sigma_y\left(1 + \frac{E}{\sigma_y}\varepsilon_p\right)^n, \quad \sigma \ge \sigma_y. \tag{11.11b}$$

At any point after yield,

$$\varepsilon = \varepsilon_y + \varepsilon_p, \tag{11.12}$$

where
  ε is the total strain
  ε_p is the nonlinear part of the total strain as shown in Figure 11.13

If we can obtain the elastic modulus E and the yield stress σ_y, and the strain-hardening exponent n, then the stress–strain relationship in Equation 11.12 can be ascertained. The main idea behind obtaining the stress–strain relationship using the indentation method is to deduce the elasto-plasticity parameters of brittle materials (E, σ_y, n from the load/unloading-deflection curves as shown in Figure 4.10). Then, by making use of the power law hardening model, the stress–strain relationship can be obtained. In what follows, we illustrate how the parameters E, σ_y, and n of elasto-plastic materials can be deduced from the load/unloading-deflection curves.

Regarding sharp indentation experiments, the load/unloading-deflection relationships can be simply expressed in terms of the Kick's Law as follows:

$$F = Ch^2, \tag{11.13}$$

where
  C is the loading coefficient curvature
  h is the indentation depth

For elasto-plastic materials that satisfy the power law hardening model, the indentation load F can be expressed as a function of the basic mechanical properties of the indented material, the indentation depth, and the elastic modulus of the indenter and its Poisson's ratio. In other words,

$$F = F(h, E, v, E_i, v_i, \sigma_y, n), \tag{11.14}$$

which can be simplified to

$$F = F(h, E^*, \sigma_y, n), \tag{11.15}$$

and rewritten as

$$F = F(h, E^*, \sigma_r, n), \tag{11.16}$$

where σ_r is the representative stress, that is the flow stress when $\varepsilon_p = \varepsilon_r$. Taking σ_r and h as the basic variables and applying the Π-theorem [10], which is a powerful and important tool for researchers to study mechanical properties of materials, Equation 11.16 can be written as

$$F = \sigma_r h^2 \Pi_1\left(\frac{E^*}{\sigma_r}, n\right), \tag{11.17}$$

from which we obtain

$$C = \frac{F}{h^2} = \sigma_r \Pi_1\left(\frac{E^*}{\sigma_r}, n\right), \tag{11.18}$$

where Π_1 is a dimensionless function.

Similarly, if we also express the unloading rate $\mathrm{d}F_u/\mathrm{d}h$ as a function of the basic mechanical properties of the indented material, the indentation depth, the elastic modulus of the indenter and its Poisson's ratio, we have

$$\frac{\mathrm{d}F_u}{\mathrm{d}h} = \frac{\mathrm{d}F_u}{\mathrm{d}h}(h, h_m, E^*, \sigma_r, n), \tag{11.19}$$

where h_m is h_{max}.

By letting E^* and h be the basic quantities when $h = h_m$, Equation 11.19 can be written as

$$\left.\frac{\mathrm{d}F_u}{\mathrm{d}h}\right|_{h_m} = E^* h_m \Pi_2\left(\frac{E^*}{\sigma_r}, n\right) \tag{11.20}$$

The unloading force F_u can also be expressed as

$$F_u = F_u(h, h_m, E^*, \sigma_r, n) = E^* h^2 \Pi_u\left(\frac{h_m}{h}, \frac{\sigma_r}{E^*}, n\right) \tag{11.21}$$

When $F_u = 0$, the unloading is complete; and it follows that $h = h_r$ (h_r, as shown in Figure 4.10, is the residual penetration depth after unloading has completed). Hence,

$$0 = \Pi_u\left(\frac{h_m}{h_r}, \frac{\sigma_r}{E^*}, n\right) \tag{11.22}$$

and

$$\frac{h_r}{h_m} = \Pi_3\left(\frac{\sigma_r}{E^*}, n\right). \tag{11.23}$$

The following two dimensionless functions are given in the paper by Dao et al. [9]:

$$\Pi_4\left(\frac{h_r}{h_m}\right) = \frac{F_{ave}}{E^*}, \tag{11.24a}$$

$$\Pi_5\left(\frac{h_r}{h_m}\right) = \frac{W_p}{W_t}, \tag{11.24b}$$

where

$F_{ave} = F_m/A_m$ is the average load

A_m is the projected area of the largest indentation pit

W_p and W_t are, respectively, the plastic work and the total work done by the load during the loading process

From Equation 11.18, we have

$$\Pi_1\left(\frac{E^*}{\sigma_r}, n\right) = \frac{C}{\sigma_r}. \tag{11.25}$$

Dao et al. [9] noted that different σ_r and ε_r may give rise to different forms of the dimensionless function Π_1. By means of finite element numerical simulation, Dao et al. [9] used 76 different combinations of parameters to calculate elastic moduli (ranging from 20 to 210 GPa), yield strengths (ranging from 30 to 3000 MPa), and various strain hardening indices (ranging from 0 to 0.5). They found that when $\varepsilon_r = 0.033$, the dimensionless function Π_1 would be independent of the strain hardening exponent n. In this case, the expression of Π_1 follows,

$$\Pi_1\left(\frac{E^*}{\sigma_{0.033}}\right) = \frac{C}{\sigma_{0.033}}. \tag{11.26}$$

They also obtained the following five explicit dimensionless functions Π_1, Π_2, Π_3, Π_4 and Π_5 [9]:

$$\Pi_1 = -1.131\left[\ln\left(\frac{E^*}{\sigma_{0.033}}\right)\right]^3 + 13.635\left[\ln\left(\frac{E^*}{\sigma_{0.033}}\right)\right]^2 - 30.594\left[\ln\left(\frac{E^*}{\sigma_{0.033}}\right)\right] + 29.267 \tag{11.27}$$

$$\Pi_2 = (-1.40557n^3 + 0.77526n^2 + 0.15830n - 0.06831)\left[\ln\left(\frac{E^*}{\sigma_{0.033}}\right)\right]^3$$
$$+(17.93006n^3 - 9.22091n^2 - 2.37733n + 0.86295)\left[\ln\left(\frac{E^*}{\sigma_{0.033}}\right)\right]^2$$
$$+(-79.99715n^3 + 40.55620n^2 + 9.00157n - 2.54543)\left[\ln\left(\frac{E^*}{\sigma_{0.033}}\right)\right]$$
$$+(122.65069n^3 - 63.88418n^2 - 9.58936n + 6.20045) \quad (11.28)$$

$$\Pi_3 = (0.010100n^2 + 0.0017639n - 0.0040837)\left[\ln\left(\frac{\sigma_{0.033}}{E^*}\right)\right]^3$$
$$+(0.14386n^2 + 0.018153n - 0.088198)\left[\ln\left(\frac{\sigma_{0.033}}{E^*}\right)\right]^2$$
$$+(0.59505n^2 + 0.034074n - 0.65417)\left[\ln\left(\frac{\sigma_{0.033}}{E^*}\right)\right]$$
$$+(0.58180n^2 - 0.088460n - 0.67290) \quad (11.29)$$

$$\Pi_4 \approx 0.268536\left(0.9952495 - \frac{h_r}{h_m}\right)^{1.1142735} \quad (11.30)$$

$$\Pi_5 = 1.61217\left\{1.13111 - 1.74756^{-1.49291\left(\frac{h_r}{h_m}\right)^{2.535334}} - 0.075187\left(\frac{h_r}{h_m}\right)^{1.135826}\right\} \quad (11.31)$$

From the aforementioned five dimensionless functions, they at first formulated the forward analytical algorithm for obtaining C, h_r (or W_p/W_t), h_m(or F_m), and $\left.\frac{dF_u}{dh}\right|_{h_m}$ or, in other words, the indentation load versus the displacement (depth of indent) curve, from the known stress–strain relationship (i.e., from the known values of E, n, σ_y, and ν). The results obtained from this forward analysis agreed very closely with the experimental indentation results. This proves that the forward analysis procedure is appropriate. On the other hand, a reverse analysis, or reverse deduction, is also feasible, C, h_r (or W_p/W_t), h_m (or F_m), and $\left.\frac{dF_u}{dh}\right|_{h_m}$, can be found from the load–displacement curve, and by which the relationship between the five dimensionless functions and the mechanical property parameters, E^*, A_m, F_{ave}, $\sigma_{0.033}$, σ_y, and n, can be established. In other words, from the load–displacement curve obtained from experiment, the values of E^*, σ_y, and n can be found. As a result, the stress–strain relationship in terms of the power law model can be obtained. Figure 11.4 gives the flow chart showing the deduction path from load–displacement curve to E^*, σ_y, and n.

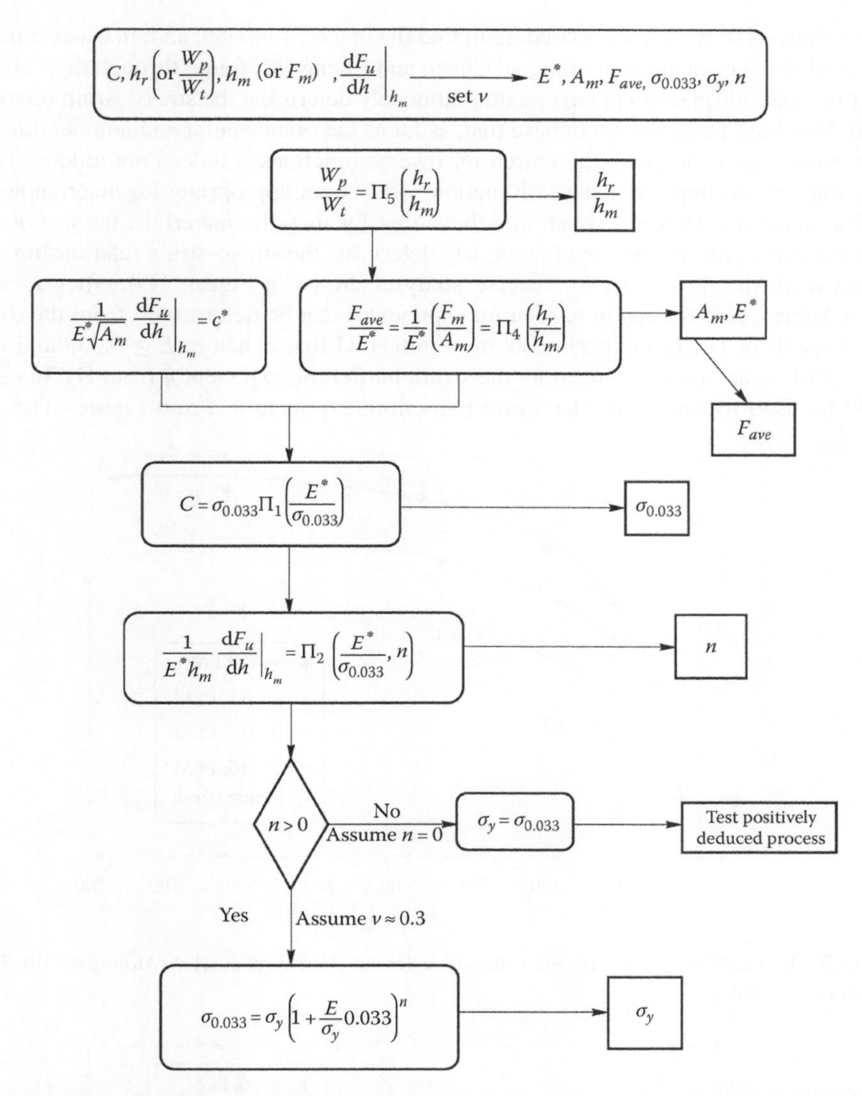

FIGURE 11.4 Reverse analysis algorithm. (From Dao, M. et al., *Acta Mater.*, 49(19), 3899, 2001; Taken from Flow Chart 2f.)

Dao et al. [9] performed both uni-axial compression and indentation experiments on two types of aluminum, 7075-T651 and 6061-T6411. In the case of the reverse analysis, the experimental *F–h* curve is used to deduce the elasto-plastic parameters of the compressed material. The results on E^* were more accurate than those obtained by Oliver and Parr [6] and Doerner and Nix [11]. Although in certain situations the mechanical property parameters obtained from their individual *F–h* curves sustain considerable errors, the results, averaged over a large number of indentation experiments, have been found to be reliable.

The results obtained from the forward analysis have been found to be consistent with experimental results. This implies that, as long as the stress–strain relationship of a material is known, there is no need to perform the indentation experiment, and the hardness of the material and its load–displacement curve can be determined directly by calculation. However, this does not mean that the reverse analysis procedure is valid (in which the stress–strain relationship can be definitely determined as long as we have the knowledge of the load–displacement curve). The reason is that the reverse analysis is in reality a procedure to find the inverse function; and, in usual situations, the inverse function to be sought is not unique. Cheng and Cheng [12] found through their analysis that the use of the load–displacement curve cannot uniquely determine the stress–strain relationship of a material. However, Dao et al. [9] believe that, as far as macromolecular and nonmetallic materials such as ceramics are concerned, the search for inverse functions is indeed not unique. The reason is mainly that the assumption made in Equation 11.12 is not appropriate for macromolecular and nonmetallic materials. Hence, Dao et al. believe that for metallic materials, the use of the load–displacement curve can still be used to uniquely determine the stress–strain relationship.

As seen from the flow chart for reverse analysis shown in Figure 11.4, when E^* and $\sigma_{0.033}$ have been determined, the strain hardening exponent n can be determined from the dimensionless functions Π_2 or Π_3. But experiments have indicated that, when $\sigma_y/E \geq 0.033$ and $n > 0.3$, it is very difficult to accurately determine the strain hardening exponent n from Π_2. In such cases, Π_3 should be used to determine the strain hardening exponent n. From Figures 11.5 and 11.6,

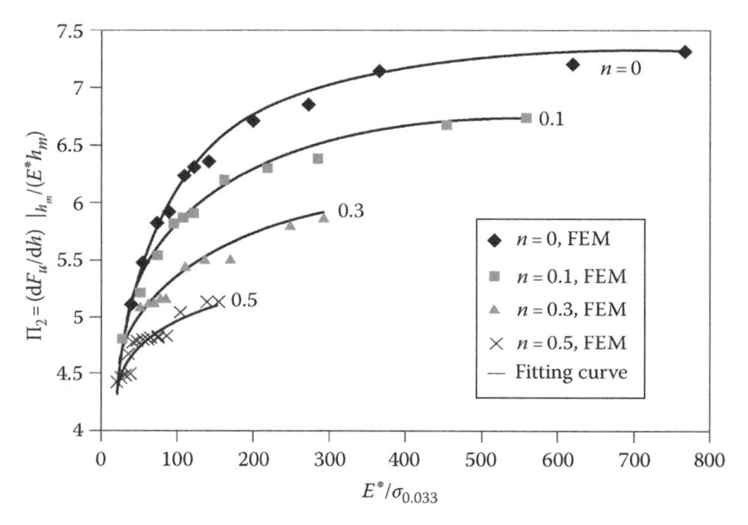

FIGURE 11.5 Π_2 vs. $E^*/\sigma_{0.033}$ for different values of n. (From Dao, M. et al., *Acta Mater.*, 49(19), 3899, 2001; Taken from Figure 10.)

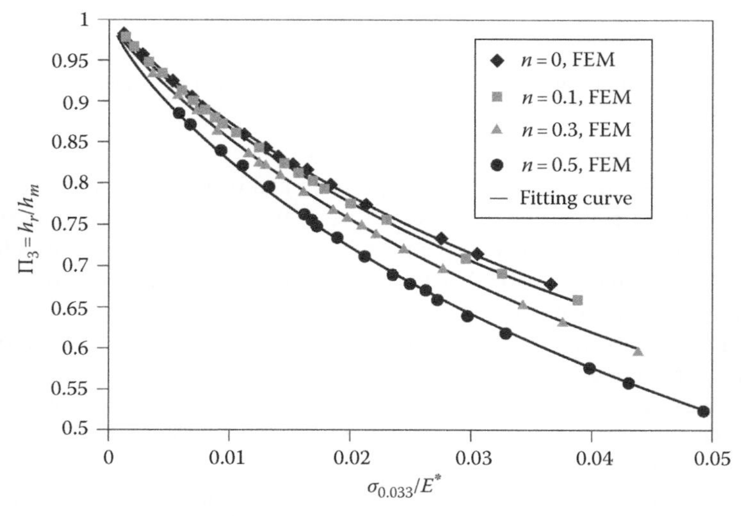

FIGURE 11.6 Π_3 vs. $\sigma_{0.033}/E^*$ for different values of n. (From Dao, M. et al., *Acta Mater.*, 49(19), 3899, 2001; Taken from Figure 10.)

it can be seen that, when $\sigma_y/E \geq 0.033$ and $n > 0.3$, the curve for $n = 0.5$ intersects with three other curves so that it becomes difficult to determine n accurately. Either one of the dimensionless functions Π_2 or Π_3 can be used to determine the value of n. The reason is that Π_2 and Π_3 are not independent of each other, but are related. Therefore, we suggest using Π_2 when $E^*/\sigma_{0.033}$ is comparatively large, and to use Π_3 when $E^*/\sigma_{0.033}$ is comparatively small, in order to determine the strain hardening exponent n.

Furthermore, Liao et al. [8] have included the effects of the substrate in their consideration, and have obtained an analytic expression for the stress–strain relationship for thin films from the F–h curve.

11.3 RESIDUAL STRESS OF THIN FILMS

11.3.1 Sources of Residual Stress

The existence of residual stress in thin films affects their quality and performance. For instance, tensile residual stress may result in stress intensification that causes the film or the interface to crack, to promote the initiation of cracks, or to spread and stretch minute cracks. On the other hand, compressive residual stress can reduce the concentration of stress internally, and enhance the anti-fatigue properties of materials [13,14]. However, excessive compressive stress can cause bubble or layer formation in thin films [15]. As far as all kinds of thin film constituted, electronic and opto-electronic components are concerned, the issue of stress is very important because it is directly related to their rate of production, stability, and reliability. Therefore, the study of residual stress in thin films is indispensable.

It is generally recognized that residual stress in thin films can be divided into two types: thermal stress and intrinsic stress. Thermal stress is caused by the difference in thermal expansion coefficients between the film and the substrate; and, therefore, is also referred to as thermal mismatch stress. The thermal expansion coefficient is the inherent property of materials, and the difference between the thermal expansion coefficients of any two materials can be considerable. Such differences are the main cause for the production of residual stress when the thin film epitaxial growth on the substrate [16]. The elastic strain corresponding to thermal stress is

$$\varepsilon_{th} = \int\left[\alpha_f(T)-\alpha_s(T)\right]\mathrm{d}T, \tag{11.32}$$

where α_f and α_s are, respectively, the thermal expansion coefficients of the film and the substrate. From Hooke's law, we obtain

$$\sigma_{th} = \frac{E_f}{1-\nu_f}\varepsilon_{th} = M_f\varepsilon_{th}, \tag{11.33}$$

where

$M_f = E_f/(1 - \nu_f)$ is the biaxial elastic modulus of the thin film

E_f and ν_f represent, respectively, the elastic modulus and the Poisson's ratio of the thin film

The cause of intrinsic stress, also referred to as eigen stress, is more complicated. At the present time, there are a number of different viewpoints. According to one viewpoint, the intrinsic stress is caused by a mismatch between crystal lattices. The crystalline mismatch produces edge dislocations, and these dislocations, in turn, give rise to a corresponding elastic stress field in their surroundings. The degree of mismatch $(a_s - a_f)/a_f$, a_f, and a_s, being the crystal constants of film and substrate, characterizes the extent of the crystalline mismatch.

In the film-substrate film/substrate material system, σ_{in}—the intrinsic stress produced in the film by the crystal constant mismatch—is obtained from Hoffman's relaxation model [17] as follows:

$$\sigma_{in} = \left(\frac{E_f}{1-\nu_f}\right)\frac{x-a}{a} = \left(\frac{E_f}{1-\nu_f}\right)\frac{\Delta}{L_g}, \tag{11.34}$$

where

a is the crystal constant of the film in the absence of residual stress
x–a is the variation in crystal constant
Δ is the grain boundary relaxation distance
L_g is the final size of the crystalline grain

In general, the intrinsic stress in a thin film is greater than its thermal stress. Cheng et al. [18] point out that, according to the conventional understanding of intrinsic stress, their order of magnitude is smaller than 1 GPa. However, experiments have shown that their levels can exceed several GPa. Recently, using the improved version of TFD (Thomas–Fermi–Dirac) theory, and the methods of the theory of elastic mechanics, Cheng et al. analyzed the production mechanism of intrinsic stress in thin films. They found that the intrinsic stress originated from the difference in electronic densities between the thin film and the substrate, and is an inevitable consequence of the continuity of electron density across the boundary. In order to make electron densities bordering both sides of the interface between film and substrate reach a certain value, a higher stress could occur at the interface. According to Cheng's theory [18], the intrinsic stress is given by

$$\sigma_{in} \cong \left(\frac{dp}{dn}\right)_f (n_{s0} - n_{f0}), \tag{11.35}$$

where

$(dp/dn)_f$ is the derivative of the film's internal pressure with respect to the electron density
n_{s0} and n_{f0} are, respectively, the electron densities at the surfaces of the film and the substrate

For a given film material, $(dp/dn)_f$ has a fixed value, while the film intrinsic stress is directly proportional to the difference between the electron densities at the surfaces of film and substrate. This theory illustrates a way of controlling a film's residual stress: through an appropriate doping to reduce the difference between electron densities at the surfaces of film and substrate, the residue stress can be reduced.

11.3.2 Measurement of Residual Stress in Thin Films [19]

11.3.2.1 Deflection Curvature Method

11.3.2.1.1 The Stoney Formula for Measuring Residual Stress [20]

Under the influence of residual stress in a thin film, its substrate will deflect. No matter how minute or tiny the deformation, the radius of curvature of the deflection can be determined by using a laser interferometer or a surface contour graph. The deflection of the substrate reflects the magnitude of residual stress in the film. The relationship between the two is given by Stoney as follows:

$$\sigma_f = \frac{E_s}{1-\nu_s}\frac{t_s^2}{6rt_f}, \tag{11.36}$$

where

σ_f is the residual stress in the film
t_f and t_s are, respectively, the thicknesses of the film and its substrate
r is the radius of curvature of the deflected film-substrate composite
E_s and ν_s are, respectively, the elastic modulus and Poisson's ratio of the substrate

While the Stoney formula has been widely used in determining residual stress in thin films, caution should be taken in ascertaining its scope of validity. The validity of Stoney's formula is based upon the following hypotheses:

- $t_f \ll t_s$; that is, the film thickness should be much smaller than the thickness of the substrate. This requirement is usually met because the difference between their thicknesses is usually very large.
- $E_f \approx E_s$; that is, the two elastic moduli should approximately be the same.
- The substrate should be homogeneous, isotropic, and linearly elastic, and should initially be free of any deflection.
- The film material should be isotropic, and its residual stress should be biaxial.
- The residual stress in the film is uniformly distributed in the direction of its thickness.
- The deflections should be small, and the film periphery has only a negligible effect on stress.

In reality, many of these hypothesized conditions cannot be completely satisfied. Hence, Stoney's formula calls for appropriate modifications.

11.3.2.1.2 The Multilayer Case [20]

In the case of multilayers—regardless of the number of thin films—if the total thickness of all films is still very small, the first of the aforementioned Stoney conditions can still be satisfied. Every time a layer of film is deposited, the residual stress in that layer alone gives rise to a bending action on substrate, which causes a change in the substrate curvature. The bending action of each and every layer accumulates in accordance with the principle of linear superposition. Hence, for an n-layer thin film, the Stoney formula can be modified to take the following form:

$$\frac{1}{r_1}+\frac{1}{r_2}+\cdots+\frac{1}{r_n}=\frac{1-\nu_s}{E_s}\frac{6}{t_s^2}(\sigma_{f1}t_{f1}+\sigma_{f2}t_{f2}+\cdots+\sigma_{fn}t_{fn}), \tag{11.37}$$

where the subscripts 1, 2, …, n of r and σ_f, respectively, refer successively to each and every layer of film, and all the σ_f denote residual stresses.

11.3.2.1.3 Thin-Film Thickness Comparable to Substrate Thickness [21]

When t_f (the film thickness) and t_s (the substrate thickness) are comparable, the nonvanishing residual stresses are $\sigma_r(r, z)$ and $-\sigma_\theta(r, z)$. The corresponding elastic strain energy density $U(r, z)$ for both the film and substrate is given by

$$U(r,z)=\frac{E}{2(1-\nu^2)}\left[\varepsilon_r^2(r,z)+\varepsilon_\theta^2(r,z)+2\nu\varepsilon_r(r,z)\varepsilon_\theta(r,z)\right], \tag{11.38}$$

where ε_r and ε_θ are the magnitudes of strain in radial and angular directions (Figure 11.7).

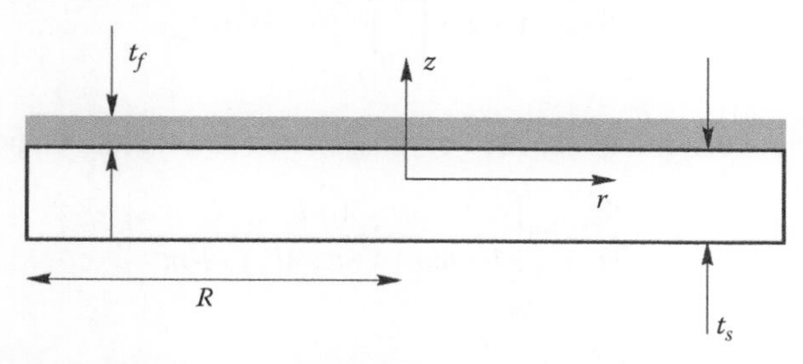

FIGURE 11.7 Cylindrical coordinate system. (From Qian, J. et al., *J. Mech. Strength*, 23(4), 393, 2001; Taken from Figure 2.)

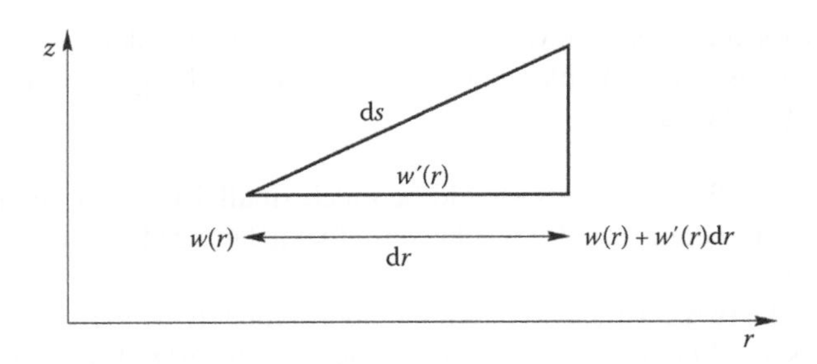

FIGURE 11.8 Strain caused by rotation of midplane of substrate in cylindrical coordinates. (From Qian, J. et al., *J. Mech. Strength*, 23(4), 393, 2001; Taken from Figure 3.)

As shown in Figure 11.8, for film, ε_r and ε_θ are given by

$$\varepsilon_r = u'(r) - zw''(r) + \varepsilon_m, \tag{11.39a}$$

$$\varepsilon_\theta = \frac{u(r)}{r} - \frac{zw'(r)}{r} + \varepsilon_m, \tag{11.39b}$$

where

- ε_m is the mismatch strain indicating the degree of mismatch
- $u(r)$ and $w(r)$ are, respectively, the displacements of the midplane of the substrate in the radial direction and in the direction of the z-axis
- $u'(r)$ and $w'(r)$ are the first derivatives of $u(r)$
- $w''(r)$ is the second derivative

The quantity $w(r)$ is also called deflection.

For small deflections,

$$u(r) = \varepsilon_0 r + \varepsilon_m, \tag{11.40a}$$

$$w(r) = \frac{1}{2}\kappa r^2, \tag{11.40b}$$

where

- ε_0 is the strain in the midplane of substrate
- κ indicates the curvature of the substrate

From Equations 11.38 through 11.40, we obtain the total strain energy V as a function of ε_0 and κ as follows:

$$V(\varepsilon_0, \kappa) = 2\pi \int_0^R \int_{-t_s/2}^{t_f + t_s/2} U(r,z) r\,dr\,dz. \tag{11.41}$$

Using the equilibrium conditions satisfied by $V(\varepsilon_0,\kappa)$, that is, $\partial V/\partial\varepsilon_0 = 0$ and $\partial V/\partial\kappa = 0$, we find

$$\kappa = \frac{6\varepsilon_m}{t_s} lm \left[\frac{1 + l}{1 + lm(4 + 6l + 4l^2) + l^4 m^2} \right], \tag{11.42}$$

where

- $l = t_f/t_s$ is the ratio between the thicknesses of film and the substrate
- $m = M_f/M_s$ is the ratio between the elastic moduli of film and substrate

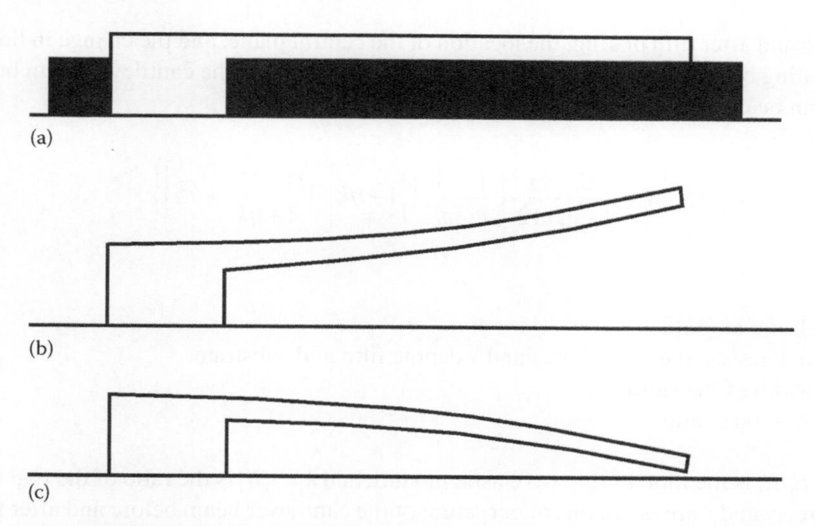

FIGURE 11.9 A microfilm with graded stress distribution in its thickness direction. (a) Microfilm structure before relief from residual stress. (b) Microfilm bending upward after relief from residual stress. (c) Microfilm bending downward after relief from residual stress. (From Qian, J. et al., *J. Mech. Strength*, 23(4), 393, 2001; Taken from Figure 4.)

When $t_f << t_s (l \rightarrow 0)$, Equation 11.42 reduces to the Stoney formula. However, when the thicknesses of film and substrate become comparable, such that $m = 1$ and $l = 0.1$, the Stoney formula in Equation 11.36 may give rise to an error of about 30%.

11.3.2.1.4 Graded Film Stress in the First Approximation

In our discussion of the Stoney formula, it has been assumed that the film is very thin, and that the residual stress in the direction of its thickness is uniformly distributed. In reality, the residual stress in thin films varies in that direction, as illustrated in Figure 11.9.

After the shaded portion of a thin layer of the film, shown in Figure 11.9a, was etched by an etching solution, the remaining portion of the film deflects as shown in Figure 11.9b or 11.9c. If the residual stress were uniformly distributed, the remaining film would only stretch or contract, but would not show deformation as in Figure 11.9.

In general, the distribution of total axial stress σ_{total} across the thickness of a thin film can be expressed as a polynomial:

$$\sigma_{total} = \sum_{k=0}^{\infty} \sigma_k \left(\frac{z}{t/2} \right)^k \tag{11.43}$$

in which z is the coordinate in the thickness direction, and t is the film thickness. In the usual first-order approximation, we have

$$\sigma = \sigma_0 \pm \frac{z}{t/2} \sigma_1, \tag{11.44}$$

where the " + " sign is for tension and the "–" sign is for compression.

11.3.2.2 Cantilever Beam Method [19,22]

The existence of stress in a film causes the film/substrate system to bend and deform to a certain extent. According to the theory of elastic mechanics, from the radii of curvature of a cantilever beam before and after film plating, the location of the central plane, and the change in flexural rigidity, the resulting bending moment at any arbitrary cross section of the cantilever beam brought forth by stress, can be determined as follows:

$$M = \frac{E_s}{1-\nu_s}\frac{bt_s{}^3}{12}\left(\frac{1}{r}-\frac{1}{r_0}\right)\left\{1+\eta k\left[3\frac{(1+k)^2}{1+\eta k}+k^2\right]\right\}, \quad (11.45)$$

where

E's are elastic moduli
t's are thicknesses, the subscripts *f* and *s* denote film and substrate
b is the width of the substrate
ν is the Poisson's ratio

Hence, $\eta = E_f/E_s$ is the ratio of the two elastic moduli, and $k = t_f/t_s$ is the ratio of the two thicknesses. Furthermore, r_0 and r are the radius of curvature of the cantilever beam before and after film plating. The aforementioned equation is derived under the condition that both the film and the substrate have the same Poisson's ratio.

Under normal situations when $t_f << t_s$, the aforementioned equation can be simplified to

$$M \approx \frac{E_s}{1-\nu_s}\frac{bt_s^3}{12}\left(\frac{1}{r}-\frac{1}{r_0}\right)(1+3\eta k). \quad (11.46)$$

Moreover, if the average stress in the film is σ when its thickness is t_f, the moment is

$$M = \frac{b\sigma}{2}t_f(t_s+t_f). \quad (11.47)$$

Because $t_f << t_s$, the aforementioned equation becomes

$$M \approx \frac{b\sigma}{2}t_f t_s. \quad (11.48)$$

From Equations 11.46 and 11.48, we obtain the expression for average stress:

$$\sigma \approx \frac{E_s t_s^2}{6(1-\nu_s)t_f}\left(1+\frac{3E_f}{E_s}\frac{t_f}{t_s}\right)\left(\frac{1}{r}-\frac{1}{r_0}\right). \quad (11.49)$$

Under the condition $t_f << t_s$, Equation 11.49 can be simplified as

$$\sigma \approx \frac{E_s t_s^2}{6(1-\nu_s)t_f}\left(\frac{1}{r}-\frac{1}{r_0}\right). \quad (11.50)$$

Before film plating, the substrate is flat, and $r_0 = +\infty$. After plating, $r \approx L^2/(2\delta)$, where L is the length of the cantilever beam and δ is the deflection of the free end of the plated cantilever beam. Hence, Equation 11.50 becomes

$$\sigma \approx \frac{E_s t_s^2}{3(1-\nu_s)L^2}\frac{\delta}{t_f}. \quad (11.51)$$

This equation is the Stoney formula. Berry revised Stoney's formula by substituting the plane strain modulus of the substrate $E_s/(1 - \nu_s^2)$ for the biaxial elastic modulus $E_s/(1 - \nu_s)$. In so doing, he obtains from Equation 11.51 the following

$$\sigma \approx \frac{E_s t_s^2}{3(1-v_s^2)L^2} \frac{\delta}{t_f}. \tag{11.52}$$

Consequently, it can be seen that the residual stress in a thin film can be found without knowing its elastic modulus.

11.3.2.3 Indentation Method

By performing identical indentation experiments on films with residual stress as well as without residual stress, and assuming [23,24] that residual stress in thin films does not alter their hardness, load–displacement curves $F - h$ and $F_0 - h_0$, as shown in Figure 11.10, can be obtained. The subscript 0 denotes the case of no residual stress. The two F versus h curves corresponds to two possible scenarios, with compressive residual stress and with tensile residual stress. By comparing the curve that has residual stress with the one without residual stress, the sign of the residual stress can be determined [25].

The actual indentation contact area under the maximum load F_{max} is given by

$$A = \left\{ \frac{\mathrm{d}F}{\mathrm{d}h} \frac{1}{C_u E^*} \right\}^2, \tag{11.53}$$

where

$E^* = \left((1-v_f^2/E_f)+(1-v_i^2/E_i)\right)^{-1}$ is the reduced modulus in Equation 11.5

E_f and ν_f are, respectively, the elastic modulus and Poisson's ratio of the film

E_i and ν_i are, respectively, the elastic modulus and Poisson's ratio of the indenter

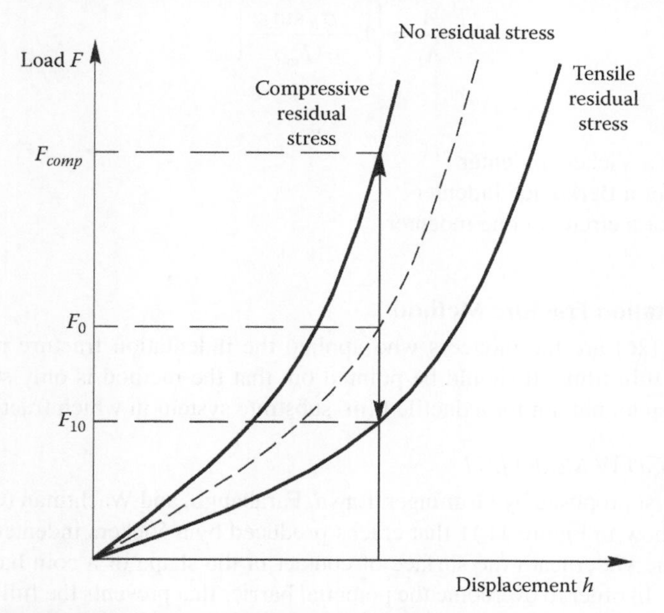

FIGURE 11.10 Using relative position of curve to determine the sign of residual stress.

For a Vickers indenter, $C_u = 1.142$, while for a Berkovich indenter $C_u = 1.167$. The average load per unit contact area is given by

$$F_{ave} = \frac{F_{max}}{A}. \tag{11.54}$$

In terms of F_0 for the load, and h_0 for the penetration depth when there is no residual stress, the effective contact area is given by

$$A_0 = \left\{ \frac{\mathrm{d}F_0}{\mathrm{d}h_0} \cdot \frac{1}{C_u E^*} \right\}^2, \tag{11.55}$$

where $(\mathrm{d}F_0/\mathrm{d}h)_0$ denotes the derivative of F at a point on the no-residual-stress curve (the curve in the middle) in Figure 11.10.

Hence,

$$\frac{A}{A_0} = \left\{ \frac{\mathrm{d}F}{\mathrm{d}h} \right\}^2 \left\{ \frac{\mathrm{d}F_0}{\mathrm{d}h_0} \right\}^2. \tag{11.56}$$

By substituting the expressions A/A_0 and F_{ave} into the following equation,

$$\frac{A}{A_0} = \left\{ 1 + \frac{\sigma_R}{F_{ave}} \right\}^{-1}, \tag{11.57}$$

the residual tensile stress σ_R can be computed. For residual compressive stress, the following equation with a minus sign is used:

$$\frac{A}{A_0} = \left\{ 1 - \frac{\sigma_R \sin\alpha}{F_{ave}} \right\}^{-1}, \tag{11.58}$$

where

α is the 22° for a Vickers indenter
α is the 24.7° for a Berkovich indenter
α is the 19.7° for a circular cone indenter

11.3.2.4 Indentation Fracture Method

Lawn and Fuller [26] are the pioneers who applied the indentation fracture method to measure residual stress in thin films. It should be pointed out that the method is only suitable for a brittle film–substrate system, but not for a ductile film–substrate system in which fractures seldom occur.

11.3.2.4.1 The GLFW Model [27]

This model was first proposed by Gruninger, Lawn, Farabaugh, and Wachtman (GLFW) in 1987. As an example, we show in Figure 11.11 that cracks produced by a Vickers indenter [28], at first, form a deformation zone underneath the surface of contact in the shape of a coin having a notch at the surface of contact. In order to overcome the potential barrier that prevents the full development of the cracks, the indenter must exert a critical load on the load recipient so that the cracks can fully develop

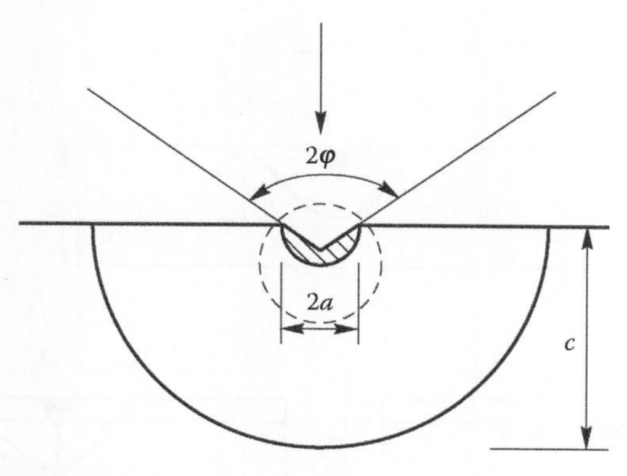

FIGURE 11.11 Vickers indentation model.

to such a depth that they spontaneously burst through the sample in radial directions, within a semi-coin-shaped region in the middle. If these cracks reach equilibrium after the completion of the loading and unloading cycle, the critical crack length at which the material ruptures can be determined by using the equation for the stress intensity factor K at the tips of cracks. In the case of a Vickers indenter, this is given by

$$K_r = \frac{\chi F}{c^{3/2}}. \tag{11.59a}$$

Under equilibrium conditions, we have

$$\frac{F}{{c_0}^{3/2}} = \frac{K_c'}{\chi} = \text{const.} \tag{11.59b}$$

In the aforementioned equations
- F is the load
- K_c' is the stress intensity factor at equilibrium
- c is the crack depth
- c_0 is the penetration depth in equilibrium condition
- χ is a dimensionless field intensity parameter

Equation 11.59b tells us that $F/{c_0}^{3/2}$ is independent of crack length.

Let us now consider the situation pertaining to Figure 11.12b, in which the substrate is plated after it has been subjected to the indentation load shown in Figure 11.12a. During the plating process, the development of residual stress in the substrate is unavoidable. Let the residual stress developed in the substrate be σ_s. This residual stress will contribute to the spread of the fracture. Let the resulting stress intensity factor attributed to σ_s be K_s. Then, according to the principle of superposition of stress fields, the stress intensity factor of the fracture system in the substrate under equilibrium conditions, or the fracture ductility K_c of the substrate material, can be expressed as

$$K = K_r + K_s = K_c. \tag{11.60}$$

In accordance with the GLFW model [27], the stress intensity factor resulting from σ_s satisfies the equation

$$K_s = \gamma \sigma_s c^{1/2}, \tag{11.61}$$

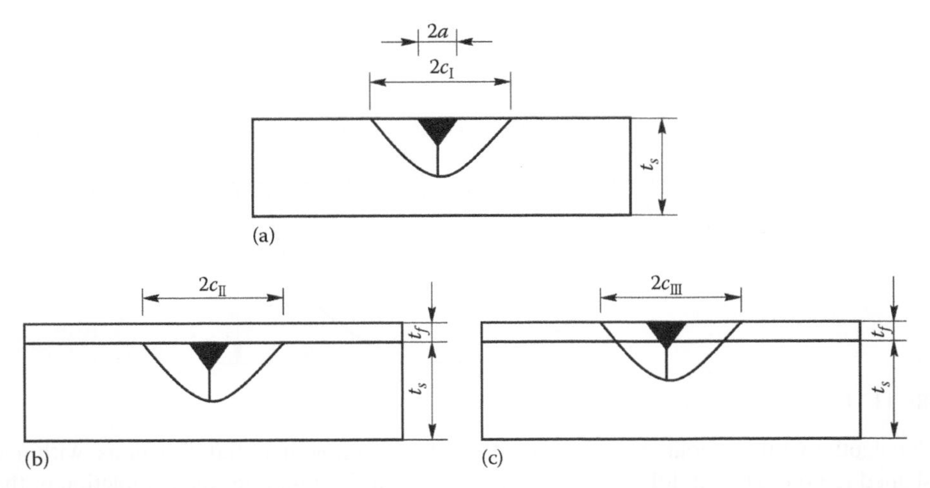

FIGURE 11.12 Three modes of Vickers indentation. (a) No film. (b) Plating after substrate is indented. (c) Indentation after plating. (Adapted from Gruninger, M.F. et al., *J. Am. Ceram. Soc.*, 70(5), 344, 1987; Figure 1.)

where γ is the one-dimensional fracture geometry factor with a value close to 1.0. Thus, Equation 11.60 becomes

$$\frac{F}{c_{\mathrm{II}}^{3/2}} = \frac{F}{c_0^{3/2}}\left[1-\left(\frac{\gamma\sigma_s}{K_c}\right)c_{\mathrm{II}}^{3/2}\right], \tag{11.62}$$

where c_{II}, as shown in Figure 11.12b, is the crack length in the substrate produced by both the indenter and the residual stresses in the film. From the aforementioned equation, it can be seen that as the size of the crack increases, the change in the value of F/c_s is related not only to the crack length c_{II}, but also to the residual stress in the substrate σ_s.

Now let us analyze what is depicted in Figure 11.12c. The figure shows the formation of cracks when a load is applied after the substrate has been plated. In this case, we have to consider not only the residual stress in the substrate developed during plating, but also the effect of residual stress in the film σ_s on the spread of cracks in the substrate. Laws and Fuller have already derived the equation pertaining to the stress intensity factor K_f [26] for the film as follows:

$$K_f = 2\gamma\sigma_f\sqrt{t_f} \tag{11.63}$$

where

t_f is the film thickness

γ is the same geometry factor as in Equation 11.61

Based upon the principle of superposition, the stress intensity factor of the fracture system in a substrate under equilibrium conditions is given by

$$K = K_r + K_s + K_f. \tag{11.64}$$

Based upon Equations 11.59 through 11.64, when the cracks have reached the equilibrium condition ($K = K_c$), the situation shown in Figure 11.12c can be represented by

$$\frac{F}{c_{\mathrm{III}}^{3/2}} = \frac{K_c - K_s - K_f}{\chi}, \tag{11.65a}$$

$$\frac{F}{c_{\mathrm{III}}^{3/2}} = \frac{F}{c_{\mathrm{I}}^{3/2}}\left[\frac{1-\gamma\sigma_s c_{\mathrm{III}}^{1/2}}{K_c} - \frac{2\gamma\sigma_f\sqrt{t_f}}{K_c}\right]. \tag{11.65b}$$

In Equation 11.65a, c_{III}, shown in Figure 11.12c, is the length of the crack in the substrate when a load is applied by the indenter on the film plated on it. Because the plated film is so thin, the entire length of the crack c_{III} can practically be seen through the film. According to Equation 11.65, as the crack length changes, the change in the magnitude of $F/c_{\mathrm{III}}^{3/2}$ is related not only to the crack length c_{II}, but also to the residual stresses σ_s and σ_f in the substrate and the film. It can also be seen from Equations 11.59, 11.62, and 11.65 that to determine the residual stresses in the film and substrate, all we need is the magnitude of $F(c)$ in the three situations of loading and crack length, depicted in Figure 11.12a through c.

Because the situation depicted in Figure 11.12b is hard to realize, we will make a simplifying assumption [26] by setting

$$\sigma_s = -\frac{t_f}{t_s}\sigma_f, \tag{11.66}$$

where t_f and t_s are, respectively, the thicknesses of film and substrate. Consequently, Equation 11.65b becomes

$$\frac{F}{c_{\mathrm{III}}^{3/2}} = \frac{F}{c_{\mathrm{I}}^{3/2}}\left[1+\frac{\gamma t_f\sigma_f}{K_c}\left(\frac{c_{\mathrm{III}}^{1/2}}{t_s} - \frac{2}{\sqrt{t_f}}\right)\right]. \tag{11.67}$$

Hence, by merely considering the situations depicted in Figure 11.12a and c, the residual stress in a film can be determined from the crack lengths under certain loading conditions and the corresponding material parameters, by using Equations 11.59 and 11.60.

11.3.2.4.2 ZCF Model

The ZCF Model was proposed by Zhang et al. [28] in 1999. Under the combined influences of the external load exerted through an indenter and the residual stress in the film, cracks can form in the film and the substrate. If the cracks resemble half-penny cracks shown in Figure 11.13, then the stress intensity factor due to the residual stress σ_f in the film at the point (r, θ) is given by

$$K_r(\theta) = \frac{2\sigma_f}{(\pi c)^{1/2}}\int_0^{t_f} dy \int_{-(c^2-y^2)^{1/2}}^{(c^2-y^2)^{1/2}} dx \frac{(c^2-x^2-y^2)^{1/2}}{c^2-2cx\cos\theta+2cy\sin\theta+x^2+y^2}, \tag{11.68}$$

where the residual stress σ_f is assumed to be uniform throughout the film.

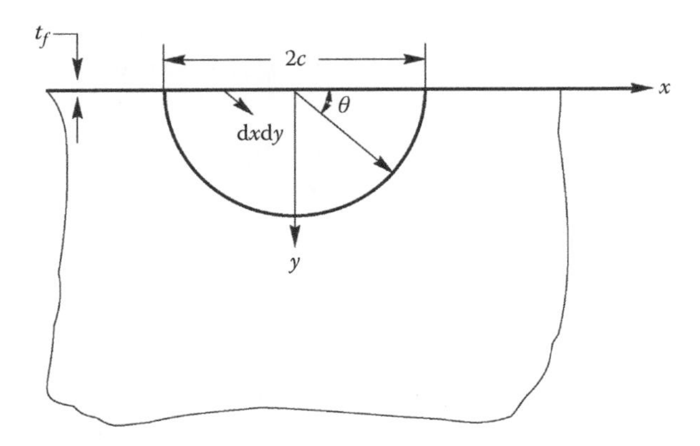

FIGURE 11.13 A half-penny crack.

From the calculation done by Zhang et al. [28] it was discovered that $K_r(\theta)$ is strongly dependent upon the angle θ. Its value is maximum at $\theta = 0$; that is, at the surface, and decreases rapidly as θ increases. Such a crack pattern, on one hand, explains why, in indentation experiments, cracks always form at the exterior surface before they gradually extend into the substrate. On the other hand, it also indicates that the existence of residual stress adversely impacts the usefulness of thin films. If we set $\theta = 0$, that is to say, cracks appear only on the surface, then Equation 11.68 reduces to Equation 11.63: $K_f = 2\,\gamma\sigma_f\,t_f^{1/2}$. In fact, whether the cracks are on the surface or within the substrate, the equilibrium condition described by Equation 11.60 is always valid. Thus, the ZCF model, unlike the GLFW model, which is valid only for $\theta = 0$, can be used to find K_r averaged over the angle θ. In other words,

$$\bar{K}_r = \frac{2}{\pi}\int_0^{\pi/2}\frac{2\sigma_f \mathrm{d}\theta}{(\pi c)^{1/2}}\int_0^{tf}\mathrm{d}y\int_{-\sqrt{c^2-y^2}}^{\sqrt{c^2-y^2}}\frac{\sqrt{c^2-x^2-y^2}}{c^2-2cx\cos\theta+x^2+y^2}\mathrm{d}x. \tag{11.69}$$

Because the film thickness t_f is very much smaller than the crack length, and $y \le t_f$, by setting $r^2 = x^2 + y^2 \approx x^2$, $x^2 - y^2 \approx x^2$, and $c^2 - y^2 \approx c^2$ in Equation 11.69, the integrand in (11.69) becomes independent of y. Since $\int \mathrm{d}y \approx t_f$, Equation 11.69 becomes

$$\bar{K}_r = \frac{4\sigma_f t_f}{\pi(\pi c)^{1/2}}\int_0^{\pi/2}\int_{-c}^{c}\frac{(c^2-x^2)}{c^2-2cx\cos\theta+x^2}\mathrm{d}x\mathrm{d}\theta = 3.545\frac{\sigma_f t_f}{c^{1/2}} = \beta\frac{\sigma_f t_f}{c^{1/2}}. \tag{11.70}$$

Under such a circumstance, the total stress intensity factor is given by

$$K = K_F + \bar{K}_r, \tag{11.71}$$

where K_F is the stress intensity factor due to the indentation load. When the equilibrium condition $K = K_c$ for the cracks is reached, Equations 11.59a, 11.70, and 11.71 give

$$\frac{F}{c^{3/2}} = \frac{K_c}{\chi} + \frac{\eta}{c^{1/2}}, \tag{11.72}$$

in which $\eta = -\delta\sigma_f\,t_f$, $\delta = \beta/\chi$. Therefore, knowing the crack length corresponding to a certain loading, and the relevant parameters, we can obtain the residual stress in the film from Equation 11.72.

In the following sections, we further discuss the GLFW and the ZCF models. In Equation 11.67 of the simplified GLFW model, the relationship between load and crack length is in the form of $F/c^{3/2} = ac^{1/2} + b$. This is the equation of a straight line of the form $y = mx + b$, where $y = F/c^{3/2}$, $x = m = \eta$ is the slope of the straight line, and $b = K_c/\chi$ is the y-intercept. However, according to Equation 11.72 of the ZCF model, the relationship between load and crack length is in the form $F/c^{3/2} = a'/c^{1/2} + b'$. It has been seen in analyzing Equation 11.68 that, if θ is set to zero—that is, if the cracks are surface cracks—Equation 11.72 of the ZCF model becomes consistent with Equation 11.67 of the GLFW model. Zhou et al. have used the aforementioned models to analyze the residual stress in PZT thin films prepared by pulsed laser deposition [29].

11.4 INTERFACE FRACTURE TOUGHNESS OF THIN FILMS

11.4.1 Types of Interface Bonding between Film and Substrate [30]

Because the interface behavior of the film layer directly determines the layer's bonding properties and its effectiveness in applications, a thorough study of all types of film bonding processes and mechanisms is a prerequisite to optimizing film ingredients, their structural design and manufacturing, and their superior quality. The bonding between film and substrate can generally be classified into the following six types:

1. *Metallurgical bonding*: A metallurgical bond is formed between film and substrate when molten film material is deposited onto a semi-molten solid substrate, which then solidifies outward along the surface, and crystallizes. It is, in essence, a metallic bond. The bonding has a high degree of strength and can withstand considerable external force or load. It does not peel off easily during application processes.
2. *Diffusing bonding*: When two solid materials in direct contact are placed in a vacuum and are heated and pressurized, interface diffusion and reaction between the two gives rise to a diffusing bond. Its salient features are the existence of graded chemical change between film and substrate, and the formation of an atomic-level compound or alloy. Boundaries formed by ion-injection techniques can be considered as a special type of diffusion bond, or sometimes, a quasi-diffusion bond, because the ions are forced into the substrate by a high-energy particle beam.
3. *Epitaxial growth bonding:* When technical conditions are appropriate, a new layer of a perfect single crystal grows in the direction of the crystallographic axis of the single crystalline substrate. This is epitaxial growth. The bond formed is called an epitaxial bond. There are two types of epitaxial processes: one is vapor phase epitaxy, such as chemical gas deposition, and the other is liquid phase epitaxy, such as electroplating. In practice, the epitaxy growth level is contingent upon the crystal type, the crystal constant of the substrate, and the epitaxial layer. Take electroplating as an example. When the two metals are of the same kind, or have almost the same lattice constant, there can be epitaxy up to a thickness of 0.1–400 nm. Since the epitaxy layer between film and substrate has the same structural orientation, the bonding strength should in principle be better. However, the actual bonding strength should depend on the type of bonding—molecular, covalent, ionic, metallic, and so on—between the single crystal layer and the substrate.
4. *Chemical bonding*: When film and substrate undergo a chemical reaction and form a stable chemical compound, a chemical bond is formed. For example, the surface of a titanium alloy is deposited with a layer of gaseous TiN and TiC. Their nitrogen and carbon atoms react with the titanium atoms in the substrate to form Ti–N and Ti–C chemical bonds. The advantage of a chemical bond is its comparatively high strength.

The disadvantage is its relatively poor ductility; such that, under the impact of load or heat, fractures easily occur because of its brittle nature.

5. *Molecular bonding*: A molecular bond is formed as a result of Van der Waal forces between the surfaces of film and substrate. The distinction of this kind of bonding is the absence of diffusion or chemical reactions. Examples of molecular bonding are bonds formed with a substrate by some gaseous deposits made by physical processes, and by organic cohered films in spreading techniques.
6. *Mechanical bonding*: A mechanical bond is formed by the mutual embedding of materials in both the film and substrate by mechanical means. The main methods for forming mechanical bonds in surface engineering are thermal spraying and cladding.

The above-mentioned text summarizes all of the typical bonding situations. In reality, the interface bonding is often a combination of the mechanisms enumerated earlier.

11.4.2 Measurement of Fracture Toughness at Interface

11.4.2.1 Tape Method [31]

The Tape Method proposed by Strong in 1935 is a method for measuring the film bonding strength at the interface. It was used to test the bonding between aluminum film and a glass substrate. The method is as follows: First, place adhesive tape onto the surface of a film. Then peel it off, determine the force exerted, and observe the residual stresses in the substrate and the thin film. From this procedure, the strength of the interface adhesive bonding between film and substrate can be determined. The tape method can only provide qualitative conclusions and is useless when the interface bonding strength exceeds the tape strength.

11.4.2.2 Stretching Method [32]

The method utilizes a pulling force perpendicular to the boundary between film and substrate to strip off the film, and relies on the stripping force to determine the adhesive strength. Specifically, a smooth, flat circular plate is glued to the film surface. Then a pulling force is applied in a perpendicular direction to strip the film from the substrate, and the stripping force is thereby determined.

11.4.2.3 Indentation Fracture Method [33,34]

Among all testing methods, the scratch method is considered to be well established and is the most widely used. Its quantitative precision is comparatively greater, and its deficiencies tend to be more manageable. The scratch method makes use of an indenter (commonly a Rockwell C diamond indenter) to make a scratch across the surface of a sample at a prescribed speed, while the force applied on the film through the indenter continuously and gradually increases to the point where the film separates. In practical situations, very little of the film along the scratch path is peeled off. The method is very convenient for determining the load at the film-substrate interface. Using this method, there are two ways to apply the perpendicular load on the indenter: stepwise and continuous. The smallest force that causes the film to peel off from the substrate is called the critical force, L_c. Zheng et al. [33,34] made use of the indentation fracture model to test and measure the fracture toughness at the interface of a PZT ferroelectric film (approximately 350 mm thick), and found it to be within the range of 0.921–35.468 J/m^2.

11.4.2.4 Shear Lag Model [35]

With regard to a plastic substrate, in the event that there are cracks in the film, whether the film will shed from substrate depends on the bonding strength and the yield strength of the substrate.

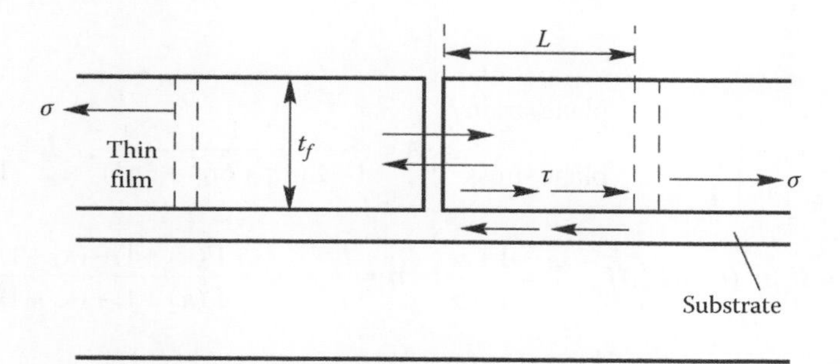

FIGURE 11.14 Schematic diagram of shear lag model. (From Zhou, Y.C. et al., *Surf. Coatings Technol.*, 157(2), 118, 2002; Taken from Figure 7.

σ_c and Y are, respectively, the fracture strength of the coating and the yield strength of the substrate. When the relative cracking strength σ_c/Y is greater than 0.2, the Shear Lag Model (Figure 11.14) can be used to obtain

$$\sigma_c = \frac{L\tau}{t_f}, \tag{11.73a}$$

and

$$P_L = \frac{1}{L} = \frac{\tau}{t_f \sigma_c}. \tag{11.73b}$$

In these two equations, L is the crack spacing; τ is the interface shear strength; t_f is the film thickness; and P_L is the crack density. The fracture toughness K_c satisfies the equation

$$K_c = \tau(\pi L)^{1/2} \tag{11.74}$$

(Note: In this section, the subscript c instead of f is used to denote film. In other words, the two words, coating and film, can be used interchangeably.)

11.4.2.5 Sh Model [36]

Suo and Hutchinson proposed two semi-infinite isotropic models for solving interface crack problems for coatings made of elastic materials. The basic solution for the models should be applicable to the following two problems: one is concerned with the interface fractures propagated by residual stress in films adhered to the substrate, and the other is the analysis and measurement of the interface fracture toughness of films.

Suo and Hutchinson pointed out that the rate of energy release by interface fractures satisfies the following equation:

$$G = \frac{c_1}{16}\left[\frac{F^2}{At_f} + \frac{M^2}{It_f^3} + 2\frac{FM}{\sqrt{AI}t_f^2}\sin\gamma\right], \tag{11.75}$$

in which

$$c_1 = \frac{\kappa_1 + 1}{\mu_1}, \quad \kappa_1 = \begin{cases} 3-4\nu & \text{plane strain} \\ \dfrac{3-\nu}{1+\nu} & \text{plane stress} \end{cases}, \quad A = \frac{1}{1+\Sigma(4\eta + 6\eta^2 + 3\eta^3)}, \quad I = \frac{1}{12(1+\Sigma\eta^3)},$$

$$\sin\gamma = 6\Sigma\eta^2(1+\eta)\sqrt{AI}, \quad \Sigma = \frac{1+\alpha}{1-\alpha}, \quad \eta = \frac{t_f}{t_s}, \quad \alpha = \frac{\Gamma(\kappa_2+1)-(\kappa_1+1)}{\Gamma(\kappa_2+1)+(\kappa_1+1)},$$

$$\beta = \frac{\Gamma(\kappa_2-1)-(\kappa_1-1)}{\Gamma(\kappa_2+1)+(\kappa_1+1)}, \quad \Gamma = \frac{\mu_1}{\mu_2},$$

where
- ν is the Poisson's ratio
- μ is the shear modulus
- t_s is the substrate thickness

The subscripts 1 and 2, respectively, denote film and substrate. F and M are the load and moment, while the relationship between stress intensity factor and rate of energy release is

$$|K| = \frac{4\cosh \pi\varepsilon}{\sqrt{c_1 + c_2}}\sqrt{G}, \tag{11.76}$$

in which $c_2 = (\kappa_2 + 1)/\mu_2$ and $\varepsilon = (1/2\pi)\ln((1-\beta)/(1+\beta))$ are the dual phase parameters. In the case of unidirectional tensile samples,

$$F = -c_1 Q, \quad M = 0. \tag{11.77}$$

If it is a four-point bending sample,

$$F = c_2\frac{F_0(L'-L)}{4bt_f} \quad M = c_3\frac{F_0(L'-L)}{4b}, \tag{11.78}$$

where
- F_0 is the external load
- b is the sample width
- L and L' are as shown in Figure 11.15

For the definition of the parameter c_3, please refer to Reference [36].

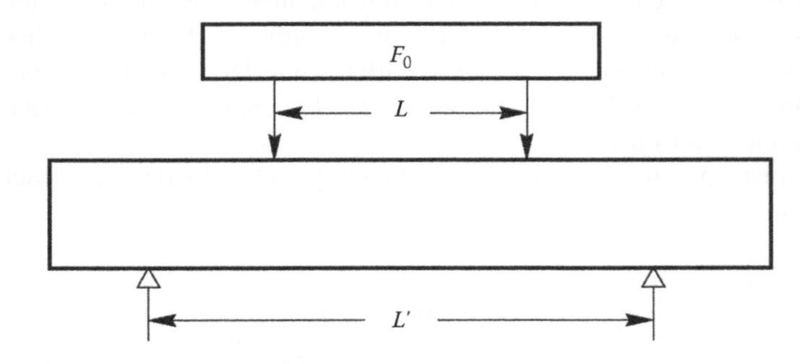

FIGURE 11.15 Schematic diagram of four-point bending.

11.4.2.6 Blister Test

Prior to preparing the test sample, puncture a hole through the flat surface of the substrate. Then deposit the film onto the substrate. Inject a liquid (pressurized oil) into the hole so that the film above the hole bulges up like a blister. As soon as the pressure reaches a critical value *p*, the film directly above and around the hole gradually expands and detaches itself from the substrate due to de-bonding, as shown in Figure 11.16. The bonding strength can then be determined, contingent upon the properties of the film, the diameter of the hole, and the liquid pressure at the time of detachment and expansion. Sample preparation in this method is rather complicated.

The schematic diagram for this method is shown in Figure 11.6. As the oil pressure p gradually increases, the cracks in the interface gradually expand and the film gradually detaches. By assuming that the detached portion of the film is an isotropic round-shaped plate of radius *a*, and by using von Karman nonlinear plate theory [37], the following overall potential energy of the detached portion of the film can be obtained:

$$U_s^{plate} = \pi D \int_0^a r\mathrm{d}r \left\{ (\nabla^2 w)^2 - \frac{2}{r}(1-\nu_1)w'w'' + \frac{12}{t_f^2}\left[\varepsilon_1^2 + 2(\nu_1 - 1)\varepsilon_2\right] - \frac{2wp}{D} \right\} b, \tag{11.79}$$

where

u and *w* denote, respectively, the displacements in the *r*- and *z*-directions in the midplane of the circular plate

D is the bending rigidity given by

$$D = \frac{E_1 h^3}{12\left(1-\nu_1^2\right)}, \tag{11.79a}$$

where

$\nabla^2 = (\mathrm{d}^2/\mathrm{d}r^2) + ((1/r)(\mathrm{d}/\mathrm{d}r))$ is the Laplace operator

E_1, ν_1, and t_f are, respectively, the elastic modulus, the Poisson's ratio, and the thickness of the plate

ε_1 and ε_2 are, respectively, the first and second invariants

By using the Euler–Lagrange variation principle, there is a coupled differential equation for the displacements *u* and *w* under fixed end boundary conditions at the center $r = 0$ and the periphery of the disk $r = a$. This differential equation enables us to solve for *u* and *w* and to obtain the one-dimensional quantities S_r^0 (film stress), and M_r^0 (the bending moment), which are given respectively by

$$S_r^0 = \frac{c_1^0 N_r\big|_{r=a}}{t_f} = \left(\frac{3p}{2\eta^3 k^2}\right)^2 \left\{ \frac{8}{k^2} + \frac{3}{2} - \left[\frac{I_0(k)}{I_1(k)}\right]^2 - \frac{2}{k}\frac{I_0(k)}{I_1(k)} \right\} \tag{11.80}$$

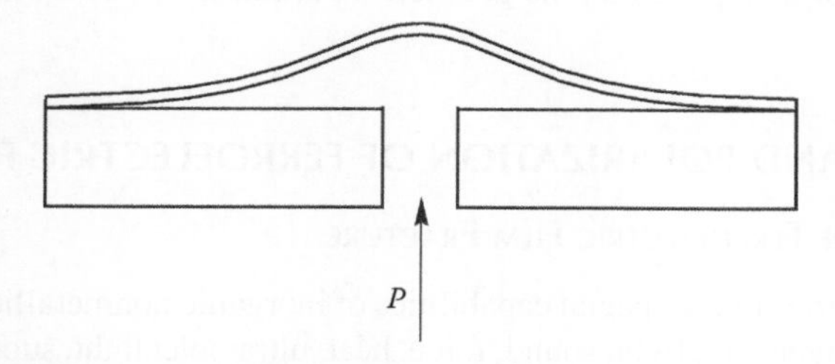

FIGURE 11.16 Schematic diagram for blister test.

$$M_r^0 = \frac{c_1^0 M_r|_{r=a}}{t_f^2} = \frac{p}{2\eta^2 k^2}\left[\frac{kI_0(k)}{I_1(k)} - 2\right] = \frac{p}{2\eta^2} f(k). \tag{11.81}$$

For the meaning of the symbols in these equations, please refer to Reference [37].

For smaller external loads, the deflection at the center of the circular plate is given by

$$w(0) = p\frac{3(1-v_1^2)a^4}{16E_1 t_f^3}. \tag{11.82}$$

In terms of the film stress S_r^0 and the bending moment M_r^0, the rate of energy release associated with interface cracks as obtained by Sue and Hutchinson is G_a (see Equation 11.75). Its dimensionless form $\bar{G}_a = (1/3)((16\eta^2/P)^2(c_1^0 G_a/t_f))$ can be expressed as follows:

$$\bar{G}_a = \frac{1}{3}\left(\frac{4\eta^2}{P}\right)^2\left[\frac{(S_r^0)^2}{A} + \frac{(M_r^0)^2}{I} - 2\frac{S_r^0 M_r^0}{\sqrt{AI}}\sin\gamma\right],$$

$$= \frac{4}{3}\frac{1}{I} f^2(k)(\lambda^2 + 1 - 2\lambda\sin\gamma) \tag{11.83}$$

where the dimensionless parameter $\lambda = \sqrt{(I/A)}\left(S_r^0/M_r^0\right)$. The other dimensionless parameters A, I, Σ, and $\sin\gamma$ are the same as those in Equation 11.75. Furthermore, the following equation, dependent only on dimensionless parameter p/η^4, can be obtained.

$$\bar{G}_a = \exp\left[-\sum_{i=1}^{9} a_i\left(\frac{3p}{128\eta^4} - \bar{a}\right)^{i-1}\right], \quad \lambda = \sqrt{\frac{I}{A}}\left[\sum_{i=1}^{7} b_i\left(\frac{3p}{128\eta^4} - \bar{b}\right)^{i-1}\right]. \tag{11.84}$$

The dimensionless parameter p/η^4 is given by

$$\frac{p}{\eta^4} = \frac{c_1^0 q}{\eta^4} = \frac{8(1-v_1^2)q}{E_1}\left(\frac{a}{t_f}\right)^4, \tag{11.85}$$

where a and b are related coefficients (see Reference [37] for their actual values). The interface crack is a mixed crack. The phase angle is used to present the relative contribution of Model I and Model II cracks, given by

$$\psi = \tan^{-1}\left[\frac{\lambda\sin\omega - \cos(\omega+\gamma)}{\lambda\cos\omega + \sin(\omega+\gamma)}\right], \tag{11.86}$$

where the magnitude of ω depends on the geometric parameter $\eta_0 = t_f/t_s$, and Dunders parameters α and β.

11.5 FRACTURE AND POLARIZATION OF FERROELECTRIC FILMS

11.5.1 Overview of Ferroelectric Film Fracture

The continuing discoveries of the special capabilities of inorganic nonmetallic materials—in connection with electricity, magnetism, light, sound, force, heat, ultraviolet light, superconductivity, osmosis, and reflection—have modernly advanced the use of these materials in technological applications.

The monographs on dielectric science edited by Yao Xi, fellow of the Chinese Academy of Sciences, are the only literature [38] published on solid inorganic compounds, especially the chemistry of oxidized metallic substances. Ferroelectric materials as functional ceramic materials have been a widely used geometric configuration in the film-substrate field, and have consequently brought forth a wide range of research [39].

Generally speaking, there are differences between the properties of thin film materials and those of bulk materials [40]. In the film making process, the most important research problem is residual stress due to the mechanical constraint imposed by the substrate. Residual stress plays a very important role in the evolving changes in the microstructure of the film. Although the extremely thin film deposited on the substrate is able to withstand a large mismatch elastic strain, once its thickness exceeds its critical value, the film–substrate system will form cracks, by generating mismatch dislocations or mismatch crystal twins [41]. This will cause surface instability, change the material's microstructure, and release energy associated with the mismatch strain [41–44]. In the case of a ferroelectric film deposited on a silicon substrate plated with metallic electrodes, the crack problem is of major importance.

Close attention has been paid to the consideration of tensile stress as a cause for film fractures [42,45,46]. When the film thickness is greater than a certain critical value, the residual tensile stress can possibly give rise to film fractures. However, when the film thickness is less than the critical value, no cracks will form. Another important problem in film research is the density of cracks, which are related to the spacing between cracks. A low crack density implies a low possibility of crack production in the microdevice. Zhao et al. [45] systematically studied the crack behavior in PZT thin films. They found that when the film thickness is less than the critical value needed for the formation of cracks, stress in the film is advantageous to the film. Due to the mechanical constraint imposed by the substrate, there is a great deal of difference between the properties of ferroelectric film materials, and the corresponding free materials in bulk. Wang and Zhang [43,47,48], using thermodynamic analysis, numerical calculations, and phase field simulation, studied the properties of ferroelectric films.

11.5.2 Characteristic Features of Cracks in Ferroelectric Films

11.5.2.1 Crack Density in Ferroelectric Films

When the optical microscope was used to test the PZT film deposited on a Pt/Ti/Si (100) substrate, using the spin coating method, it was discovered that cracks had formed in the thick film along the [100] direction of a silicon wafer. Figure 11.17a and b show, respectively, typical cracks formed in a 0.93 μm film and in a 1.12 μm film. The crack numbers of the latter are more than that of the former.

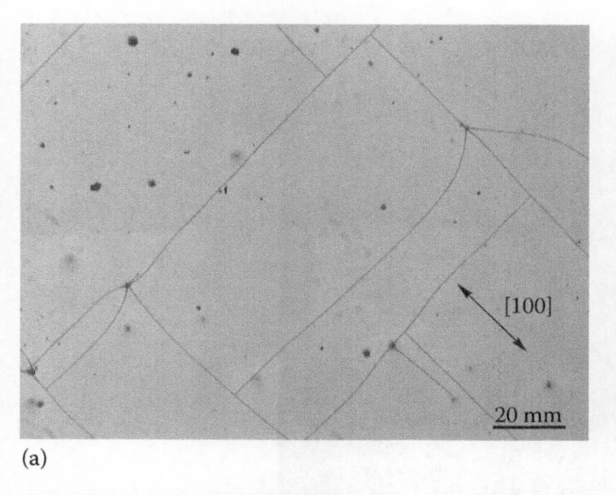

(a)

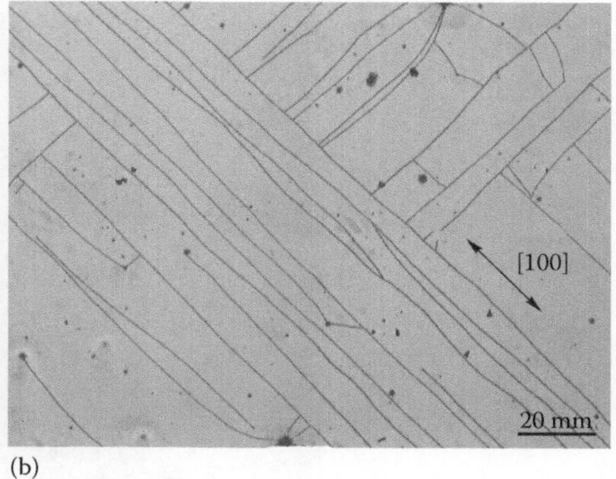

(b)

FIGURE 11.17 Cracks in thin films. (a) Crack density 1.00 mm/mm^2, 0.93 μm thick PZT thin film. (b) Crack density 2.37 mm/mm^2, 1.12 μm thick PZT thin film.

In order to provide a quantitative description of crack numbers, let us calculate the crack length per unit area of the film surface. To do so, magnify 100-fold a 225 mm^2 area on the surface of each film. By measuring the crack lengths shown in each photograph, the crack density for each thin film can be determined.

Figure 11.18 shows how crack density varies with the thickness of film. It can be concluded that when the film thickness is less than 0.7588 μm, no cracks are formed (crack density = 0). For films with thickness greater than 0.7963 μm, the crack density increases sharply as the film thickness increases. Experimental observations have clearly shown a critical thickness t_{fc} for a crack-containing PZT film deposited on a Pt/Ti/Si substrate.

From Figure 11.19, an image taken by the scanning electron microscope (SEM), it can be seen that cracks in the film terminate at the boundary between the PZT film and the metallic layer. In the PZT film with a thickness of 0.903 μm, the crack tips become blunt, and the cracks have displaced and increased their separation by 0.1 μm. The plastic deformation of tough metallic layers can blunt

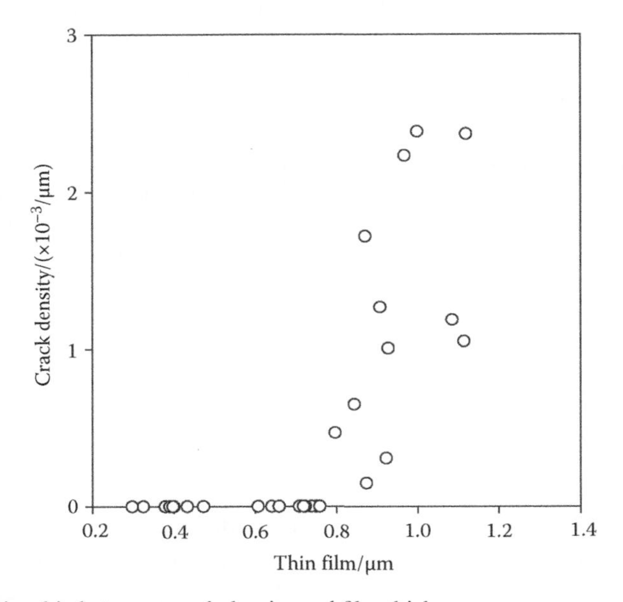

FIGURE 11.18 Relationship between crack density and film thickness.

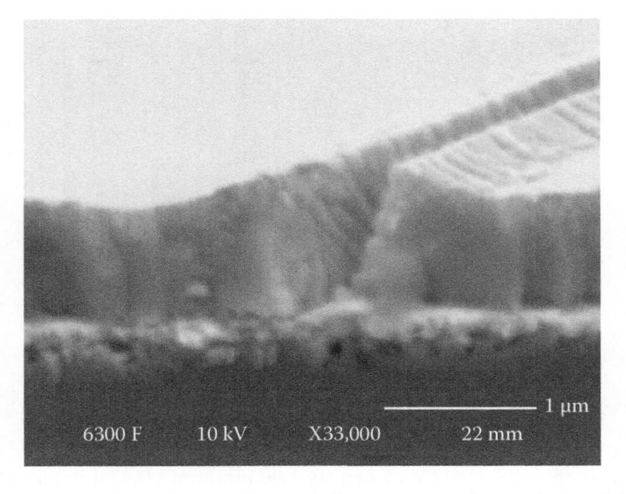

FIGURE 11.19 SEM image showing the opening and displacement of a crack in a 0.903-μm PZT film.

the crack tips and release residual stress, which is tantamount to a weakening of the driving force behind crack infiltration; so that the cracks in the film are unable to cross the boundaries. In accordance with their experimental observations, Zhao et al. proposed a hybrid elasto-plastic crack shear-lag model, as discussed in the next section, for predicting the crack behavior of brittle thin films with an underlying tough substrate.

11.5.2.2 Compound Elasto-Plastic Crack Shear-Lag Model

We make use of the schematic diagram in Figure 11.20 to explain the hybrid elasto-plastic crack shear-lag model.

The relevant one-dimensional parameters are

$$s=\frac{l}{t_f},\quad \xi=\frac{t_s}{t_f},\quad \chi=\frac{r}{t_f},\quad \gamma=\frac{\tau}{\sigma},$$

$$\eta=\xi\sqrt{\frac{1}{(1-\nu_s)\left(1+\left(E_s^*/E_f^*\right)\xi\right)}},\quad R=\frac{\Gamma_f E_f^*}{\pi\sigma^2 t_f} \tag{11.87}$$

where

l is the periodic crack spacing
r is a coordinate variable
$E^* = E/(1-\nu^2)$
t, E, and ν are respectively the thickness, elastic modulus, and Poisson's ratio of the film
σ is the residual stress in the film
Γ_f is the rate of energy release at interface
R is the crack resistant number

The subscripts f and s, respectively, denote film and substrate. For a set of fixed parameters, making the change in total energy a minimum is tantamount to requiring an identical spacing between any of the cracks. Generally speaking, a large value of the crack-resistant factor R, corresponds to a bigger spacing between the cracks. This is similar to the elastic case [44]. Here, there is an interface crack resistant number R_c, and when $R > R_c$, there can be no cracks in the film. The crack-resistant factor is determined under the following hypothesis: For the case of a

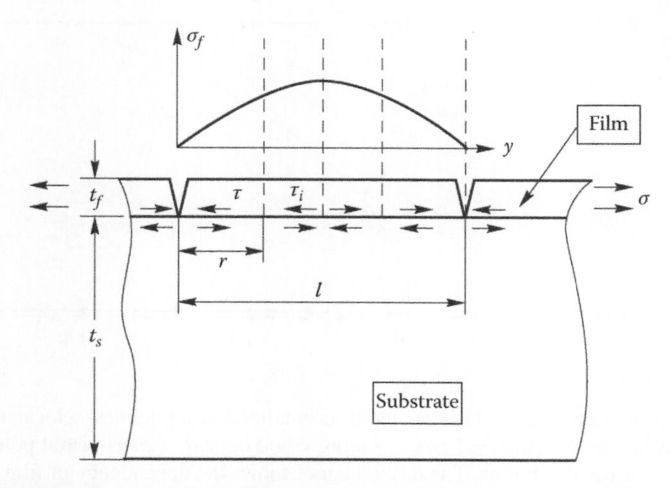

FIGURE 11.20 Schematic diagram for the elasto-plastic compound shear-lag model.

single crack, when the crack depth is the same as the film thickness, the change in total energy of the overall system is zero. At the same time, the single crack case must satisfy the condition that the crack spacing l tends to be infinity, or $s \to \infty$. Then we have

$$R_c = \frac{1}{\pi}\left[(1+\gamma\eta)\eta + \frac{1}{3}\gamma^2\left(\frac{1}{\gamma}-\eta\right)^3 - 2(1-\gamma\eta)\gamma\eta^2\right]. \tag{11.88}$$

If the substrate is sufficiently thick such that $\eta \to \infty$, then Equation 11.88 becomes

$$R_c = \frac{1}{3\pi\gamma}. \tag{11.89}$$

Equation 11.89 is consistent with the result for large-size yield of a single crack obtained by Hu and Evans [49]. It should be pointed out that they did not consider the rate of elastic energy.

When γ is relatively small, such as when $\gamma \leq 1/100$, a wide range of yield conditions is satisfied. If the yield region is not sufficiently large, the released elastic strain energy due to the expansion of cracks should be counted into the total change in energy. In the film–substrate system shown in Figure 11.21, the substrate thickness t_s = 525 μm. By using the nano-indentation method, it was found that the elastic modulus of film is E_f = 59.09 GPs and the Poisson's ratio is ν_f = 0.23 approximately. Numerical calculations showed that when γ = 1/350, the predicted crack density should be in close agreement with the experimental result as indicated by the solid curves shown in Figure 11.21. Furthermore, the solid curves demonstrate the existence of a transition point. When the film thickness t_f slightly exceeds or is approximately equal to the critical thickness t_{fc}, that is, when $t_f/t_{fc} \approx 1$, the normalized crack density curve $1/s = t_f/l$ jumps sharply from 0 for a single crack (for which case

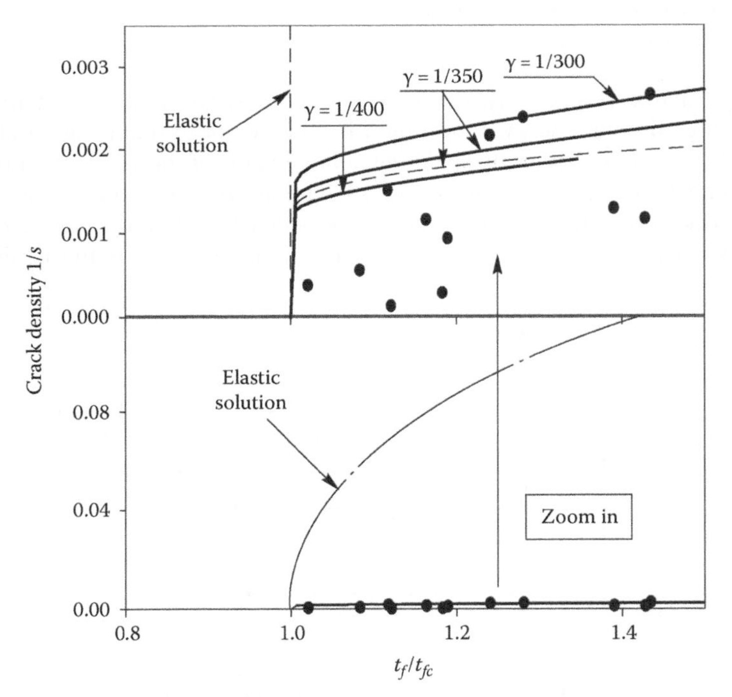

FIGURE 11.21 Plot of normalized crack spacing vs. normalized film thickness. Normalized crack density = $1/s$, s being the dimensionless normalized crack spacing. Solid dots are experimental points. Solid curves are plotted under constant stress conditions. The dashed curve shows the dependency of film stress on film thickness. Dashed line and curve were given by the elasticity solutions.

$l = \infty$ and $1/s = 0$), to 0.001 for $l \approx 1000\, t_{fc}$. The point $1/s = 0.001$ is referred to as the transition point because after this point the curve levels off and rises slowly.

The exact location of the transition point depends on the value of γ. That is, the greater the value of γ, the higher the transition point. To obtain a better agreement with the experimental data, the value of γ should be kept lower than 1/350. However, upon using lower values of γ in calculations, very small numbers may emerge in the calculation process, and cause numerical errors. As a comparison, elasticity solutions based upon the same data are obtained by the method introduced in Reference [44], and are plotted in Figure 11.21. When the normalized crack spacing exceeds 50 (because the interactions between cracks are negligibly small in elasticity solutions), such solutions are unable to make predictions about the behavior of cracks with lower crack densities and larger crack spacing.

11.5.3 Polarization of Epitaxial Ferroelectric Thin Film under Nonequally Biaxial Misfit Strains

11.5.3.1 Effects of Misfit Strains on the Phase Diagram of Epitaxial Ferroelectric Thin Films

Recent experimental results [50–52] show that if the ferroelectric thin film grows in a square fashion, the misfit strains along one crystallographic axis will be different from those on another crystallographic axis. This phenomenon is called nonequally biaxial misfit strains. Nonequally biaxial misfit strains may grant new properties to the material and lead to new directions in designing electronic devices.

Wang and Zhang [43] studied epitaxial ferroelectric thin film grown in a square fashion, assuming that residual stress exists only in the thin film, and that the basis is an unstressed state. Paraelectric is treated as unstressed. Denote the strain with a Voigt matrix and apply rectangular coordinates to the x_3 axis, which is vertical to the thin film/basis interface. In this way, nonequally biaxial misfit strains can be defined as $e_1 = (b - a_0)/b$ and $e_2 = (c - a_0)/c$, in which b and c are crystal constants of the basis, and a_0 is the crystal constant of the paraelectric thin film in the unstressed state. Assume that $e_1 \neq 0$ and $e_2 \neq 0$, and that all other strains are equal to zero due to the very thin film. As shown in Chapter 10, Table 10.3, Wang and Zhang [43] obtained the Helmholtz free energy of a ferroelectric thin film under nonequally biaxial misfit strains. By analyzing the Helmholtz free energy, the equilibrium phase in the thin film must include:

1. a_1 phase $(p_1 \neq 0,\ p_2 = p_3 = 0)$;
2. a_2 phase $(p_2 \neq 0,\ p_1 = p_3 = 0)$;
3. c phase $(p_1 = p_2 = 0,\ p_3 \neq 0)$;
4. a_1c phase $(p_1 \neq 0,\ p_2 = 0,\ p_3 \neq 0)$;
5. a_2c phase $(p_1 = 0,\ p_2 \neq 0,\ p_3 \neq 0)$;
6. a_1a_2 phase $(p_1 \neq 0,\ p_2 \neq 0,\ p_3 = 0)$; and
7. r phase $(p_1 \neq 0,\ p_2 \neq 0,\ p_3 \neq 0)$.

Figure 11.22a and b show the room temperature phase diagram of epitaxial $BaTiO_3$(BT) and $PbTiO_3$(PT) under misfit strains, respectively. Figure 11.22a, in a BT thin film, when the misfit strain increases from −0.008 to 0.008, the r phase (monocline) does not exist in the diagram. In addition, the other six phases, three orthogonal (a_1a_2, a_1c, a_2c) and three tetragonal phases (c, a_1, a_2), appear in different regions on the phase diagram. A thermodynamic analysis indicated that the a_1a_2 phase to the a_1c, a_2c phase, and the a_1c phase to the a_2c phase is the first kind of phase transition, while all other phase transfers belong to the second kind. The a_1 and a_2 tetragonal phases exist only when one of the two nonequally biaxial misfit strains is tensional and the other is compression. When both directions of the biaxial strains are in compression, the c tetragonal phase exists. The a_1a_2 monocline

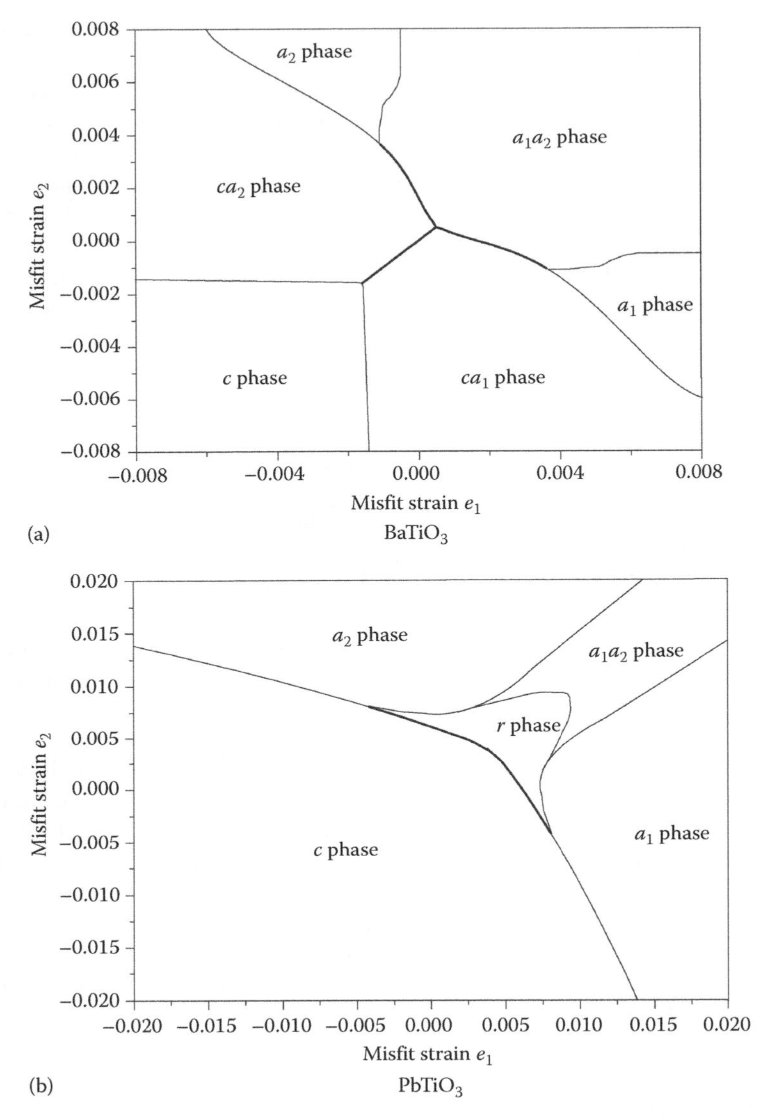

FIGURE 11.22 Misfit strain phase diagrams of single-domain ferroelectric thin films. (a) $BaTiO_3$ thin film. (b) $PbTiO_3$ thin film. The thick and thin lines represent phase transformation of first and second kind.

phase exists only when both direction of the misfit strains are under tension. In Figure 11.22b, the PT thin film, when misfit strains increase from −0.02 to 0.02, no a_1c or a_2c phases exist in the phase diagram. The other five phases r, r, a_1a_2, c, a_1, and a_2, appear in different regions on the diagram. Moreover, the a_1 and a_2 tetragonal phases exist only when one of the two nonequally biaxial misfit strains is tensional, and the other is compression. In order to eliminate the a_1 and a_2 tetragonal phases in BT and PT thin films, one of the strains in the nonequally biaxial misfit strains must be compressive. Because of this, applying nonequally biaxial misfit strains through the basis is inadequate to obtain a steady planar tetragonal phase.

The actual phase diagram should be three dimensional. The present two-dimensional temperature–strain phase diagram was obtained with one of the misfit strains fixed. Figure 11.23a and b show the temperature–strain phase diagram of BT and PT thin films at $e_1 = 0.005$. The phase transformation from the paraelectric to the ferroelectric phase takes place at $T_c = \max[T_1, T_2, T_3]$,

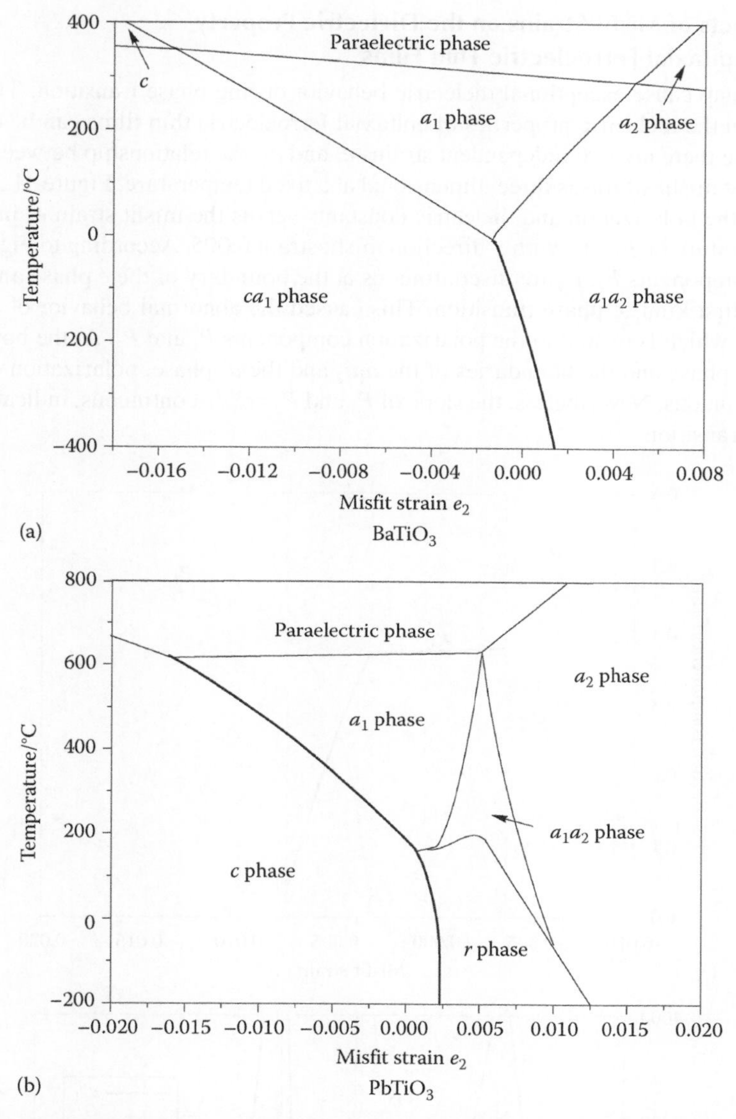

FIGURE 11.23 Temperature–misfit strain phase diagram of ferroelectric thin films. (a) $BaTiO_3$ thin film. (b) $PbTiO_3$ thin film. The thick and thin lines represent phase transformation of the first and second kind.

where T_1, T_2, and T_3 represent the phase transition temperature from paraelectric to a_1, a_2, and c, respectively. The paraelectric–ferroelectric phase transition temperature $T_c(e_2, e_1 = 0.005)$ of the BT and PT thin films has a steady state in the a_1 phase, which is distinct from the phase transition temperature of the seagull shape under nonequally biaxial misfit strain conditions [53]. For a BT thin film under equally biaxial strain conditions, the paraelectric phase could change to a c, aa (P1 = P2 ≠ 0, P3 = 0), or an r ferroelectric phase. Under nonequally biaxial misfit strain $e_2(e_1 = 0.005)$, the paraelectric phase may change to the c, a_1c, a_1, a_1a_2, or the a_2 ferroelectric phase. For a PT thin film under equally biaxial strain conditions, the paraelectric phase could change to the c or aa ferroelectric phase. When nonequally biaxial misfit strain conditions $e_2(e_1 = 0.005)$ are applied to a PT thin film, the paraelectric phase could change to the c, a_1, a_1a_2, or one of the a_2 ferroelectric phases. One dominant characteristic of the nonequally biaxial misfit strain condition is the formation of tetragonal phases a_1 and a_2, which do not exist under equally biaxial strain conditions [53].

11.5.3.2 Effects of Misfit Strains on the Dielectric Property of Epitaxial Ferroelectric Thin Films

Misfit strains may cause exceptional dielectric behavior during phase transition. The influence of misfit strains on the dielectric properties of epitaxial ferroelectric thin films can be analyzed using (10.44). Because there are two independent strains e_1 and e_2, the relationship between the dielectric constant and the misfit strains is three dimensional at a fixed temperature. Figure 11.24a and b show the relation of the polarization and dielectric constants versus the misfit strain e_2 in a $PbTiO_3$ thin film at a temperature of 300°C, with x-direction misfit strain 0.005. According to Figure 11.24a, the polarization components P_1, P_3 are discontinuous at the boundary of the c phase and the a_1 phase, indicating the first kind of phase transition. This caused the abnormal behavior of ε_{11}, ε_{33} and the jumping of ε_{22}, which is related to the polarization components P_1 and P_3. At the boundaries of the a_1a_2 and the a_1 phase and the boundaries of the a_1a_2 and the a_2 phase, polarization components P_1 and P_2 are continuous. Nevertheless, the slope of P_1 and P_2 are discontinuous, indicating the second type of phase transition.

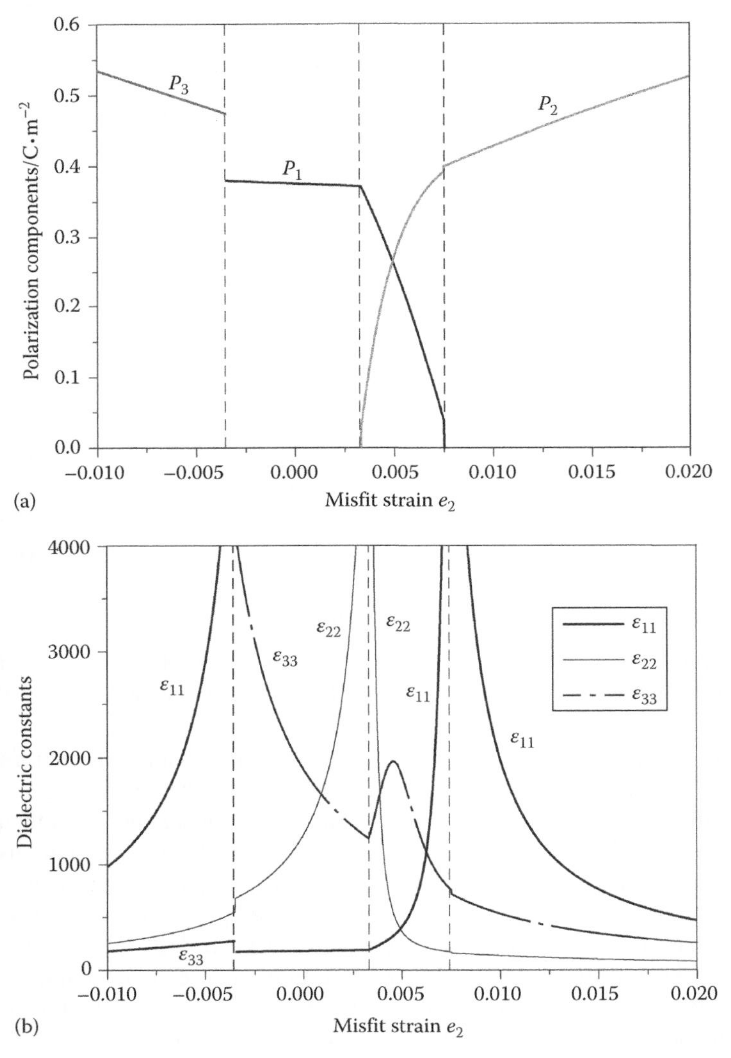

FIGURE 11.24 At 300°C, x-direction misfit strain 0.005, in $PbTiO_3$ ferroelectric thin film. (a) Polarization component against misfit strain. (b) Dielectric constant against misfit strain.

The discontinuity of P_2 at the boundary of the a_1a_2 phase and the a_1 phase, and the P_2's approaching zero at the boundary of a_1a_2 phase, caused the dielectric constant ε_{22} to converge to infinity at the boundaries. Moreover, the discontinuity of the slope of polarization component P_1 at the a_1a_2 and a_1 boundaries results in a peak value of ε_{33} at the boundaries. In the monocline phase a_1a_2, the nonequally biaxial misfit strains $e_2 = 0.005 = e_1$ causes a local maximum of dielectric constant at $P_1 = P_2$. Similarly, at the boundary of the a_1a_2 and the a_2 phases, the low value and discontinuity of polarization constant P_1 leads to dielectric constant ε_{11} approaching infinity. Noticeable differences between dielectric constants along two lattice directions were observed in experiments. This phenomenon was detected in a (Pb,Sr)TiO_3 thin film deposited on $NdGaO_3$(NGO), manifesting the effects of nonequally biaxial misfit strains on the dielectric constant [50].

11.5.4 Effects of Depolarization on Polarization State of Epitaxial Ferroelectric Thin Films

The noncompensation electricity at the surface and interface of ferroelectric thin films and nonuniform distribution of polarization in the films may cause a depolarization field. Surface electric charge is noncompensational under open circuit boundary conditions. However, under short-circuit boundary conditions, the surface charge is fully compensational. Thus, a depolarization field will only be caused by a nonuniform distribution of polarization. The nonuniform distribution of polarization could be categorized as the so-called thin film inherent surface effect. This effect can be described using the extrapolation length of Landau–Devonshire theory, where the extrapolation length represents the distance from the zero-polarization surface of the thin film [54]. In order to describe inhomogeneous polarization distribution, the free energy of the thin film must include the polarization gradient energy. Polarization of the equilibrium state could be calculated by solving the Euler–Lagrange equation, which is obtained by solving for the minimum of chemical free energy. It is almost impossible to analytically solve the Euler–Lagrange equation in a unidimensional situation. In most cases, numerical methods are applied. Glinchuk [55] studied the three dimensional phase diagram of ferroelectric thin films after depolarization. They introduced supplemental surface polarization and considered short-circuit boundary conditions and equally biaxial misfit strains [55]. As previously mentioned, under open circuit boundary conditions, surface electrons of the thin film leads to a large depolarization field, which further affects the equilibrium polarization state of the ferroelectric thin film. Equally biaxial misfit strains will cause a distinct ferroelectric thin film polarization state from nonequally biaxial misfit strains. Wang and Zhang [47] studied ferroelectric thin film polarization under nonequally biaxial misfit strains at both open and short-circuit boundary conditions. The following describes a brief introduction to their findings.

Figure 11.25 is the misfit strain–misfit strain phase diagram when extrapolation length $\delta_1 = \delta_3 =$ 3 nm. The solid line and the dash-dot line represent the phase diagram with and without a depolarization field, respectively. Both phase diagrams have a monocline phase r, two orthogonal phases a_1c and a_2c, and a tetragonal phase c. The two diagrams share some similarities. However, the depolarization field shifts the phase boundaries and reduces the area that a certain phase occupies outside the surface. In general, the value of extrapolation lengths δ_1 and δ_3 are unequal. The influence of the extrapolation length on the polarization state can be studied by varying δ_1 and δ_3. For example, let $\delta_1 = 3$ nm, $\delta_3 = 4$ nm; under short-circuit boundary conditions, the phase diagram is Figure 11.26. In comparison to Figure 11.25, the c phase and the r phase occupy smaller areas in Figure 11.26. When $\delta_1 = 4$ nm and $\delta_3 = 3$ nm, only one c phase exists. These results indicate that the extrapolation length can have a significant effect on the equilibrium polarization state. One of the main tasks at present is to evaluate r, which can be calculated using the first principle, or measured via better experimental methods.

The depolarization field and misfit strains are two important factors concerning the properties of ferroelectric thin films. Current research indicates that if the tensile stress applied on the thin film exceeds a critical value at a fixed thickness, or the thickness of the thin film exceeds a critical thickness,

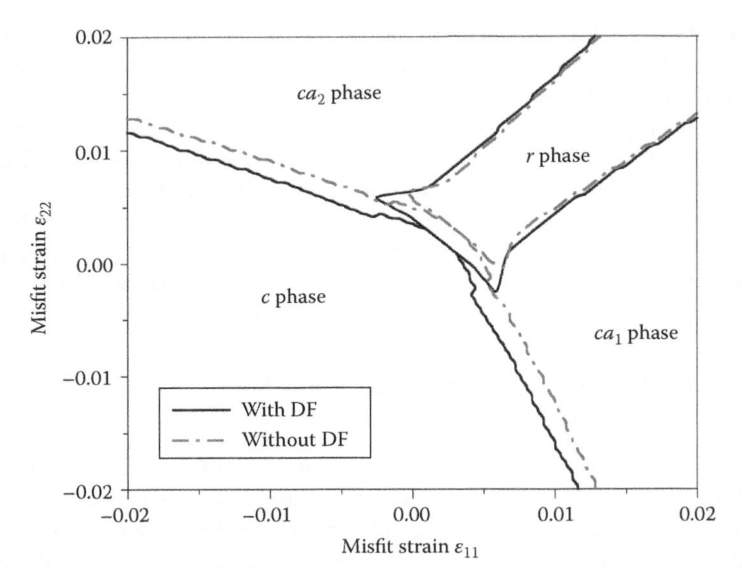

FIGURE 11.25 Misfit strain phase diagram when extrapolation length ($\delta_1 = \delta_3 = 3$ nm) at short-circuit boundary condition.

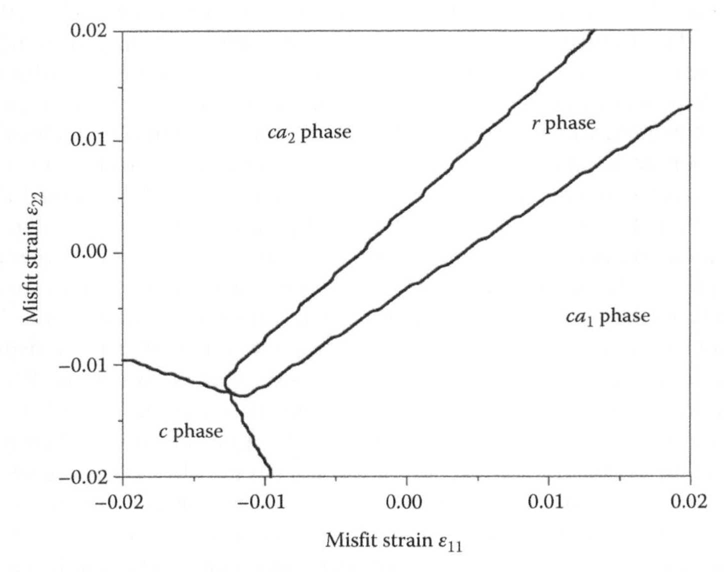

FIGURE 11.26 Misfit strain phase diagram when extrapolation length $\delta_1 = 3$ nm, $\delta_3 = 4$ nm at short-circuit boundary condition.

cracks will develop inside the thin film. Under critical conditions, membrane stress could be adopted to optimize the polarization structure and enhance the performance of ferroelectric thin films.

11.6 FLEXURE OF DUCTILE THIN FILMS

11.6.1 Concepts of Ductile Thin Films

New developments in electronics primarily come from enhancing the running speed and the integrity of circuits to lower the electric power consumption and enable a larger area for display systems. In recent years, research has mainly been focused on seeking the development of a new method of loading high-performance circuits and discovering unconventional backing materials with special wave form factors [56], such as the supple plastic basis used by paper monitors and optical scanners [57–61], the sphere dome support for focal-plane arrays [62,63], as well as the guard shield at the surface of the integrating robot's sensor spring [64,65]. When electronic materials are shaped to thin films and placed on a thin basis laminar [66–71], or on the near neutral mechanical plane of the lamination basis [72,73], these materials appear to have good flexibility. In this case, the strain induced during the curving process could be better kept under the standard magnitude of break (~1%). The integrated tensile property is a more challenging and inadequate property of a component. It requires the component to suffer curving and stretching, or reach a curving limit under active conditions. Other devices may require the material to be wound-up onto a support at a complex form while staying active. In these systems, deformation on the circuit scale has extended the fracture limit of almost every known electronic material, particularly well-developed and stably functioning electronic materials. This problem may be solved by linking extensible wires between these electronic devices sustained by rigidly isolated islands [74–79]. Some meaningful results could be obtained through this method, although it is more suitable to active electric devices under relatively low coverage fraction.

Huang Yonggang [80] posted another method which involves acquiring the tensile property from nanometer-scale, high-quality monocrystalline silicon with periodically undulated geometry. This structure could adapt to huge compression and tension caused by variations in amplitude and wavelength, avoiding potential destructive strain of the material itself. By directly integrating this monocrystalline silicon with electrolyte materials, dopant model, electrode and metal thin film, high-quality stretchable electronic devices may be produced.

11.6.2 Preparation of Undulated Monocrystalline Silicon Ribbon on Elastic Substrate

Figure 11.27 describes one process of preparing an undulated monocrystalline silicon ribbon grown on an elastic substrate. It begins by using lithography to place a corrosion-resistance layer on the SOI (Si-on-insulator) chip, which is followed by engraving the silicon at the top using lithography. After removing the corrosion-resistance layer with acetone and using concentrated hydrofluoric acid to engrave the SiO_2 layer, band-shaped silicon will appear from the substrate. The ends of the bands are connected with the silicon chip to prevent them from being washed out during the engraving. The width (5–50 μm) and length (~15 mm) limit the dimension of these bands and the crown silicon thickness (20–320 nm) above the SOI chip determines the height of these bands. Next, a plane elastic basis (PDMS, 1–3 mm thick) can easily be stretched and placed onto the bands according to shape size. Then strip the PDMS off such that the band-shaped silicon on the SOI surface remains attached to the PDMS. Releasing the strain inside the PDMS will cause surface deformation and produce a distinct wavy geometric shape on the surface of the PDMS and the silicon (see Figure 11.28a and b). The geometric shape simulates sinusoid curves with a period of 5–50 μm and an amplitude of 100 nm–1.5 μm, depending on the thickness of the silicon and the strain previously assigned to the PDMS. For a given system—namely, inside an area less than several centimeter square—the period and amplitude of the wave are coherent. The planes between bands and the phase difference between two adjacent bands indicate no strong connection between these bands. Figure 11.28c shows the micro-Raman measurement of silicon peaks with the horizontal axis representing the distance along one undulated band. These results indicate the condition of stress distribution.

11.6.3 Analysis of Flexure of Ductile Thin Film

The flexure behavior of static undulated materials is consistent with the inhomogeneous analysis of the initial buckling mechanism of a homogeneous thin high modulus layer on a semi-infinite low module substrate [81,82]:

$$\lambda_0 = \frac{\pi h}{\sqrt{\varepsilon_c}}, \quad A_0 = h\sqrt{\frac{\varepsilon_{pre}}{\varepsilon_c} - 1}. \tag{11.90}$$

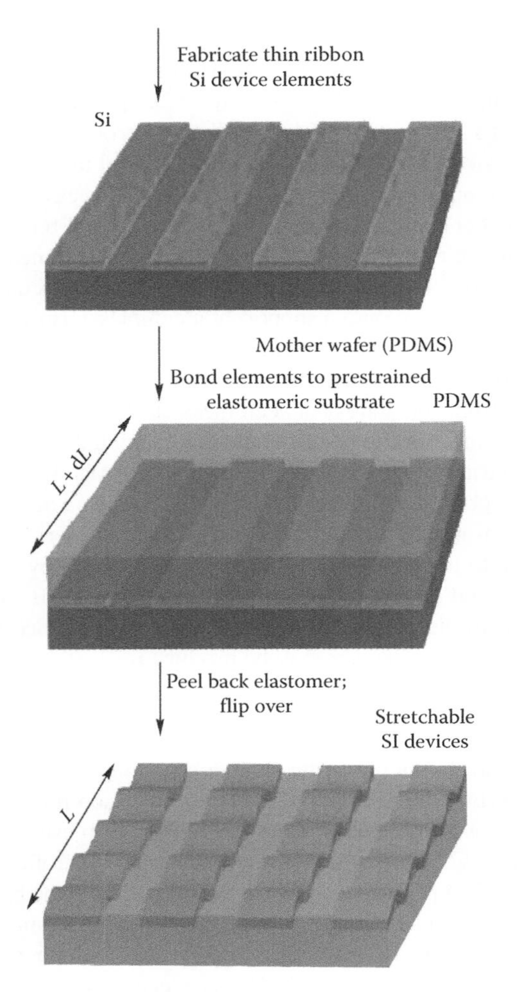

FIGURE 11.27 Schematic illustration of the process of preparing an undulated monocrystalline silicon ribbon grown on an elastic substrate.

In Equation 11.90, $\varepsilon_c = 0.52[E_{PDMS}(1-\nu_{Si}^2)/E_{Si}(1-\nu_{PDMS}^2)]^{2/3}$ represents the flexure critical strain and ε_{pre} represents the magnitude of the previously applied strain. λ_0 is the wavelength with amplitude A_0 and h is the thickness of Si. Assume E is Young's modulus and ν is the Poisson's ratio, referring to PDMS and Si by subscript, respectively. This execution method could retain many characteristics of a wavy structure.

In Figure 11.28d, for instance, given pre-assigned strain, both oscillation wavelength and amplitude show a linear relation with the thickness of silicon. The wavelength does not depend on the pre-assigned strain. In addition, using the data from the document [83,84] about the mechanical parameters of Si and PDMS, the calculated oscillation amplitude and wavelength have a 10% deviation from the measured values. The ratio of the Si band's effective length to their actual length should provide us with their strain and yield value, which are approximately equal to the pre-assigned strain on the PDMS (up to 3.5%). When the waveform attains its extreme values, the maximum strain of Si could be calculated from the thickness and radius of curvature of the $\kappa h/2$ band in the formula. When a wave form exists, κ stands for the radius of curvature in the strain system, and the critical strain should be smaller than the fringe deflection maximum strain. In Figure 11.28, the maximum strain of Si is 0.36(±0.8)%, which is more than two times less than the strain of the bands.

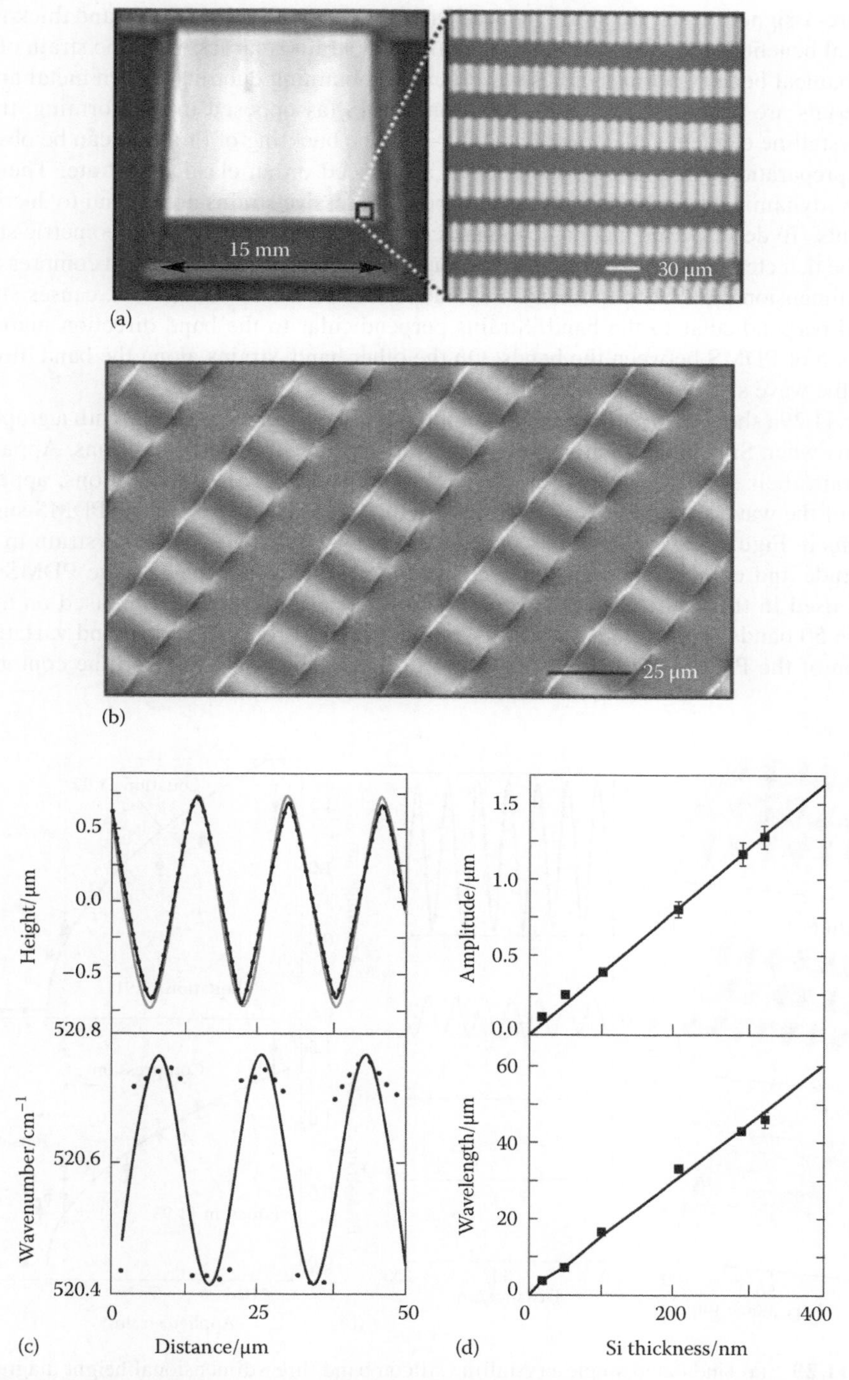

FIGURE 11.28 (a) Optical diagram of the large-scale arrays of single crystalline silicon bands on PDMS. (b) Tilt viewing angle scanning electron micrographs of four undulated Si bands in (a). The wavelength and amplitude of the wave structure across array directions are highly consistent. (c) The AFM (upper) and micro-Raman (lower) measurement of silicon surface's height and wavenumber. The horizontal axis represents the distance along one strip of band. The curves on the graph fit the results. (d) Function of the oscillation amplitude (upper) and wavelength (lower) of waveform Si band against Si thickness. These results are obtained by setting pre-strain on the PDMS.

Given a pre-assigned strain, the strain induced in Si has the same value for all band thicknesses. The mechanical benefit of the synthesis in the Si maximum strain is far less than the strain of the band. The mechanical benefit of this synthesis is crucial to obtaining ductility. When metal and electrolyte materials are evaporated or spin-coated on PDMS (as opposed to yop-forming, transferred, single-crystalline element and components) [85–87], the buckling of thin film can be observed.

After preparation, the undulated structures are placed on an elastic substrate. The undulated structures' dynamic responses toward tensile and compressive strains are crucial to ductile electric components. To describe the mechanical characteristics of this process, the geometric structure of Si could be detected using AFM, because the load AFM exerts on PDMS could compress or stretch it to the dimension parallel to the bands. This load, due to the Poisson effect, causes strains both along and perpendicular to the band. Strains perpendicular to the band direction mainly lead to deformation of PDMS between the bands. On the other hand, strains along the band direction can adjust to the wave structure of bands.

Figure 11.29a shows a three-dimensional topographical diagram together with a graph showing waveforms when Si is under compressed, unperturbed, and stretched conditions. Apparently, the bands retain their sinusoidal shape during deformation. Under these conditions, approximately one-half of the wave structure is confined within the force-free region of the PDMS surface. The two graphs in Figure 11.29b show the relationship between the peak value of strain in Si as well as amplitude and compression and tensile strain, due to external load on the PDMS substrate. The data used in the figure are related to AFM statistical data, which are based on a survey of more than 50 bands. These external strains are determined from the end-to-end variations in the dimension of the PDMS substrate. Both direct measurements by AFM and the contour integral

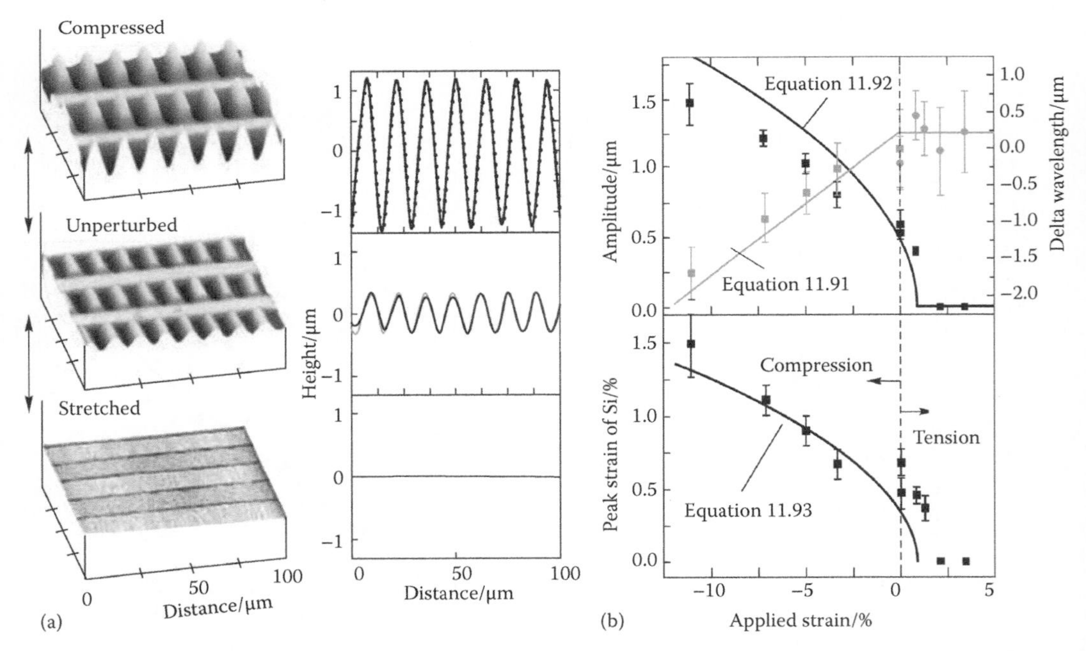

FIGURE 11.29 (a) Undulated single-crystalline silicon band three-dimensional height diagram and surface diagram on PDMS substrate, corresponding to PDMS suffer along length direction, −7% compression strain (upper figure), unperturbed (middle figure), and 4.7% tension strain (bottom figure). All three measurements were taken at similar position. (b) The average oscillation amplitude (black) and wavelength change (gray) of waveform Si band against applied strain on PDMS substrate (upper figure). As for wavelength measurement, compression (square) and tension (circle) were used on different basis. Bottom figure is the function of Si peak strain against applied strain. The curve in was obtained through calculation without fitting parameter.

related to sinusoidal waves indicate that, from the perspective of measurement, externally applied strain is equivalent to strain in the band. The results reveal two kinds of physical response of undulated bands toward external strain conditions. Under the tensile condition, the wave grows in a nonintuitive manner. The wavelength experiences no change under external strain, which is coincident with the mechanism of post-buckling. On the other hand, strain will cause a corresponding amplitude change. In this case, with the PDMS stretched, the strain of Si is reduced and even converges to 0% when the external strain equals the pre-assigned strain. In contrast, under the compressive condition, there is a reduction in wavelength and an increase in amplitude as the external load grows larger. This mechanical response is similar to folding bellows, which has a different behavior under stretching. As the radius of curvature at peak and valley is reduced during compression, the strain in Si increases with external strain. Nevertheless, the increase rate and magnitude of the Si strain are much smaller than the strain in the bands, as shown in Figure 11.29b. These characteristics enable them to have ductile behavior.

Under a strain system, the geometric configuration of the wave structure is consistent with the full response of a single crystalline Si band, which could be quantificationally described by equations. The equation is given by wavelength λ, λ_0 under a condition of initial flexure, and applied strain $\varepsilon_{applied}$:

$$\lambda = \begin{cases} \lambda_0 & \text{for tension} \\ \lambda_0(1+\varepsilon_{applied}) & \text{for compression} \end{cases} \tag{11.91}$$

At this point, the compression and tension conditions lead to asymmetry. For example, in the compression process, the tiny reversible peels off from bumps on PDMS and Si. This condition—together with systems without asymmetric behavior—when tension or compression is applied, the oscillation amplitude can be given by an independent expression

$$A = \sqrt{A_0^2 - h^2 \frac{\varepsilon_{applied}}{\varepsilon_c}} = h\sqrt{\frac{\varepsilon_{pre} - \varepsilon_{applied}}{\varepsilon_c} - 1}. \tag{11.92}$$

In the formula, A_0 is the value at the initial buckling state, which is applicable to a suitable strain (less than 10% or 15%). This expression explains the result from the experiment without a fitting parameter, as shown in Figure 11.29a. When an undulated structure forms under tension or compression strains, the maximum of Si will be given by mainly the deflection term

$$\varepsilon_{Si}^{peak} = 2\varepsilon_c\sqrt{\frac{\varepsilon_{pre} - \varepsilon_{applied}}{\varepsilon_c} - 1}. \tag{11.93}$$

This equation agrees well with the strain measured from the curvature in Figure 11.29b. The analysis formula is helpful in defining the range of applied strain, thus maintaining the present condition of the whole system, and preventing Si from fracturing. For 0.9% pre-strain, if the breaking strain of Si is 2% (the situation of compression or stretching), the scope can be from negative 27%–2.9%. By controlling pre-assigned strain, the scope of the strain (30%) can achieve balance with expected compression or tension deformation. For example, a 3.5% pre-strain (the maximum value that can be detected) could reach a scope from negative 24%–5.5%. This sort of calculations require the assumption that even though the deformation is maximized, the applied strain equals the strain in the band. The experiment shows that these estimated values are frequently oversized. The reason may be that PDMS can reduce to strain beyond both ends of the band and among the bands, which prevents the applied strain from fully transferring to the bands.

11.6.4 Application of Ductile Thin Films

Huang and Rogers [88] have prepared functional stretchable components. The following is an introduction to metal contact and the traditional method used to prepare an electrolyte layer. The diode and transistor are produced by two or three terminal devices. This procedure has laid a solid foundation for the advancing functionality of circuits. Integrated circuit components have a dual transmission process. Use the SOI chip on the nondeformational PDMS plate first, then on the substrate with a pre-assigned strain. This will produce a takeout device used to detect waveforms of the contact metal.

Figure 11.30a and b display the *p–n* surface diode's optic pictures and electrical responses. In the relative dispersion of the data, the electric characteristics of the component do not change under compression or stretching. Bending of the curve mainly comes from the contacting probe. These *p–n* junction diodes could be applied as photo detectors (in a reverse bias situation) or as photovoltaic

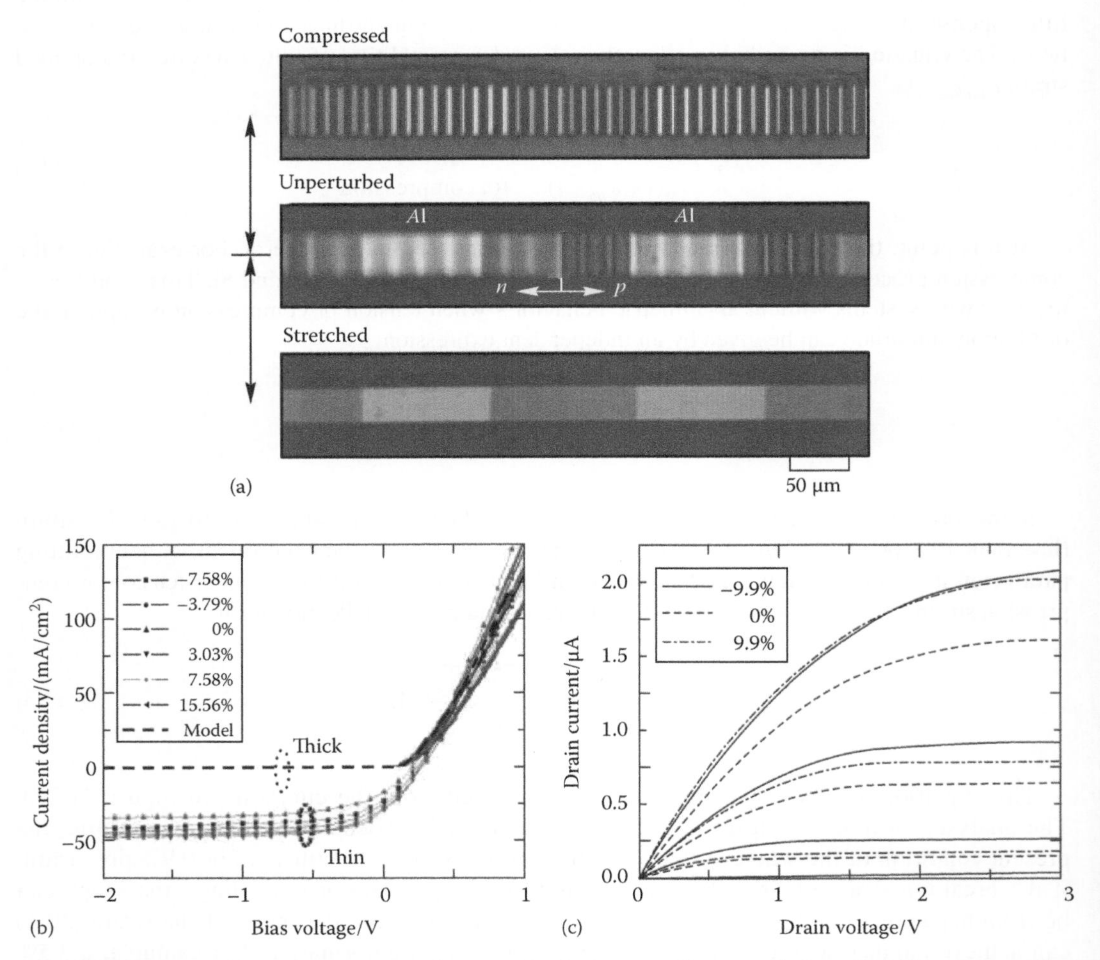

FIGURE 11.30 (a) The optical diagram of stretchable single crystalline *p–n* diode on PDMS basis, at applied strain of −11% (upper), 0% (middle), and 11% (bottom). *A*1 region is connected to thin *A*1 electrode. Slightly dark and dark regions are connected with *n*-type and *p*-type doping region, respectively. (b) The relation between current density and bias voltage of single crystalline Si *p–n* diode under different applied strains. The "thick" in the figure represents no-light condition and "thin" stands for lighting conditions. (c) Current–voltage relation of stretchable Schottky barrier Si MOSFET measured at applied strain of 9.9%, 0%, and 9.9% (trigger voltage varies from 0 to −5 V with step of 1 V).

devices. In addition, these p–n junctions can be used as ordinary rectifier units. When reverse bias voltage is −1 V, the photoelectric current density is ~35 mA/cm^2. In the forward bias condition, the short-circuit density and the open circuit current are 17 mA/cm^2 and 0.2 V, respectively. From this the space factor is derived as 0.3. The shape of this response is consistent with the model (see the solid line in Figure 11.30b). Even after redundant compression and tension for hundreds of repetitions, the performance of this device undergoes no significant change. Figure 11.30c describes the current–voltage curve of a stretchable, wavy, Si Schottky, barrier metal-oxide semiconductor, field-effect transistor (MOSFET), whose production process is similar to p–n junction diode, making an oxide gate from a 40 nm thick, integrated SiO_2 thin layer. Under the same technological conditions, the device parameter obtained from the profile transistor can be compared with those from SOI components. In comparison to the p–n diode, these profile transistors can be compressed or stretched to a large deformation and will recover to their original condition without damage to the component, or any variation of their characteristic electrical properties. Regarding the diode and transistor, the deformation that exceeds the end of the component on PDMS leads to strain in the bands, which is smaller than the applied strain. The complete stretch identity is a combination of the stretch property of the component, and deformation of the PDMS. For compressive strain larger than observed value, the PDMS will experience bending, which makes measurement increasingly difficult. Under a relatively large tensile strain, bands will suffer either breakage or slipping. The Si thickness, the band length, and the PDMS bonding strength all influence whether the device will remain perfect.

The Si metal-oxide semiconductor field-effect transistor, and the p–n diode, are only two of the large number of realizable waveform electric components. A complete circuit board or a thin Si metal board could also form a single or a dual axial wave structure. Except for the unique mechanical property of waveform structures, various kinds of strain coupling appear in many semi-conductor components. The effects on their electrical properties could create new opportunities in component structure design. These new structures endow the strain with mechanically modifiable periodic variation, which leads to unconventional electrical responses. This area is definitely a very promising research direction for the future.

EXERCISES

11.1 Explain the definition of thin film, and categories of thin films.
11.2 Access relevant books and understand the applications of thin films.
11.3 Derive Equation 11.1.
11.4 Derive Equation 11.7.
11.5 Briefly describe what mechanical property the indentation method measures, and explain why.
11.6 Explain the reason why the biaxial elastic modulus of a thin film is $M_f = E/(1-\nu_f)$.
11.7 Prove the Stoney formula (11.36).
11.8 Prove the Stoney formula for the multilayer case (11.37).
11.9 Derive Equations 11.79 through 11.81.
11.10 Given that residual stress exists in a certain thin film, the indentation method gives a depth of 200 nm, $A = 2 \times 10^6$ nm^2, $A_0 = 1\times10^6$ nm^2, and $H = 3$ GPa. Please calculate σ_R. What if $A = 1\times10^6$ nm^2, and $A_0 = 2\times10^6$ nm^2? (Berkovich indenter)
11.11 What is the effect of residual stress on the performance of a thin film?
11.12 What are the types of interface bonding between film and substrate?
11.13 Read through reference file [35] and learn more about the application of the scratch method in determining the fracture toughness of thin films.
11.14 How are samples prepared in the bubble method, which is used to measure the surface fracture toughness of thin films?
11.15 What is the difference between the indentation method used to determine thin film hardness, and the traditional hardness testing method?

REFERENCES

1. Freund L. B. and Suresh S. *Thin Film Materials: Stress, Defect Formation, and Surface Evolution*. Cambridge: Cambridge University Press, 2003.
2. Chen G. H. and Deng J. X. *New Electronic Thin Film Materials*. Beijing: Chemical Industry Press, 2002.
3. Ohring M. *Materials Science of Thin Films*. Singapore: Elsevier Pte Ltd, 2006.
4. Nix W. D. Mechanical properties of thin films. *Metall Mater Trans A*, 1989, 20: 2217–2245.
5. Ren F. Z., Zhou G. S., Zhao W. Z. et al. Beam three-point bending measurement of thin film elastic modulus. *Rare Metal Mat Eng*, 2004, 33(1): 109–112.
6. Oliver W. C. and Pharr G. M. An improved technique for determining hardness and elastic modulus using load and displacement sensing indentation experiments. *J Mater Res*, 1992, 7(6): 1564–1583.
7. Bhushan B. *Handbook of Micro/Nanotribology*. New York: CRC Press Inc, 1995.
8. Liao Y. G., Zhou Y. C., Huang Y. L. et al. Measuring elastic-plastic properties of thin films on elastic-plastic substrates by sharp indentation. *Mech Mater*, 2009, 41(3): 308–318.
9. Dao M., Chollacoop N. et al. Computation modeling of the forward and reverse problems in instrumented sharp indentation. *Acta Mater*, 2001, 49(19): 3899–3918.
10. СеДов, Л;.И;., *Similar Methods in Mechanics and Dimension Theory*. Beijing: Science Press, 1982, pp. 1–24.
11. Doener M. F. and Nix W. D. A method for interpreting the data from depth-sensing indentation instruments. *J Mater Res*, 1986, 1: 601–609.
12. Cheng Y. T. and Cheng C. M. Can stress-strain relationships be obtained from indentation curves using conical and pyramidal indenters? *J Mater Res*, 1999, 14(9): 3493–3496.
13. Knuyt G., Lauwerens W., Stals L. M. A unified theoretical model for tensile and compressive residual film stress. *Thin Solid Films*, 2000, 370(1): 232–237.
14. Chen C. J. and Lin K. L. Internal stress and adhesion of amorphous Ni-Cu-Palloy on aluminum. *Thin Solid Films*, 2000, 370: 232–237.
15. Andersen P., Moske M., Dyrbye K. et al. Stress formation and relaxation in amorphous Ta-Cr films. *Thin Solid Films*, 1999, 340: 205–209
16. Qian J., Zhao Y. P., Zhu R. Z. et al. Analysis of residual stress gradient in MEMS multi-layer structure. *Int J Nonlin Sci Num Simul*, 2002, 3(3/4): 727–730.
17. Pauleau Y. Generation and evolution of residual stresses in physical vapour-deposited thin films. *Vacuum*, 2001, 61: 175–181.
18. Cheng K. J. and Cheng S. Y. Analysis and computation of the internal stress in thin films. *Prog Nat Sci,* 1998, 8(6): 679–689.
19. Chen L. Q., Zhao M. H., and Zhang T. Y., Thin film mechanical testing techniques. *J Mech Strength*, 2001, 23(4): 413–442
20. Stoney G. G. The tension of metallic films deposited by electrolysis. London: *Proc Roy Soc*, 1909, 82(A553): 172–175.
21. Qian J., Liu C., Zhang D. C., and Zhao Y. B. The problem of the residual stress in the microelectronic mechanical systems. *J Mech Strength*(100 MEMS tribute album), 2001, 23(4): 393–401.
22. Yuan F. R. and Wu S. L. *Residual Stress Measurement and Calculation*. Hunan, China: Hunan University Press, 1987.
23. Tsui T. Y., Oliver W. C., and Pharr G. M. Influences of stress on the measurement of mechanical properties using nanoindentation: Part I. Experimental studies in an aluminum alloy. *J Mater Res*, 1996, 11(3): 752–759.
24. Bolshakov A., Oliver W. C., and Pharr G. M. Influences of stress on the measurement of mechanical properties using nanoindentation: Part II. Finite element simulations. *J Mater Res*, 1996, 11(3): 760–768.
25. Suresh S. and Giannakopoulos A. E. A new method for estimating residual stresses by instrumented sharp indentation. *Acta Mater*, 1998, 46(16): 5755–5767.
26. Lawn B. R. and Fuller E. R. Measurement of thin-layer surface stresses by indentation fracture. *J Mater Sci*, 1984, 19(12): 4061–4067.
27. Gruninger M. F., Brian R. L., Edward N. F. et al. Measurement of residual stresses in coatings brittle substrates by indentation fracture. *J Am Ceram Soc*, 1987, 70(5): 344–348.
28. Zhang T. Y., Chen L. Q., and Fu R. Measurement of residual stresses in thin films deposited on silicon wafers by indentation fracture. *Acta Mater*, 1999, 47(14): 3869–3878.
29. Zhou Y. C., Yang Z. Y., and Zheng X. J. Residual stress in PZT thin films prepared by pulsed laser deposition. *Surf Coat Tech*, 2003, 162(2–3): 202–211.

30. Xu B. S., Zhu S. H. et al. Surface engineering theory and technology. Beijing: National Defence Industry Press, 1999.
31. Strong J. On the cleaning of surfaces. *Rev Sci Instrum*, 1935, 6: 97–98.
32. Ma F. and Cai W. X. Interfacial bond strength of membrane-based characterization and evaluation. *Surf Technol*, 2001, 30(5): 5–19.
33. Zheng X. J., Zhou Y. C., Liu J. M. et al. Use of the nanomechanical fracture-testing for determining interfacial adhesion of PZT ferroelectrics thin films. *Surf Coat Technol*, 2003, 176: 67–74.
34. Zheng X. J., Zhou Y. C., and Li J. Y. Nano-indentation fracture test of $Pb(Zr_{0.52}Ti_{0.48})O_3$ ferroelectric thin films. *Acta Mater*, 2003, 51: 3985–3997.
35. Zhou Y. C., Tonomori T., Yoshida A. et al. Fracture characteristic of thermal barrier coatings after tensile and bending tests. *Surf Coat Technol*, 2002, 157(2–3): 118–127.
36. Suo Z. G. and Hutchinson W. Interface crack between two elastic layers. *Int J Fract*, 1990, 43(1): 1–18.
37. Zhou Y. C., Hashida T., and Jian C. Y. Determination of interface fracture toughness in thermal barrier coating system by blister test. *J Eng Mater Tech, ASME*, 2003, 125(2): 176–182.
38. Smith (USA). Editor: Yao X. *Classic Dielectric Science Books*. Xian, China: Xi'an Jiaotong University Press, 2006.
39. Jiang D. L., Li L. T., Ouyang S. W. et al. *Inorganic Non-Metallic Materials Engineering*. Beijing: Chemical Industry Press, 2005.
40. Fang D. N., Wan Y. P., Feng X. et al. Deformation and fracture of functional ferromagnetics. *Appl Mech Rev*, 2008, 61: 020803–020811.
41. Liu L. L., Zhang Y. S., and Zhang T. Y. Strain relaxation in heteroepitaxial films by misfit twinning. I. critical thickness. *J Appl Phys*, 2007, 101(6): 063501.
42. Gao H. J. and Nix W. D. Surface roughness of heteroepitaxial thin films. *Annu Rev Mater Sci*, 1999, 29: 173–209.
43. Wang J. and Zhang T. Y. Effects of nonequally biaxial misfit strains on the phase diagram and dielectric properties of epitaxial ferroelectric thin films. *Appl Phys Lett*, 2005, 86(19): 192905.
44. Zhang T. Y. and Zhao M. H. Equilibrium depth and spacing of cracks in a tensile residual stressed thin film deposited on a brittle substrate. *Eng Fract Mech*, 2002, 69(5): 589–596.
45. Zhao M. H., Fu R., Lu D. X. et al. Critical thickness for cracking of $Pb(Zr_{0.53}Ti_{0.47})O_3$ thin films deposited on Pt/Ti/Si(100) substrates. *Acta Mate*, 2002, 50(17): 4241–4254.
46. Thouless M. D., Olsson E., and Gupta A. Cracking of brittle films on an elastic substrate. *Acta Metall Mater*, 1992, 40: 1287–1292.
47. Wang J. and Zhang T. Y. Influence of depolarization field on polarization states in epitaxial ferroelectric thin films with nonequally biaxial misfit strains. *Phys Rev B*, 2008, 77(1): 014104.
48. Wang J. and Zhang T. Y. Size effects in epitaxial ferroelectric islands and thin films. *Phys Rev B*, 2006, 73(14): 144107.
49. Hu M. S. and Evans A. G. The cracking and decohesion of thin films on substrates. *Acta Metall*, 1989, 37: 917–925.
50. Lin Y., Liu S. W., Chen C. L. et al. Anisotropic in-plane strains and dielectric properties in $(Pb,Sr)TiO_3$ thin films on $NdGaO_3$ substrates. *Appl Phys Lett*, 2004, 84(4): 577–579.
51. Lee H. N. and Hesse D. Anisotropic ferroelectric properties of epitaxially twinned $Bi_{3.25}La_{0.75}Ti_3O_{12}$ thin films grown with three different orientations. *Appl Phys Lett*, 2002, 80(6): 1040–1042.
52. Garga A., Barber Z. H., Dawber M. et al. Orientation dependence of ferroelectric properties of pulsed-laser-ablated $Bi_{4-x}Nd_xTi_3O_{12}$ films. *Appl Phys Lett*, 2003, 83(12): 2414–2416.
53. Pertsev N. A., Zembilgotov A. G., and Tagantsev A. K. Effect of mechanical boundary conditions on phase diagrams of epitaxial ferroelectric thin films. *Phys Rev Lett*, 1998, 80(9): 1988–1991.
54. Hu Z. S., Tang M. H., Wang J. B. et al. Effect of extrapolation length on the phase transformation of epitaxial ferroelectric thin films. *Phys B*, 2008, 403: 3700–3704.
55. Glinchuk M. D., Morozovska A. N., and Eliseev E. A. Ferroelectric thin films phase diagrams with self-polarized phase and electret state. *J Appl Phys*, 2006, 99(11): 114102.
56. Forrest S. R. The path to ubiquitous and low-cost organic electronic appliances on plastic. *Nature*, 2004, 428: 911–918.
57. Rogers J. A., Bao Z., Baldwin K. et al. Paper-like electronic displays: Large-area rubber-stamped plastic sheets of electronics and microencapsulated electrophoretic inks. *Proc Natl Acad Sci*, 2001, 98: 4835–4840.
58. Jacobs H. O., Tao A. R., Schwartz A. et al. Fabrication of a cylindrical display by patterned assembly. *Science*, 2002, 296: 323–325.

59. Huitema H. E. A., Gelinck G. H., Puttenet J. V. D. et al. Plastic transistors in active-matrix displays. *Nature*, 2001, 414: 599.
60. Sheraw C. D., Zhou L., Huang J. R. et al. Organic thin-film transistor-driven polymer-dispersed liquid crystal displays on flexible polymeric substrates. *Appl Phys Lett*, 2002, 80: 1088.
61. Chen Y., Au J., Kazlas P. et al. Electronic paper: Flexible active-matrix electronic ink display. *Nature*, 2003, 423: 136.
62. Jin H. C., Abelson J. R., Erhardt M. K., et al. Soft lithographic fabrication of an image sensor array on a curved substrate. *J Vac Sci Technol B*, 2004, 22: 2548.
63. Hsu P. H. I., Huang M., Gleskova H. et al, Effects of mechanical strain on TFTs on spherical domes. *IEEE Trans Elect Dev*, 2004, 51: 371–377.
64. Someya T., Sekitani T., Iba S. et al. A large-area, flexible pressure sensor matrix with organic field-effect transistors for artificial skin applications. *Proc Natl Acad Sci*, 2004, 101: 9966–9970.
65. Lim H. C., Schulkin B., Pulickal M. J. et al. Flexible membrane pressure sensor. *Sens Act A*, 2005, 119: 332–335.
66. Vandeputte J. Mechanical resistance of a single-crystal silicon wafer. U.S. Patent, 2003, 6(580): 151.
67. Sekitani T., Kato Y., Iba S., et al. Bending experiment on pentacene field-effect transistors on plastic films. *Appl Phys Lett*, 2005, 86: 073511.
68. Menard E., Nuzzo R. G., and Rogers J. A. Bendable single crystal silicon thin film transistors formed by printing on plastic substrates. *Appl Phys Lett*, 2005, 86: 093507.
69. Gleskova H., Hsu P. I., Xi Z, et al. Field-effect mobility of amorphous silicon thin-film transistors under strain. *J Noncryst Solids*, 2004, 338: 732–735.
70. Hur S. H., Park O. O., Rogers J. A. Extreme bendability of single-walled carbon nanotube networks transferred from high-temperature growth substrates to plastic and their use in thin-film transistors. *Appl Phys Lett*, 2005, 86: 243502.
71. Duan X. F. High-performance thin-film transistors using semiconductor nanowires and nanoribbons. *Nature*, 2003, 425: 274.
72. Suo Z., Ma E. Y., Gleskova H. et al. Mechanics of rollable and foldable film-on-foil electronics. *Appl Phys Lett*, 1999, 74: 1177.
73. Loo Y. L., Someya T., Baldwin K. W. et al. Soft, conformable electrical contacts for organic semiconductors: High-resolution plastic circuits by lamination. *Proc Natl Acad Sci*, 2002, 99: 10252–10256.
74. Someya T., Kato Y., Sekitani T. et al. Conformable, flexible, large-area networks of pressure and thermal sensors with organic transistor active matrixes. *Proc Natl Acad Sci*, 2005, 102: 12321.
75. Kim S. K., Lin C. C., Lei X., et al. Crosstalk reduction in mixed-signal 3-D integrated circuits with inter-device layer ground planes. *Proc IEEE*, 2005, 93: 1459.
76. Lacour S. P., Wagner S., Huang Z. et al. Stretchable gold conductors on elastomeric substrates. *Appl Phys Lett*, 2003, 82: 2404.
77. Gray D. S., Tien J., and Chen C. S. High-conductivity elastomeric electronics. *Adv Mater*, 2004, 16: 393.
78. Faez R., Gazotti W. A., and Paoli M. A. D. An elastomeric conductor based on polyaniline prepared by mechanical mixing. *Polymer*, 1999, 40: 5497.
79. Marquette C. A. and Blum L. J. Conducting elastomer surface texturing: A path to electrode spotting: Application to the biochip production. *Biosens Bioelectron*, 2004, 20: 197–203.
80. Khang D. Y., Jiang H. Q., Huang Y. et al. A stretchable form of single-crystal silicon for high-performance electronics on rubber substrates. *Science*, 2006, 311: 208.
81. Chen X. and Hutchinson J. W. Herringbone buckling patterns of compressed thin films on compliant substrates. *J Appl Mech*, 2004, 71: 597.
82. Huang Z. Y., Hong W., and Suo Z. Nonlinear analyses of wrinkles in a film bonded to a compliant substrate. *J Mech Phys Solids*, 2005, 53: 2101.
83. INSPEC. *Properties of Silicon*. New York: Institution of Electrical Engineers, 1988.
84. Bietsch A. and Michel B. Conformal contact and pattern stability of stamps used for soft lithography. *J Appl Phys*, 2000, 88: 4310.
85. Bowden N. et al. Spontaneous formation of ordered structures in thin films of metals supported on an elastomeric polymer. *Nature*, 1998, 393: 146.
86. Huck W. T. S. et al. Ordering of spontaneously formed buckles on planar surfaces. *Langmuir*, 2000, 16: 3497.
87. Stafford C. M. et al. A buckling-based metrology for measuring the elastic moduli of polymeric thin films. *Nat Mater*, 2004, 3: 545.
88. Kim D. H., Ahn J. H., Choi W. M. et al. Stretchable and foldable silicon integrated circuits. *Science*, 2008, 320: 507–511.

12 Mechanical Properties of Polymer Materials

Polymers are an important category of engineering materials, and are widely used in a variety of areas such as mechanical, construction, automotive, electrical appliances, light industry, electronics, and aerospace. Because of their unique molecular chain structures and aggregation structures, polymers have different physical properties compared with other materials, especially their mechanical properties. The mechanical properties of a polymer strongly depend on the conditions of temperature, loading time, deformation rate, and load frequency. Usually—depending on different temperatures and the time of observation—a polymer shows different mechanical behaviors including the glassy state, the viscoelastic state, the high-elastic state (hyperelastic or rubbery state), and the viscous flow state. As a structural material, the main features of a polymer's mechanical properties are its viscoelasticity and its hyperelasticity. This chapter will introduce viscoelasticity and its mechanical models, the hyperelasticity of polymer materials, and the yield and pattern of the brittle–ductile transition of polymer materials.

12.1 VISCOELASTICITY OF HIGH POLYMER

In understanding substances and materials, human beings first became familiar with two types of materials: elastic solids and viscous fluids. Elastic solids have specific volumes and configurations, and their stress state and deformation are time-independent when subject to static loads. Elastic solids can be fully restored after the external force is removed. The work done by external forces in the elastic deformation process is not only stored in the form of elastic potential energy, but is also completely released in the process of unloading. On the contrary, viscous fluids do not have determined volumes and configurations, and their shapes depend on the container. Under external forces, they continually deform over time, resulting in an irreversible flow. When deformation movement occurs, the adjacent fluid layers will cause internal friction that consumes energy. In general, polymer materials are neither attributable to the elastic solid, nor attributable to the viscous fluid. They often have the characteristics of both elastic solids and viscous fluids, and comprehensively show the two different mechanisms of elastic and viscous deformations. This property of substances is called viscoelasticity [1]. The macroscopic phenomenological description of the viscoelasticity of materials focuses on the mechanical behavior of materials, and its relevance to time, loading rate, frequency, and temperature.

12.1.1 Stress Relaxation and Strain Rate Effect

Under certain loads, the strain or stress of elastic solids is a constant value, and does not change over time. For an ideal viscous fluid, deformation increases at a constant strain rate over time. Viscoelastic materials under certain stresses will continue to produce deformation, and under certain strain conditions their stress amplitudes will reduce over time.

12.1.1.1 Stress Relaxation

Under a constant strain, the phenomenon or process of stress decreasing over time is known as stress relaxation. Figure 12.1 shows the general stress relaxation. At the beginning, the stress quickly

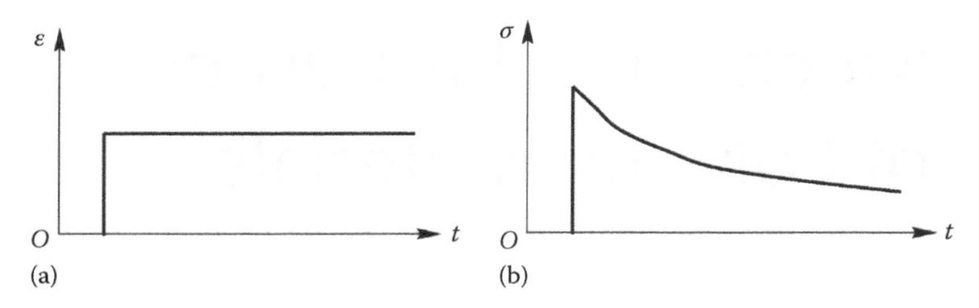

FIGURE 12.1 Stress relaxation. (a) Strain vs. time. (b) Stress vs. time.

decays, and then gradually reduces and tends to a constant value. Thus, under the conditions of a certain strain, a material whose stress quickly reduces and finally tends to zero is called a viscoelastic fluid, while a material whose stress after a longer period of time decreases to a constant value is called a viscoelastic solid.

12.1.1.2 Strain Rate Effect

In addition to creep (see Chapter 7), stress relaxation, and anelastic phenomena, the loading rate effect is also an important feature of a viscoelastic material. In order to analyze the correlation between the viscoelasticity of materials and the loading rate, we must study the stress response under different strain rates, or the impact of different stress rates on the strain. When the strain rate increases, the modulus, the yield stress, and the fracture strength of the material normally increase, but the ultimate elongation decreases (see Figure 12.2). This effect in a polymer is more significant than in a metal, and the effect for a flexible polymer is more significant than for a rigid polymer. Figure 12.3 shows the stress–strain curve [1] of a high-density polyethylene (HDPE) under different strain rates, and illustrates the impact of strain rate on the yield stress of materials.

12.1.2 Frequency-Dependent Properties

Creep, stress relaxation, and the strain rate effect describe viscoelastic behavior under the quasi-static load over a period of time. However, the applied load on many viscoelastic materials and their structures alternates over time. Due to viscous effects, materials can become frequency-dependent, resulting in energy dissipation—one of the important properties of viscoelasticity [1].

When an elastic solid is subject to a sinusoidal (cosinusoidal) stress, the strain and stress change in phase, in the form of a sinusoid (cosinusoid). There is no energy loss at this time, as shown in Figure 12.4a. For an ideal viscous fluid, we can see from the stress–strain rate relationship $\sigma = \eta\dot{\varepsilon}$ that the strain lag phase is $\pi/2$ and the lag time is $\pi/2\omega$, where ω is the angular frequency (see Figure 12.4b). For the general linear viscoelastic body, the strain response under harmonic stress is

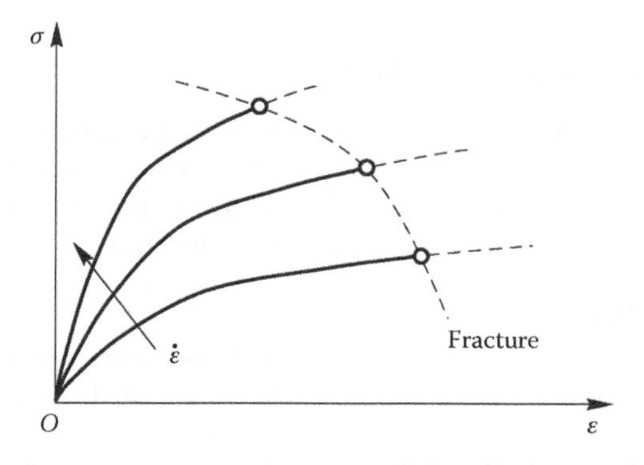

FIGURE 12.2 Effect of strain rate to stress–strain curves of viscoelastic materials.

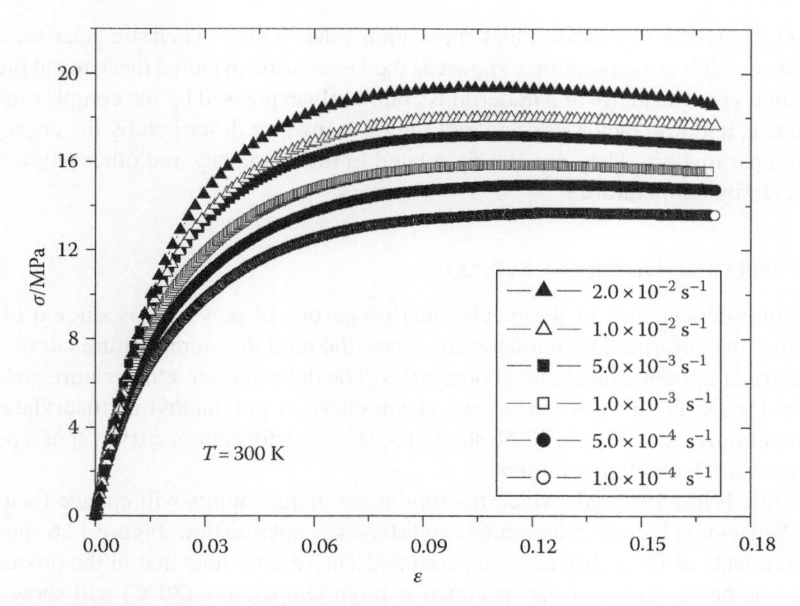

FIGURE 12.3 Stress–strain curves of HDPE under different strain rates. (From Yang, T.-Q. et al., *Theory of Viscoelasticity and Application*, Science Press, Beijing, 2004.)

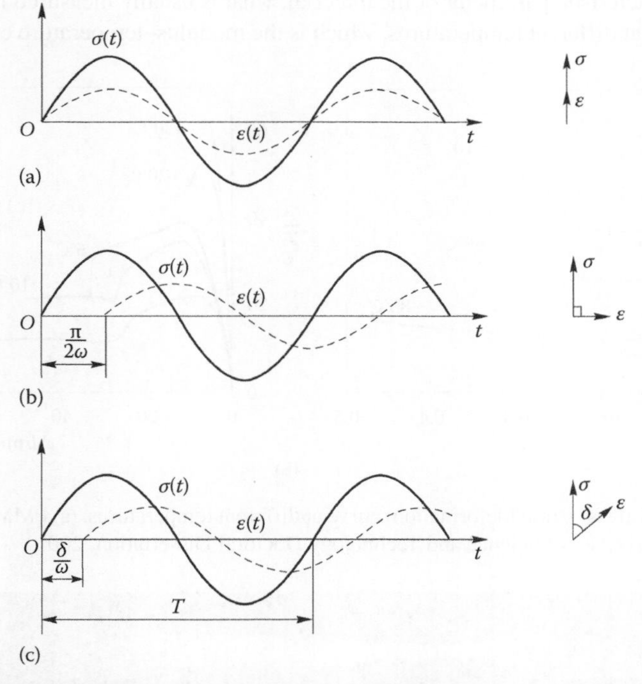

FIGURE 12.4 Strain response under alternating stress. (a) Elastic solid. (b) Ideal viscous fluid. (c) General linear viscoelastic body. (From Yang, T.-Q. et al., *Theory of Viscoelasticity and Application,* Science Press, Beijing, 2004.)

between elastic solids and viscous fluids. If we use δ to represent the strain lag phase, then we get $0 < \delta < \pi/2$, and the lag time is δ/ω, as shown in Figure 12.4c.

Heat generation in all kinds of running tires and transmission parts, and the damping effects of vibration reduction or soundproof materials and their structures are closely related to the viscoelastic properties of materials. Therefore, the study of viscoelastic behavior under steady-state resonance conditions has important theoretical significance and application value. This viscoelastic behavior under steady-state resonance conditions is sometimes known as the viscoelastic dynamic mechanical property.

The dynamic viscoelasticity of a material is commonly expressed by the complex modulus, complex compliance, and loss factor (lag angle of tangent $\tan\delta$), or described by the energy dissipation and its related parameters. They are closely related to the frequency and other physical properties, and are affected by temperature.

12.1.3 Temperature-Dependent Properties

The temperature dependence of the mechanical properties of polymers is studied in a variety of ways, including the deformation–temperature curve, the modulus–temperature curve, and the temperature spectra of dynamic mechanical properties. The deformation–temperature curve is the simplest method. Figure 12.5a shows the stress–strain curve of polymethyl methacrylate (PMMA) at different temperatures, and Figure 12.5b shows the stress–deformation curve [2] of a polypropylene (PP) specimen at different temperatures.

It is clear that brittle materials which fracture at low temperatures will change their deformation and fracture behavior at higher temperatures, and show viscoplastic flow. Figure 12.6 shows the tensile necking experiments of PP at different temperatures. The results show that in the process of necking development, the necking area of the specimen at room temperature (30°C) will show some serious stress-whitening phenomena. When the temperature rises to 50°C, stress-whitening mitigates; and at 100°C, the experiment shows no stress-whitening phenomena in the necking area [2]. Because deformation is not the characteristic parameter of the material, what is usually measured is the variation of the specimen modulus at different temperatures, which is the modulus–temperature curve.

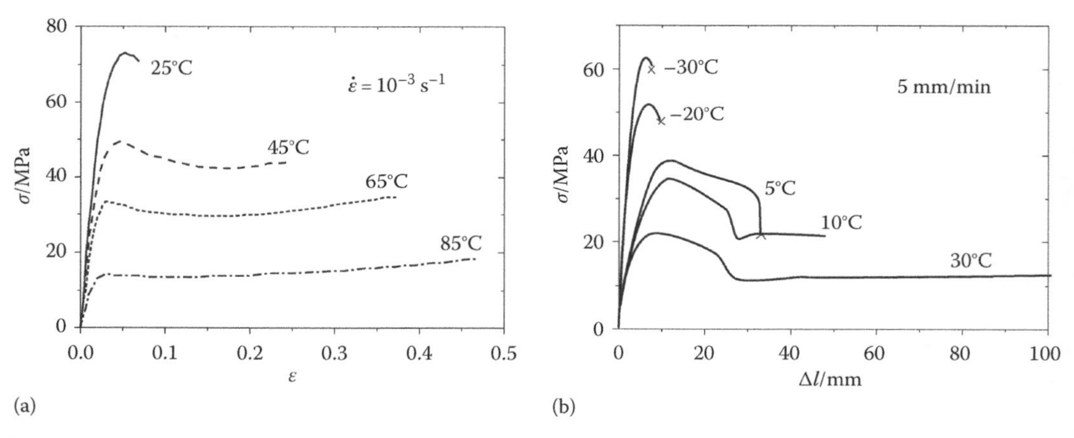

FIGURE 12.5 The stress–strain (deformation) curve at different temperatures. (a) PMMA. (b) PP. (From Luo, W.-B., Huazhong University of Science and Technology Doctoral Dissertation, 2001.)

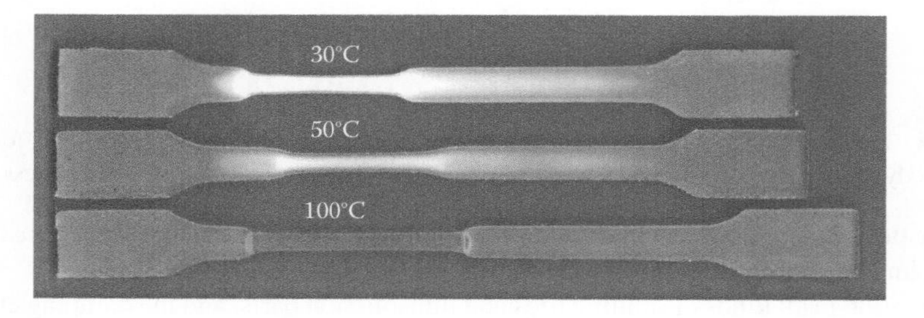

FIGURE 12.6 The stress-whitening phenomenon of the necking area of the PP specimen at different temperatures. (From Luo, W.-B., Huazhong University of Science and Technology Doctoral Dissertation, 2001.)

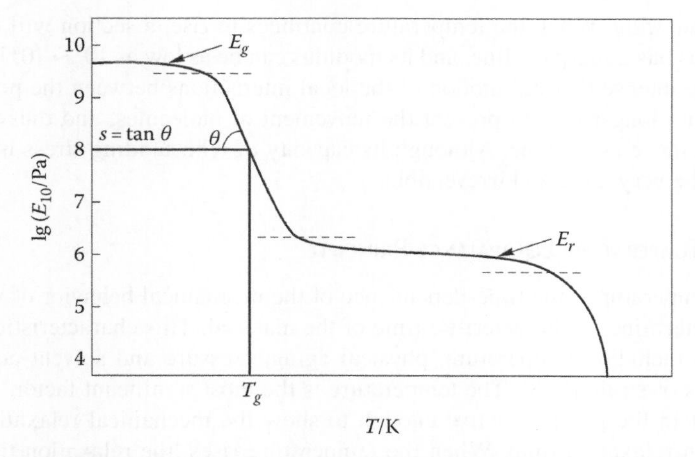

FIGURE 12.7 The typical curve of the elastic modulus of amorphous polymers with temperature change.

Figure 12.7 is the typical amorphous polymer modulus–temperature curve, which describes the temperature dependence of the modulus under the timing conditions. The main method of static determination is the tensile stress relaxation test, taking the modulus $E_{10} = \sigma(t = 10\text{s}) / \varepsilon_0$ at time $t = 10$ s as the characteristic parameters. Typical $E_{10} - T$ curves consist of four sections, corresponding to the four states of the mechanical properties of polymers: (1) the glassy state, (2) the viscoelastic state, (3) the rubbery state, and (4) the viscous flow state. The basic characteristics of each material state, and the corresponding molecular motion conditions, can be summarized as follows [1].

1. *Glassy state*: When the temperature is below the glass transition temperature (T_g), the molecular chain of the material and the movement of its chain segments are in the "frozen" state. The movement of the chain segments changes their conformation through internal rotation (or the spread of the movement of the chain segments from one location to another) in less than 10 s, even if there are only a few. This applies only for thermal vibration in a fixed position. The $E_{10} - T$ curve has a corresponding plateau, taking the modulus E_g as a characteristic parameter. The modulus is up to 10^9 Pa orders of magnitude. At this time, the material has hard and brittle mechanical properties, and is in the glassy state. As the temperature increases, the amplitude of the thermal motion increases. The molecular chain loosens due to repulsion. A secondary relaxation transition, corresponding to the individual chain movement on the side chain or main chain, might be expected. However, there is little impact on both the entire molecular chain conformation and the resulting modulus changes.
2. *Viscoelastic state*: When the temperature rises and approaches T_g, the material enters into the glass transition zone, and the modulus can be reduced by three orders of magnitude. At this time, the material has ductility. There are two characteristic parameters representing this section: one is the corresponding transition temperature T_g of the modulus $E_{10} = 10^8$ Pa, and the other is the negative slope $s = \tan\theta$ of the curve at the turning point. In this section, although the overall movement of the molecular chain is still not possible, the free volume will gradually increase (at temperature T_g, free volume is about 2.5%) and the chain segments start to change from short-range diffusion to greater freedom of movement due to the exacerbated thermal vibration.
3. *Rubbery state*: As the temperature continues to rise, the material enters into the rubbery state, or called the high-elastic state. Its modulus is generally stable at 10^6 Pa orders of magnitude, and its value E_r can be used as a characteristic parameter. In this section, the number of chain segments with short-range diffusion increases to 20–50, the diffusion movement accelerates, and the relaxation time is $\ll$10 s. The movement of the chain segments constantly changes the conformation, but local interactions between the polymer chains still exist (van der Waals cross-linking), thus making the overall long-range movement of the molecular chain more difficult.

4. *Viscous flow state*: When the temperature continues to rise, a section will emerge where the modulus has a sharp decline, and its modulus can be as low as $10^5 \sim 10^4$ Pa. This is due to the more intense thermal motion of the local interactions between the polymer chains, which are no longer able to prevent the movement of molecules, and thus the molecular chains can move as a whole. Although its capacity of withstanding stress is very low, the strain can be very large and irreversible.

12.1.4 Time–Temperature Equivalence Principle

Under certain temperatures, the time-dependence of the mechanical behavior of viscoelastic polymer materials determines a characteristic time of the material. This characteristic time is affected by many factors including temperature, physical aging, pressure and solvent concentration, and the level of stress or strain [3–8]. The temperature is the most significant factor. In order to make a movement unit in the polymer active enough to show the mechanical relaxation phenomenon, it needs a certain relaxation time. When the temperature rises, the relaxation time is shortened. Therefore, the same mechanical relaxation phenomenon can appear either at a higher temperature and a shorter load time (or observed in a short period of time), or at a lower temperature and a longer load time (or observed in a longer period of time). This is because at a lower temperature, the relaxation time of the molecular movement is longer.

The impact of applied loads on polymers may not be observed, or takes a very long time before it is observed. Therefore, if the temperature is increased and the relaxation time is shortened, it can be observed within a short period of time on its mechanical response. Under alternating loads, the interaction time is equivalent to the reciprocal of the effect of frequency, so that reducing the frequency represents an increase of the load time (extension of the observation time). This also manifests the mechanical relaxation phenomenon, which originally cannot keep up with the response. Therefore, the extension of time (or lowering of frequency), and the increase of temperature, has an equivalent impact on molecular movement, and thus an equivalent impact on the viscoelastic behavior of polymers [9]. That is, changing the temperature and changing the length of time is equivalent, and is known as the time–temperature equivalence principle.

According to the time–temperature equivalence principle, the stress relaxation modulus curves of some materials at different temperatures can translate along the time axis and be superimposed together, constituting the relaxation modulus master curve at a reference temperature (see Figure 12.8).

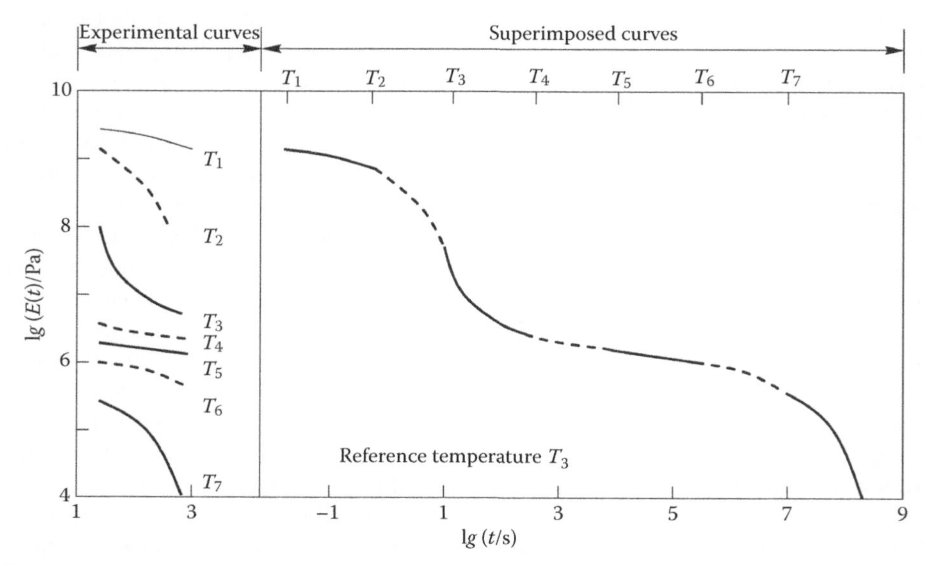

FIGURE 12.8 The constructed schematic of a relaxation modulus master curve.

Typically, when the temperature is low, it can only come to a section of the modulus–time curve in a short period of time. For example, it takes a longer time for the section of the T_1 temperature curve on the left side of Figure 12.8 to continually receive the modulus in the T_1 temperature–time relationship. If the temperature is increased, it can come to another section of the curve at the T_2 temperature in a relatively short time. If we shift this dotted line to the right, you can see the two curves of temperatures T_1 and T_2 completely overlap in the same part of the modulus. This means that the short-time modulus at the T_2 temperature is the result of a long period of continued observation at the T_1 temperature. Such a superimposed curve would indicate the modulus of a reference temperature–time curve. For the reference temperature corresponding to the section of the modulus–time curve, its corresponding time scale should be identical with the actual observation time. See Figure 12.8, which shows the experimental curve and the superimposed curve at temperature T_3.

If you constantly maintain the shape of the curve, and move the curve corresponding to the temperature T along the time axis to superimpose the curve corresponding to the temperature T_0, the amount needed to move is denoted by ϕ_T, and the time–temperature equivalence is given by

$$E(T,t) = E\left(\frac{T_0,t}{\phi_T}\right). \tag{12.1}$$

It is shown that the modulus at temperature T and at time t can be expressed by the value at temperature T_0 and at time t/ϕ_T, where ϕ_T is called the temperature shift factor, T is the test temperature, and T_0 is the reference temperature. Clearly, for the relaxation modulus, when $T > T_0$, $\phi_T < 1$; and when $T < T_0$, $\phi_T > 1$; similarly for the creep compliance J, the storage modulus E' of the dynamic mechanical tests, and the storage compliance J'. According to the time–temperature equivalence [9], we have

$$J(T,t) = J\left(\frac{T_0,t}{\phi_T}\right),$$

$$E'(T,\omega) = E'(T_0,\omega\phi_T),$$

$$J'(T,\omega) = J'(T_0,\omega\phi_T).$$

Discussed here is the simple behavior of the thermal flow of the polymer (the thermal-rheology), which is the change in viscoelastic materials caused by temperature differences, and is equivalent to the result caused by a horizontal shift of the logarithmic time coordinate. Figure 12.9 shows that

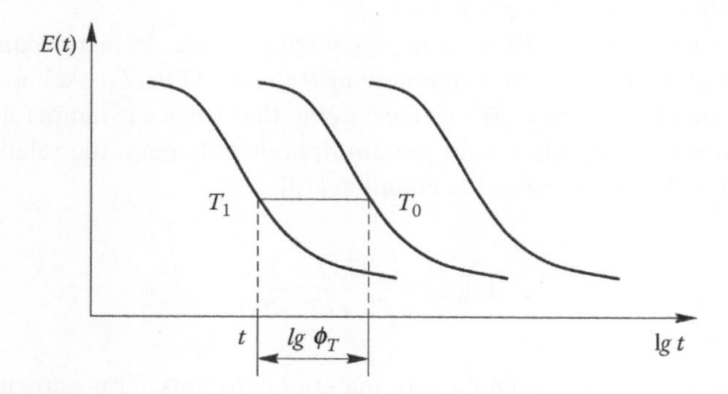

FIGURE 12.9 The schematic diagram of time–temperature translation.

the relaxation modulus $E(t)$ at an increased temperature is equivalent to the curve shifted along the direction of decreasing logarithmic time at the reference temperature T_0. Therefore, we may write

$$E(T,t) = E\left(\frac{T_0, t}{\phi_T}\right) = t\Lambda(T_0, \log t - \log\phi_T). \tag{12.2}$$

The first equation explains that the temperature change is equivalent to the time change, and the second equation describes the change of time scale, reflecting the translation amount of the value of the logarithmic coordinate.

Strictly speaking, due to the change of temperature, the modulus itself and the material density also change, and the modulus changes with the amount of material contained in the unit volume. Considering these factors, the earlier time–temperature conversion relationship requires modification by the factor $\rho T/\rho_0 T_0$, where ρ and ρ_0 are the densities of the material under temperatures T and T_0, respectively. This correction is equivalent to the vertical shift of the curve in the stress relaxation modulus curve. After such corrections, the temperature dependence of the viscoelasticity of materials can be expressed by the time–temperature equivalence principle as follows:

$$E(T,t) = \frac{\rho T}{\rho_0 T_0} E\left(\frac{T_0, t}{\phi_T}\right),$$

$$J(T,t) = \frac{\rho_0 T_0}{\rho T} J\left(\frac{T_0, t}{\phi_T}\right),$$

$$E'(T,\omega) = \frac{\rho T}{\rho_0 T_0} E'(T_0, \omega\phi_T).$$

The application of the time–temperature equivalence principle could greatly simplify polymer viscoelasticity testing. In order to describe the time (or frequency) and temperature dependence of the mechanical behavior of polymers, the relationship $E(T,t) \sim \log t \sim T$, or $E(T,\omega) \sim \log\omega \sim T$, must be addressed.

There are two independent variables (T,t) and (T,ω), and thus the graphical representation is a three-dimensional curved surface. The application of the time–temperature equivalence principle linking the two variables (T,t) and (T,ω) together reduces the independent variables to one. This independent variable, with the temperature shift factor ϕ_T, can simplify the original three-dimensional representation of a curved surface to a plane curve. In addition, the use of the time–temperature equivalence principle may reduce the time needed for the test through the method of raising the temperature and making the translation. This method of accelerating the test is very important in the long-term study of material performance.

The temperature shift factor $\phi_T(T)$ is a function of temperature. In fact, it can be expressed as the relaxation time ratio of the material at different temperatures $\tau(T)/\tau(T_0)$. According to experimental results, Williams, Landel, and Ferry (WLF) have found that when the temperature is near the glass transformation temperature, for almost all the amorphous polymers, the relationship between the shift factor $\log\phi_T$ and $(T - T_0)$ satisfies the equation [10]

$$\log\phi_T = \frac{-C_1(T - T_0)}{C_2 + (T - T_0)}. \tag{12.3}$$

This is the WLF equation, where C_1 and C_2 are material constants. This equation usually applies at a temperature range from T_g to T_g + 50°C. If we take the glass transition temperature as a reference temperature, then $C_1 = 17.44$, and $C_2 = 51.6$. The WLF equation can be derived from the free volume theory, and has been widely applied in the study of temperature dependencies in the viscoelastic mechanical properties of polymers.

12.2 MECHANICAL MODELS OF VISCOELASTIC BEHAVIOR OF POLYMERS

The viscoelastic mechanical properties of polymers are usually expressed by time-dependent material functions. The differential and integral types are two main categories for describing the viscoelastic mechanical behavior of the stress–strain–time relationship. This section introduces the model descriptions, Boltzmann superposition principle, and material functions of linear viscoelastic mechanical properties; and discusses the constitutive equations of one-dimensional differential and integral types. For more detailed discussions of the theory of linear viscoelasticity and the research advances in nonlinear viscoelasticity, refer to References [1,11].

12.2.1 A Simple Description of Viscoelastic Mechanical Behavior

The differential constitutive equations were widely used in the early development of viscoelasticity. In order to understand the mechanical relaxation behavior of polymers and other viscoelastic materials, researchers often used discrete elastic components and viscous components, namely springs and dampers (dashpots), as well as their different combinations, to describe a polymer's linear viscoelastic mechanical behavior. Through these mechanical models, we can obtain the differential stress–strain–time relationship of materials.

12.2.1.1 Basic Components

The elastic component is represented by a spring, as shown in Figure 12.10, subject to Hooke's law; and the stress and strain are independent of time:

$$\sigma = E\varepsilon, \tag{12.4a}$$

or

$$\tau = G\gamma, \tag{12.4b}$$

where σ, τ, ε, and γ are the normal stress, the shear stress, the normal strain, and the shear strain respectively; and E and G are the elastic modulus and shear modulus, which are constants. The stress–strain ratio relationship of the spring does not change over time, showing the instantaneous elastic deformation and recovery.

Viscous components, namely dampers, sometimes referred to as dashpots, are subject to Newton's law of viscosity (see Figure 12.11):

$$\tau = \eta_1\dot{\gamma}, \tag{12.5a}$$

FIGURE 12.10 Elastic component.

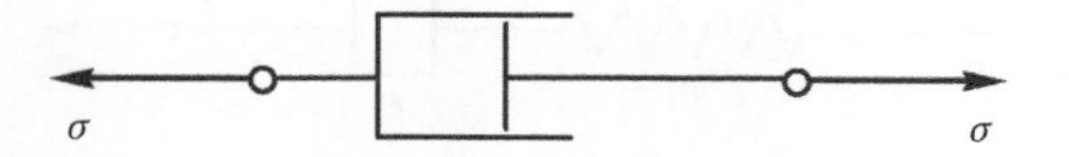

FIGURE 12.11 Viscous component.

or

$$\sigma = \eta \dot{\varepsilon}, \tag{12.5b}$$

where η or η_1 is the viscosity coefficient, and $\dot{\varepsilon} = d\varepsilon/dt$ is the strain rate. The rheological features of dampers can be explained by their quasi-static responses to constant stress or strain. Here, the quasi-static response means that the loading rate suddenly imposed on the object does not stir up a dynamic response.

Through a series or parallel construction of the two basic elastic and viscous components, a variety of viscoelastic material models can be constituted. In parallel, the stress of the whole model is the sum of the stress of each basic component, and the deformation, or deformation rate, of each basic component is the same. In series, the deformation, or deformation rate, of the entire model is the sum of the deformation, or deformation rate, of each basic component; and the stress of each basic component is equal, and equal to the total stress. The simplest combination model is a binary model, constituted by the series or parallel of a spring element and a damper element; that is, the Maxwell model and the Kelvin model.

12.2.1.2 Maxwell Model

The Maxwell model is composed of an elastic component and a viscous component in series, as shown in Figure 12.12. Assuming that the model is under the effect of stress $\sigma(t)$, and that the strain of the spring and the damper is ε_1 and ε_2, respectively, the total strain is

$$\varepsilon = \varepsilon_1 + \varepsilon_2.$$

Using Equations 12.4a and 12.5b, we can obtain

$$\dot{\varepsilon} = \frac{\dot{\sigma}}{E} + \frac{\sigma}{\eta}. \tag{12.6a}$$

Equation 12.6a can be rewritten as

$$\sigma + p_1 \dot{\sigma} = q_1 \dot{\varepsilon}, \tag{12.6b}$$

where the model parameters $p_1 = \eta/E$ and $q_1 = \eta$ represent the material's constants.

Equation 12.6a and b is the stress–strain–time relationship of the Maxwell model; it is a differential-type constitutive equation. If the material constant is known, we can use Equation 12.6a and b to analyze the process and phenomenon of creep, creep recovery, and stress relaxation.

1. *Creep.* Under a step stress σ_0, the total strain of the Maxwell model is the sum of the strain of the spring and the strain of the damper. That is,

$$\varepsilon(t) = \frac{\sigma_0}{E} + \frac{\sigma_0}{\eta} t. \tag{12.7}$$

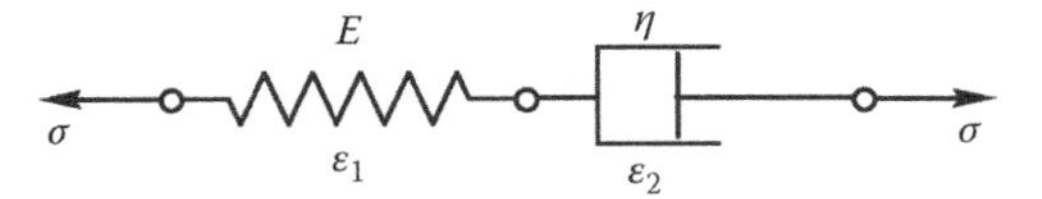

FIGURE 12.12 Maxwell model.

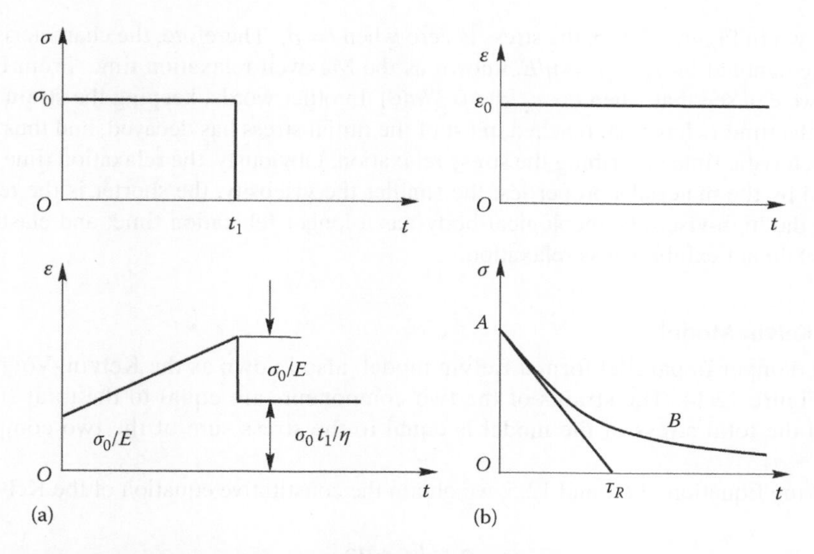

FIGURE 12.13 (a) Creep and (b) stress relaxation of the Maxwell body.

This equation can be derived using the differential Equation 12.6a. To integrate Equation 12.6a after substituting $\dot{\sigma} = 0$ when $t > 0$, we obtain $\varepsilon(t) = \sigma_0 t/\eta + C$, where C is the integral constant. From the initial conditions of instantaneous elasticity $\varepsilon(0^+) = \sigma_0/E$ at $t = 0^+$, we obtain $C = \sigma_0/E$. Thus, under the action of a suddenly applied constant stress σ_0, the strain–time relation of the Maxwell model is Equation 12.7. It shows that after instantaneous elastic deformation, the strain of the Maxwell body is increased linearly over time, as shown in Figure 12.13a. Under the action of a certain stress, the material produces continually increasing deformation, which is the characteristic of a fluid. Therefore, the materials represented by the Maxwell model are often referred to as Maxwell fluids.

2. *Creep recovery*. If the external force is removed at time $t = t_1$, then the original steady-state flow under the action of σ_0 is terminated, and the elastic deformation immediately disappears. That is, the instantaneous elastic recovery is σ_0/E, and the permanent deformation remaining in the material is $\sigma_0 t_1/\eta$, as shown in Figure 12.13a.
3. *Stress relaxation*. Under the action of $\varepsilon(t) = \varepsilon_0 H(t)$, $\dot{\varepsilon} = 0$. When $t > 0$, Equation 12.6a becomes a homogeneous ordinary differential equation, whose solution is $\sigma = Ce^{-t/p_1}$. Under the initial conditions where $t = 0$ and $\sigma(0^+) = E\varepsilon_0$, we can find the constant C and obtain the stress as

$$\sigma = E\varepsilon_0 e^{-t/p_1} \tag{12.8}$$

where $p_1 = \eta/E$. Equation 12.8 describes the stress relaxation process of the Maxwell model, as shown by the curve AB in Figure 12.13b: suddenly applied strain will have instantaneous stress response value $E\varepsilon_0$. Under the action of constant strain ε_0, the stress continually decreases; and with an infinite increase in time, the stress gradually decays to zero. The stress change rate of the relaxation process is

$$\dot{\sigma} = -\frac{\sigma(0)}{p_1} e^{-t/p_1}.$$

Clearly, the change rate (absolute value) in the beginning of the stress relaxation is the maximum. That is, when $t = 0^+$, $\dot{\sigma}(0) = -\sigma(0)/p_1$. If the stress is in accordance with this ratio and changes over time, it is expressed as $\sigma(t) = \sigma(0) - \sigma(0)t / p_1$. That is, for a straight line as shown in Figure 12.13b, the stress is zero when $t = p_1$. Therefore, the characteristic time can be denoted by $\tau_R = p_1 = \eta/E$, known as the Maxwell relaxation time. From Equation 12.8 we can see that when $t = \tau_R$, $\sigma = 0.37\sigma(0)$. In other words, keeping the strain value ε_0 until the time τ_R has been reached, most of the initial stress has decayed, and thus τ_R is the characteristic time describing the stress relaxation. Obviously, the relaxation time is determined by the material's properties: the smaller the viscosity, the shorter is the relaxation time; the high-viscosity rheological body has a longer relaxation time; and elastic solids ($\eta \to \infty$) do not exhibit stress relaxation.

12.2.1.3 Kelvin Model

Spring and damper in parallel form a Kelvin model, also known as the Kelvin–Voigt model, as shown in Figure 12.14. The strains of the two components are equal to the total strain of the model, and the total stress of the model is equal to the stress sum of the two components, so that $\sigma = \sigma_1 + \sigma_2$.

Considering Equations 12.4 and 12.5, we obtain the constitutive equation of the Kelvin model.

$$\sigma = E\varepsilon + \eta\dot{\varepsilon}, \tag{12.9a}$$

or

$$\sigma = q_0\varepsilon + q_1\dot{\varepsilon}, \tag{12.9b}$$

where the model parameters are $q_0 = E$, $q_1 = \eta$. The viscoelasticity reflected by Equation 12.9 is as follows:

1. *Creep*. Under the action of constant stress, according to the differential Equation 12.9a, we have

$$\varepsilon(t) = Ce^{-t/\tau_d} + \frac{\sigma_0}{E},$$

where $\tau_d = \eta/E$. The initial condition of the creep is $t = 0$, $\varepsilon(0) = 0$. The reason is that, if $\varepsilon(0^+)$ is a certain value due to $\varepsilon(0^-) = 0$, we will have $\dot{\varepsilon} \to \infty$ at $t = 0$. This cannot be tolerated by Equation 12.9, and thus $\varepsilon(0^+) = 0$. From the earlier equation, the integral constant $C = -\sigma_0/E$ can be obtained. Thus, the creep expression of the Kelvin model is

$$\varepsilon(t) = \frac{\sigma_0}{E}\left(1 - e^{-\frac{t}{\tau_d}}\right). \tag{12.10}$$

Clearly, the strain gradually increases over time. When $t \to \infty$, $\varepsilon \to \sigma_0/E$, the material is a kind of elastic solid. Therefore, sometimes materials represented by the Kelvin model are

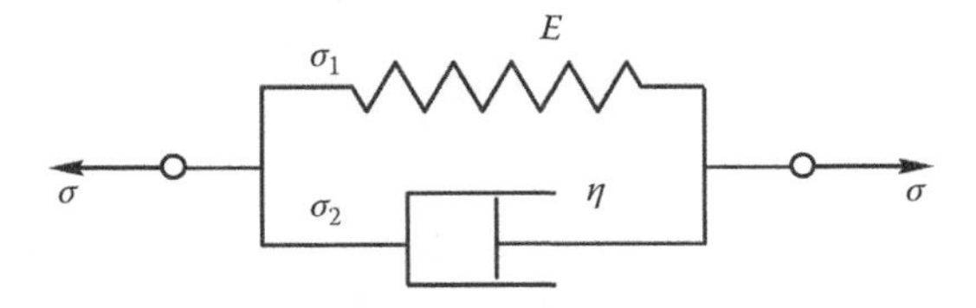

FIGURE 12.14 Kelvin model.

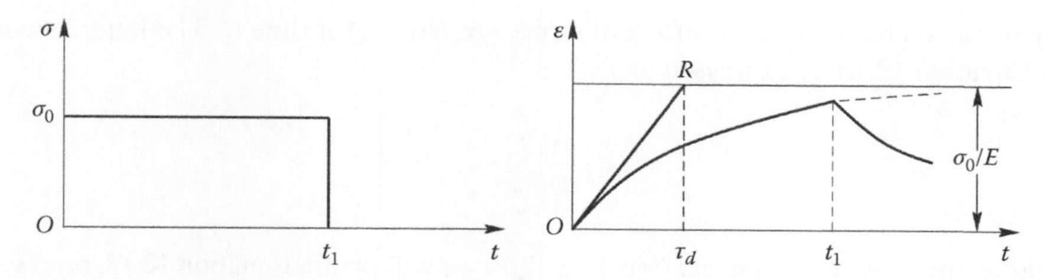

FIGURE 12.15 Creep and recovery in the Kelvin model.

called Kelvin solids. However, the Kelvin solid has no instantaneous elasticity. Instead, deformation occurs according to the change rate $\dot{\varepsilon}(t) = \sigma_0 e^{-t/\tau_d}/\eta$, and the strain gradually tends to the asymptotic value σ_0/E over time. The initial strain rate is $\dot{\varepsilon}(0) = \sigma_0/\eta$. If the deformation occurs according to this strain rate (*OR* shown in Figure 12.15), then when $t = \tau_d = \eta/E$, the strain reaches σ_0/E. Therefore, $\tau_d = \eta/E$ is often referred to as the retardation time of the Kelvin body.

2. *Creep recovery.* Equation 12.10 gives the strain value under the effect of stress σ_0 at any time t. When $t = t_1$, we have

$$\varepsilon(t_1) = \frac{\sigma_0}{E}\left(1 - e^{-\frac{t_1}{\tau_d}}\right). \tag{12.11}$$

If σ_0 is removed at time $t = t_1$, the strain begins to recover according to the values shown in Equation 12.11. When $t \geq t_1$, from Equation 12.9a, we can obtain the equation $E\varepsilon + \eta\dot{\varepsilon} = 0$, which describes the recovery process of the Kelvin model. Its solution is

$$\varepsilon(t) = C_1 e^{-\frac{t}{\tau_d}}, \qquad t \geq t_1. \tag{12.12}$$

Consider Equation 12.11, the initial conditions of recovery, and use the continuous conditions of the strain at $t = t_1$. We obtain

$$C_1 e^{-\frac{t_1}{\tau_d}} = \frac{\sigma_0}{E}\left(1 - e^{-\frac{t_1}{\tau_d}}\right).$$

Thus, substituting the derived C_1 into Equation 12.12 we obtain the strain–time relations of the recovery process:

$$\varepsilon(t) = \frac{\sigma_0}{E}\left(e^{\frac{t_1}{\tau_d}} - 1\right) e^{-\frac{t_1}{\tau_d}}. \tag{12.13}$$

Equation 12.13 describes the recovery process at time $t \geq t_1$, after removing σ_0 at time $t = t_1$. Clearly, when $t \to \infty$ then $\varepsilon \to 0$, which reflects the characteristics of elastic solids. In fact, Equation 12.13 expresses the strain under the action of the stress $\sigma(t) = \sigma_0 H(t) - \sigma_0 H(t - t_1)$, which can be obtained from the superposition of two creep processes at different times. (Here, the function $H(x)$ is defined as 1 when $x > 0$, and 0 when $x < 0$.) That is, in the creep equation expressed by Equation 12.10, we superimpose a generated creep under the effect of stress $-\sigma_0 H(t-t_1)$ at time t_1. The latter, according to Equation 12.10, is expressed as

$$\varepsilon(t) = \frac{-\sigma_0}{E}\left[1 - e^{-\frac{(t-t_1)}{\tau_d}}\right].$$

Adding the above equation and Equation 12.10 we will obtain Equation 12.13, representing the strain–time relationship of the recovery process.

It is noteworthy that in the previous analysis, although the stress is zero at time $t > t_1$, the strain of the material is not zero. The strain is time- and load history-dependent, indicating that the material has memory.

3. *Stress relaxation*. The Kelvin model cannot reflect the stress relaxation process, because the deformation of dampers takes time. There must be a strain rate $\dot{\varepsilon}$, and then the stress σ. So, when the strain is maintained at a constant ε_0, the damper is free of applied forces. The spring withstands all of the stresses. On the other hand, under the action of the step strain, $\dot{\varepsilon}(t) = \varepsilon_0 \delta(t)$, and from the stress–strain relations, we obtain

$$\sigma(t) = E\varepsilon_0 H(t) + \eta\varepsilon_0 \delta(t), \tag{12.14}$$

where the first item on the right-hand side of the equation represents the stress withstood by the spring. The second item in the equation represents the infinite stress pulse at $t = 0$. Hence the sudden strain ε_0 at $t = 0$, which does not make sense for the Kelvin model.

From this analysis we can see that the Maxwell model and the Kelvin model are both the simplest parameter viscoelastic models. The Maxwell model is able to show the phenomenon of stress relaxation, but not creep; only steady-state flow. The Kelvin model can reflect the creep process, but not the stress relaxation. At the same time, the stress relaxation or creep process reflected by these two basic models has only one exponential time function. It is unsuitable for expressing the more complex rheological process of polymers and other materials. Therefore, in order to better describe the viscoelastic behavior of the actual material, we often use other models composed of more basic components.

12.2.2 Creep Compliance and Relaxation Modulus

The creep or relaxation process of the basic model shows that the strain or stress response is a function of time, which reflects the viscoelastic behavior of materials subject to a simple load. Therefore, two important material functions can be defined: the creep function and the relaxation function, also known as the creep compliance and the relaxation modulus.

When a linear viscoelastic material is under the action of $\sigma(t) = \sigma_0 H(t)$, the strain response change over time can be expressed as

$$\varepsilon(t) = J(t)\sigma_0, \tag{12.15}$$

where $J(t)$ is called the creep compliance. It indicates that the strain value at time t under the action of the unit stress is a function increasing monotonically with time. From Equations 12.7 and 12.10, we can obtain the creep compliance of some basic models.

$$\text{For the Maxwell body,} \quad J(t) = \frac{1}{E} + \frac{t}{\eta}.$$

$$\text{For the Kelvin body,} \quad J(t) = \frac{1}{E}\left(1 - e^{-\frac{t}{\tau_d}}\right).$$

When studying the stress relaxation, the stress response after the action of a constant strain ε_0 can be expressed as

$$\sigma(t) = Y(t)\varepsilon_0, \tag{12.16}$$

where $Y(t)$ is called the relaxation modulus. It indicates that the stress under the effect of the unit strain is a function decreasing with increasing time. The relaxation functions of the previously described models can be derived, respectively, from Equations 12.8 and 12.14. As a special case, the relaxation modulus of elastic solids and viscous fluids are, respectively, E and $\eta\delta(t)$.

12.2.3 One-Dimensional Differential-Type Constitutive Relations

The material function of the simple model described earlier only contains an exponential function, which is often insufficient to describe the viscoelastic behavior of the actual material, and thus requires a model with a combination of more components. For several Maxwell units in series, their constitutive relations are identical with the constitutive form of the Maxwell body, showing the same viscoelastic properties as a single Maxwell model. Similarly, when a number of Kelvin units are in parallel, we obtain the same constitutive equation as (12.9), expressing the viscoelastic mechanical behavior in a single Kelvin model. Therefore, what a number of Maxwell units in series, or Kelvin units in parallel, describe, is still the simplest material model.

The model constituted by multiple Maxwell units in parallel, or Kelvin units in series, can express more complicated material properties, and describe the general viscoelastic mechanical behavior. This is the generalized Maxwell model and the generalized Kelvin model shown in Figure 12.16. The latter is sometimes also known as the Kelvin chain.

Using this general model, we can derive a one-dimensional linear viscoelastic differential-type constitutive equation. For example, suppose the strain of the ith Kelvin unit in the Kelvin chain is ε_i, and its spring elastic modulus and damper viscosity coefficient are, respectively, E_i and η_i. Thus, by Equation 12.9, we can obtain $\sigma = E_i\varepsilon_i + \eta_i\dot{\varepsilon}_i$, and express the strain by the differential operator D:

$$\varepsilon_i = \frac{\sigma}{(E_i + \eta_i D)}.$$

The total strain of the generalized Kelvin model by n Kelvin units is then

$$\varepsilon = \sum_{i=1}^{n} \varepsilon_i = \sum_{i=1}^{n} \frac{\sigma}{(E_i + \eta_i D)}.$$

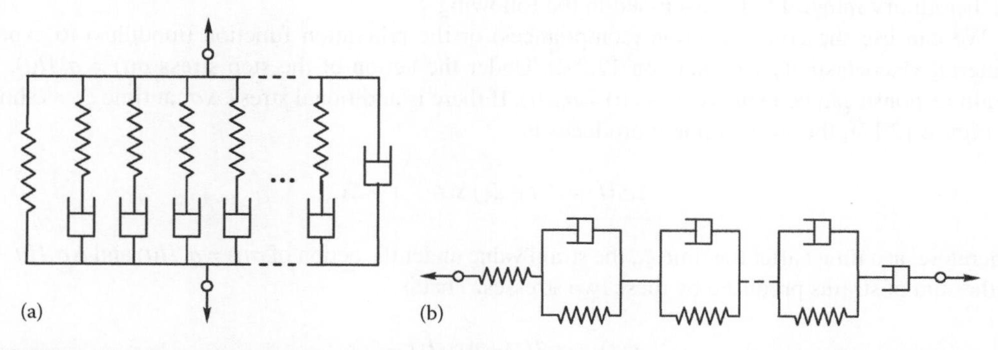

FIGURE 12.16 General viscoelastic model. (a) Generalized Maxwell model. (b) Generalized Kelvin model.

After refining the expansion of this equation, we can get the constitutive equation of a general model,

$$p_0\sigma + p_1\dot{\sigma} + p_2\ddot{\sigma} + p_3\dddot{\sigma} + \cdots = q_0\varepsilon + q_1\dot{\varepsilon} + q_2\ddot{\varepsilon} + q_3\dddot{\varepsilon} + \cdots, \tag{12.17a}$$

which can be rewritten as

$$\sum_{k=0}^{m} p_k \frac{\mathrm{d}^k\sigma}{\mathrm{d}t^k} = \sum_{k=0}^{n} q_k \frac{\mathrm{d}^k\varepsilon}{\mathrm{d}t^k} \qquad n \geq m, \tag{12.17b}$$

or

$$\boldsymbol{P}\sigma = \boldsymbol{Q}\varepsilon, \tag{12.17c}$$

where the differential operator is

$$\boldsymbol{P} = \sum_{k=0}^{m} p_k \frac{\mathrm{d}^k}{\mathrm{d}t^k}; \ \mathbf{Q} = \sum_{k=0}^{n} q_k \frac{\mathrm{d}^k}{\mathrm{d}t^k}.$$

Here, the effect operator is expressed in bold. Equation 12.17 is the general one-dimensional linear viscoelastic differential-type constitutive equation. p_k and q_k are constants determined by the material properties, usually $p_0 = 1$. The basic components, and the constitutive relation of the basic model described previously, are all special circumstances of Equation 12.17a through c. For example, the spring stress–strain relationship is found by taking the first items on the left and right sides of Equation 12.17a through c.

12.2.4 One-Dimensional Integral-Type Constitutive Relations and the Boltzmann Superposition Principle

The linear viscoelastic stress–strain–time relationship in the form of the differential operator is more convenient for expressing the material's properties when using a model constituted by springs and dampers, and is also desirable in the solution of certain problems. However, we often use the integral-type constitutive relation through the material function, in order to be more specific about expressing the viscoelastic behavior of materials and conducive to the actual test; to better describe the memory properties of materials and the response process of objects under loading; for ease of considering the factors of material aging and temperature effects; and to give greater flexibility in application.

In linear viscoelastic problems, the total effect of multiple causes is equal to the sum of the effect of each separate cause. This is the basis and essence of the Boltzmann superposition principle, and the hereditary integral to be discussed in the following.

We can use the creep function (compliances) or the relaxation function (modulus) to express material viscoelasticity (see Section 12.2.2). Under the action of the step stress $\sigma(t) = \sigma_0 H(t)$, the strain response can be expressed as $\varepsilon(t) = \sigma_0 J(t)$. If there is additional stress $\Delta\sigma_1$ at time ζ_1, as shown in Figure 12.17a, the strain value it produces is

$$\Delta\varepsilon_1(t) = J(t-\zeta_1)\Delta\sigma_1, \quad t > \zeta_1.$$

Therefore, at a time t after the time ζ_1, the strain value under the action of $\sigma(t) = \sigma_0 H(t)$ and $\Delta\sigma_1 H(t-\zeta_1)$ is the sum of strains produced by these two stresses. That is,

$$\varepsilon(t) = \sigma_0 J(t) + \Delta\sigma_1 J(t-\zeta_1).$$

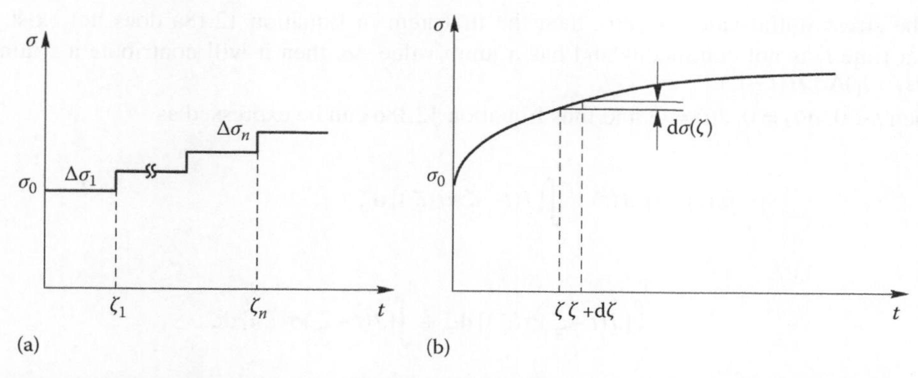

FIGURE 12.17 Superposition principle diagram. (a) Step-by-step. (b) Smoothly increasing stress.

Similarly, if at time ζ_i there are n stress increments in sequence (as shown in Figure 12.17a) acting on an object, the total strain after the time ζ_n is

$$\varepsilon(t) = \sigma_0 J(t) + \sum_{i=1}^{n} \Delta\sigma_i J(t - \zeta_i), \quad t > \zeta_n.$$

This is the Boltzmann superposition principle.

The general loading process is relatively complicated, but it can be seen as the superposition of many forces. Suppose the stress acting on an object is a continuous differentiable function (as shown in Figure 12.17b). We can break it down into the action of $\sigma_0 H(t)$ and numerous very small stresses $d\sigma(\zeta)H(t - \zeta)$, among which

$$d\sigma(\zeta) = \left.\frac{d\sigma}{dt}\right|_{t=\zeta} d\zeta = \frac{d\sigma(\zeta)}{d\zeta} d\zeta.$$

Thus, at time t, the strain response can be written as

$$\varepsilon(t) = \sigma_0 J(t) + \int_{0^+}^{t} J(t - \zeta) \frac{d\sigma(\zeta)}{d\zeta} d\zeta. \tag{12.18a}$$

This is the integral expression of the Boltzmann superposition principle, also known as the inherited integral, or the hereditary integral.

By the partial integration of the second item on the right side of Equation 12.18a, we can substitute it back to Equation 12.18a and obtain

$$\varepsilon(t) = J(0)\sigma(t) + \int_{0^+}^{t} \sigma(\zeta) \frac{dJ(t - \zeta)}{d(t - \zeta)} d\zeta. \tag{12.18b}$$

It can be seen that the strain expressed in Equation 12.18a is the strain caused by the initial stress value σ_0 plus the strain response produced during the process of stress changes. The strain in Equation 12.18b is expressed as the sum of the strain value produced by the stress at time t, and the creep caused by the stress history. These two equations are completely equivalent.

If the stress initial value is zero, then the first item in Equation 12.18a does not exist. If the stress at time t_1 is not continuous and has a jump value $\Delta\sigma$, then it will contribute a strain value $\Delta\varepsilon = J(t-t_1)\Delta\sigma H(t-t_1)$.

When $t < 0$, $\sigma(t) = 0$, $J(t) = 0$, and thus Equation 12.18a can be expressed as

$$\varepsilon(t) = \sigma_0 J(t) + \int_{0^+}^{t} [J(t-\zeta)\dot{\sigma}(\zeta)]\,\mathrm{d}\zeta$$

$$= \int_{-\infty}^{0^-} [J(t-\zeta)\dot{\sigma}(\zeta)]\,\mathrm{d}\zeta + \int_{0^-}^{0^+} [J(t-\zeta)\dot{\sigma}(\zeta)]\,\mathrm{d}\zeta$$

$$+ \int_{0^+}^{t} [J(t-\zeta)\dot{\sigma}(\zeta)]\,\mathrm{d}\zeta + \int_{t}^{\infty} [J(t-\zeta)\dot{\sigma}(\zeta)]\,\mathrm{d}\zeta.$$

Therefore, Equation 12.18a can also be rewritten as

$$\varepsilon(t) = \int_{-\infty}^{t} J(t-\zeta)\dot{\sigma}(\zeta)\,\mathrm{d}\zeta, \tag{12.18c}$$

or

$$\varepsilon(t) = \int_{-\infty}^{\infty} J(t-\zeta)\dot{\sigma}(\zeta)\,\mathrm{d}\zeta. \tag{12.18d}$$

A viscoelastic integral-type constitutive relation can be written in the abbreviated form of Stieltjes convolution:

$$\varepsilon(t) = J(t) * \mathrm{d}\sigma(t) = \sigma(t) * \mathrm{d}J(t). \tag{12.18e}$$

Equations 12.18a through e are integral constitutive equations. They are creep-type constitutive relations. Having a known material creep function and a given stress $\sigma(t)$, we can obtain the strain response or the material creep process from those equations. However, the creep we are talking about here is not just simple creep under a constant stress.

In a similar way, suppose the external action on the objects is the strain $\varepsilon(t)$ changing over time, and the relaxation modulus function of the material is $Y(t)$. We can then obtain the stress response formula based on the superposition principle.

$$\sigma(t) = \varepsilon_0 Y(t) + \int_{0^+}^{t} Y(t-\zeta)\frac{\mathrm{d}\varepsilon(\zeta)}{\mathrm{d}\zeta}\,\mathrm{d}\zeta. \tag{12.19a}$$

$$\sigma(t) = Y(0)\varepsilon(t) + \int_{0^+}^{t} \varepsilon(\zeta)\frac{\mathrm{d}Y(t-\zeta)}{\mathrm{d}(t-\zeta)}\,\mathrm{d}\zeta. \tag{12.19b}$$

$$\sigma(t) = \int_{-\infty}^{t} Y(t-\zeta)\dot{\varepsilon}(\zeta)\,\mathrm{d}\zeta. \tag{12.19c}$$

$$\sigma(t) = \int_{-\infty}^{\infty} Y(t-\zeta)\dot{\varepsilon}(\zeta)\mathrm{d}\zeta. \tag{12.19d}$$

$$\sigma(t) = Y(t) * \mathrm{d}\varepsilon(t) = \varepsilon(t) * \mathrm{d}Y(t). \tag{12.19e}$$

These are one-dimensional relaxation-type constitutive equations.

Equations 12.18a through e and 12.19a through e describe the linear viscoelastic stress–strain–time relation, which is a one-dimensional integral-type constitutive equation. It is noteworthy that the integral-type constitutive relation and the differential-type constitutive relation are consistent. For the same material, both should show the same physical property relations, but expressed in different forms. Knowing a linear viscoelastic material function, we can then write the integral-type or differential-type constitutive relation. The following are examples.

If a given material relaxation function is

$$Y(t) = q_0 + Ae^{-t/p_1}, \tag{a}$$

where $A = (q_1/p_1) - q_0$, the integral-type constitutive relation can be expressed as

$$\sigma(t) = \int_{-\infty}^{t} Y(t-\zeta)\dot{\varepsilon}(\zeta)\mathrm{d}\zeta = q_0\varepsilon(t) + Ae^{-t/p_1}\int_{0^-}^{t} e^{\zeta/p_1}\,\dot{\varepsilon}(\zeta)\mathrm{d}\zeta. \tag{b}$$

Using this equation to get the derivative with respect to time, we can obtain

$$\dot{\sigma}(t) = q_0\dot{\varepsilon}(t) + A\dot{\varepsilon}(t) + \frac{\left[q_0\varepsilon(t) - \sigma(t)\right]}{p_1}.$$

It can be rewritten as

$$\sigma + p_1\dot{\sigma} = q_0\varepsilon + q_1\dot{\varepsilon}. \tag{c}$$

Equations a through c thus represent the same kind of standard linear solid.

12.3 HYPERELASTICITY OF POLYMERS

A polymer has a unique mechanical state, called the high-elastic state at temperatures above its glass transition. The mechanical property of a polymer in the high-elastic state, known as rubber elasticity, is a significant characteristic that distinguishes the polymer from other materials. Different from the general elasticity of other materials, high elasticity is essentially the entropic elasticity. The high elasticity of a polymer mainly shows the following characteristics [12]: (1) the high elasticity of the polymer has large reversible elastic deformation, up to 1000%, while the elastic deformation of a common metal material does not exceed 1%; (2) the high elasticity of the polymer has a small elastic modulus, generally 10^5 Pa, while the elastic modulus of the metal material is generally up to 10^{11} Pa; (3) the high elastic modulus of the polymer increases with increasing temperature, while the elastic modulus of the metal material decreases with increasing temperature; and (4) the temperature of the high-elastic material increases when undergoing rapid stretching (adiabatic process); however, the metal material does not.

12.3.1 Thermodynamic Analysis of High Elasticity

If we use a rubber specimen as a thermodynamic system, its environment is the change of external force, pressure, and temperature. Given the equilibrium thermodynamic analysis on this system, we can understand the response of the rubber to a variety of deformations (stretching, compression, shear, and so on), as well as its relationship with factors such as temperature, pressure, and volume changes.

For simplicity, we can take the uniaxial tension as an example. The initial length of the rubber specimen is l_0. With one end fixed, a pulling force f at the other end is applied along the length of the specimen, so that the specimen is stretched dl. According to the first law of thermodynamics, the internal energy U of the system changes as follows:

$$\mathrm{d}U = \mathrm{d}Q - \mathrm{d}W,$$

where dQ is the thermal energy that the system obtains from the environment, and dW is the external work the system has done. If the isothermal process is reversible, then by the second law of thermodynamics, we have

$$\mathrm{d}Q = T\mathrm{d}S.$$

The total external work the system has done includes the work done by the volume change of the specimen $p\mathrm{d}V$ and the work done by the external force $f\mathrm{d}l$. That is,

$$\mathrm{d}W = p\mathrm{d}V - f\mathrm{d}l.$$

Thus,

$$\mathrm{d}U = T\mathrm{d}S - p\mathrm{d}V + f\mathrm{d}l. \tag{12.20}$$

During the process of stretching, the volume of the rubber is almost constant (dV = 0), so that

$$\mathrm{d}U = T\mathrm{d}S + f\mathrm{d}l. \tag{12.21}$$

Therefore,

$$f = \left.\frac{\partial U}{\partial l}\right|_{T,V} - T\left.\frac{\partial S}{\partial l}\right|_{T,V}. \tag{12.22}$$

The rubber specimen under the action of the external force makes the internal energy of the rubber change with the elongation; but it also makes the entropy of the rubber change with the elongation. That is to say, the tension of the rubber is caused by the internal energy and entropy changes during deformation. Because $T\left.(\partial S/\partial l)\right|_{T,V}$ cannot be measured directly, we can use the Gibbs free energy G and the Helmholtz free energy H to make it measurable.

$$G = A + PV = U + PV - TS.$$

Making a total differential of this equation, we obtain

$$\mathrm{d}G = \mathrm{d}U + p\mathrm{d}V + V\mathrm{d}p - T\mathrm{d}S - S\mathrm{d}T.$$

Substituting Equation 12.20 into this equation, it may be rewritten as

$$\mathrm{d}G = f\mathrm{d}l + V\mathrm{d}p - S\mathrm{d}T.$$

As a result, we have

$$f = \left.\frac{\partial G}{\partial l}\right|_{T,p}, \quad S = -\left.\frac{\partial G}{\partial T}\right|_{l,p}. \tag{12.23}$$

Using this equation, the unmeasured $(\partial S/\partial l)|_{T,V}$ can be converted into a measurable quantity:

$$\left.\frac{\partial S}{\partial l}\right|_{T,V} = \left[\frac{\partial}{\partial l}\left(\left.\frac{\partial G}{\partial T}\right|_{l,p}\right)\right]_{T,V} = -\left[\frac{\partial}{\partial T}\left(\left.\frac{\partial G}{\partial l}\right|_{T,p}\right)\right]_{l,V} = -\left.\frac{\partial f}{\partial T}\right|_{l,V}. \tag{12.24}$$

This can be substituted into Equation 12.22, which will give us

$$f = \left.\frac{\partial U}{\partial l}\right|_{T,V} + T\left.\frac{\partial f}{\partial T}\right|_{l,V}. \tag{12.25}$$

This is the mechanical state equation of the rubber specimen under isovolumic condition. Accordingly, from the temperature-dependent relationship of the tension, we can derive the internal energy and entropy changes when the length of the specimen changes.

At different temperatures, the rubber specimen is stretched to different predetermined lengths. After quite a long period of stress relaxation, it makes a balanced tension–temperature curve, as shown in Figure 12.18. Thus, for the entire elongation deformation, the curve becomes a linear relationship. Clearly, the slope of the line is $\left.\frac{\partial f}{\partial T}\right|_{l,V}$ and the intercept is $\left.\frac{\partial U}{\partial l}\right|_{T,V}$. It was found that almost all straight lines are through the origin of the coordinates; that is, $\left.\frac{\partial U}{\partial l}\right|_{T,V} = 0$. This explains that during

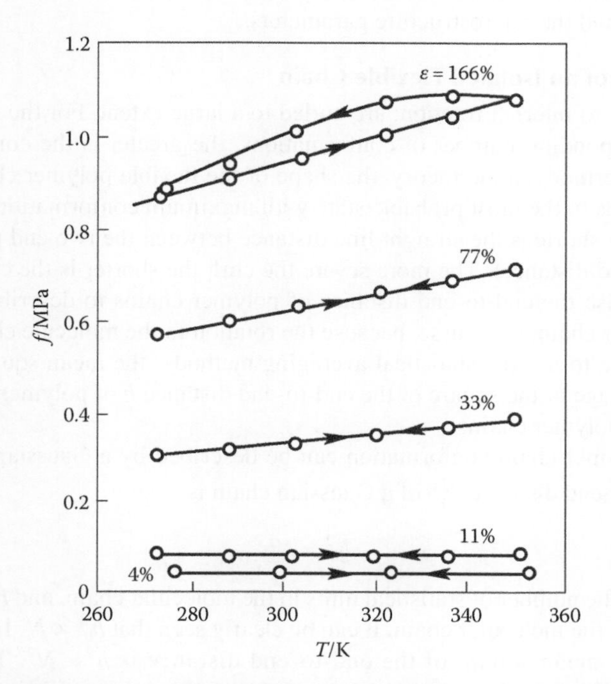

FIGURE 12.18 The relationship of balanced tension and temperature of the rubber specimen in constant elongation.

the rubber tensile process, the internal energy is almost constant. High elasticity is mainly caused by the change in entropy; so in essence, the rubber's high elasticity is a kind of entropy elasticity.

At this time,

$$f = T\left.\frac{\partial f}{\partial T}\right|_{l,V} = -T\left.\frac{\partial S}{\partial l}\right|_{T,V}. \tag{12.26}$$

It should be pointed out that recent research has shown that in addition to the contribution of entropy, the internal energy of high elasticity also makes contributions, which account for about 10%. Nevertheless, the essential characteristics of high elasticity given by the thermodynamic analysis are still valid.

During the rubber tensile process, the internal energy is almost the same. Under isovolumic condition, from Equation 12.21 we can obtain

$$f\mathrm{d}l = -T\mathrm{d}S = -\mathrm{d}Q.$$

Thus, when the rubber specimen is stretched, $\mathrm{d}l > 0$, and so $\mathrm{d}Q < 0$; and the system is exothermic. On the contrary, when the rubber specimen is compressed, $\mathrm{d}l < 0$. But because $f < 0$, $\mathrm{d}Q < 0$; and the system is still exothermic, which is one of the deformation characteristics of high-elastic polymers.

12.3.2 Statistical Theories of High-Elastic Deformation

Thermodynamics analyzes the entropic nature of high elasticity in rubber-type materials. The deformation process of rubber is mainly due to the change in entropy, and the internal energy is only a minor contribution. However, thermodynamic analysis only gives the relationship between macroscopic quantities. Only with the help of statistical theory can we obtain the quantitative expression, and the change of entropy in deformations of the polymer chain through the microstructure parameters, and further establish the relationship between the macromechanical parameters (such as elastic modulus) and the microstructure parameters.

12.3.2.1 Entropy of an Isolated Flexible Chain

Polymer chains, due to internal rotation, are curled to a large extent. For the molecular chain curl, the more the corresponding number of conformations, the greater is the conformational entropy. According to the thermodynamic theory, the shape of the flexible polymer chain at a certain temperature always tends to the most probable state with maximum conformational entropy.

Curl significantly shortens the straight-line distance between the two end points of the polymer chain (the end-to-end distance). The more severe the curl, the shorter is the end-to-end distance h. Therefore, we can use the end-to-end distance of polymer chains to describe or characterize the shape of the polymer chain. Of course, because the rotation in the molecule changes the conformation, it is appropriate to use the statistical averaging method—the mean square of the end-to-end distance $\overline{h^2}$ (the average of the square of the end-to-end distance h of polymer chains)—to describe the average size of polymer chains.

Suppose the polymer chain conformation can be described by a Gaussian function. The mean square of the end-to-end distance $\overline{h_0^2}$ of a Gaussian chain is

$$\overline{h_0^2} = Nb^2,$$

where N represents the number of statistical units in the molecular chain, and b represents the length of statistical units in the molecular chain. It can be clearly seen that $\overline{h_0^2} \propto N$. If the polymer chain is fully extended, the mean square of the end-to-end distance is $\overline{h^2} \propto N^2$. Therefore, we obtain $\overline{h_{ext}^2}/\overline{h_0^2} = N$. Usually, the number of statistical units N in the polymer chain is large, so $\overline{h_{ext}^2}/\overline{h_0^2}$ is large, which is why the rubber may have a large deformation.

If one end of the isolated flexible Gaussian chain is fixed at the origin of the coordinate system and a pulling force f is applied at the other end, and the length of the Gaussian chain is l, then the probability of the end point at the point (x, y, z) is

$$\omega(x,y,z)=\left(\frac{\beta}{\sqrt{\pi}}\right)^3 e^{-\beta^2(x^2+y^2+z^2)}=\left(\frac{\beta}{\sqrt{\pi}}\right)^3 e^{-\beta^2 l^2}, \quad (12.27)$$

where $\beta^2 = 3/2Nb^2$. By the entropy of the Boltzmann relation $S = k\ln\Omega$, and when Ω and ω are proportional, then the conformational entropy of this Gaussian chain is

$$S = C - k\beta^2(x^2+y^2+z^2) = C - k\beta^2 l^2,$$

where C is a constant. Thus, under isothermal and isovolumic conditions, the tension f is

$$f = -T\left.\frac{\partial S}{\partial l}\right|_{T,V} = 2kT\beta^2 l = \frac{3kT}{Nb^2} l. \quad (12.28)$$

Clearly, the tension f and the temperature T are proportional. The required force f for stretching to l grows linearly with T; that is, the modulus increases with increasing temperature.

12.3.2.2 Entropy Changes in Deformation of Cross-Linked Rubber Network

The real cross-linked rubber network is very complicated. We will simplify it as an ideal cross-linked model, making the following four assumptions:

1. Four chains constitute each cross-linked point that is randomly distributed;
2. The molecular chain between the adjacent cross-linked points is a Gaussian chain, and its end-to-end distance satisfies the Gaussian distribution;
3. The total number of isotropic cross-linked conformations constituted by these Gaussian chains is equal to the sum of the conformation of each individual network chain;
4. Cross-linked points are fixed at the average position before and after deformation, and the deformation is an affine deformation, as shown in Figure 12.19.

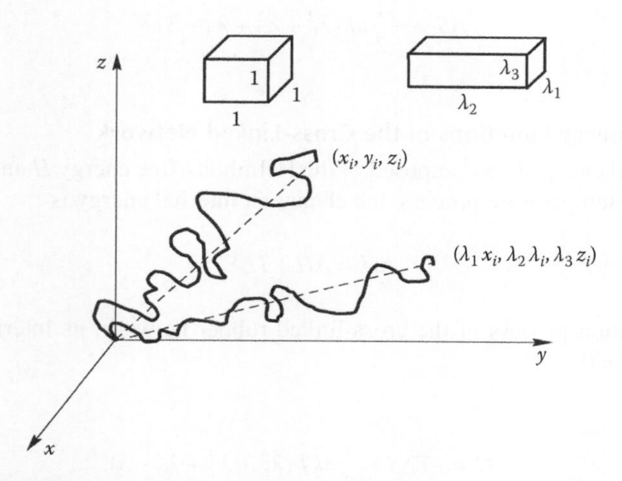

FIGURE 12.19 Affine deformation of the ith network chain.

Accordingly, if the elongation ratio of the deformation of the rubber specimen in the x-direction is λ_1, and in the y-direction is λ_2, and in the z-direction is λ_3, then, after the deformation, one end of the cross-linked chain will change from the initial position (x,y,z) to $(\lambda_1 x, \lambda_2 y, \lambda_3 z)$. Before the deformation, the conformational entropy of the ith network chain is $S_{iu} = C - k\beta^2(x_i^2 + y_i^2 + z_i^2)$, and after deformation it is $S_{id} = C - k\beta^2(\lambda_1^2 x_i^2 + \lambda_2^2 y_i^2 + \lambda_3^2 z_i^2)$. Therefore, the entropy change of the ith network chain due to deformation becomes

$$\Delta S_i = S_{id} - S_{iu} = -k\beta^2\left[(\lambda_1^2 - 1)x_i^2 + (\lambda_2^2 - 1)y_i^2 + (\lambda_3^2 - 1)z_i^2\right].$$

Assuming that the number of network chains contained in the cross-linked network in the unit volume is n, the total entropy change in unit volume of the specimen becomes

$$\Delta S = -k\beta^2 \sum_{i=1}^{n}\left[(\lambda_1^2 - 1)x_i^2 + (\lambda_2^2 - 1)y_i^2 + (\lambda_3^2 - 1)z_i^2\right].$$

Because the end-to-end distance of each network chain is not identical, we use their average value. Then

$$\Delta S = -nk\beta^2\left[(\lambda_1^2 - 1)\overline{x}^2 + (\lambda_2^2 - 1)\overline{y}^2 + (\lambda_3^2 - 1)\overline{z}^2\right].$$

For the isotropic cross-linked network before deformation, we have $\overline{x}^2 = \overline{y}^2 = \overline{z}^2 = \frac{1}{3}\overline{h}^2$.

Thus,

$$\Delta S = -\frac{1}{3} nk\beta^2\overline{h}^2\left[(\lambda_1^2 - 1) + (\lambda_2^2 - 1) + (\lambda_3^2 - 1)\right].$$

Substituting $\beta^2 = 3/2Nb^2$ and $\overline{h}^2 = \overline{h_0^2} = Nb^2$ into this equation, we obtain

$$\Delta S = -\frac{1}{2} nk(\lambda_1^2 + \lambda_2^2 + \lambda_3^2 - 3). \tag{12.29}$$

12.3.2.3 Strain Energy Functions of the Cross-Linked Network

The system's internal energy U is composed of the Helmholtz free energy H and the binding energy TS. For the constant temperature process, the change in internal energy is

$$\Delta U = \Delta H + T\Delta S.$$

During the deformation process of the cross-linked rubber network, its internal energy is almost constant, and so $\Delta U = 0$.

Thus,

$$\Delta H = -T\Delta S = \frac{1}{2} nkT(\lambda_1^2 + \lambda_2^2 + \lambda_3^2 - 3).$$

For the constant temperature process, the increase of the system's Helmholtz free energy ΔH is equal to the work W done to the system. Therefore,

$$W = \Delta H = \frac{1}{2}nkT(\lambda_1^2 + \lambda_2^2 + \lambda_3^2 - 3).$$

Define $G = nkT$ as the shear modulus of the rubber. Then

$$W = \frac{G}{2}(\lambda_1^2 + \lambda_2^2 + \lambda_3^2 - 3). \tag{12.30}$$

This is called the strain energy function, representing the stored energy per unit volume of rubber under the action of external forces during the deformation process.

12.3.3 Stress–Strain Relationship of High-Elastic Material

From the strain energy function W, we can obtain the stress–strain relationship of cross-linked rubber. Letting the volume be constant during deformation, for the unit cube we have

$$\lambda_1\lambda_2\lambda_3 = 1. \tag{12.31}$$

For uniaxial tension, set the stretch ratio as $\lambda_1 = \lambda$. Then $\lambda_2 = \lambda_3 = 1/\sqrt{\lambda}$. At this time, the strain energy function is

$$W = \frac{G}{2}\left(\lambda^2 + \frac{2}{\lambda} - 3\right).$$

Under constant temperature and isovolumic conditions, $dW = fdl$. Thus, the external force per unit volume is

$$f = \left.\frac{dW}{dl}\right|_{T,V} = \left.\frac{\partial W}{\partial \lambda}\right|_{T,V}\left.\frac{\partial \lambda}{\partial l}\right|_{T,V} = \frac{G}{l_0}\left(\lambda - \frac{1}{\lambda^2}\right).$$

Setting the initial cross-sectional area to A_0, we obtain $V_0 = A_0 l_0$ and the nominal tensile stress is

$$\sigma = \frac{fV_0}{A_0} = \frac{fA_0 l_0}{A_0} = fl_0 = G\left(\lambda - \frac{1}{\lambda^2}\right). \tag{12.32}$$

The true tensile stress is

$$\sigma^* = \lambda\sigma = G\left(\lambda^2 - \frac{1}{\lambda}\right). \tag{12.33}$$

For general elastic solids, the stress–strain relationship of uniaxial tension satisfies Hooke's law:

$$\sigma = E\varepsilon = E\frac{l - l_0}{l_0} = E(\lambda - 1). \tag{12.34}$$

Obviously, Equation 12.32 is different from Equation 12.34, showing that the rubber stress–strain relationship does not satisfy Hooke's law. Equation 12.33 is called the neo-Hookean Law. However, because $\lambda = 1 + \varepsilon$, we obtain

$$\frac{1}{\lambda^2} = \frac{1}{(1+\varepsilon)^2} = 1 - 2\varepsilon + 3\varepsilon^2 - 4\varepsilon^3 + \cdots .$$

When ε is very small, we can ignore the higher-order terms, and thus $(1/\lambda^2) = 1 - 2\varepsilon$. Equation 12.32 can be approximately written as

$$\sigma = G\left(\lambda - \frac{1}{\lambda^2}\right) = G(1+\varepsilon-1+2\varepsilon) = 3G\varepsilon = 3G(\lambda-1).$$

Assuming that the volume is constant, and taking Poisson's ratio μ as 0.5, the shear modulus $G = E/[2(1+\mu)] = E/3$. Substituting this into the previous equation, we have $\sigma = E(\lambda - 1)$. It can be seen that only when the deformation is very small does the deformation of the rubber satisfy Hooke's Law.

It is noteworthy that (because $G = nkT$ and $E = 3G$) when the number of network chains n in the unit volume increases, elastic modulus and shear modulus of rubber materials increases. When the temperature increases, elastic modulus and shear modulus of rubber materials also increase. This is one of the characteristics of the high elasticity of rubber materials.

12.3.4 Phenomenological Theories for Large High-Elastic Deformation

According to the theory of elasticity in Chapter 1 of this book, the strain energy of isotropic elastic materials can be expressed as the function of three strain invariants I_1, I_2, and I_3. The three strain invariants can be determined by the main stretch ratio λ_1, λ_2, and λ_3:

$$I_1 = \lambda_1^2 + \lambda_2^2 + \lambda_3^2, \quad I_2 = \lambda_1^2\lambda_2^2 + \lambda_2^2\lambda_3^2 + \lambda_1^2\lambda_3^2, \quad I_3 = \lambda_1^2\lambda_2^2\lambda_3^2. \tag{12.35}$$

Usually, the rubber is considered as incompressible; that is, $I_3 = 1$, and I_3 does not contribute to the strain energy function. In this case, Rivlin recommended that the strain energy function be expressed by the following formula [13]:

$$W = W(I_1, I_2) = \sum_{i,j=0}^{\infty} C_{ij}(I_1 - 3)^i (I_2 - 3)^j, \tag{12.36}$$

where C represents the material parameters and $C_{00} = 0$.

If, in Equation 12.36, we only take ($i = 1, j = 0$); that is, we take C_{10} only, then

$$W = C_{10}(I_1 - 3) = C_{10}(\lambda_1^2 + \lambda_2^2 + \lambda_3^2 - 3). \tag{12.37}$$

Setting $C_{10} = G/2$, Equation 12.37 is exactly the same as Equation 12.30.

If, in Equation 12.36, we take ($i = 1, j = 0$) and ($i = 0, j = 1$), then

$$W = C_{10}(I_1 - 3) + C_{01}(I_2 - 3). \tag{12.38}$$

This is the Mooney function [14], which is one of the main strain energy functions in phenomenological theories for large high-elastic deformation.

In addition, Yeoh proposed the following strain energy function [15]:

$$W = C_{10}(I_1 - 3) + C_{20}(I_1 - 3)^2 + C_{30}(I_1 - 3)^3. \tag{12.39}$$

After determining the strain energy function, by derivation similar to that in Section 12.3.3, we can obtain the stress–strain relationship of rubber materials.

12.4 YIELDING AND FRACTURE OF POLYMERS

12.4.1 Plastic Yielding of Polymers

It is generally accepted that polymer yielding may occur by shear yielding, crazing, or a combination of both [16–18]. In shear yielding, the material volume is essentially unchanged, while the crazing process exhibits partial plastic deformation with a large number of scattered voids between microfibrils.

12.4.1.1 Stress Analysis of Uniaxial Tensile Yielding

For an equal cross-sectional specimen with a cross-sectional area A_0, the stress on the cross section is $\sigma_0 = F/A_0$, when subjected to a uniaxial tensile force as shown in Figure 12.20. The area of the inclined plane with a bevel angle α is $A_\alpha = A_0/\cos\alpha$. The tensile F acting on the inclined plane A_α can be decomposed into two mutually perpendicular force components F_n and F_s. $F_n = F\cos\alpha$, $F_s = F\sin\alpha$; therefore, the normal stress and the shear stress acting on the inclined cross section are, respectively,

$$\sigma_\alpha = \frac{F_n}{A_\alpha} = \sigma_0 \cos^2 \alpha. \tag{12.40}$$

$$\tau_\alpha = \frac{F_s}{A_\alpha} = \left(\frac{\sigma_0}{2}\right) \sin 2\alpha. \tag{12.41}$$

These equations show that the normal stress and the shear stress on any inclined section of the specimen are only related to the normal stress on the cross section, and the bevel angle of the

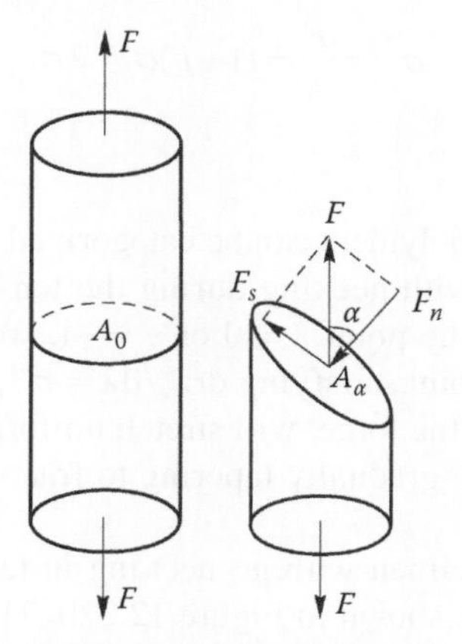

FIGURE 12.20 Mechanical analyses on the inclined cross section.

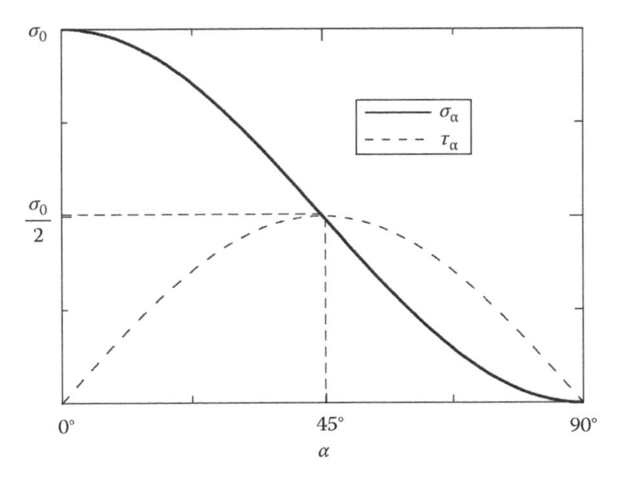

FIGURE 12.21 Variations of normal stress and shear stress with the tilt angle of the inclined section.

inclined section of the specimen. As shown in Figure 12.21, the normal stress reaches its maximum value σ_0 when acting on the cross section, while the shear stress reaches its maximum value $\sigma_0/2$ when acting on the inclined cross section at the bevel angle of 45°. According to the reciprocal law of shear stresses, the shear stress simultaneously reaches the maximum value on the inclined section at the tilt angle of 135°.

Different materials vary in their ability to resist damage and fracture. In general, when ductile materials are subjected to a tensile force, the maximum shear stress on the inclined section is the first to reach the shear strength of the material, and the shear slip deformation zone, or cross-cutting shear zone, occurs at the location at an angle of 45° with the stretching direction. The formation and development of shear bands causes necking of the specimen, until it extends to the entire specimen. For brittle materials, the normal stress exceeds the tensile strength before the maximum shear stress reaches the shear strength, and specimen fracture occurs without yielding.

The stress used in the previous analyses is usually engineering stress, and the cross-sectional area of the specimen changes during stretching. Therefore, the true stress and the engineering stress acting on the specimen are quite different.

Assume that the volume is constant when deformation of the specimen occurs ($A_0 l_0 = Al$), and define the elongation ratio as $\lambda = l/l_0 = 1+\varepsilon$. Then the actual force area is $A = A_0 l_0/l = A_0/(1+\varepsilon)$, and the true stress σ^* is

$$\sigma^* = \frac{F}{A} = (1+\varepsilon)\sigma = \lambda\sigma. \tag{12.42}$$

Clearly, σ^* is greater than σ.

The true stress–strain curve of polymers can be categorized into three types.

The first is for the specimens with necking during the tensile process. Its $\sigma^* \sim \varepsilon$ relationship is shown in Figure 12.22a. From the point $\lambda = 0$ or $\varepsilon = -1$, we can draw a tangent to the $\sigma^* \sim \varepsilon$ curve. On the curve there is one point satisfying $d\sigma^*/d\lambda = \sigma^*/\lambda$, and this point is the yield point. Polymers, after being subjected to the force, will stretch uniformly to this point and begin to neck. This is followed by a narrow neck, gradually tapering to fracture. This mapping method is called Considère mapping.

The second type is for the specimen with no necking in tension, but uniform elongation with increasing load. Its $\sigma^* \sim \varepsilon$ curve is shown in Figure 12.22b. This curve is concave, and we cannot draw a tangent to the curve at $\lambda = 0$. $d\sigma^*/d\lambda$ is always greater than σ^*/λ.

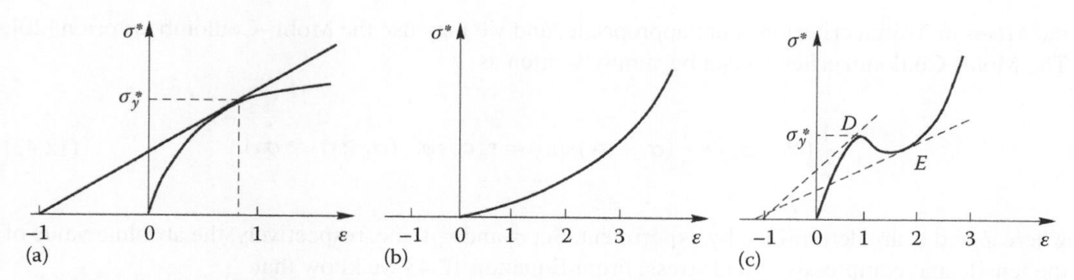

FIGURE 12.22 Three typical types of true stress–strain curve. Specimen (a) with necking, (b) without necking, and (c) with necking and cold drawing.

The third type is the $\sigma^* \sim \varepsilon$ curve on which there are two points that may satisfy $\mathrm{d}\sigma^*/\mathrm{d}\lambda = \sigma^*/\lambda$, as shown in Figure 12.22c. $\sigma = \sigma^*/\lambda$ reaches maximum value at point D, which is the yield point. If stretching continues, σ^*/λ decreases along the curve until point E. After the tangency point E, the material is in the stage of cold drawing in the strain hardening, until its fracture. This is a type of necking and cold drawing polymer.

12.4.1.2 Yield Criteria under Complex Stress State

Two important yield criteria have been introduced in Chapter 1 of this textbook: the Tresca criterion and the Mises criterion. The Tresca criterion states that when the maximum shear stress reaches the critical value, the material begins to yield. The Mises criterion holds that when the second invariant of the deviatoric stress tensor, the octahedral shear stress, or the distortion energy reaches the critical value, the material begins yielding. For polymer materials, the Tresca criterion and the Mises criterion are approximately effective within a certain range. However, the tensile and compressive yield stresses of polymers are usually not the same: this is called the Bauschinger effect. Whitney and Andrews have done research on the yield behavior of polystyrene, polymethyl methacrylate, polycarbonate, and polyvinyl formaldehyde [19], as shown in Figure 12.23. The figure shows that

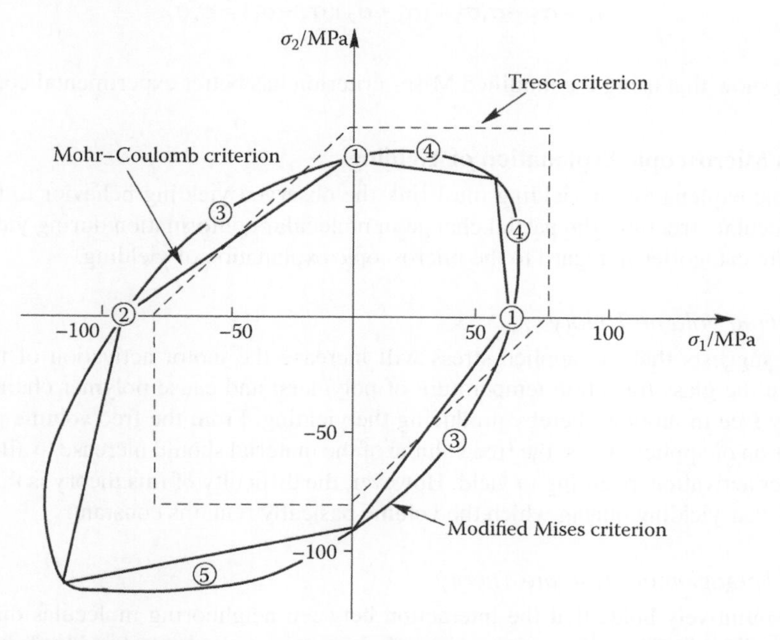

FIGURE 12.23 The yield of polystyrene within the $\sigma1$–$\sigma2$ stress plane: 1, Uniaxial tension; 2, uniaxial compression; 3, torsional shear; 4, biaxial tension; 5, biaxial compression. (From Whitney, W. and Andrews, R.D., *J. Polym. Sci., Part C*, 16, 2981, 1967.)

the Mises or Tresca criterion is not appropriate, and we may use the Mohr–Coulomb criterion [20]. The Mohr–Coulomb criterion can be simply written as

$$\frac{1}{2}(\sigma_1-\sigma_3)+\frac{1}{2}(\sigma_1+\sigma_3)\sin\phi=\tau_c\cos\phi,\quad(\sigma_1\geq\sigma_2\geq\sigma_3), \tag{12.43}$$

where ϕ and τ_c are determined by experiment. Set σ_t and σ_c to be, respectively, the absolute value of the tensile and compressive yield stress. From Equation 12.43 we know that

$$\sin\phi=\frac{(\sigma_c-\sigma_t)}{(\sigma_c+\sigma t)},\quad\tau_c=\frac{\sqrt{\sigma_c\sigma_t}}{2}.$$

The Mohr–Coulomb criterion can be rewritten as

$$\frac{\sigma_1}{\sigma_t}-\frac{\sigma_3}{\sigma_c}=1\,,\quad m\sigma_1-\sigma_3=\sigma_c,\quad(\sigma_1\geq\sigma_2\geq\sigma_3), \tag{12.44}$$

where

$$m=\frac{\sigma_c}{\sigma_t}=\frac{1+\sin\phi}{1-\sin\phi}.$$

Another criterion that may be used is the modified Mises criterion, which is the Mohr–Coulomb asymmetric hexagonal external oval. That is,

$$\sigma_1^2+\sigma_2^2-\sigma_1\sigma_2+(\sigma_1+\sigma_2)(\sigma_c-\sigma_t)=\sigma_c\sigma_t. \tag{12.45}$$

Experiments show that using the modified Mises criterion has better experimental consistency.

12.4.1.3 A Microscopic Explanation of Yielding

A microscopic explanation of yielding must link the observed yielding behavior to the polymer's internal molecular structure (the partial change in molecular conformation during yielding). There are three main categories in regard to the microscopic explanation of yielding.

12.4.1.3.1 Free Volume Theory

This theory suggests that the applied stress will increase the motor activation of the molecular chain, reduce the glass transition temperature of polymers, and cause polymer chain segments to become fully free in motion, thereby producing the yielding. From the free volume point of view, under the action of applied stress, the free volume of the material should increase to allow the chains to have higher activation, resulting in yield. However, the difficulty of this theory is that it is hard to explain the shear yielding during which the volume basically remains constant.

12.4.1.3.2 Entanglement Fracture Theory

This theory intuitively holds that the interaction between neighboring molecules during yielding changes, including the breakdown of a variety of geometric entanglements and sub-valence forces. This theory gives a good explanation for the phenomenon of post-yield strain-softening of the material.

12.4.1.3.3 Argon Theory

This theory suggests that the resistance of the plastic deformation of polymers occurs mainly from intermolecular interactions, and the deformation is the generation of a dual distortion of molecules. This theory can describe the plastic flow of glassy polymers within the range from absolute 0 K to near T_g. Moreover, this theory also explains that the strain hardening after material yielding is caused by entropic resistance of the molecular rearrangement.

The Argon theory considers yielding as the process of activation rate. It is based on the concept given by Eyring et al. [22] that stress promotes the microscopic mechanism of thermally activated plastic deformation, and assumes that the deformation of polymers is caused by the entire chain-like molecule, or part of it, moving over the barrier. The external force σ causing the barrier has a symmetrical linear displacement, making the strain thermal activation transfer, and causing partial molecular conformation across a large energy barrier, tending in the direction of the applied stress. Under the effect of stress σ, the change speed of the strain is proportional to the net flow in this direction. We then get

$$\frac{\mathrm{d}r}{\mathrm{d}t} = \dot{r} = \dot{r}_0 \exp\left(-\frac{\Delta H}{RT}\right)\sinh\left(\frac{v\sigma}{RT}\right), \tag{12.46}$$

where $\dot{r}_0$ is the constant factor, ΔH is the activation energy, and v is the activation volume, representing the volume of the chain segment of a certain size of polymer. In order to produce plastic deformation, this chain segment should undergo the entire movement.

Based on the viewpoint that applied stress causes the molecules to flow, and according to Eyring's concept that increasing stress causes the internal viscosity to decrease [23], the so-called yield stress is the stress value when the plastic strain speed making the viscosity decreased to deformation is exactly equal to the plastic strain speed $\dot{r}$ given by Eyring's equation.

12.4.1.4 Factors Affecting the Yielding of Polymers

The yield point previously described is the critical point of the material, producing plastic deformation. In addition to the molecular weight and its distribution, the material-forming technologies, the strain rate, the temperature, and the hydrostatic stress are the main factors affecting the material's yielding behavior [21].

1. *Dependence of yield stress on strain rate*: Because polymers are viscoelastic in nature, the movement of their molecular chains is strongly dependent on the loading time. Therefore, the yield stress of polymer materials also has a great dependence on the strain rate. When the strain rate increases, the corresponding yield stress also increases. In general, the yield stress σ_Y and the logarithm of the strain rate show a linearly increasing relationship. That is, $\sigma_Y = A + B\log\dot{\varepsilon}$, where A and B are empirical constants. Therefore, we can use the slope of the line $\mathrm{d}\sigma_Y/\mathrm{d}\log\dot{\varepsilon}$ to represent the extent to which the yield stress is dependent on the strain rate.
2. *Dependence of yield stress on temperature*: All of the mechanical properties of polymer materials have a strong dependence on temperature. Similarly, both the yield stress and the yield strain of the polymer are strongly dependent on temperature. On the low temperature side, the yield stress terminates at the brittle–ductile transition temperature. Below this transition temperature, the polymer becomes brittle, and there is no yield point. On the high temperature side, the glass transition temperature of polymers limits the yield stress. At the glass transition temperature, the yield stress of the polymer tends to zero. Between the brittle–ductile transition temperature and the glass transition temperature, the yield stress of polymers decreases linearly with increasing temperature, as shown in Figure 12.24.

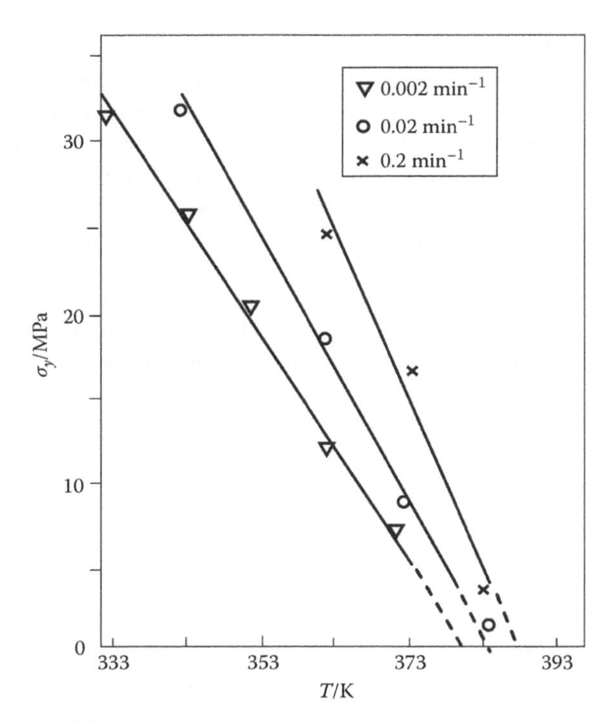

FIGURE 12.24 The yield stress of polymethyl methacrylate varies with temperatures at different strain rates. (From Langford, G. et al., *The Material Research Lab. Research Report No. R63–49*, MIT School of Engineering, Cambridge, MA, 1963.)

3. *Influence of hydrostatic stress on yield stress*: The impact of hydrostatic stress on the yield of polymers is obvious. It has been proven that when hydrostatic stress increases to several hundred MPa, the yield stress of the polymer will significantly increase. Researches on the relationship between the yield stress and the hydrostatic pressure have resulted in the following conclusions: (1) For all crystalline and amorphous polymers, the dependence of yield stress on pressure is approximately a linear relationship; that is, when the pressure increases, the yield stress increases. (2) The yield stress–pressure dependence for crystalline polymers is greater than that for amorphous polymers. (3) The impact of pressure on the yield stress of low-modulus polymers is greater than that of polymers with high modulus.

12.4.2 Crazing of Glassy Polymers

Crazing and crazing damage are unique phenomena for polymers. From the mechanical state point of view, the crazing which causes material damage is the precursor to macroscopic fracture. On the other hand, in a multiphase polymer, crazing consumes a significant amount of energy used for crack propagation, and thus constrains crack propagation and improves the toughness of the material, which is one of the toughening mechanisms of polymers. To correctly understand the crazing phenomenon and explore the basic law of initiation, growth and breakdown are the keys to understanding the deformation and fracture of polymer materials [17,18,25–32].

12.4.2.1 Mesoscopic Structure and Morphology of Craze

Under certain thermal and mechanical conditions, the surface, the interior, or the crack tip region of polymer materials are likely to produce crazes. Crazing has unique structural features, and through a variety of experimental techniques we can study the structure and morphology of crazes.

Optical methods use optical interferometry to measure the refraction index, estimated the volume percent of the hole in the silver lines, and determine the cross-sectional shape of the craze. However, this method applies only to transparent specimens. Moreover, the displacement of the crack tip opening and the thickness of the craze should have an equal order of magnitude with the wavelength of light. The refraction index of materials in the crazing area should be quite different from the refraction index of the bulk material.

X-ray and electron diffraction methods can measure the size of the microvoids and microfibrils in the craze, and can also characterize the orientation of the craze fibril.

Transmission electron microscopy (TEM) and scanning electron microscopy (SEM) methods can directly observe the craze structure of thin films and bulk materials.

The literature has extensively described the typical craze structure [17,18,26], and consensus has been formed on the following points: (a) crazing is generally caused by tensile stresses; (b) the craze consists of highly oriented microfibrils and voids; and (c) the volume fraction of microvoids can reach 50%–80%, the diameter of the microvoid is about 10–20 nm, and the average craze fibril diameter and fibril spacing is about 6–45 nm. All these values are affected by the polymer molecular structure, ambient temperature, loading rate, and stress level.

In the past, it was thought that craze fibrils were independent of each other, and arranged in parallel. The craze fibril is an intermittent phase, the void is a continuous phase, and the craze fibril is dispersed in the continuous void phase. However, low-angle electron diffraction (LAED) analysis shows [18] that (i) some cross-tie fibrils appear quasi-periodically, connecting the main fibrils, and (ii) the main craze fibrils are oriented at a certain angle with the direction of principal stress. The presence of cross-tie fibrils makes the craze fibrils become the continuous phase, and thus the craze is a complex fibril network structure, as shown in Figure 12.25. The cross-tie fibrils

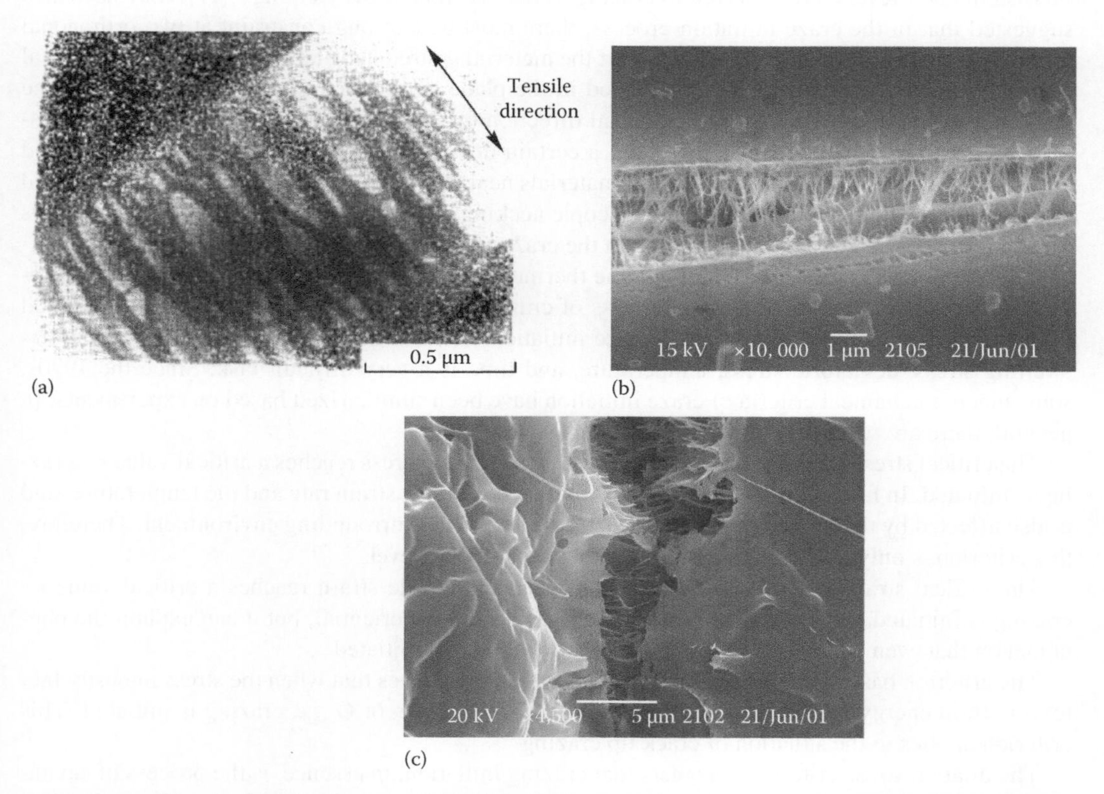

FIGURE 12.25 Craze structures. (a) TEM images of the craze structure in a PS thin film. SEM images of (b) surface and (c) internal crazing in a PP specimen.

lend the craze some load-carrying capacity in the direction perpendicular to the main fibrils, and the shear stress can be transferred between the craze fibrils. The cross-tie fibril is created in the craze-widening process. Some of the entangled polymer strands which bridge the two main fibrils in the active zone are not broken or disentangled during fibrillation; when several such strands pile up locally, the craze/bulk interface bypasses the pile-up.

Figure 12.25a shows TEM images of the craze structure in a thin film of polystyrene (PS). Figure 12.25b and c show SEM images of surface crazing and internal crazing in a PP specimen [2]. These images show that the internal craze structure is substantially similar for both an amorphous polymer such as PS and a semicrystalline polymer such as PP. Usually, the craze fibril in semicrystalline polymers is finer than in amorphous polymers, and the thickness of a single craze observed in semicrystalline polymers is 2–10 times greater than that in amorphous polymers. Crazes are short and irregular when they form above the glass transition temperature in a crystalline polymer composed of many smaller spherulites, as in polyethylene (PE) at room temperature. This is because their growth is limited to a small number of spherulites and is significantly affected by the local stress direction and the local spherulite structure. In a semicrystalline polymer, the amount of crazing below the glass transition temperature is small, but the length is longer than those formed above the glass transition temperature. They grow along the direction perpendicular to the larger principal stress, regardless of the structure of spherulites, just as in an amorphous polymer, and their microstructures are still similar to the crazes formed above the glass transition temperature.

12.4.2.2 Craze Initiation

The craze structure shows that its formation is accompanied by local cavitation and fibrillation along the direction of deformation. In order to differ from the shear yielding pore forming effect, Sternstein and Meyers have referred to crazing as the "normal stress yielding." [17] They have also suggested that in the craze initiation process, there must be a strong constraint at two orthogonal directions (e.g., x-direction and z-direction) if the material is stretched along the y-direction. Internal crazing is liable to occur in materials loaded in the plane strain condition. However, because the stress constraint does not exist in the normal direction on the surface, it does not meet the necessary condition to cause crazing. Therefore, a certain deformation first occurs on the surface, and crazing emerges at the internal location of materials near the surface deformation zone. This initial deformation on the surface might be microscopic necking caused by surface cracks or defects. This local micro-shear plays an important role in the craze initiation process.

The core issues on craze initiation are the thermo mechanical condition or criterion, the microscopic mechanism, and the dynamic process of craze initiation. Craze initiation has been studied by many researchers, but developing a craze initiation criterion involving multiple factors such as swelling stress, deviatoric stress, temperature, and time is a very difficult task. Since the 1970s, some macromechanical criteria of craze initiation have been summarized based on experiments. In general, there are several categories as will be discussed [17].

The critical stress criterion considers that when the tensile stress reaches a critical value σ_c, crazing is initiated. In fact, the craze initiation stress depends on the strain rate and the temperature, and is also affected by the impact of molecular orientation and the surrounding environment. Therefore, this criterion is only useful experientially at the macroscopic level.

The critical strain criterion prescribes that when the tensile strain reaches a critical value ε_c, crazing is initiated. This criterion is also macroscopic and experiential, but it can explain the phenomenon that even when the stress is relaxed, crazing is still initiated.

The criterion based on fracture mechanics parameters believes that when the stress intensity factor, or strain energy release rate, reaches its critical value K_{craze} or G_{craze}, crazing is initiated. This criterion applies to the situation of crack tip crazing.

The dilation stress criterion considers that crazing initiation, in essence, is the process of cavitation. This criterion believes that the crazing initiation criterion should include the effect of hydrostatic tensile stress.

Argon and others considered the impact of a material's inhomogeneity on crazing nucleation. They believe that crazing initiation is the thermal activation process [34] promoted by stress concentration. It consists of two basic stages:

a. The locally concentrated shear stress enables the material to develop microscopic shear bands through thermal activation. When the development of microshear bands is obstructed, they can overcome the material's surface energy and form microvoids.
b. When the number density of microvoids reaches a critical value, the stress field they have caused will interact and, accordingly, the microvoids will expand, accompanied by strain softening and cold drawing of materials between the adjacent voids. Using the plastic expansion of microvoids as the crazing condition, Argon obtained the craze initiation time [34], which is the first explanation for time dependency of crazing. Regarding the cavitation, there are also different views. Kawogoe and Kitagawa [35] suggest that microvoids are caused by elastic expansion of microcracks formed by the interaction of two intersected microshear bands. Gent believes that the local dilation stress causes the decrease in the material's glass transition temperature T_g, and that therefore materials are subject to cavitation and crazing [36].

12.4.2.3 Craze Growth

The growth of crazes comes in two forms: the craze tip advances and craze thickening.

1. *Craze tip advance*: The craze tip advance model proposed by Argon et al., based on the Taylor meniscus instability mechanism, has become widely accepted [18,26]. The mechanism suggests that there is a narrow wedge region at the craze tip, within which—due to strain softening—the material forms a fluid-like layer. The craze tip experiences unstable finger-like extensions in this layer as shown in Figure 12.26. This model has been confirmed in many polymer materials. The degree of tip stress concentration and the internal void spacing indicated by this model are consistent with experiment results.
2. *Craze thickening*: During the process of craze tip advances, the thickness of the craze increases simultaneously. The craze-thickening mechanism has been a focus of research in recent years.

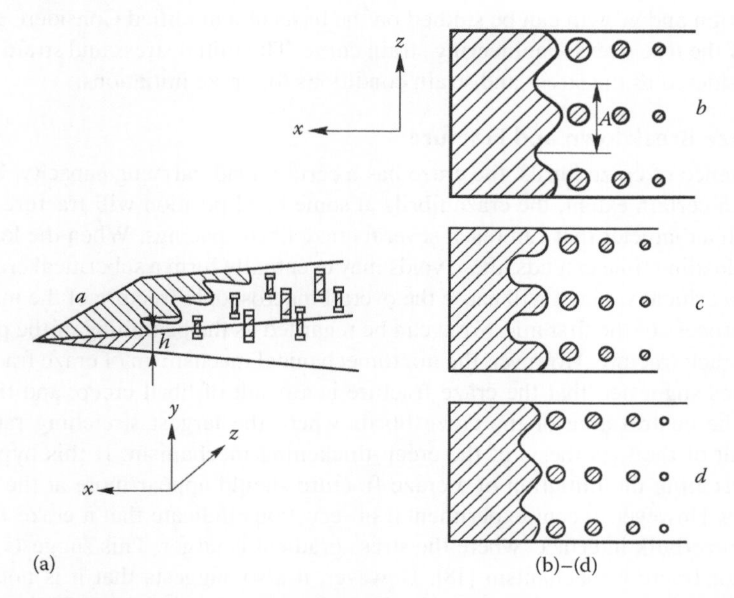

FIGURE 12.26 Schematic of the meniscus instability mechanism for craze tip advance. (a) Three-dimensional view. (b)–(d) Sections along the vertical y-direction.

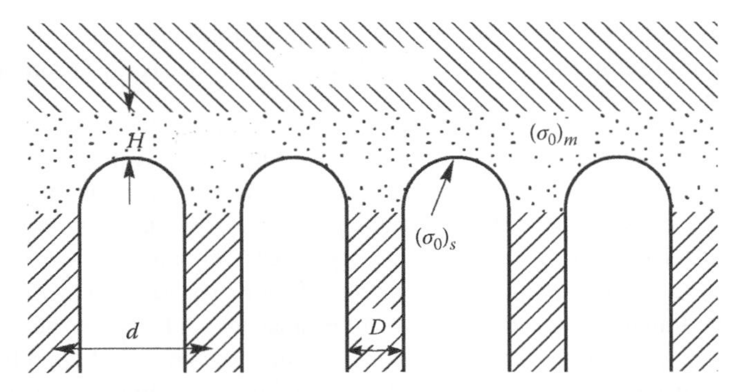

FIGURE 12.27 Schematic of the interface drawing mechanism for craze thickening. (From Kausch, H.H., *Adv. Polym. Sci. 91/92*, Springer-Verlag, Berlin, 1990.)

There are two possible mechanisms for craze thickening: (a) the creep of craze fibrils; and (b) substances in the active zone of the bulk/craze interface are gradually transformed into fibrils. Although some studies suggest that the creep of craze fibrils is the main mechanism, recent calculations and analyses indicate that it is not enough to only consider creep as a mechanism for the thickening of stress-crazing. Rather, the cold-drawing of the active zone of the interface may be the major mechanism. Experimental results show that the craze fibrils in the active zone stretch with a constant stretch ratio λ, and the craze fibrils fracture almost at the craze/bulk interface. These are reasons to be more inclined to accept the second mechanism. This mechanism suggests that there is a strain-softening layer at the interface between the oriented crazes and the bulk material, called the active zone. The thickness of this layer depends on the local strain rate and temperature; it is roughly the same as the diameter of the craze fibrils, and it can be observed using a variant of the gold decoration technique. Under stress, the molecular chain in the active zone is disentangled or fractured, and continuously drawn into the craze fibrils, increasing the length of the craze fibrils and increasing craze thickness, as shown in Figure 12.27, here, D is the fibril diameter, d is the distance between fibrils, H is the thickness of the active zone. Recent studies suggest that a continuous micronecking takes place in concurrence with surface drawing during craze fibrillation. Therefore, the craze initiation and growth can be studied on the basis of a modified Considère graphic method with the help of the true stress–engineering strain curve. The initial stress and strain for micronecking can be considered as the stress and strain conditions for craze initiations.

12.4.2.4 Craze Breakdown and Fracture

Due to the presence of craze fibrils, the craze has a certain load-carrying capacity. When the craze is thickened to a certain extent, the craze fibrils at some local position will fracture, and thus form microvoids with a diameter that can reach several craze fibril spacings. When the load continues to increase or the loading time extends, these voids may eventually form a subcritical crack. If the crack length reaches a critical value, it will cause the overall macroscopic fracture of the material. Clearly, the formation process of the first microvoid can be regarded as the beginning of the process of craze breakdown or crack fracture. However, the micromechanical mechanism of craze fracture is unclear.

Early theories suggested that the craze fracture is a result of fibril creep, and that the fracture location is at the central area of the craze fibrils where the largest stretching ratio exists. This is a direct result of the hypothesis of the creep-thickening mechanism. If this hypothesis is true, the voids representing the initiation of a craze fracture should appear more at the central area of the craze fibrils. However, recent experimental observations indicate that a craze fracture usually occurs at the craze/bulk interface, where the stress gradient is larger. This suggests that creep may not be the major fracture mechanism [18]. However, it also suggests that it is not enough to say that the fibril fracture process relies only on the molecular chain scission. Disentanglement also plays a very important role. Craze fracture is caused by the integrated effect of disentanglement and molecular chain scission at the craze/bulk interface. The fibril fracture model, which considers both chain scission and disentanglement, has obtained results in good agreement with experiments. Molecular weight and the effective entanglement density have greater impact on craze fracture. Studies have shown that the stability of craze fibrils rises when the molecular weight and effective entanglement density of materials increases. In addition, recent studies have considered the cross-tie fibril effects. Assuming that the crazing region is an anisotropic elastic media, Brown [28] and Hui [29,30] have correlated the macroscopic fracture toughness, G_c, with the microscopic areal chain density of entanglements, Σ, the force required to break a craze fibril, f_b, as well as the craze microstructure parameters, such as the average diameter and the stretching ratio of craze fibrils. This attempt provides a possible way for designing the toughness of a material [37].

12.4.3 Strength and Brittle–Ductile Transition of Polymers

Compared with inorganic nonmetallic materials, one of the biggest advantages of polymer materials is their inherent toughness: before fracture they can absorb a large amount of mechanical energy. However, the inherent toughness of polymer materials is not always revealed in practical applications. As mentioned before, the mechanical properties of polymers are dependent on load rate and temperature. Changing the load method or the shape and size of specimens can cause the material to transition from ductile fracture to brittle fracture. In engineering, we always try to avoid the brittle fracture. Therefore, we must understand both the brittle fracture and the ductile fracture process of the polymer material, and master the law of brittle–ductile transition to ensure the material against brittle fracture.

There is no exact definition for the brittleness and toughness of materials. It is generally believed that brittleness, in essence, is always related to the material's elastic response. Specimen deformation is uniform before the brittle fracture point. Cracks run rapidly through the plane perpendicular to the stress direction and cause the overall fracture of the specimen. Before fracture, the specimen does not show obvious plastic deformation, and the corresponding stress–strain relationship is basically linear. The strain value to fracture is less than 5%; therefore, the energy required for fracture is not large. Ductile fracture usually has a larger and nonuniform deformation. The fracture surface often shows irreversible plastic deformation. The corresponding stress–strain relationship is nonlinear. Moreover, before the fracture point, the curve slope can be zero or even negative. Therefore, the fracture energy consumed is large. The fracture surface morphology and the fracture energy are the most important indicators that distinguish brittle fracture from ductile fracture.

Crazing is the precursor to the brittle fracture of glassy polymers. The fracture of craze fibrils usually results in the formation and extension of cracks, and eventually leads to the overall fracture of the material or structure.

It is generally believed that brittle fracture and plastic yielding are two separate processes. The relationship of brittle fracture stress σ_B and the plastic yield stress σ_Y, with the strain rate, is quite different from that with temperature [9,12,21], as shown in Figure 12.28.

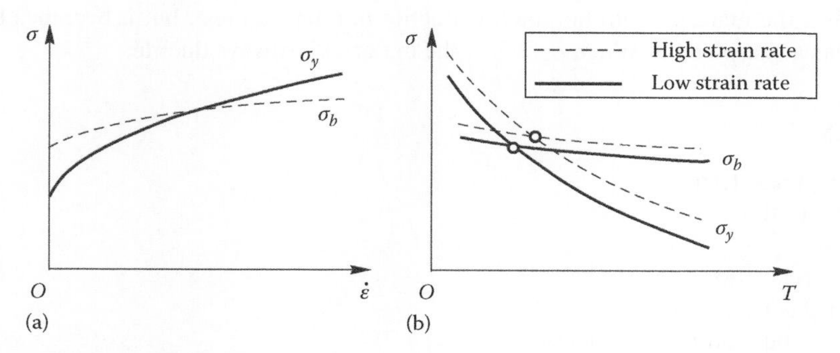

FIGURE 12.28 Brittle–ductile transition.

At a certain temperature and strain rate, when the applied stress reaches σ_B or σ_Y (whichever is lower), the corresponding brittle fracture or plastic yielding occurs. Obviously, the intersection of the curve $\sigma_B \sim T$ (or $\sigma_B \sim \dot{\varepsilon}$) and $\sigma_Y \sim T$ (or $\sigma_Y \sim \dot{\varepsilon}$) should be the brittle–ductile transition point. When the temperature is higher than the brittle–ductile transition temperature, the material is always ductile. Meanwhile, at a certain temperature, the material is ductile in the low strain rate, and brittle fracture occurs at the high strain rate. The brittle–ductile transition of polymer materials is affected by many factors. When the strain rate increases, the brittle–ductile transition point moves toward higher temperatures. Compared with brittle fracture stress, plastic yield stress is affected more by the strain rate and the temperature, and these two factors have a smaller impact on brittle fracture stress.

In addition, other factors also affect the brittle–ductile transition of a material [21]:

1. *Molecular weight.* Molecular weight has no direct impact on the yield stress of a material. The increase in the number average molecular weight M_n increases the tensile strength σ_B in polymers. Within a certain range, σ_B and M_n have the following approximate relationship:

$$\sigma_B = A - \frac{B}{M_n},$$

where A and B are material constants.
2. *Side groups.* Vincent's studies have shown that rigid side groups increase both the yield stress and the brittle fracture stress, while flexible side groups decrease both the yield stress and the brittle fracture stress. Therefore, there is no apparent regularity for the impact of side groups on the brittle–ductile transition.
3. *Cross-linking.* In general, cross-linking increases the yield stress of polymers, and thus increases the material's brittleness and moves the brittle–ductile transition toward higher temperatures, but there is little impact on tensile strength.
4. *Plasticization.* Adding an appropriate amount of plasticizer in the polymer material can reduce the probability of brittle fracture of the material, and the brittle–ductile transition moves toward lower temperatures. The added plasticizer causes a reduction of the material's yield stress greater than the reduction of the material's brittle fracture stress.
5. *Molecular orientation.* Molecular orientation will result in anisotropy of the mechanical properties of materials. In general, fracture stress and yield stress are dependent on the direction of the applied stress, but fracture stress is more dependent on the direction of the stress than yield stress.
6. *Notch.* A sharp notch will make the fracture of polymers change from ductile to brittle. In an infinite solid, plastic constraints caused by the sharp and deep notch can raise the yield stress about three times.

In general, for fracture stress σ_B and yield stress σ_Y, when $\sigma_B < \sigma_Y$, the material is brittle. When $\sigma_Y < \sigma_B < 3\sigma_Y$, the material with no notch is ductile in a tension test, but it becomes brittle if the specimen has a sharp notch. When $\sigma_B > 3\sigma_Y$, the material is always ductile.

EXERCISES

12.1 Define these terms:
Viscoelasticity
Creep
Stress relaxation
High elasticity
Time–temperature equivalence
The Boltzmann superposition principle

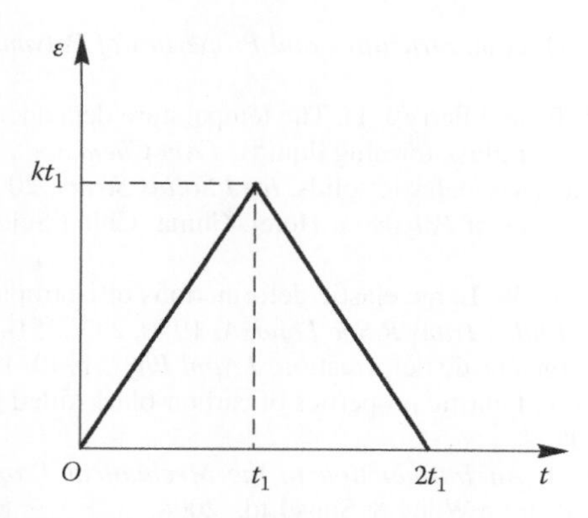

FIGURE 12.29 Strain history in Exercise 12.3.

12.2 What are the characteristics of the high elasticity of polymer materials?

12.3 Let the relaxation modulus function of a polymer material be $E(t) = E_1 e^{-t/\tau_1} + E_2 e^{-t/\tau_2}$. Calculate the stress after $2t_1$ if a material undergoes a strain history as shown in Figure 12.29.

12.4 It is known that the relaxation modulus function of a linear viscoelastic material is $E(t) = E_0 e^{-t/\tau}$. Write the creep compliance function J(*t*).

12.5 Considering the WLF equation, when selecting different temperatures as reference temperatures, the equation is in the same form, and only the values of the parameters C_1 and C_2 in the equation are different. Take $T_0 = T_g$, and set the universal parameters $C_1 = C_1^* = 17.44$, $C_2 = C_2^* = 51.6$. Find the material parameters C_1 and C_2 when $T_0 = T_g + 50°\text{C}$.

12.6 Polystyrene is creeping under a fixed stress, and it is known that the glass transition temperature (T_g) of the materials is 358 K.
How much faster is the creep at a temperature of 423 K than at a temperature of 393 K?

12.7 A rubber band keeps its elongation constant. At a temperature of 273 K, the force is 450 N. How much is the force at a temperature of 323 K?

12.8 The Mooney–Rivlin strain energy function is $W = C_{10}(I_1 - 3) + C_{01}(I_2 - 3)$. Derive the stress–strain relationship of the rubber under uniaxial tension; that is, the Mooney–Rivlin equation.

12.9 What are the factors that influence the brittle–ductile transition of polymer materials?

12.10 What are the main criteria for craze initiation? Outline the mechanism of craze growth.

REFERENCES

1. Yang T. Q., Luo W. B., Xu P. et al. *Theory and Application of Viscoelasticity*. Beijing: Science Press, 2004 (in Chinese).
2. Luo W. B. Studies on deformation-induced heat effect, nonlinear viscoelastic behavior and crazing in polymers [Doctoral Dissertation], Huazhong University of Science and Technology, 2001 (in Chinese).
3. Adolf D. B., Chambers R. S., Flemming J. et al. Potential energy clock model: Justification and challenging predictions. *J Rheol,* 2007, 51(3): 517–540.
4. Losi G. U. and Knauss W. G. Free volume theory and nonlinear thermoviscoelasticity. *Polym Eng Sci,* 1992, 32(8): 542–557.
5. O'Connell P. A. and McKenna G. B. Large deformation response of polycarbonate: Time-temperature, time-aging time, and time-strain superposition. *Polym Eng Sci*, 1997, 37(9): 1485–1495.
6. Lai J. and Bakker A. Analysis of the non-linear creep of high-density polyethylene. *Polymer,* 1995, 36(1): 93–99.
7. Luo W. B., Yang T. Q., and An Q. L. Time-temperature-stress equivalence and its application to nonlinear viscoelastic materials. *Acta Mech Solida Sin,* 2001, 14(3): 195–199.
8. Akinay A. E. and Brostow W. Long-term service performance of polymeric materials from short-term tests: Prediction of the stress shift factor from a minimum of data. *Polymer*, 2001, 42(10): 4527–4532.

9. Ma D. C., He P. S., Xu C. D. et al. *Structures and Properties of Polymers*. (2nd edn.). Beijing: Science Press, 1999 (in Chinese).
10. Williams M. L., Landel R. F., and Ferry J. D. The temperature dependence of relaxation mechanisms in amorphous polymers and other glass-forming liquids. *J Am Chem Soc.*, 1955, 77: 3701–3707.
11. Schapery R. A. The nonlinear viscoelastic solids. *Int J Solids Struct,* 2000, 37(1–2): 359–366.
12. He P. S. *Mechanical Properties of Polymers*. Hefei, China: China Science and Technology University Press, 1997 (in Chinese).
13. Rivilin R. S. and Saunders D. W. Large elastic deformations of isotropic materials VII. Experiments on the deformation of rubber. *Philos Trans R Soc Lond, A,* 1950, 243: 251–288.
14. Mooney M. A. theory of large elastic deformation. *J Appl Phys*, 1940, 11: 582–592.
15. Yeoh O. H. Characterization of elastic properties of carbon-black-filled rubber vulcanizates. *Rubb Chem Technol*, 1990, 63: 792–805.
16. Ward I. M. and Sweeney J. *An Introduction to the Mechanical Properties of Solid Polymers*. (2nd edn.).West Sussex, England: John Wiley & Sons Ltd., 2004.
17. Narisawa I. and Yee A. F. Crazing and fracture of polymers. In: Thomas E. L. *Structure and Properties of Polymers, Materials Science and Technology, A Comprehensive Treatment, Vol. 12*. Weinheim, Germany: VCH Publication. 1993, pp. 698–765.
18. Kausch H. H. Crazing in polymers. *Advances in Polymer Science 91/92*. Berlin: Springer-Verlag, 1990.
19. Whitney W. and Andrews R. D. Yielding of glassy polymers: Volume effects. *J Polym Sci, Part C*, 1967, 16: 2981–2990.
20. Coulomb C. A. Essai sur une application des règles de maximis et minimis à quelques problèmes de statique relatifs à l'architecture. *Mémoire prèsentè à l'Acadèmie Royale des Sciences Paris*. Paris: De l'Imprimerie Royale, 1776: pp. 343–382.
21. Gao J. G. and Li Y. X. *Polymer Materials*. Beijing: Chemical Industry Press, 2002 (in Chinese).
22. Krausz A. S. and Eyring H. *Deformation of Kinetics*. New York: Wiley-Interscience, 1975.
23. Eyring H. Viscosity, plasticity, and diffusion as examples of absolute reaction rates. *J Chem Phys.*, 1936, 4: 283–291.
24. Langford G., Whitney W., and Andrews R. D. *The Material Research Lab*. Research Report No. R63-49. MIT School of Engineering, Cambridge, MA, 1963.
25. Li Q., He Z. R., and Song M. S. Review on the statistical mechanics of meso-damage and fracture for glassy polymers. *Adv Mech,* 1995, 25(4): 451–470 (in Chinese).
26. Luo W. B., Yang T. Q., and Zhang P. An extensive review on meso-damage evolution in polymers. *Adv Mech,* 2001, 31(2): 264–275 (in Chinese).
27. Wang T. J., Yin Z. N., and Wang J. G. Mechanics for crazing in glassy polymers—A review, *Adv Mech,* 2007, 37 (1): 48–66 (in Chinese).
28. Brown H. R. A molecular interpretation of the toughness of glassy polymers. *Macromolecules,* 1991, 24: 2752–2756.
29. Sha Y., Hui C. Y., Ruina A. et al. Detailed simulation of craze fibril failure at a crack tip in a glassy polymer. *Acta Mater,* 1997, 45(9): 3555–3563.
30. Hui C. Y. and Kramer E. J. Molecular weight dependence of the fracture toughness of glassy polymers arising from crack propagation through a craze. *Polym Eng Sci,* 1995, 35(5): 419–425.
31. Luo W. B., Yang T. Q., and Wang X. Y. Time-dependent craze zone growth at a crack tip in polymer solids. *Polymer*, 2004, 45(10): 3519–3525.
32. Luo W. B. and Yang T. Q. Crack tip damage and crazing in polymers under loading. *Acta Mech Sin,* 2003, 35(5): 553–560 (in Chinese).
33. Luo W. B. and Liu W. X. Incubation time to crazing in stressed poly (methyl methacrylate). *Polym Test,* 2007, 26(3): 413–418.
34. Argon A. S. and Hannoosh J. G. Initiation of crazes in polystyrene. *Phil Mag*, 1977, 36(5): 1195–1216.
35. Kawogoe M. and Kitagawa M. Craze initiation in poly (methyl-methacrylate) under biaxial stress. *J Polym Sci Polym Phys*, 1981, 19: 1423–1433.
36. Argon A. S. and Salama M. M. Growth of crazes in glassy polymers. *Phil Mag,* 1977, 36(5): 1217–1234.
37. Luo W. B. and Yang T. Q. Crazing toughening design of brittle polymer solids. *Mater Sci Eng,* 2002, 20(3): 422–424 (in Chinese).

13 Ceramics and the Mechanical Properties of Ceramic Coating Materials

A space shuttle reentering the atmosphere withstands enormous atmospheric pressure and intense high-temperature friction. Without protective measures, the metal shell of the space shuttle would be ablated and melted. Therefore, the surface of the metal shell of the space shuttle must be covered by thermal tiles and an insulation liner. But what is the material of these thermal tiles? Why does it take a leading role?

In home improvement or in an office space, people like to cover the floor with tiles or marble; these materials are not only beautiful, but also durable. What materials have good resistance to wear and tear? Why are ceramic materials chosen for the repair of human bones and teeth? Why does the porcelain rice bowl, or other ceramic tableware, easily break when it hits the ground, but tableware made of metal materials remains unharmed?

Ceramic materials are important, and are used in a lot of applications. In daily life, we use ceramics composed of materials of different types and different manufacturing technologies, and use their special mechanical, physical, and chemical properties to solve specific problems. Ceramic materials are indispensable to human life and manufacturing, and the range of application of their products is widespread throughout various areas in the economy. The development of ceramic materials has grown from the simple to the complex, from coarse to fine, from nonglazed to glazed, and from low temperature to high temperature. Historically, the development, technological improvement, and the scope of ceramics have changed. This chapter will introduce the basic concepts, the microstructure, and the main characteristics of ceramic materials, and then introduce the relevant knowledge and mechanical properties of high-performance ceramic coating materials.

13.1 OVERVIEW OF CERAMIC MATERIALS

13.1.1 Concepts of Ceramic Materials

Ceramic materials were the first man-made materials. Traditional ceramics, household ceramics, construction ceramics, and electrical porcelain all use clays and other natural and mineral raw materials, after the process of crushing, molding, and sintering to produce containers and other vessels. The main raw material of ceramics is a silicate mineral, and is attributed to silicate materials. The development of manufacturing and the advance of technology require the full use of the mechanical, physical, and chemical properties of ceramic materials. The use of high-purity and ultrafine synthetic materials, the precise control of chemical composition, and special processing technologies produce the fine structure and excellent mechanical and thermal properties of ceramic materials. For example, high-temperature ceramics, wear-resistant ceramics, piezoelectric ceramics, and high-thermal conductivity ceramics are all used in integrated circuit boards, and are collectively referred to as special ceramics.

Modern ceramic materials have been developed from traditional ceramics, and are a new class of ceramics with the distinctive characteristics from traditional ceramics. The modern ceramic material—the inorganic nonmetallic material—is a general class of crystalline or amorphous inorganic nonmetallic materials, comprising metal and nonmetallic elements, or simple substances possessing the binding properties of covalent bonding, ionic valence bonding, or mixed valence bonding. It includes not only a variety of oxides, composite oxides, and a variety of silicates, but also carbides, silicides, nitrides, borides, and intermetallic compounds. Moreover, metal ceramics, and simple substance inorganic materials such as diamonds, graphite, and monocrystalline silicon, are all included in the category of ceramics. Thus, ceramics have become a large material family with a variety of types and crystal seeds, and, with metal and polymer organic materials, constitute three contemporary solid engineering materials. Therefore, we can say that the study of ceramic materials is both an old and a young subject [1].

13.1.2 Characteristics of Ceramic Materials

It is well known that the chemical bond of metallic materials (pure metals and alloys) is primarily the metallic bond, which is constituted by the positive metal ion and the electron cloud. Because the metal bond has no direction, metals have good plastic deformation capacity. As for ceramic materials in inorganic nonmetallic compounds, the chemical bonding is covalent bonding, complex ionic valence bonding, and the coexistence of the two—mixed valence bonding. These chemical bonds have strong directionality and high binding energy. Therefore, the plasticity and toughness properties of ceramic materials are much lower than those of metals. They are very sensitive to defects and have less reliable strength.

The Weibull modulus is commonly used to characterize their strength uniformity. Factors such as manufacturing technology, porosity, inclusions, grain boundaries, grain structure, and uniformity have significant effects on the mechanical properties of ceramic materials. They are all the fatal weakness of ceramic materials. However, it is because of these chemical bonds that ceramic materials exhibit greater performance than metal materials in certain areas: (a) high hardness, which determines wear resistance; (b) high melting point, which determines outstanding heat resistance; and (c) high chemical stability, which determines excellent corrosion resistance. Although ceramic materials exhibit these special qualities, their fatal flaw—brittleness—limits their practical application. Therefore, the toughening of ceramic materials has become the core issue of ceramic material research worldwide [1,2].

13.1.3 Microstructure of Ceramic Materials

Ceramic materials are usually polycrystalline, and their microstructures include phase distributions, grain sizes and shapes, pore sizes and distributions, impurity defects, and grain boundaries. Ceramic materials are characterized by the crystal phase, the glass phase, and the gas phase. The crystalline phase is a major component of ceramic materials and determines their physical and chemical properties. The glass phase is the amorphous low-melting-point solid phase, enabling the binding of crystalline phases, the filling of pores, and the reduction of sintering temperature effects. In the gas phase, pores are an inevitable phenomenon, because the accumulation of particles of raw materials is not dense enough in the manufacturing of ceramic materials. The presence of pores will reduce chemical corrosion resistance, electrical insulation, and the mechanical properties of ceramic materials, but will help to improve their thermal insulation, soundproofing, and their wavelength absorption properties. When the porosity increases, the density of the ceramic material is reduced, decreasing its strength and hardness. If the glass phase distribution is in the main phase interface, the strength of ceramic materials at high temperatures drops, making them prone to plastic deformation. Conducting heat treatment for the sintered sample of ceramics makes the grain boundary glass phase recrystallize or enter the crystalline phase and become a solid solution. This can significantly increase the strength of ceramic materials at high temperatures [2].

Although ceramic crystal structures are similar to those of metal, what distinguishes ceramics from metals is that the ceramic structure does not have a large number of free electrons.

This is because ceramics have ionic crystals (MgO, Al_2O_3, etc.) or covalent crystals (SiC, SiN_4, etc.), which are mainly composed of ionic bonds or covalent bonds. Oxide and silicate structures are the two most important types of ceramic crystal structures. Their common characteristics are that the bond is mainly ionic, containing a certain proportion of covalent bonding; and the composition is determined so that they can be represented by a determined molecular formula. Therefore, different types, different particle sizes, different numbers, and different shapes and distributions of the main crystalline phase, the amorphous phase, and the gas phase comprise countless types of ceramic materials with different physical and chemical properties.

13.1.4 Thermophysical Properties of Ceramic Materials

Ceramic material density refers to the ratio of the mass of ceramic materials in air to the mass of the same volume of water at 4°C. The relative density of the vast majority of ceramic materials is in the range of 2.5–4.0, and its size is related to the chemical composition and crystal structure of the material. The melting point (T_m) of the material is a reflection of the strength of the binding force between atoms that maintain the crystal structure. The stronger the binding force, the more stable is the thermal vibration of atoms, the higher the temperature that the crystal structure can maintain, and thus the higher the melting point will be. Because the ceramic crystal is composed of covalent bonds and ionic bonds, the bonding force between lattice particles is strong. High energy is required to convert crystalline structures showing a regular arrangement to the amorphous (molten) state with disordered arrangement. Therefore, the melting point of ceramic materials is generally high, and their high temperature resistance is one of the outstanding characteristics of ceramic materials.

In simple substance materials, carbonaceous materials have the highest melting point; in ceramic materials, carbides have the highest melting point, and a large number of carbides with high melting points possess the NaCl crystal structure. The thermal expansion coefficient is directly related to the repulsion between atoms and the interatomic bond energy in the material. The material with strong bond energy generally has a relatively small thermal expansion coefficient, and vice versa. The melting point of a material is one of the characteristics of its bonding strength. Therefore, materials with high melting points have smaller thermal expansion coefficients.

Generally speaking, the thermal expansion coefficient of ceramic materials is very small, which further reflects their good thermal stability and volume stability. What are the electrical properties of ceramic materials? Because of the action of capacitors and ceramic insulation, it is thought that ceramics are insulators only. In fact, the chemical bond and the crystal structure of ceramic materials are far more complicated than metal materials. Therefore, their electrochemical properties are varied, and include conductive ceramics, semiconductor ceramics, insulation or dielectric ceramics, and superconducting ceramics.

13.2 MECHANICAL PROPERTIES OF CERAMIC MATERIALS

Because the chemical bond of ceramic materials is primarily the ionic bond and the covalent bond, the bonding is solid and the directionality is obvious. Compared with metal materials, in general, ceramic materials have more complicated crystal structures and a smaller surface energy. Therefore, the ceramic material's strength, hardness, elastic modulus, wear resistance, corrosion resistance, and heat resistance are superior to those of metal. However, the plasticity, toughness, workability, thermal shock resistance, and the reliability of use of ceramic materials are not as good as metal materials.

13.2.1 Elastic Deformation of Ceramic Materials

In general, materials under the effect of static tensile loads should experience elastic deformation, plastic deformation, and fracture in three stages, which can usually be expressed by the stress–strain curve,

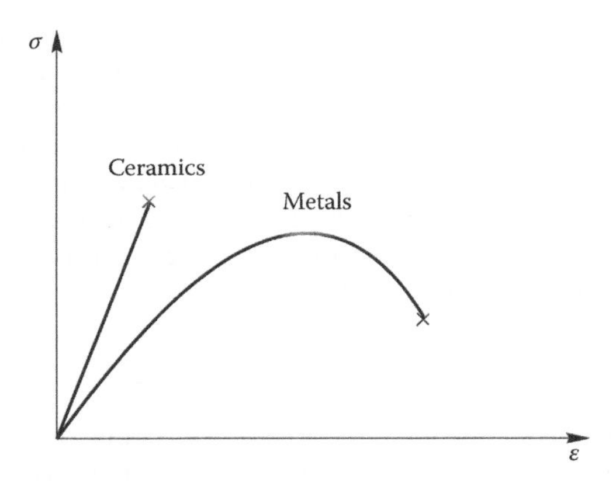

FIGURE 13.1 Comparison of stress–strain curves of ceramic and metal materials under tensile loads. (From Jin, Z.H. et al., *Engineering Ceramic Materials*, Xi'an Jiaotong University Press, Shaanxi, China, 2000; Taken from Figure 5.1.)

shown in Figure 13.1. For all kinds of metal materials, in varying degrees, there exists a plastic deformation phase before fracture. The vast majority of ceramic materials under the effect of tensile or bending loads at room temperature will not have plastic deformation. That is, immediately after the elastic deformation phase, brittle fracture characteristics are shown. This is because the elastic deformation of ceramic materials actually results from a very small displacement of the distance between atoms, generated by the equilibrium position under the external force. This critical value allowed by the small displacement between the atoms is very small. When exceeding this value, bond cleavage is generated (ceramics at room temperature) or atomic slip plastic deformation is generated (ceramics at high temperature).

The stress–strain relationship of the material's elastic deformation phase obeys Hooke's law: $\sigma = E\varepsilon$. The slope of the linear part of the σ–ε curve (the elastic modulus E) is the important performance indicator that describes the mechanical behavior of the elastic deformation phase. The physical meaning of the elastic modulus E is the stress required for the material to produce unit strain. The size of the elastic modulus reflects the size of the bonding force between atoms of the material. The greater the E value, the higher is the bonding strength of the material.

13.2.2 Plastic Deformation of Ceramic Materials

According to Chapter 1, plastic deformation refers to stress exceeding the yield point when the material is bearing force, as the material continues to deform without fracture. After the applied force is removed, the material retains some residual permanent deformation that cannot be restored. The size of plasticity of the material is typically measured by tension and elongation, or cross-sectional shrinkage of the compression test. There are two basic ways plastic deformations occur in general crystalline structures: sliding and twinning. At higher temperatures, plastic deformation may occur by grain boundary sliding or it may be rheological. Obvious plastic deformation of metal materials at room temperature mainly comes from sliding generated by the movement of the dislocation. The corresponding grain boundary sliding, or flowing, generally occurs at higher temperatures (~0.3–0.5 T_m). For ceramic materials at room temperature, there is no or little plastic deformation; the main reason is that there is very little sliding in the system. Ceramic materials are generally ionic bonds or covalent bonds, with a significant directionality. When ions with the same charge meet, a great repulsion occurs. The more complex the crystal structure, the more difficult it is to meet these conditions. Hence, only a very small number of simple ceramic crystal structures, such as MgO, KCl, and KBr (NaCl-type structures) have plasticity at room temperature. Normally, ceramic materials, due to the complexity of their crystal structures at room temperature, have no plasticity. Furthermore, ceramic materials generally show a polycrystalline state, and sliding in polycrystalline is more difficult than in single crystals. Because of the confusion of the grain orientation in polycrystalline structures, the sliding of each grain is subject to the constraint of the surrounding grains and grain boundaries, and thus sliding occurs with difficulty. At the grain boundary, due to the accumulation of dislocations, stress concentrations that result in microcracks occur, which also limit the continuance of plastic deformation.

13.2.3 Superplastic Deformation of Ceramic Materials

Nonferrous metals or alloys with fine grain structure, at the appropriate temperature and strain rate (slow), will show abnormal plastic deformation rates, or superplasticity. Superplasticity will also occur in ceramic materials, under the condition of appropriate organizations and deformations. The only difference is that the temperature of superplasticity in ceramics is much higher than that in metals. The superplasticity of ceramics specifically refers to the material under the effect of certain temperatures and stresses, and shows a very high plastic deformation rate. Its tensile deformation rate can be up to several hundred percent. Currently, there are two types [1,3] of superplasticity in ceramic materials:

1. *Transformation superplasticity*: This is superplastic behavior caused by the phase transition due to temperature change that passes the phase transition point. For example, tetragonal zirconia polycrystalline (TZP) is the most typical superplastic ceramic.
2. *Structural superplasticity*: This is a uniform, fine-grained material (<1 μm) of grains with equiaxial shapes, which shows superplastic behavior under the effect of stresses. It is also known as fine-grained superplasticity.

As far as technology is concerned, structural superplasticity is more important. Therefore, equiaxial fine-grained ceramic materials are able to display different levels of superplasticity. For example, Si_3N_4 and SiC materials are considered to be typical brittle materials; if the deformation is more than 3%, the material will be destroyed. However, fine-grained Si_3N_4–SiC multiphase materials (20 wt% SiC) produced by using a special technology are able to reach 150% of their tensile deformation capacity under conditions of 1600°C and strain rate $\dot{\varepsilon}_0 = 4\times10^{-5}s^{-1}$, demonstrating significant superplasticity [4–6]. In addition, for cubic phase, stabilized zirconia materials doped with 5 wt% SiO_2 and 8 wt% Y_2O_3 (8 mol% Y_2O_3 cubic stabilized of ZrO_2, referred to as 8Y-CSZ) are able to reach 520% and 190% of their tensile deformation capacity under the condition of a 1430°C high-temperature environment and a constant strain rate ($\dot{\varepsilon}_0 = 1.0\times10^{-4}s^{-1}$ and $\dot{\varepsilon}_0 = 1.0\times10^{-3}s^{-1}$), as shown in Figure 13.2 [7].

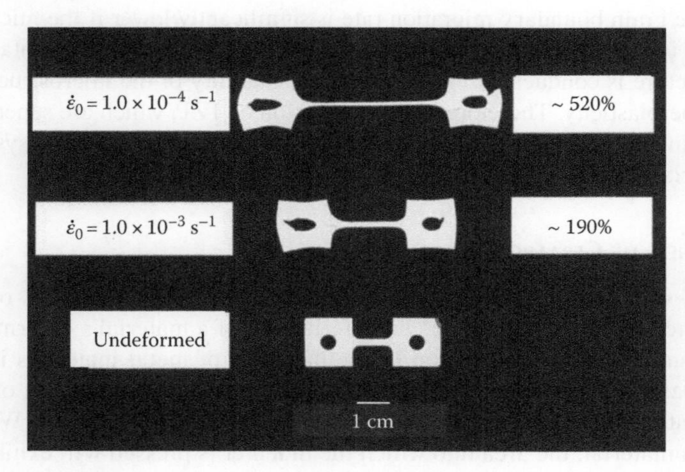

FIGURE 13.2 Macromorphology observations of 8Y-CSZ before and after superplastic deformation. (From Dillon, R.P. et al., *Scr. Mater.*, 50, 1441, 2004; Taken from Figure 3.)

Recent studies have shown that ceramic materials at high temperatures are able to display superplasticity with the following conditions: fine grain size and the range of critical size is about 200–500 nm; grains are equiaxial; the second phase dispersed distribution can inhibit the matrix grain growth at high temperatures; and in intergranular liquid phases or amorphous phases. Typical ceramic materials with superplastic characteristics are powders with Al_2O_3 and ZrO_2 produced by the chemical co-precipitation method. Sintering at about 1250°C after the formation, we can obtain the sinters with a relative density of about 98%. With this type of ceramic at 1250°C and 3.5×10^{-2} s^{-1} strain rate, the maximum deformation can reach up to 400%. The superplasticity of ceramic materials is related to grain boundary sliding or grain boundary liquid phase flowing. Similar to metals, the superplastic flow of ceramic materials is controlled by diffusion processes, and the strain rate [8–10] is

$$\dot{\varepsilon} = A\left(\frac{1}{d}\right)^{p} \sigma^{n} \exp\left(-\frac{Q}{RT}\right), \tag{13.1}$$

where $\dot{\varepsilon}$ represents the strain rate, d is the grain size, σ is the flow stress, p is the exponent related to the grain size, n is the stress exponent, Q is the plastic deformation activation energy, R is the gas constant, and T is the thermodynamic temperature. Within the optimum superplastic region, the value of n is approximately 2.0. When the purity of ceramic materials is low, the n value is lower, $n = 1.9$; when the purity is high, $n = 3.2$.

Current research shows that factors affecting the superplasticity of ceramic materials are mainly grain sizes, the stability of the microstructure, strain rates, and deformation temperatures. The first two are internal factors that affect superplasticity, and the latter two are external factors that affect superplasticity. Grain size and grain boundary properties are the main factors affecting the superplasticity of ceramic materials. The mechanism of fine-grained superplasticity is grain boundary sliding. The smaller the grain size and the greater the grain boundary becomes, the more likely it is that grain boundary sliding will occur at high temperatures. This results in greater deformations that show high superplasticity.

During the process of superplastic deformation, before and after the deformation, the ultrafine grain does not remain unchanged. Due to the combined effects of strain and high temperatures, the microstructure of materials displays the phenomenon of grain growth. Studies have shown that the grain of $Al_2O_3 + 0.1W_B\%$ MgO ceramics shows significant growth after superplastic deformation. The grain growth speed also rises with the increase of tensile strain, thus affecting superplastic performance. The grain boundary migration rate is significantly lower if the microstructure of the ceramic material is a two-phase structure at a constant high temperature. Maintaining the original fine-grained structure is conducive to maintaining the stability of the microstructure, and demonstrating good superplasticity. Therefore, the organization of TZP, which are superplastic materials, is often selected in the two-phase region of the cubic system and the tetragonal system; or, selecting multiphase materials [11,12] such as TZP–Al_2O_3, TZP–SiO_2, and TZP–MgO.

13.2.4 Hardness of Ceramic Materials

From Chapter 4, we know that hardness is one of the important parameters of the mechanical properties of materials. Hardness is the characterization of a material's deformation capacity to resist local pressures. The determination of the hardness of metal materials is a measurement of the surface plastic deformation. Therefore, the hardness and the strength of metal materials have a direct relationship. However, ceramic materials are brittle materials. While determining the hardness of a material, the area into which the indenter is pressed will exhibit pseudo-plastic deformation, including compression and shear composite destruction. Therefore, for ceramic materials, it is difficult to determine a direct correspondence between hardness and strength.

However, high hardness and good wear resistance are excellent characteristics of ceramic materials, and hardness and wear resistance are closely related.

The determination of the hardness of ceramic materials does have some convenient features: (1) it follows the metal hardness testing methods; (2) experimental methods and equipment are simple, and the specimen is small and economical; (3) hardness as the physical parameter of the material itself may obtain stable values; (4) while the Vickers hardness is measured, the fracture toughness can also be measured. Therefore, in the evaluation of the mechanical properties of ceramic materials, the determination of hardness is the most widely used method for the assessment. Commonly used hardness testing methods for ceramic materials are Vickers hardness, microhardness, and Mohs hardness [1].

This is different from metal hardness; the high-temperature hardness of the ceramic material is an important indicator of its mechanical properties [2]. High-temperature hardness measurements are based on the Vickers hardness method and the microhardness method. The determination of high-temperature Vickers hardness must consider the heating method, the determination of sample temperatures, the holding time, temperature changes of the indenter, the protective atmosphere, and other factors. Strictly speaking, the temperature of the specimen should be the temperature near the surface where the indenter was pressed into the specimen. However, we must take into account the temperature and the thermal conductivity of the indenter itself, and the internal temperature distribution of the specimen. By controlling the uniformity of the specimen temperature, and selecting the appropriate preheating method for the indenter, we ensure the stability of the measured hardness data. Moreover, a diamond indenter is easily oxidized when heating at high temperatures. Therefore, vacuum or inert gas is used for protection and to ensure good control of the temperature distribution in both vacuum and atmosphere.

When the indenter has no heating devices, it should be kept near the surface of the specimen under testing until the temperature of the indenter is close to the specimen temperature, then continuously hit a number of indentations. Begin the test when the diagonal line is stable. Experiments have shown that we can see the impact of the pressed indenter on the temperature drop only from the first two indentations. Subsequently, we can obtain stable indentation diagonal line length. From Chapters 4 and 7 of this book we know that within the hold time during the high-temperature testing process, maintained loads will show creep properties. Therefore, high-temperature hardness is the high-temperature performance test method commonly used for ceramic materials.

13.2.5 Wear Resistance of Ceramic Materials

Wear resistance refers to the ability of materials to resist dual pieces of friction, or abrasive wear. Friction and wear occur on the surface of crystals or materials, and are affected by friction pair matches, the abrasive nature (type, particle size, shape, proportion, and distribution), the state of lubrication, pressure, temperature, atmosphere, and other factors. These involve complex systems engineering. In all operating machinery, there is relative movement between the various parts and components. That is, sliding friction or rolling friction, or the coexistence of both frictions, occurs. Under the effect of friction a series of mechanical, physical, and chemical interactions occur, thus resulting in mechanical surface size changes and material losses. This phenomenon is known as wear and tear.

Ceramic wear occurs due to the sliding motion between two surfaces, or between two particles between surfaces, and may also be generated from particle collisions, leading to surface cracking and peeling. Therefore, the amount of wear of ceramic materials is related to the smoothness of the contact surface, or the particle size, and the positive pressure of the wear surface. The smoother the surface, the smaller is the sliding movement between the particles. Wear rate rises with the increase of impact angle and relative velocity. In addition, ceramics wear is related to the relative material hardness, strength, elastic modulus, density, and environmental factors [1].

13.3 FRACTURE TOUGHNESS AND TESTING METHODS OF CERAMIC MATERIALS

13.3.1 Static Toughness of Ceramic Materials

The static toughness of ceramic materials—the work absorbed per unit volume of a material prior to fracture—can be calculated according to the following formula:

$$W = \frac{\sigma_i^2}{2E}. \tag{13.2}$$

The fracture strength of ceramic materials is not higher than the yield strength of steel, but its elastic modulus is higher than that of steel. Therefore, the static toughness of ceramic materials is very low. Because ceramic materials are brittle materials, the crack propagation resistance of a cracked ceramic specimen is the surface energy 2γ required for the formation of two new surfaces. If the surface energy value γ is known, the fracture toughness of ceramic materials can be estimated in accordance with the following formula:

$$K_{IC} = \left[\frac{2E\gamma}{(1-\nu^2)}\right]^{1/2}. \tag{13.3}$$

In addition, because the plasticity of the crack tip of ceramic materials is much smaller than the crack length, linear elastic mechanics can be used to study the problems of crack propagation and fracture. The crack propagation of ceramic materials belongs to the open crack model, that is, the type *I* crack. The tensile or flexural yield strength of ceramic materials and metal materials does not vary considerably. However, fracture toughness values, reflecting the material's crack propagation resistance, vary widely. The fracture of metal materials must absorb a large amount of plastic deformation energy, and the plastic deformation energy can be larger than the surface energy by several orders of magnitude. Therefore, the fracture toughness of ceramic materials is lower than that of metal materials by one to two orders of magnitude: from 2 or 3 $\text{MPa}\cdot\sqrt{\text{m}}$ [1] up to 12 to 15 $\text{MPa}\cdot\sqrt{\text{m}}$.

13.3.2 Fracture Toughness Testing Methods of Ceramic Materials [2]

To determine the fracture toughness of materials, the specimen must be pre-cracked. However, it is very difficult to produce a pre-cracked specimen and the success rate is very low, because ceramic materials are brittle, and the range of ΔK ($\Delta\sigma$) is very narrow. Even so, a number of ceramic fracture toughness testing methods have been developed; including the single-edge notched beam (SENB) method, the double cantilever beam (DCB) method, the double-twist (DT) method, the short bar/rod (SB/SR) method, and the indentation method (IM). Because the SB/SR method opens a V-shaped incision, it is also known as the chevron notch (CN) method. However, because each method has its advantages and disadvantages, there is no recognized test standard as yet. Figure 13.3 shows specimen shapes and loading methods for a variety of fracture toughness testing methods. The most widely used are the SENB method and the IM method. We will focus on the SENB method and the IM method.

13.3.2.1 The SENB Method

The SENB method is similar to the method of three-point bend specimens measuring K_{IC} in metals. The difference is that this method uses an SENB instead of a pre-cracked specimen. It can also use the four-point bending loading method. When using the three-point bending or four-point bending loading method, K_I can be calculated by the following formula:

$$K_I = Y \times \frac{3PL}{2bW^2} \times \sqrt{a}, \tag{13.4}$$

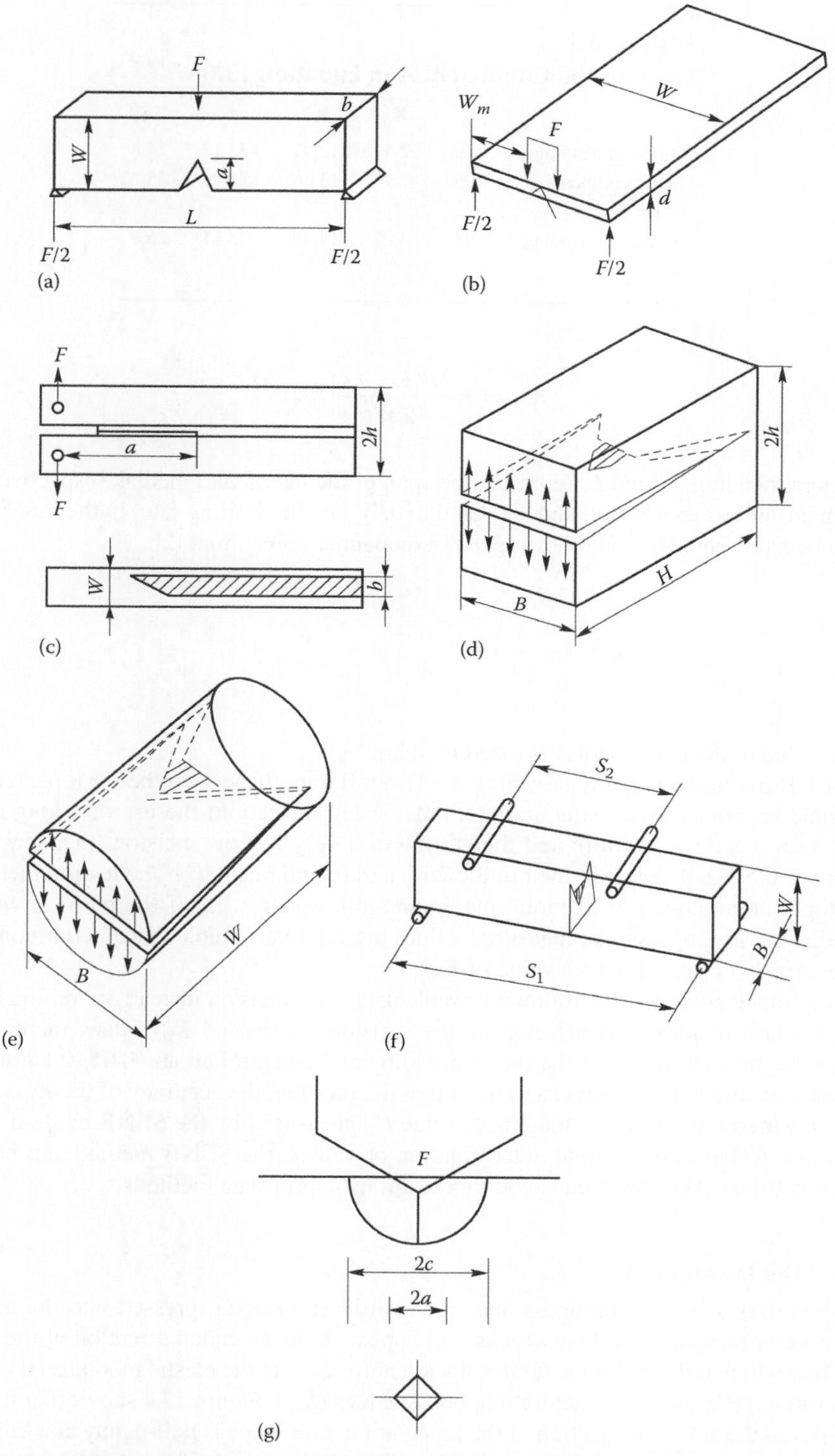

FIGURE 13.3 Different specimen shapes and test methods used to measure the fracture toughness of ceramic materials. (a) SENB specimen, (b) DT specimen, (c) DCB specimen, (d) SB specimen, (e) SR specimen, (f) CN specimen, and (g) IM specimen. (From Zhou, Y., *Ceramic Materials*, 2nd edn., Science Press, Beijing, 2004; Taken from Figures 10.37, 10.39, 10.43, and 10.53.)

TABLE 13.1
Value of the Coefficient *A* in Equation 13.6

	A_0	A_1	A_2	A_3	A_4
Four-point bending	1.99	−2.47	12.97	−23.17	24.8
Three-point bending $L/W = 8$	1.96	−2.75	13.66	−23.98	25.22
Three-point bending $L/W = 4$	1.93	−3.07	14.53	−25.11	25.8

$$K_I = Y \times \frac{3P(L_1 - L_2)}{2bW^2} \times \sqrt{a}. \tag{13.5}$$

For four-point bending, L_1 and L_2 represent the span of the lateral and medial, respectively. Y represents a dimensionless coefficient, and is related to a/W and the loading rate. In the range $0 \leq a/W \leq 0.6$, it can be represented by the following a/W exponential polynomial [2]:

$$Y = A_0 + A_1 \times \frac{a}{W} + A_2\left(\frac{a}{W}\right)^2 + A_3\left(\frac{a}{W}\right)^3 + A_4\left(\frac{a}{W}\right)^4, \tag{13.6}$$

where the value of the coefficient A is listed in Table 13.1.

The SENB method has many advantages: (1) small data dispersion; better repeatability; relatively simple specimen processing and methods of determination; the use of a long rectangular specimen (2 × 4 × 36 – 40 mm); and the cutting of a very narrow incision (incision width ≤0.2 mm, depth of $0.5W \pm 0.1$ mm) in the middle with a diamond blade. (2) The testing method is suitable for high temperatures or different media and atmospheres. (3) If the incision width can be controlled at 0.2 mm or less, the measured values are relatively stable, have better comparability, and are relatively close to the true value of K_{IC}.

However, this method has the following problems: it uses incision instead of pre-cracking; therefore, the fracture toughness is affected by the incision width, and K_{IC} values increase with the increase of the incision width. If the incision width can be controlled at ~0.05–0.1 mm or less, or can precast a certain length of cracks to the top of the incision, the accuracy of the K_{IC} value should improve. In general, the fracture toughness value K_{IC} measured by the SENB method is relatively high. In order to have an accurate determination of values, the SENB method can be used, and ceramic prefabricated cracks can also be successful by appropriate methods.

13.3.2.2 The IM Method

This method uses a Vickers hardness or a microhardness indenter, pressed into the polished surface of the ceramic specimen. Four cracks will appear in the extended direction of the indentation diagonal line which will determine (a) the crack length (2c), (b) the elastic modulus (E) and Vickers hardness values (HV), and (c) fracture toughness values (K_{IC}). Figure 13.4 shows that there are two crack shapes in the side cross section of the indentation: one shape is half-penny cracking, the other is Palmqvist cracking. The calculation formula will be different for different types of cracks. There are many formulas reported in the literature using the indentation method to calculate K_{IC}. Even for the same types of crack, there are also several different proposals for calculating equations. Here, we only show the formula proposed by Niihara [13]:

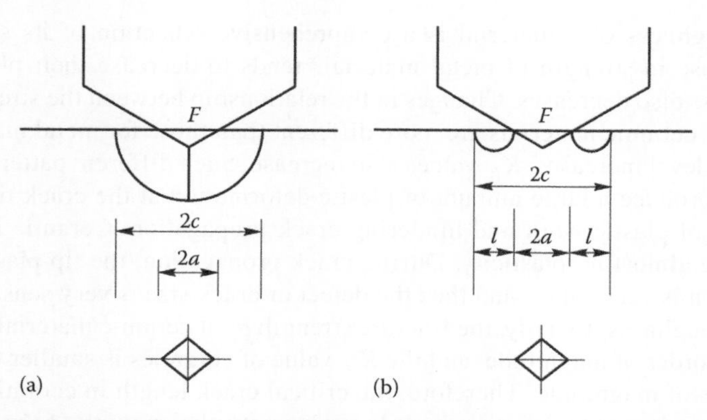

FIGURE 13.4 Two types of crack form in indentation tests. (a) Half-penny crack. (b) Palmqvist crack. (From Zhou, Y., *Ceramic Materials*, 2nd edn., Science Press, Beijing, 2004; Taken from Figures 10.53 and 10.54.)

Half-penny cracks:

$$K_{IC} = 2.109H^{0.6}E^{0.4}a^2c^{-1.5}e \quad (c/a \geq 2.5). \tag{13.7}$$

Palmqvist cracks:

$$K_{IC} = 0.572H^{0.6}E^{0.4}ac^{-1.5} \quad (0.25 \leq l/a \leq 2.5). \tag{13.8}$$

The biggest advantage of the indentation method is that it is not necessary to prepare specialized specimens. A small specimen can be used for testing. When testing the Vickers hardness, the K_{IC} value can be obtained at the same time. This method is therefore both simple and feasible. However, this method is effective only for materials that can produce good indentation cracks. The formation of cracks is mainly an effect of residual stresses, while residual stresses resulting from the plastic zone around the indentation do not match the elastic matrix. Therefore, this method does not allow the material at the lower part of the indenter to undergo the phase transformations or volume densification phenomenon during the loading process. Moreover, the indentation surface must not have any fragmentation phenomenon.

Recently, Lugovy and others have discussed in detail the impact of residual stress on the fracture toughness of Si_3N_4 laminate ceramic materials [14]. In addition, the stress and strain fields around the indentation are very complicated, and so far we have not obtained an exact solution for fracture mechanics, only an approximate solution. Along with the different properties of materials, this method will lead to considerable errors, and, especially when the material shows phase transformation toughening effects, the error is greater. In addition, the length of four corner cracks will change due to the effect of residual stress around the indentation. If placed at different times after the indentation crack, the crack length will change, affecting the accuracy of the measured values.

In order to reduce the error using the indentation method in the determination of the ceramic fracture toughness value of phase transformation toughening, the following amendments are proposed, based on Equations 13.7 and 13.8:

For half-penny cracks,

$$K_{IC}^{C} = (1 + 0.0083V_fE_c)2.109H^{0.6}E^{0.4}a^2c^{-1.5} \quad (c/a \geq 2.5). \tag{13.9}$$

For Palmqvist cracks,

$$K_{IC}^{C} = (1 + 0.0083V_fE_c)0.572H^{0.6}E^{0.4}ac^{-1.5} \quad (0.25 \leq l/a \leq 2.5). \tag{13.10}$$

The fracture toughness of a material is a comprehensive reflection of its strength and plasticity. The increase in strength of metal materials tends to decrease their plasticity, and their fracture toughness also decreases. Changes in the relationship between the strength and the fracture toughness of ceramic materials are more different than those for metal materials. When the ceramic strength level increases, K_{IC} values also increase. Such different patterns of change from metal materials produce a large amount of plastic deformation at the crack tip before fracture, consuming a lot of plastic work and hindering crack propagation. Ceramic materials at room temperature have almost no plasticity. During crack propagation, the tip plastic zone is small, work consumption is very small, and thus the defect or crack size is very sensitive to the impact of strength and toughness. Usually, the fracture strength σ_b of ceramic materials and metal materials is the same order of magnitude, and the K_{IC} value of ceramics is smaller than for metals by one to two orders of magnitude. Therefore, the critical crack length in ceramics is smaller than in metals by 2–4 orders of magnitude. In order to improve the strength of the ceramic material, it is better to reduce its internal defects and cracks. Selecting ultrafine powder raw materials and using hot pressing or hot isostatic pressing technologies can reduce the number and size of defects and cracks in the ceramic material. Employing ZrO_2 phase transformation toughening, microcrack toughening, and the second phase or fiber toughening can also increase the crack propagation resistance; and these are also effective measures to improve the strength and toughness of ceramic materials.

13.4 STRENGTH OF CERAMIC MATERIALS

The strength of a material refers to the maximum stress when the material is destroyed under certain loads. Because of its exceptional binding properties and crystal structures, ceramic materials, immediately after elastic deformation, experience brittle fracture without the plastic deformation phase. Therefore, ceramic materials are different from metal materials, and only have fracture strength σ_f (as shown in Figure 13.1). Before σ_f, there exists the yield strength σ_s and tensile strength σ_b in metals. Therefore, poor plastic toughness is the fatal weakness of ceramic materials, and is also the major obstacle affecting the engineering application of ceramic materials. Because ceramic materials have no plasticity, ceramic strength mainly refers to the fracture strength σ_f. Experiments have shown that the actual strength of ceramic materials is lower than the theoretical value by one to two orders of magnitude. Only the actual strength of whiskers and fibers is close to the theoretical value. The reason for the huge difference can be explained by the Griffith crack strength theory: ceramic materials often contain microcracks.

13.4.1 FLEXURAL STRENGTH OF CERAMIC MATERIALS

The bending test is the major experimental method used to assess the strength of ceramic materials. The bending strength of ceramic materials is also known as flexural strength. Flexural strength refers to the maximum stress of the tensile surface fracture when the rectangular interface is under bending stress. Loading methods can be divided into three-point bending and four-point bending, as shown in Figure 13.5. The length is $L_T \geq 36$ mm, the width is b = mm, the thickness is $h = 3$ mm, the span is $L = 30 \pm 0.5$ mm, $l = 10 \pm 0.5$ mm, and the ballasting head is $R_1 = 2.0$–5.0 mm, $R_2 = 2.0$–3.0 mm. In the bending test, load at the displacement speed of 0.5 mm/min, find the maximum fracture load, and then use the following bending strength to calculate the following.

Three-point bending:

$$\sigma_{f3} = \frac{3FL}{2bh^2}, \tag{13.11}$$

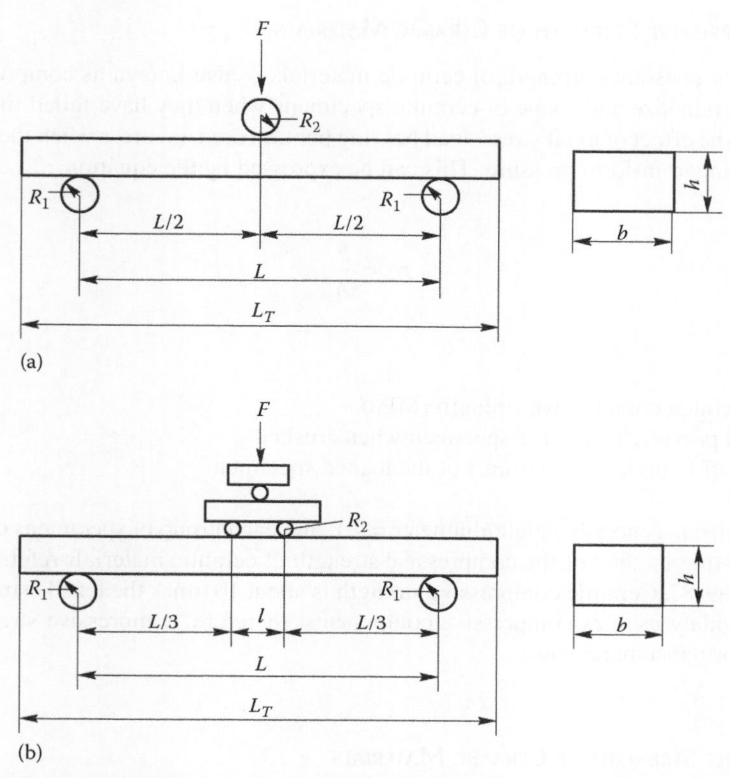

FIGURE 13.5 Schematic of specimen size and experimental setup of (a) three- or (b) four-point bending tests. (From Zhou, Y., *Ceramic Materials*, 2nd edn., Science Press, Beijing, 2004; Taken from Figure 10.19.)

Four-point bending:

$$\sigma_{f4} = \frac{3F(L-l)}{2bh^2}, \tag{13.12}$$

where
- σ_f is the moment strength
- F is the maximum load when the specimen fractures
- L is the distance between the specimen bearings
- l is the distance between the indenters
- b is the width of the specimen

The range of the maximum bending moment for the four-point bending experiment is wide. The stress state is close to the actual component of the service state, and it is more practical. At room temperature, the ceramic material does not yield, and brittle fracture occurs often if the state of deformation is small (0.01%). When the temperature is raised to a certain extent (about 1000°C), most ceramic materials are transformed from brittle into semi-brittle. Before fracture, varying degrees of plastic deformation will occur. Excellent high-temperature structural ceramic materials can maintain their strength at a higher temperature (~1000°C–1200°C).

13.4.2 Compressive Strength of Ceramic Materials

The compression resistance strength of ceramic materials is also known as compressive strength. It refers to a certain size and shape of ceramic specimens when they have failed the specified test machine under the effect of axial stress, load bearing per unit area, or stress when the ceramic material is crushed under uniform pressure. This can be expressed by the equation

$$\sigma_c = \frac{F}{A}, \tag{13.13}$$

where

σ_c is the specimen compressive strength (MPa)
F is the total pressure (N) of the specimen when crushed
A is the cut-off boundary area (mm^2) of the loaded specimen

The specimen size is generally height/diameter = 2:1, and each group of specimens comprises more than 10. The testing method of the compressive strength of ceramic materials refers to the national standard GB8489-87. Ceramic compressive strength is about 10 times the tensile strength. Ceramic materials are widely used as compressive components; therefore, compressive strength is also an important performance parameter.

13.4.3 Tensile Strength of Ceramic Materials

The design of ceramic parts commonly uses its tensile strength values as a criterion. Ceramic materials, due to their brittleness, easily fracture at gripping parts in the tensile test. In addition, the axes of the fixture and the specimen are inconsistent, generating additional moment, and thus the true tensile strength of ceramic materials often cannot be determined. In order to ensure correct tensile testing of ceramic materials, lot of work must be performed on the specimen and the chuck design. For example, in parallel chuck, a rubber pad is added to fix a lamellar specimen, which prevents specimen fracture at the gripping part. The elastic deformation of the specimen is used to reduce the additional moment.

13.4.4 Major Factors Affecting Strength of Ceramic Materials

The brittleness of the ceramic material itself is based on the type of its chemical bonds. Ceramic crystals mainly contain ionic bonds and covalent bonds with a strong direction. The majority of crystals have complicated structures, their average interatomic distance is wide, and thus the surface energy is small. Compared with metal materials, the start of the slip system barely exists at room temperatures, and thus the slip dislocation and the proliferation of dislocations occur with difficulty. Thus, it easily causes brittle fracture due to stress concentration resulting from surface or internal defects. This is the reason for the brittleness of ceramic materials, and also the reason for the bigger dispersion of strength values. This book will focus on the influencing patterns of organizational factors on the strength of single-phase polycrystalline ceramics. Factors affecting the strength of ceramic materials include microstructures, the shape and size of internal defects, the size and shape of the specimen itself, the strain rate, environmental factors (temperature, humidity, pH), the force state and stress state, and so on. We will now discuss several main influencing factors.

1. The influence of microstructures on the strength of ceramic materials.
   The microstructure of ceramics mainly includes grain sizes, morphologies, and orientations; pore sizes; shapes and distributions; the nature, size, and distribution of second-phase impurities; grain boundary phase compositions, structures, and morphologies; and sizes, densities, and shapes of the crack. Their formation is related to manufacturing technologies of ceramic materials. The relationship between the strength of ceramic materials and porosity can be expressed by the following formula:

$$\sigma_c = \frac{F}{A}, \tag{13.14}$$

   where σ_f is the strength of ceramic materials having pores, σ_o is the strength of ceramic materials without pores, p is the porosity, and b is a constant for the related material. We know from Equation 13.14 that the strength of ceramic materials decreases with the increase of porosity. The reason is due to the presence of pores, reducing the cross-sectional area of the loading, and leading to an increase of the actual stress. On the other hand, because pores cause stress concentration, the strength decreases. In addition, the elastic modulus and fracture energy, with the change of porosity, also affect the intensity value.
   The influence of the grain boundary phase on the strength of ceramic materials is mainly because most ceramic materials, when sintered, require a sintering aid. Therefore, the formation of a certain amount of low-melting-point grain boundary phases promotes densification. The composition, nature, and properties (thickness) of the grain boundary phase have a significant effect on the strength. The presence of a grain boundary glass phase has an adverse effect on the strength. Therefore, heat treatment is needed to cause crystallization, and minimize the brittle glass phase as much as possible. The grain boundary phase is best to stop crack propagation, and release the effect of the stress field at the crack tip.
2. The impact of temperature on the strength of ceramic materials.
   One of the most significant characteristics of a ceramic material is that its high-temperature strength is much higher than that of metals. The temperature of automotive gas engines is expected to rise to 370°C, and Ni, Cr, and Co superalloys cannot withstand such working temperatures. However, ceramic materials such as Si_3N_4 and SiC are very promising. Most ceramic materials have better high-temperature performance. Usually, below 800°C, temperature has little effect on the strength of ceramic materials. High-temperature performance of ceramic materials with ionic bonds is lower than ceramics with covalent bonds. At the lower temperature range, the destruction of ceramics is a brittle fracture, without plastic deformation. The ultimate strain is also very small, and very sensitive to minor defects. In the high-temperature zone, ceramic materials can incur minor plastic deformations before fracture, a large increase in ultimate strain, and have a small amount of elastoplastic behavior. At this time, the sensitivity of strength to defects changes greatly, producing a dividing line between the low-temperature zone and the high-temperature zone of the ceramic material properties change. This is known as the brittle–ductile transition temperature. The brittle–ductile transition temperature is related not only to the chemical composition of the material, but also to the material's microstructure, grain boundary impurities, and especially the glass phase ingredients. Most of the strength of ceramic materials at high temperature declines with increasing temperature. Radovic and others studied the elastic material parameters (elastic modulus and Poisson's ratio) and the wear resistance trends and features in the ~300–1573 K temperature range, using different polycrystalline ceramic materials (including Ti_3SiC_2, Ti_3GeC_2, $Ti_3Si_{0.5}A_{l0.5}C_2$, and Ti_2AlC). They found that the elastic parameter of the various types of ceramic materials almost shows a linear decrease with increasing temperature. For example, the elastic modulus of a Ti_2AlC ceramic material at room temperature is 340 GPa; however, under a temperature of 1573 K, its elastic modulus is reduced to 119 GPa [15].

13.5 THERMAL SHOCK RESISTANCE OF CERAMIC MATERIALS

Ceramic materials have the advantages of high-temperature resistance, wear resistance, and corrosion resistance. However, their poor plasticity and toughness, their intolerance of impacts, and difficulty in processing are fatal disadvantages. Designers should avoid bearing mechanical stress and impact stress, but engineering ceramic components, especially high-temperature structural components, are mainly used in the high-temperature field. Thermal shock stress, caused by drastic temperature changes, is large. Therefore, if the material has plasticity, it is able to abate the stress peak, ease stress concentration, absorb the impact work, and prevent crack initiation and propagation. In addition, ceramic materials generally have poor thermal conductivity, and the stress gradient caused by temperature changes is large. Therefore, thermal shock fracture and damage are one major form of failure of ceramic engineering materials, but they are also an important indicator for performance evaluation in the use of engineering ceramics.

So-called thermal shock resistance is the ability of a material to bear sudden changes in temperature without damage. Thermal shock damage is divided into two categories: (1) instantaneous fractures caused by thermal shock, known as thermal shock fractures; (2) under the effect of the thermal shock cycle, cracks in the material first appear, followed by crack propagation, resulting in lower material strength, and ultimately thermal shock damage.

Thermal shock resistance of a material is the overall performance of its mechanical properties and thermal properties. The thermal shock resistance of ceramic materials is not only affected by geometrical factors and the impact of environmental media, but also depends on the strength and fracture toughness of the material. Thermal stress caused by a variety of thermal environments, and the corresponding stress intensity factor, is the driving force behind thermal shock damage. When a material's inherent strength is not enough to resist thermal stress caused by the thermal shock temperature difference, it will result in instantaneous thermal shock fracture of the material. Formation and propagation of cracks occur when the strain energy stored in the material, caused by the thermal stress, is sufficient to allow the newly added surface energy to undergo crack nucleation and propagation. Along with repeated heating, cooling, and crack propagation, the strength is drastically reduced, and the parts may become partially flaked or cracked, which is the process of thermal shock damage. Thermal shock resistance of ceramic materials is usually represented by the thermal shock resistance parameters.

13.5.1 Thermal Shock Resistance Fracture of Ceramic Materials

The theory of thermal shock fracture is based on the theory of thermoelastics, using the equilibrium condition between thermal stress and the inherent material strength as a criterion of thermal shock fracture; that is, ($\sigma_H \geq \sigma_f$). When thermal shock stress caused by rapid change of temperature (ΔT) exceeds the inherent strength of the material, instantaneous fracture occurs, and is known as thermal shock fracture. Because transient thermal stress caused by thermal shock is much larger than the thermal stress under normal circumstances, it acts on an object with great speed and impact. This phenomenon is known as thermal shock. For specimens without boundary constraints, thermal stress occurs due to transient, uneven distribution of the surface, and the internal temperature field of specimens. When the specimen is under the effect of a quench temperature, at the initial moment, the surface contraction rate is $\alpha \propto \Delta T$ and the inner layer is not cooling and contracting. Therefore, the surface layer is subject to a pull (tensile) force from the inner layer, while the inner layer is subject to a compressive stress from the surface. This tensile stress arising from the rapid cooling of the surface of the material is expressed as

$$\sigma_t = \frac{E\alpha\Delta T}{1-\nu}, \tag{13.15}$$

where E, α, and ν are the material elastic modulus, expansion coefficient, and the Poisson's ratio respectively.

In general, when surface thermal stress reaches the inherent strength of the material, it is considered to be a critical state, and the critical temperature ΔT_c is the thermal shock coefficient (R). From Equation 13.15 we obtain

$$R = \Delta T_c = \frac{1-\nu}{E\alpha}\sigma_f, \tag{13.16}$$

where σ_f is the fracture strength.

For slow heating and cooling of ceramic materials, the parameter of the thermal shock fracture resistance is

$$R^I = \frac{K(1-\nu)}{E\alpha}\sigma_f = KR, \tag{13.17}$$

where K is thermal conductivity. Studies have shown that ceramic materials have high thermal shock fracture when they have high strength and thermal conductivity, and a low coefficient of thermal expansion, Young's modulus, Poisson's ratio, coefficient of thermal radiation, and viscosity. In addition, appropriately reducing the material density and the molar heat capacity will also help to improve the thermal shock resistance of ceramic materials.

13.5.2 Thermal Shock Resistance Damage of Ceramic Materials

Thermal shock damage of a material refers to the thermal damage process under thermal shock stresses, starting from material cracking and spalling, until fragmentation or overall fracture. Ceramic materials in the manufacturing process inevitably will have the presence of (large or small) varying amounts of microcracks and pores. The occurrence of crack nucleation in a thermal shock environment does not always lead to material fracture. Thermal shock damage theory, based on the theory of fracture mechanics, analyzes the dynamic process of crack nucleation, expansion, and inhibition of materials under the condition of temperature changes, and uses the equilibrium condition between the thermoelastic strain energy W and the fracture energy of materials U, as a criterion for thermal shock damage ($W \geq U$). Formation and propagation of cracks occur when the strain energy stored in the material, caused by thermal stress, is sufficient to cover the newly added surface energy required for crack nucleation and expansion. Setting the radius of a heated sphere as r, at the maximum thermal stress, we can consider the temperature distribution inside the sphere as approximately parabolic. When the thermal stress of the center of the ball is equivalent to the material's fracture strength σ_f, the total elastic strain energy reserved in the sphere is

$$W = \frac{4\pi r^3 \sigma_f^2 (1-\nu)}{3nE}, \tag{13.18}$$

where n is the geometric factor. For simplicity, we assume that at the fracture, all of the thermal elastic strain can be converted to effective surface energy of N cracks (the average area is A): $W = 2NA\gamma_f$. Then we substitute it into Equation 13.18, and obtain the average area of crack propagation as

$$A = \frac{2\pi r^3 \sigma_f^2 (1-\nu)}{3nEN\gamma_f}. \tag{13.19}$$

Therefore, the ratio of the average crack area to the maximum cross-sectional area of the ball is

$$\frac{A}{\pi r^2} = \frac{2\sigma_f^2(1-\nu)}{3nE\gamma_f} \cdot \frac{r}{N}. \tag{13.20}$$

Thus, the larger the sphere, the greater the relative crack area becomes, the bigger is the thermal shock crack, and the smaller the relative crack area becomes. The relative crack area is a measurement of the extent of the component's thermal shock damage. The smaller the $A/\pi r^2$, the greater is the capacity for thermal shock damage resistance. Take the reciprocal and exclude the geometric factor related to the specimen shape, and we can then define the parameter of thermal shock damage resistance of ceramic materials:

$$R^{II} = \frac{E\gamma_f}{(1-\nu)\sigma_f^2}. \tag{13.21}$$

When a series of components with generally equal γ_f are compared, γ_f can be regarded as constant. Therefore, another thermal shock damage parameter can be derived from Equation 13.21:

$$R^{III} = \frac{E}{(1-\nu)\sigma_f^2}. \tag{13.22}$$

If $K_{IC} = (2E\gamma_f)^{1/2}$ is substituted into Equation 13.21, we obtain

$$R^{IV} = \frac{(K_{IC}/\sigma_f)^2}{(1-\nu)}. \tag{13.23}$$

From these equations we can see that materials with better thermal shock damage resistance should have the highest possible elastic modulus and fracture surface energy, and the lowest possible strength. It is easy to see that these requirements are contrary to that of high thermal shock fracture resistance. In other words, in order to improve thermal shock damage resistance, we should, as much as possible, improve the fracture toughness of the material and reduce the strength of the material. In fact, a ceramic material inevitably has the presence of (large or small) varying amounts of microcracks and pores, and in a thermal shock environment, they do not always result in the immediate fracture of the material. For example, a thermal shock crack in a nondense ceramic with porosity of 10%–20% tends to be inhibited by pores. Here, the presence of pores not only plays the role of passivating the crack tip and reducing stress concentration, but also reduces the thermal conductivity rate and contributes to thermal insulation. On the contrary, a dense high-strength ceramic material under thermal stress is prone to bursting.

Thermal shock damage of ceramic materials is primarily reflected in their strength attenuation. Under normal circumstances, attenuation of a ceramic material's residual strength under thermal shock reflects the material's thermal shock resistance. The most common method of thermal shock is dropping (quenching) a ceramic specimen directly from high temperature to room temperature water (water cooling) or into the air (air cooling). Then, we test its strength attenuation or find the critical temperature difference where the strength does not significantly decline. In engineering applications, failure analysis of ceramic components is very important. If failure of materials is caused mainly by thermal shock fracture, such as high-strength and dense fine ceramics, crack initiation plays a dominant role. In order to prevent thermal failure, we should improve the thermal shock fracture resistance, that is, be committed to improving the material's strength, and reducing the elastic modulus and the expansion coefficient. If the primary reason leading to thermal shock failure is thermal shock damage, then crack propagation plays a major role, such as in nondense ceramics (industrial SiC kiln, ceramic heat storage, and ceramic high-temperature filters). In this case, we should seek to improve fracture toughness and reduce the strength [16,17].

13.6 CREEP OF CERAMIC MATERIALS

In Chapter 7, we described in detail the related concepts and features of creep. For brittle ceramic materials at room temperature, this generally occurs as brittle fractures without, or with very little, plastic deformation. However, along with the increase of temperature and the extension of time, ceramic materials will, to some extent, show the capacity for plastic deformation and creep deformation.

13.6.1 Creep Mechanisms in Ceramic Materials

The creep mechanism of metal materials as described in Chapter 7 can basically be applied to ceramic materials. At high temperatures, intragranular dislocation slip and dislocation climb, and grain boundary sliding and migration, alternate as strengthening and softening alternations. However, ceramic materials have their own characteristics. The creep mechanism has two mechanisms: the grain boundary mechanism (diffusion creep) and the lattice mechanism (dislocation mechanism). The grain boundary mechanism is related to the process of polycrystalline creep. The lattice mechanism not only controls the creep behavior of single crystals, but is also able to dominate the creep process of polycrystallines. Because engineering ceramics are polycrystalline materials, the motion of dislocations in ceramic crystals must overcome the higher stress. Here, we will focus on the mechanism of a ceramic material's grain boundaries [1,2,20]. The grain boundary of ceramic materials may be a microcrystalline grain boundary without a second phase, and may also be a second-phase material in the liquid phase of high temperature. Due to the difference of grain boundary phases, the microscopic process of creep is different.

13.6.1.1 Vacancy Diffusion Flow (Diffusion Creep)

New types of inorganic nonmetallic engineering materials, such as pure oxide (e.g., Al_2O_3 and WO_2) crystallizing into nitrides, are almost free of second-phase grain boundary materials. Thus, vacancy diffusion is an important mechanism to control high-temperature creep in the above-mentioned ceramic materials. The mechanism considers that creep is the process of directional vacancy diffusion, as shown in Figure 13.6 [18]. Set an isolated tetragonal grain in the polycrystalline. When it is subjected to shear stresses, the vacancy concentration on the grain boundaries in tension *AB* and

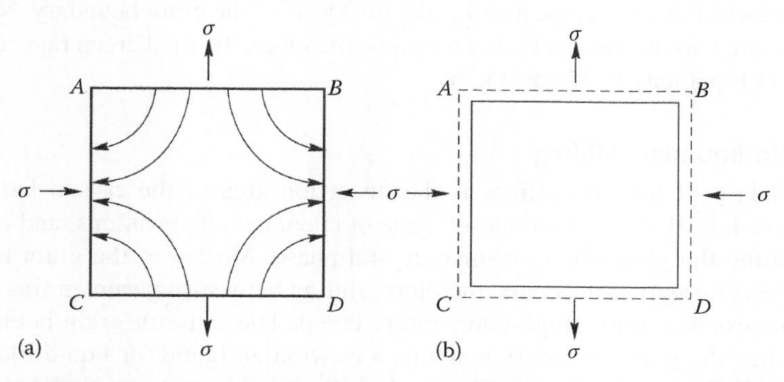

FIGURE 13.6 Vacancy flow induced by applied shear stress. (a) Vacancy flow through the grain. (b) Vacancy flow along the grain. (From Jin, Z.H. et al., *Engineering Ceramic Materials*, Xi'an Jiaotong University Press, Shaanxi, China, 2000; Taken from Figures 5-29.)

CD is higher than the equilibrium value, while the vacancy concentration on the grain boundaries in compression AC and BD is lower than the equilibrium value. Therefore, poor vacancy concentration in a grain causes vacancy flow through the grain (Figure 13.6a), or along the grain (Figure 13.6b). The directional flow of vacancies is equivalent to atomic flow in the opposite direction, causing grain elongation along the direction of tensile stress. At the same time, the vacancy on the AC and BD borders increases, and the vacancy on the AB and CD borders reduces. The tensile stress on the grain boundary causes vacancy concentration increase to

$$C = C_o \times \exp\left(\frac{\sigma V}{KT}\right), \tag{13.24}$$

where

V is the vacancy volume, which is equivalent to the atomic volume b^3

C_o is the equilibrium concentration (i.e., the number of vacancies an atom/unit cell can contain)

Compressive stresses reduce the vacancy concentration to $C = C_o \times \exp(-\sigma V/KT)$. The deformation caused by a vacancy leaving or an atom entering the grain boundary—that is, the strain caused by an atom per unit area of the grain boundaries AC and BD diffusing to the AB and CD grain boundaries—is the phenomenon of stress-induced vacancy diffusion. Nabarro and Herring calculated the grain diffusion creep rate, expressed as [19,20]

$$\dot{\varepsilon} = \frac{B_1 V D_g \sigma}{KTd^2}, \tag{13.25}$$

where d is the grain size, B_1 is the coefficient depending on the grain shape and stress state (for those grains subject to uniform tensile stress and fully relaxed $B_1 \approx 13$), D_g is the lattice diffusion coefficient, T is the thermodynamic temperature, K is the Boltzman constant, and σ is the applied stress.

If diffusion occurs along the grain boundary, that is, the vacancy flow along the grain model (Figure 13.6b), the Coble diffusion creep rate is calculated as follows:

$$\dot{\varepsilon} = \frac{B_2 \sigma D_b \delta V}{KTd^3}, \tag{13.26}$$

where B_2 is the coefficient depending on the grain shape and stress state. In general, $B_2 = 47$; D_b is the grain boundary diffusion coefficient; and δ is the thickness of the grain boundary. Nabarro/Herring and Coble creep are two independent rate processes; therefore, the total creep rate can be expressed by the addition of Equations 13.25 and 13.26.

13.6.1.2 Grain Boundary Sliding

The grain boundary of polycrystallines is the distortion area of the crystal lattice, which has defects and traps. It is also the microuneven zone of chemical compositions, and collects impurities easily, forming the glass phase or microcrystal phase. Moreover, the grain boundary is the place where stress concentration occurs; therefore, sliding between adjacent grains is an important microscopic process of ceramic high-temperature creep. The cause of grain boundary sliding is plastic flow. When the grain boundary contains a Newtonian liquid, or liquid-like second phase materials, the diffusion coefficient of the liquid-phase grain boundary is related to the thermal activation of the second-phase material. If the liquid-phase layer has an appropriate thickness, and the extent of grain irregularity on both sides of the grain boundary does not hinder the shear process, the creep rate has the characteristics of Newtonian viscosity and is subject to the control of the grain boundary separation rate under tensile stress. If the extent of grain boundary irregularity is more serious, and the grain boundary layer thickness is thinner, the creep rate is a non-Newtonian viscous flow, and the creep rate should be related to the formation of holes at the grain boundary, and the crack growth at the three-intersection-point area. Therefore, plastic flow is only a partial reason for grain boundary sliding. This type of grain boundary sliding mechanism is due to the slip and climb motion of dislocations along the grain boundaries, or near the grain boundary, under high-temperature conditions.

13.6.2 Analysis of Creep Testing Examples of Ceramic Materials

Recently, Radovic and others studied the tensile creep properties [21] of Ti_3SiC ceramic materials at temperatures within the range of ~1000°C–1200°C, and stress within the range of 10–100 MPa. In the second stage of this ceramic material creep, the creep rate is the smallest and can be expressed by the following equation:

$$\dot{\varepsilon}_{min} = \varepsilon_0 \exp(17 \pm 1)\left(\frac{\sigma}{\sigma_0}\right)^{2.0\pm0.1} \exp\left(\frac{-458 \pm 12}{RT}\right), \tag{13.27}$$

where $\sigma_0 = 1$ MPa, $\varepsilon_0 = 1\ s^{-1}$, and the creep failure time is $t_f(s) = \exp(-2 \pm 0.3)\dot{\varepsilon}_{min}^{-1}$.

Experimental results have shown that dislocation creep is the main failure mechanism.

In the creep process, a huge internal stress occurs due to the high plastic anisotropy of Ti_3SiC ceramic materials.

Internal stress is the result of competition between continually accumulating and releasing internal stress in the ceramic material.

Under high-temperature environments or low strain rate conditions, the internal stress of ceramic materials can dissipate by itself, causing the material to show toughness behavior. In the third stage of creep, the common effects of dislocation creep and the formation of voids and cracks affect the creep behavior of ceramic materials.

A coarse-grained Ti_3SiC ceramic material has a lower creep rate than a fine-grained Ti_3SiC ceramic material, which means that it has a longer lifetime.

Carrère and others studied the high-temperature creep behavior of SiC/Si–BC composite materials in a 1200°C high-temperature environment [22]. First, they used the chemical vapor infiltration (CVI) method to prepare a two-dimensional SiC/Si–BC composite ceramic material, whose microstructure was shown in Figure 13.7. The schematic diagram for the high-temperature creep experiment setup is shown in Figure 13.8. The device includes a high-temperature furnace, an electrical and mechanical tensile testing device, and a temperature measurement device. The entire heating device is in a closed cavity, which can be filled with argon or nitrogen as a protective gas to prevent the oxidation of ceramic materials in high-temperature environments, which will affect the creep experiment results. When studying creep–oxide interactions, it is not necessary to use the protective gas. They first studied the creep behavior of SiC/Si–BC ceramic composite materials in static fatigue experiments. However, even under very long saturated conditions, the creep behavior of ceramic materials did not incur the creep second stage, and the creep rate of the ceramic material gradually decreased, as shown in Figure 13.9. In addition, they observed (with an optical microscope) sectional morphology changes after the static fatigue and cyclic fatigue experiments for the SiC/Si–BC ceramic composite material. They are shown in Figures 13.10 and 13.11.

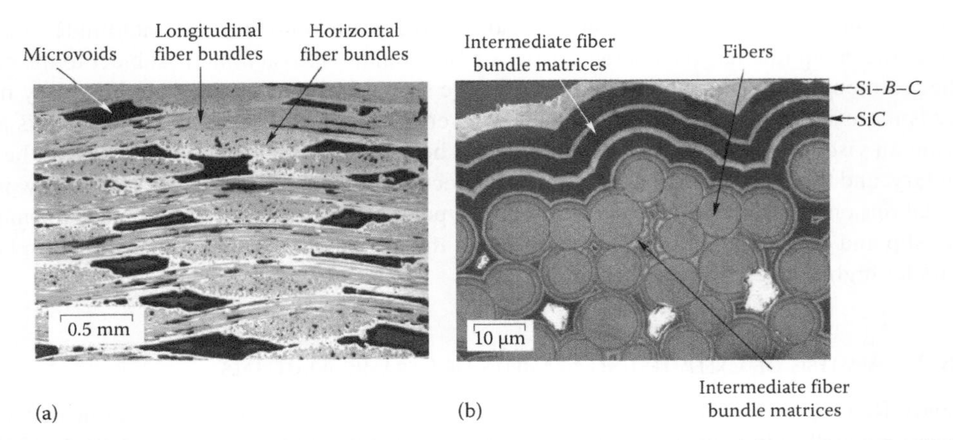

FIGURE 13.7 Microstructure of SiC/Si–BC composite with a self-healing multilayered matrix. (a) SEM micrograph of the structure of the Si–B–C composite. (b) SEM micrograph of the structure of the multilayer matrix. (From Carrère, P. and Lamon, J., *J. Eur. Ceram. Soc.*, 23, 1105, 2003; Taken from Figure 1.)

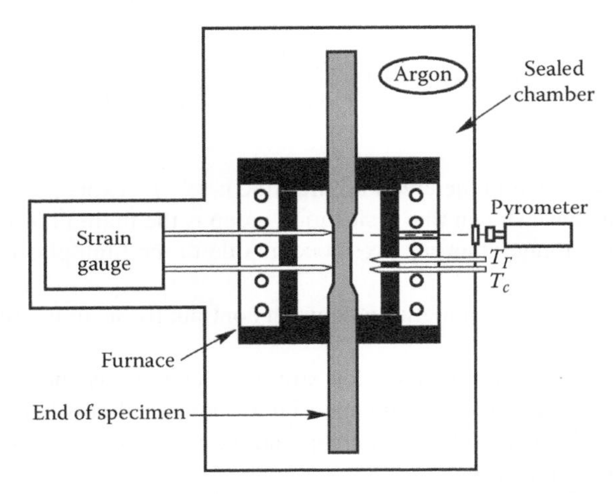

FIGURE. 13.8 Diagram of the creep experimental setup. (From Carrère, P. and Lamon, J., *J. Eur. Ceram. Soc.*, 23, 1105, 2003; Taken from Figure 2.)

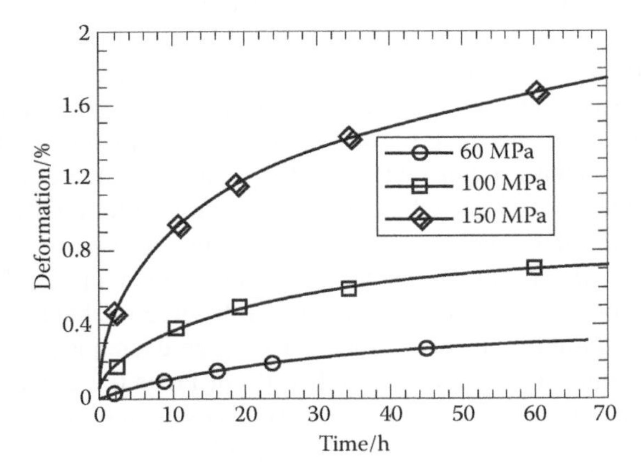

FIGURE 13.9 The creep curve of ceramic materials in the static fatigue experiments under different stress conditions, where $\varepsilon_0 = 0.8\%$, and the temperature is 1200°C. (From Carrère, P. and Lamon, J., *J. Eur. Ceram. Soc.*, 23, 1105, 2003; Taken from Figure 4.)

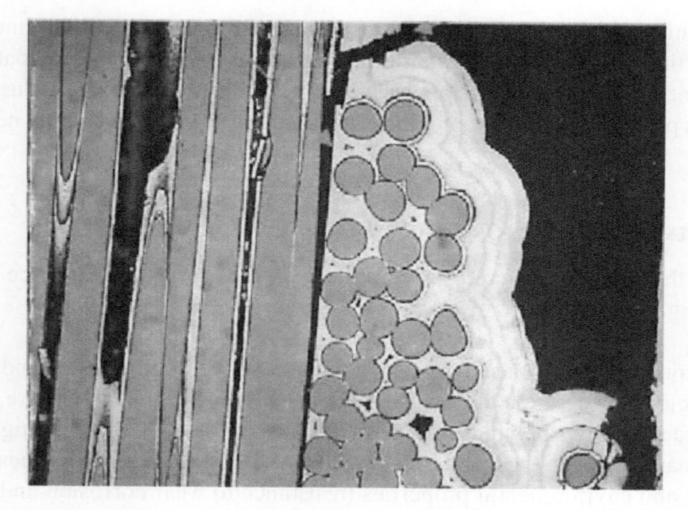

FIGURE 13.10 Optical micrograph of a SiC/Si–BC composite initially damaged under $\varepsilon_0 = 0.2\%$ and tested during 100 h in static fatigue under 100 MPa at 1200°C. (From Carrère, P. and Lamon, J., *J. Eur. Ceram. Soc.*, 23, 1105, 2003; Taken from Figure 9.)

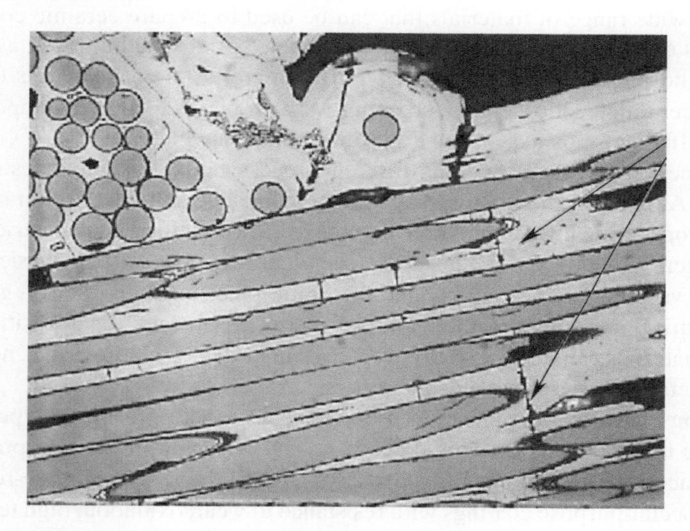

FIGURE 13.11 Optical micrograph of a SiC/Si–BC composite initially damaged under $\varepsilon_0 = 0.25\%$ and tested during 100 h in static fatigue under 150 MPa at 1200°C. (From Carrère, P. and Lamon, J., *J. Eur. Ceram. Soc.*, 23, 1105, 2003; Taken from Figure 12.)

13.7 OVERVIEW OF HIGH-PERFORMANCE CERAMIC COATING MATERIALS

The previous sections in this chapter have mainly described the structural characteristics, physical and chemical properties, and the relevant mechanical properties of ceramic materials. Researchers have continued to explore and study how to take full advantage of their excellent properties. With the development of the aerospace, shipbuilding, industrial chemicals, and automobile manufacturing industries, researchers gradually recognized that preparing powders of ceramic materials into ceramic coatings not only expands their application scope, but also solves many important practical considerations. High-performance ceramic coating (also called fine ceramic coatings, advanced ceramic coatings, or new ceramic coatings) is a general term for a wide range of nonmetallic inorganic coatings that have followed organic resin coatings, and metal and alloy coatings. With the evolution of cutting-edge technology in the aerospace, electronics, and military industries in the last half-century, particularly from the 1990s, these high-performance ceramic coatings experienced rapid development. In the developed countries, it is becoming an emerging industry in the twenty-first century. It is possible that we are moving toward an era of even developing new materials from clay and stone!

13.7.1 Features of High-Performance Ceramic Coatings [23,24]

Compared with the integral structure of ceramic materials, high-performance ceramic coatings have the following features:

1. They can combine metal materials' strong toughness, workability, and conductivity of heat and electricity with ceramic materials' good resistance to high temperature, wear, and corrosion systematically, showing their different advantages. Meanwhile, high-performance ceramic coatings meet the needs for products with structural properties (strength and toughness) and environmental properties (resistance to wear corrosion and high temperature), as well as those with special functions (infrared radiation, wavelength absorption, thermal sensitivity, and photosensitivity), so as to realize an ideal composite material structure.
2. There is a wide range of materials that can be used to prepare ceramic coatings, including various oxides and composite oxides, carbides, borides, and nitrides; as well as metal ceramics and intermetallic compounds. Furthermore, different materials (ceramics with ceramics, ceramics with metals, and ceramics with plastics) can be composed and combined in different ratios to produce diversified composite materials.
3. Preparing ceramic coatings features different process methods, small investment, and high flexibility. At present, the main process methods include solid deposition (thermal spraying, self-propagating high-temperature synthesis (SHS) method, and electric spark surface reinforcement method), vapor deposition (chemical vapor deposition, physical vapor deposition, and vacuum ion deposition), and the liquid-phase method (the sol–gel method, the electrochemical deposition method, and the chemical autocatalysis deposition method).
4. Ceramic materials can deposit on different base materials, including steel, nonferrous metals, rare metals, glass, resin plate, and others.
5. The coatings have wide application scope. Because there are many types of materials that can be used to prepare ceramic coatings, and different preparation processes can be employed according to the specific needs, we can obtain various surface-reinforced coatings and special-purpose coatings with resistance to wear, oxidation, high temperature and radiation, thermal isolation, electric insulation, and sensitive elements.

Ceramic coatings also have many other advantages. However, they also have inherent disadvantages, which include:

1. Poor plastic deformation, susceptibility to stress concentration and crack, poor resistance to thermal shock and fatigue, as well as brittleness.
2. Ceramic coating materials differ greatly from metal materials in their thermophysical properties (such as coefficient of expansion and thermal conductivity), so different stress states may appear in their use, leading to reduced service life.
3. The bonding between ceramic coatings and the base materials is mainly mechanical embedment or molecular force, so the adhesive strength is not high enough. As a result, ceramic coatings can't be applied in working conditions with high stress, impact, and fatigue. Ceramic coatings can only remedy or add functions to the base structural materials, not replace them.

13.7.2 High-Performance Ceramic Coating—Thermal Barrier Coating [25,26]

This section will focus on the description of an important high-performance thermal insulation ceramic coating—thermal barrier coatings (TBCs), also called thermal shielding coating. It is one of the key technologies in the modern aero engine. The basic principle is that these coatings reflect the properties of ceramic materials, such as high melting point, low thermal conductivity, low vapor pressure, low radiance, and high reflectivity. With air plasma-sprayed (APS) or electron beam physical vapor deposition (EB-PVD) technology [23,24], ceramic powder is sprayed or deposited on the high-temperature alloy surface of the aero engine's hot end components (such as the liner, afterburner, turbine blade, and rocket nozzle) to prepare one layer of heat insulation ceramic coating. The high-temperature components (alloys) are isolated from the high-temperature gas to reduce their working temperatures, which would protect them from high-temperature corrosion and erosion in service, and greatly prolong their operational life.

13.7.2.1 Air Plasma-Sprayed (APS)

This method is a surface treatment process developed in the 1950s. It is the main method of producing TBC currently. This technology mainly includes air plasma spraying, argon-shrouded plasma spraying, vacuum plasma spraying, and water-stable plasma spraying. The thermal spraying is a surface treatment method in which the spraying material (powder, wire, or bar) is fed into the high-temperature and high-speed flame or plasma jet sprayed by a gun nozzle to make heat it quickly and is sprayed on the pretreated base material surface at a high speed in a molten or semi-molten form; when the molten particles collide with the base, energy transformation, deformation, spreading, distribution, and wetting occur, after which it is cooled (fast at 10^6 K/s), curdled, and stacked to form a coating. The powder gas flame spraying method is shown in Figure 13.12.

The coating is a lamellar structure stacked by a number of deformation particles in a waveform. It is inevitable that some pores and cavities, as well as oxide inclusions, exist among the particles in the coating, as shown in Figure 13.13. The typical thermal spraying coating consists of deformed particles, unmolten particles, oxides or impurities, air holes, and bonding interfaces. They are different in microscopic size, shape, quantity, distribution, and grain orientation. In a typical multiphase inhomogeneous coating structure, the material between the ceramic coating and the base is the transition layer, which has the functions of bonding and antioxidation, to protect the base materials from corrosion.

Before producing high-quality ceramic coatings, the spraying system's parameters must be optimized and the entire production process should be monitored carefully to ensure a good regeneration rate of the coatings.

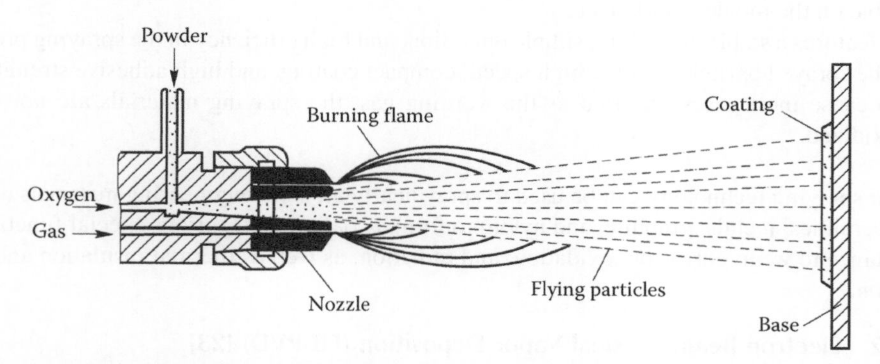

FIGURE 13.12 Schematic of the powder gas flame spraying technique. (From Wu, Z.J., *Thermal Spraying Technology and Its Application*, China Machine Press, Beijing, 2006; Taken from Figures 2-12.)

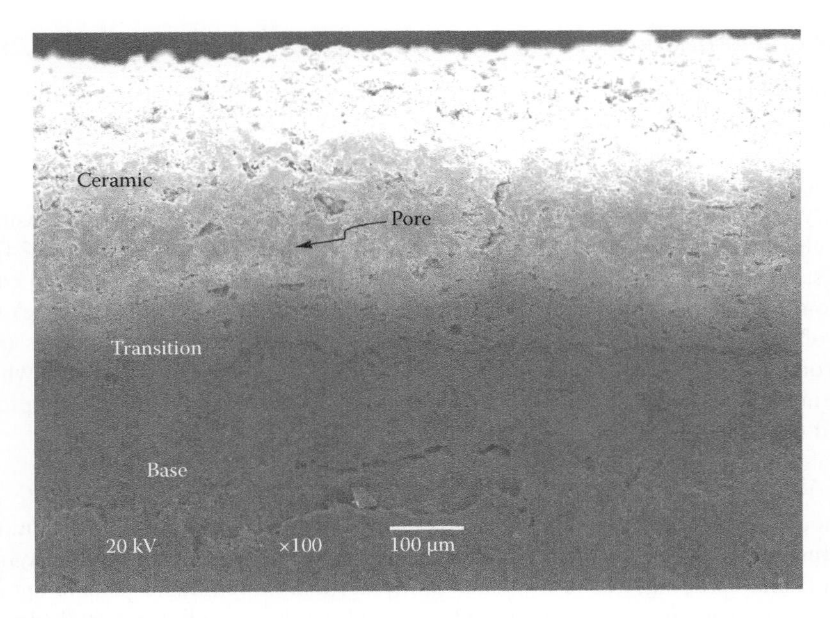

FIGURE 13.13 SEM observation of typical cross-sectional microstructure of APS TBCs.

For the APS ceramic coating process, the plasma gas must meet the following requirements [23]:

1. Its corrosivity to the gun's cathodic and anodic materials must be as low as possible to extend the gun's service life.
2. In the spraying process, to avoid any adverse effect on the coating, the plasma gas should not react with the sprayed coating materials.
3. It must be decomposed and ionized easily, so arcing becomes easy.
4. It should have sufficient enthalpy to ensure that the plasma flame has a very high temperature.
5. In the continuous spraying process, the arc flame should be stable. At present, the common plasma gases include Ar, N_2, and H_2.

The APS method has the following advantages:

1. The heat source has a high temperature, so it is suitable for spraying high-melting materials.
2. The thermal impact on the base is low in the spraying process, so surface spraying is possible on the molded workpiece.
3. It features a stable procedure, simple operation, and high efficiency in the spraying process.
4. The sprayed particles realize high speed, compact coating, and high adhesive strength.
5. Because inert gases are used as the working gas, the spraying materials are not easily oxidized.

Thermal spraying technology can be used to spray nearly all solid engineering materials (e.g., hard alloys, ceramics, metals, graphite, and nylon) to form a coating with various special functions such as resistance to wear, corrosion, oxidation, and radiation, as well as thermal insulation and electric insulation.

13.7.2.2 Electron Beam Physical Vapor Deposition (EB-PVD) [23]

This method uses a high-energy electron beam generated by a high-energy electron gun (200 kW) to melt the material in the crucible and evaporate it into vapor, and then deposit it on the base surface by using the physical vapor deposition to form a ceramic coating. The entire process is carried out in a vacuum chamber. The general process is as follows: vacuum-pump the working chamber to 2 Pa; keep the workpiece base temperature at ~960°C–980°C in the heating method; start the high-energy electron gun to generate a high-energy electron beam; melt the material in the crucible; and evaporate it into vapor. The vapor will deposit on the heated metal base under the action of the strong electric field, and will grow quickly at high temperature, as shown in Figure 13.14. Its evaporation rate is high (~100–150 μm/min). The formed coating is compact, and its chemical composition can be controlled precisely.

In recent years, TBCs have developed from a double-layer structure to a graded structure. Based on Zinsmeister's theory, [27] when the electron beam evaporation power is constant, the evaporation rate of all components in the alloy is a function of the vapor pressure ratio and the component ratio. Single-source multicomponent evaporation technology can be used to prepare the graded TBC, with continuous composition change, on the metal adhesive layer. After high-temperature posttreatment, diffusion occurs between the bonding layer and the ceramic layer, so as to eliminate the internal interface and effectively reduce thermal stress. Its construction is a columnar crystal structure perpendicular to the base surface, as shown in Figure 13.15. Different from the APS lamellar or equiaxed crystal structure, the columnar crystal structure has high resistance to shear stress

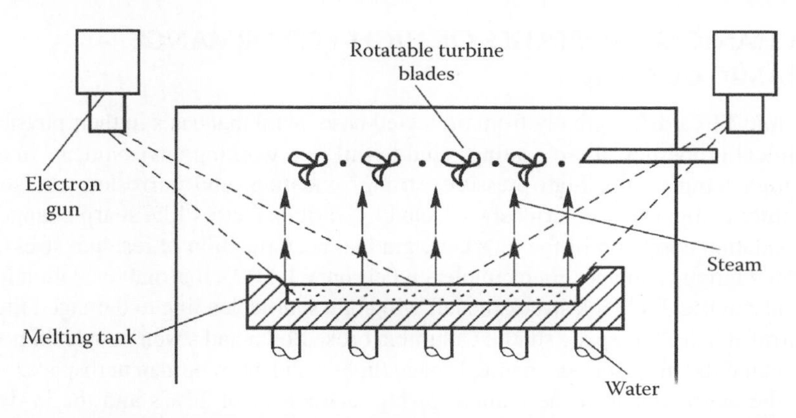

FIGURE 13.14 Schematic of EB-PVD technique. (From Deng, S.J., *High-Performance Ceramic Coatings*, Chemical Industry Press, Beijing, China, 2004; Taken from Figures 9-6.)

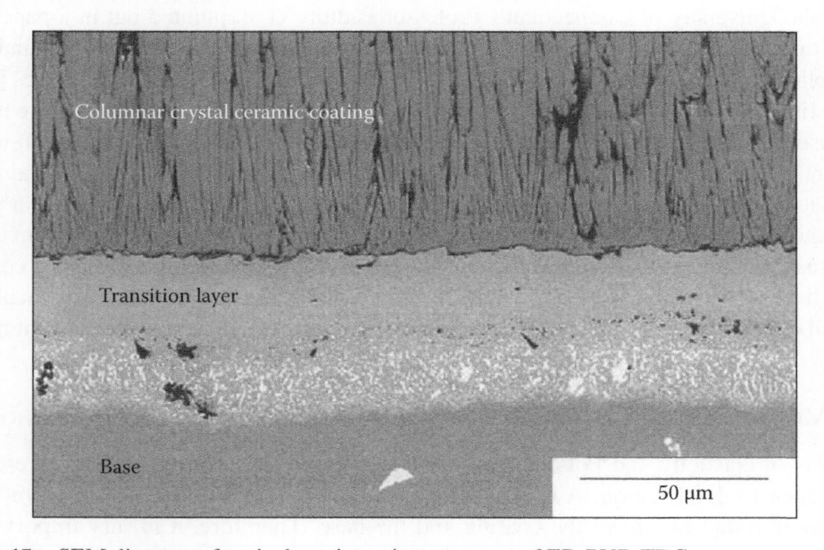

FIGURE 13.15 SEM diagram of typical section microstructure of EB-PVD TBC.

and bending stress generated by high-speed gas erosion. Hence, it greatly improves the TBC's resistance to failing in a direction parallel to the interface. The coating is more compact and has stronger resistance to oxidation and corrosion.
Studies show that EB-PVD technology has the following advantages:

1. The columnar crystal structure leads to an EB-PVD coating with higher strain tolerance, and its thermal life cycle is eight times that of plasma TBC.
2. The coating is more compact and has better resistance to oxidation and corrosion.
3. The coating interface is mostly chemical bonding, so the bonding force obviously goes up.
4. The surface cleanness is higher, which is helpful in keeping the blade's aerodynamic properties.
5. Fewer coating parameters must be controlled, so that changing the structure of the ceramic layer can be controlled.

With these advantages, EB-PVD technology is used to produce hot end components (such as TBC on an aero engine's blade) working in rough environments.

13.8 MECHANICAL PROPERTIES OF HIGH-PERFORMANCE CERAMIC COATINGS

Because ceramic TBCs differ greatly from protected base metal materials in their physical, mechanical, and chemical properties, these coatings tend to fail in a working environment. In a tough environment of high temperature, high pressure, strong oxidation, and corrosion, the service life of high-temperature components is seriously affected by various factors like sharp change of temperature grade, oxidation corrosion of the interface, gradual accumulation of residual stress in the material, and thermal fatigue damage. For example, on February 1, 2003, thermal insulation foam material falling from an external fuel tank struck the left wing of a shuttle, leading to damaged thermal insulation and a harmful hole. The space shuttle Columbia broke down and seven people on board lost their lives. This became the most serious manned space flight accident in human aerospace.

However, the development of the manufacturing technology of TBCs and the in-depth study of their failure mechanism [28–35] have gradually led to the recognition that TBCs can work with a relatively low safety factor, thereby showing their potential and saving a large quantity of energy and materials. University of Connecticut's Professor Padture et al. pointed out in a paper published in *Science* that the TBC system is the most complex structure in all coating systems, and is also the coating applied on high-temperature components in aero engines and industrial turbines [33]. Due to the complexity and diversity of the TBC's application environment and interface failure mechanism, different institutes and researchers have obtained different results. Particularly, in a high-temperature environment or a high-temperature grade, studies on the interface failure between ceramic coatings and metals (e.g., ceramic coating buckling and peel-off) and its mechanism are not clear yet, and we can't accurately predict a TBC's peel-off and its service life. Therefore, researchers from many countries pay great attention to the study of the mechanical properties of high-performance ceramic coatings. They have analyzed the TBC's interface failure mechanism and relevant mechanical properties in a detailed manner by using theoretical calculation, test verification, and finite element modeling.

13.8.1 Measurement of Coating's Modulus of Elasticity and Poisson's Ratio

The modulus of elasticity and Poisson's ratio are two important properties that represent a material's resistance to deformation. A TBC has residual stress, which plays an important role in the adhesive strength and failure of the coating and the base. Therefore, it is very important to measure the coating's resistance to deformation (i.e., the modulus of elasticity and the Poisson's ratio) under residual stress. However, it is hard to measure these two indicators. At present, there is no general criterion. We describe a cantilever beam specimen measurement method as an example.

The so-called modulus of elasticity refers to the ratio between the unidirectional positive stress σ and the strain along its direction ε, or $E = \sigma/\varepsilon$. The Poisson's ratio refers to the ratio between the elastic rod's lateral strain caused by the unidirectional positive stress ε_1 and the absolute value of the axial strain ε, or $\nu = |\varepsilon_1/\varepsilon|$. From the earlier definition, we know that as long as a certain load is applied on the coating specimen, and its strain value is measured by the strain gauge, the coating's modulus of elasticity and Poisson's ratio can be calculated and obtained.

Figure 13.16 shows the tester for measuring an APS coating's modulus of elasticity and Poisson ratio, using a cantilever beam specimen. Because ceramic coating materials have small strain under certain stresses, their modulus of elasticity is large.

13.8.2 Measurement of TBC's Interface Adhesive Strength [34]

The tensile method is used to measure the adhesive strength of TBCs. The test specimen is shown in Figure 13.17. After sandblasting treatment, the specimen base is sprayed with the transition layer material to a thickness of ~100 μm, then sprayed with the ceramic coating to a thickness of ~200–400 μm. The base material is a nickel superalloy or stainless steel with a thickness of 1.5 or 3.0 mm.

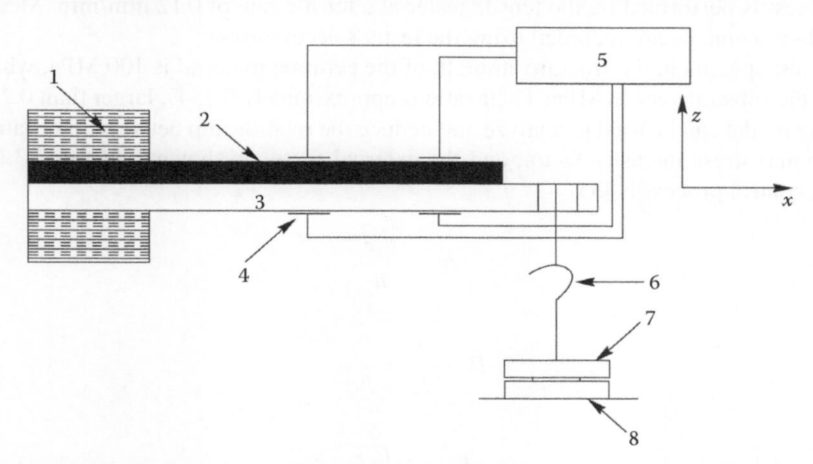

FIGURE 13.16 Schematic of cantilever beam experimental setup of Young's modulus and Poisson's ratio of APS TBCs. (1) Bench clamp; (2) coating; (3) substrate; (4) strain gauge; (5) multichannels strain display; (6) drawbar; (7) nominal load; (8) load-bearing tray. (From Deng, S.J., *High-Performance Ceramic Coatings*, Chemical Industry Press, Beijing, 2004; Taken from Figures 6-9.)

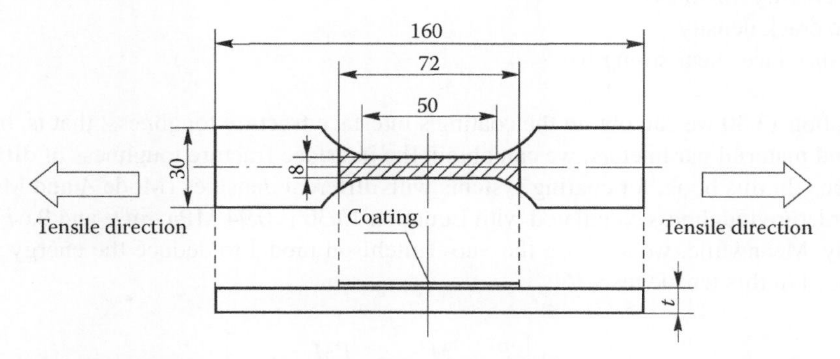

FIGURE 13.17 Schematic of TBCs sample for tensile tests, t is the substrate thickness. (From Zhou, Y.C., Tonomori, T., Yoshida, A. et al., *Surf. Coat. Technol.*, 157(2–3), 118, 2002; Taken from Figure 1.)

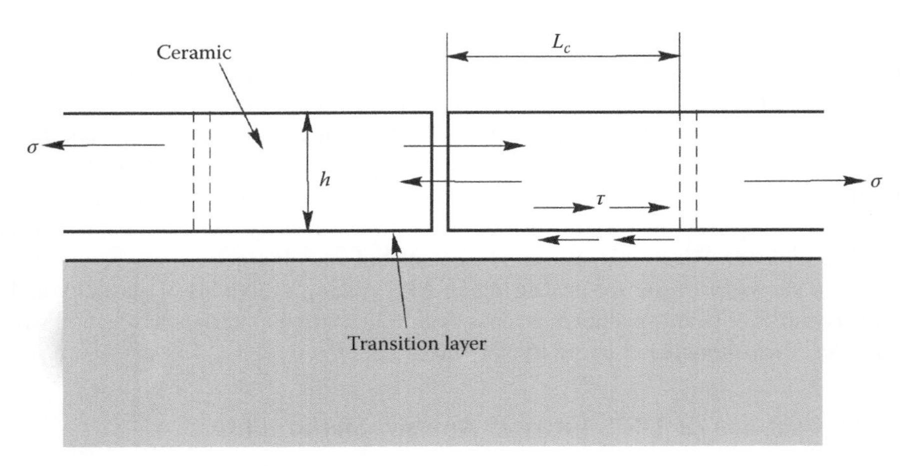

FIGURE 13.18 TBC's interface failure analyzed by shear lag model. (From Hu, M.S. and Evans, A.G., *Acta Metall.*, 37(3), 917, 1989; Zhou, Y.C., Tonomori, T., Yoshida, A., et al. *Surf. Coat. Technol.*, 157(2–3), 118, 2002; Taken from Figure 7.)

The tensile test is performed on the tensile tester at a tensile rate of 0.12 mm/min. Meanwhile, the loads and displacements are recorded using the tester's accessories.

For this test specimen, the fracture strength of the ceramic material is 100 MPa, while the yield strength of the substrate is 426 MPa. Their ratio is approximately 0.2347, larger than 0.2. Therefore, the shear lag model can be used to analyze and deduce the relationship between the coating's energy release rate and stress intensity factor, and the external force, as shown in Figure 13.18. The corresponding control process is [35]

$$\sigma_c = \frac{L_c \tau}{h}, \tag{13.28}$$

$$P_L = \frac{1}{L_c} = \frac{\tau}{h\sigma_c}, \tag{13.29}$$

$$K_c = \tau\sqrt{\pi L_c}, \tag{13.30}$$

where

L_c is the crack width
h is the coating thickness
P_L is the crack density
τ is the interface shear strength

From Equation 13.30 we can obtain the coating's interface fracture toughness; that is, based on the test data and material parameters, we can obtain the interface fracture toughness of different types of specimens. In this book, for coating systems with different densities (Mode *A* and Model *B*), the interface fracture toughness calculated with Equation 13.30 is 0.94 MPa $\cdot$ m$^{1/2}$ and 0.67 MPa $\cdot$ m$^{1/2}$, respectively. Meanwhile, we also use the Suo–Hutchison model to deduce the energy release rate equation used in this tensile test [36]:

$$G = \frac{c_1}{16}\left[\frac{P^2}{Ah} + \frac{M^2}{Ih^3} + 2\frac{PM}{\sqrt{AI}h^2}\sin\gamma\right]. \tag{13.31}$$

As the transition layer material's parameters are close to those of the substrate (SUS304 stainless steel) material, we consider them to be basically the same layer materials. Hence, the mixed stress intensity factor can be written as follows:

$$|K| = \frac{4\cosh \pi\varepsilon}{\sqrt{c_1 + c_2}}\sqrt{G}. \tag{13.32}$$

Therefore, in the tensile test here, the stress intensity factor of the TBC with interface crack can be equivalently deemed to be caused by the external force and bending moment below [36]:

$$P_1 = 0, \quad M_1 = 0, \quad P_3 = -Q, \quad M_3 = 0 \tag{13.33}$$

Hence, the equivalent external force and bending moment are

$$P = C_1 Q, \qquad M = 0. \tag{13.34}$$

For the relevant parameters of Equation 13.34, refer to References [34,36]. In our tests, the influence of factors—such as the substrate's thickness and the transition layer's surface roughness—on the ceramic coating's adhesive strength is taken into consideration. After the calculation, we can obtain the relationship between the TBC sample's strain and the interface stress intensity factor in the tensile test, as shown in Figure 13.19. The stress intensity factor shown in the figure is the value when the coating thoroughly peels off from the substrate. It is actually the TBC interface's fracture toughness (failure toughness). We find that, for TBC samples with different densities (Mode *A* and Model *B*), the change range of their interface's fracture toughness is ~1.05–1.27 and ~1.0–1.17 MPa · $m^{1/2}$, respectively. Its value is identical with the calculation result of the previous shear model.
In the tensile test, the coating's failure process is summarized as follows:

The coating crack appears, perpendicular to the coating surface in the tensile direction, as shown in Figure 13.20a. As the tensile load continues to go up, some cracks grow toward the coating interface, while some cracks grow with an angle of 45° to the interface. When it reaches the coating and

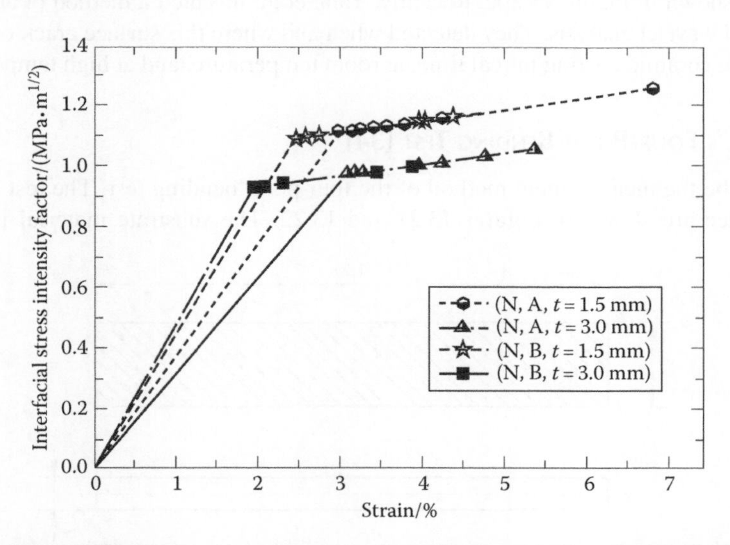

FIGURE 13.19 Relationship between interface stress intensity factor and applied strain in a double TBC system under tensions. (From Zhou, Y.C., Tonomori, T., Yoshida, A. et al. *Surf. Coat. Technol.*, 157(2–3), 118, 2002; Taken from Figure 8.)

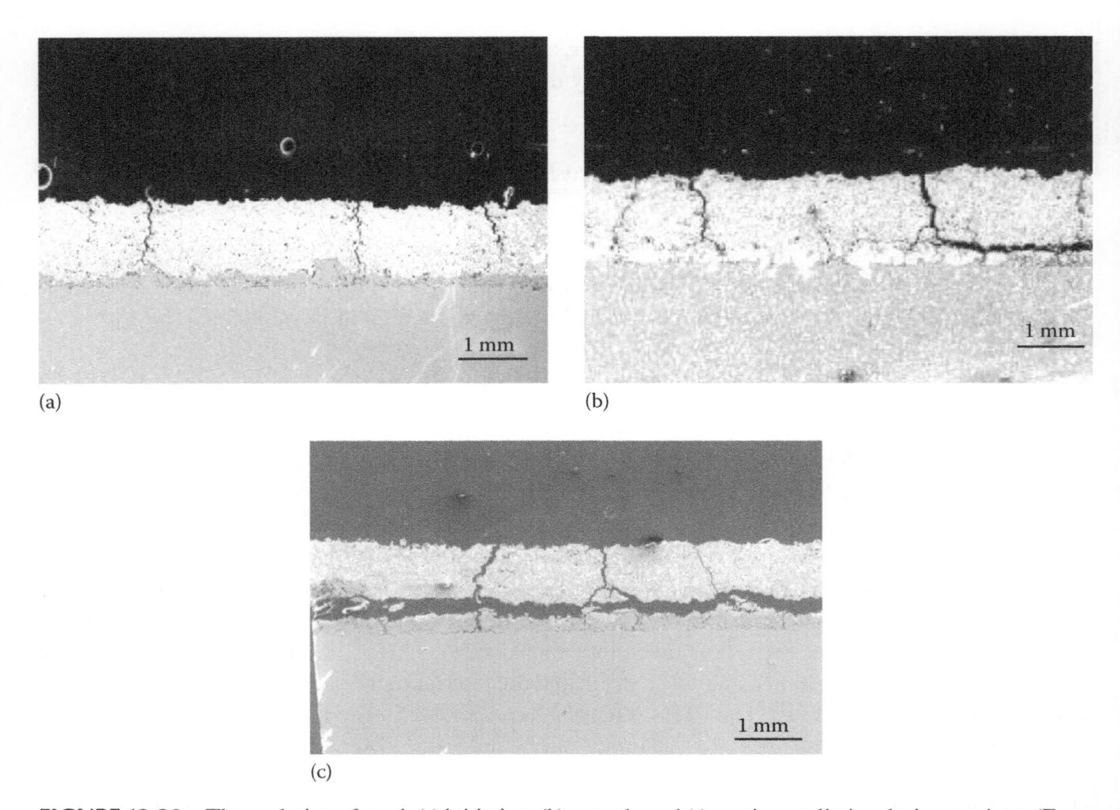

FIGURE 13.20 The evolution of crack (a) initiation, (b) growth, and (c) coating spallation during tensions. (From Zhou, Y.C., Tonomori, T., Yoshida, A. et al., *Surf. Coat. Technol.*, 157(2–3), 118, 2002; Taken from Figure 6).

transition layer interface, the interface cracks at different locations join along the interface, forming larger interface cracks, as shown in Figure 13.20b. As the tensile load continues to increase, the interface cracks at different locations continue to join up on the interface. Meanwhile, some coatings begin to peel off, as shown in Figure 13.20c. Recently, Yang et al. invented a method of acoustic emission technology and wavelet analysis. They detected when and where this surface crack or interface crack appeared on the ceramic coating in real time, at room temperature, and at high temperature [37].

13.8.3 TBC's Four-Point Bending Test [34]

We will describe the measurement method of the four-point bending test. The test sample size and specimen device are shown in Figures 13.21 and 13.22. The substrate material is stainless steel

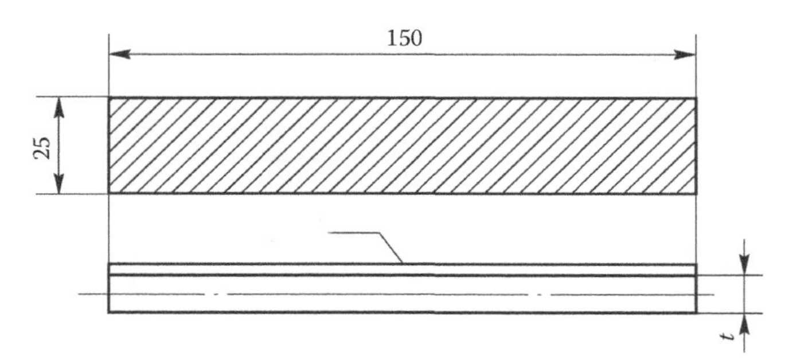

FIGURE 13.21 Schematic of sample size used for four-point bending tests. (From Zhou, Y.C., Tonomori, T., Yoshida, A. et al., *Surf. Coat. Technol.*, 157(2–3), 118, 2002; Taken from Figure 2.)

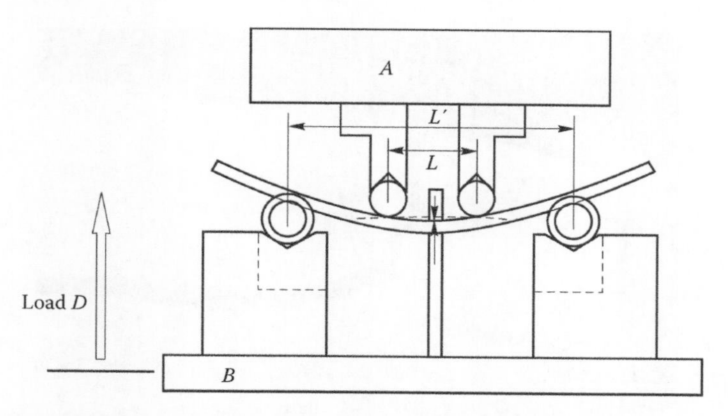

FIGURE 13.22 Schematic of four-point bending experimental equipment. (From Zhou, Y.C., Tonomori, T., Yoshida, A. et al., *Surf. Coat. Technol.*, 157(2–3), 118, 2002; Taken from Figure 3.)

with a thickness of 1.5, 3.0, and 5.0 mm, respectively. The transition layer is 100 μm thick and the ceramic coating is 440 μm thick. The sample's total length is 150 mm, where $L' = 90$ mm, which represents the distance between the lower two support centers in the four-point bending tester; and $L = 30$ mm, which represents the distance between the upper two indenter centers in the tester. In the test process, the laser displacement testing system is used to monitor the bending deflection of the sample's center part, so as to obtain the relationship between the load and the center deflection.

In different substrate thickness conditions, we can obtain the curve of the load value and the strain. From this curve, we find that the larger the substrate thickness t, the harder the specimen is deformed. Before coating failure and peel-off, a long elastic deformation process occurs on the sample. By using the Suo–Hutchinson theoretical model [36] and test data, Zhou et al. used a four-point bending test to obtain the coating interface's SIF value. The result is shown in Figure 13.23 [34]. The interface's SIF value increases as the bending load goes up. When the coating has thoroughly peeled off from the substrate, the SIF value is equal to the material interface's fracture toughness. From the figure we also find that the interface fracture toughness of this batch of samples is within ~4.26–7.21 $\text{MPa} \cdot \sqrt{\text{m}}$.

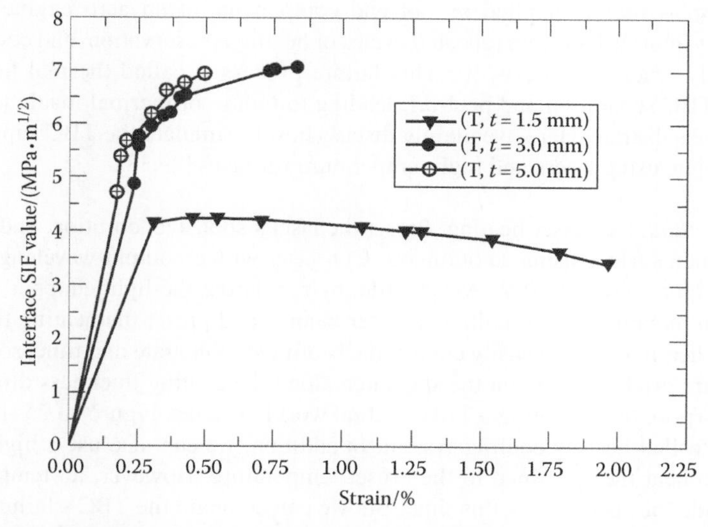

FIGURE 13.23 The relationship between interface SIF and strain of APS TBCs under four-point bending tests. (From Zhou, Y.C., Tonomori, T., Yoshida, A. et al., *Surf. Coat. Technol.*, 157(2–3), 118, 2002; Taken from Figure 11.)

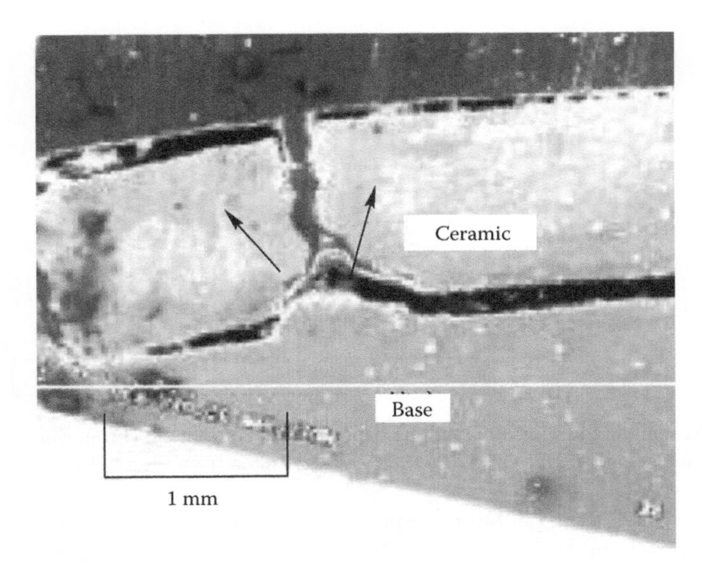

FIGURE 13.24 SEM observations of interface spallation between the coating and substrate. (From Zhou, Y.C., Tonomori, T., Yoshida, A. et al., *Surf. Coat. Technol.*, 157(2–3), 118, 2002; Taken from Figure 13.)

Meanwhile, we find that for samples with thick substrates, their interface fracture toughness has a larger value. In the test, we observe that as the load goes up, cracks perpendicular to the interface appear first inside the ceramic coating. Cracks along this direction keep growing and branching. Finally, they gradually become parallel to the interface at the joint between the ceramic coating and the transition layer. When the load continues to increase, interface cracks at different locations gradually grow and join, forming larger interface cracks. Finally, the ceramic coating is separated from the substrate, as shown in Figure 13.24.

13.8.4 TBC's Thermal Fatigue Test

The TBC system is mainly applied on hot end components of an aero engine's turbine blade. Generally, this system suffers from repeated cycles of heating, preservation, and cooling. Its ceramic coating gradually cracks and peels off. This failure process is called thermal fatigue failure. Its direct result is TBC separation and peel-off, leading to failure of thermal insulation and corrosion protection for the substrate. Here, we mainly discuss how to simulate the TBC's interface oxidation and thermal fatigue using lasers and high-temperature furnaces [29,38].

1. *Testing method.* CO_2 laser heating: Preset the laser's spot size, duration, and density. Heat the specimen surface using a continuous CO_2 laser with an output wavelength of 10.6 μm and an output power of 50 W. Meanwhile, by regulating the light chopper with pore, we can obtain monopulse and multipulse laser beams, and preset the heating time. By using CO_2 laser heating, we can easily control the heating/cooling rate and time, so that a certain temperature grade appears on the specimen along the coating thickness direction, realizing the purpose of simulating a TBC's actual working status. Figure 13.25 shows the computer-controlled heating/cooling system. In addition, we can also use a high-temperature furnace to heat the specimen to the preset temperature. However, no temperature grade exists inside the specimen at this time. But we can compare the TBC's failure characteristics by using the laser heating method.
2. *Nondestructive measurement and observation.* The temperature on the ceramic coating surface and the substrate surface is measured by a thermal infrared imager and a

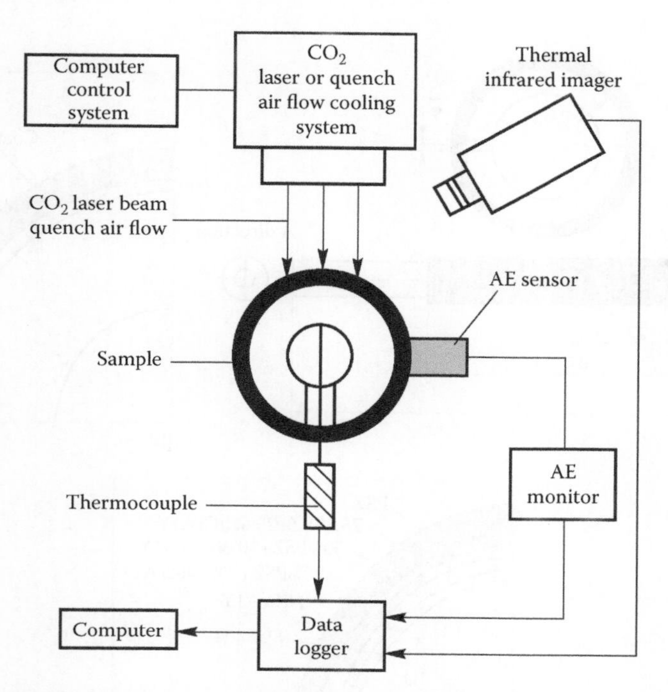

FIGURE 13.25 Schematic of laser heating experimental equipment of TBC system. (From Zhou, Y.C., Hashida, T., *Int. J. Fatigue*, 24(2–4), 407, 2002; Taken from Figure 1.)

thermocouple, respectively. In the heating process, an acoustic emission (AE) detector is used to detect the microcrack growth process. The specimen is a cylindrical drum. Its middle part is the laser-heating zone. The AE sensor should be kept far away from the laser radiation zone. The AE signal is detected by using a piezoelectric transducer with a resonant frequency close to 1 MHz. The square of the signal voltage's peak value is used as an energy measurement method. The signal sent from the transducer is amplified to 80 db. The band frequency passing the filter is ~5–500 Hz. AC impedance spectroscopy is used to measure the interface oxidation reaction, and the physical parameter change between the middle transition layer and the ceramic coating. The thickness and micropattern of thermal growth oxidation (TGO) between the middle transition layer and the ceramic coating can be observed and analyzed, using scanning electron microscopy (SEM).

3. *Specimen*. In the test, two coating material specimens are made. One is a common double-layer coating system (nonfunctionally graded material [non-FGM]). A partially stabilized zirconia (PSZ) layer (8%Y_2O_3–ZrO_2) is sprayed on the NiCrAlY middle transition layer using the APS method. The substrate material is SUS304 stainless steel with a thermo-elastic property close to an Ni-based superalloy material. The other one is an FGM coating. The FGM coating is a five-layer coating made up of PSZ and NiCrAlY. Figure 13.26 shows the sectional view of the cylindrical specimen's non-FGM and FGM coating, and the specimen's geometric size.

   For the non-FGM specimen, the coating thickness and transition layer thickness are, respectively, 0.35 and 0.15 mm. For the FGM specimen, the PSZ coating thickness is 0.35 mm and the other layers' thickness is 0.1 mm.
4. *Test results*. In the non-FGM and FGM coating, SEM observation shows that the laser results in coating failure in two forms: vertical (or surface crack) and interface peel-off. The vertical crack and interface peel-off of non-FGM and FGM TBCs caused by laser heating are shown in Figure 13.27. After analysis of the SEM photo, we find that the interface crack

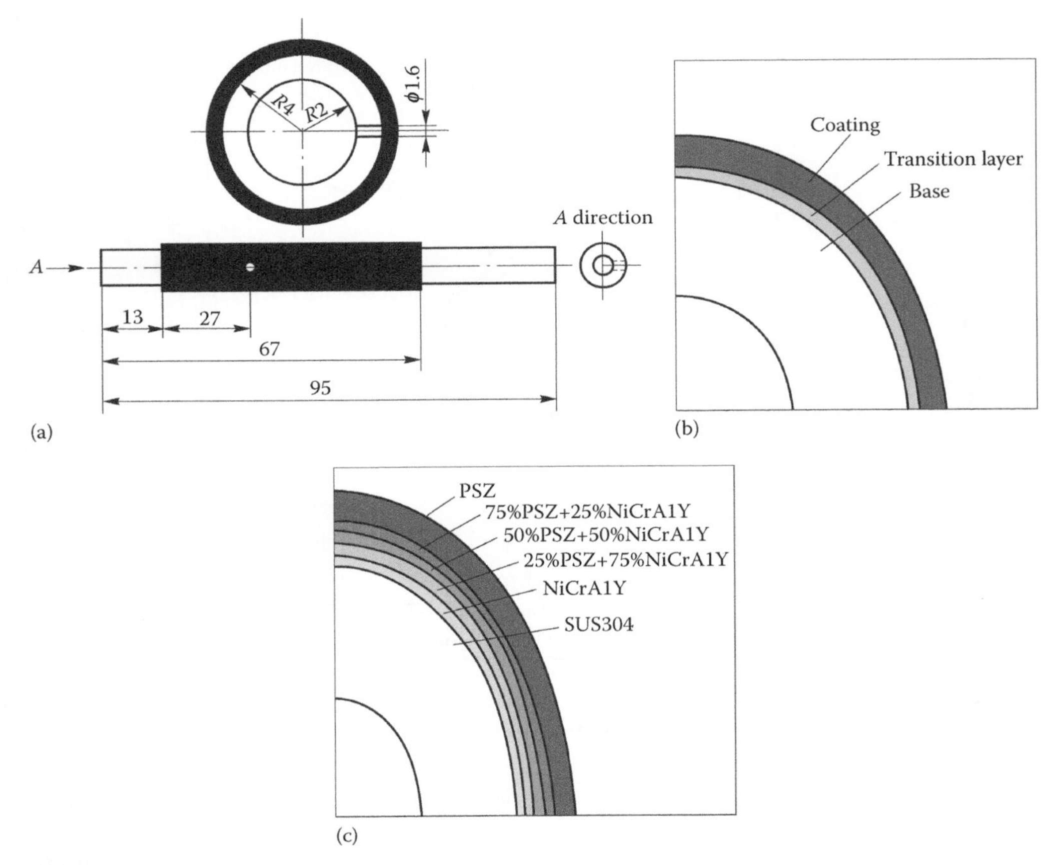

FIGURE 13.26 TBC sample and structure used in thermal fatigue tests. (a) Sample's geometric size. (b) Sectional view of non-FGM sample. (c) Sectional view of FGM sample. (From Zhou, Y.C. and Hashida, T., *Int. J. Fatigue*, 24(2–4), 407, 2002; Taken from Figure 2.)

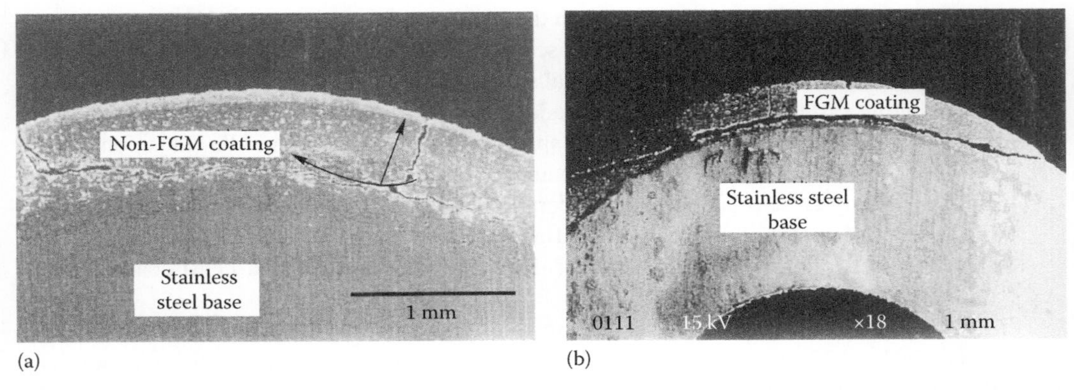

FIGURE 13.27 Vertical crack and interface delamination of non-FGM and FGM TBC caused by laser heating. (a) Failure characteristics of non-FGM coating after six thermal fatigue cycles (thermal cycle time: 70 s, the maximum temperature of coating and base surface: 1200°C and 600°C). (b) Failure characteristics of FGM coating after six thermal fatigue cycles (laser power: 34 W). (From Zhou, Y.C. and Hashida, T., *Int. J. Fatigue*, 24(2–4): 407, 2002; Taken from Figures 3 and 4.)

in the non-FGM coating always occurs inside the NiCrAlY layer, the yttrium-stabilized zirconia (YSZ) layer, while the interface crack in the FGM coating generally exists in two different interfaces: (1) the interface of the PSZ layer and the 75% PSZ/25% NiCrAlY layer, and (2) the interface of the 75% PSZ/25% NiCrAlY and 50% PSZ/50% NiCrAlY. TBC final failure is characterized by coating peel-off.

In addition, from the temperature history curve and detected AE result in the laser thermal fatigue test, we find that the AE signal is usually recorded in the heating or cooling process. An AE signal can even be detected in the post-cooling process. The temperature grade at this time is zero. Voyel et al. also observed the interface peel-off phenomenon in the cooling process [39]. By comparing AE characteristics and SEM observation results, we find that the initially observed AE signal is caused by the vertical crack's initiation and growth. The high-energy AE signal, appearing after the first recorded AE signal, is caused by the interface crack's initiation and growth. This means that the vertical crack appears more easily than the interface crack. In addition, when FGM coatings and non-FGM coatings are in the same test condition, the FGM coating is not peeled off thoroughly, while the non-FGM coating is peeled off thoroughly.

13.8.5 TBC's Buckling Failure Test [40]

In actual application, TBC components usually work in a high-temperature, high-pressure, strong oxidation environment. Such environments easily result in interface oxidation on the TBC material, forming a TGO layer. The thickness of this oxidation layer keeps increasing. The existence of TGO further leads to microcracking on the TBC interface, reducing the interface's adhesive strength. On the other hand, because the TBC system material parameters don't match, residual stress appears in the thermal cycle process. The residual stress, gradually accumulating, makes the interface crack grow, finally leading to buckling and peel-off failure occurring between the ceramic coating and the substrate. Here, we mainly analyze the TBC buckling failure test condition and failure process.

First, prepare the TBC sample with interface defects and perform the compression test at different high temperatures. Observe the TBC buckling failure and study the TBC buckling failure factors.

13.8.5.1 Specimen Preparation

Because the ceramic material is a typical brittle material, it is nearly impossible to make a defect or a buried crack crossing the width direction at the undamaged TBC sample interface. Therefore, we must try to make a buried crack crossing the width direction in the sample preparation process, in order to simulate the interface crack occurring in an actual application process. The sample preparation method is shown in Figure 13.28. First, cut the substrate into a specimen of $40 \times 5 \times 5$ mm^3 with a wire cutter. In the APS process, spray a 0.1 mm thick transition layer on the metal substrate. Then, cover both ends of the specimen with a metal sheet. On the middle part, maintain a space with a length of *a* (*a* values are, respectively, 1, 2, 5, 8, and 10 mm). Spray an Al_2O_3 coating, whose thickness should not exceed 10 μm. Then, remove the mask and spray the ceramic layer. For a specimen prepared in this method, the adhesive strength of the corresponding interface part is much weaker than that of other locations.

Therefore, we can essentially consider that this location zone is an initial delamination or crack, appropriate for studying the specimen's buckling failure behavior. Finally, we perform a thermal cycle test on the specimen, and make an oxidation layer at the interface between the ceramic coating and the transition layer, to simulate an actual failure condition. The substrate material used in this test is SUS304 stainless-steel alloy. NiCrAlY alloy is used as the bonding layer, and 8 wt%Y_2O_3–ZrO_2 powder is used as the ceramic layer material. The basic parameters are listed in Table 13.2.

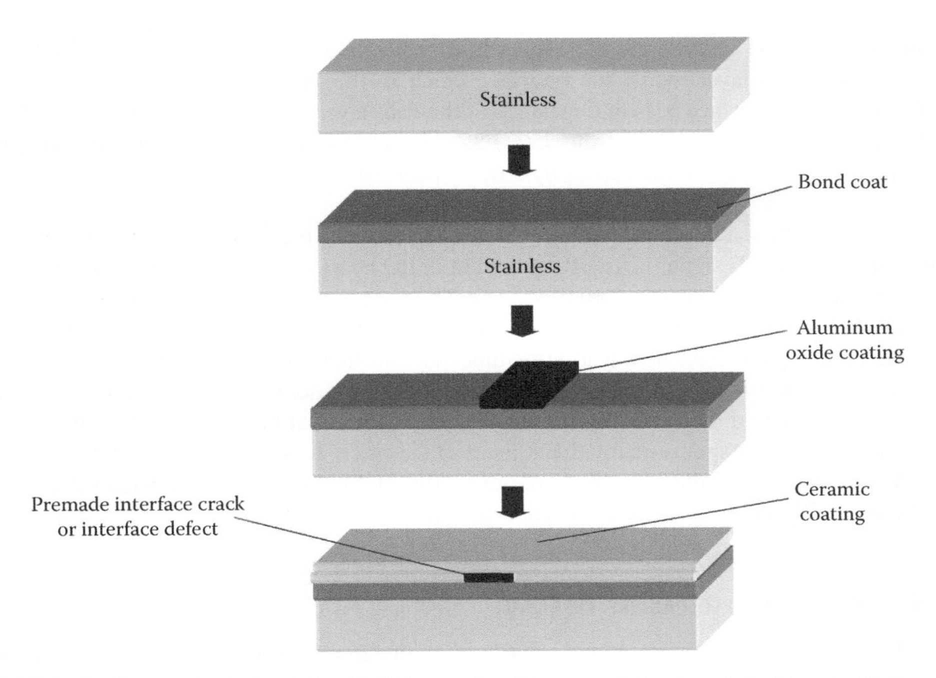

FIGURE 13.28 Preparation schematic of TBCs sample with a preexisting interface delamination.

TABLE 13.2
TBC Material's Physical Properties

Material	Density (g/cm³)	Melting Point (°C)	Thermal Expansion Coefficient (× 10⁻⁶/°C)	Heat Conduction Coefficient (W/mK)	Modulus of Elasticity (GPa)
8 wt%-ZrO_2	6.2	2700	11	0.6–1.1	22.5–49.6
NiCrAlY	7.7	1400	13.3	6.7	126–168

13.8.5.2 Tester

The testers are shown in Figure 13.29. The loading device is a microcomputer-controlled universal electronic tester. This system consists of three parts: a main device, an EDC100 control system, and a computer control system. It is a fully digital and graphics precision instrument. It can accurately measure the relevant data of materials in the tensile, compression, and bending condition, such as load, deformation, and stress–strain relationships. The heating device is an oxyacetylene flame gun. By observing the temperature value on the digital instrument, we can control the distance from the nozzle to the coating surface, and thereby control the surface temperature on the specimen coating. The temperature measurement device is a thermocouple. The probes of two thermocouples are respectively fixed on the ceramic surface and the substrate surface to measure their temperature and record the temperature gradient. Meanwhile, a high-powered observation instrument is used to monitor and record the sample's thermal buckling failure process.

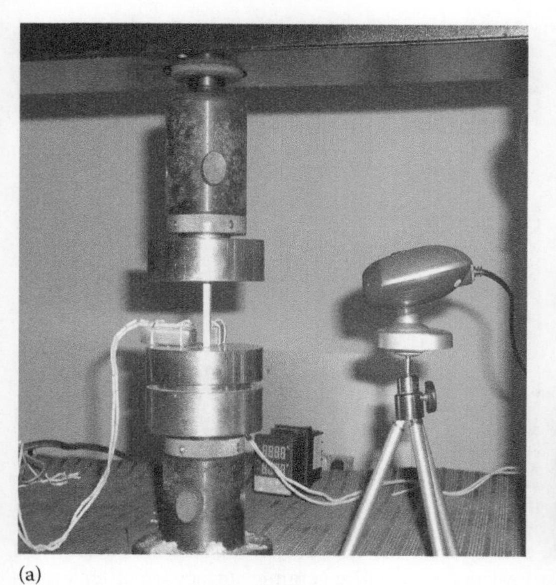

(a) (b)

FIGURE 13.29 Experimental equipment of thermomechanical buckling failure of TBC system. (a) Temperature measurements on the ceramic surface and substrate by thermocouple. (b) Temperature measurements of the substrate bottom surface.

1. Testing steps: Estimate the critical buckling load P_{max} in advance.
2. Adjust the retaining ring of the tester's limit lever to limit its travel, and protect the instrument and specimen.
3. Load the workpiece, keeping it in a vertical and aligned state, to prevent eccentric bending failure.
4. Set the test parameters.
5. Operate the tester to compress the sample.
6. When the compression load is equal to one preset value $P(P \approx P_{max}/3)$, keep it at a constant value.
7. Heat the ceramic layer surface with its interface crack evenly using the oxyacetylene flame, and measure the temperature of the substrate and ceramic layer using the thermocouple. Perform dynamic observation and take photos using the digital camera and video camera, as shown in Figure 13.30.
8. When the buckling phenomenon is observed, stop loading and heating.
9. Save the test data.
10. Lift the indenter of the universal tensile testing machine and take out the test sample.
11. Complete the test.

13.8.5.3 Test Results

Under the joint action of heat and force, buckling failure occurs on the ceramic coating. By controlling test conditions, we observe the ceramic material's buckling failure process. Testing shows that the factors affecting TBC buckling failure mainly include the ceramic coating surface's temperature, the ceramic coating's thickness, the premade interface's defect length (L), and the pre-applied compression load's value. The length of the premade interface defect or delamination is the key factor. Buckling failure occurs on the ceramic coating materials mainly in two forms: center buckling failure and edge buckling failure. Figure 13.31 shows that buckling failure occurs on the TBC sample with interface defects, under the joint action of heat and force. From the figure we see that, in the sample's premade defect zone, some ceramic coatings not only suffer from buckling failure, but are also peeled

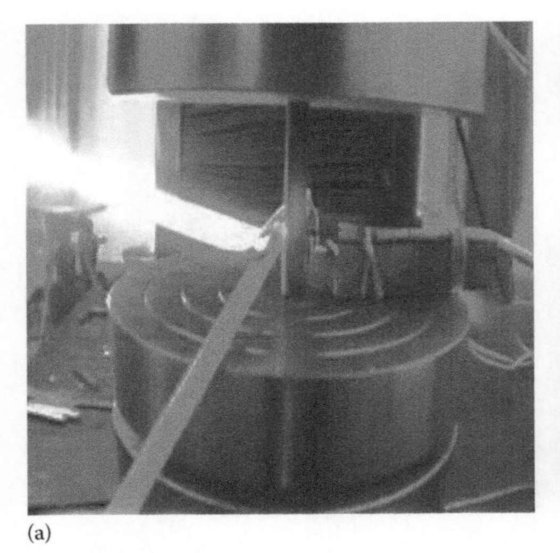

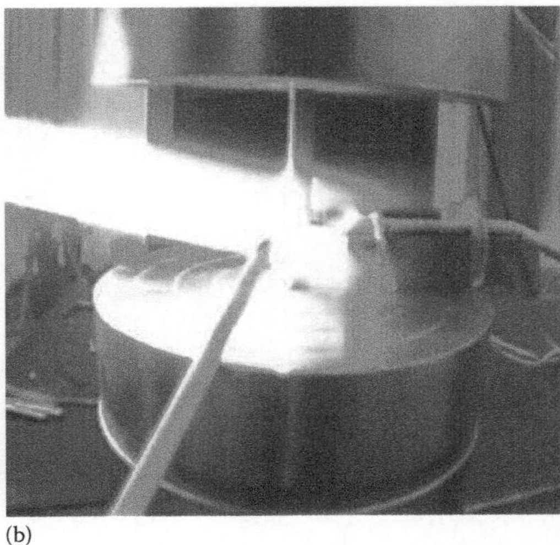

(a) (b)

FIGURE 13.30 In the test process, heat the sample and measure the temperature of ceramic surface and substrate using thermocouple. (a) When heating begins, warm up the entire ceramic surface; (b) after warm-up, heat the sample's defect using an oxyacetylene flame.

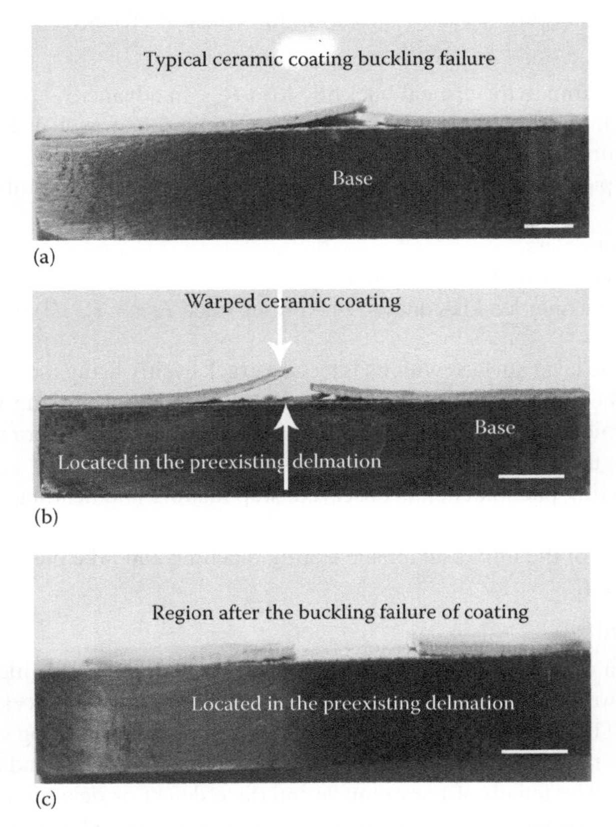

FIGURE 13.31 Digital images of thermomechanical buckling failure of APS TBCs. (a) Digital images of thermomechanical buckling failure of APS TBCs under Condition 1. (b) Digital images of thermomechanical buckling failure of APS TBCs under Condition 2. (c) Digital images of thermomechanical buckling failure of APS TBCs under Condition 3. (From Mao, W.G., Dai, C.Y., Zhou, Y.C. et al., *Surf. Coat. Technol.*, 201, 6217, 2007; Taken from Figures 4 and 8.)

off from the substrate thoroughly, exposing the substrate materials. In Figure 13.31a,b, and c, buckling failure occurs on the ceramic coating under the following conditions:

1. Pre-applied compression load P = 2000 N, interface defect length L = 10 mm, ceramic surface temperature T_1 = 1370°C, and the substrate's lower surface temperature T_2 = 985°C.
2. Pre-applied compression load P = 2000 N, interface defect length L = 10 mm, ceramic surface temperature T_1 = 1280°C, and the substrate's lower surface temperature T_2 = 915°C.
3. Pre-applied compression load P = 2000 N, interface defect length L = 8 mm, ceramic surface temperature T_1 = 277°C, and the substrate's lower surface temperature T_2 = 214°C.

To identify the specific location of the buckling failure on the APS TBC system, we perform SEM and energy-dispersive x-ray (EDX) observation analysis on the sample after buckling failure. Figure 13.32 shows SEM observation of the sample in Figure 13.31a. We find that, in the place where the buckling failure occurs, there is still one layer of thin ceramic coating material on the substrate. EDX composition analysis on the peeled ceramic coating fragment's lower surface further proves this point. The material of the residual part is ZrO_2. The transition layer material is not present. Finally, we carry out pattern observation and analysis on the peeled ceramic coating fragment after buckling failure, as shown in Figure 13.33. From the figure we see that the ceramic coating micropattern is the existence of many cracks and holes after the fracture. Irregular patterns appear after brittle fracture occurs. This is the typical failure pattern of APS TBC sample peel-off.

Because most chemical bonds of ceramic materials are ionic bonds and covalent bonds, they bind firmly and have obvious directivity. Compared with common metals, their crystal structure is complex, with low surface energy. Therefore, ceramic materials feature high strength, hardness, and modulus of elasticity, as well as good resistance to wear and corrosion. They have been widely applied in industrial fields such as aerospace, electronics, information, energy, petrochemical, metallurgy, and textiles. However, because ceramic materials have poor plastic toughness, low resistance to impact, and high processing difficulty, brittle fracture often occurs in their application. Because ceramic coating materials don't match the substrate materials in their characteristics, residual stress and interface cracks appear easily in the application process, leading to delamination, blistering, and peel-off on the interface of the ceramic coating and the substrate. Therefore, the interface issue of ceramic coating material systems

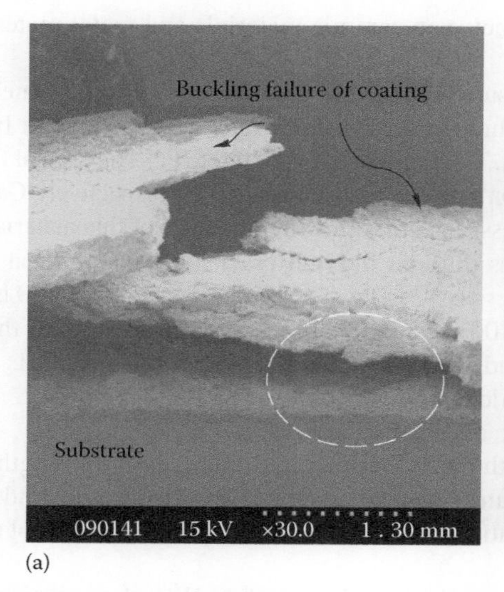

(a)

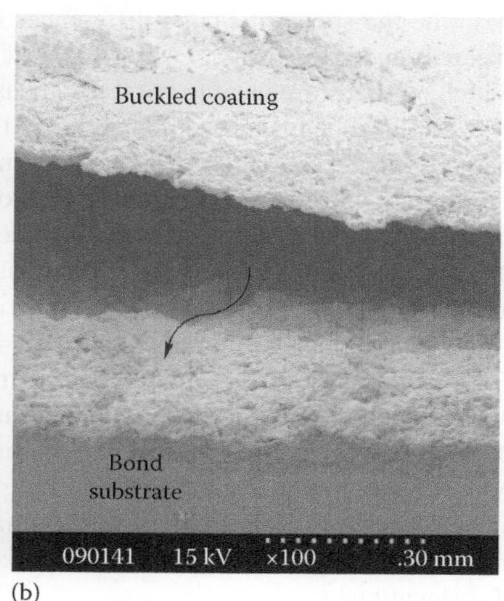

(b)

FIGURE 13.32 TBC's typical buckling failure phenomenon in the compression test of heat and force. (a) TBC's buckling failure SEM observation in the center zone. (b) Micropattern of the white dotted line zone in (a) after zoom-in. (From Mao, W.G., Dai, C.Y., Zhou, Y.C. et al., *Surf. Coat. Technol.*, 201, 6217, 2007; Taken from Figure 4.)

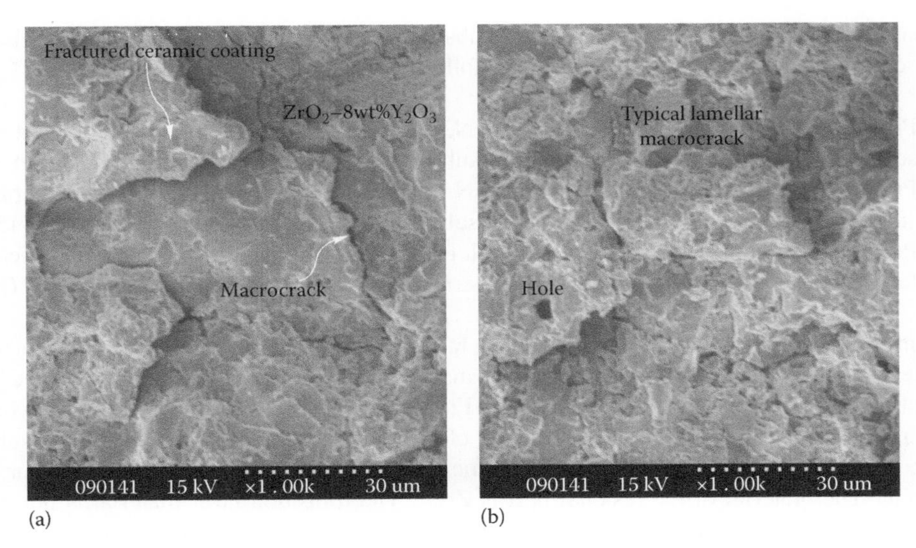

FIGURE 13.33 SEM observations of buckling failure of the ceramic coating. (a) The SEM image of the bottom surface of the spalled ceramic coating. (b) The SEM image of the exposed substrate surface. (From Mao, W.G., Dai, C.Y., Zhou, Y.C. et al., *Surf. Coat. Technol.*, 201, 6217, 2007; Taken from Figure 4.)

is the target of engineers and researchers, and should be solved quickly. Many countries have invested resources (including manpower, materials, and money) to study the interface failure mechanism and reliability forecasting of the ceramic coating system, in order to create better prospects for application.

EXERCISES

13.1 Explain the following terms: ceramic, ceramic coating, fracture toughness, thermal shock fracture, thermal shock damage, and thermal barrier coating.

13.2 Describe the differences between ceramic materials and metal materials in elastic deformation, plastic deformation, and fracture.

13.3 Describe the similarities and differences between ceramic materials and metal materials in the creep mechanism.

13.4 Which methods are used to correctly measure the fracture toughness of ceramic materials?

13.5 One ZrO_2 notched beam with a thickness-through surface crack fails in the four-point bending method (Mode I crack). The height is 10 mm. The fracture stress is 100 MPa. The critical fracture size is 1 mm. Its Young's modulus and Poisson's ratio are 200 GPa and 0.30, respectively. Calculate the fracture surface energy, fracture toughness, and crack growth resistance of this material.

13.6 One Al_2O_3 notched plate with a thickness-through internal crack fails in the even tensile condition (Mode I crack). The fracture stress is 12 MPa. The critical crack size is 60 mm. Its Young's modulus and Poisson's ratio are 400 GPa and 0.30, respectively. Calculate the fracture surface energy, fracture toughness, and crack growth resistance of this material.

13.7 Which factors affect the strength of ceramic materials?

13.8 Describe the types and features of ceramic material fatigue.

13.9 Why is a ceramic material's tensile strength much lower than its compression strength?

13.10 In metal materials, plastic deformation mainly depends on the slip of dislocation and climbing. However, in ceramic material deformation, why is attention not paid to the role of the slip of dislocation?

13.11 Ceramic materials are generally deemed as typical brittle materials. Why do ceramic materials have superplastic behavior?

13.12 Explain why the strength of ceramic materials has high dispersion.

REFERENCES

1. Askeland D. R. and Phulé P. P. *Essentials of Materials Science and Engineering*. Beijing: Tsinghua University Press, 2005.
2. Zhou Y. *Ceramic Materials* (2nd ed.). Beijing: Science Press, 2004.
3. Besterci M. Preparation, microstructure and properties of Al–Al4C3 system produced by mechanical alloying. *Mater Design*, 2006, 27(5): 416–421.
4. Gasch M. J., Wan J., and Mukherjee A. K. Preparation of a Si3 N4/SiC nanocomposite by high-pressure sintering of polymer precursor derived powders. *Scr Mater*, 2001, 45(9): 1063–1068.
5. Sharif A. A. and Mecartney M. L. Superplasticity in cubic yttria-stabilized zirconia with intergranular silica. *Acta Mater,* 2003, 51: 1633–1639.
6. Xu X., Nishimura T., Hirosaki N., Xie R. J. et al. Superplastic deformation of nano-sized silicon nitride ceramics. *Acta Mater*, 2006, 54: 255–262.
7. Dillon R. P., Sosa S. S., and Mecartney M. L. Achieving tensile superplasticity in 8 mol% Y_2O_3 cubic stabilized ZrO_2 through the addition of intergranular silica. *Scr Mater*, 2004, 50: 1441–1444.
8. Hiraga K., Kim B. N., Morita K. et al. High-strain-rate superplasticity in oxide ceramics. *Sci Technol Adv Mat*, 2007, 8(8): 578–587.
9. Morita K., Hiraga K., Kim B. N. et al. High-strain-rate superplastic flow in tetragonal ZrO_2 polycrystal enhanced by the dispersion of 30 vol.% $MgAl_2O_4$ spinel particles. *Acta Mater*, 2007, 55(13): 4517–4526.
10. Morita K., Hiraga K., and Kim B. N. Effect of minor SiO_2 addition on the creep behavior of superplastic tetragonal ZrO_2. *Acta Mater*, 2004, 52(11): 3355–3364.
11. Kamiya S., Motohashi Y., Harjo S. et al. Ion-plating of TiN on superplastically deformed 3Y-TZP ceramic and 3Y-TZP/Al_2O_3 composite. *Surf Coat Technol*, 2003, 169(2): 464–467.
12. Cina B. and Eldror I. Bonding of stabilised zirconia (Y-TZP) by means of nano Y-TZP particles. *Mater Sci Eng A*, 2001, 301(2): 187–195.
13. Niihara K., Morena R., and Hasselman D. P. H. Evaluation of K_{IC} of brittle solids by the indentation method with low crack-to-indent ratios. *J Mater Sci Lett*, 1982, 1: 13–16.
14. Lugovy M., Slyunyayev V., and Orlovskay N. Apparent failure toughness of Si3N4-based laminates with residual compressive or tensile stresses in surface layers. *Acta Mater*, 2005, 53: 289–296.
15. Radovic M., Barsoum M., Barsoum W. et al. On the elastic properties and mechanical damping of Ti_3SiC_2, Ti_3GeC_2, $Ti_3Si_{0.5}Al_{0.5}C_2$ and Ti_2AlC in the 300–1573 K temperature range. *Acta Mater*, 2006, 54(10): 2757–2767.
16. Posarac M., Dimitrijevic M., and Volkov-Husovic T. Determination of thermal shock resistance of silicon carbide/cordierite composite material using nondestructive test methods. *J Eur Ceram Soc*, 2008, 28(6): 1275–1278.
17. Si T. Z., Liu N., Zhang Q. A., and You X. Q. Thermal shock fatigue behavior of TiC/Al_2O_3 composite ceramics. *Rare Metals*, 2008, 27(3): 308–314.
18. Jin Z. H., Gao J. Q., and Qiao G. J. *Engineering Ceramic Materials*. Shaanxi, China: Xi'an Jiaotong University Press, 2000.
19. Herring C. Diffusional viscosity of a polycrystalline solid. *J Appl Phys*, 1950, 21: 437–445.
20. Nabarro F. R. N. Restriction of prismatic punching to a limited class of crystals. *Phys Rev*, 1950, 79: 894–899.
21. Radovic M., Barsoum M. W., Raghy T. et al. Tensile creep of coarse-grained Ti_3C in the 1000°C~1200°C temperature range. *J Alloy Compd*, 2003, 361:299–312.
22. Carrère P. and Lamon J. Creep behavior of a SiC/Si-B-C composite with a self-healing multilayered matrix. *J Eur Ceram Soc.*, 2003, 23: 1105–1114.
23. Deng S. J. *High-Performance Ceramic Coatings*. Beijing: Chemical Industry Press, 2004.
24. Wu Z. J. *Thermal Spraying Technology and Its Application*. Beijing: China Machine Press, 2006.
25. Li J. B. and Zhou Y. C. *New Materials Science and Application Technology*. Beijing: Tsinghua University Press, 2004.
26. Cao X. Q. *Thermal Barrier Coating Materials*. Beijing: Science Press, 2007.
27. Guo H. B., Xu H. B., Gong S. K., and Liu F. S. Failure mechanism of gradient thermal barrier coating subjected to thermal cycling, *Acta Metall Sin*, 2001, 37(2): 151–155.
28. Zhou Y. C. and Hashida T. Coupled effects of temperature graded and oxidation on the thermal stress in thermal barrier coating system. *Int J Solids and Struct*, 2001, 38(24–25): 4235–4264.
29. Zhou Y. C. and Hashida T. Thermal fatigue failure induced by delamination in thermal barrier ceramic coating. *Int J Fatigue*, 2002, 24(2–4): 407–417.
30. Mao W. G. and Zhou Y. C. Modeling of residual stresses variation with thermal cycling in thermal barrier coatings. *Mech Mater*, 2006, 38: 1118–1127.

31. Evans A. G., He M. Y., and Hutchinson J. W. Mechanics-based scaling laws for the durability of thermal barrier coating. *Prog Mater Sci*, 2001, 46: 249–271.
32. Evans A. G., Mumm D. R., and Hutchinson J. W. Mechanisms controlling the durability of thermal barrier coatings. *Prog Mater Sci*, 2001, 46: 505–553.
33. Padture N. P., Gell M., and Jordan E. H. Thermal barrier coatings for gas-turbine engine applications. *Science*, 2002, 296: 280–284.
34. Zhou Y. C., Tonomori T., Yoshida A. et al. Failure characteristics of thermal barrier coatings after tensile and bending tests. *Surf Coat Technol*, 2002, 157(2–3): 118–127.
35. Hu M. S. and Evans A. G. The cracking and decohesion of thin films on ductile substrates. *Acta Metall*, 1989, 37(3): 917–925.
36. Suo Z. G. and Hutchinson J. W. Interface crack between two elastic layers. *Int J Fract*, 1990, 43: 1–18.
37. Yang L. Research on non-destructive evaluation of TBC oxidation, damage and fracture, [Xiangtan University Doctoral Dissertation]. Hunan, China, 2007.
38. Zhou Y. C. and Hashida T. Thermal fatigue in thermal barrier coating. *JSME Int J*, 2002, A45(1): 57–64.
39. Voyer J., Gitzhofer F., and Boulos M. I. Study of the performance of TBC under thermal cycling conditions using an acoustic emission rig. *J Therm Spray Tech*, 1998, 7: 181–190.
40. Mao W. G., Dai C. Y., Zhou Y. C. et al. An experimental investigation on thermo-mechanical buckling delamination failure characteristic of air plasma sprayed thermal barrier coatings. *Surf Coat Technol*, 2007, 201: 6217–6227.

14 Mechanical Properties of Composite Materials

Why is a racing bike much lighter than an ordinary bike, yet more durable? Why has the loading capability of a modern flight been increasing while the weight of the aircraft has been reduced? This is because more composite materials have been used in a racing bike, or in a modern airplane, instead of steel and other metallic materials. The reason that composite materials have high performance—materials such as glass fiber-reinforced plastic (GRP) and carbon fiber-reinforced plastic composites—is due to their structural complexity.

In recent years, there have been increasing demands in the development of advanced materials. Composite materials have been one of the key players due to their combination of two different material components, each with its own advantages. In fact, many natural materials, including biological materials, are composite materials that make our lives easier and more colorful. So far, composite materials have been widely used in the fields of aerospace, energy, transportation, and architecture, as well as in industrial machinery, the biomedical industry, and in sport. We could even say that when we entered into the twenty-first century, mankind also entered the era of composite materials. With the development and application of composite materials, the field of composite materials mechanics was created, and has been extensively developed [1,2].

In this chapter, the definition, characteristics, and development of composite materials will be introduced. Then we will focus on their mechanical properties, including elasticity, strength, and the fracture properties of two typical composite materials: fiber-reinforced composite materials and particle-enhanced composite materials. Finally, new trends in the development of composite materials will be discussed.

14.1 INTRODUCTION TO COMPOSITE MATERIALS

14.1.1 What Are Composite Materials? [1,2]

A compound material is defined as a solid-state material that consists of two or more different phases of distinct physical and chemical properties, with one acting as a substrate and another dispersed into it. The substrate is normally a continuous phase. The dispersed phase is also called the strength enhancement phase. The composite material takes advantage of the two phases, while overcoming their disadvantages. Composite materials have been used in the cost-effective manufacture of components with complex structures and higher performance.

The use of composite materials in daily life dates back to ancient times. People at the time realized that it was beneficial to make a material composed of several different materials. From the study of physical historical evidence and literature, it was concluded that about 7000 years ago, people in Xi'an Banpo village had used composite materials composed of grass mixed with mud in their house walls and bricks. They realized that the mixed material had properties that mud and grass lacked separately. This was thought to be the first time human beings began to use composite materials. About 4000 years ago, lacquer wares began to be used. This is a typical fiber-reinforced composite material consisting of hemp and textiles, as enhanced phases are placed on top of the mold layer, using raw lacquer as an adhesive. The surface made of lacquer is not only bright and clean but also has good antiaging performance. The lacquer ware of Jianzhen Buda in Yangzhou Pingshan's hall is still in good condition, even though it was made about 1000 years ago. Another typical example of an application of

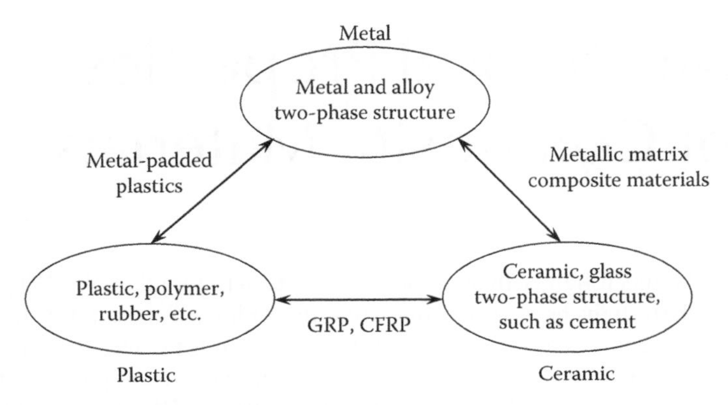

FIGURE 14.1 Relationship between different engineering materials.

composite materials can be found in ancient China, when people made their bows using laminated materials consisting of bamboo and steel, in order to gain high module and high strength. In ancient Egypt, people made a material similar to the plywood we have today, by reconstruction of sliced wooden planks. This type of material has high strength and anti-distortion properties. These examples show that human beings have known that composite materials have enhanced properties, compared to any of its components singly, and have used them in daily life for a long time. Since the middle of the twentieth century—with the development of modern science and technology, especially in chemistry, mechanics, mechanicals, metallurgy, and ceramics—a new materials system called composite materials has been created, which is no longer limited to natural composite materials. Composite materials have been playing an increasing role in science and technology.

Figure 14.1 shows the relationship between different engineering materials. Steels are typical and widely used metallic materials. They are composite materials and consist of normally soft metallic matrixes and dispersed hard ceramic phases for strength enhancement. These ceramic phases can have different shapes: some are needle like, some platelike, some are spherical, and some have irregular shapes. A polymer is normally composed of two phases with a strength-enhanced or a toughness-enhanced phase dispersed into a matrix.

Composites are also found in nature. Wood is an example of a composite because it is made of cellulose and lignin. The cellulose fibers in wood are held together by lignin. These fibers can be found in cotton and thread, but it's the bonding power of lignin in wood that makes it much tougher. Another natural composite is rock and sand, which are materials used in concrete. Rock is just smaller rocks held together, and sand is made of small grains. Bones, teeth, and shells are another category of natural composite materials that consist of ceramic hard phases and natural organic substrates.

Concrete is a typical example of a ceramic matrix composite material, which is composed of sand grains dispersed in a cement matrix [3].

So far, a variety of composite materials have been developed. There are several ways to sort out these composite materials [4–6]. One way to sort out composite materials is based on their application. Composite materials can be grouped into structural composite materials and functional composite materials.

A composite material is called a functional composite material if its application is based on its physical, chemical, and biological properties. A composite is called a structural composite material if its application is based on its mechanical properties. Composite materials are also classified into different groups by the shape of the reinforcement phases. There are particle-reinforced, plate-reinforced, fiber-reinforced, and layered composite materials. The shapes of the reinforced phases are shown in Figure 14.2. Composite materials are also classified according to their matrix materials, and include metallic matrix composite materials, ceramic matrix composite materials, polymer matrix composite materials, and carbon matrix composite materials.

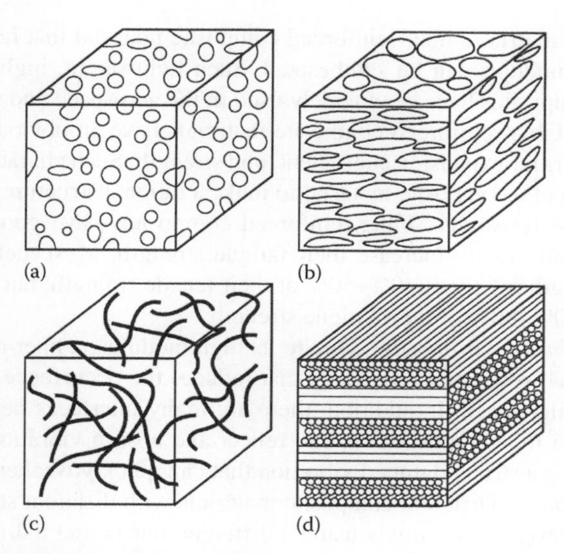

FIGURE 14.2 Shapes of the reinforcement phases in composite materials. (a) Pellet-reinforced composite materials. (b) Plate-reinforced composite materials. (c) Fiber-reinforced composite materials. (d) Layered composite materials.

14.1.2 Characteristics of Traditional and Composite Materials

Because there are big differences in properties among the metals, plastics, and ceramics, we cannot say which material is superior to another by simply listing their properties in a simple table. Nevertheless, we can still describe the advantages and disadvantages of the materials qualitatively [3].

1. A plastic material normally has a low density, short-term chemical stability, and good bearing conditions performance. It is also easy to manufacture into different shapes. However, it has bad thermostability and low mechanical strength.
2. Most ceramics also have low densities, but have extremely high thermostability, good anticorrosive, abrasion, and attrition performance. However, ceramics are brittle and that makes their manufacture difficult.
3. Most metallic materials have relatively high densities. But some light alloys—such as magnesium alloys, aluminum alloys, and beryllium alloys—have similar densities to those of plastics. Metallic materials normally have excellent thermostability and good anticorrosive properties. They also have great mechanical properties and are easy to manufacture.

As mentioned earlier, each type of material has both advantages and disadvantages. The design of a composite material usually considers the combination of the advantages of both the matrix and the reinforcement phase. Therefore, the composite material has superior properties compared to most traditional materials. The properties of a composite material are determined by its matrix property and the property and distribution of its reinforcement phase, which can be summarized as follows [7]:

1. *High strength and modulus*: The salient properties of a composite material are its high strength and high modulus (high strength/density and modulus/density, respectively.) For example, a carbon fiber only has a density 1.80 g/cm^3, but a strength of 3700–5500 MPa, and a modulus of 550 GPa. A boron fiber and an SiC fiber have densities of 2.5–3.40 g/cm^3 and they have a modulus of 350–450 MPa. If such a fiber is

added into a matrix, a fiber-reinforced composite material that has ultrahigh strength and a high modulus can be synthesized. Such lightweight, high-strength, and high-modulus composites have been widely used in the aerospace and space industries.

2. *Anisotropy*: Fiber-reinforced composite materials have an obvious anisotropic elastic constant, thermal-expansion coefficient, and strength. Such characteristics can be used in the design of a composite material to make a rational structure.
3. *Good fatigue resistance*: Fiber-reinforced composites have good crack propagation properties that greatly increase their fatigue strength. Most metallic materials have fatigue strengths of about 40%–50% of their tensile strength, but composite materials may reach 70%–80% of their fatigue strength.
4. *Good antivibration properties*: Due to its high modulus, a fiber-reinforced composite has a high self-resonance frequency that reduces the occurrence of brittle fracture at regular loading speed. In addition, there are many interfaces between the fibers and the matrix in the composite that can reflect and absorb vibration waves. Hence, the composite has a strong damped vibration that can quickly weaken any vibrations.
5. *Design flexibility*: Different composite materials with different structures and properties can be designed by simply using a different matrix and a different reinforcement phase, and varying the type, amount, and arrangement of the matrix.

The reason these composite materials have high strengths and moduli is that the reinforcement phases have shared most of the load to the materials. The reinforcement phases play a very important role in the performance of the materials. For example, when glass fibers are added into a polystyrene plastic matrix, its tensile strength will increase from 600 to 1000 MPa, and the elasticity modulus will increase from 3000 to 8000 MPa. Moreover, its impact strength at −40°C will increase up to 10 times.

In summary, along with the increasing improvement of their properties and the growing reduction in their cost, the application of composite materials to a variety of industrial applications will be increasingly extended [7].

14.1.3 Reinforcement Phases (Fibers and Particles) and Metallic Matrix

The reinforcement phases can be of different geometric shapes, including granular (zero-dimensional), fiber-shaped (one-dimensional), lamellar-shaped (two-dimensional), and fiber net-shaped (three-dimensional). We will introduce the commonly used granular reinforcement and fiber phases. Interested readers can refer to the references and learn more about the other two types of reinforcements [5–7].

Fiber reinforcement phases include glass fiber, carbon fiber, and silicon carbide fiber. Glass fibers, similar to other inorganic fibers, have high strength but low modulus. The properties of glass fibers can be improved by modifying their chemical composition. So far, the most used glass fibers in composite materials are alumino-boronsilicate glass fibers, or E-glass fibers. Another type of glass fiber, the S-glass fiber (also called the R-glass fiber in France), is a special aluminum silicate–magnesium glass fiber having high tensile strength and high thermostability. However, its high cost limits its application.

Carbon fiber (graphite fiber) is usually obtained through the oxidation and thermal decomposition of organic textile fibers. Commercialized carbon fibers are always treated by chemical or electrochemical methods in order to improve their surface adhesive properties.

Granular reinforcement phases are normally ceramic or graphite, nonmetallic particles such as SiC, B_4C, TiC, Al_2O_3, mullite ($3Al_2O_3$–$2SiO_2$), Si_3N_4, TiB_2, and fine diamond particles. Because these particles have high strength, high modulus, good heat and wear resistance, and high-temperature properties, these particles are also called rigid particles or ceramic reinforcement phases. The size of these particles is normally 0.1–100 μm. The properties of several commonly used particle reinforcements are shown in Table 14.1.

TABLE 14.1
Properties of Several Commonly Used Particulate Reinforcements

Particulates	Density/ g cm^{-3}	Thermal-Expansion Coefficient/10^{-1-6} °C	Thermal Conductivity/ W (Cm °C)-one	Hardness/ MPa	Bending Strength/ MPa	Modulus of Elasticity/ GPa
Silicon carbide (SiC)	3.21	4.8	1.8	2700	400–500	427
Boron carbide $(B)_4C$	2.52	5.73	—	2700	300–500	360–460
Titanium carbide (TiC)	4.92	7.40	—	2600	500	—
Alumina (Al_2O_3)	—	9.00	—	—	—	—
Silicon nitride (Si_3N_4)	3.2	2.5–3.2	0.3–0.7	—	900	330
Mullite $(3Al_2O_3\text{-}2SiO_2)$	3.17	4.2	—	3250	1200	—
Titanium di-boride (TiB_2)	4.5	—	—	—	—	—

It has been shown that the shape of ceramic particles plays an important role in determining the properties of particle-reinforced metallic matrix composite (PMMC) materials. Figure 14.3 shows scanning electron microscopy (SEM) images of three kinds of particles: (a) angular Sic particles, (b) spherical Al_2O_3 particles synthesized through atomization, and (c) sol–gel methods, respectively [4].

For most PMMC materials, the matrix bears the main load with particle reinforcements that normally have irregular shapes dispersed homogeneously inside the matrix. Because the particles can block the movement of dislocations in the matrix, the strength and modulus of PMMC materials can be greatly improved, although the materials tend to be very brittle. The microstructures of PMMC materials are inhomogeneous even though they are thought to be homogeneously isotropic at the macroscale. Due to the existence of many defects and microcracks in the materials, the microstructures of the materials are far more continuous.

The metallic matrix composite materials are one type of widely used composites. These composites can be classified as aluminum matrix, titanium matrix, and magnesium matrix composites. The titanium matrix composites have not only high strength and modulus, but also good antioxidation and high-temperature properties. The magnesium matrix composites have good thermostability.

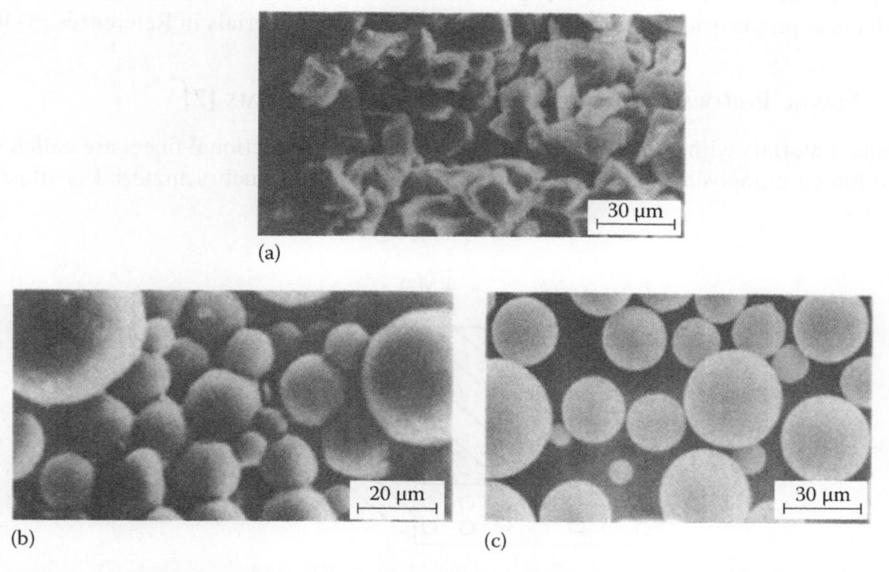

FIGURE 14.3 (a) Irregular-shaped SiC particles. (b) Sol–gel-synthesized spherical Al_2O_3 particles. (c) Atomization synthesized spherical Al_2O_3 particles.

TABLE 14.2
Compositions and Properties of the Commonly Used Al Alloys in PMMC Materials

Grades	China / US (Corresponding Trademark)	LY12 / 2024	LD2 / 6061	LD10 / 2014	LC4 / 7075	ZL101 / A356	ZL104 / A360
Composition	Cu	3.8–4.9	0.2–0.6	3.8–4.9	1.4–2.0	<0.2	<0.3
	Mg	1.2–1.8	0.45–0.9	0.4–0.8	1.8–2.8	0.2–0.4	0.17–0.3
	Mn	0.4–1.0	0.15–0.35	0.4–1.0	0.2–0.6	<0.5	0.2–0.5
	Si	<0.5	0.5–1.2	0.6–1.2	<0.5	6.0–8.0	8.0–10.0
	Zn	—	—	—	5.0–7.0	—	—
	Al	Balance	Balance	Balance	Balance	Balance	Balance
Property	Tensile strength (σ *b*/MPa)	500	320	480	600	230	230
	Elongation ratio (δ/%)	10–13	16	12	12	1	2
	Density (ρ/g cm^{-3})	2.80	2.69	2.80	2.85	–	–
	Thermal-expansion coefficient (~20–300°C)/10^{-1-6}°C	24.8	25.5	24.5	26	24.5	23.5

Source: Chen, H. et al., *Modern Composite Materials*, China Substances Press, Beijing, 1998.

Nevertheless, the aluminum matrix composites, which are cheaper than titanium matrix composites, have been the most widely used for manufacturing different products. The composition and properties of commonly used aluminum alloys in PMMC materials are listed in Table 14.2.

14.2 MECHANICAL PROPERTIES OF FIBER-REINFORCED COMPOSITE MATERIALS

The mechanical properties of uniaxial long-fiber-reinforced composite materials will be introduced, including their elastic performance, strength, and failure characteristics. Interested readers can find the mechanical properties of short-fiber-reinforced composite materials in References [6–10].

14.2.1 Elastic Performance of Uniaxial Composite Materials [7]

Composite materials with continuously parallel arranged unidirectional fibers are called uniaxial fiber-reinforced composite materials. The typical structure of such a material is illustrated in Figure 14.4.

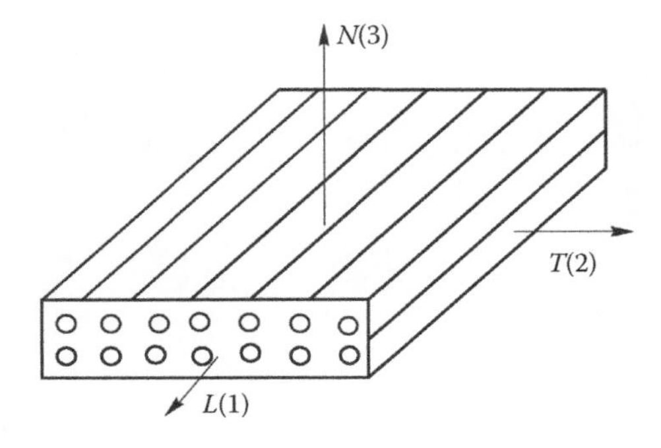

FIGURE 14.4 Layered structure of a uniaxial composite material.

To best describe composite materials, three symmetrical planes perpendicular to each other are commonly defined. Thus, three directions are defined: the direction parallel to the fiber direction is called the longitudinal direction, L(1), and the two other directions perpendicular to L(1) are called the transverse directions T(2) and N(3). In the longitudinal direction, the mechanical property is different from the other two directions T(2) and N(3), but the two transverse directions have the same mechanical properties [7].

The strength and modulus of uniaxial fiber-reinforced composite materials vary from direction to direction. There are five independent parameters for describing the strength of these composites: (1) longitudinal tensile strength, (2) transverse tensile strength, (3) longitudinal compressive strength, (4) transverse compressive strength, and (5) shear strength. Only four independent parameters are defined to describe the moduli of composite materials: the longitudinal modulus, the transverse modulus, the Poisson's ratio, and the shear modulus. The strength and modulus of a composite material are dependent on the characteristics of its constitutional components, the direction, and volume fraction of the reinforcement.

14.2.1.1 Longitudinal Elastic Modulus

It is convenient to make a simple model of a fiber-reinforced composite material, in order to calculate its longitudinal elastic modulus. The composite can be treated as two elastic bodies coexisting together, with regular shapes and distributions as shown in Figure 14.5.

The matrix will transfer the tensile force to the fibers if the fibers exist in the matrix continuously and homogeneously and are parallel to each other; if the textile fibers and the matrix are intimately connected; and if the textile fibers, the matrix, and the composite material have the same tensile strain, according to the following formula:

$$F = F_f + F_m = \sigma_f A_f + \sigma_m A_m, \tag{14.1}$$

$$A_c = A_f + A_m, \tag{14.2}$$

$$V_f = \frac{A_f}{A_c}, \quad V_m = \frac{A_m}{A_c}, \quad V_f + V_m = 1. \tag{14.3}$$

In the formula, A_c, A_f, and A_m are the cross-sectional areas of the composite, the fiber, and the matrix, respectively; V_f and V_m are the volume fractions of the fiber and the matrix; and σ_f and σ_m are the loading stresses of the fiber and the matrix, respectively. (In the following chapters, c, f, and m are denoted as the composite, the fiber, and the matrix, and L and T represent the longitudinal and transverse directions of the composite.)

The average tension stress that the composite materials receive is

$$\sigma_{cL} = \sigma_f V_f + \sigma_m V_m. \tag{14.4}$$

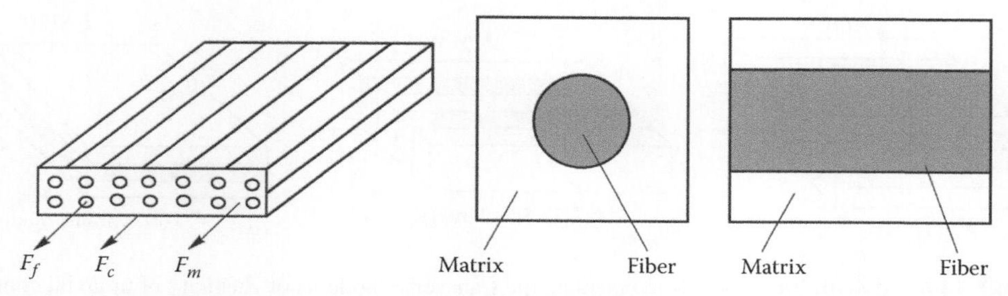

FIGURE 14.5 Schematic drawing of uniaxial composite materials.

Because the fiber and the matrix are in the elastic deformation range, according to Hooke's law,

$$\sigma_f = E_f \varepsilon_f \quad \sigma_m = E_m \varepsilon_m \quad \sigma_{cL} = E_c L \varepsilon_{cL}. \tag{14.5}$$

In the formula, ε_{cL}, ε_f, and ε_m are the longitudinal strain of the composite, the strain of the fiber, and the strain of the matrix, respectively. Because we assumed that $\varepsilon_{cL} = \varepsilon_f = \varepsilon_m$, the modulus of the elasticity of the uniaxial composite material can be calculated by the following equation:

$$E_{cL} = E_f V_f + E_m V_m = E_f V_f + E_m (1 - V_f). \tag{14.6}$$

This is also called the Mixing Rule of the composite. In practice, the actual value slightly deviates from the calculated one, due to buckling and the irregular arrangement of the fibers, and the weak binding at the interface. Thus, a correction coefficient K is added and the equation is then rewritten as

$$E_{cL} = K\left[E_f V_f + E_m (1 - V_f)\right]. \tag{14.7}$$

14.2.1.2 Transverse Modulus of Elasticity

It is more difficult to calculate the transverse modulus of elasticity, and there is also much error. The uniaxial composite materials are classified into two different model structures: Type I and Type II. In the Type I structure, there are fewer fibers and the fiber and the matrix are connected serially. The fiber and the matrix will share the same stress. This is called the serial model. In the Type II structure, many of the fibers are bundled together and dispersed in the matrix intimately, with some thin matrix included. This part of the matrix will have the same strain as the fibers. This is also called the parallel model. Figure 14.6 shows the two modeled structures.

According to the series model (Figure 14.7), under loading, the transverse elongation of a composite material is equal to the sum of the transverse elongations of the fiber and the matrix:

$$\Delta L_{cT} = \Delta L_{fT} + \Delta L_{mT}. \tag{14.8}$$

According to Hooke's Law, the lateral stress of the composite material is

$$\sigma_{cT} = E_{cT}^{I} \varepsilon_{cT} = E_{cT}^{I} \frac{\Delta L_{cT}}{L_{cT}}. \tag{14.9}$$

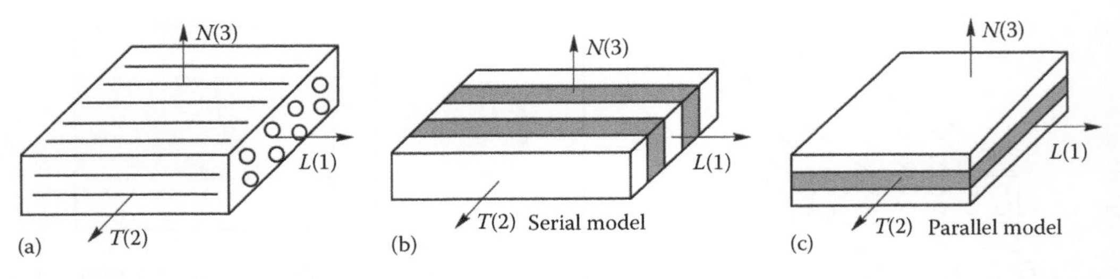

FIGURE 14.6 Two different models to calculate the transverse modulus of elasticity of uniaxial composite materials. (a) Fiber orientation model. (b) Serial model. (c) Parallel model.

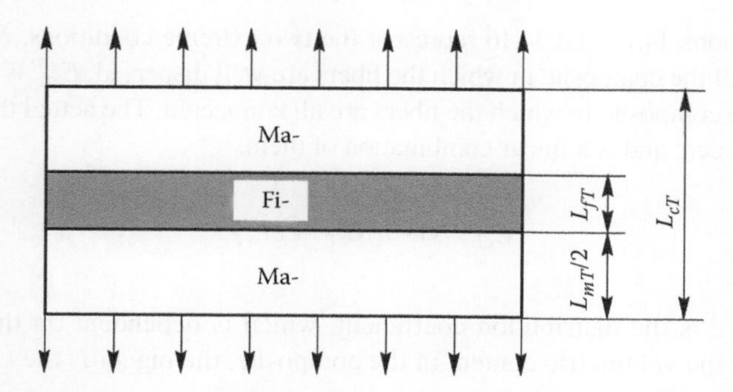

FIGURE 14.7 Serial model for the calculation of the modulus of elasticity of uniaxial composite materials.

The fiber lateral stress is

$$\sigma_{fT} = E_{fT}\varepsilon_{fT} = E_{fT}\frac{\Delta L_{fT}}{L_{fT}}. \tag{14.10}$$

The matrix lateral stress is

$$\sigma_{mT} = E_{mT}\varepsilon_{mT} = E_{mT}\frac{\Delta L_{mT}}{L_{mT}}. \tag{14.11}$$

Then,

$$\frac{\sigma_{cT}L_{cT}}{E_{cT}^{I}} = \frac{\sigma_{fT}L_{fT}}{E_{fT}} + \frac{\sigma_{mT}L_{mT}}{E_{mT}}, \tag{14.12}$$

because

$$\frac{L_{fT}}{L_{cT}} = V_f, \quad \frac{L_{mT}}{L_{cT}} = V_m. \tag{14.13}$$

Therefore,

$$\frac{\sigma_{cT}}{E_{cT}^{I}} = \frac{\sigma_{fT}V_f}{E_{fT}} + \frac{\sigma_{mT}V_m}{E_{mT}}. \tag{14.14}$$

According to the assumption $\sigma_{cT} = \sigma_{fT} = \sigma_{mT}$, therefore,

$$\frac{1}{E_{cT}^{I}} = \frac{V_f}{E_{fT}} + \frac{V_m}{E_{mT}}. \tag{14.15}$$

Because the parallel model is the same as the model for the calculation of the longitudinal module of elasticity, therefore,

$$E_{cT}^{II} = E_{fT}V_f + E_{mT}V_m. \tag{14.16}$$

Obviously, Equations 14.15 and 14.16 represent the two extreme conditions. E_{cT}^{I} is the minimum transverse modulus of the composite in which the fibers are well dispersed. E_{cT}^{II} is the maximum transverse modulus of the composite in which the fibers are all connected. The actual transverse modulus of elasticity lies in between, and is a linear combination of them:

$$E_{cT} = (1-c)E_{cT}^{I} + cE_{cT}^{II}. \tag{14.17}$$

In the equation, c is the distribution coefficient, which is dependent on the fiber's volumetric content. The higher the volumetric content in the composite, the bigger is the c value.

14.2.1.3 Shear Modulus

To calculate the shear modulus of fiber-reinforced composite materials, the composite can be sorted into two different model structures. In the model I structure, the fiber and the matrix are connected in series, along the axial direction. When a torque is applied to the modeled structure, a pure shear stress is loaded. The fiber and the matrix share the same shear stress, which results in different shear strain due to their different shear moduli. Therefore, the model I structure is called the equal-stress model. In the model II structure, the fibers and the matrix are connected in parallel, along the axial direction; that is, the fiber is surrounded by the matrix. The fiber and the matrix, under applied torque, produce the same shear strain, but the resulting shear stresses are different. Thus, model II is called the equal strain model. Figure 14.8 schematically illustrates the two different model structures.

In model I, the cylindrical composite produces a shear strain γ under an applied torque. If a straight line is drawn on the surface of the cylinder along the cylindrical direction, after the shear deformation, the line will be twisted and position a will be at a'. The circumferential displacement of position a will be the sum of the displacements of both the fiber and the matrix:

$$\gamma_c l_c = \gamma_f l_f + \gamma_m l_m. \tag{14.18}$$

In the range of elastic deformation, it obeys Hooke's law:

$$\gamma_c = \frac{\tau_c}{G_c^I}, \quad \gamma_f = \frac{\tau_f}{G_f}, \quad \gamma_m = \frac{\tau_m}{G_m}. \tag{14.19}$$

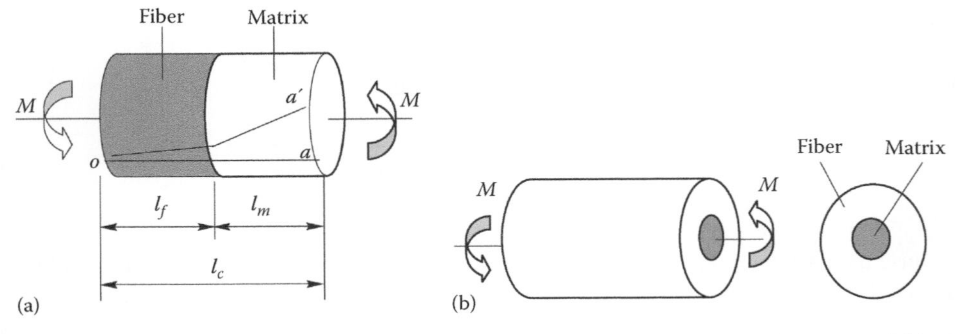

FIGURE 14.8 Model structures of a fiber-reinforced composite material. (a) The equal stress model (model I). (b) The equal strain model (model II).

Here, G_c^I, G_f, and G_m are the shear modulus of the composite, the fiber, and the matrix respectively; τ_c, τ_f, and τ_m are the shear stress of the composite material, the fiber, and the matrix, respectively; and γ_c, γ_f, and γ_c are the shear strain of the composite material, the fiber, and the matrix, respectively.

Because $l_f/l_c = V_f, l_m/l_c = V_m$ and $\tau_c = \tau_f = \tau_m$, therefore

$$\frac{l}{G_c^I} = \frac{V_f}{G_f} + \frac{V_m}{G_m}. \tag{14.20}$$

For model II, under an allied torque, forces loaded on the fiber and the matrix will be different. In the cross-sectional direction, the total torque will be

$$M_c = M_f + M_m. \tag{14.21}$$

Suppose τ_c is the average shear stress in the cross-sectional direction, and A_c and R_c are the cross-sectional area and the radius, respectively. Then

$$M_c = \tau_c A_c R_c. \tag{14.22}$$

Similarly, the torque of the fiber is

$$M_f = \tau_f A_f R_f, \tag{14.23}$$

and the torque of the matrix is

$$M_m = \tau_m A_m R_m, \tag{14.24}$$

where A_f, A_m, R_f, and R_m are the cross-sectional area and the radius of the fiber and the matrix, respectively. Because the matrix is very thin in model II, we suppose that $R_c = R_f = R_m$. Therefore,

$$\tau_f A_f + \tau_m A_m = \tau_c A_c. \tag{14.25}$$

Suppose that $\tau_c = \gamma_c G_c^{II}, \tau_f = \gamma_f G_f, \tau_m = \gamma_m G_m$ and that $\gamma_c = \gamma_f = \gamma_m$. According to Hooke's law,

$$G_c^{II} = G_f V_f + G_m V_m. \tag{14.26}$$

Similarly, Equations 14.20 and 14.26 represent two extreme conditions. G_c^I is the upper limit value of the shear modulus of the composite, whereas G_c^{II} is the lower limit of the value. In practice, the commonly used form is their linear combination:

$$G_c = (1-c)G_c^I + cG_c^{II}. \tag{14.27}$$

14.2.1.4 Poisson's Ratio

A uniaxial composite material is orthotropic. Thus, Poisson effects on the composite are different along its longitudinal and transverse directions, and there are two Poisson's ratios for the material. When a tensile force is applied to the fiber direction of the composite, it will shrink along its transverse direction. The ratio between its longitudinal strain and its transverse (lateral) strain is called the longitudinal Poisson's ratio (v_{LT}):

$$v_{LT} = \frac{-\varepsilon_{cT}}{\varepsilon_{cL}}. \tag{14.28}$$

In the equation, ε_{cT} and ε_{cL} are the composite material's transverse strain and longitudinal strain, respectively.

The longitudinal Poisson's ratio v_{LT} can be obtained in a simple way, based on a similar model in which the longitudinal tensile is considered as shown in Figure 14.9. For the longitudinal deformation, it is assumed that

$$\varepsilon_{cL} = \varepsilon_{fL} = \varepsilon_{mL}. \tag{14.29}$$

The transverse strain of the composite is a sum of the transverse strains of the fiber and the matrix:

$$\varepsilon_{cT} L_{cT} = \varepsilon_{fT} L_{fT} + \varepsilon_{mT} L_{mT}. \tag{14.30}$$

Consider Equation 14.13:

$$\varepsilon_{cT} = \varepsilon_{fT} V_f + \varepsilon_{mT} V_m. \tag{14.31}$$

Consider Equations 14.29, 14.31, and 14.28. Then

$$v_{LT} = -\frac{\varepsilon_{mT}}{\varepsilon_{mL}} V_m - \frac{\varepsilon_{fT}}{\varepsilon_{fL}} V_f. \tag{14.32}$$

According to the definitions of the fiber Poisson's ratio and the matrix Poisson's ratio, we may write:

$$v_{LT} = v_f V_f + v_m V_m. \tag{14.33}$$

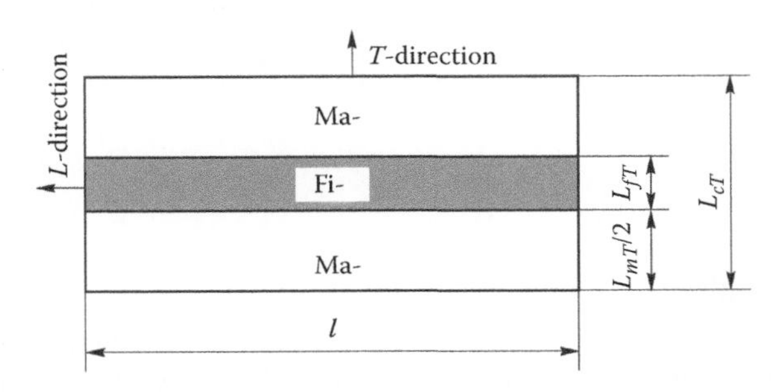

FIGURE 14.9 Model for computation of Poisson's ratio of a uniaxial composite material.

When an elastic tensile force is applied to the direction perpendicular to the fiber direction of the composite, the ratio of its longitudinal strain and lateral strain is called the transverse Poisson's ratio:

$$V_{TL} = -\frac{\varepsilon_{cL}}{\varepsilon_{cT}}. \tag{14.34}$$

Because a uniaxial composite material is an orthotropic elastic body, the relationship between the Poisson's ratios and its modulus of elasticity obeys Maxwell's law:

$$V_{TL} = \nu_{LT}\frac{E_{cT}}{E_{cL}}. \tag{14.35}$$

14.2.2 Tensile Strength of Uniaxial Composite Materials [7]

The straining process of a uniaxial composite material under tensile loading can be sorted into four stages: in Stage (I), both the fibers and the matrix experience elastic strain; in Stage (II), the fibers are still in elastic strain but the matrix is at inelastic strain; in Stage (III), both the fibers and the matrix are at inelastic strain; in Stage (IV), the fibers break first, along with the composite material fracture, as shown in Figure 14.10.

Normally, a composite material with glass fibers, carbon fibers, boron fibers, or ceramic fibers as reinforcements, and a thermosetting resin as the matrix, only experiences Stage I and Stage IV. A metal-based or a thermoplastic resin-based composite material experiences Stages I, II, and IV. For those composite materials with brittle fibers as reinforcements, there is no Stage III. However, composite materials with toughness fibers can experience Stage III.

In Stage I, both the fibers and the matrix are at the elastic strain states, as well as the composite material with $\varepsilon_c = \varepsilon_f = \varepsilon_m$. Considering Equations 14.4 and 14.5, we have

$$\sigma_{cL} = E_f\varepsilon_f V_f + E_m\varepsilon_m(1-V_f). \tag{14.36}$$

The ratio of the load applied to the fibers and that applied to the matrix is

$$\frac{F_f}{F_m} = \frac{E_f\varepsilon_f V_f}{E_m\varepsilon_m(1-V_f)} = \frac{E_f V_f}{E_m(1-V_f)}. \tag{14.37}$$

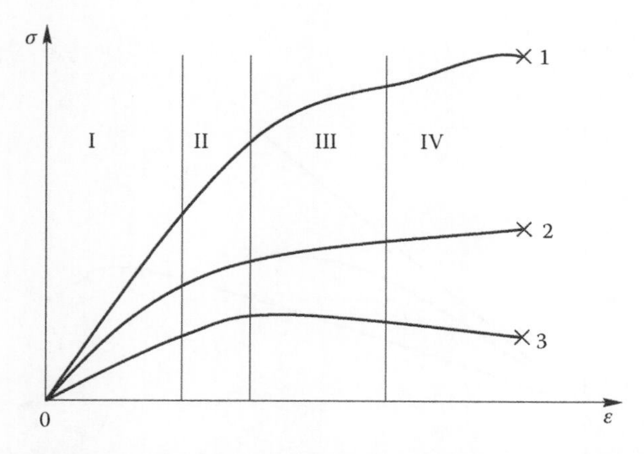

FIGURE 14.10 Four stages of materials under a tensile loading: 1, fiber; 2, composite material; 3, matrix.

For a fiber-reinforced composite material with a constant volume fraction (V_f) of the fibers, the larger the E_f/E_m, the more load the fibers will bear. This means that the fibers have strong enhancement. This is usually why fibers with high strength and modulus are commonly used in fiber-reinforced composite materials. In a fiber-reinforced composite material with a constant E_f/E_m, the bigger the V_f, the greater is the contribution from the fibers to the strength of the material. However, this does not mean that the more fibers are added, the stronger the materials will be. In fact, when the fiber volume fraction $V_f > 80\%$, the strength of the composite decreases. This is because there are not enough matrix materials to wrap the fibers, which causes gaps between the fibers and the matrix. In general, most fiber-reinforced composite materials have a V_f within a range of 30–60%.

During Stage II of a material straining, the fiber is still at the elastic stage, but the matrix has undergone plastic deformation. At this stage, the stress of the composite material can be described as

$$\sigma_{cL}(\varepsilon) = \sigma_f(\varepsilon)V_f + \sigma_m(\varepsilon)V_m. \tag{14.38}$$

At this stage, the load on the composite is mainly borne by the fiber, and with the increase in deformation of the matrix, the loading on the fiber is becoming greater. Once the loading is greater than the tensile strength of the fiber, the fiber will break. Because the matrix alone cannot bear the entire loading, the composite will then fracture. The tensile strength of the composite material will be

$$\sigma_{cLu} = \sigma_{fu}V_f + \sigma_m^*(1 - V_f). \tag{14.39}$$

Here, σ_m^* is the stress of the matrix when its strain is equal to that of the fiber when it is broken. Two conditions must be met when this equation is used: (1) the fiber should be in the elastic state when a load is applied; and (2) the total strain of the matrix should be larger than that of the fibers when they are broken. Figure 14.11 shows the strain–stress curves of the fiber, matrix, and the composite under a tensile loading, in which ε_{my} is the yield strain of the matrix, and ε_{fu} and ε_{mu} are the fracture strains of the fiber and the matrix, respectively.

The enhancement of the fiber can only be applied when the tensile strength of the composite material is larger than that of the matrix, which means that Equation 14.40 must be satisfied:

$$\sigma_{cLu} = \sigma_{fu}V_f + \sigma_m^*(1 - V_f) \geq \sigma_{mu}. \tag{14.40}$$

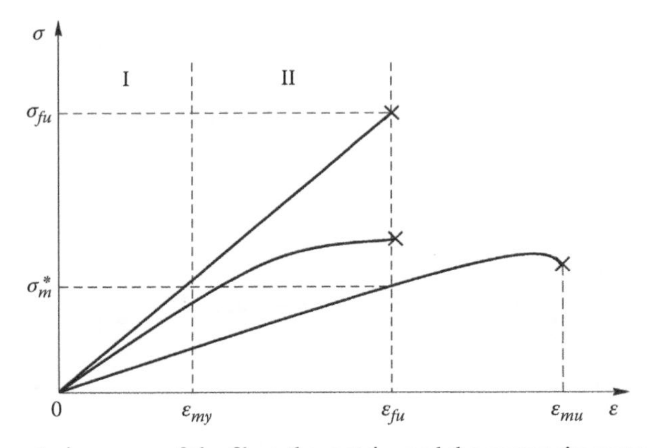

FIGURE 14.11 Stress–strain curves of the fiber, the matrix, and the composite material.

In this equation, σ_{mu} is the tensile strength of the matrix. Equation 14.40 has also defined the critical fiber volume fraction V_{fcr} in a composite material. In order to achieve the fiber reinforcement effect, the actual volume fraction of the fiber should be greater than V_{fcr}:

$$V_{fcr} = \frac{\sigma_{mu} - \sigma_m^*}{\sigma_{fu} - \sigma_m^*}. \quad (14.41)$$

Obviously, when the strength of the fiber is much larger than that of the matrix, V_{fcr} is smaller. When the matrix and the fiber have similar strength, V_{fcr} is bigger. Hence, to reach the same enhancement effect, the need is less for stronger fibers, while the need is greater for weak fibers. For a fiber-reinforced resin composite material, the volume fraction of the fiber V_{fcr} is smaller because the fibers are always stronger than the resin matrix. For example, for carbon fiber-reinforced epoxy resin-based composite materials, the fracture strains of the carbon fiber (ε_{fu}) and the matrix (ε_{mu}) are 0.5% and 2%, respectively, and their tensile strengths are $\sigma_{fu} = 2100\,\text{MPa}, \sigma_{mu} = 80\,\text{MPa}, \sigma_m^* = 26.5\,\text{MPa}$, and so $V_{fcr} = 2.6\%$. In fact, it is not applicable for such a composite material because, in practice, the volume fraction of the fiber is far bigger than this number in composite materials. Nevertheless, it is applicable for fiber-reinforced metal-based composite materials. Usually, the strength of a fiber is about 2000 MPa. For Ni-based and stainless-steel-based composite materials, V_{fcr} will be 13% and 15%, respectively.

14.2.3 Failure Characteristics of Fiber-Reinforced Composite Materials [7]

Composite materials are different from traditional metallic materials, due to the special microstructure of their fracture modes, and many other factors. However, the fracture process of a composite material is also composed of the formation of a crack and its propagation. The crack can be formed either in the material synthesis or in its use. The sources of the crack are from microscopic defects like voids, fiber ends, and lamination. When a load is applied to the composite material, the crack will propagate forward once the stress at the crack tip has reached its critical value. During propagation, the crack tip may merge with other damage regions like the fiber fracture, the deformation and fracture region in the matrix, and the separation zone between the matrix and the fibers. This results in the extension of the damage region, until macroscopic fracture occurs as shown in Figure 14.12. Thus, the fracture process results from the accumulation of damages of several types.

Figure 14.13 illustrates the changes in the number of fractured fibers in an inorganic fiber-reinforced composite material under loading. It clearly indicates that with an increase in loading, the number of fractured fibers increases rapidly, resulting in the quick fracture of the material. Such a fracture mechanism usually applies to composite materials with low interfacial bond strengths [7–10].

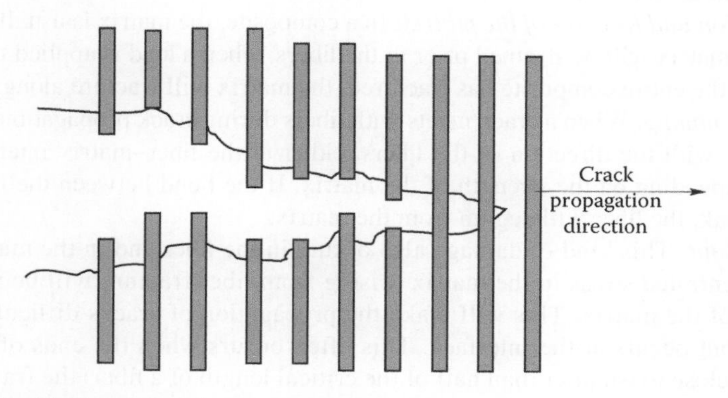

FIGURE 14.12 Model of a crack tip in a composite material.

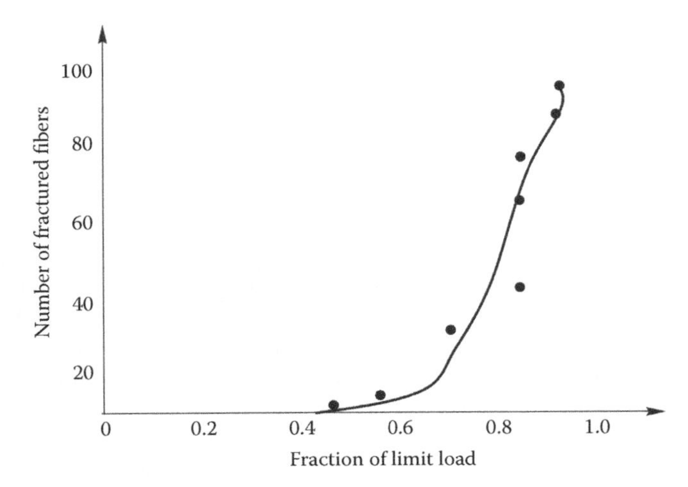

FIGURE 14.13 Relationship between the number of breakage fibers and the limit load.

It should be mentioned that there is also another fracture mechanism of fiber-reinforced composite materials: the nonaccumulation mechanism. This is a brittle fracture resulting from the breaking of one fiber, which causes the entire composite to fracture. Most metal-based composite materials that have high interfacial bond strengths obey such a mechanism. When fracture occurs there is no pull-off of fibers from the matrix.

There are three different types of such fractures: (a) cascading failure occurs when one fiber is broken, resulting in the breaking of a nearby fiber (due to the accumulation of stress on it) and so on, until the entire composite fractures; (b) brittle adhesive fracture refers to the fracture of the entire composite due to the fracture of the matrix, resulting from fractured fibers which adhere to it; and (c) the weakest component mechanism, which is the fracture that results from broken fibers that are strongly bonded to the matrix.

In practice, the fracture of a composite material is more complex. It is usually a process that combines both the accumulation and nonaccumulation types of fracture. All of the mentioned types of fractures can occur under static tensile, shock loading, or alternative loading. It is important to understand the basic concepts of fracture in order to understand the mechanical behavior of composite materials under different loading conditions.

1. *Fiber fracture*: Fibers with their directions perpendicular to the direction of crack tip propagation will fracture once the strain has reached its fracture strain. At the early stage of composite loading, some fibers will fracture. With the increase in loading, the number of fractured fibers increases.
2. *Deformation and fracture of the matrix*: In a composite, the matrix is usually a soft phase. Thus, the matrix will be strained prior to the fibers, when a load is applied to the composite. When the entire composite has fractured, the matrix will fracture along with it.
3. *Fiber degumming*: When a crack meets with fibers during crack propagation, it will propagate along with the direction of the fibers, either at the fiber–matrix interface or in the matrix, depending on the strength of the matrix. If the bond between the matrix and the fiber is weak, the fiber will degum from the matrix.
4. *Fiber pullout*: This kind of damage also occurs in the fiber and in the matrix interface. The concentrated stress in the matrix arising from fiber fracture will be relaxed due to the yield of the matrix. This will make the propagation of cracks difficult. As a result, fiber pullout occurs at the interface. This often occurs when the ends of the fractured fibers are close to (smaller than half of the critical length of a fiber) the fractured surface of the composite.

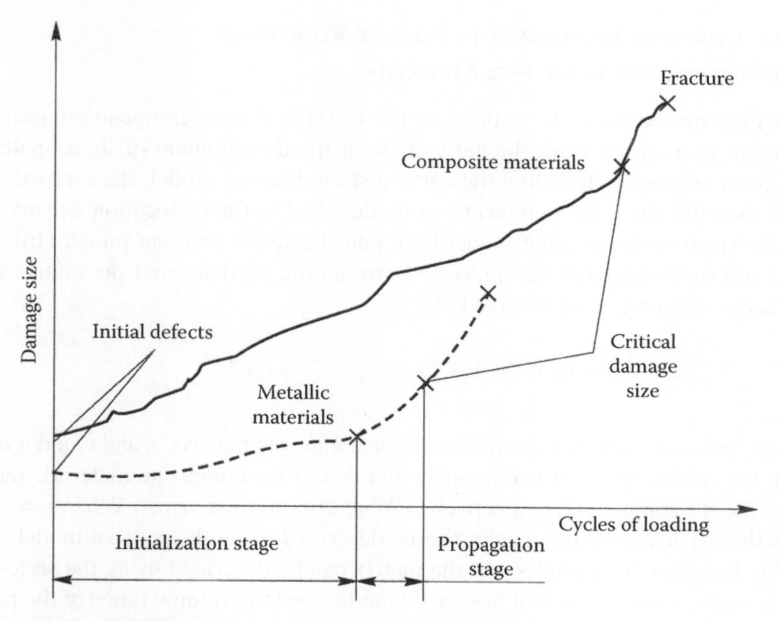

FIGURE 14.14 Fatigue characteristics of composite and metallic materials.

5. *Lamination crack*: This fracture normally occurs in laminated composite materials. When a crack propagates through one layer of the composite, the crack tip will be blocked by nearby layers. Because of large shear stress in the neighboring matrix, the crack will begin to propagate along the interface, parallel to the laminated layer structure. This results in the formation of the lamination crack.

Fiber-reinforced composite materials are widely used to make parts in the aviation industry, in astronautics, in automobiles, and in the power industry, by alternating load function components. Therefore, more and more research has been undertaken on their fatigue behavior. In general, the antifatigue properties of composite materials are better than for metallic materials. Figure 14.14 outlines the fatigue characteristic of the two different types of materials.

From Figure 14.14 we can see that the damage scale of a composite at the initial stage of fracture is larger than that of the metallic materials. However, composite materials have better fracture toughness and lower notch sensitivity. Thus, they have longer fatigue life than metallic materials. Composite materials also have a large critical damage size. In addition, unlike the unpredictable fatigue of metallic materials, the fatigue of composite materials is cumulative and thus predictable.

14.3 MECHANICAL PROPERTIES OF PARTICLE-REINFORCED COMPOSITE MATERIALS

The most commonly used particle-reinforced composite materials are the PMMCs. Extensive research has been performed on the properties of PMMC materials. In fact, theoretical work on the mechanical properties of these composites was conducted before their appearance. For example, the mixed-mechanics model proposed by Eshelby in 1957 has been one of the most widely used models on PMMC mechanical properties [11]. The following section will introduce the enhancement mechanism, tensile strength, fatigue failure, laser thermal shock, and thermal fatigue failure properties of PMMC materials.

14.3.1 Strengthening Mechanisms in Particle-Reinforced Metallic Matrix Composite Materials

The strengthening mechanisms of particles in the metallic matrix composite materials have been studied for many years, even from the early stage of the development of these materials. Several models have been proposed, including the earliest strengthening model, the mix rule of composite materials, [5] then the shear lag enhancement model [12,13], the dislocation density enhancement theory [14], the strain-strengthening model [15], and the strain gradient model [16]. According to the mix rule, and supposing that the interface between the particles and the matrix is perfect, the following general relation can be obtained [5]:

$$y_c{}^n = v_1 x_1{}^n + v_2 x_2{}^n + \cdots + v_i x_i{}^n. \tag{14.42}$$

Here, y_c represents the material parameters of the composite materials; and v_i and v_i are the volume ratio between the volume of the ith component and that of the composite material, and the material parameters of the ith component, respectively. (What does n mean here?) When $n = 1$, the relationship between the particles and the matrix can be described using the parallel model. When $n = -1$, the relationship between the particles and the matrix can be described using the series model. If we use E_p and E_m, and v_p and v_m represent the elastic moduli and the volume ratios of the particle and the matrix, respectively, then the elastic modulus of the composite material is

$$E_c = v_p E_p + v_m E_m. \tag{14.43}$$

Because the elastic modulus of the particle is much larger than that of the matrix, the rigidity of the composite material is enhanced.

According to the shear lag enhancement model, it is thought that the strengthening mechanism in a PMMC material occurs mainly through shear action across the interface between the particle and the matrix, which induces the load to be applied to the particle, transferred from the matrix. As a result, the composite material is enhanced [12,13]. For composite materials with a good interface connection, the interface shear stress is proportional to the strength of the matrix, and obeys the following relationship:

$$\sigma_c = \sigma_p(1-a)\rho + \sigma_m(1-\alpha). \tag{14.44}$$

Here, σ_c, σ_p, and σ_m are the strengths of the composite, the particle, and the matrix, respectively. α is a parameter related to the size of the particle.

Nardone et al. [14] proposed a revised shear lag model and thought that the enhancement is due more to loading being applied to the particle than to the matrix:

$$\sigma_c = \sigma_m(1 + f_1). \tag{14.45}$$

Here, f_l is a strengthening factor which is determined by the volume fraction, size, and shape of the particle.

Regarding the dislocation density enhancement model proposed by Arsenault et al. [17] strength enhancement is due to the increase in dislocation density in the matrix, upon loading being applied to the composite. Ramakrishnan et al. [18] proposed the composite sphere model. According to this model, the particle only experiences elastic deformation when a load is applied to the composite, because of its large elastic modulus. It is therefore called the elastic region; whereas the matrix near the particle experiences plastic deformation, and is thus called the plastic region. The matrix far away from the particle also experiences elastic deformation, and it is therefore called the elastic region of the matrix. The following equation can be used to describe the enhancement:

$$\sigma_0^c = \sigma_0^m(1+f_d)(1+f_L). \tag{14.46}$$

In the equation, f_d is the dislocation enhancement factor and f_L is the particle-bearing enhancement factor.

Cheng and Wang [16] systematically studied the strain gradient strengthening in PMMC, and concluded that strengthening is due to the existence of a large strain gradient in the composite. In addition, they numerically simulated the effect of the volume fraction, size, aspect ratio, elastic modulus, and matrix materials on composite materials with ellipsoidal particles. They found that in order for composite materials to exhibit a better strengthening effect, the following factors were required: (a) a larger volume fraction of the particle, (b) a larger difference in elastic modulus of the matrix and the particle and the matrix hardening index, and (c) a larger aspect ratio. In addition, the characteristic size of the composite material and the size of the particle had tremendous effects on strengthening.

The strength of a PMMC material is first dependent on the matrix strength. The higher the matrix strength, the higher is the strength of the composite. Second, the strength of the composite also depends on the volume fraction of the particles. Figure 14.15 displays the relationships between the tensile strength of two SiC-reinforced aluminum matrix composite materials, and the volume fraction of SiC [19]. Figure 14.15 clearly indicates that the tensile strength of the composite increases with the increase of the volume fraction of the SiC particles. Moreover, it also shows that composite materials with different matrices have different fracture strengths.

The strength of composite materials can also vary from different heat treatments. Hwu et al. [20] investigated the dependence of the strength of an Al/SiC-6061 alloy on the parameters of different processes. They found that the material has its highest strength when heat-treated at the eutectic temperature, and when its bending strength is increased using a water-quenching treatment, due to the reduction in the average particle size.

In summary, the strengthening mechanisms in PMMC materials can be attributed to the following three factors: (1) the composite material is enhanced when more loading is applied to the particle than is applied to the matrix; (2) the introduction of particles increases the dislocation

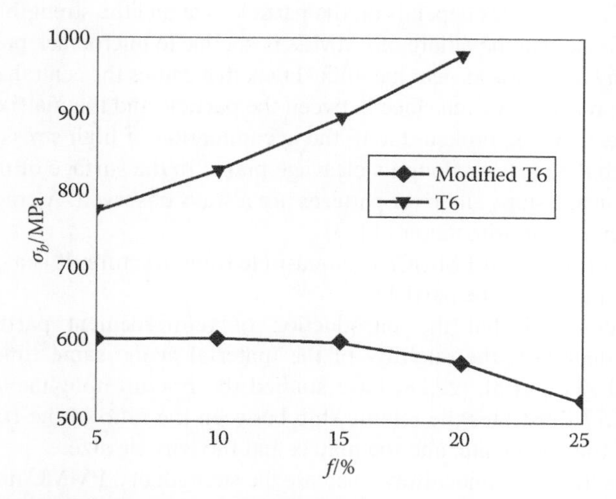

FIGURE 14.15. Relationships between the tensile strengths of two SiC-reinforced Al matrix composites and the volume fractions of the SiC particles.

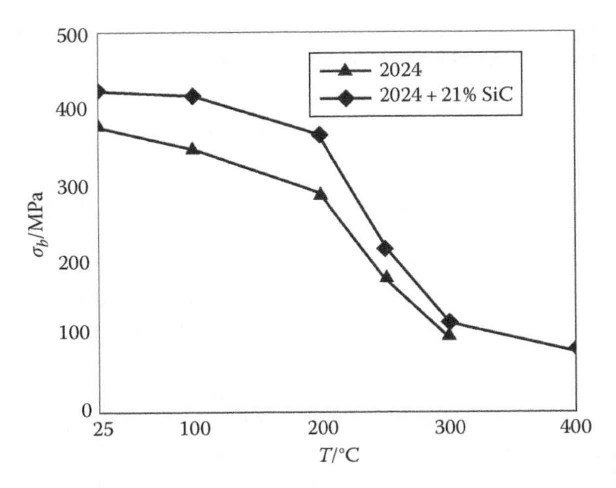

FIGURE 14.16 Relationship between the tensile strength and temperature of the matrix and the composite.

density and thus enhances the composite; (3) enhancements can also be introduced by treatment of the composite, to increase the strain gradient effect in the material.

It should also be mentioned that the strength of PMMC materials varies also with temperature. Figure 14.16 displays the relationship between the strength of a SiC-reinforced Al-based composite and the temperature. From Figure 14.16 we can see that the strength of the composite decreases with the increase in temperature.

14.3.2 Tensile and Fatigue Failure of PMMC Materials

The failure mechanism of PMMCs is very complicated. PMMC materials under different loading and with different microstructures experience failure differently. Extensive studies have been done on materials under different failure conditions. In the study of tensile testing of PMMCs at room temperature, Quan Gaofeng et al. found that large particles in the composite are easily destroyed under loading, which induces the generation of microcracks [21]. Brechet et al. [22] found the generation of microcracks dependent on the distribution of particles in the matrix. For a PMMC with nonuniform distributed particles, much damage occurs at the particle-clustering area. Wang et al. found that the degree of damage of the particles depends on the particle size and the strength of the matrix [23].

Moreover, it was found that the failure of PMMCs is not due to microcrack propagation. Instead it is the generation of many microcracks near the initial crack that causes the centralization of damage [24].

When a crack has reached the interface between the particle and the matrix in the stage of crack propagation, a particle will be broken due to the accumulation of high stress at the crack tip. The crack will go through the particle along its cleavage plane. In the surface of the fracture cross section of a PMMC, cleavage steps and river patterns are always observed. A crack can also propagate along the interface in a composite material [13].

So, in general, the fracture of PMMCs is a quasi-cleavage fracture. It is a ductile fracture to the matrix, but a brittle fracture to the particle.

It should be mentioned that the introduction of reinforcement particles that strengthen the composite also decreases the ductility of the material at the same time. Leggoe et al. [25] Zhao et al. [26] and Zhou et al. [27,28] have studied the fracture toughness property of PMMC materials. Figure 14.17 illustrates the relationship between the ratio of the fracture toughness and the yield strength of the composite, and the matrix and the particle size.

Figure 14.17 tell us that we cannot always increase the strength of a PMMC material by reducing the particle size. There is an optimized particle size for the enhancement. Zhao et al. [26] reported that the reduction in the composite's fracture toughness is mainly due to the crystal mismatch between the particle

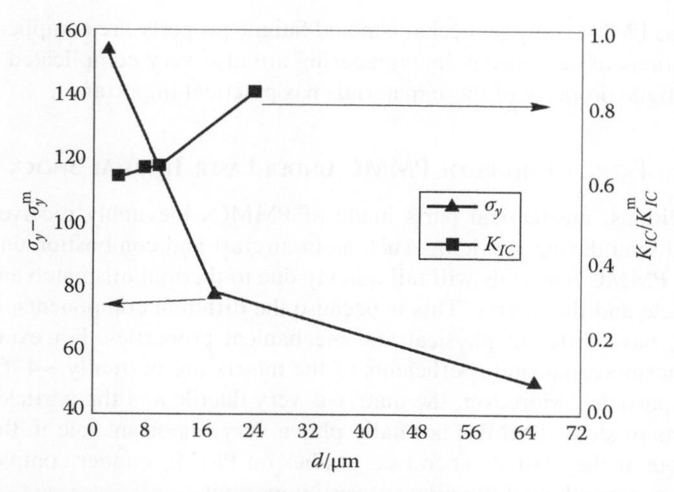

FIGURE 14.17 Changes in the ratio between the PMMC fracture toughness and the matrix fracture toughness and their ratio of the yield with the particle size.

and the matrix, which causes the creation of high dislocation density. Fracture toughness is dependent on the particle size and the Poisson's ratio, and is not sensitive to the volume fraction of the particle.

The fatigue property has been extensively studied, and is one of the most important properties of PMMC materials, because it defines the safety issue in the application of these materials. There are two aspects that need to be considered to understand the fatigue mechanism: (1) the microstructure effect: the effect of the microstructure on the fatigue; and (2) the force effect: the effect of the stress and strain of the particle, the matrix, and the interface on fatigue. The fatigue process consists of three stages: (1) the initial damage, (2) the generation of microcracks, and (3) the propagation of the cracks.

The effect of a microscopic structure on the fatigue failure of PMMC materials has been extensively studied. Various researches have been conducted on the effects of the types and properties of the matrix [29], the interface conditions, [30,31] as well as the property, size, volume fraction, and distribution of particles on the fatigue property [32,33]. Moreover, much theoretical and experimental research has been performed that has investigated the fatigue lifetime of these materials [34,35].

Figure 14.18 shows the relationship between the fatigue life of the matrix and the composite, and the strain amplitude of 8090Al + 15%SiC composite materials [34].

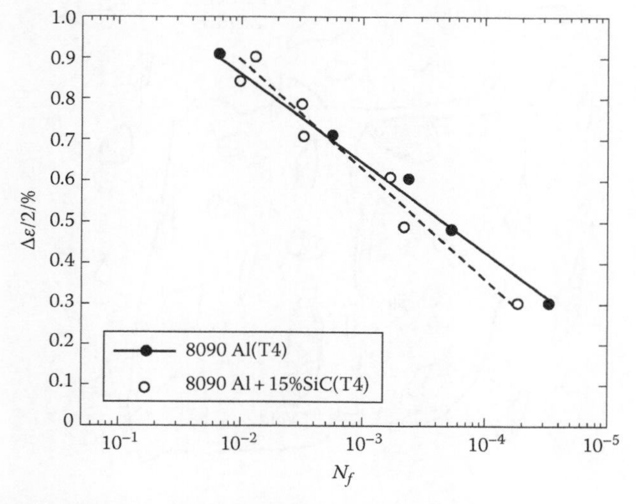

FIGURE 14.18 Relationship between fatigue life and strain amplitude.

In summary, the PMMC fatigue mechanism and fatigue property are complicated. Furthermore, the loading conditions of composites in engineering are also very complicated. Hence, an understanding of the fatigue property of these materials has practical importance.

14.3.3 Thermal Fatigue Failure of PMMC under Laser Thermal Shock

Under most conditions, mechanical parts made of PMMCs inevitably receive both mechanical force and thermal load during servicing, such as in aircraft and combustion engines. Under such tough conditions, PMMC materials will fail quickly due to thermal mismatch and strain mismatch between the particle and the matrix. This is because the different components, such as the matrix and the particles, have different physical and mechanical properties. For example, the thermo-conductive and thermo-expansion coefficients of the matrix are normally ~4–6 times larger than those of ceramic particles. Moreover, the matrix is very ductile and the particles are very fragile. Furthermore, parts made of PMMCs normally play a very important role in the structure, and it would be disastrous if they failed. Therefore, studies on PMMCs under complex heat-mechanics response and under strength load function are very important.

Traditionally, a thermal fatigue experiment is performed in an oven via a quenching method. Recently, laser technology—an advanced technique for the study of the thermal properties of materials—has been introduced in thermal shock and fatigue experiments. Using the new technology has many advantages over the traditional one. Because the energy, the working time, the frequency, the heating position, and the size of the laser can be easily adjusted, the heating speed, the time, and the frequency can be adjusted, and the heating temperature and strain can be measured. In addition, the use of lasers makes it convenient for mechanical loading. Hence, lasers are suitable when studying the thermal force effect on materials under complicated loading conditions.

Zhou Yichun and Long Tuguo [36–40] systematically studied the thermal fatigue failure effect in PMMC materials under laser thermal shock load. They found that there are three fatigue modes: (a) the generation of voids and microcracks in the matrix, (b) interfacial debonding, and (c) particle damage. Figure 14.19 illustrates the three modes.

The fatigue failure processes are composed of three major stages. Stage I is the generation of microcracks and voids. Stage II is the formation of main cracks through the coalescence of microcracks and voids. Stage III is the fast propagation of main cracks, resulting in material failure. Figure 14.20 shows the microstructure of the crack tip, from which we can see that there is no damage to the

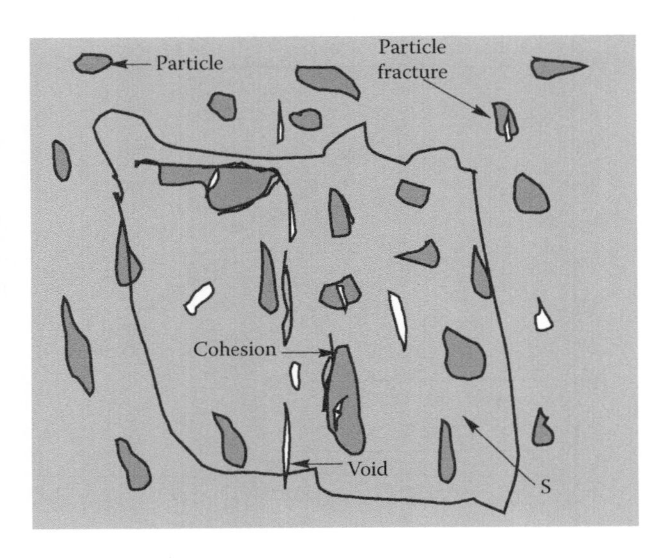

FIGURE 14.19 PMMC fatigue modes.

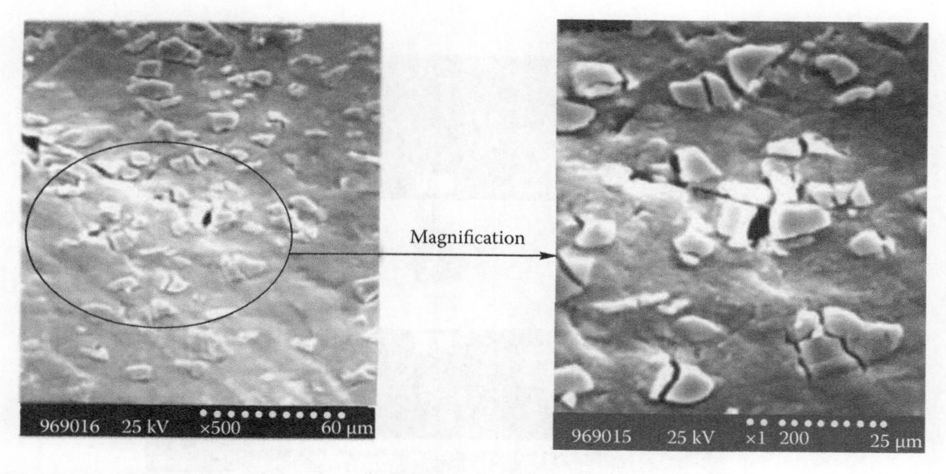

FIGURE 14.20 Microstructure of crack tips.

nearby matrix during propagation of the crack. Instead, the particle at the tip is fractured and the crack propagates through the fractured particle. Interested readers can refer to the work by Zhou Yichun and Long Tuguo [36–40].

It was found that nonobvious destruction phenomena occurred when a mechanical load of σ_{max} = 100 MPa was applied to the material. When a laser with an energy density of $E_J \leq 8$ J/cm^2 is introduced, there is still no obvious damage occurrence. With the gradual increase of laser energy density, the destruction grows more and more serious until the part fails at an energy density of E_J = 45 J/cm^2. Similar experiments were also performed with mechanical loadings at σ_{max} = 0, 50, 150, 200, 240, and 400 MPa, respectively. The occurrence of damage was defined by the observation of microcracks and voids, or the fracture of particles in the SEM images taken at a magnification of 500. E_J and σ_{max} under such conditions are defined as damage thresholds. When the main cracks have gone through the entire part, the material is considered to have totally failed. E_J and σ_{max} under these conditions are defined as fracture thresholds. The relationship between the energy densities E_J and the mechanical loadings σ_{max} is

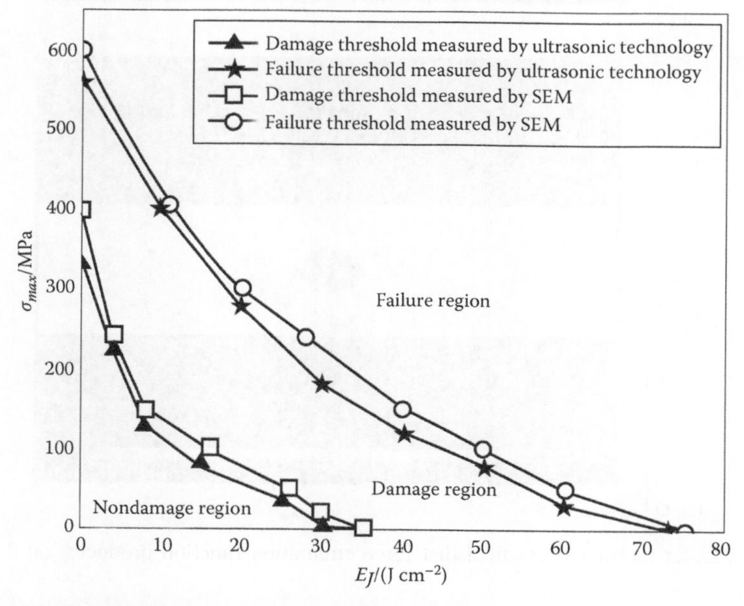

FIGURE 14.21 Damage threshold and fracture threshold values.

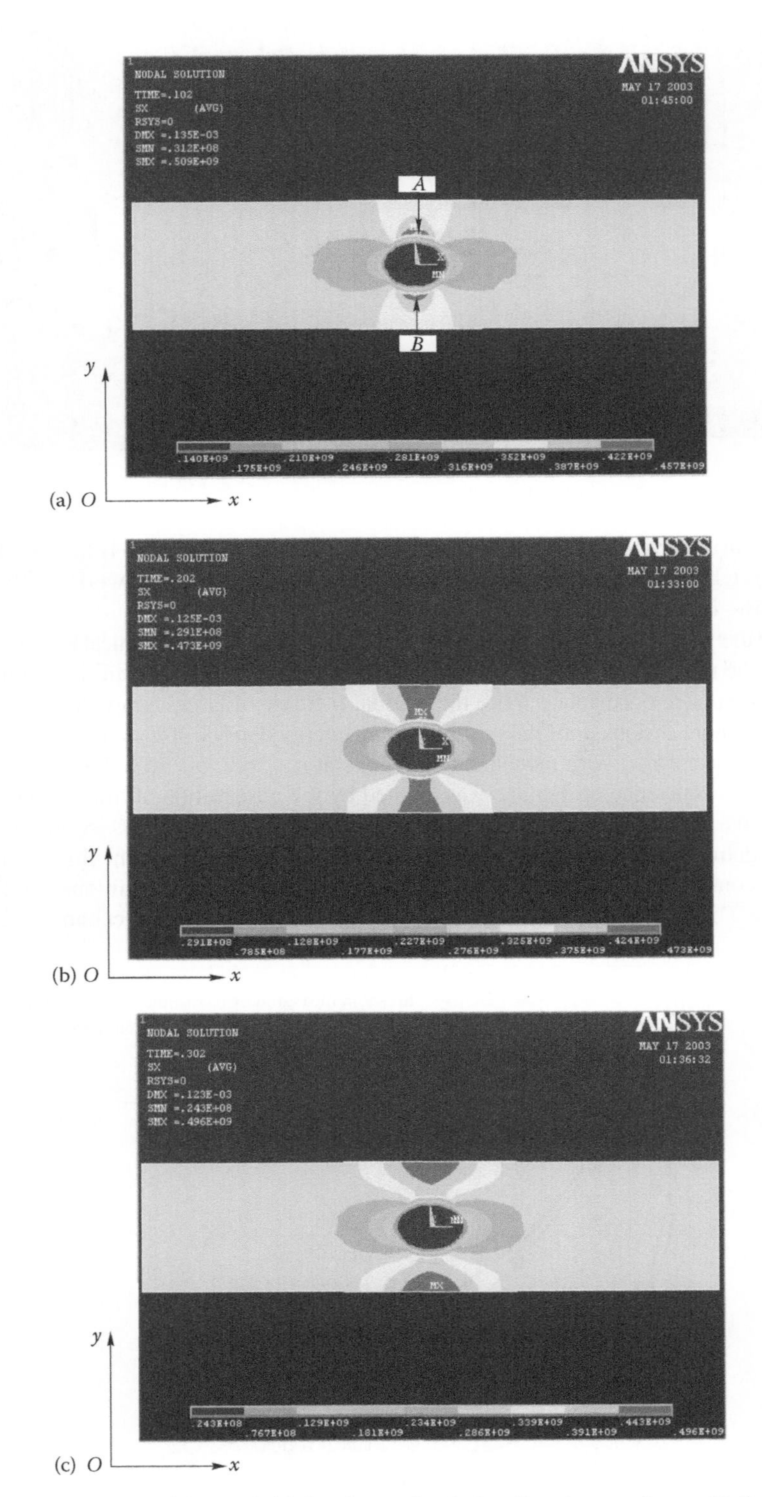

FIGURE 14.22 Laser thermal stress field that force circulation function produces. (a) 2 ms. (b) 102 ms. (c) 202 ms.

summarized in Figure 14.21. The E_J and σ_{max} fracture thresholds of the PMMC, under the laser thermal and mechanical loadings, were obtained through the ultrasonic wave test method. Figure 14.21 shows the damage threshold and fracture threshold values of the materials under different tests. The E_J–σ_{max} plane can be divided into three zones: the undamaged zone, the damage zone, and the fracture zone.

Theoretical work using the finite element method has also been performed on the dynamic behavior of PMMC materials under laser thermal and mechanical force loadings. The results are shown in Figure 14.22. Figure 14.22 indicates that under the action of thermal stress and tensile stress induced by laser and mechanical loading, the maximum positive stress σ_x along the x-direction appears in the regions A and B near the laser-irradiated area, as shown in Figure 14.22a. Therefore, the crack will be generated in region A and region B.

Due to the existence of tensile stress, the stress in the laser-irradiated area is also a tensile stress, but it is much smaller than in the nearby areas. Nevertheless, cracks generated in regions A and B will still propagate to the laser-irradiated area, because the temperature in this area is much higher than in the nearby areas. At high temperatures, the strength of the PMMC materials will significantly decrease. On the other hand, even if there are no large temperature changes in the areas outside regions A and B, and thus no large changes in the material's properties, a large number of cracks will be generated in the outside areas, due to the existence of high tensile stress. See Figure 14.22b and Figure 14.22c, which are consistent with experimental results.

14.4 APPLICATIONS AND PROSPECTS OF THE DEVELOPMENT OF COMPOSITE MATERIALS

14.4.1 Applications of Composite Materials

Along with the development of science and technology, composite materials have been widely used in aerospace, the automobile industry, civil engineering, sporting goods, and in other areas as well [3].

1. Applications in the aeronautics and space industry
   In both fixed-wing and unfixed-wing aircraft, many parts—including load/unload-bearing parts—are made of composite materials. For example, in many commercial and military aircraft, light carbon and glass fiber-reinforced composite materials have been used for making honeycomb-like laminated structural parts, including cabin doors, wing beams, speed brakes, in the horizontal tail structure, fuel tanks, auxiliary oil tanks, cabin walls, floors, helicopter rotor blades, propellers, high-pressure gas containers, antenna caps, nose cones, landing gear doors, gas rectifiers, contra-propellers, in engine compartments (particularly jet engine compartments), turbine engine bypasses, seat parts, and aircraft table boards. Many lightweight aircraft try to use as many lightweight composite materials as possible. Carbon fiber-reinforced carbon matrix composite materials have been widely used in aircraft, for example. They are used in parts that operate at high temperatures, for example, in Concord brake discs, rocket spray nozzles, and modules for humans. They are also used in static parts of jet engines. Rocket engine shells and rocket launchers are normally made of reinforced composite materials. One important and interesting application of composite materials is their application in aircraft metallic structure repair.
2. Applications in the automobile industry
   In the automobile industry, the number of lightweight automobiles is increasing due to their high energy and engine efficiencies. The key to reducing the weight of a car is to reduce its structural weight. If enough weight can be eliminated, a smaller engine can be used, and thus the car will be more energy-efficient. The primary composite materials currently used in the automobile industry are glass fiber-reinforced composite materials (GRPs), and not carbon fiber-reinforced composites. This is due to the high prices of carbon fibers and aramid. Nevertheless, the price of GRPs still cannot compete with traditional

pressed steel plates, which limits their application. Nowadays, many cars and trucks use molded parts for their bodies, and composite materials for their panels and doors: for example, in various automobile bodies, gauge boards, bumpers, and finishing materials. In the automobile industry, there is also interest in fabricating anti-vibration parts using high energy-absorbing composite materials. For example, composite materials have been used in automotive reed plates and in transmission rods for trucks. The automobile reed and truck joystick also uses composite materials. According to reports, GRPs have been used as vehicle spoke strips and air feeders. In addition, if aluminum oxide fibers are used, the heat resistance of the materials will be enhanced.

3. Applications in civil engineering
   GRP composite materials have been used in civil engineering. In their construction, the GRP is usually shaped as a hyperbolic structure and folded plate structure to overcome its low elasticity coefficient. In addition, thin GRP is semitransparent, which makes it ideal for construction, even though some architects still suspect the sustainability of use of alkali-resistant glass fiber-reinforced composites. Glass fiber-reinforced cements (GRCs) consisting of alkali-resistant glass fibers have been increasingly used in construction. Stable polymer-reinforced cements that can be used under different conditions can be made from high-strength polymer fibers and polymer nets. Even though the introduction of polymer fibers into normal cements significantly increases their cost, GRC composite materials have been widely used in many lightweight concrete structures: for example, shell structures, folded plate structures, cover sheets, decoration "carving" boards (similar to the boards in the Liverpool Rome Cathedral gates), storage containers, convey pipes, frame and anti-water structures, water storage tanks for resident and commercial uses, concrete frames, and small parts in chain bridges, and so on. In addition, lightweight composites have been used in some structures like partitions.
4. Applications in sporting goods
   There has been much attention paid to applications of composite materials in sporting goods. The earliest use of composite materials in sporting goods is in the pole vault jump bar. Nowadays, composite materials are widely used in tennis rackets, board rackets, golf clubs, fishing poles, boat oars, archery instruments, dugout canoes, transmission devices, surfboards, skis, ski poles, bicycles, and in instruments used by lifeguards.

   Moreover, composite materials have also obtained broad application [3] in bioengineering, electrical engineering, and in the shipbuilding industry.

14.4.2 Developmental Trends in Composite Materials [41]

Although composite materials are widely used, they are still in the vigorous development stage. At present, the trend of their development mainly focuses on the following aspects [41].

1. Capturing overwhelming market share in certain specific product areas such as middle- or high-grade sport goods. One example might be fishing rods made of carbon fiber-reinforced composites. However, much effort needs to be put into the application of composite materials in the automobile and electronic industries, in architecture, textiles, and chemical engineering, and in the shipbuilding industry. In addition, composite materials are necessary in the patching, reinforcement, and reconstruction of existing architecture and structures such as high-rise buildings, tunnels, culverts, dams, bridges, and historic buildings, and even in lignin fishing boats.
2. Developing new and cost-efficient composite materials to replace some widely used traditional materials, such as wood, steel, and iron. Developing thermoplasticity composite materials and polymer-reinforced composite materials using polypropylene (PP), polyamide (PA), polycarbonate (PC), polyether ether alkone (PEEK), polyether sulfone (PES), and polyphenyl sulfide (PPS, gathering diphenyl sulfide).

3. Exploring new functional composite materials—like home position synthesis (home position growth) compound materials, nano-compound materials, tiny view condition compound materials, and so on—that solve the interface weak link in macroscopic compound materials and enhance the plasticizing effect.
4. Strengthening the optimization of design research, and forms to design expert systems, that give full play to the composition effect.
5. Studying thermoplasticity polymer base composite materials.

In summary, along with the progress of science and technology, composite materials will certainly undergo further development. The mechanical properties of composite materials are important; therefore, the development of composite materials has also introduced new areas of study to solid-state mechanics. Rapid development of composite material mechanics will certainly continue.

EXERCISES

14.1 Give definitions of the following concepts: (1) the volume fraction threshold of fibers; (2) the minimum volume fraction of fibers; (3) strength and modulus; (4) uniaxial composite materials' longitudinal and transverse Poisson's ratios.

14.2 What is a composite material? How is a composite material sorted?

14.3 Summarize the characteristics of particle-reinforced metal matrix composites.

14.4 List the enhancement mechanisms of particle-reinforced metal matrix composites.

14.5 List the characteristics of fiber-reinforced composite materials. What are the behaviors of the matrix and the fibers in a forced loaded composite?

14.6 Under which condition can the mixing rule of the composite be applied to calculate the properties? Under which condition can the paralleling rule of the composite be applied to calculate the properties? What is the difference between the two rules?

14.7 Which factors can affect the strength of fiber-reinforced composite materials?

14.8 List the characteristics of fatigue properties of a composite material.

14.9 List all the factors that can affect the fatigue properties of a composite material.

14.10 What is the difference in the impact properties of metallic materials and metal matrix composite materials?

14.11 What is the fracture mechanism for fiber-reinforced composite materials?

14.12 What are the trends in the current development of composite materials?

14.13 In this chapter, we assumed that the interfaces between the matrix and the reinforcement are the geometrical interfaces in mathematics. But in practice, the interfaces consist of a transit layer. How you do carry on the theoretical analysis and numerical simulation?

14.14 How can we benefit from the study of the failure mechanism of fiber-reinforced metal matrix composites, when studying other composite materials?

14.15 How can we benefit from the study of the failure mechanism of fiber-reinforced metal matrix composites when designing other, similar, composite materials?

REFERENCES

1. Wu R. J. *Composite Materials*. Tianjin, China: Tianjin University Press, 2000.
2. Chen H. H., Deng H. J., and Li M. *Modern Composite Materials*. Beijing: China Substances Press, 1998.
3. Harris B. *Engineering Composite Materials*. Translated by Chen Xiangbao, Zhang Baoyan. Beijing: Chemical Industry Press, 2004.
4. Clyne T. W. and Withers P. J. *An Introduction to Metal Matrix Composites*. Translated by Yu Yongning, Fang Zhigang. Beijing: Metallurgical Industry Press, 1996.
5. Zhang J. and Zhang N. G. *Mechanics and Applications of New Composite Materials*. Beijing: Beijing University of Aeronautics and Astronautics Press, 1993.

6. Wo D. Z. *Encyclopedia of Composite Materials*. Beijing: Chemical Industry Press, 2000.
7. Shu D. L. *Mechanical Properties of Engineering Materials*. Beijing: Mechanical Industry Press, 2003.
8. Qiao S. R. *Mesomechanics of Composite Materials*. Xi'an, China: Northwestern Polytechnical University Press, 1997.
9. Shen G. L. *Composite Materials Mechanics*. Beijing: Tsinghua University Press, 1996.
10. Luo Z. D. and Wang Z. M. *Progress in Composite Materials Mechanics*. Beijing: Beijing University Press, 1992.
11. Eshelby J. D. The determination of the elastic field of an ellipsoidal inclusion and related problems. *Proc Roy Soc A*, 1957, 241(2): 376–396.
12. Ma Z. Y., Wu S. J., and Luo M. Dispersed particles and SiC particle reinforced Al matrix composite materials, I. preparation and microscopic structure. *Acta Metall Sin*, 1994, 30(1): 27–32.
13. Ma Z. Y., Wu S. J., and Luo M. Dispersed particles and SiC particle reinforced Al matrix composite materials, II. Performance and fracture. *Acta Metall Sin*, 1994, 30(1): 33–38.
14. Nardone V. C. and Prewo K. M. The strength of the particle rein-forced metal matrix composite. *Scr Metall*, 1986, 20(2): 68–87.
15. Wang L. N. and Wang J. Strengthening mechanism of particle reinforced metal matrix composite materials. *J Southeast Univ*, 1995, 25(1): 77–82.
16. Chen S. H. and Wang T. C. Size effects in the particle-reinforced metal-matrix composites. *Acta Mech*, 2002, 157(1): 113–127.
17. Arsenault R. J., Wang L., and Feng C. R. Strengthening of composites due to microstructural changes in the matrix. *Acta Matall Mater*, 1991, 39(1): 47–57.
18. Ramakrishnan N. An analytical study on strengthening of particulate reinforced metal matrix composites. *Acta Mater*, 1996, 44 (2): 69–77.
19. Nair V., Tien J. K., and Bates R. C. SiC-reinforced aluminum metal matrix composites. *Int Metals Rev*, 1985, 30(6): 275–290.
20. Hwu B. K. Effects of process parameters on the properties of squeeze-cast SiCp-6061/Al metal matrix composite. *Mater Sci Eng*, 1996, 207(1): 135–141.
21. Quan G. F., Song Y. J., and Tu M. J. Fracture proper-ties of metal matrix composites. *J Xi'an Jiaotong Univ*, 1995, 29(2): 84–89.
22. Brechet Y. and Embury J. D. Damage initiation in metal matrix composites. *Metall Mater*, 1995, 39(8): 1781–1786.
23. Wang B. SiC particle cracking in powder metallurgy processed aluminum matrix composite materials. *Metall Mater Trans A*, 1995, 26(12): 2457–2467.
24. Kim Y. H., Lee S., and Kim N. T. Fracture mechanisms of a 2024Al matrix composite reinforced with SiC whiskers. *Metall Mater Trans A*, 1992, 23(1): 68–80.
25. Leggoe J. W. Crack tip damage development and growth resistance in particulate reinforced metal matrix composites. *Eng Fract Mech*, 1996, 53(6): 873–895.
26. Zhao D. and Tuler F. R. Effect of particle size on fracture toughness in metal matrix composites. *Eng Fract Mech*, 1994, 47(3): 303–308.
27. Zhou Y. C., Zhu Z. M., and Duan Z. P. Rheological-thermal fracture by laser beam thermal shock. *Proceedings of IUTAM Symposium on Rheology of Bodies with Defects*, Edited by R. Wang, Kluwer Academic Publishers, Dordrecht, 1998, pp. 56–65.
28. Zhou Y. C., Long S. G., and Duan Z. P. Laser-induced rheological thermal failure in ceramic reinforced metal matrix composites. *The Third Pacific Rim International on Advanced Material and Progress*, Vol. 2, edited by M. A. Imam, *The Minerals Metals and Materials Society*, Warrendale, PA, 1998, pp. 1995–2000.
29. Han N. L., Yang J. M., and Wang Z. G. Role of real matrix strain in low cycle fatigue life of a SiC particulate reinforced aluminum composite. *Scr Mater*, 2000, 43(9): 801–805.
30. Ghorbel E. Interface degradation in metal-matrix composites under cyclic thermo mechanical loading. *Compos Sci Technol*, 1997, 57(8): 1045–1056.
31. Mogilevsky R., Soni K. K., and Chabala J. M. Reactions at the matrix/reinforcement interface in aluminum alloy matrix composites. *Mater Sci Eng A*, 1995, 191(1–2): 209–222.
32. Srivatsan T. S., Meslet A.-H., Petraroli M., Hotton B., and Lam P. C. Influence of silicon carbide particulate reinforcement on quasi static and cyclic fatigue fracture behavior of 6061 aluminum alloy composites. *Mater Sci Eng A*, 2002, 325(1–2): 202–214.
33. Hartmann O., Kemnitzer M., and Biermann H. Influence of reinforcement morphology and matrix strength of metal-matrix composites on the cyclic deformation and fatigue behavior. *Int J Fatigue*, 2002, 24(2–4): 215–221.

34. Llorca J. Fatigue of particle-and whisker-reinforced metal-matrix composites. *Prog Mater Sci*, 2002, 47(3): 283–353.
35. Ding H. Z., Hartmann O., and Biermann H. Modeling low-cycle fatigue life of particulate-reinforced metal-matrix composites. *Mater Sci Eng A*, (2002), 333(1–2): 295–305.
36. Long S. G. and Zhou Y. C. Thermal fatigue of particle reinforced metal matrix composite induced by laser heating and mechanical load. *Compos Sci Technol*, 2005, 65(9)1391–1400.
37. Zhou Y. C. and Long S. G. Thermal failure mechanism and failure threshold of SiC particle reinforced metal matrix composites induced by laser beam. *Mech Mater*, 2003, 35(10): 1003–1020.
38. Long S. G. and Zhou Y. C. Determination of damage parameter in particle reinforced metal matrix composite by ultrasonic method. *J Mater Sci Lett*, 2003, 22(2): 911–913.
39. Zhou Y. C. and Long S. G. Thermal damage and fracture of particulate reinforced metal matrix composites induced by laser beam. *J Eng Mater Trans ASME*, 2001, 123(2): 251–260.
40. Long S. G. and Zhou Y. C. Laser induced damage and fracture of metal matrix composite through coupling of heat and force. *JSM*, 2000, 21(3): 277–281.
41. Hao Y. K. and Xiao J. Y. *High Performance Composite Materials*. Beijing: Chemical Industry Press, 2004.

Index

H

I

N

O

P

Q

R

T

U

V

W

X

Y

For Product Safety Concerns and Information please contact our
EU representative GPSR@taylorandfrancis.com Taylor & Francis
Verlag GmbH, Kaufingerstraße 24, 80331 München, Germany